Finite Mathematics

Finite Mathematics

André L. Yandl
Seattle University

Brooks/Cole Publishing Company
Pacific Grove, California

In memory of my parents, Louis and Rosette Yandl

Brooks/Cole Publishing Company
A Division of Wadsworth, Inc.

Printed in the United States of America
10 9 8 7 6 5 4 3 2 1

Library of Congress Cataloging-in-Publication Data
Yandl, André L.
    Finite mathematics / André L. Yandl.
      p.   cm.
    Includes index.
    ISBN 0-534-14190-0
    1. Mathematics.   I. Title.
  QA39.2.Y33   1990   90-44982
  510—dc20   CIP

Sponsoring Editor: *Jeremy Hayhurst*
Consulting Editor: *Robert Wisner*
Project Development Editor: *Patricia P. Gadban*
Signing Representative: *Karen Buttles*
Editorial Assistants: *Nancy Champlin, Gay Bond*
Production Editor: *Timothy A. Phillips*
Manuscript Editor: *Margaret Chang*
Permissions Editor: *Carline Haga*
Interior and Cover Design: *Katherine Minerva*
Cover Photo: *Lee Hocker*
Art Coordinator: *Lisa Torri*
Interior Illustration: *Lori Heckelman*
Typesetting: *Jonathan Peck Typographers, Ltd.*
Cover Printing: *Lehigh Press*
Printing and Binding: *Arcata Graphics/Fairfield*

Preface

This book is the tangible result of my thirty-four years of teaching experience at Seattle University. For many of those years I have taught a two-quarter mathematics sequence to business students. That experience is the foundation and, I hope, the strength of this text.

I have seen a lot of changes in courses and curricula in those years of teaching. Indeed, from 1968 to 1970 I was a member of the MAA's CUPM committee chaired by Ralph Boas to examine the national undergraduate curricula and make recommendations for improving those curricula at all levels. I had the opportunity to serve on two interesting panels, that for mathematics in the two-year colleges, and the parallel committee on basic mathematics. I am pleased to recall that many of the needs of today's students and their instructors were anticipated then and recommendations made accordingly. My hope is that the best of those ideas have been implemented.

In the twenty intervening years, "finite" mathematics topics have become consolidated into a coherent program which has obvious utility for students in business and managerial science. Finite mathematics—the application of sets, matrices, some linear algebra, probability, and statistics to practical problems—is now standard in most business, economics, and sometimes social science curricula.

Alas, changes in curricula have not provided any more time for classroom instruction, and courses necessarily emphasize some of these topics to the exclusion of others. This text is no exception, and it will be necessary to select the topics that best fit your syllabus. A chart showing the interdependence of chapters is given on page ix. I am hopeful that even the material not explicitly covered in class will be useful to some students in some way, perhaps as motivation for exploring more mathematics that is useful in the business world, or as reference material for future courses.

Philosophy and Rationale

I am well aware that most students resist actually reading their texts, preferring to start by working the exercises assigned in class, resorting to examples in the section where necessary, and, as a last resort, reading from the beginning of the section to the end. I hope that the efforts spent to make this material readable and comprehensible will convince many students that a better strategy is to read before and after class, to read with pencil and paper at hand, and to study theorems to see precisely what they say (and do not say) before attempting the exercises.

I strongly believe that a textbook and an instructor should complement one another. Either the text should provide necessary details with the instructor providing a clear outline and reinforcement for the important points, or the text should be a brief outline in notelike form to which the instructor adds detail and explanation. Textbooks that take the latter point of view are preferred by some instructors who believe that a minimum of verbiage is desirable in order to avoid confusing students. These texts are also short and preferred by publishers because the page length, and therefore cost of producing the book, is minimized. They look accessible, are easy to carry, and seem to minimize the effort required by the course. Many instructors are seduced.

I believe that a far better approach is to let the instructor guide the student toward the mathematical priorities while letting the text provide details, reinforcement, and rationale as well as be a repository of good examples and exercises. My teaching experience tells me that short, overly concise, "notelike" texts do most students a disservice. When students use books with little or no detail they appear to be successful because they are able to memorize how to do the problems from the templates provided; however they do not retain what they've learned because they never understood the reasoning behind the procedures. I was fortunate to find a publisher who agreed with my views.

I have tried to explain each topic in enough detail so that students are able to learn to follow mathematical arguments without fear. Whenever appropriate, I have attempted to appeal to their visual intuition (see, for example, page 256, the use of the "squirrel on the roof" in motivating the simplex method).

Structure and Coverage

The basic structure and content of this book is comparable to other finite mathematics texts, but there are enough differences to justify some comments. (**1**) This text stresses algebra concepts earlier and more thoroughly than most texts for a simple reason. Many students, especially mature returning students, have forgotten their previous mathematics and need to be reminded of what they once knew. They are unable to succeed without mastery of basic algebra, whether it is formally covered in the course, given as assigned reading, or used just for reference when needed. (**2**) Students typically use texts as a source of worked examples to help solve the problems they are assigned, and it is here that I have spent much effort in writing. Many of my examples were created as a result of explaining concepts for which students requested extra help. It seems that most instructors do not have the opportunity to work enough examples in class, and it is here that a text can and should provide support. I have provided an abundance of those that I feel are most illustrative of the concepts and most helpful to students. Most examples end with a reference to an exercise in the text at the end of the section. These references serve two purposes. Students are able to immediately apply the technique and test whether or not they understand the example given, and they are able to match certain types of exercises with corresponding examples so that the student may refer to that example when doing assigned exercises. (**3**) Some students, even those successful in other courses, are not successful in mathematics despite spending a lot of time studying. I hope to make my students more efficient and proficient in approaching mathematics by providing some strategies to help them succeed. In this text, this objective is addressed by the introductory section titled "Getting Started." Students should be urged, even required, to read it. The main emphasis is on helping students to understand why mathematical methods work, and provide some insight to a sensible *method* of studying mathematics.

Optional Sections

With all the material inherent in this course, it is necessary to explain why the sections designated optional are present. There are three reasons for the optional sections. Many of them deal with material that students have almost certainly encountered but have forgotten. Other optional sections are present because I believe they are valuable (from my teaching experience). Still other optional sections are designed for students with more interest and aptitude, because these students tend to want to know *why* techniques work the way they do. All of the optional topics may be omitted without loss of continuity.

Pedagogy and Design

Most students in business programs require a great deal of assistance in learning mathematics. I believe that the text optimizes students' chances of mastering the material in the following ways: (**1**) Each chapter concludes with a careful chapter summary that can refresh students' memory and boost their confidence. These summaries include lists of important symbols and terms, a summary of the chapter discussion, and sample examination exercises. (**2**) Step-by-step procedures for solving important problems are listed on the endpapers of the text for easy reference. (**3**) Icons such as the "slippery road" and "3 × 5 cards" , alert students to possible pitfalls and the material which should be committed to memory. (**4**) Definitions, theorems, and basic formulas are boxed, labeled, and identified by the 3 × 5 card icon, . I always encourage my students to make and use these flash cards for this and future courses. These icon-flagged statements are reproduced with examples on real 3 × 5 cards and are available in the accompanying student study guide for the text. (**5**) Some discussion in the text is labeled "A Closer Look" and delineated with a grey screen. These portions of the text give a deeper understanding of the material (see, for example, pages 95, 309–312) but may be omitted without loss of continuity. (**6**) Examples reference exercises that can and should be worked immediately after the student has understood the example. (**7**) Exercises are numerous and every effort has been made to be as realistic as possible without making them overly difficult. Starred (*) exercises are indicated for one of the following reasons: (a) they require material from one of the optional sections, (b) they are more theoretical or difficult, or (c) they are most appropriate for better-prepared students. (**8**) For most new concepts the algebraic manipulations are as simple as possible because the emphasis should be on understanding, not applying, the concept. However, many students at this level need much more exposure to and experience with algebra, and therefore some of the examples are explicitly designed to challenge and help sharpen their algebraic skills.

Calculator Usage

One of the obvious changes in thirty-four years of teaching is the relatively recent emergence of calculators in the classroom. I embrace them. Many of the text's exercises deliberately involve large numbers to encourage the use of calculators. Similarly, simple reference tables have been minimized. Before the widespread use of calculators students had to spend considerable time doing simple arithmetic and hand calculation. Calculators have saved students much of this time and I believe that it can be profitably invested in actually reading their texts and mastering concepts. Of course, students should refer to their owner's manuals before using their calculators, but many have lost their manuals or acquired their calculators secondhand. Accordingly, a short prescription for using calculators is given in "Getting Started."

Chapters in Brief

Chapter 1 gives a very brief introduction of sets including Venn diagrams. It also reviews equations in one variable and inequalities as well as the function concept. The exponential and logarithm functions are discussed because they are needed in Chapter 5 on mathematics of finance. Those who do not plan to cover that chapter may omit these two sections. Section 1.8 on sigma notation is included because many business faculty like students to be proficient in using it as soon as possible.

Chapter 2 is devoted to systems of linear equations and inequalities. The treatment is standard but the presentation is aimed at preparing the student for matrices and solving systems of linear equations by row reduction. The last section includes some important business applications.

Chapter 3 begins the study of matrices. I discuss cryptography in addition to the more usual business applications for two reasons. It provides an opportunity to demonstrate how a simple mathematical tool can be used effectively in a practical situation, and students usually enjoy the intrigue of coding and uncoding messages. Markov chains are included here because not all instructors have the luxury of including probability in their syllabi.

Chapter 4 introduces linear programming via the geometric approach. The simplex method is discussed in more detail than most books attempt at this level. I introduce the three-dimensional coordinate system because I want to explain geometrically *why* the simplex method works. I also feel that it is useful not to restrict the discussion to the plane, especially when the big *M* method is discussed.

Sequences and mathematics of finance are discussed in Chapter 5. Although I want to show the students how to use standard formulas to solve annuity problems, I strongly encourage them to solve the problems using only two basic formulas. I believe that they better understand the unifying ideas this way. I encourage them to use their calculators by restricting the use of annuity tables to two pages.

Chapter 6 contains basic probability material, including Bayes' formula, the binomial distribution, and more on the Markov chains.

Elementary theory of games and graphs is presented in Chapter 7 because it is a natural extension of the treatment of probability, linear programming, and matrices given earlier. It also introduces the student to a variety of useful applications of mathematics to business.

Topic Dependencies

There is much more material in this book than can be covered in one semester or two-quarter course. The emphasis can be on linear algebra, linear programming, or probability and decision theory. The nature of the course and/or the instructor's preference will dictate the selection of topics. The following chart shows the interdependency of sections.

Note: A blue arrow indicates that Topic A is prerequisite to Topic B. A black arrow indicates that Topic A *should* be covered before Topic B but may be omitted when students are well prepared.

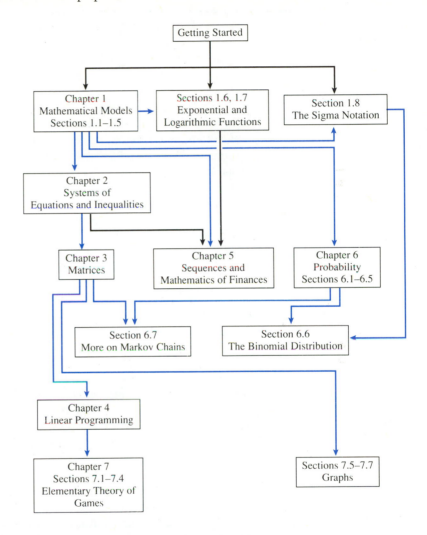

Accompanying Texts

The material in this text is accompanied by two separate volumes, *Introduction to Mathematical Analysis*, Brooks/Cole, 1991, and *Applied Calculus*, Brooks/Cole, 1991. The first text is a combined volume for two-semester courses in finite and calculus mathematics.

A Student Solutions Manual is available for this and the separate volumes giving complete, worked solutions to all the odd-numbered exercises in the text(s). Another solutions manual for the instructor gives complete solutions to all even-numbered exercises as well as sample examination questions.

A separate Study Guide includes flash 3×5 cards, objectives, and other aids for learning. Computerized testing for both IBM family and Apple Macintosh computers is also available to adopting schools. [Contact your Wadsworth-Brooks/Cole sales representative, or call 1-800-876-2350, for more details.]

Preface to Students

Many of the teaching ideas used in this book originated with conversations I have had with my students. I usually leave home early to miss the traffic and arrive on campus at 7:00 A.M. I often go to the school cafeteria for a cup of tea and almost always some of my students join me, sometimes when they're *cramming* for an exam and *anxious* and sometimes when they're just curious about material from a lecture. These numerous hours of one-on-one or one-on-few conversations have helped me to better understand the many difficulties some students have. Often, the real problem is simple: they haven't done enough work. But I've also found that many students are not afraid of doing a little work if it helps them understand *why* they are being taught this material and *why* things work as they do. I have also found that students (like you) *can be genuinely enthusiastic* about mathematics *when they experience success*. My goal is to aid your teacher in conveying that experience of success to you. To aid in this process, this text includes some topics and methods that often do not appear in other books for this course. In general, I encourage you to use whichever method gives a correct answer and is understandable for you. This does not mean that we will stress "monkey-see, monkey-do." You will have to do some work in this course, but the result should be more satisfying to you than if you had blindly followed a method without understanding it and why it works.

Many of the examples you'll encounter were originally made up during my conversations with students. I selected those that seemed to stir their interest the most, such as the problem on basketball playoffs (see Example 10, page 407). I made a real effort to explain the simplex method, geometrically and visually, because I know that my students learn it best that way. Many of my students are surprised to learn that the best strategy for studying mathematics may be quite different from that which has helped them succeed in other disciplines. I have included some hints on how to study mathematics in the preliminary section titled "Getting Started." I urge you to carefully read that section before you begin with the core material.

Very often, students come to my office with a question about an example in a textbook. Almost always the question is: "How did the author get from this step to that step?" Usually I suggest that they take a pencil and paper and try to work out the details omitted by the author. Most of the examples given here have very detailed solutions so you should not find it too difficult to see how to arrive at each step. It is always good practice, however, to study with paper and pencil handy. For example, in a "story" or word problem, I may show you how to get the equation, and give you the answer without giving all the steps to the solution. In such a case, it is best for you to try to solve the equation anyway to convince yourself that you are able to do so. If you cannot solve it, go back to the discussion on how to solve this type of equation, then, read the corresponding examples and try again.

I feel very strongly that you should understand *why* procedures that you are asked to follow work the way they do. Certainly you will enjoy studying more if you actually *understand* what is being done. With this approach, you will better remember what you learned and you will be able to use your acquired knowledge when needed, both in this course and in the real world.

I have tried to help you differentiate between useful explanations and essential facts by putting all formulas, definitions, theorems, and procedures for solving each type of problem in colored boxes. On the inside cover of the book, you will find a list of the procedures used to solve certain types of problems, for easy reference. You will find the 3 × 5 icon beside a statement important enough for you to memorize. Some parts of the text have been labeled "A Closer Look" and appear in grey color screens. Your instructor will advise you about these sections.

It is likely that your instructor will not have time to cover all of the material, especially the examples, in class. Therefore, it is *essential that you read* each section you've been assigned *carefully*. It is also essential that you test your understanding of the material as you proceed. To help you do this, I refer you to an exercise at the end of the section at the end of a worked example. Immediately after studying the material and reading an example, try to *do the exercise without looking at the corresponding example*. I have referred you only to odd-numbered exercises so that you may check your answers in the back of the book. If you are not able to solve the exercise, go back and study the corresponding discussion, read the example, and try again. Remember that if your answer doesn't match the answer in the book exactly, you may still be right and further simplification of the algebra will yield my answer. Do not proceed to the next topic until you succeed in doing the exercise correctly. The symbol ■ signifies the end of the example if no exercise is indicated. The same symbol is used to indicate the end of a proof.

I truly believe that in my thirty-four years of teaching, I have seen *all* the types of errors that students can possibly make when doing elementary mathematics. I have tried to warn you *not* to make these mistakes whenever possible. These warnings can be found in the text, or in the footnotes, and are indicated in the margin by the icon ⬦, which you can think of as a "slippery road ahead" sign.

In addition, I have purposely included throughout the text some exercises that involve large numbers to encourage you to use your calculator. If you are not proficient in using a calculator, read the appropriate material in the "Getting Started" section, read your owner's manual, and practice as often as possible. The time you will save can be used to think about the concepts you are learning.

At the beginning of each of the chapter review sections, I have listed the new terms and symbols that appear in the chapter. Make sure that you are familiar with the meaning of each of these terms. Following the list is an extensive summary of the chapter. These summaries are detailed because I believe that a summary should contain almost all the really important information. To really test your knowledge, mentally construct your own summary before reading those given in the text. Sample exam questions are given after the summaries. Some of these are straightforward; others are quite challenging. The types of problems that you will have when you take an exam depend, of course, on your instructor. However, if you can do all of the sample problems that appear in the review sections, you can be confident you have mastered the material.

Finally, I wish you the very best success in this course and in your future endeavors.

Acknowledgments

I thank the following people whose criticism at the different stages of the manuscript was invaluable to me in developing and improving the book:

Patricia M. Bannatine, *Marquette University*; Theresa J. Barz, *St. John's University*; Joseph Brody, *Concordia University*; Lynn Cleary, *University of Maryland*; John E. Gilbert, *University of Texas, Austin*; Madelyn T. Gould, *DeKalb University*; Bennette R. Harris, *University of Wisconsin, Whitewater*; Judith H. Hector, *Walters State Community College*; Alec Ingraham, *New Hampshire College*; Gerald Leibowitz, *University of Connecticut*; Lawrence P. Maher, *North Texas State University*; H. T. Mathews, *University of Tennessee, Knoxville*; John Morrison, *University of Delaware*; William C. Ramaley, *Ft. Lewis College*; Gordon Shilling, *University of Texas, Arlington*; S. J. Stack, *Wilfrid Laurier University*; Louis M. Weiner, *Northeastern Illinois University*; and Carroll G. Wells, *Western Kentucky University*.

During the writing of this book, I have received help and encouragement from many colleagues and friends. I wish to acknowledge a debt to my former dean, Terry van der Werff, who first suggested that I write a book. His interest and support greatly influenced my writing. I also express my appreciation to the Seattle University sabbatical and summer faculty fellowship committees for their support.

I am indebted to my colleagues: Mary B. Ehlers, who worked all of the exercises; Sr. Kathleen Sullivan, who helped prepare the study guide; C. C. Chang, Wynne A. Guy, Janet Mills, Ahmad Mirbagheri, Carl Swenson, Burnett Toskey, and Alan Troy; who gave me much support during the many years of writing. I owe many thanks to Kristi Weir of Bellevue Community College's Department of Economics for the many informative conversations we had on the teaching of mathematics to economics students while she was on the faculty of Seattle University. I am also indebted to my students, who in the past thirty-four years have helped me develop the ideas incorporated into this book.

I also wish to express my deepest appreciation to the staff of Brooks/Cole Publishing Company: Acquisitions Editor, Jeremy Hayhurst, for his sponsorship, and advice and support throughout the production of the book; Pat Gadban, Senior Project Development Editor, for the careful editing of the early draft of the manuscript and many helpful suggestions for improvement; Professor Robert Wisner of New Mexico State University, Mathematics Series Editor, for his careful reading of the manuscript and constructive criticism; Tim Phillips, Production Editor, for coordinating all the tasks and providing me with needed assistance and enthusiastic support; Faith Stoddard, Assistant Mathematics Editor, for coordinating the production of the accompanying solutions manuals and the study guide; Katherine Minerva, Senior Designer, for a beautiful and effective interior and cover design; Nancy Champlin, Editorial Assistant, for providing editorial support for the text, solutions manuals, and study guide; and last but not least, William M. Roberts, President, and Michael V. Needham, past President, Craig F. Barth, Vice President, and the entire Brooks/Cole staff, for the support that they gave me and for making me feel so welcome during my visits to Pacific Grove.

Finally, my thanks go to my wife, Shirley, and to my three sons Steven, Michael, and Kris, and their families, for their patience during the last few years, while much of my time was spent writing instead of with them.

André L. Yandl

Contents

Getting Started

Philosophy and Rationale

The purpose of having this course as an integral part of your program of study is to help you understand the use of mathematics in subsequent courses such as statistics, probability, and economics. It is important that you understand the basic concepts rather than simply memorize some examples in order to pass the exams. Many of you may find mathematics difficult. Nevertheless, if you study correctly, you can improve your understanding of the subject significantly. In this introduction you will learn how to study mathematics, how to use this book effectively, and how to use a calculator. What follows is an alternative to memorization through the understanding of mathematical processes.

Getting Started

It is true that to work successfully with mathematics you must know many definitions, theorems, and formulas. However, it is not necessary or desirable to memorize every detail. In fact, it is possible to recall a great amount of information from relatively few memorized facts. *The trick is to look for patterns that occur frequently in mathematics.*

Patterns

In basic algebra courses, most of you encountered the following formulas:

$$(a + b)^2 = a^2 + 2ab + b^2 \tag{1}$$
$$(a - b)^2 = a^2 - 2ab + b^2 \tag{2}$$
$$(a + b)^3 = a^3 + 3a^2b + 3ab^2 + b^3 \tag{3}$$
$$(a - b)^3 = a^3 - 3a^2b + 3ab^2 - b^3 \tag{4}$$
$$a^2 - b^2 = (a - b)(a + b) \tag{5}$$
$$a^3 - b^3 = (a - b)(a^2 + ab + b^2) \tag{6}$$

Many of you learned all of these, remembered two or three of them a few weeks after completion of the course, and then later promptly forgot all of them!

There is a way to remember formulas more effectively. Consider the last two formulas. The left sides are in the form $a^n - b^n$, where n is a positive integer. In both cases, $a - b$ is a factor on the right side. This is the case not only for $n = 2$ and $n = 3$, but for all positive integers n. Therefore, if you are asked to factor $a^5 - b^5$, you would know that $a - b$ is one of the factors. However, how do you get the other factor? You could divide $a^5 - b^5$ (the left side) by $a - b$ (the factor you know) to obtain $a^4 + a^3b + a^2b^2 + ab^3 + b^4$. Thus,

$$a^5 - b^5 = (a - b)(a^4 + a^3b + a^2b^2 + ab^3 + b^4)$$

Recall that the degree of a term is the sum of all exponents assigned to the variables in that term. For example, the degree of $3x^2y^3$ is 5 since $2 + 3 = 5$. If n is a positive integer, one factor of $a^n - b^n$ is $a - b$. The other factor consists of the sum of all terms of degree $n - 1$ in the variables a and b, with nonnegative integer exponents and with coefficient 1.

EXAMPLE 1   Factor $a^7 - b^7$.

Solution   The first factor is $a - b$. Each term of the second factor is of degree 6. Hence, the terms are a^6, a^5b, a^4b^2, a^3b^3, a^2b^4, ab^5, and b^6. The factorization is

$$a^7 - b^7 = (a - b)(a^6 + a^5b + a^4b^2 + a^3b^3 + a^2b^4 + ab^5 + b^6). \qquad \blacksquare$$

We have been looking at the formulas (5) and (6). Now consider (1) and (3). In both, the left side is in the form $(a + b)^n$, where n is a positive integer. In the expansion of $(a + b)^2$, there are three terms of degree 2; while in the expansion of $(a + b)^3$, there are four terms of degree 3. The pattern we see is that in the expansion of $(a + b)^n$, where n is a positive integer, there are $n + 1$ terms of degree n. Thus, in the expansion of $(a + b)^4$, there are five terms of degree 4. Without their coefficients, the terms are a^4, a^3b, a^2b^2, ab^3, and b^4.

There are several ways to find the appropriate coefficients, known as the *binomial coefficients*. One way is to represent the coefficients as an array known as *Pascal's triangle*. (See Figure 1.)

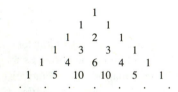

FIGURE 1

In each row of this triangular array the first and last entries are 1s. Each of the other entries is the sum of the two nearest entries in the row just above. Thus, the first three elements of the next row in Figure 1 would be 1 (the border element), 6 (since $1 + 5 = 6$), and 15 (since $5 + 10 = 15$). To write the expansion of $(a + b)^4$, find the row of Pascal's triangle whose second entry is 4. The entries for this row are 1 4 6 4 1. Thus,

$$(a + b)^4 = a^4 + 4a^3b + 6a^2b^2 + 4ab^3 + b^4$$

Although Pascal's triangle is an interesting way of obtaining the coefficients, it is not very efficient. Trying to find the 51 coefficients in the expansion of $(a + b)^{50}$ by writing 51 rows of Pascal's triangle would be both time- and space-consuming. However, there is an easier way to find the binomial coefficients using the following method:

1. In the expansion of $(a + b)^n$, the first term is a^n and may be written $1a^nb^0$.

2. Having obtained one term of the expansion, the coefficient of the next term can be determined as follows. Multiply the coefficient of the term just obtained by the exponent assigned to a in that term and divide the result by one more than the exponent assigned to b in that term. In other words, if $ka^i b^j$ is a term of the expansion, the coefficient of the next term is $\frac{ki}{j+1}$. In fact, the next term will be $\frac{ki}{j+1}a^{i-1}b^{j+1}$ and the last term will be b^n.

EXAMPLE 2   Write the expansion of $(a + b)^5$.

Solution   Without coefficients, the terms are a^5, a^4b, a^3b^2, a^2b^3, ab^4, and b^5. Note that the first term is a^5 and may be written $1a^5b^0$. Thus, the coefficient of the second term is 5, since $\frac{(1)(5)}{0+1} = 5$, and the second term is $5a^4b^1$. The next coefficient is 10, since $\frac{(5)(4)}{1+1} = 10$, and the third term is $10a^3b^2$. The other three coefficients are similarly obtained to complete the expansion:

$$(a + b)^5 = a^5 + 5a^4b + 10a^3b^2 + 10a^2b^3 + 5ab^4 + b^5$$ ∎

Now consider formulas (2) and (4). The left sides are in the form $(a - b)^n$. Here, we note that $(a - b) = (a + (-b))$ and, starting with the expansion of $(a + b)^n$, we replace every b by $-b$.

EXAMPLE 3   Write the expansion of $(a - b)^5$.

Solution   Start with the expansion of $(a + b)^5$ obtained in Example 2 and replace every b by $-b$. Thus,

$$(a - b)^5 = a^5 + 5a^4(-b) + 10a^3(-b)^2 + 10a^2(-b)^3 + 5a(-b)^4 + (-b)^5$$
$$= a^5 - 5a^4b + 10a^3b^2 - 10a^2b^3 + 5ab^4 - b^5$$ ∎

Notice the symmetry in the row of coefficients for the expansion of $(a + b)^n$. The first and last coefficients are both 1. The second and next-to-last coefficients are equal, and so forth.

Here is an example of another pattern. The squares of the first seven nonnegative integers are 0, 1, 4, 9, 16, 25, and 36. The differences of successive terms are $1 - 0 = 1$, $4 - 1 = 3$, $9 - 4 = 5$, $16 - 9 = 7$, $25 - 16 = 9$, and $36 - 25 = 11$. Observe that 1, 3, 5, 7, 9, and 11 are the first six odd integers. It is easy to show that this pattern will continue. To this end, note that if n is any nonnegative integer, then

$$(n + 1)^2 - n^2 = n^2 + 2n + 1 - n^2 = 2n + 1$$

This pattern can be used to sketch the graph of $y = x^2$. Start at the point $(0, 0)$. Move to the right 1, and up 1, to obtain $(1, 1)$. Move to the right 1 and up 3, to obtain $(2, 4)$. Move to the right 1 and up 5, to obtain $(3, 9)$. Move to the right 1 and up 7, to obtain $(4, 16)$. Move to the right 1 and up 9, to obtain $(5, 25)$, and so on. (See Figure 2.)

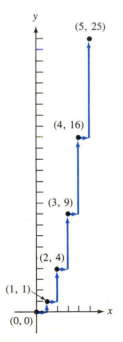

FIGURE 2

To introduce the next illustration, we need the following definition.

> **DEFINITION 1:**   An *Arithmetic Progression* is a sequence of numbers where the difference between any two consecutive terms of the sequence is constant. For example, 2, 5, 8, 11, 14, 17, 20, . . . is an arithmetic progression since $5 - 2 = 8 - 5 = 11 - 8 = 14 - 11 = 17 - 14 = 20 - 17 = \ldots$.

Suppose that you must find the sum of the first ten terms of this arithmetic progression. Of course, you could simply add them. However, for cases in which there are many terms it is best to use formulas. One of these is

$$a + (a + d) + (a + 2d) + \cdots + (a + (n - 1)d) = \frac{[2a + (n - 1)d]n}{2}$$

This formula need not be memorized; there is a better way to remember how to do this. Let us go back to finding the sum of the first ten terms of the progression 2, 5, 8, 11, 14, ...

We first write

$$S = 2 + 5 + 8 + 11 + 14 + 17 + 20 + 23 + 26 + 29$$

We can also perform the addition in reverse order,

$$S = 29 + 26 + 23 + 20 + 17 + 14 + 11 + 8 + 5 + 2$$

Adding these two equalities, we obtain

$$2S = (2 + 29) + (5 + 26) + (8 + 23) + (11 + 20) + (14 + 17) + (17 + 14)$$
$$+ (20 + 11) + (23 + 8) + (26 + 5) + (29 + 2) = 31 + 31 + 31 + 31 + 31$$
$$+ 31 + 31 + 31 + 31 + 31 = (31)(10)$$

Thus,

$$S = \left(\frac{31}{2}\right)(10) = 155$$

But, $\frac{31}{2} = \frac{2 + 29}{2}$. So, $\frac{31}{2}$ is the average of the first and last terms of the sum we evaluated; also, 10 is the number of terms. Therefore, *to find the sum of the first n terms of an arithmetic progression, we calculate the average of the first and nth terms and multiply this average by the number of terms*. Most of you will find this much easier to remember than the formula.

Mnemonic Devices

In addition to noticing patterns, it is often useful to think up a "trick" or mnemonic device* to recall a fact we need to know. For example, the graph of the equation $y = ax^2 + bx + c$, where a, b, and c are constants and $a \neq 0$, is a parabola. It is concave up if $a > 0$ and concave down if $a < 0$. (See Figure 3.)

---

*A mnemonic device is any device or "trick" that helps you remember—such as a string around your finger or a rhyme ("30 days has September . . .").

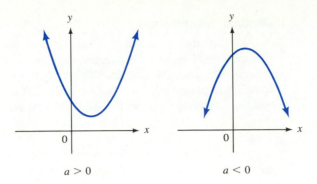

FIGURE 3

$a > 0$ $a < 0$

Some students remember the shapes of these graphs by associating the graph on the left with a smiling face and the graph on the right with a sad face. (See Figure 4.) (Of course, it is usually best to think up your own mnemonic device, since otherwise it may be just as difficult to remember the trick as it is to remember the fact.)

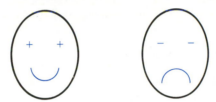

FIGURE 4

Geometric Interpretations

Another idea that many students find effective in studying mathematics is to draw, whenever possible, a geometric interpretation of the problem at hand. As an illustration, we use the formula $(a + b)^2 = a^2 + 2ab + b^2$. In Figure 5, we have drawn two squares: one with side a units, the other with side $(a + b)$ units. We see from the figure that the area of the larger square is equal to the area of a square with side a units, plus the area of a square with side b units, plus the areas of two rectangles with sides a units by b units. The areas are a^2, b^2, and $2ab$ square units, respectively. Thus, the formula $(a + b)^2 = a^2 + 2ab + b^2$ is illustrated geometrically.

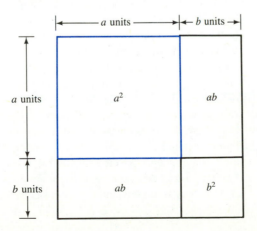

FIGURE 5

Individual Relationships

You may also relate a concept to a familiar situation. For example, when we multiply a sum of several terms by a number, it is necessary to multiply each term by that number. On the other hand, when we multiply a product of several factors by a number, it is sufficient to multiply only one factor by that number. You could relate this to the following situation. Suppose that you had two part-time jobs. Your total income is the sum of the incomes you get from each job. If you need to double your total income, you can accomplish this by doubling *each* of the two incomes. On the other hand, if you have only one part-time job, your weekly income is found by multiplying your hourly wage by the number of hours you work in a week. If you must double your income, you may accomplish this by either doubling the number of hours you work each week, or by talking your employer into doubling your hourly wage. If you relate mathematical concepts to real-life situations as we have just illustrated, you will find that you can understand more mathematical concepts than you expected.

In addition to looking for patterns, thinking up mnemonic devices, trying to give geometric interpretations whenever feasible, and relating mathematical concepts to your own experiences, you must also acquire good study habits.

Suggestions for Good Study Habits

1. Study daily. Two hours each day is better than 14 hours on the weekend.
2. Have regular review periods.
3. Schedule study sessions so that they either immediately precede or follow the lecture.
4. Go over each set of lecture notes as soon as possible after class and outline the main ideas. It is useful to copy key facts, definitions, theorems, and formulas on 3×5 cards. These cards may be used not only for study, but also for review in subsequent courses for which the present course is a prerequisite.
5. Concentrate and think clearly during the lectures.
6. Do your studying *before* you start your written assignment.
7. Read slowly and do not go to a new concept until you clearly understand the previous one.
8. Many students spend too much time "studying" and not enough time "thinking." Keep asking yourself why a certain step was taken by the professor in the lecture or by the author in the text. Once you understand why a step is taken, you will be more likely to take it yourself when it is needed.
9. Learn what can be done and what cannot be done in mathematics. There are rules, just like in a game. No one would try to play poker, or baseball, or chess without learning the rules. Do the same with mathematics!
10. Attend class regularly. This is especially important in mathematics.
11. Have occasional discussions with the professor.
12. Each weekend, write a letter to someone explaining, in your own terms, the basic concepts covered during the week. You need not send the letter. The purpose of this exercise is to reflect on what was done during that week. A good way to understand a concept is to try to explain it to someone else.

How to Use a Calculator

Most arithmetic calculations that you will encounter in this book should be done using a calculator. The time saved is better spent on studying the concepts and trying to understand the applications. Therefore, this section includes a brief introduction to calculators. Since there are many different brands, you are urged to consult the owner's manual because there may be some variation in the way each brand handles a specific problem.

Calculators are manufactured for many different types of users. For the person who carries a calculator to the local store simply to compare prices, a four-function calculator, perhaps one with memory, may be sufficient. On the other hand, a student majoring in a scientific field may need a scientific calculator or a programmable calculator. There are also special-purpose calculators available to handle problems peculiar to statistics, business, and so on. You will have to decide which type you are more likely to need.

Calculators are classified according to the type of logic they use in doing calculations. Basically, there are three types of logic used by calculators.

A calculator using arithmetic logic works from left to right and performs the operations in the order in which they appear. For example, an arithmetic calculator will give the answer 20 for $3 + 7 \times 2$ because it will add the 3 and 7 first to obtain 10, then multiply the 10 by 2 to get 20.

In algebra, however, the convention is to perform multiplication before addition, which is the way a calculator using algebraic logic works. Thus, an algebraic calculator will yield the answer 17 for the problem above. It will multiply the 7 by 2 first to get 14, and then it will add 3 to this result to get 17.

Finally, there are calculators that use reverse Polish notation (RPN) logic. These can be identified by the $\boxed{\text{ENTER}}$ key. When we use this type of calculator the numbers are entered first, then the operation symbols are entered.

In algebra, we have

$$5 + 7 \times 3 = 26$$

because we perform the multiplication first to get 21 and then we add this result to 5 to get 26. We have illustrated the sequence of key strokes that must be used with each type of calculator to obtain the answer 26.

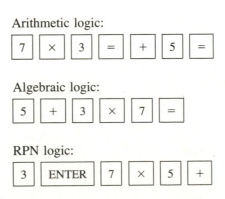

Arithmetic logic:

$\boxed{7}$ $\boxed{\times}$ $\boxed{3}$ $\boxed{=}$ $\boxed{+}$ $\boxed{5}$ $\boxed{=}$

Algebraic logic:

$\boxed{5}$ $\boxed{+}$ $\boxed{3}$ $\boxed{\times}$ $\boxed{7}$ $\boxed{=}$

RPN logic:

$\boxed{3}$ $\boxed{\text{ENTER}}$ $\boxed{7}$ $\boxed{\times}$ $\boxed{5}$ $\boxed{+}$

Several examples are given to illustrate the use of calculators. In these examples, a scientific calculator that uses algebraic logic is used. Those of you who own calculators that use RPN logic should work the same examples consulting your owner's manuals, if necessary.

EXAMPLE 4   Use a calculator to perform the following addition problem: 43 + 29.

Solution   First make sure that the calculator is on, or if it is already on, that it is cleared by pushing the key labeled ON/C . The display should show 0. Proceed as follows:

ENTER	PRESS	DISPLAY
43		43
	+	43
29		29
	=	72

The sum is 72.   ■

If a wrong number is entered, it is not necessary to start over. The calculator should have a key labeled CE , which means clear entry. For example, in the foregoing problem if 39 is entered instead of the desired 29, we can clear the 39 by using the clear entry key. We should note that a subtotal is displayed each time the + or − key is pushed. The sequence of steps is the following:

ENTER	PRESS	DISPLAY
43		43
	+	43
39		39
	CE	0
29		29
	=	72

EXAMPLE 5   Perform the following:

$$2.45 + 3.4 + 7 - 5.32$$

Solution

ENTER	PRESS	DISPLAY	ENTER	PRESS	DISPLAY
2.45		2.45	7		7
	+	2.45		−	12.85
3.4		3.4	5.32		5.32
	+	5.85		=	7.53

The answer is 7.53. ■

Some calculators provide parentheses so that certain operations may be grouped.

EXAMPLE 6 Calculate the value of $\dfrac{9^2 + 3}{50 - 6^2}$.

Solution Since the values of the numerator and the denominator have to be determined and the results divided, we use parentheses as shown in the following expression:

$$\frac{(9^2 + 3)}{(50 - 6^2)}$$

We now display the sequence of steps using the calculator.

ENTER	PRESS	DISPLAY	ENTER	PRESS	DISPLAY
	ON/C	0		(	0
	(	0	50		50
9		9		−	50
	x^2	81	6		6
	+	81		x^2	36
3		3		)	14
	)	84		=	6
	÷	84			

The answer is 6. ■

Annuities will be discussed in detail in Chapter 5. The amount A dollars of an annuity is the future value of all the payments, plus the compound interest on each of these payments. It is given by the formula

$$A = R\left[\frac{(1 + i)^n - 1}{i}\right]$$

where R dollars is the payment per period, i is the rate of interest per period, and n is the number of payments. The next example illustrates how to use this formula.

EXAMPLE 7   I.M. Thrift deposits \$200 at the end of every three months in a bank paying 8% interest compounded quarterly. What will the balance in the account be after 15 years?

Solution   Using the formula

$$A = R\left[\frac{(1 + i)^n - 1}{i}\right]$$

with $R = 200$, $i = 0.02$ (since $\frac{.08}{4} = .02$), and $n = 60$ (since $15(4) = 60$), we get $A = 200[(1 + .02)^{60} - 1]/.02$. We now use the calculator to find the value of A.

ENTER	PRESS	DISPLAY
	ON/C	0
200		200
	×	200
	(	0
1.02		1.02
	y^x	1.02
60		60
	−	3.28103079
1		1
	)	2.28103079
	÷	456.206158
.02		.02
	=	22810.3079

The amount in the account will be \$22,810.31, rounded to two decimal places. ■

Calculators and Significant Digits*

One simple method for specifying the accuracy of a number is to state how many **significant digits** it has. The significant digits in a number are the ones from the first nonzero digit to the last nonzero digit (reading from left to right). Thus 1074 has four significant digits, 1070 has three, 1100 has two, and 1000 has one significant digit.

This rule may sometimes lead to ambiguities. For example, if a distance is 200 km to the nearest kilometer, then the number 200 really has three significant digits, not just one. This ambiguity is avoided if we use *scientific notation*—that is, if we express the number as a multiple of a power of 10:

$$2.00 \times 10^2$$

Number	Significant digits
12,300	3
3000	1
3000.58	6
0.03416	4
4200	2
4.2×10^3	2
4.20×10^3	3

When working with approximate values, students often make the mistake of giving a final answer with **more** significant digits than the original data. This is incorrect because you cannot "create" precision by using a calculator. The final result can be no more accurate than the measurements given in the problem. For example, suppose we are told that the two shorter sides of a right triangle are measured to be 1.25 and 2.33 inches long. By the Pythagorean Theorem, we find, using a calculator, that the hypotenuse has length

$$\sqrt{1.23^2 + 2.33^2} \approx 2.644125564 \text{ in.}$$

But since the given lengths were expressed to three significant digits, the answer cannot be any more accurate. We can therefore say only that the hypotenuse is 2.64 in. long, rounding to the nearest hundredth.

In general, the final answer should be expressed with the same accuracy as the *least-*accurate measurement given in the statement of the problem. The following rules make this principle more precise.

Rules for Working with Approximate Data

1. When multiplying or dividing, round off the final result so that it has as many *significant digits* as the given value with the fewest number of significant digits.

---

*Reprinted with permission from Stewart/Redlin/Watson–Mathematics for Calculus, 1989

2. When adding or subtracting, round off the final result so that it has its last significant digit in the *decimal place* in which the least-accurate given value has its last significant digit.
3. When taking powers or roots, round off the final result so that it has the same number of *significant digits* as the given value.

As an example, suppose that a rectangular table top is measured to be 122.64 in. by 37.3 in. We express its area and perimeter as follows:

$$\text{Area} = \text{length} \times \text{width} = 122.64 \times 37.3 \approx 4570 \text{ in.}^2 \quad \text{(three significant digits)}$$

$$\text{Perimeter} = 2(\text{length} + \text{width}) = 2(122.64 + 37.3) \approx 319.9 \text{ in.} \quad \text{(tenths digit)}$$

Note that in the formula for the perimeter, the value 2 is an exact value, not an approximate measurement. It therefore does not affect the accuracy of the final result. In general, if a problem involves only exact values, we may express the final answer with as many significant digits as we wish.

Note also that to make the final result as accurate as possible, you should wait until the last step to round off your answer. If necessary, use the memory feature of your calculator to retain the results of intermediate calculations.

Exercises

In Exercises 1–10, write the given expression as a product. Each expression is in the form $a^n - b^n$, where n is a positive integer.

1. $x^7 - y^7$

2. $16x^4 - 81y^4$

3. $8x^3 - 125y^3$

4. $\dfrac{x^2}{4} - 9y^2$

5. $32a^5 - 243c^5$

6. $x^6 - 64$

7. $x^3 + y^3$ (*Hint:* $x^3 + y^3 = x^3 - (-y^3).$)

8. $243x^5 + 32y^5$

9. $\dfrac{x^3}{8} + 27y^3$

10. $32 - a^5$

11. Write the row of Pascal's triangle that corresponds to $n = 6$.

12. Same as Exercise 11 for $n = 7$.

For Exercises 13–22, write the expansion of the given expressions.

13. $(x + y)^6$

14. $(x - y)^6$

15. $(x + y)^7$

16. $(x - y)^7$

17. $(2x + 3y)^3$

18. $(2x - 3y)^3$

19. $(3a - 2b)^4$

20. $\left(\dfrac{x}{2} + \dfrac{y}{3}\right)^5$

21. $\left(2x + \dfrac{y}{2}\right)^8$

22. $\left(2x - \dfrac{y}{2}\right)^8$

In Exercises 23–32, a term of the expansion of $(a + b)^n$ for some positive integer n is given.

a. Find the value of n.
b. Find the next term.
c. Find the preceding term.

23. $495a^8b^4$

24. $286a^3b^{10}$

25. $1001a^4b^{10}$

26. $5005a^6b^9$

27. $6188a^5b^{12}$

28. $3876a^4b^{15}$

29. $4845a^{16}b^4$

30. $12{,}870a^8b^8$

31. $6435a^7b^8$

32. $715a^9b^4$

33. The first term of an arithmetic progression is 3 and the 50th term is 101. What is the sum of the first 50 terms of the progression?

34. The first term of an arithmetic progression is -4 and the 100th term is 293. What is the sum of the first 100 terms of the progression?

35. What is the sum of the first 50 positive integers? (*Hint:* Note that 1, 2, 3, 4, 5, . . . is an arithmetic progression.)

36. What is the sum of the first 850 positive integers?

37. Compare the sums
$1^3 + 2^3 + 3^3 + \cdots + n^3$ and $1 + 2 + 3 + \cdots + n$
for the following values of *n*:

   a. $n = 2$ **b.** $n = 3$

   c. $n = 4$ **d.** $n = 5$

   e. $n = 6$ **f.** $n = 7$

Do you notice a pattern? Recalling that
$1 + 2 + 3 + \cdots + n = \frac{(n + 1)}{2}n$,
guess a formula for
$1^3 + 2^3 + 3^3 + \cdots + n^3$.

38. Given a sequence of numbers $a_1, a_2, a_3, \ldots$, the sum $a_1 + \cdots + a_n$ is called the *nth partial sum of the sequence*.
   a. Write the first 20 odd positive integers.
   b. Find the 20 partial sums. Do you notice a pattern?

Using a calculator, find the value of each of the following expressions.

39. $53(47) + 36(72)$

40. $23.4(64.3) + 17.2(41.7)$

41. $72(23.5) + 64(90.2) - 13.3(14.7)$

42. $[17.3(82.5) + 36.4(17.9)]/82.3$

43. $[23.5(45.3) - 17.6(34.8)]/17.5$

44. $[47.6(19.8) + 18.3(23.4)]/(17.4 + 35.6)$

45. $[59.5(23.6) + 27.5(41.7)]/(19.8 + 64.2)$

46. $[58.3(39.6) + 32.4(37.5)]/(23.6 + 17.4)$

47. $45(13.2 + 17.8) - 16(14.9 + 13.1)$

48. $27(24.6 + 42.9) + 25(23.4 + 19.1)$

49. $32^2 + 53^2$ **50.** $41^2 + 67^2$

51. $45^2 - 78^2$ **52.** $3^4 + 5^3$

53. $5.2^5 - 7.4^4$ **54.** $21.1^4 + 35.3^3$

55. $(31.2^2 + 14.7^3)6.2^3$

56. $(3.5^2 - 3.4^4)(3.4^2 - 5.2^3)$

57. $(8.3^2 + 63.2^3)(51.3^2 + 63.2^3 - 17.4^4)$

58. $(1.05^{21} + 1.02^{17})(2.1 + 3.2^5)$

59. $(1.04^{40} - 1.03^{38})(1.02^{12} - 23^3)$

60. $\dfrac{2^5 + 3^6}{4^3 + 6^2}$ **61.** $\dfrac{5.1^5 + 6.2^4}{3.2^2 + 4.1^3}$

62. $\dfrac{3.2^6 + 2.7^4}{5.2^4 - 2.6^3}$ **63.** $\dfrac{32^3 + .43^4}{5.2^2 + 4.5^4}$

64. $\dfrac{3.2^6 + 4.3^6}{3.14^2 + .6^4}$ **65.** $\dfrac{4^{12} + 1.02^{42}}{1.04^{25} - 13.2}$

66. $\dfrac{3.4^{10} + 1.05^{35}}{1.03^{15} - 1.01^{16} + 23.72}$

67. $\dfrac{1.6^{20} + 1.05^{30}}{1.015^{20} + 1.025^{10} - 2313.2}$

If P dollars earns interest at a rate of i (written as a decimal) per time period and if interest is compounded at the end of each of n periods, then the balance in the account will be $P(1 + i)^n$. Use this formula in working Exercises 68–71.

68. If \$500 is invested at 6%, what is the amount after 15 years if interest is compounded **a.** annually, **b.** semiannually, **c.** quarterly, and **d.** monthly? Write your answer to the nearest cent.

69. If \$3500 is invested at 8%, what is the amount after 15 years if interest is compounded **a.** annually, **b.** semiannually, **c.** quarterly, and **d.** monthly? Write your answer to the nearest cent.

70. Beth wishes to invest \$750 for a period of 5 years. Bank A pays 5.85% interest compounded monthly and bank B pays 6% interest compounded semiannually. Which bank should she select? Assuming she selects the bank that will pay the highest return, what will her balance be at the end of 5 years?

71. Gregory wishes to invest \$500 for a period of 8 years. Bank A pays 6.95% interest compounded monthly and bank B pays 7% interest compounded quarterly. Which bank should he select? Assuming he selects the bank that will pay the highest return, what will his balance be at the end of 8 years?

Use the formula presented in Example 7 in working Exercises 72–75.

72. Mr. Planner deposits \$500 at the end of each month in a bank paying 6% interest compounded monthly. What will the balance in the account be after 12 years?

73. Ms. Wise deposits \$350 every 6 months in a bank paying 9% interest compounded semiannually. What will the balance in the account be after 20 years?

74. Miss Prudente deposits the same amount at the end of every 3 months in a bank paying 8% interest compounded quarterly. She has $30,200.99 in her account after 10 years. How much did she deposit each time?

75. J. De Largent deposits the same sum at the end of each month in a Montreal bank paying 9% interest compounded monthly. He has $77,313.47 in his account after 12 years. How much did he deposit at the end of each month?

76. Use a calculator to find 231(9) + 507(5). When the answer is displayed, turn your calculator upside down to find out how you should not feel while driving.

77. If you wish to know who one of the largest customers of the Arab countries is, perform the following function on your calculator and turn the calculator upside down before you read it: 78.25469(5) + 15(21.3).

Mathematical Models

1

1.1 Elementary Introduction to Sets

When we use mathematics to solve a problem in any field, the first step is to construct a mathematical model for the problem at hand. For example, when solving a story problem in basic algebra, it is necessary to write an equation (the mathematical model) which represents the conditions stated in the problem. Many applications of mathematics involve collections of objects; these collections are called *sets*. Mathematical models in such cases will, therefore, involve sets. It is important that you have some knowledge of elementary facts about sets and about operations involving sets. This section provides some basic terminology and notation which will be used frequently in later sections.

Set Description

> **DEFINITION 1.1:**   A *set* is a collection of objects.

To describe a set, we usually state a necessary and sufficient property* that an object must have to be a member of that set. For example, the set of all smart people is not well defined; one person may be declared smart by one teacher and slow by another. On the other hand, the set of the first five counting numbers is well defined.

In general, we use capital letters to denote sets and small letters to denote members of sets. If A is the name of a set and b is the name of an object, the sentence "b is an element of A" is abbreviated $b \in A$. Similarly, $b \notin A$ is an abbreviation for "b is not an element of A." The terms *member* and *element* are synonymous.

When a set has a small number of elements, we can describe it by displaying the names of its members, listing each only once, separating consecutive ones by commas and putting the list within braces. This technique is often called the *roster method*. For example, the set whose members are the first three counting numbers is described as follows:

$$\{1, 2, 3\}$$

A collection with no members at all is also a set. It is called *the empty set* (it can be shown that there is only one such set). The symbol $\emptyset$ (or $\{\ \ \}$) is used to denote the empty set.

---

*If p and q are properties, and q is satisfied whenever p is satisfied, we say that p is sufficient for q. On the other hand, if q must be true for p to be true, we say that q is necessary for p. For example, "having wings" is a necessary property to "be a normal bird," but it is not sufficient (bats and bees have wings also). However, "having feathers" is a property that is both necessary and sufficient to "be a normal bird" in the animal world. As another example, to "be a woman" is a sufficient condition to "be a human," but it is not necessary (men are human beings also!).

In algebra we often use letters to represent numbers. For example, you have seen statements such as

$$(a + b)^2 = a^2 + 2ab + b^2$$
$$3x + 6 = 7 - 10x$$

and

$$x + y = y + x$$

At first, statements such as these may seem strange. However, a moment of reflection should reveal that even in the statement $2 + 8 = 4 + 6$, the 2, 8, 4, and 6 are symbols which represent some specific numbers. On the other hand, in the statement

$$x + 5 = 2x + 1$$

x may be regarded as a symbol that does not denote a specific number; it may be replaced by any one member from a certain set of numbers. In general, such a symbol is called a *variable*. In any discussion in which a variable is involved, we must have a set so that we may replace the variable by the name of some member of that set. This set is called the *replacement set* for the variable. For example, if we consider the statement $x + 5 = 2x + 1$ with replacement set $\{3, 4, 5\}$, we may replace x by any of the numbers 3, 4, or 5. If we replace x by 3, $2x$ must be replaced by 6 and we get

$$3 + 5 = 6 + 1$$

which is false. If we replace x by 4, $2x$ must be replaced by 8 and we get

$$4 + 5 = 8 + 1$$

which is true. Hence, 4 is a *solution* of the sentence. If x is replaced by 5, again we get a false statement.

 Be careful not to replace x by one number on one side of the equal sign and by a different number on the other side.

 DEFINITION 1.2:   If a sentence involves a variable, it is called an *open sentence*.

Often, the replacement set which should accompany the open sentence is not given, but its membership is understood from context. For example, the replacement set for the open sentence "x is the president of General Motors Co." is understood to be a set of names of people.

When a set has many members, the roster method is not a very practical way to describe it. It is common practice to use an open sentence to describe the set whose members are exactly those solutions of the open sentence. We use the following procedure:

If $p(x)$ (read "p of x") is an open sentence in the variable x, the symbol

$$\{x \mid p(x)\}$$

(read "the set of all x's such that $p(x)$") describes the set of solutions of the open sentence. Of course, the replacement set should be described if it is not clear from context.

EXAMPLE 1 List five members of the set $\{x \mid x$ is, or has been, a professional basketball player$\}$.

Solution Jerry West, Kareem Abdul Jabbar, Elgin Baylor, Jack Sikma, and Ralph Sampson are five members of the set. There are, of course, many others.
Do Exercise 21. ■

EXAMPLE 2 Describe the set $\{3, 4\}$ using an open sentence.

Solution $\{3, 4\} = \{y \mid y$ is a whole number and $2 < y < 5\}$, or

$$\{3, 4\} = \{t \mid (t - 3)(t - 4) = 0\}$$ ■

EXAMPLE 3 Let $A = \{u \mid u$ is a real number and $2 < u < 5\}$. Which of the following statements is true?

a. $\pi \in A$ **b.** $\dfrac{10}{3} \in A$ **c.** $6 \in A$

Solution **a.** $\pi \in A$ is true since π is a real number larger than 2 and less than 5.

b. $\dfrac{10}{3} \in A$ is true since $\dfrac{10}{3}$ is a rational (hence real) number which is greater than 2 and less than 5.

c. $6 \in A$ is false since 6 is not less than 5.

Do Exercise 31. ■

Elementary Set Algebra

> **DEFINITION 1.3:** Suppose that A and B are sets and that each element of A is also an element of B, we then say that A is a *subset* of B. We use the symbol $A \subseteq B$ to denote this.

For example, if $A = \{1, 2, 3\}$ and $B = \{0, 1, 2, 3, 4\}$, then $A \subseteq B$ (read: "A is a subset of B") since each element of A is also in B. According to the definition, it is clear that if A is any set, then $A \subseteq A$. Also if A, B, and C are sets, such that $A \subseteq B$ and $B \subseteq C$, then $A \subseteq C$.

> **DEFINITION 1.4:** If the set A is a subset of the set B, but B has at least one element which is not in A, then we say that A is a *proper subset* of B and we write $A \subset B$.

It can be shown that if A is an arbitrary set, the empty set is a subset of A. Symbolically, if A is any set, $\emptyset \subseteq A$.

EXAMPLE 4   Let $A = \{a, b, c\}$. List all the subsets of A.

Solution   The subsets of A are $\emptyset$, $\{a\}$, $\{b\}$, $\{c\}$, $\{a, b\}$, $\{a, c\}$, $\{b, c\}$, and A.
Do Exercise 37.   ■

In general, it is true that if a set has n elements, then it has 2^n subsets. Set A has three elements and eight subsets: $2^3 = 8$. There is an interesting relationship between the number of subsets of a set with n members and the coefficients in the expansion of $(a + b)^n$. In the preceding example, we have shown that the set $\{a, b, c\}$ has *one* subset with no element (the empty set), *three* subsets each with one member, *three* subsets each with two members, and *one* subset with three members (the set itself). In the "Getting Started" section we saw that the coefficients of the expansion of $(a + b)^3$ are 1, 3, 3, and 1. We also saw that the coefficients of the expansion of $(a + b)^5$ are 1, 5, 10, 10, 5, and 1. It is also true that a set with five members has *one* subset with no member, *five* subsets each with one member, *ten* subsets each with two members, *ten* subsets each with three members, *five* subsets each with four members, and *one* subset with five members. As expected, $1 + 5 + 10 + 10 + 5 + 1 = 32 = 2^5$, the total number of subsets. (See Exercise 39.)

We will now discuss several ways to derive new sets from given sets.

DEFINITION 1.5:   The set whose members are the elements under consideration in a given discussion is called the *universe*.

For example, in an algebra course, the universe may be the set of complex numbers; while in a sociology course, the universe may be the set of all people.

DEFINITION 1.6:   If A is a set and U is the universe, the set which consists of all members of U which are not in A is called the *complement of A* (with respect to U) and is denoted A'.

For example, if U is the set of all counting numbers and A is the set of even numbers, then A' is the set of odd numbers.

DEFINITION 1.7:   The *union* of the sets A and B is the set of all elements that are in A, or in B, or in both. We use the symbol $A \cup B$ to denote this set.

That is,

$$A \cup B = \{x \mid x \in A, \text{ or } x \in B, \text{ or both}\}$$

DEFINITION 1.8: The *intersection* of the sets A and B is the set of all elements that belong to both A and B. We use the symbol $A \cap B$ to denote this set.

In other words,

$$A \cap B = \{x \mid x \in A \text{ and } x \in B\}$$

EXAMPLE 5 Let $A = \{1, 2, 3\}$ and $B = \{2, 3, 4, 5\}$. Find $A \cup B$ and $A \cap B$.

Solution $A \cup B = \{1, 2, 3, 4, 5\}$ and $A \cap B = \{2, 3\}$.
Do Exercise 53. ■

Verify that if A and B are any sets, then $A \cap B \subseteq A \cup B$.

EXAMPLE 6 Let the universe be the set $\{i, j, k, l, m, n, o, p, q, r\}$, $R = \{j, l, n, o, q, r\}$, $S = \{i, j, q, r\}$, $T = \{j, m, o\}$, and $V = \emptyset$. Find:

  a. R' **b.** V' **c.** $(R \cup T)'$

  d. $R' \cap T'$ **e.** $R \cap S$ **f.** $R \cap T$

  g. $R \cap (S \cup T)$ **h.** $(R \cap S) \cup (R \cap T)$

Solution   **a.** $R' = \{j, l, n, o, q, r\}' = \{i, k, m, p\}$

  b. $V' = \emptyset' = \{i, j, k, l, m, n, o, p, q, r\}$

  c. $(R \cup T)' = (\{j, l, n, o, q, r\} \cup \{j, m, o\})' = \{j, l, m, n, o, q, r\}' = \{i, k, p\}$

  d. $R' \cap T' = \{i, k, m, p\} \cap \{i, k, l, n, p, q, r\} = \{i, k, p\}$

  e. $R \cap S = \{j, l, n, o, q, r\} \cap \{i, j, q, r\} = \{j, q, r\}$

  f. $R \cap T = \{j, l, n, o, q, r\} \cap \{j, m, o\} = \{j, o\}$

  g. $R \cap (S \cup T) = \{j, l, n, o, q, r\} \cap (\{i, j, q, r\} \cup \{j, m, o\}) =$
    $\{j, l, n, o, q, r\} \cap \{i, j, m, o, q, r\} = \{j, o, q, r\}$

  h. $(R \cap S) \cup (R \cap T) = \{j, q, r\} \cup \{j, o\} = \{j, o, q, r\}$

Do Exercises 49, 57, and 59. ■

You should compare the answers in **c** and **d**, and in **g** and **h** of the preceding example.

DEFINITION 1.9: If two sets have no common elements, their intersection is the empty set. In that case, we say that the two sets are *disjoint*.

Exercise Set 1.1

In Exercises 1–10, use the roster method to describe the given collections.

1. The first six counting numbers.

2. The first ten counting numbers.

3. The first counting number.

4. The last five letters of the English alphabet.

5. The states in the United States with names beginning with the letter "W."

6. All counting numbers between 15 and 20.

7. All letters of the English alphabet between b and f.

8. All counting numbers between 3 and 5.

9. The presidents of the United States who succeeded John F. Kennedy.

10. The vowels of the English alphabet.

In Exercises 11–20, an open sentence and a replacement set for the variable are given. In each case, determine the solution set. Recall that each solution must be a member of the replacement set.

11. $2x + 3 = x + 10$; $\{5, 6, 7, 8\}$

12. $x + 5 = 2x - 1$; $\{4, 5, 6, 7, 8\}$

13. $\dfrac{x}{2} = x - 3$; $\{4, 5, 6\}$

14. $x^2 + 5x - 6 = 0$; $\{0, 1, 2\}$

15. $3x + 10 = 5x + 6$; $\{1, 2, 3\}$

16. $5x - 6 = 2x + 3$; $\{1, 2, 3, 4\}$

17. $4x + 1 = 5x - 4$; $\{2, 6, 11\}$

18. $3x + 7 = 5x + 1$; $\{1, 2, 3, 4\}$

19. $6 - 3x = 5x - 2$; $\{0, 1, 2\}$

20. $x^3 - 8 = 0$; $\{0, 2, 5\}$

In Exercises 21–30, a set is described. In each case, list three members of the given set.

21. $\{x \mid x$ has been a president of the United States$\}$

22. $\{x \mid x$ is a professional football player$\}$

23. $\{y \mid y$ is a whole number larger than 2 and less than 5$\}$

24. $\{t \mid t$ is a real number larger than 5$\}$

25. $\{z \mid z$ is a counting number$\}$

26. $\{u \mid (u - 1)(u - 3)(u - 5)(u - 7) = 0\}$. (*Hint:* Recall that a product of real numbers is 0 if, and only if, at least one of the factors is 0.)

27. $\{x \mid x$ is a whole number less than 5$\}$

28. $\{x \mid x$ is an American movie actor$\}$

29. $\{y \mid y$ is a television personality$\}$

30. $\{s \mid s$ is a professional baseball player$\}$

In Exercises 31–35, indicate whether the given statement is true or false.

31. $2 \in \{x \mid x$ is a whole number greater than 3$\}$

32. $5 \in \{x \mid (x - 1)(x - 5)(x - 7) = 0\}$

33. $\pi \in \{y \mid y$ is a real number greater than 2$\}$

34. $\dfrac{9}{2} \in \{t \mid t$ is a positive real number$\}$

35. $m \in \{x \mid x$ is a vowel in the English alphabet$\}$

36. Let $A = \{1\}$. How many subsets does this set have? List all the subsets.

37. Same as Exercise 36 for $A = \{1, 2\}$.

38. Same as Exercise 36 for $A = \{1, 2, 3, 4\}$.

39. Same as Exercise 36 for $A = \{1, 2, 3, 4, 5\}$.

40. How many proper subsets does the set $\{1\}$ have?

41. Same as Exercise 40 for the set $\{1, 2\}$.

42. Same as Exercise 40 for the set $\{1, 2, 3\}$.

43. Same as Exercise 40 for the set $\{1, 2, 3, 4\}$.

44. Same as Exercise 40 for the set $\{1, 2, 3, 4, 5\}$.

45. If a set has n elements, how many proper subsets does it have?

46. If a set has four elements, how many subsets does it have

    a. each with no element?

    b. each with one element?

    c. each with two elements?

    d. each with three elements?

    e. each with four elements?

47. If a set has six elements, how many subsets does it have

    a. each with no element?

    b. each with one element?

    c. each with two elements?

    d. each with three elements?

    e. each with four elements?

    f. each with five elements?

    g. each with six elements?

48. Let S be a set with n members and let k be an integer between 0 and n. Explain why the number of subsets, each with k elements, is the same as the number of subsets each with $n - k$ elements.

In Exercises 49–60 the universe is the set $\{0, 1, 2, 3, 4, 5, 6, 7, 8, 9\}$, $A = \{1, 2, 3, 4, 5, 6\}$, $B = \{2, 5\}$, $C = \{3, 4, 8\}$, *and* $D = \emptyset$.

49. Find A'.

50. Find B'.

51. Find C'.

52. Find D'.

53. Find $A \cup B$ and $A \cap B$.

54. Find $A \cup C$ and $A \cap C$.

55. Find $B \cup C$ and $B \cap C$.

56. Find $A \cap (B \cup C)$.

57. Find $(A \cap B)'$ and $A' \cup B'$.

58. Find $(B \cup C)'$ and $B' \cap C'$.

59. Find $(A \cap C)'$ and $A' \cup C'$.

60. Find $A \cup (B \cap C)$ and $(A \cup B) \cap (A \cup C)$.

1.2 Venn Diagrams

In this section we illustrate the concepts of the preceding section geometrically. We also show how sets can be used to solve certain survey problems.

Description of Venn Diagrams

The geometric illustration is made using a *Venn diagram*. If the universe for a certain discussion is set U, we represent that set by a rectangle. In general, each set in that discussion is represented by points on a plane bounded by one or more simple closed curves. Circles, ellipses, squares, and triangles are examples of simple closed curves. (See Figure 1.1a.) On occasion, it is useful to represent a single set using several simple closed curves.

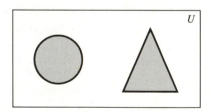

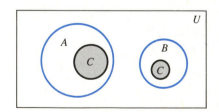

C is represented by two disjoint circles

FIGURE 1.1

            (a)   (b)

For example, we have represented two disjoint sets A and B with two separate circles. We want to represent a third set C and to emphasize that C has points in common with both A and B, but it has no points outside of either A or B. In that case, we represent C by the union of two circles. (See Figure 1.1b.) The only stipulation is that all configurations representing sets must lie within the rectangle which represents the universe.

We should understand that there are infinitely many points inside a simple closed curve, while there may be only a finite number of elements in the set represented by that simple closed curve. This creates no difficulty, because a Venn diagram is often used only to illustrate the relationship between sets, and not their sizes. For example, in Figure 1.2a we illustrate the fact that the set A is a subset of the set B, in Figure 1.2b we show that sets A and B are disjoint, while Figure 1.2c displays the set which is the intersection of the sets A and B.

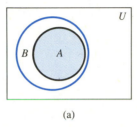
(a)

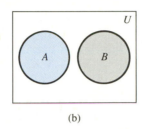
(b)

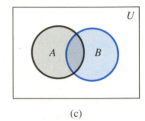
(c)

FIGURE 1.2

If it is important to know how many elements are in a set, we place a number within the simple closed curve which represents that set. Since we usually have several sets represented in a Venn diagram, we have several simple closed curves and there is often a number within more than one of these. Therefore, the following convention is used: Any number placed in a Venn diagram gives the number of elements in the smallest of all the sets which are represented by the simple closed curves which enclose that number. As an illustration, three sets A, B, and C are represented by circles in Figure 1.3. The number 5 is within several simple closed curves, each of which represents a set. The smallest of these sets is the set of all points which belong to both A and B, but which do not belong to C. Hence, we know that there are five elements in the intersection of the sets A, B, and C'. The number 10 in the upper right corner is in the interior of the rectangle representing the universe and is in the exterior of each of the three circles. So the number 10 indicates that there are ten elements in the universe that are not in any of the sets A, B, or C.

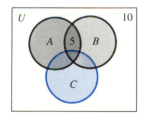

FIGURE 1.3

EXAMPLE 1 Use a Venn diagram to illustrate the fact that if A and B are arbitrary sets, then

$$(A \cup B)' = A' \cap B'$$

(See also Example 6(c, d) and Exercise 58 in the preceding section.)

Solution In Figure 1.4a the shaded region illustrates $A \cup B$, in Figure 1.4b it illustrates $(A \cup B)'$, in Figures 1.4c, 1.4d, and 1.4e, the shaded regions represent A', B', and $A' \cap B'$, respectively. Comparing the shaded regions in Figures 1.4b and 1.4e, we conclude that these regions are the same. Hence, we have

$$(A \cup B)' = A' \cap B'$$

Do Exercise 7.

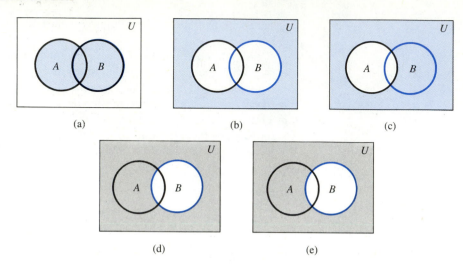

FIGURE 1.4

(a)   (b)   (c)

(d)   (e)   ■

The foregoing equality is one of *De Morgan's Laws*. It can be proved rigorously, but we shall restrict our argument to its geometric illustration given in Example 1.

Solving Survey Problems

Venn diagrams can be used in solving survey problems as illustrated in the following two examples.

EXAMPLE 2   A large university is planning to build a physical education complex and wants to know how to divide space for racquetball, squash, and tennis courts. To that end, 250 students were selected randomly and it was found that among these,

    55 play racquetball;
    25 play squash;
    65 play tennis;
    15 play racquetball and squash;
    10 play tennis and squash;
    25 play tennis and racquetball; and
     5 play all three games.

Use this data and a Venn diagram to answer the following questions.

a. How many of the students surveyed play none of the three games?
b. How many of them play only racquetball?
c. How many play racquetball and tennis but not squash?

Solution   Let R, S, and T be the sets of students surveyed who play racquetball, squash, and tennis, respectively. Let U (the universe) be the set of all students surveyed. We know that U has 250 members; R, S, and T have 55, 25, and 65 members, respectively. Further, $R \cap S$ has 15 members, $T \cap S$ has 10 members, $T \cap R$ has 25 members, and $R \cap S \cap T$ has 5 members. Since this latter set is the smallest of the sets we are considering, we begin with it and place a 5 in the region that is within all three circles. (See Figure 1.5a.) We know that there are 25 members in $T \cap R$. Hence, there are

25 members within the circles representing T and R, respectively. But five of them are already accounted for because they are in $R \cap T \cap S$. Thus, we place a 20 in the region that is within the circles representing T and R, but outside the circle representing S. (See Figure 1.5b.) The reason for each of the other entires should now be clear. (See Figure 1.5c.)

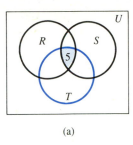
(a)

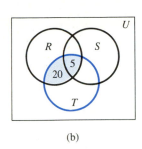
(b)

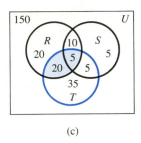
(c)

FIGURE 1.5

a. Adding all the numbers which have been placed within at least one circle, we get 100. Thus, 100 of the students play at least one of the three games. Therefore, 150 of the students surveyed play none of the three games since $250 - 100 = 150$.

b. The number placed within the circle representing R but outside the other two circles is 20. We conclude that 20 of the students surveyed play only racquetball.

c. The number placed in the region within the circles representing R and T but outside the other circle is also 20. Thus, 20 of the students surveyed play racquetball and tennis but not squash.

Do Exercise 13.

EXAMPLE 3   A company hired a college student to survey a sample of the student population at a certain eastern university. The student claims that he selected 200 students at random and obtained the following results. Among the 200 students surveyed,

    120 drink Coke;
    100 drink Seven-Up;
    115 drink Pepsi;
     55 drink Coke and Seven-Up;
     60 drink Coke and Pepsi;
     50 drink Seven-Up and Pepsi; and
     35 drink all three.

The manager of the project received a report that the hired student was seen spending a great deal of time in the cafeteria socializing and that perhaps he had made up the data. The manager, who had taken a mathematics for business students course while she was in college, used a Venn diagram to check the consistency of the data. Shortly after that, she fired the student. What was her reason?

Solution   Let C, S, and P denote the sets of students who reportedly drink Coke, Seven-Up, and Pepsi, respectively. Then the sets C, S, P, $C \cap S$, $C \cap P$, $S \cap P$, and $C \cap S \cap P$ should have 120, 100, 115, 55, 60, 50, and 35 members, respectively. We begin by placing a 35 in the region that is within the three circles representing the three sets,

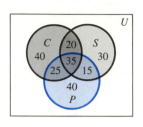

because that region represents the intersection of the three sets. (See Figure 1.6.) We then place a 20 in the region that is within the circles representing C and S, but outside the circle representing P. The justification for this entry is that we must have 55 elements in the region within the circles representing C and S, but 35 are already accounted for in the region within all three circles and $55 - 35 = 20$. We continue in this manner and place numbers in all the appropriate regions. Adding all the entries in Figure 1.6, we get

$$40 + 20 + 30 + 25 + 35 + 15 + 40 = 205$$

Since the hired student claims that he surveyed only 200 students, either he has made up the data or his work was performed carelessly. Hence, he deserves to be fired!

FIGURE 1.6

Do Exercise 19. ■

Exercise Set 1.2

In Exercises 1–6, A, B, and C are sets. Use a Venn diagram to illustrate the given set.

1. $A \cap (B \cup C)$

2. $A' \cap (B \cup C)$

3. $A \cup (B' \cap C)$

4. $(A \cap B) \cup C$

5. $(A \cap B)' \cup C'$

6. $B' \cup (A \cap B)$

In Exercises 7–12, use a Venn diagram to illustrate geometrically the truth of the given statement.

7. $(A \cap B)' = A' \cup B'$

8. $(A \cap B) \subseteq A$

9. $A \subseteq (A \cup B)$

10. $A \cap (B \cup C) = (A \cap B) \cup (A \cap C)$

11. $A \cup (B \cap C) = (A \cup B) \cap (A \cup C)$

12. $(A \cap B) \subseteq (A \cup B)$

13. Suppose that 350 students at a large western university were surveyed with the following results. Among these students,

33 read Scientific American;
40 read National Geographic;
43 read Playboy;
20 read National Geographic and Playboy;
15 read Scientific American and Playboy;
19 read Scientific American and National Geographic; and
7 read all three magazines.

Use a Venn diagram to find how many of the students surveyed

a. read only Scientific American.

b. read Scientific American and Playboy but do not read National Geographic.

c. read none of the three magazines.

14. A survey of 400 people in a small town in eastern Washington provided the following results. Among these 400 people,

38 drink coffee;
48 drink tea;
39 drink mineral water;
15 drink both coffee and tea;
13 drink both coffee and mineral water;
18 drink both tea and mineral water; and
5 drink all three.

Using a Venn diagram, find how many of the people surveyed drink

a. mineral water but neither coffee nor tea.

b. tea and coffee but not mineral water.

c. none of the three.

15. At the end of their freshman year, 200 randomly selected students at a small western university were surveyed. They were asked whether they had taken English, mathematics, or philosophy during the year. The following results were recorded:

20 had taken none of the three subjects;
50 had taken only English;
30 had taken only mathematics;
20 had taken only philosophy;
10 had taken English and philosophy but not mathematics;
40 had taken English and mathematics but not philosophy; and
15 had taken mathematics and philosophy but not English.

Using a Venn diagram, determine how many among the students surveyed took

a. all three subjects.
b. English and mathematics.
c. English.
d. mathematics.
e. philosophy.

16. A Safety Inspection Center selected 350 cars randomly and checked their tires, brakes, and headlights. It was found that

43 had defective tires;
45 had defective brakes;
24 had defective lights;
23 had defective tires and brakes;
17 had defective brakes and lights;
12 had defective tires and lights; and
10 had defective tires, lights, and brakes.

Using a Venn diagram, determine how many of the cars inspected had

a. only defective tires.
b. defective tires and brakes but safe lights.
c. safe brakes, lights, and tires.

17. The director of the Sports program at a small private university randomly selected 250 students and collected the following data. Among these 250 students,

53 smoked;
78 drank alcoholic beverages;
58 played some sport;
25 smoked and played some sport;
30 drank alcoholic beverages and played some sport;
28 smoked and drank alcoholic beverages; and
10 did all three.

Using a Venn diagram, determine how many of the 250 students surveyed

a. play some sport but do not smoke or drink alcoholic beverages.
b. play some sport and drink some alcoholic beverages but do not smoke.
c. do none of the three.

18. A clinic hired a college student to survey a sample of the student population at a certain western university. The student claims that she selected 300 of her peers at random and obtained the following results. Among the 300 students surveyed,

134 eat red meat;
183 eat fish;
151 eat poultry;
63 eat red meat and fish;
50 eat red meat and poultry;
63 eat fish and poultry; and
25 eat all three.

The manager of the project saw that the hired student was spending a lot of time in the cafeteria socializing and he worried that the student had made up the data. The manager used a Venn diagram to check the consistency of the data. Shortly after that, he fired the student. What was his reason?

19. The manager of a large store noticed that one of the young clerks was working hard and thought to promote him to assistant manager. As a test, she asked him to survey 230 customers and he came up with the following data. Among the 230 people who entered the store,

173 were males;
193 made a purchase;
113 used a credit card;
123 were males who made a purchase; and
63 males used a credit card.

The manager used the data and a Venn diagram to determine how many, among the people who entered the store,

a. were males who did not make a purchase.
b. were males who paid cash for their purchases.
c. were females who used credit cards.
d. were females who paid cash for their purchases.

She then decided that the young clerk was not ready for the promotion. What were her answers to the four calculations and why did she decide not to promote the young clerk?

1.3 Equations in One Variable

Equivalent Equations

DEFINITION 1.10: An open sentence in a variable that is a statement of equality is called an *equation in one variable*.

For example,

$$2x + 9 = 13x - 24$$

and

$$y^2 - 5y + 6 = 0$$

are equations in one variable. In general, with each equation in one variable we must have a replacement set. That is, we must have a set of numbers to use as replacements for the variable. The replacement set for a variable is usually not described. In this book, however, unless otherwise specified, it is the *largest* set of real numbers that may be used as replacements for the variable so that each replacement yields quantities that are both defined and real. For example, the replacement set for the equation

$$\frac{3}{x - 2} + \frac{x + 4}{2x + 1} = 4$$

is the set of all real numbers except 2 and $\frac{-1}{2}$, since replacing x by either of these two numbers would yield a division by 0, which is not defined. As another example, the replacement set for the equation

$$\sqrt{x - 3} = 2x - 21$$

is the set of all real numbers greater than or equal to 3, because replacing x by a number less than 3 would yield a negative quantity under the radical sign and square roots of negative numbers are not real numbers.

> **DEFINITION 1.11:**   A *solution* of an equation is a member of the replacement set which yields a true statement when it is used as a replacement for the variable.

For example, the number 3 is a solution of the equation

$$2x + 9 = 13x - 24$$

since

$$2(3) + 9 = 13(3) - 24$$

is true. Also, -2 and 5 are solutions of the equation

$$y^2 - 3y - 10 = 0$$

since

$$(-2)^2 - 3(-2) - 10 = 0 \text{ and } 5^2 - 3(5) - 10 = 0$$

are both true.

The set of all solutions of an equation is called the *solution set* of the equation. It is important to note that the solution set of an equation depends on the replacement set. For example, the solution set of the equation

$$x^2 + 1 = 0$$

is empty if the replacement set is the set of real numbers, but has two members if the replacement set is the set of complex numbers.

> **DEFINITION 1.12:** If the solution set of an equation is equal to its replacement set, the equation is called an *identity*; otherwise, it is called a *conditional equation*.

For example,

$$(x + 1)^2 = x^2 + 2x + 1$$

is an identity, but

$$3x + 2 = 5x - 12$$

is a conditional equation.

> **DEFINITION 1.13:** Two equations whose replacement sets are the same are said to be *equivalent* if, and only if, their solution sets are equal. The process of finding the solution set of an equation is called *solving the equation*.

In general, when given an equation, we attempt to form a sequence of equivalent equations until we get an equation which is simple enough so that its solution set is found by inspection. Since the last equation is equivalent to the original one, its solution set is also the solution set of the original equation. For example, the following is a sequence of equivalent equations,

$$5x - 11 = 3x - 3$$
$$2x - 11 = -3$$
$$2x = 8$$
$$x = 4$$

Obviously, $\{4\}$ is the solution set of the last equation. Therefore, $\{4\}$ is also the solution set of the equation

$$5x - 11 = 3x - 3$$

> If an equation is given, an equivalent equation will be found by doing any one of the following:
>
> 1. Add the same polynomial to both sides.
> 2. Subtract the same polynomial from both sides.
> 3. Multiply both sides by the same nonzero number.
> 4. Divide both sides by the same nonzero number.

Be aware that although any polynomial may be added to, or subtracted from, both sides of an equation, you should not multiply or divide both sides of an equation by a quantity involving the variable. Consider, for example, the following equation:

$$(x - 1)(x - 3) = 0$$

It is obvious that its solution set is $\{1, 3\}$. If we divide both sides by $(x - 1)$, we obtain the equation

$$x - 3 = 0$$

with solution set $\{3\}$. We have lost one solution. The new equation is *defective* relative to the previous one. Now, if we multiply both sides of the equation

$$(x - 1)(x - 3) = 0$$

by $(x - 7)$, we obtain the equation

$$(x - 1)(x - 3)(x - 7) = 0$$

whose solution set is $\{1, 3, 7\}$. We have gained one solution. The new equation is *redundant* with respect to the previous one.

Linear Equations in One Variable

An equation that can be written in the form

$$ax + b = cx + d$$

where a, b, c, and d are real numbers and either a or c does not equal 0, is called a *linear equation in x*. This type of equation is easy to solve as is shown in the next two examples.

EXAMPLE 1   Solve the equation $3x + 5 = 7x - 11$.

Solution   Subtract 5 from both sides of the equation.

$$3x = 7x - 16$$

Now subtract $7x$ from both sides of the new equation.

$$-4x = -16$$

Dividing both sides of this last equation by -4 yields

$$x = 4$$

The solution set of the last equation is obviously $\{4\}$. Thus, the solution set of the equation $3x + 5 = 7x - 11$ is also $\{4\}$.
Do Exercise 7.   ■

EXAMPLE 2   Solve the equation

$$\frac{x + 12}{2} = \frac{2x + 10}{3} + 3$$

Solution   First multiply both sides of the equation by 6 in order to eliminate fractions, obtaining

$$3x + 36 = (4x + 20) + 18$$

Adding $-4x - 36$ to both sides we get

$$-x = 2$$

Multiplying both sides of the last equation by -1 yields

$$x = -2$$

The solution set of the last equation, and of the original equation, is $\{-2\}$.
Do Exercise 13.   ■

Quadratic Equations in One Variable

If a, b, and c are real numbers with $a \neq 0$, an equation that can be written in the form

$$ax^2 + bx + c = 0$$

is called a *quadratic equation*. As we pointed out in the preceding section, you have undoubtedly learned to solve such equations, as well as linear equations, in previous courses. We shall, however, give several examples to serve as a quick review of the available techniques.

EXAMPLE 3   Solve the equation

$$2x^2 + 5x - 18 = 0$$

by factoring.

Solution   The left side of the equation may be factored as follows:

$$(2x + 9)(x - 2) = 0$$

Using the fact that a product of two real numbers is equal to 0 if, and only if, at least one of the two numbers is equal to 0, we get

$$2x + 9 = 0 \quad \text{or} \quad x - 2 = 0$$

The solution set of the foregoing open sentence is easily seen to be $\left\{\frac{-9}{2}, 2\right\}$. Hence, the solution set of the original quadratic equation is also $\left\{\frac{-9}{2}, 2\right\}$.
Do Exercise 15.   ■

EXAMPLE 4   Solve the quadratic equation

$$3x^2 - 12x + 9 = 0$$

by completing squares.

Solution   Dividing both sides by 3 to change the leading coefficient to 1, we get

$$x^2 - 4x + 3 = 0$$

We then subtract 3 from both sides

$$x^2 - 4x = -3$$

The square of half the coefficient of x is 4, since $4 = \left(\frac{-4}{2}\right)^2$. Thus, we add 4 to both sides, so that the left side will be a perfect square. (See Exercise 56.) We obtain

$$x^2 - 4x + 4 = -3 + 4$$

which may be written

$$(x - 2)^2 = 1$$

This equation is equivalent to the open sentence

$$x - 2 = 1 \quad \text{or} \quad x - 2 = -1$$

Thus,

$$x = 3 \quad \text{or} \quad x = 1$$

It follows that the solution set of the given equation is $\{1, 3\}$.
Do Exercise 19.   ■

Using exactly the same steps as we did in the preceding example, we can show that if a, b, and c are real numbers with $a \neq 0$, the equations

$$ax^2 + bx + c = 0$$

and

$$x = \frac{-b \pm \sqrt{b^2 - 4ac}}{2a}$$

are equivalent. This last equation is called the *quadratic formula*.

EXAMPLE 5   Solve the equation of Example 4 using the quadratic formula.

Solution   For the equation of Example 4, $a = 3$, $b = -12$, and $c = 9$. Thus, the equation is equivalent to

$$x = \frac{-(-12) \pm \sqrt{(-12)^2 - 4(3)(9)}}{2(3)}$$

$$= \frac{12 \pm \sqrt{36}}{6}$$

$$= \frac{12 \pm 6}{6}$$

That is,

$$x = \frac{12 + 6}{6} = 3 \quad \text{or} \quad x = \frac{12 - 6}{6} = 1$$

Therefore, the solution set of the equation is $\{1, 3\}$.
Do Exercise 25.   ◢

Equations Involving Radicals

We have said earlier that multiplying both sides of an equation by a quantity which involves the variable may introduce extraneous solutions. However, it is sometimes necessary to do so as illustrated in the following example.

EXAMPLE 6 Solve the equation

$$\sqrt{x - 1} = x - 3$$

Solution To eliminate the radical on the left side of the equation, we must square both sides of the equation. In doing so, we multiply by a quantity involving the variable, which may introduce extraneous solutions. We obtain

$$x - 1 = x^2 - 6x + 9$$

This equation is equivalent to

$$x^2 - 7x + 10 = 0$$

which may be written

$$(x - 5)(x - 2) = 0$$

It is easy to see that the solution set of the last equation is $\{2, 5\}$. However, this last equation may be redundant with respect to the original one. Hence, we *must* check the validity of the solutions in the original equation. Replacing x by 2 in the given equation, we get

$$\sqrt{2 - 1} = 2 - 3, \quad \text{or} \quad 1 = -1$$

which is false. Hence, 2 is an extraneous solution and we reject it. Replacing x by 5 in the original equation, we get

$$\sqrt{5 - 1} = 5 - 3, \quad \text{or} \quad 2 = 2$$

which is true. Hence 5 is a solution and the solution set of the given equation is $\{5\}$. We can be sure that there are no other solutions since the solution set of the equation $\sqrt{x - 1} = x - 3$ must be a subset of the solution set of the equation $x - 1 = x^2 - 6x + 9$.

Do Exercise 31.   ■

Story Problems

EXAMPLE 7 We wish to enclose a rectangular region measuring 15,000 square feet with 400 feet of fence. One side of the area is along an existing wall and does not require fencing. What should the dimensions of the rectangle be?

Solution Let one side perpendicular to the wall be x feet long. Since there are two sides perpendicular to the wall, $2x$ feet of fence will be used for these two sides. Hence, there will be $(400 - 2x)$ feet of fence available for the third side. Therefore, the length of the side parallel to the wall should be $(400 - 2x)$ feet. The area of the rectangle will be $x(400 - 2x)$ square feet. Since the area is 15,000 square feet, we write

$$x(400 - 2x) = 15,000$$

This equation may be written

$$2x^2 - 400x + 15,000 = 0$$

and

$$2(x - 50)(x - 150) = 0$$

The solution set of the equation is {50, 150}. Thus, we could have the two sides perpendicular to the wall be 50 feet each and the side parallel to the wall be 300 feet long. Or, we could have the two sides perpendicular to the wall be 150 feet each and the side parallel to the wall be 100 feet long.

Clearly, in the first case the area is 15,000 square feet (since $300 \cdot 50 = 15,000$) and the length of the fence is 400 feet (since $50 + 300 + 50 = 400$). These are the conditions stated in the original problem. In the other case, the area is also 15,000 square feet (since $150 \cdot 100 = 15,000$) and the length of the fence is 400 feet since $(150 + 100 + 150 = 400)$. Thus, both solutions are correct.

Do Exercise 37. ■

The preceding is a *word problem*. In general, a written or verbal statement expressing some condition or conditions of equality that exist among some quantities, of which at least one quantity is unknown, is called a word (or story) problem.

General Procedure for Solving Word or "Story" Problems

Step 1. Read the problem carefully, and write a short summary of the problem on scratch paper if necessary.

Step 2. Write down the particular idea or formula to be used in the problem. For example, total cost of production = fixed cost + cost per unit · number of units produced.

Step 3. Ask yourself "What are the quantities involved that are unknown?"

Step 4. Represent one of these quantities by some symbol, say x, and be specific in writing down in what units the quantities are to be expressed. For example, to say "let x be the weight of . . ." is not enough. Is it x pounds? x ounces? x kilograms?

Often, there is more than one way to set up a problem. A different choice of unknown quantity for the variable will lead to a different, possibly simpler, equation. You will learn to make the better choice with practice. Right now, the solution of the problem does not depend on what you choose to call x. Therefore, it seems natural at the beginning to call x the unknown that is to be found.

Step 5. Write all unknown quantities in terms of x (or whatever symbol is used in Step 4).

Step 6. Form an equation that, according to the statement of the problem, expresses the relation between the unknown quantities introduced in Step 5 and the known quantities of the problem.

Step 7. Solve the equation. Go back and find the corresponding values of the other relevant quantities. (See Step 5.)

Step 8. Check your results.

To check the correctness of the solution you must verify that the results satisfy the conditions stated in the original problem. If you check only that the results are solutions to the equation obtained in Step 6, then you have only checked the correctness of the work in solving the equation—you have not checked the accuracy of the analysis of the original problem to see if the solution fits.

We conclude this section with several applications.

EXAMPLE 8   A waiter earns a basic salary of $500 per month plus tips. On the average, he makes $20 per hour in tips. How many hours must he work each month to earn $2900?

Solution   Let x be the number of hours he works per month. His average monthly take in tips is $20x$. Adding this to his basic salary, we get $(20x + 500)$. But this must equal $2900. Thus,

$$20x + 500 = 2900$$

It is easy to see that the solution of this equation is 120. Therefore, the waiter must work 120 hours to earn $2900 per month. You should check the correctness of this result to the original problem.   ∎

EXAMPLE 9   Mr. J. De Largent has $45,000 to invest. He wants to earn at least $3450 in interest per year. He has a choice of investing in industrial bonds paying 9% per year, investing in safer government bonds at 6% per year, or investing in a combination of the two. How should he invest his money to minimize his risk and yet accomplish his goal?

Solution   If he wanted to be absolutely safe, he could invest all of his money in government bonds. However, he would earn only $2700, since $45,000\left(\frac{6}{100}\right) = 2700$. Let x be the maximum amount he can invest in government bonds and still accomplish his goal. Then, he can invest $(45,000 - x)$ in the more risky industrial bonds. He will earn $x\left(\frac{6}{100}\right)$ dollars from the government bonds and $(45,000 - x)\left(\frac{6}{100}\right)$ dollars from the industrial bonds. Thus, his yearly earnings will be $\left[\frac{6x}{100} + \frac{(45,000 - x)9}{100}\right]$. Since he wishes to earn at least $3450, we have the equation:

$$\frac{6x}{100} + \frac{(45,000 - x)9}{100} = 3450$$

Multiplying both sides of this equation by 100 and solving for x, we obtain

$$6x + (45,000 - x)9 = 345,000$$

$$-3x = -60,000, \text{ and } x = \frac{-60,000}{-3} = 20,000$$

Hence, he should invest $20,000 in government bonds and $25,000 in industrial bonds (since $45,000 - 20,000 = 25,000$). You should verify that this strategy will enable him to reach his goal and yet minimize his risk.   ∎

EXAMPLE 10   A motel in Reno, Nevada has 80 rooms. The manager knows that all the rooms will be rented if she charges $40 a room per day. She also knows that for each $7 increase in rent, one less room will be rented and, because of the competition, no room should be rented for more than $120. It costs $5 per day per occupied room for cleaning. On a given day, how much was the rent per room if the profit on that day was $5250? This is a larger profit than if all 80 rooms are rented at $40 per room.

Solution   Suppose each room is rented for $$x$. Then, the increase per room is $$(x - 40)$ and the number of $7 increases is $\frac{x - 40}{7}$. Therefore, the number of rented rooms is $\left(80 - \frac{x - 40}{7}\right)$. Since it costs $5 to clean each rented room, the profit per rented room is $$(x - 5)$. The total profit is

$$\text{(number of rooms rented)} \cdot \text{(profit per room)}$$

or,

$$\$\left[80 - \frac{x - 40}{7}\right](x - 5)$$

But on that day the total profit is $5250. So, we have

$$\left[80 - \frac{x - 40}{7}\right](x - 5) = 5250$$

Multiplying both sides of this equation by 7 and solving for x we obtain

$$[560 - (x - 40)](x - 5) = 36{,}750$$
$$(600 - x)(x - 5) = 36{,}750$$
$$x^2 - 605x + 39{,}750 = 0$$
$$(x - 75)(x - 530) = 0$$

Thus, $x = 75$ or $x = 530$. We reject the larger of the two solutions since the rent should not exceed $120. We conclude that on the day the profit was $5250, the rent per room was $75. You are urged to check the correctness of this result. ∎

EXAMPLE 11   The manager of a barbershop knows that if he charges $$p$ per haircut, the number of customers he will get per day is x, where x and p are related as follows:

$$p = 24 - .25x$$

The average daily cost of running the barbershop when x customers get haircuts is $$(100 + 5x)$. How many customers had a haircut in the shop on a day when the profit was $260?

Solution   In general, the profit is equal to the revenue minus the cost. Also, the revenue is equal to the price per unit multiplied by the number of units. In this case, the revenue per day is $$px$. But,

$$px = (24 - .25x)x = 24x - .25x^2$$

The cost is $$(100 + 5x)$. Thus, the profit is $$[(24x - .25x^2) - (100 + 5x)]$. Since the profit made on the day in question was $260,

$$[(24x - .25x^2) - (100 + 5x)] = 260$$

This equation may be solved as follows:

$$-.25x^2 + 19x - 360 = 0$$

Multiplying both sides by -4 we obtain

$$x^2 - 76x + 1440 = 0$$
$$(x - 40)(x - 36) = 0$$
$$x = 40 \quad \text{or} \quad x = 36$$

Thus, the shop will realize a $260 profit on a day when the number of customers getting haircuts is either 36 or 40. However, when the number of customers is 36, the price per haircut will be $15 since $24 - .25(36) = 15$ and when the number of customers is 40, the price per haircut will be $14 since $24 - .25(40) = 14$. Verify that both answers are correct by using them in the original problem. ∎

EXAMPLE 12   When the price of a certain brand of dog food is $p per case, the quantity (in thousands) the consumers are willing to buy is D and the quantity (also in thousands) the producers are willing to supply is S, where S, D, and p are related as follows:

$$D = 200 - \frac{1}{2}p^2 \quad \text{and} \quad S = \frac{23}{3}p + \frac{1}{4}p^2$$

Find the price per case for which the demand D and supply S are equal.

Solution   We start with the equation $D = S$. Thus,

$$200 - \frac{1}{2}p^2 = \frac{23}{3}p + \frac{1}{4}p^2$$

We first multiply both sides of the equation by 12 and then solve for p as follows:

$$2400 - 6p^2 = 92p + 3p^2$$
$$9p^2 + 92p - 2400 = 0$$
$$(p - 12)(9p + 200) = 0$$

Thus, $p = 12$ or $p = \frac{-200}{9}$. Obviously, the negative solution is not applicable. Therefore, the price per case for which supply and demand are equal is $12. This is called the *equilibrium price*. This concept will be studied more extensively later.
Do Exercise 41. ∎

Exercise Set 1.3

In Exercises 1–5, find the replacement set of the equation but do not solve the equation.

1. $\dfrac{x - 3}{x + 2} + \dfrac{5}{x - 3} = \dfrac{3}{x + 5}$

2. $\dfrac{4}{y^2 - 4} - \dfrac{y}{y + 7} = \dfrac{y - 4}{y - 3}$

3. $\dfrac{z + 3}{z + 1} + \dfrac{z - 3}{z - 1} = \dfrac{z + 5}{z + 6} - \dfrac{z - 7}{z + 1}$

4. $\dfrac{x - 1}{x^2 + 5x - 6} = \dfrac{x + 1}{x^2 + 3x - 10}$

5. $\sqrt{x - 4} + x + 1 = \sqrt{x + 3}$

In Exercises 6–35, solve the given equations.

6. $2x + 10 = 5x - 11$

7. $5 - 3y = 2y + 20$

8. $5x - 6 = 3x + 10$

9. $2x - (3 + 5x) = 4x - 17$

10. $2 - 3(x + 4) = 5(2 - x) - 14$

11. $3y - (2 - 4y) = 2(y + 3) - 18$

12. $\dfrac{z + 4}{2} + 2 = \dfrac{2z + 11}{3}$

13. $\dfrac{x + 3}{5} - \dfrac{x - 8}{3} = \dfrac{x + 4}{2}$

14. $y - \dfrac{3}{2} = \dfrac{y + 3}{3} + \dfrac{3}{2}$

15. $x^2 - 6x + 5 = 0$ **16.** $y^2 + 3y - 18 = 0$

17. $z^2 + 7z - 18 = 0$ **18.** $x^2 - 3x - 10 = 0$

19. $y^2 + 10y - 75 = 0$ **20.** $-z^2 + 3z + 18 = 0$

21. $(x - 1)(x + 2) = (x + 3)(x - 3) + 13$

22. $(y - 2)(y + 3) = (y + 3)(y + 1)$

23. $2z^2 + 3z - (z + 1)(z - 1) = 11$

24. $3x^2 - 5x - 28 = 0$ **25.** $-5x^2 + 7x + 196 = 0$

26. $4y^2 + 3y - 76 = 0$ **27.** $-3z^2 - 11z + 60 = 0$

28. $6 + \dfrac{2}{x - 3} = x + 4$ **29.** $\dfrac{6}{x + 1} + \dfrac{5}{x + 3} = 3$

30. $\sqrt{x - 3} + x = 5$ **31.** $\sqrt{x + 1} + 1 = x - 4$

32. $\sqrt{2x - 1} + 2 = 3x - 10$

33. $\sqrt{x + 6} + 3 = \sqrt{27 + x}$

34. $\sqrt{3x + 7} + x = 7$

35. $2x - \sqrt{5x - 1} = 3x - 5$

36. A man invested part of $25,000 at 11% and the remainder at 6%. His income in one year from these two investments was the same as if he had invested the whole sum at 9%. How much did he invest at each rate?

37. A woman invested part of $40,000 at 12% and the remainder at 8%. Her income in one year from these two investments was the same as if she had invested the whole sum at 9%. How much did she invest at each rate?

38. If the price of a brand of mineral water is $p per case, the quantity (in thousands) the consumers are willing to buy is D and the quantity (also in thousands) the producers are willing to supply is S, where S, D, and p are related as follows:

$$D = 675 - \frac{1}{3}p^2 \quad \text{and} \quad S = 19.5p + \frac{2}{3}p^2$$

Find the price per case for which the demand D and supply S are equal.

39. When the price of a certain racquetball racquet is $p, the quantity (in hundreds) the consumers are willing to buy is D and the quantity (also in hundreds) the producers are willing to supply is S, where S, D, and p are related as follows:

$$D = 1600 - \frac{1}{4}p^2 \quad \text{and} \quad S = \frac{37}{11}p + \frac{1}{2}p^2$$

Find the price per racquet for which the demand D and supply S are equal.

40. The Board of Trustees of a small private university met to discuss an increase in tuition for the following year. The tuition for the past year was $150 per credit hour and the total number of credits taught were 180,000. It is known that for each $1 increase per credit hour, the total number of credits taught will decrease by 800. The trustees established that the total revenue from tuition must increase by $604,800 and that the tuition should not increase by more than 10%. How much will the university charge per credit hour the following year?

41. The manager of an apple orchard in Wenatchee, Washington, knows that if 30 trees are planted per acre, the average yield per tree will be 475 apples. He also knows that for each additional tree per acre, the average yield per tree is expected to decrease by 7. Because of local regulations, the number of trees per acre cannot exceed 50. If the total yield per acre is 16,318 apples, how many trees have been planted on each acre?

42. The manager of a wine shop knows that if she charges $p per bottle of a certain wine and if she keeps the price below $20 per bottle, the number of bottles she will sell per month is x, where x and p are related as follows:

$$p = 40 - .2x$$

The cost of selling x bottles of this particular wine in a month is $C, where C and x are related as follows:

$$C = 20 + 6x$$

How many bottles of this wine were sold during a certain month if the profit during that month was $580? What was the price per bottle?

43. The racquetball pro at an Athletic Club knows that if he charges $p per hour for private lessons he will teach on the average x hours per month as long as he does not charge more than $75 per hour, where x and p are related as follows:

$$p = 120 - 1.5x$$

Whenever he teaches x hours in a month, the club charges him $(100 + 5x)$. How many hours did he teach in a month if his profit was $1264? How much did he charge per hour?

44. A grocer has two kinds of chocolate. The first kind sells at $4.20 a pound, the second kind at $3.70 a pound. How much should he use of each kind to get 50 pounds of a mixture that he could sell at $4 a pound?

45. A grocer has two kinds of nuts. The first kind sells at $5.10 a pound, the second kind at $4.50 a pound. How much should she use of each kind to get 120 pounds of a mixture that she could sell at $4.90 a pound?

46. A van radiator contains 21 quarts of a solution that is 30% antifreeze. How much of the solution should be drained out and replaced by pure antifreeze in order to get 21 quarts of a solution that is 40% antifreeze?

47. A truck radiator with a 30-quart capacity has been prepared for fall driving with 3 quarts of antifreeze, but winter driving requires a solution that is 20% antifreeze. How many quarts of the fall solution should be withdrawn and replaced by pure antifreeze?

48. A man has $20,000 invested in three bonds that pay different rates of interest. He has three times as much invested in a 6% bond as he has in a 8% bond, and the remainder is invested in a 9% bond. His annual interest from these three investments is $1380. How much did he invest in each bond?

49. A garage owner buys a certain number of spark plugs at a cost of three for $2. He sells half of them at $.80 a piece and, to attract business, he puts the other half on sale at $.60 a piece. When they are sold, he has realized a $10 profit on the spark plugs. How many of them did he sell?

50. We wish to enclose 7800 square feet in a rectangular region against an existing wall. The side against the wall does not require fencing. If we use 250 feet of fence, what are the dimensions of the rectangle?

51. Betty has just been promoted to the position of vice president and her office staff decided to have a party to celebrate. John was given $84 to buy ginger ale and cranberry juice to prepare 18 gallons of punch. If the ginger ale and cranberry juice cost $4 and $6 per gallon, respectively, and John spent all the money given to him, how much ginger ale was in 1 gallon of punch?

52. The distance from the ground to an object thrown vertically upward with an initial velocity of 256 ft/sec is given by the formula $h = 256t - 16t^2$, where t is the time in seconds and h is the distance in feet above the ground t seconds after the throw. In how many seconds will the object return to the ground?

53. A square sheet of metal is used to construct a tray by cutting four corners and turning up the four flaps. If the tray has a volume of 18,000 cubic centimeters and is 5 centimeters high, what were the dimensions of the original sheet of metal?

54. The distance between Seattle and Spokane is 315 miles. Ralph and Norma leave Seattle at the same time. However, Norma arrives in Spokane 1 hour and 10 minutes before Ralph with an average speed that was 9 miles per hour faster than his. How fast was each traveling?

55. A homeowner wants to add a 500-square-foot rectangular room to his house against a wall 35 feet long. The cost of building the exterior walls is $120 per linear foot, and the cost of removing part of the existing wall is $40 per linear foot. If the total cost is $9200, what are the dimensions of the room?

56. Suppose that h is a nonzero real number. What real number should be added to $x^2 + hx$ to get a quantity that may be written $(x + k)^2$? (*Hint:* Expand $(x + k)^2$ and compare the first two terms of the expansion to $x^2 + hx$.)

1.4 Inequalities

In this section we discuss inequalities in one variable. First, we must recall the meaning of $<$ (less than) and $>$ (greater than) and their fundamental properties.

Properties of Inequalities

The following properties of real numbers are needed.

1. If c is an arbitrary real number, exactly one of the following is true: **a.** c is positive; **b.** $-c$ is positive; **c.** $c = 0$.
2. The sum and the product of any two positive real numbers are positive real numbers.

> **DEFINITION 1.14:** If a and b are real numbers and $b - a$ is positive, we say that a is less than b and we write $a < b$.
>
> In that case, we also say that b is greater than a and write $b > a$.
>
> If $a < b$, or $a = b$, we write $a \leq b$ (read: "a is less than or equal to b"). *Greater than or equal to* is defined similarly.

The Basic Properties of Inequalities

In the following statements, a, b, c, and d denote real numbers.

1. Precisely one of the following statements is true at one time: $a < b$, $b < a$, $a = b$.
2. $a < b$ ($a \leq b$) if, and only if, $a + c < b + c$ ($a + c \leq b + c$). (Adding the same number to both sides of an inequality preserves the inequality.)
3. If $0 < c$, then $a < b$ ($a \leq b$) if, and only if, $ac < bc$ ($ac \leq bc$). (Multiplying both sides of an inequality by the same positive number preserves the inequality.)
4. If $c < 0$, then $a < b$ ($a \leq b$) if, and only if, $ac > bc$ ($ac \geq bc$). (Multiplying both sides of an inequality by a negative number reverses the inequality—that is, reverses the inequality symbol.)
5. If $a < b$ and $b < c$, then $a < c$. Also if $a \leq b$ and $b \leq c$ then $a \leq c$.
6. If $0 < a < b$, then $\frac{1}{b} < \frac{1}{a}$.
7. If $a < b$ and $c < d$, then $a + c < b + d$. Also, if $a \leq b$ and $c \leq d$, then $a + c \leq b + d$.*

In the following example, we prove two of the properties listed above. The proofs of the remaining properties are left as exercises.

EXAMPLE 1
1. Prove that if a, b, and c are real numbers, then $a < b$ and $b < c$ implies $a < c$.
2. Prove that if a, b, and c are real numbers, then $a < b$ and $0 < c$ implies $ac < bc$.

Solution
1. By definition, $a < b$ if, and only if, $b - a$ is positive. Similarly, $b < c$ if, and only if, $c - b$ is positive. Thus, $(b - a) + (c - b)$ is positive. But, $(b - a) + (c - b) = c - a$. Therefore, $c - a$ is positive and we conclude that $a < c$.
2. Since $a < b$, $b - a$ is positive, and since $0 < c$, c is positive. Thus, $(b - a)c$ is positive. However, $(b - a)c = bc - ac$. Therefore, $bc - ac$ is positive, and so $ac < bc$.

Do Exercise 41. ∎

*Note that Property 7 allows us to add corresponding sides of inequalities. However, *it is wrong to subtract corresponding sides of inequalities*, as is shown by the following: $10 < 11$ is true, $2 < 5$ is also true; however, $10 - 2 < 11 - 5$ is false since $8 > 6$.

Linear Inequalities

> **DEFINITION 1.15:** We say that two open sentences with same replacement set are *equivalent* whenever they have the same solution set.

As it is with equations, solving a linear inequality means finding its solution set. The procedures used to solve a linear inequality are the same as those used to solve equations with one important exception. When we multiply (or divide) both sides of an inequality by a negative number, we must reverse the inequality symbol.

EXAMPLE 2 Solve the inequality $5x + 2 < 2x + 11$.

Solution Use Property 2 and subtract 2 from each side of

$$5x + 2 < 2x + 11$$

to obtain

$$5x < 2x + 9$$

Now subtract $2x$ from each side to get

$$3x < 9$$

Multiplying both sides by $\frac{1}{3}$ (by Property 3), we get the equivalent inequality

$$x < 3$$

The solution set is the set of all real numbers less than 3.
Do Exercise 1.   ■

Sometimes we encounter statements such as $a < b$ and $b < c$. This particular statement is abbreviated as $a < b < c$. We may also write $a \leq b \leq c$ instead of $a \leq b$ and $b \leq c$. The meaning of statements such as $a < b \leq c$ should now be clear. The solution sets of inequalities often are special sets, as described in the following.

> **DEFINITION 1.16:** Let a and b be two real numbers such that $a < b$.
> **a.** The set $\{x \mid x \text{ is a real number and } a \leq x \leq b\}$ is called a *closed interval* and is denoted $[a, b]$, where we use two square brackets to indicate that the end points a and b belong to that interval.
> **b.** The set $\{x \mid x \text{ is a real number and } a < x < b\}$ is called an *open interval* and is denoted (a, b), where we use two parentheses to indicate that the end points a and b do not belong to that interval.
> **c.** The sets $\{x \mid x \text{ is a real number and } a \leq x < b\}$ and $\{x \mid x \text{ is a real number and } a < x \leq b\}$ are called *half-open intervals* and are denoted $[a, b)$ and $(a, b]$, respectively. The intervals $[a, b]$, (a, b), $[a, b)$, and $(a, b]$ are *bounded intervals*.
> **d.** The sets $\{x \mid x \text{ is a real number and } x < a\}$, $\{x \mid x \text{ is a real number and } x \leq a\}$, $\{x \mid x \text{ is a real number and } x > a\}$, and $\{x \mid x \text{ is a real number}$

and $x \geq a$} are denoted $(-\infty, a)$, $(-\infty, a]$, (a, ∞), and $[a, \infty)$, respectively, and are called *unbounded intervals*—closed or open depending on whether the end point a is or is not in the set.

The symbols $-\infty$ and ∞ are called minus infinity and infinity, respectively. You should not think of them as numbers. The solution set of the inequality of Example 2 is the unbounded open interval $(-\infty, 3)$.

Double Inequalities

EXAMPLE 3   Solve the double inequality $2x + 1 < 3x + 2 < 38 - 6x$.

Solution   We write the double inequality as follows:

$$2x + 1 < 3x + 2 \quad \text{and} \quad 3x + 2 < 38 - 6x$$

Using Properties of Inequalities 1–7, we obtain the following sequence of equivalent open sentences:

$$2x + 1 + (-1) < 3x + 2 + (-1) \quad \text{and} \quad 3x + 2 + (-2) < 38 - 6x + (-2)$$
$$2x < 3x + 1 \quad \text{and} \quad 3x < 36 - 6x$$
$$2x + (-3x) < 3x + 1 + (-3x) \quad \text{and} \quad 3x + 6x < 36 - 6x + 6x$$
$$-x < 1 \quad \text{and} \quad 9x < 36$$
$$(-1)(-x) > (-1)(1) \quad \text{and} \quad \frac{1}{9}(9x) < \frac{1}{9}(36)$$
$$x > -1 \quad \text{and} \quad x < 4$$

That is, $-1 < x < 4$. Since the last and first inequalities are equivalent, they have identical solution sets. Thus, the solution set of the original inequality is the open interval $(-1, 4)$.   ■

It is often helpful to have a graphic representation of the solution set of an inequality. If S is a set of real numbers, its graph is the set of all points on a number line whose coordinates are members of the set S. For example, Figure 1.7 displays the solution set of Example 3.

The heavy line shows the points on the graph. We have placed a small open circle around the end points -1 and 4 to indicate that they are *not* on the graph.

Do Exercise 7.

FIGURE 1.7

Using the fact that the product of two real numbers is negative if, and only if, one of the factors is positive and the other is negative, we obtain the following sequence of equivalent inequalities:

$$(x + 6)(x - 1) < 0$$
$$(x + 6 > 0 \text{ and } x - 1 < 0) \quad \text{or} \quad (x + 6 < 0 \text{ and } x - 1 > 0)$$
$$(x > -6 \text{ and } x < 1) \quad \text{or} \quad (x < -6 \text{ and } x > 1)$$

Since no real number is less than -6 and greater than 1, the solution set of the given inequality is

$$\{x \mid x \text{ is a real number and } -6 < x < 1\}$$

This set is the open interval $(-6, 1)$. Its graph is displayed in Figure 1.8.
Do Exercise 11.   ∎

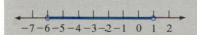

FIGURE 1.8

Inequalities similar to that of Example 4 often involve more than one factor. It is much simpler to obtain the solution sets of such inequalities graphically, by displaying the sign of each factor (positive or negative) under a "number line" for values of x. We can then use the fact that a product of nonzero real numbers is negative if, and only if, the number of negative factors is odd. We could have solved Example 4 using the following display:

x		-6		1	
$x + 6$	$-$	0	$+$	$+$	$+$
$x - 1$	$-$		$-$	0	$+$
$(x + 6)(x - 1)$	$+$	0	$-$	0	$+$

In this table, the first line represents values of x. We entered the number -6 and 1 because $x + 6 = 0$ when $x = -6$, and $x - 1 = 0$ when $x = 1$. The second line represents the values of $x + 6$. We entered a 0 under the -6 since $x + 6 = 0$ when $x = -6$. We displayed a minus sign to the left to indicate that $x + 6$ is negative when x is less than -6. We also displayed some plus signs to the right to indicate that $x + 6$ is positive when x is greater than -6. We proceeded similarly for $x - 1$ on the third line. The fourth line represents values of the product $(x + 6)(x - 1)$ which are positive when $x + 6$ and $x - 1$ are both positive or both negative, and negative when these factors have opposite signs. We also entered 0 below -6 and 1 because the product is 0 when x is replaced by either of these two numbers. Since we are solving the inequality $x^2 + 5x - 6 < 0$, we wish to find values of x for which this product is negative. From the table we see that this will occur whenever $-6 < x < 1$. Therefore, the solution set is the open interval $(-6, 1)$.

To use the method of the last example, one side of the inequality *must* be 0. We illustrate with the following:

EXAMPLE 5   Solve the inequality $\dfrac{2x^2 + 4x - 10}{x + 3} \leq x - 2.$

Solution   We subtract $x - 2$ from both sides to get

$$\frac{2x^2 + 4x - 10}{x + 3} - (x - 2) \leq 0$$

From this, we get

$$\frac{2x^2 + 4x - 10 - (x - 2)(x + 3)}{x + 3} \leq 0$$

which simplifies to

$$\frac{x^2 + 3x - 4}{x + 3} \leq 0$$

or to

$$\frac{(x + 4)(x - 1)}{x + 3} \leq 0$$

The solution set can now be obtained.

x			-4		-3		1	
$x + 4$	$-$	0	$+$	$+$	$+$	$+$	$+$	$+$
$x - 1$	$-$	$-$	$-$	$-$	$-$	$-$	0	$+$
$x + 3$	$-$	$-$	$-$	0	$+$	$+$	$+$	$+$
$\dfrac{(x + 4)(x - 1)}{x + 3}$	$-$	0	$+$	U	$-$		0	$+$

We entered a U in the bottom row under the -3 because $(x + 4)(x - 1)/(x + 3)$ is undefined when $x = -3$. The solution set is $(-\infty, -4] \cup (-3, 1]$. The graph of the solution set is shown in Figure 1.9. The small filled-in circles around the end points -4 and 1 indicate that these two numbers *are* in the solution set.

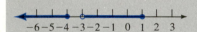

$-6\ -5\ -4\ -3\ -2\ -1\ \ 0\ \ 1\ \ 2\ \ 3$

FIGURE 1.9

Do Exercise 35.   ■

In solving inequalities in the form $\frac{f(x)}{g(x)} < 0$, where $f(x)$ and $g(x)$ are polynomials, it is necessary to factor $f(x)$ and $g(x)$ which sometimes may be difficult. In this regard, it is useful to recall the following.

1. If $f(x)$ is a polynomial, then $x - a$ is a factor of $f(x)$ if, and only if, $f(a) = 0$. In particular, if r_1 and r_2 are solutions of the equation $ax^2 + bx + c = 0$, with $a \neq 0$, then $ax^2 + bx + c = a(x - r_1)(x - r_2)$.
2. The solutions of a quadratic equation can always be found using the quadratic formula.
3. If the solutions of the equation $ax^2 + bx + c = 0$ are not real (that is, $b^2 - 4ac < 0$), then $ax^2 + bx + c$ is either positive for all values of x, or negative for all values of x.

EXAMPLE 6   Solve the inequality $\dfrac{8 - x^3}{3x^2 - 6x - 6} \geq 0$.

Solution First write

$$8 - x^3 = 2^3 - x^3 = (2 - x)(4 + 2x + x^2)$$

and observe that $x^2 + 2x + 4 = 0$ has no real solution (since $2^2 - 4 \cdot 1 \cdot 4 < 0$) and $x^2 + 2x + 4$ is positive when $x = 0$. Thus, $x^2 + 2x + 4 > 0$ for all values of x.

The equation $3x^2 - 6x - 6 = 0$ may be solved using the quadratic formula. We obtain

$$x = \frac{-(-6) \pm \sqrt{(-6)^2 - 4(3)(-6)}}{2(3)}$$

$$= \frac{6 \pm \sqrt{108}}{6} = \frac{6 \pm 6\sqrt{3}}{6} = 1 \pm \sqrt{3}$$

Hence,

$$3x^2 - 6x - 6 = 3(x - (1 + \sqrt{3}))(x - (1 - \sqrt{3}))$$

The original inequality may now be written,

$$\frac{(2 - x)(x^2 + 2x + 4)}{3(x - (1 + \sqrt{3}))(x - (1 - \sqrt{3}))} \geq 0$$

We will use the following table to obtain the solution set.

x				$1-\sqrt{3}$		2		$1+\sqrt{3}$
$2 - x$	+	+	+	0	−	−	−
$x^2 + 2x + 4$	+	+	+	+	+	+	+
$x - (1 + \sqrt{3})$	−	−	−	−	−	0	+
$x - (1 - \sqrt{3})$	−	0	+	+	+	0	+
$\dfrac{(x - 2)(x^2 + 2x + 4)}{3(x - (1 + \sqrt{3}))(x - (1 - \sqrt{3}))}$	+	U	−	0	+	U	−

We locate the plus signs and the number 0 on the last line of the table and find the solutions directly above on the first line. The solution set is $(-\infty, 1-\sqrt{3}) \cup [2, 1+\sqrt{3})$. We could have omitted the factor $x^2 + 2x + 4$ in the table, as was the factor 3, because its values are always positive and therefore do not affect the sign of the expression in the bottom line. However, when a factor does not change sign but is negative for all values of x, it *must* be included in the table.
Do Exercise 39. ■

Compact Solution of Nonlinear Inequalities

It is possible to use the methods of the preceding examples by displaying only the first and last lines of the table. Note the following properties.

1. If $f(x)$ is a polynomial and the real solutions of the equation $f(x) = 0$ are r_1, $r_2, \cdots, r_k$, then $f(x)$ can be factored as

$$f(x) = (x - r_1)^{n_1}(x - r_2)^{n_2} \cdots (x - r_k)^{n_k}g(x)$$

where $n_1, n_2, \cdots, n_k$ are positive integers, and $g(x)$ is a polynomial which has no real roots.

2. As the values of x increase and pass through r_1, the values of $f(x)$ will change sign only if n_i is odd. It is easy to see this because when n_i is even, $(x - r_i)^{n_i} > 0$ when $x \neq r_i$, and therefore that factor does not affect the sign of $f(x)$.

We illustrate how these may be used to simplify the solution in the next example.

EXAMPLE 7   Solve the inequality $\dfrac{(x - 2)^4(x + 1)^5(x^2 + 1)}{(x - 3)^7(-x^2 + 3x - 9)} > 0$.

Solution   Let

$$f(x) = \frac{(x - 2)^4(x + 1)^5(x^2 + 1)}{(x - 3)^7(-x^2 + 3x - 9)}$$

Obviously, the only real solutions of the equation $(x - 2)^4(x + 1)^5(x^2 + 1) = 0$ are -1 and 2, and 3 is the only real solution of $(x - 3)^7(-x^2 + 3x - 9) = 0$. Thus, $f(-1) = f(2) = 0$ and $f(3)$ is not defined. The exponent in $(x - 2)^4$ is even, thus the sign of the values of the left side of the inequality will not change as x increases through 2. However, in $(x + 1)^5$ and $(x - 3)^7$ the exponents are odd. Thus, the sign of the values of the left side of the inequality will change as x increases through -1 and will change again as x increases through 3.

   In the first line of the table, we will display only the values -1, 2, and 3. However, we will place -1 and 3 within small squares to remind us that the sign of the values of $f(x)$ must change as x increases through these numbers. The table will have only two lines. Under the -1 and under the 2 we will enter 0 in the second line. Under the 3 we will enter a U. We must determine the sign of the value of $f(x)$ for some value of x. The simplest way is to find $f(0)$. We have

$$f(0) = \frac{(0 - 2)^4(0 + 1)^5(0^2 + 1)}{(0 - 3)^7(-0^2 + 3(0) - 9)} = \frac{(-2)^4(1)^5(1)}{(-3)^7(-9)}$$

We need not calculate the exact value of $f(0)$ because we are interested only in knowing whether it is positive or negative. The value of $(-2)^4(1)^5(1)/(-3)^7(-9)$ is positive since the number of negative factors is even $(4 + 7 + 1 = 12)$. Since 0 is between -1 and 2 on the first line, we enter some plus signs between these numbers on the second line. Recalling that the signs of the values of $f(x)$ change at -1 and at 3, we enter minus signs on the second line, to the left of -1 and to the right of 3 and plus signs between 2 and 3. The solution set can now be found.

x			$\boxed{-1}$		2		$\boxed{3}$	
$\dfrac{(x - 2)^4(x + 1)^5(x^2 + 1)}{(x - 3)^7(-x^2 + 3x - 9)}$	$-$	$-$	0	$+$	$+\ 0\ +$	$+$	U	$-\ -$

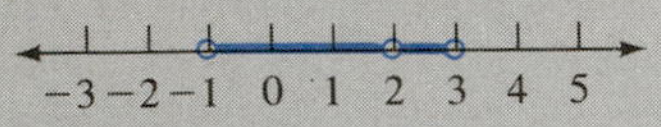

FIGURE 1.10

The solution set is $(-1, 2) \cup (2, 3)$. Its graph is displayed in Figure 1.10.

Do Exercise 55.

Caution Against Common Errors

Before giving you story problems involving inequalities, we caution you against making some common errors.

1. In solving inequalities such as

$$\frac{(x - 3)(x + 2)}{(x - 5)} \leq 0$$

   students often multiply both sides of the inequality by $x - 5$ to obtain the inequality $(x - 3)(x + 2) \leq 0$, which is free of fractions. However, $\frac{(x - 3)(x + 2)}{(x - 5)} \leq 0$ and $(x - 3)(x + 2) \leq 0$ are not equivalent. This is because, when we multiply both sides of an inequality by a negative number, we must reverse the inequality symbol. However, when both sides of the inequality are multiplied by $x - 5$, that quantity may be positive or negative, depending on the value of x. It is important to remember in solving inequalities not to multiply or divide both sides by a quantity involving a variable; unless, of course, we know that the quantity has the same sign for all values of x. For example, we know that $x^2 + 1$ is positive for all values of x. Consequently, multiplying both sides of an inequality by $x^2 + 1$ would yield an equivalent inequality.

2. You must also be careful in combining double inequalities. If a and b are real numbers with $a < b$, it is correct to write the statement $x > a$ *and* $x < b$ in the more compact form $a < x < b$. Note that this open sentence uses the word "and." Another example, the set $(-\infty, -1) \cup (2, \infty)$, whose graph is displayed in Figure 1.11, is defined by the open sentence $x < -1$ *or* $x > 2$. (Notice the word "or.") You cannot write that sentence in the form $-1 > x > 2$ because it would mean $-1 > x$ *and* $x > 2$ and would imply the false statement $-1 > 2$.

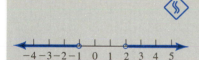

FIGURE 1.11

3. If you use the methods of Examples 4, 5, and 6, you must be careful where you display the plus and minus signs on each line. If the factor is in the form $x - a$, 0 should be written under the a, plus signs should be displayed to the right of 0, and minus signs to the left. However, if the factor is in the form $a - x$, again 0 should be written under the a, but plus signs should be displayed to the *left* of 0, and minus signs to the *right*. (See the second line in the table of Example 6.) We suggest that you actually substitute a value of x in $x - a$, or in $a - x$, to determine the correct entries on the line.

4. Finally, when you use the method of Example 7, you must be very careful to determine the sign of the value of the left side of the inequality for some value of x accurately. Every entry on the bottom line depends on that result. We suggest that as a check you use at least two distinct values of x.

Applications

We conclude this section with some story problems involving inequalities.

EXAMPLE 8   Pierre and Suzie have jobs paying $12 and $8 per hour, respectively. They have agreed that Pierre would work 20 more hours than Suzie does each week until she finishes her work for an MBA. They must earn at least $540 to meet their obligations. What is the minimum number of hours that Suzie should work each week?

Solution   Let x be the number of hours that Suzie works each week. Then Pierre is working $(x + 20)$ hours. Thus, each week Suzie earns $\$8x$ while Pierre earns $\$12(x + 20)$. Their total income per week is $[8x + 12(x + 2)]$. Since they need at least $\$540$ per week, their income should be greater than or equal to $\$540$. Consequently,

$$8x + 12(x + 20) \geq 540$$
$$20x \geq 300$$
$$x \geq 15$$

Thus, Suzie should work at least 15 hours per week.
Do Exercise 45.

EXAMPLE 9   The market research department of some company has established that if the company's new product is to be put on the market at a price of $\$p$ per unit, consumers would be likely to buy q units per day where $q = 315 - 9p$. The production department established that the cost of producing q units per day is $\$C(q)$ where $C(q) = 1125 + 5q$. When the product is put on the market, what should the price per unit be in order for the company to experience a daily profit?

Solution   The company will experience a daily profit if each day the revenue is greater than the cost. That is, if $R > C$. But the revenue is obtained by multiplying the number of units sold by the price per unit. So

$$R > C$$

But the revenue is obtained by multiplying the number of units sold by the price per unit. So

$$R = qp = (315 - 9p)p = -9p^2 + 315p$$

Since we want to compare R and C, we must express the cost C in terms of p. Thus,

$$C = 1125 + 5q = 1125 + 5(315 - 9p) = 2700 - 45p$$

We must have

$$-9p^2 + 315p > 2700 - 45p$$

This inequality is equivalent to

$$p^2 - 40p + 300 < 0$$

which can be written

$$(p - 10)(p - 30) < 0$$

We now use the following table to find the solution set of the inequality.

p		0	10		30	
$(p - 10)(p - 30)$		+	0	−	0	+

From the table, we see that $(p - 10)(p - 30) < 0$, whenever $10 < p < 30$. Thus, the company will make a profit if the price per unit is somewhere between $10 and $30. If the price per unit is exactly $10, or exactly $30, the profit will be 0. At these prices, the company breaks even. Later, we shall be able to determine the price $$p$ which maximizes the daily profit.

Do Exercise 47.   ■

Exercise Set 1.4

Solve the following inequalities and sketch the graphs of their solution sets.

1. $2x + 13 < 5x + 28$

2. $3x - 4 \geq 7x + 12$

3. $6x + 5 \leq 2x + 33$

4. $2 - 3x > 5x - 22$

5. $\dfrac{x}{2} + \dfrac{1}{3} < \dfrac{3x}{5} - \dfrac{2}{7}$

6. $5 + 4x \leq 5x + 3 \leq 2x + 18$

7. $3x + 1 < 6x + 4 < 4x + 10$

8. $9x + 10 \leq 12x + 16 < 8x + 16$

9. $15 + 4x < 8x + 3 < 6x + 5$

10. $x^2 + 6x - 7 < 0$

11. $x^2 + 8x - 20 \leq 0$

12. $2x^2 + 2x - 24 > 0$

13. $x^2 - 3x - 10 \leq 0$

14. $x^3 - 27 > 0$

15. $8x^3 - 27 < 0$

16. $x^4 - 16 > 0$

17. $16x^4 - 81 \leq 0$

18. $x^2 + 4x - 6 < 0$

19. $3x^2 + 24x - 9 > 0$

20. $x^2 + 5x + 7 < 0$

21. $\dfrac{x^2 + 3x - 4}{x + 5} > 0$

22. $\dfrac{x^2 + x - 6}{x - 7} < 0$

23. $\dfrac{x^2 - 5x + 6}{5 - x} > 0$

24. $\dfrac{x^2 + x - 3}{x^2 - 25} < 0$

25. $\dfrac{x^2 - 16}{x^3 - 8} > 0$

26. $\dfrac{x^2 + 2x + 4}{x^2 - 2x + 5} < 0$

27. $\dfrac{x^2 + 3x + 8}{x + 5} < 0$

28. $\dfrac{x^2 - 4x + 10}{4 - x} > 0$

29. $\dfrac{2x^2 + 4x - 4}{x + 1} < 0$

30. $\dfrac{5x + 1}{x + 5} < 5 - x$

31. $\dfrac{x^2 + 3x - 6}{x + 3} > 0$

32. $\dfrac{3x^2 + 5x + 1}{5 - 2x} < 0$

33. $\dfrac{3x^2 - 6x + 2}{x^2 - 5} \leq 0$

34. $\dfrac{x^2 + 3x + 9}{4x^2 - 8x + 3} \leq 0$

35. $\dfrac{x^2 + 5x + 2}{x + 5} > 2x - 3$

36. $\dfrac{2x^2 - 3x + 7}{3 - x} < 5x + 2$

37. $\dfrac{3x + 2}{x - 3} > \dfrac{6x + 4}{x - 2}$

38. $\dfrac{2x + 7}{x - 3} > \dfrac{3x - 4}{x + 3}$

39. $\dfrac{-x^2 + 3x - 9}{x^2 + 3x + 2} < 0$

40. $\dfrac{-x^2 + 2x - 6}{2x^2 + 3x - 7} > 0$

*41. Prove that if a, b, and c are real numbers, $a < b$ and $c < 0$, then $bc < ac$.

*42. Prove that if $0 < a < b$, then $\dfrac{1}{b} < \dfrac{1}{a}$.

43. A student tried to solve the inequality $x + 2 < 5x + 10$ as follows:

$$x + 2 < 5x + 10$$
$$x + 2 < 5(x + 2)$$
$$1 < 5$$

He concluded that the solution set was the set of all real numbers. Was he correct? Explain.

44. A student wishes to earn an "A" grade in her business calculus course. The professor has stated that an average of at least 92 is required for an "A" grade. The student received grades of 89, 93, 90, and 94 on the first four exams. What minimum score must she receive on the fifth and final exam in order to earn an "A"?

45. Roy and Betty have agreed that he will work 10 more hours per week than she does so that she can complete her college education. She earns $7 per hour on her part-time job, while he earns $6 per hour on his. They have calculated that they need at least $385 per week to meet their obligations. What is the minimum number of hours that Betty should work per week?

46. Company X rents a 1987 Chevrolet Citation for $20 per day plus $.12 per mile, while company Y rents the same model car for $14 per day plus $.15 per mile. How many miles per day should a renter drive so that renting from company X would be to his advantage?

47. Some company has established that if it prices its new calculator at p per unit, consumers would be likely to buy q units per day where

$$q = 252 - 7p$$

The production department established that the cost of producing q calculators per day is $C(q)$ where

$$C(q) = 1092 + 4q$$

What should the price be per calculator for the company to experience a daily profit?

48. The market research department of the AZ Company has established that if the company's new talking doll is priced at p per doll, consumers would be likely to buy q dolls per month where

$$q = 1541 - 23p$$

The production department established that the cost of producing q dolls per month is $C(q)$ where

$$C(q) = 14{,}467 + 13q$$

When the doll is put on the market, what should the price be per doll for the company to experience a monthly profit?

49. The Baldhead Company has established that if the company's new hair dryer is priced at p per unit, consumers would be likely to buy q units per month where

$$q = 935 - 17p$$

The production department established that the cost of producing q dryers per month is $C(q)$ where

$$C(q) = 5100 + 15q$$

When the hair dryer is put on the market, what should the price be per unit for the company to experience a monthly profit?

50. A manufacturer can manufacture and sell q thousand units of a commodity at a price of p per unit and a cost of C thousand dollars, where p and C are given by

$$p = 20 - .5q$$

and

$$C = 36 + 6q$$

How many units of the commodity should be produced and sold in order for the profit to be at least $60,000?

51. Same as Exercise 50 where

$$p = 30 - .25q$$

and

$$C = 200 + 5q$$

and the profit should be at least $400,000.

52. A van radiator contains 21 quarts of a solution which is 30% antifreeze. What amount of solution should be drained out and replaced by pure antifreeze in order to get 21 quarts of a solution that is at least 40% but not more than 55% antifreeze?

53. A car radiator contains 15 quarts of a solution which is 25% antifreeze. What amount of solution should be drained out and replaced by pure antifreeze in order to get 15 quarts of a solution that is at least 35% but not more than 45% antifreeze?

54. Solve

$$\frac{(x - 2)^6(x + 3)^7(x^2 + x + 1)}{(x + 1)^3(x^2 + 6x + 10)^5} < 0$$

using the method of Example 7.

55. Solve

$$\frac{(x + 3)^3(x - 5)^7(x^2 + 2x + 6)}{(x + 5)^3(x^2 + 8x + 20)^5} < 0$$

using the method of Example 7.

56. Solve

$$\frac{(x - 1)^9(x + 8)^6(x^2 + 3x + 4)}{(x + 3)^5(x^2 + 5x + 10)^5} > 0$$

using the method of Example 7.

1.5 Functions and Graphs

When we say that the distance traveled by an object in a fixed amount of time is a *function* of its speed, we mean that the distance traveled is dependent on the speed of the object and, therefore, if the speed is known, the distance traveled can be determined. Speed and distance are both variable quantities. In a formula such as $d = 5s$, d and *s* are both variables and for each value of *s* (the speed), the corresponding value of *d* can be found (the distance traveled in 5 units of time).

> **DEFINITION 1.17:** In general, a *function* from a set S to a set T is a rule that assigns to each element x of set S a unique element of set T. The set S is called the *domain* of the function. If f is the name of a function, then the unique element in T corresponding to an element x in S is denoted $f(x)$ (read: "f of x") and is called the *image* of x. The set $\{f(x) \mid x \in S\}$ is called the *range* of the function. It is the set of all possible values of $f(x)$ in T.

There are several ways to describe functions. The most common is to use a formula.

EXAMPLE 1   The function h from the set of integers to the set of nonnegative integers is defined by the equation

$$h(x) = x^2$$

a. Find $h(-2)$, $h(0)$, $h(3)$, and $h(2.4)$.
b. What is the range of the function h?

Solution   **a.** $h(-2) = (-2)^2 = 4$, $h(0) = 0^2 = 0$, $h(3) = 3^2 = 9$, and $h(2.4)$ is not defined since 2.4 is not an integer and hence is not in the domain of h.
b. The range of the function h is the set of all square integers $\{0, 1, 4, 9, 16, 25, 36, \ldots\}$.
Do Exercise 5. ■

Independent and Dependent Variables

> **DEFINITION 1.18:** If f is the name of a function and we write $y = f(x)$, x is called the *independent variable* and y is called the *dependent variable*.

The replacement set for the independent variable is the domain of the function. In most of our work, we shall give only the rule defining the function. The domain and range will be sets of real numbers. In particular, unless otherwise specified, the domain will be the largest set of real numbers that can be used as a replacement set for the independent variable so that the corresponding values of the dependent variable are defined and are real numbers.

EXAMPLE 2   Let the function g be defined by

$$y = g(x) = \frac{2}{x - 3}$$

Find the domain of g.

Solution   Replacing x by 3 would yield a division by 0, which is not defined. Hence, 3 is not in the domain of g. If x is replaced by any other real number, the corresponding value of y will be a real number. Thus, the domain of g is the set of all real numbers except the number 3.
Do Exercise 7.   ■

EXAMPLE 3   Let the function f be defined by

$$y = f(x) = \frac{x - 3}{\sqrt{x + 5}}$$

Find the domain of f.

Solution   To find the value of y, we must divide by $\sqrt{x + 5}$. Thus, this expression must be real and nonzero. Since the square root of a negative number is not a real number, $x + 5 > 0$. This inequality is equivalent to $-5 < x$. Therefore, the solution set and the domain of the function is $(-5, \infty)$.
Do Exercise 11.   ■

Graphing Functions

Often, names for pairs of objects are listed in order. For example, in a telephone directory, one can find the name of a resident listed first and then the corresponding phone number. In general, (a, b) denotes an ordered pair where a is considered the first member of the pair, and b is considered its second member; however, a and b need not be distinct. For example, $(2, 2)$ is an ordered pair. Recall that there is a one-to-one correspondence between the set of all ordered pairs of real numbers and a Cartesian coordinate plane. If the ordered pair (a, b) corresponds to a point P, then P is said to be the *graph* of (a, b) and (a, b) are the *coordinates* of the point P. If a point Q has coordinates (c, d), it is customary to refer to it as the point (c, d). We also say that c is the *abscissa* of Q and d is its *ordinate*. If the domain and range of a function f are sets of real numbers, then the graph of f is the set of all points $(x, f(x))$ where x is in the domain of f.

EXAMPLE 4   Let f be the function from the set $\{-2, 1, 3, 4\}$ to the set of real numbers defined by $f(x) = 2x - 3$. Sketch the graph of f.

Solution   Since $f(-2) = 2(-2) - 3 = -7$, the point whose coordinates are $(-2, -7)$ is on the graph of f. Similarly, $f(1) = 2(1) - 3 = -1$. Thus, the point with coordinates $(1, -1)$ is also on the graph of f. It should now be clear that the points with coordinates $(3, 3)$ and $(4, 5)$ are also points of the graph of f. The graph of f is shown in Figure 1.12. Note that the domain and the graph of a function have the same number of elements.
Do Exercise 21.

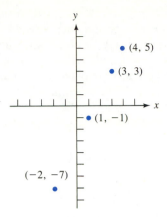

FIGURE 1.12

Linear Functions

Recall that if m and b are any two real constants, the function defined by

$$y = f(x) = mx + b$$

is called a *linear function*. The graph of a linear function is a straight line and the real number m is called the *slope* of that line. The number b is often called the *y-intercept* of the line because the point with coordinates $(0, b)$ is the intersection of the line with the y-axis. To find the *x-intercept*, simply replace y by 0 in the equation $y = mx + b$ and solve for x. The following facts about linear functions should be familiar.

> **Facts About Linear Functions**
>
> 1. If the points with coordinates (x_1, y_1) and (x_2, y_2) are on a line, the slope m of the line is
>
> $$m = \frac{y_2 - y_1}{x_2 - x_1}$$
>
> provided that $x_2 - x_1 \neq 0$. In the case of $x_2 - x_1 = 0$, the line is vertical. Its slope is not defined and the equation of such a line is $x = c$, where c is the abscissa of the point of intersection of the line with the x-axis.
> 2. If two lines have slopes m_1 and m_2, respectively, then the two lines are parallel if, and only if, $m_1 = m_2$. The lines are perpendicular if, and only if, $m_1 m_2 = -1$.
> 3. If a line is parallel to the x-axis, its slope is 0. Replacing m by 0 in the equation of a straight line, we get $y = b$. Thus, if a line is parallel to the x-axis and passes through the point $(0, b)$, its equation is $y = b$.

A typical problem you will encounter is having to find the equation of a line given some information about that line. Though there are several ways to do this, we suggest the following method.

To find the equation of a straight line, select a point arbitrarily on the line and call its coordinates (x, y). Using given information, find the slope of the line in two ways, at least one of which makes use of the point (x, y). Get the equation of the line by setting the two answers obtained for the slope equal to each other. Of course, this equation can often be simplified.

EXAMPLE 5 Find the equation of each of these lines:

  a. The line passing through the points $(-1, 3)$ and $(2, 9)$.
  b. The line passing through the point $(-2, 5)$ and parallel to the line of part **a**.
  c. The line having x-intercept 3 and perpendicular to the graph of $3x + 2y = 12$.
  (See Figures 1.13 and 1.14.)

Solution   **a.** Select a point arbitrarily on the line and let its coordinates be (x, y). Since the points with coordinates (x, y) and $(-1, 3)$ are on the line, the slope is given by

$$m = \frac{y - 3}{x - (-1)} = \frac{y - 3}{x + 1}$$

Using the fact that the points $(-1, 3)$ and $(2, 9)$ are also on the line, we find that the slope is 2 since $\frac{9 - 3}{2 - (-1)} = \frac{6}{3} = 2$. Setting the two answers equal, we obtain

$$\frac{y - 3}{x + 1} = 2$$

This may be written

$$y - 3 = 2(x + 1)$$

This is called the *point-slope* form of the line. The equation can also be written

$$y = 2x + 5$$

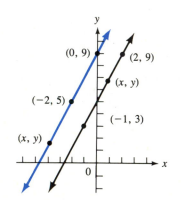

FIGURE 1.13

which is called the *slope-intercept* form of the line.

If we replace y by 0 in the equation of the line, then

$$0 = 2x + 5$$

which yields $x = \frac{-5}{2}$. Hence, the point $\left(\frac{-5}{2}, 0\right)$ is the intersection of the line with the x-axis. Therefore, the x-intercept is $\frac{-5}{2}$.

  b. Since the line of part **a** has slope 2, any line parallel to it will also have slope 2. Let (x, y) be an arbitrary point on the line. Then the points $(-2, 5)$ and (x, y) are on the line and the slope is $\frac{y - 5}{x - (-2)}$. Since the slope is 2, we write

$$\frac{y - 5}{x + 2} = 2$$

This can be written

$$y = 2x + 9$$

c. Since the line has x-intercept 3, the point $(3, 0)$ is on the line. Let (x, y) be an arbitrary point on the line. Then the slope of the line is

$$m = \frac{y - 0}{x - 3} = \frac{y}{x - 3}$$

The equation $3x + 2y = 12$, may be written

$$y = \left(\frac{-3}{2}\right)x + 6$$

Therefore, the graph of that equation has slope $\frac{-3}{2}$. If a line perpendicular to that graph has slope m, we must have $\left(\frac{-3}{2}\right)m = -1$. Thus, $m = \frac{2}{3}$. Setting the two answers we obtained for m equal to each other, we get

$$\frac{y}{x - 3} = \frac{2}{3}$$

This equation can be written in the point-slope form as

$$y - 0 = \frac{2}{3}(x - 3)$$

and in slope-intercept form as

$$y = \left(\frac{2}{3}\right)x - 2$$

Do Exercises 23, 27, and 31. ■

Quadratic Functions

Recall from basic algebra that if a, b, and c are real numbers with $a \neq 0$, the function defined by

$$y = f(x) = ax^2 + bx + c$$

is a *quadratic function*. The graph of a quadratic function is a *parabola*, concave up if $a > 0$ (see Figure 1.15a) and concave down if $a < 0$ (see Figure 1.15b). The *vertex*

FIGURE 1.14

FIGURE 1.15

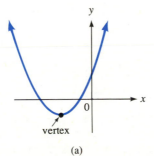

(a)

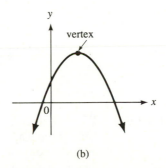

(b)

of the parabola is its lowest point when $a > 0$ and its highest point when $a < 0$. The abscissa of the vertex is the number $\frac{-b}{2a}$ and the parabola is symmetric with respect to a vertical line passing through the vertex.

EXAMPLE 6   The function f is defined by

$$y = f(x) = -2x^2 + 8x + 1$$

Sketch its graph.

Solution   The graph is a parabola which is concave down since $-2 < 0$. Therefore, the vertex will be the highest point of the graph. Its abscissa is 2 since $\frac{-8}{2(-2)} = 2$. To find the ordinate of the vertex, we calculate the value of y when $x = 2$. We find $y = -2(2^2) + 8(2) + 1 = 9$. We conclude that the vertex has coordinates $(2, 9)$. Hence the parabola is symmetric with respect to the vertical line whose equation is $x = 2$; therefore, we need sketch only the half of the parabola to the left of that line. The other half is obtained by symmetry. We calculate the values of y when x is replaced successively by -1, 0, and 1. The results are tabulated as

x	-1	0	1
y	-9	1	7

Therefore, the points $(-1, -9)$, $(0, 1)$, $(1, 7)$, and $(2, 9)$ are on the graph of f. We plot these points and draw a smooth curve through them. (See Figure 1.16a.) The other half of the graph results from symmetry. (See Figure 1.16b.)
Do Exercise 33.

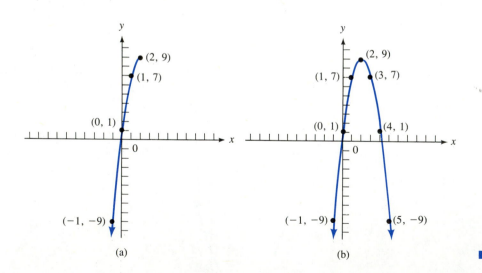

FIGURE 1.16   (a)   (b)

EXAMPLE 7   We wish to enclose a rectangular region against an existing wall. No fence is required against the wall. Fencing materials cost $20 per foot for the side parallel to the wall and $15 per foot for the other two sides. Find the dimensions of the largest area that can be enclosed for $840.

Solution   Let x feet be the length of each side perpendicular to the wall. The total length of these two sides is $2x$ feet, and the cost of materials for these two sides is \$30x since $15(2x) = 30x$. Thus, $840 - 30x$ dollars is available for the side parallel to the wall. Since \$20 a foot is to be spent on that side, its length is $\frac{840 - 30x}{20}$ feet. Thus, the area A square feet of the enclosed region is

$$A = f(x) = \left(\frac{840 - 30x}{20}\right)x = \frac{-3}{2}x^2 + 42x$$

The graph of this equation is a parabola which is concave down since $\frac{-3}{2}$, the coefficient of x^2, is negative. The vertex is the highest point of the graph and the y-coordinate of the vertex gives the maximum value of A. The x-coordinate of the vertex is 14 since in the foregoing equation

$$b = 42, a = \frac{-3}{2}, \text{ and } \frac{-b}{2a} = \frac{-42}{2\left(\frac{-3}{2}\right)} = \frac{-42}{-3} = 14$$

Thus, the sides perpendicular to the wall must be 14 feet. The other side must be 21 feet since $\frac{840 - 30(14)}{20} = 21$.

Do Exercise 45. ■

Here is another example.

EXAMPLE 8   The racquetball pro at an athletic club knows that if she charges \$40 an hour for private lessons she will teach 45 hours a month, but if she charges only \$20 an hour she will teach 110 hours. The club charges the pro $50 + 12x$ dollars for the use of x hours of court time during a month.

   a. Find the price \$$p$ per hour of private lesson in terms of the number of x hours the pro is teaching per month, assuming a linear relationship between p and x.
   b. Find an equation giving the monthly revenue \$$R$ in terms of x.
   c. How many hours should the pro teach each month to maximize her profit?

Solution   **a.** The graph of the equation will be a line. Let (x, p) be an arbitrary point on the line. The points $(45, 40)$ and $(110, 20)$ are on the line. Hence, the slope of the line can be obtained in the following ways:

$$m = \frac{p - 40}{x - 45} \quad \text{and} \quad m = \frac{40 - 20}{45 - 110} = \frac{-4}{13}$$

Thus,

$$\frac{p - 40}{x - 45} = \frac{-4}{13}$$

This equation may be written

$$p = \frac{-4}{13}x + \frac{700}{13}$$

b. The revenue R is obtained by multiplying the number of x hours taught by the price p per hour. Therefore,

$$R = xp = x\left(\frac{-4}{13}x + \frac{700}{13}\right), \text{ so } R = \frac{-4}{13}x^2 + \frac{700}{13}x$$

c. Since the club charges the pro $50 + 12x$ dollars to teach x hours in a month, the monthly profit P is given by

$$P = \frac{-4}{13}x^2 + \frac{700}{13}x - (50 + 12x) = \frac{-4}{13}x^2 + \frac{544}{13}x - 50$$

The graph of this equation is a parabola that is concave down since $\frac{-4}{13}$, the coefficient of x^2, is negative. The vertex is the highest point of the graph and its y-coordinate gives the maximum value of the monthly profit. To find the x-coordinate of the vertex, note that $a = \frac{-4}{13}$ and $b = \frac{544}{13}$. The x-coordinate of the vertex is 68 since

$$\frac{-\dfrac{544}{13}}{2\left(\dfrac{-4}{13}\right)} = 68$$

Thus, the pro should teach 68 hours per month to maximize her profit.
Do Exercise 49. ■

Functions Defined by Multiple Formulas

We conclude this section with a few remarks on functions. In our work, most of the functions will be defined by formulas. Sometimes, however, we need more than one formula to define a function, as illustrated in the following example.

EXAMPLE 9   Betty has a job paying $10 per hour. She gets paid time and a half for overtime (time over 40 hours per week). If she works x hours during a certain week, express her earnings $I(x)$ in terms of x.

Solution   If $0 \le x \le 40$, then clearly $I(x) = 10x$. However, if $40 \le x \le 168$, Betty will get paid $400 for the first 40 hours, plus the overtime pay. Note that she has worked $x - 40$ hours of overtime for which she receives $15(x - 40)$ since her overtime pay is $15 per hour. In that case, $I(x) = 400 + 15(x - 40) = 15x - 200$. We express our result as

$$I(x) = \begin{cases} 10x, & \text{if } 0 \le x \le 40 \\ 15x - 200, & \text{if } 40 < x \le 168 \end{cases}$$

Do Exercise 41. ■

Exercise Set 1.5

*In Exercises 1–6, two sets S and T are given and a function f from S to T is described. In each case, find **a.** the domain of f and **b.** f(x) for the given values of x.*

1. $S = \{0, 1, 2, 3, 4\}$, $T = \{-2, -1, 0, 2, 3, 4, 5, 6, 7, 8, 9\}$, and f is defined by $f(x) = 2x$; $x = 1$, $x = 3$, $x = 4$.

2. $S = \{-2, -1, 1, 2, 3, 4\}$, $T = \{-7, -4, 2, 5, 8, 11\}$, and f is defined by $f(x) = 3x - 1$; $x = -1$, $x = 2$, $x = 4$.

3. $S = \{0, 1, 2, 3, 4\}$, $T = \{0, 1, 7, 8, 10, 27, 30, 64, 79\}$, and f is defined by $f(x) = x^3$; $x = 0$, $x = 2$, $x = 4$.

4. S is the set of positive integers, T is the set of rational numbers, and f is defined by $f(x) = \frac{x}{(x+1)}$; $x = 2$, $x = 6$, $x = 9$.

5. S is the set of real numbers, T is the set of real numbers, and f is defined by $f(x) = x^2 + 1$; $x = -2$, $x = 2$, $x = 6$.

6. S is the set of positive integers, T is the set of prime numbers, and $f(x)$ is the smallest prime number larger than x; $x = 4$, $x = 8$, $x = 14$.

In Exercises 7–18, an equation defines a function. Find the domain of each.

7. $f(x) = \dfrac{2}{x - 4}$

8. $g(x) = \dfrac{x + 1}{x + 3}$

9. $h(x) = \dfrac{x - 2}{x^2 - 16}$

10. $f(x) = \sqrt{x + 3}$

11. $h(x) = \dfrac{5 + 2x}{\sqrt{x - 4}}$

12. $g(x) = \sqrt{x - 8}$

13. Find $f(5)$, $f(0)$, $f(-3)$ where f is the function of Exercise 7.

14. Find $g(-2)$, $g(0)$, $g\left(\frac{1}{2}\right)$ where g is the function of Exercise 8.

15. Find $h(-2)$, $h(1)$, $h(5)$ where h is the function of Exercise 9.

16. Find $f(3)$, $f(5)$, $f(-8)$ where f is the function of Exercise 10.

17. Find $g(9)$, $g(12)$, $g(5)$ where g is the function of Exercise 12.

18. Find $h(5)$, $h(8)$, $h(1)$ where h is the function of Exercise 11.

19. Plot the graph of the function f of Exercise 2.

20. Plot the graph of the function f of Exercise 3.

21. Plot the graph of the function f of Exercise 1.

In Exercises 22–32, a straight line is described. Find its equation.

22. The line passing through the points $(3, 4)$ and $(5, 2)$.

23. The line passing through the points $(-5, -2)$ and $(-3, 6)$.

24. The line passing through the points $(-4, -6)$ and $(-4, 8)$.

25. The line passing through the points $(3, 2)$ and $(-13, 2)$.

26. The line passing through the point $(2, 5)$ and parallel to the line of Exercise 1.

27. The line passing through the point $(3, 8)$ and parallel to the line of Exercise 2.

28. The line passing through the point $(2, 7)$ and parallel to the line of Exercise 3.

29. The line passing through the point $(3, 9)$ and perpendicular to the line of Exercise 4.

30. The line passing through the point $(12, -15)$ and perpendicular to the line whose equation is $2x + 5y = 25$.

31. The line with x-intercept -7 and perpendicular to the graph of the equation $12x - 6y = 36$.

32. The line with x-intercept 5 and y-intercept -4.

In Exercises 33–38, sketch the graph of the given function showing the vertex and symmetry.

33. The function f defined by $f(x) = x^2 + 10x - 4$.

34. The function g defined by $g(x) = 2x^2 - 12x + 7$.

35. The function h defined by $h(x) = 3x^2 - 18x + 5$.

36. The function f defined by $f(x) = -x^2 + 4x - 3$.

37. The function g defined by $g(x) = -4x^2 + 24x - 9$.

38. The function h defined by $h(x) = -.5x^2 - 3x + 4$.

39. **a.** Let g be the absolute value function defined by $g(x) = |x|$. Sketch its graph.
    b. Let $h(x) = |x - 2|$. Sketch the graph of the function h.

40. Roy has a job paying \$12 per hour. He gets paid time and a half for overtime (time over 40 hours per week). If he works x hours during a week, express his earnings \$$I(x)$ in terms of x. Sketch the graph of the function I.

41. Ralph has a job paying \$14 per hour for the first 40 hours of the week. He gets time and a half for the next 8 hours and double time after that. If he works x hours during a

week, express his earnings $I(x)$ in terms of x. Sketch the graph of the function I.

42. The owner of a dude ranch knows that if he charges $40 per day to rent a horse, 30 horses will be rented; and if he charges $25 a day per horse, 60 horses will be rented.

   a. Find an equation relating the price p per horse per day and the number x of horses rented per day, assuming this equation is linear.
   b. Find an equation giving the revenue R in terms of the number x of horses rented.
   c. If the daily cost to have x horses is $(90 + 10x)$, how many horses should the owner have to maximize his daily profit? What is the maximum daily profit?

43. A store's market research department found that if the store priced a new electric razor at $45, 90 razors would be sold per month. On the other hand, 135 would be sold per month if the price were $30. The cost to the store to sell x razors per month is $(200 + 21x)$.

   a. Find an equation relating the price p per razor and the number x of razors sold per month, assuming that this equation is linear.
   b. Find an equation giving the revenue R in terms of the number x of razors sold in a month.
   c. How many razors should be sold to maximize the monthly profit? What is the maximum monthly profit?

44. We wish to enclose a rectangular region against an existing wall. No fence is required along the wall. The cost of fencing materials for one of the sides perpendicular to the wall is $12 per linear foot, while the cost for the other two sides to be fenced is $9 per linear foot. What are the dimensions of the maximum area that can be enclosed if $966 is spent on fencing materials?

45. A farmer wishes to enclose a rectangular region. Fencing materials for one side will cost $13 per linear foot, while the other three sides will cost $8 per linear foot. Give the

dimensions of the largest area that can be enclosed if $840 is available for fencing.

46. The members of the Oregon Ducklings, a Little League baseball team, have been collecting aluminum cans to help pay for the cost of equipment. They have 180 pounds of aluminum, for which they can get $.80 a pound now. If they continue to collect, they can get 6 more pounds per day, but the price per pound will decrease $.02 per day. Knowing that they can make only one transaction, how many days should they wait to maximize their income? What is the maximum income?

47. The Lebanese University of New York charged $160 per credit hour in the 1988–1989 academic year and students registered for a total of 225,000 credit hours. The president of the university wanted to raise tuition. Studies showed him that an increase of x dollars per credit hour would cause a decrease of $1250x$ in the number of credit hours taught. He correctly calculated the increase that would create the maximum revenue from tuition in the next academic year. What was his answer?

48. The owner of an apple orchard knows that if 26 apple trees are planted on 1 acre of land, each tree will yield an average of 400 apples. For each additional tree per acre, the average yield per tree will decrease by 8. How many trees should be planted per acre to maximize the yield?

49. The manager of a sports equipment shop knows that 10 table tennis paddles will be sold per week if the price is $15 per paddle, while 30 paddles will be sold a week if the price is $5 per paddle.

   a. Find an equation to give the price p per paddle in terms of the number of x paddles sold each week, assuming that the equation is linear.
   b. Find an equation giving the revenue R in terms of x.
   c. If the cost of selling x paddles a week is $(25 + 4x)$, how many paddles should be sold each week to maximize the profit?

1.6 The Exponential Function

The Laws of Exponents

For each positive integer n and each real number a, a^n denotes the product $a \cdot a \cdot a \cdot \cdots \cdot a$ (n factors), and is called the *nth power of a*. The number a is called the *base* and the positive integer n is called the *exponent*. We will extend this definition and give a meaning to a^x where x is any real number.

If a and b are real numbers, and m and n are positive integers, then the following are easily verified:

1. $a^m \cdot a^n = a^{m+n}$
2. $\dfrac{a^m}{a^n} = a^{m-n}$ (provided $a \neq 0$ and $m > n$)
3. $(a^m)^n = a^{mn}$
4. $(ab)^m = a^m b^m$
5. $\left(\dfrac{a}{b}\right)^m = \dfrac{a^m}{b^m}$ (provided $b \neq 0$)

These five equalities are often called the *laws of exponents*.

We need to define a^x in such a way that the laws of exponents always hold. Begin with

$$\frac{a^m}{a^n} = a^{m-n} \text{ (provided } a \neq 0 \text{ and } m > n)$$

and remove the restriction $m > n$. If $m = n$, we have

$$\frac{a^m}{a^m} = a^{m-m} \text{ (provided } a \neq 0)$$

which simplifies to

$$1 = a^0$$

Thus, we define $a^0 = 1$ whenever $a \neq 0$.

Let n be a positive integer and $a \neq 0$. Replace m by $-n$ in

$$a^m \cdot a^n = a^{m+n}$$

to get

$$a^{-n} \cdot a^n = a^{-n+n} = a^0 = 1$$

Thus,

$$a^{-n} = \frac{1}{a^n}$$

Thus, we define $a^{-n} = 1/a^n$ where n is any positive integer and a is any nonzero real number.

If n is a positive integer, and a and b are real numbers such that $a^n = b$, we say that a is a *square root* of b when $n = 2$, a *cube root* of b when $n = 3$, and an *nth root* of b when $n > 3$.

In more advanced courses it can be shown that any nonzero complex number has exactly n nth roots. For example, -2 and 2 are the two square roots of 4 since $(-2)^2 = 2^2 = 4$. However, in general, at least $n - 2$ of the n nth roots of a number are not real. In this book, unless otherwise specified, we will consider only nth roots of nonnegative real numbers and call the nonnegative nth root of a nonnegative real number its *principal nth root*. For example, the principal fourth root of 16 is 2 since $2^4 = 16$. Note that -2 is also a fourth root of 16 and the other two fourth roots of

16 are imaginary numbers. If $a \geq 0$ and n is an integer greater than 1, the principal nth root of a is denoted $\sqrt[n]{a}$. If $n = 2$, we write $\sqrt{a}$ instead of $\sqrt[2]{a}$. Note also that if x is any real number, x^2 is nonnegative so that $\sqrt{x^2}$ is a real number. However, if x is a negative number, then $\sqrt{x^2} \neq x$ since the principal square root of x^2 is nonnegative. Thus, we write

$$\sqrt{x^2} = |x|$$

for all real numbers x.

Suppose now that n is an integer greater than 1 and a is a nonnegative real number. Replace m by $\frac{1}{n}$ in

$$(a^m)^n = a^{mn}$$

to get

$$(a^{1/n})^n = a^{1/n \cdot n} = a^1 = a$$

Therefore, $a^{1/n}$ must be the principal nth root of a. We write

$$a^{1/n} = \sqrt[n]{a}$$

Thus,

$$(a^{1/n})^m = (\sqrt[n]{a})^m$$

and if the third law of exponents is to hold,

$$a^{m/n} = (\sqrt[n]{a})^m$$

DEFINITION 1.19:   If a is a nonnegative real number, and m and n are positive integers with $n > 1$, then

$$a^{m/n} = \sqrt[n]{a^m} = (\sqrt[n]{a})^{m\,*}$$

In more advanced calculus texts, a formal definition is given for a^x where a is a nonnegative real number and x is any real number, rational or irrational. This definition is beyond the scope of this text. However, the important fact is that the laws of exponents hold in all cases.

If a and b are any two nonnegative real numbers and x and y are any two real numbers, then

$$a^x \cdot a^y = a^{x+y}$$

$$\frac{a^x}{a^y} = a^{x-y}$$

$$(a^x)^y = a^{xy}$$

---

*The definition of $a^{m/n}$ is also valid when a is negative and n is odd.

$$(ab)^x = a^x \cdot b^x$$

$$\left(\frac{a}{b}\right)^x = \frac{a^x}{b^x}$$

Of course, we need to make the necessary restrictions to avoid division by 0 and to avoid 0^0, which is not defined.

The Exponential Function with Base *b*

We are now ready to define a new type of function.

> **DEFINITION 1.20:**   If b is a positive real number and $b \neq 1$, the *exponential function* with base b is defined by
>
> $$f(x) = b^x$$
>
> for all real numbers x.

As stated earlier, we require the base b to be nonnegative to avoid imaginary numbers (for example $(-3)^{1/2}$ is imaginary). We also require that base b be different from 0 and from 1, since either of these two numbers used as a base would yield a constant function.

EXAMPLE 1   Sketch the graph of the exponential function with base 2.

Solution   We have $y = 2^x$ for all real numbers x. Replace x successively by $-3, -2, -1, 0, 1, 2,$ and 3 and calculate y. The results are

x	-3	-2	-1	0	1	2	3
y	$\dfrac{1}{8}$	$\dfrac{1}{4}$	$\dfrac{1}{2}$	1	2	4	8

Thus, the points $\left(-3, \frac{1}{8}\right), \left(-2, \frac{1}{4}\right), \left(-1, \frac{1}{2}\right), (0, 1), (1, 2), (2, 4),$ and $(3, 8)$ are on the graph of the function. We plot these seven points and draw a smooth curve through them to get a sketch of the graph of $= 2^x$. (See Figure 1.17.)
Do Exercise 11.

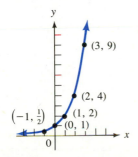

FIGURE 1.17

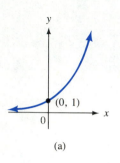

(a)

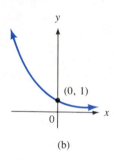

(b)

FIGURE 1.18

Properties of the Exponential Function with Base b, Where b is an Arbitrary Positive Number and b ≠ 1

1. The domain is the set of all real numbers.
2. The range is the set of all positive real numbers.
3. The graph of $y = b^x$ passes through the point $(0, 1)$ since $b^0 = 1$.
4. If $b > 1$, the exponential function with base b is an increasing function. That is, $b^{x_1} < b^{x_2}$ whenever $x_1 < x_2$. (See Figure 1.18a.).
5. If $b < 1$, the exponential function with base b is a decreasing function. That is, $b^{x_1} > b^{x_2}$ whenever $x_1 < x_2$. (See Figure 1.18b.)
6. The graph of $y = b^x$ can be drawn without lifting the pencil. That is, the exponential function with base b is a continuous function.
7. $b^{x+y} = b^x \cdot b^y$.
8. $b^{x-y} = \dfrac{b^x}{b^y}$.
9. $b^{xy} = (b^x)^y$.

The expression $\left(1 + \frac{1}{n}\right)^n$, where n is a positive integer, occurs frequently in mathematics. As n increases without bound, this expression approaches a useful irrational number. This number is denoted e and its value is approximately 2.71828. The proof that the expression $\left(1 + \frac{1}{n}\right)^n$ approaches a specific number as the value of n increases without bound requires concepts not yet covered at this level. However, the value of the expression $\left(1 + \frac{1}{n}\right)^n$ can be obtained for any value of n. For $n = 10$, using a calculator we get

ENTER	PRESS	DISPLAY
	ON/C	0
	(	0
1		1
	+	1
1		1
	÷	1
10		10
	)	1.1
	y^x	1.1
10		10
	=	2.59374246

You may gain some insight for what happens by completing this table.

n	1	10	100	1000	10,000	100,000
$\left(1 + \dfrac{1}{n}\right)^n$	2.0000	2.5937	2.7048	2.7169		

DEFINITION 1.21: The exponential function with base e is often called the *natural exponential function*.

This function is so useful that most calculators have a function key e^x.

Applications

Exponential functions are important in mathematics. In this section, we give two illustrations of how they occur in a natural way. The first includes a brief discussion of compound interest and the second is an elementary treatment of population growth and decay, and exponential depreciation.

Compound Interest

Suppose \$$P$ is deposited in a bank that pays interest at a stated annual rate r, written as a decimal. For example, if the annual rate of interest is 7%, then $r = 0.07$. Suppose further that there are k interest periods of equal lengths per year. Then, the interest for each period is added to the principal at the end of that period and this interest, as well as the principal, earns interest during the following periods. In such cases, we say that the interest is *compounded*. The rate of interest per conversion period is i, where $i = \frac{r}{k}$. Let \$$A_n$ be the amount in the bank account at the end of the nth conversion period. We shall derive a formula for A_n in terms of the principal P, the rate of interest per period i, and the number of conversion periods n.

During the first period, the interest earned is \$$P \cdot i$. Thus,

$$A_1 = P + P \cdot i = P(1 + i)$$

The interest earned during the second period is \$$A_1 \cdot i$. Therefore,

$$A_2 = A_1 + A_1 i = A_1(1 + i)$$
$$= P(1 + i)(1 + 1) = P(1 + i)^2$$

The interest earned during the third period is \$$A_2 \cdot i$. It follows that

$$A_3 = A_2 + A_2 i = A_2(1 + i)$$
$$= P(1 + i)^2(1 + i) = P(1 + i)^3$$

Continuing in this way,

$$A_n = P(1 + i)^n$$

EXAMPLE 2   Suppose that $500 is deposited at 8% annual interest, compounded quarterly. Find the amount in the account 5 years later. What is the total interest earned?

Solution   We use the formula

$$A_n = P(1 + i)^n$$

with $P = 500$, $i = 0.02$ (since $i = \frac{r}{k} = \frac{0.08}{4} = 0.02$), and $n = 20$ (since $n = kt = 4 \cdot 5 = 20$). We get

$$A_{20} = 500(1 + 0.02)^{20}$$
$$= 500(1.02)^{20} \, {}^*$$
$$\doteq 500(1.4859474) = 742.9737 {}^{**}$$

To the nearest cent, the amount after 5 years is $742.97. The interest is $242.97 since $742.97 - 500 = 242.97$.
Do Exercise 19.   ■

Nominal and Effective Rates

DEFINITION 1.22:   The advertised yearly rate of interest is often called the *nominal interest rate*, while the actual percentage by which an account grows during each year is called the *effective interest rate*.

---

*The value of $(1.02)^{20}$ may be found using tables or the y^x key on a scientific calculator. We refer you to any book of mathematical tables. Find the page giving values of $(1 + i)^n$. To find $(1.02)^{20}$ look for 2% across the top and 20 (for 20 conversion periods) down the side. Where the column and row intersect you should find 1.485947, which is the approximate value of $(1.02)^{20}$. In the example, we used a calculator as follows:

ENTER	PRESS	DISPLAY
	ON/C	0.
1.02		1.02
	y^x	1.02
20		20.
	×	1.4859474
500		500
	=	742.9737

**The symbol $\doteq$ means "approximately equal to."

EXAMPLE 3 A bank advertises a nominal interest rate of 6% compounded monthly. What is the effective interest rate?

Solution For convenience, assume that $100 is deposited in the bank and use the formula

$$A_n = P(1 + i)^n$$

with $P = 100$, $i = 0.005$ $\left(\text{since } \frac{0.06}{12} = 0.005\right)$, and $n = 12$. We obtain

$$A_{12} = 100(1 + 0.005)^{12}$$

$$= 100(1.005)^{12} \doteq 100(1.06167781) = 106.167781$$

Therefore, the interest earned during 1 year when $100 is deposited is, to the nearest cent, $6.17. The effective interest rate is approximately 6.17%.
Do Exercise 25. ∎

Let us write the formula for A_n in terms of the nominal interest rate r, the number of conversion periods per year k, and the number of years t. We get

$$A_n = P(1 + i)^n = P\left(1 + \frac{r}{k}\right)^{kt}$$

Let $m = \frac{k}{r}$ so that $\frac{1}{m} = \frac{r}{k}$ and $k = mr$. Then, the formula may be written

$$A_n = P\left(1 + \frac{1}{m}\right)^{mrt}$$

and, using one of the laws of exponents,

$$A_n = P\left[\left(1 + \frac{1}{m}\right)^m\right]^{rt}$$

Since $m = \frac{k}{r}$, for a fixed nominal interest rate r, the value of m increases as the number k of conversion periods per year increases. Note that the value of $\left(1 + \frac{1}{m}\right)^m$ increases as the value of m increases. (See Exercise 47.) Therefore, a bank that compounds interest frequently will attract more customers than a bank offering the same nominal interest rate compounded less frequently.

What happens to A_n as the number of conversion periods increases without bound? That is, what will the formula be if interest is *compounded continuously*? We stated earlier that the value of $\left(1 + \frac{1}{m}\right)^m$ approaches the number e as the value of m increases without bound. It follows that the value of $P\left[\left(1 + \frac{1}{m}\right)^m\right]^{rt}$ approaches Pe^{rt} as the value of m increases without bound. From this we can derive the formula for interest compounded continuously. If $P is invested at a nominal interest rate r and interest is compounded continuously, then the balance after t years is

$$A(t) = Pe^{rt}$$

EXAMPLE 4   Suppose \$5000 is invested at a nominal interest rate of 6% compounded continuously. Find the value of that investment after 10 years.

Solution   Use the formula

$$A(t) = Pe^{rt}$$

with $P = 5000$, $r = 0.06$, and $t = 10$. Using the e^x function key on a calculator we find that

$$A(10) = 5000e^{(0.06)10} = 5000e^{0.6}$$

$$\doteq 5000(1.8221188) = 9110.594002$$

To the nearest cent, the answer is \$9110.59.
Do Exercise 29.   ∎

Population Growth and Exponential Depreciation

Suppose that the initial size of a population is P_0 and that the increase in its size in a unit of time is a certain percentage of its initial size. Then, as in the formula for compound interest, we find that after t units of time, the size $P(t)$ of the population is given by

$$P(t) = P_0(1 + r)^t$$

where r, expressed as a decimal, is the rate of growth of the population per unit of time.

EXAMPLE 5   At the beginning of 1975 the population of the earth was about 4 billion. Assuming population growth at 2% per year, predict the approximate size of the earth's population at the end of the year 2000.

Solution   The number of years from the beginning of 1975 to the end of 2000 is 26. Thus, we use

$$P(t) = P_0(1 + r)^t$$

with $P_0 = 4$, $r = 0.02$, and $t = 26$. We get

$$P(26) = 4(1 + 0.02)^{26}$$

$$= 4(1.02)^{26}$$

$$\doteq 4(1.67341811)$$

$$= 6.69367244$$

Thus, at the end of the year 2000, the population should be approximately 6.7 billion.
Do Exercise 31.   ∎

Using techniques of calculus, it can be shown that if the rate of change of the size of a population at instant t is proportional to the size of the population at that instant, then the size $P(t)$ is given by

$$P(t) = P_0e^{kt}$$

where P_0 is the size of the population at instant 0, k is a constant that depends on the population, and t is the number of time units. Since e and k are both constants, we can write b for e^k and obtain the formula

$$P(t) = P_0 b^t$$

since $e^{kt} = (e^k)^t = b^t$.

In the formula $P(t) = P_0(1 + r)^t$ written earlier, we can also replace $1 + r$ with b and get

$$P(t) = P_0 b^t$$

Note that both formulas are identical, although the base b represents different quantities in each case.

Since the size of the population at instant t is a constant multiple of the value of the exponential function with base b at t, we say that the population is *growing exponentially*. Recalling that the exponential function with base b is increasing when $b > 1$ and decreasing when $b < 1$, we see that the population experiences an exponential growth when $b > 1$ and an exponential decay whenever $b < 1$.

EXAMPLE 6   The number of bacteria in a certain culture is 160,000 at 1:00 P.M. At 3:00 P.M. the same day, the count is 320,000. Assuming that the population grows exponentially, what will the count be at 7:00 P.M. that day?

Solution   We use $P(t) = P_0 b^t$, where P_0 is 160,000 (since this was the initial count) and t is in hours. It is given that $P(2) = 320,000$ (2 hours elapsed from 1:00 to 3:00 P.M.), and we are asked to find $P(6)$. To find the value of b, we proceed as follows:

$$P(2) = 160,000 b^2$$

Thus,

$$320,000 = 160,000 b^2$$

It follows that

$$2 = b^2$$

and

$$b = 2^{1/2}$$

Hence,

$$P(t) = 160,000(2^{1/2})^t$$

Therefore,

$$P(6) = 160,000(2^{1/2})^6$$
$$= 160,000(2^3)$$
$$= 1,280,000$$

At 7:00 P.M. the number of bacteria in the culture will be 1,280,000.
Do Exercise 35.

> **DEFINITION 1.23:** If a substance decays exponentially, its *half-life* is the amount of time it takes for its size to be half of the initial amount.

EXAMPLE 7 The half-life of a certain radioactive substance is 25 years. If 20 grams of that substance was present in 1985, how much will remain in the year 2025?

Solution We use $P(t) = P_0 b^t$, where $P_0 = 20$ since 20 grams was the initial amount in 1985. We know that $P(25) = 10$ because the half-life of the substance is 25 years. Using this data, we get

$$P(25) = 20b^{25} = 10$$

It follows that

$$b^{25} = 0.5$$

and

$$b = (0.5)^{1/25}$$

Hence,

$$P(t) = 20[(0.5)^{1/25}]^t$$

The time elapsed from 1985 to 2025 is 40 years. Thus,

$$P(40) = 20[(0.5)^{1/25}]^{40}$$
$$= 20(0.5)^{1.6}$$
$$\doteq 20(0.329876978)$$
$$= 6.59753956$$

Approximately 6.6 grams of the substance will be left in the year 2025.
Do Exercise 37. ■

The following is an example of *exponential depreciation*.

EXAMPLE 8 The value of a building that cost its owner \$90,000 is declining 20% per year. What will the building be worth at the end of **a.** 1 year? **b.** 2 years? **c.** 3 years? **d.** t years?

Solution Let V_t be the value of the building after t years.

a. The value of the building declined 20% the first year. Thus, at the end of the first year, the value of the building is 80% of the original value. Thus,

$$V_1 = 90,000(.8) = 72,000$$

The value of the building is \$72,000 after 1 year.

b. At the end of the second year, the value of the building is 80% of V_1. That is,

$$V_2 = V_1(.8) = [90,000(.8)](.8) = 90,000(.8)^2 = 57,600$$

At the end of the second year the building is worth \$57,600.

c. At the end of the third year, the value of the building is 80% of V_2. That is,

$$V_3 = V_2(.8) = [90,000(.8)^2](.8) = 90,000(.8)^3$$

$$= 46,080$$

At the end of the third year, the building is worth \$46,080.

d. Continuing in the same manner, it is easy to see that at the end of t years, the value of the building is given by

$$V_t = 90,000(.8)^t$$

Do Exercise 43.

Exercise Set 1.6

In Exercises 1–10, simplify the given expression.

1. $5(8^{2/3})(9^{3/2})$

2. $-2(16^{3/4})(25^{5/2})$

3. $7(27^{4/3})(49^{3/2})$

4. $\dfrac{-9(8^{5/3})(81^{5/4})}{4^{5/2}}$

5. $\dfrac{(16^{-5/4})(8^{7/3})}{27^{-5/3}}$

6. $\sqrt[3]{27} \cdot \sqrt[4]{256}$

7. $\sqrt[5]{59049} \cdot (\sqrt[3]{4096})^2$

8. $\dfrac{\sqrt[4]{512} \cdot \sqrt[3]{32}}{\sqrt[6]{128}}$

9. $\dfrac{\sqrt[3]{81} \cdot \sqrt[5]{729}}{(\sqrt[5]{2187})^6}$

10. $\sqrt[4]{32} \cdot 7^0 \cdot 4^{5/2}$

11. Sketch the graph of the exponential function with base 3.

12. Sketch the graph of the exponential function with base 5.

13. Sketch the graph of the exponential function with base $\frac{1}{2}$.

14. Sketch the graph of the exponential function with base $\frac{1}{3}$.

15. Sketch the graph of the exponential function with base $\frac{1}{4}$.

16. Let the function f be defined by $f(x) = 3x$ and let g be the exponential function with base 2. Sketch the graph of the composition $g \circ f$ and that of the composition $f \circ g$.

17. Let the function g be defined by $g(x) = 2x - 3$ and let h be the exponential function with base 3. Sketch the graph of the composition $h \circ g$ and that of the composition $g \circ h$.

18. Suppose that \$1000 is deposited in a bank that pays 6% compounded monthly. Find the amount in that account (to the nearest cent) 7 years later. What is the compound interest earned during the 7 years?

19. Suppose that \$5000 is deposited in a bank that pays 8% compounded quarterly. Find the amount in that account (to the nearest cent) 9 years later. What is the compound interest earned during the 9 years?

20. Suppose that \$3000 is deposited at 6% compounded quarterly. Find the amount in that account 4 years later. What is the interest earned?

21. Suppose that \$8000 is deposited at 9% compounded monthly. Find the amount in that account 28 months later. What is the interest earned?

22. You have \$5000 to deposit in a bank. Bank A pays 7.9% interest compounded monthly, while bank B pays 8% interest compounded quarterly. Which bank would you choose?

23. You have \$3000 to invest. Banks A, B, and C pay 7.95% interest compounded daily, 8% interest compounded monthly, and 8.1% interest compounded quarterly, respectively. Which bank would you choose?

24. You have \$8000 to invest. Banks A, B, and C pay 9% interest compounded daily, 9.05% interest compounded monthly, and 9.1% interest compounded quarterly, respectively. Which bank would you choose?

25. A bank advertises a nominal interest rate of 8% compounded quarterly. What is the effective interest rate?

26. A bank advertises a nominal interest rate of 9% compounded monthly. What is the effective interest rate?

27. A bank advertises a nominal interest rate of 7% compounded daily. What is the effective interest rate?

28. Suppose that \$8000 is invested at a nominal interest rate of 8% compounded continuously. Find the value of that investment after 12 years.

29. Suppose that \$4000 is invested at a nominal interest rate of 9% compounded continuously. Find the value of that investment after 15 years.

30. Suppose that \$7500 is invested at a nominal interest rate of 7% compounded continuously. Find the value of that investment after 25 years.

31. The population on an island was estimated to be 75,000 at the beginning of 1986. Assuming that the population is growing at a rate of 2.5% per year, predict the approximate size of the population at the end of 1995.

32. The population on an island was estimated to be 90,000 at the beginning of 1987. Assuming that the population is growing at the rate of 1.5% per year, predict the approximate size of the population at the end of 1998.

33. The population on an island was estimated to be 250,000 at the beginning of 1983. Assuming that the population is growing at the rate of 3.5% per year, predict the approximate size of the population at the end of 2000.

In Exercises 34–39, assume that the population grows, or decays, exponentially.

34. The number of bacteria in a culture is 250,000 at 2:00 A.M. At 5:00 A.M. the same day, the count is 325,000. What will the count be at 2:00 P.M. that day?

35. The number of bacteria in a culture is 150,000 at 3:00 A.M. on Monday. At 2:00 A.M. Tuesday, the count is 250,000. What will the count be at 5:00 P.M. Thursday?

36. The number of wild rabbits on an island was estimated to be 150,000 at the beginning of 1985. The estimate at the beginning of 1987 was 225,000. Predict the approximate size of the rabbit population at the beginning of 1995.

37. Carbon 14 has a half-life of 5730 years. If there were 5 grams of carbon 14 initially, how much will there be after 17,190 years?

38. Cobolt 57 has a half-life of 270 days. If there were 8 grams of cobolt 57 initially, how much will there be after 1080 days?

39. Hydrogen 3 (or Tritium) has a half-life of 12.3 years. Suppose there were 3 grams of hydrogen 3 at the beginning of 1975. How much will there be at the end of the year 2000?

40. A secretarial school has established that on the average, if t weeks is the amount of time one student has attended the school and $W(t)$ is the number of words per minute that a student is able to type, then $W(t) = 90(1 - e^{-0.28t})$. Find the number of words per minute a student should type, on the average, after attending that school for a period of **a.** 3 weeks, **b.** 6 weeks, and **c.** 12 weeks.

41. Same as Exercise 40 if $W(t) = 110(1 - e^{-0.2t})$.

42. On the average, should a student attend the secretarial school of Exercise 40, or that of Exercise 41, if that student can go to school only for a period of **a.** 2 weeks, **b.** 5 weeks, and **c.** 8 weeks?

43. The original value of a truck is $30,000. If the value of the truck is declining 15% per year, what will the truck be worth at the end of **a.** 2 years? **b.** 4 years? **c.** 8 years?

44. The value of a machine that cost a roofing company $120,000 is declining 18% per year. What will the machine be worth at the end of **a.** 3 years? **b.** 6 years? **c.** 12 years?

45. The original value of a car is $15,000. If the value of the car is declining 18% per year, what will the car be worth at the end of **a.** 4 years? **b.** 6 years? **c.** 8 years?

46. The value of a boat that cost its owner $60,000 is declining 9% per year. What will the boat be worth at the end of **a.** 2 years? **b.** 5 years? **c.** 15 years?

47. Using a calculator, fill in the blanks in the following table and verify that your entries on the bottom line are increasing as the values of m increase.

m	2	5	15	50	150	500	1500
$\left(1 + \dfrac{1}{m}\right)^m$							

1.7 The Logarithmic Function

The Logarithmic Function with Base *b*

The exponential function with base b was defined in the preceding section where we stated that this function is decreasing in the case $0 < b < 1$ and increasing in the case $b > 1$. It follows that if $x_1 \neq x_2$ then $b^{x_1} \neq b^{x_1}$. Consequently, if we write $x = b^y$ for each value of x we have one, and only one value of y. So we have a new function called the *logarithmic function with base b* and denoted $\log_b$.

> **DEFINITION 1.24:** $y = \log_b x$ if, and only if, $x = b^y$

(It is customary to write $\log_b x$ instead of the usual functional notation $\log_b(x)$.)

It is useful to think of the logarithm of a positive number x to a base b ($b > 0$, $b \neq 1$) as the exponent to which b must be raised to obtain x.

It is clear that the domain of the logarithmic function is the range of the exponential function and vice versa. Thus,

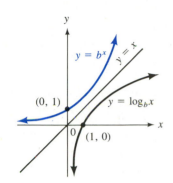

FIGURE 1.19

1. The domain of $\log_b$ is the set of all positive real numbers since it is equal to the range of the exponential function with base b.
2. The range of $\log_b$ is the set of all real numbers since it is equal to the domain of the exponential function with base b.
3. $b^{\log_b x} = x$ for every positive real number x. To see this, let $f(x) = b^x$, so that $f^{-1}(x) = \log_b x$ and $b^{\log_b x} = f(\log_b x) = f(f^{-1}(x)) = x$.
4. $\log_b b^x = x$ for every real number x. To see this, let $f(x) = b^x$, so that $f^{-1}(x) = \log_b x$ and $\log_b b^x = f^{-1}(b^x) = f^{-1}(f(x)) = x$.
5. The graphs of $y = \log_b x$ and $y = b^x$ are symmetric with respect to the graph of $y = x$. (See Figure 1.19.)
6. The graph of $y = \log_b x$ passes through the point $(1, 0)$ since the graph of $y = b^x$ passes through the point $(0, 1)$.
7. $\log_b b = 1$ since $b^1 = b$ and therefore, $\log_b b = \log_b b^1 = 1$.

Laws of Logarithms

There are, in addition to the preceding properties, the *laws of logarithms*.

> **The Laws of Logarithms**
>
> 1. $\log_b xy = \log_b|x| + \log_b|y|$, whenever xy is a positive real number.
> 2. $\log_b \dfrac{x}{y} = \log_b|x| - \log_b|y|$, whenever $\dfrac{x}{y}$ is a positive real number.
> 3. $\log_b x^y = y\log_b x$ for all positive real numbers x and arbitrary real numbers y.
> 4. $\log_b a = \dfrac{\log_c a}{\log_c b}$ for all positive real numbers a, b, and c with b and c not equal to 1.

We shall prove only the first of the four laws; the proofs of the other three are left as exercises.

Proof: If x and y are real numbers and $xy > 0$, using Property 7 of the exponential function, we get

$$b^{(\log_b|x| + \log_b|y|)} = b^{\log_b|x|} \cdot b^{\log_b|y|}$$

$$= |x|\,|y| = |xy| = xy, \text{ since } xy > 0$$

Thus,

$$\log_b xy = \log_b b^{(\log_b|x| + \log_b|y|)}$$
$$= \log_b|x| + \log_b|y|$$

The proof of the first law is now complete. Note that we used the fact that $b^{\log_b z} = z$ for any positive real number z and that $\log_b b^z = z$ for any real number z.

Common and Natural Logarithmic Functions

Although we may use any positive real number not equal to 1 as a base of a logarithmic function, the two numbers used most frequently as bases are 10 and e. The logarithmic function with base 10 is called the *common logarithmic function*, and we write $\log x$ instead of $\log_{10} x$.* The logarithmic function with base e is called the *natural logarithmic function* and we write $\ln x$ instead of $\log_e x$. It follows that
a. $\log 10 = 1$, since $\log 10 = \log_{10} 10 = 1$ (by Property 7 of the function $\log_b$), and
b. $\ln e = 1$, since $\ln e = \log_e e = 1$ (also by Property 7 of the function $\log_b$).

We will use the natural logarithmic function almost exclusively. However, we will give some examples using both logarithmic functions so that you can become familiar with the basic laws of logarithms.

EXAMPLE 1 Given that $\log 2 \doteq .3010$ and $\log 3 \doteq .4771$, and that $\ln 2 \doteq .6931$ and $\ln 3 \doteq 1.0986$, find the following:

a. $\log 8$ b. $\log 12$ c. $\ln \sqrt[3]{2}$ d. $\ln\left(\dfrac{9}{8}\right)$

e. $\log 5$ f. $\ln(3e^4)$ g. $\log_2 3$ h. $\log_5 12$

Solution
a. Since $8 = 2^3$, $\log 8 = \log 2^3 = 3\log 2 \doteq 3(.3010) = .9030$.
b. Observe first that $12 = 3 \cdot 4 = 3 \cdot 2^2$. Therefore, $\log 12 = \log(3 \cdot 2^2) = \log 3 + \log 2^2 = \log 3 + 2\log 2 \doteq .4771 + 2(.3010) = .4771 + .6020 = 1.0791$.
c. Recall that $\sqrt[3]{2} = 2^{1/3}$. Hence, $\ln \sqrt[3]{2} = \ln 2^{1/3} = \left(\frac{1}{3}\right)\ln 2 \doteq \left(\frac{1}{3}\right)(.6931) = .2310$.
d. We write $\frac{9}{8} = 3^2/2^3$. Thus, $\ln \frac{9}{8} = \ln (3^2/2^3) = \ln 3^2 - \ln 2^3 = 2\ln 3 - 3\ln 2 \doteq 2(1.0986) - 3(.6931) = .1179$.

e. Confronted with the task of finding $\log 5$ when only $\log 2$ and $\log 3$ are given, students are often tempted to start with $2 + 3 = 5$. However, this does not lead anywhere because there is no law to give the value of $\log(x + y)$ in terms

---

*Prior to the electronic calculator, common logarithms were used for computational purposes. The laws of logarithms illustrate the fact that a logarithmic function changes a product to a sum, a quotient to a difference, and a power to a product. Values of $\ln x$ and $\log x$ may be obtained using a calculator. Tables of values of common and natural logarithms are given in appendices of most standard mathematics books. Since it is no longer fashionable to use logarithms to perform calculations, we will not illustrate how this is done. If you are interested, you may find examples of such calculations in many mathematics books published prior to 1970.

of log x and of log y. Instead, start with $5 = \frac{10}{2}$. Thus, $\log 5 = \log\left(\frac{10}{2}\right) = \log 10 - \log 2 \doteq 1 - .3010 = .6990$.

f. We have $\ln(3e^4) = \ln 3 + \ln e^4 = \ln 3 + 4 \ln e \doteq 1.0986 + 4(1) = 1.0986 + 4 = 5.0986$.

g. Using Law 4, we write $\log_2 3 = \frac{\log 3}{\log 2} \doteq \frac{.4771}{.3010} \doteq 1.5850$. Note that we also could have used the natural logarithmic function to write $\log_2 3 = \frac{\ln 3}{\ln 2} \doteq \frac{1.0986}{.6931} \doteq 1.5850$. As expected, we obtained the same value for $\log_2 3$.

h. We again use Law 4 to write $\log_5 12 = \frac{\log 12}{\log 5}$. We found in part **b** that $\log 12 = 1.0791$ and in part **e** that $\log 5 = .6990$. Thus, $\log_5 12 \doteq \frac{1.0791}{.6990} \doteq 1.54377$.

Do Exercises 1 and 9.* ∎

EXAMPLE 2 Solve the equation $\log_4(3x - 2) = 3$.

Solution Since the domain of the function $\log_4$ is the set of positive real numbers, we must have $3x - 2 > 0$. That is, $x > \frac{2}{3}$. By definition, $\log_4(3x - 2) = 3$ is equivalent to $4^3 = 3x - 2$. Consequently, $4^{\log_4 z} = z$ for each positive real number z. Using this fact, and the fact that the exponential function is one-to-one, we conclude that the following equations are equivalent:

$$\log_4(3x - 2) = 3$$
$$4^{\log_4(3x-2)} = 4^3$$
$$3x - 2 = 4^3$$
$$3x = 66$$
$$x = 22$$

Since 22 is larger than $\frac{2}{3}$, it is the solution of the equation $\log_4(3x - 2) = 3$.
Do Exercise 11. ∎

 When solving equations involving one or more functions of the unknown, check that the answers belong to the domain(s) of the function(s).

EXAMPLE 3 Solve the equation $\log_2(x - 3) = 1 - \log_2(x - 4)$.

Solution Note that both $x - 3$ and $x - 4$ must be positive. Thus x must be greater than 3 and 4. A solution will be feasible only if it is greater than 4. The equation can be written

$$\log_2(x - 3) + \log_2(x - 4) = 1$$

and so

$$\log_2(x - 3)(x - 4) = 1$$

This equation is equivalent to

$$2^{\log_2(x-3)(x-4)} = 2^1$$

---

*Most of the values of the logarithmic functions in tables, or with a calculator, are approximate values. You should, therefore, be aware that the answers in the examples and the exercises are approximations, usually correct to four decimal places.

and to

$$(x - 3)(x - 4) = 2$$

That is,

$$x^2 - 7x + 10 = 0$$

The solutions are easily found by factoring. They are 2 and 5. However, only 5 is greater than 4. Thus, the only solution of the initial equation is 5.
Do Exercise 17. ■

Solving for Unknown Exponents

Logarithms are often used to solve equations where the unknown appears as an exponent, as illustrated in the following example.

EXAMPLE 4 Solve the equation $3^{2x} = 2^{x+1}$.

Solution We may use either of the two logarithmic functions since we were given their values at 2 and at 3 in Example 1. Since a logarithmic function is one-to-one, we have

$$3^{2x} = 2^{x+1}$$

if, and only if,

$$\ln 3^{2x} = \ln 2^{x+1}$$

This equation is equivalent to

$$2x \ln 3 = (x + 1)\ln 2$$

Thus,

$$2x(1.0986) \doteq (x + 1)(.6931)$$

We have replaced the exponential equation with a linear equation that can easily be solved. We have

$$(2.1972)x \doteq (.6931)x + .6931$$

From this, we get

$$(1.5041)x \doteq .6931$$

Thus,

$$x \doteq \frac{.6931}{1.5041} \doteq .4608$$

The solution of the exponential equation is approximately .4608.
Do Exercise 27. ■

We are now ready to give other examples of compound interest and population growth problems.

EXAMPLE 5   A bank pays 8% interest compounded quarterly. How long will it take for a deposit to double in value?

Solution   Use the formula $A = P(1 + i)^n$. Replace i by .02 $\left(\text{since } \frac{.08}{4} = .02\right)$, A by $2P$ (since the deposit is to double in value), and then solve for n. We get

$$2P = P(1 + .02)^n$$

That is,

$$2 = 1.02^n$$

This is equivalent to

$$\ln 2 = \ln 1.02^n$$

Thus,

$$\ln 2 = n \ln 1.02$$

and

$$n = \frac{\ln 2}{\ln 1.02} \doteq \frac{.6931}{.0198} \doteq 35$$

It will take approximately 35 quarters, or $8\frac{3}{4}$ years, for a deposit to double.
Do Exercise 29.   ∎

EXAMPLE 6   We initially have 3 grams of sodium-24; 7.5 hours later, 2.12 grams remain. Calculate the half-life of sodium-24.

Solution   Use the formula $P(t) = P_0 b^t$. It is given that $P_0 = 3$ and $P(7.5) = 2.12$. We are asked to find the value of t for which $P(t) = P_0/2 = 1.5$. We have

$$P(7.5) = 3b^{7.5}$$

Thus,

$$2.12 = 3b^{7.5}$$

and

$$b^{7.5} = \frac{2.12}{3} \doteq .70667$$

Using the laws of exponents, we obtain

$$(b^{7.5})^{1/7.5} \doteq .70667^{1/7.5}$$

and

$$b^1 = b \doteq .70667^{1/7.5}$$

It follows that

$$P(t) \doteq 3(.70667^{t/7.5})$$

Replacing $P(t)$ by 1.5, we get

$$1.5 \doteq 3(.70667^{t/7.5})$$

Thus,

$$.5 \doteq .70667^{t/7.5}$$

and

$$\ln .5 \doteq \ln .70667^{t/7.5} \doteq \frac{t}{7.5} \ln .70667$$

It follows that

$$t \doteq \frac{(7.5) \ln .5}{\ln .70667}$$

$$\doteq \frac{(7.5)(-.6931)}{-.3472} \doteq 14.97$$

The half-life of sodium-24 is approximately 15 hours.
Do Exercise 41.   ■

Exercise Set 1.7

In Exercises 1–10, use the following: log 2 $\doteq$ .3010, log 3 $\doteq$ .4771, ln 2 $\doteq$ .6931, and ln 3 $\doteq$ 1.0986 to find:

1. a. $\log 64$ **b.** $\log 36$ **c.** $\log 54$

2. a. $\log 1.5$ **b.** $\log .75$ **c.** $\log \dfrac{27}{16}$

3. a. $\log \sqrt[5]{4}$ **b.** $\log \sqrt[4]{27}$ **c.** $\log \sqrt[7]{.75}$

4. a. $\log 25$ **b.** $\log 200$ **c.** $\log .03$

5. a. $\log_3 2$ **b.** $\log_4 27$ **c.** $\log_9 64$

6. a. $\ln 64$ **b.** $\ln 36$ **c.** $\ln 54$

7. a. $\ln 1.5$ **b.** $\ln .75$ **c.** $\ln \dfrac{27}{16}$

8. a. $\ln \sqrt[5]{4}$ **b.** $\ln \sqrt[4]{27}$ **c.** $\ln \sqrt[7]{.75}$

9. a. $\ln 5e^3$ **b.** $\ln \sqrt[3]{e^2}$ **c.** $\ln \dfrac{27}{e}$

10. a. $\log_9 8$ **b.** $\log_8 243$ **c.** $\log_{.5} 27$

In Exercises 11–28, solve the given equation.

11. $\log_3(2x + 3) = 4$

12. $\log_2(5x - 6) = 3$

13. $\log_5(5 - 3x) = 2$

14. $\log_3(2x - 5) = 2 + \log_3(3x - 7)$

15. $\log_2(5 - 6x) = 3 + \log_2(3x - 7)$

16. $\log(x + 4) = 1 - \log(x + 1)$

17. $\log_{12}(x - 2) = 1 - \log_{12}(x - 3)$

18. $\log_2(x + 7) = 4 - \log_2(x + 1)$

19. $\log_2(x + 7) = 2 - \log_2(x + 4)$

20. $\ln(x - e) - 2 = \ln 6 - \ln(x - 2e)$

21. $\log|x + 4| = 1 - \log|x + 1|$

22. $\log_2|x + 7| = 2 - \log_2|x + 4|$

23. $\log_x(x + 12) = 2$ **24.** $\log_x(2x + 15) = 2$

25. $\log_x(3 + x - 3x^2) = 3$ **26.** $e^{3x-4} = 2^{2x+1}$

27. $3^{2x-1} = 2e^{3x+5}$ **28.** $3^{x/2} = 2(10^{3x})$

29. A bank pays 8% interest compounded quarterly. How long will it take for a deposit to triple in value?

30. A bank pays 6% interest compounded monthly. How long will it take for a deposit to double in value?

31. A bank pays 6% interest compounded monthly. How long will it take for a deposit to triple in value?

32. A bank pays 7% interest compounded continuously. How long will it take for a deposit to double in value?

33. A bank pays 8% interest compounded continuously. How long will it take for a deposit to triple in value?

34. The population of the earth was estimated to be 4 billion at the beginning of 1975. Assuming that the population grows 2% per year, when can we expect the population to be approximately **a.** 6 billion? **b.** 8 billion?

35. On an island, it was estimated that there were 20,000 wild rabbits at the beginning of 1987. Assuming that the rabbit population grows 30% per year, when can we expect the population to be 60,000?

36. A secretarial school has established that on the average, if t weeks is the amount of time one student has attended the school and $W(t)$ is the number of words per minute that student is able to type, then $W(t) = 90(1 - e^{-0.15t})$. Approximately how many weeks must a student attend the school to be able to type 70 words per minute?

37. Same as Exercise 36 with $W(t) = 110(1 - e^{-0.2t})$ and the student being able to type 80 words per minute.

In Exercises 38–41, assume that the population grows, or decays, exponentially.

38. In 1980, the population on an island was estimated to be 70,000; in 1985, the estimate was 80,000. When can we expect the population to be approximately 95,000?

39. The population of a city was 150,000 in 1970 and 195,000 in 1985. When can we expect the population to be approximately 225,000?

40. If the half-life of mercury 197 is 65 hours, how long does it take mercury 197 to lose three fourths of its mass?

41. The half-life of radium is 1690 years. How long does it take radium to lose one fourth of its mass?

*42. Prove that if x and y are real numbers, and $\frac{x}{y} > 0$, then

$$\log_b \frac{x}{y} = \log_b |x| - \log_b |y|.$$

*43. Prove that for all positive real numbers x and arbitrary real numbers y,

$$\log_b x^y = y \log_b x.$$

*44. Prove that for all positive real numbers a, b, and c with b and $c \neq 1$: $\log_b a = \log_c a / \log_c b$.

*45. Note that the equality of Exercise *44 may be written $(\log_c b)(\log_b a) = \log_c a$. Prove the following generalization. For any positive real numbers a, b, c, and d with a, b, and $c \neq 1$: $(\log_a b)(\log_b c)(\log_c d) = \log_a d$. (*Hint:* Let $\log_a b = x$, $\log_b c = y$, $\log_c d = z$, $\log_a d = w$, and note that $\log_a b = x$ is equivalent to $a^x = b$, and so on.) State a further generalization.

46. Suppose that money is invested at $n\%$ interest. Let x be the number of years it will take for the investment to double.

   a. Find a formula giving x in terms of n.
   b. Using a calculator, find the values of the product xn for $n = 5$, $n = 6$, $n = 7$, $n = 8$, $n = 9$, and $n = 10$.
   c. Based on your results in part **b**, find a positive integer K such that the following rule may be stated: "money invested at $n\%$ interest will double in approximately $\frac{K}{n}$ years."

47. Use the rule obtained in Exercise 46c to approximate how long it will take for money invested at 5.5% to double.

48. Use the rule obtained in Exercise 46c to approximate how long it will take for money invested at 6.8% to double.

1.8 The Sigma Notation

In mathematics, it is often desirable to compactly write a sum with many terms. For example, the expression

$$1 + 2^2 + 3^2 + \cdots + 50^2$$

may be used to write the sum of the squares of the first 50 positive integers. However, this type of notation is often not practical. For example, the individual terms may not be clearly defined from the first few terms. There is another way of expressing such sums. The symbol commonly used for this purpose is the capital sigma (Σ) of the Greek alphabet. Once this new notation becomes familiar, you will find not only that much labor is saved in writing sums, but also that the sums are clearer.

We now explain how this symbol is used. Suppose that A is a finite set and f is a function with domain A and range B where B is a set of real numbers. Then for each x in A, there is one, and only one, real number $f(x)$. The symbol

$$\sum_{x \in A} f(x)$$

denotes the sum whose terms are these $f(x)$'s. We stress that the terms of that sum need not be distinct and that *there are exactly as many terms in the sum as there are elements in A*. Also, if A is a set consisting of n consecutive integers, with a and b its smallest and largest members, respectively, then we use the symbol $\sum_{i=a}^{b} f(i)$ instead of $\sum_{x \in A} f(x)$.

EXAMPLE 1   Let $A = \{2, 4, 6, 8\}$ and let f be the function with domain A and defined by $f(x) = x^2$. Evaluate $\sum_{x \in A} f(x)$.

Solution   Note that $f(2) = 2^2 = 4$, $f(4) = 4^2 = 16$, $f(6) = 6^2 = 36$, and $f(8) = 8^2 = 64$. Thus,

$$\sum_{x \in A} f(x) = \sum_{x \in A} x^2 = 4 + 16 + 36 + 64 = 120$$

Do Exercise 1.   ■

EXAMPLE 2   Let $B = \{2, 3, 4, 5, 6\}$ and let g be the function with domain B and defined by $g(i) = 3i - 1$. Find $\sum_{i=2}^{6} g(i)$.

Solution   Note that $g(2) = 3(2) - 1 = 5$, $g(3) = 3(3) - 1 = 8$, $g(4) = 3(4) - 1 = 11$, $g(5) = 3(5) - 1 = 14$, and $g(6) = 3(6) - 1 = 17$. It follows that

$$\sum_{i=2}^{6} g(i) = \sum_{i=2}^{6} (3i - 1) = 5 + 8 + 11 + 14 + 17 = 55$$

Do Exercise 9.   ■

EXAMPLE 3   Evaluate $\sum_{i=1}^{7} 3$.

Solution   In this case we have $A = \{1, 2, 3, 4, 5, 6, 7\}$ and the function with domain A is the constant function defined by $f(i) = 3$. Thus, $f(1) = f(2) = f(3) = f(4) = f(5) = f(6) = f(7) = 3$, and

$$\sum_{i=1}^{7} 3 = \sum_{i=1}^{7} f(i) = 3 + 3 + 3 + 3 + 3 + 3 + 3 = 21$$

Do Exercise 23.   ■

In general, if n is a positive integer and k is a constant, using the same steps as in the preceding example, we can show that

$$\sum_{i=1}^{n} k = nk$$

EXAMPLE 4   Express the sum $1 + 3 + 5 + 7 + 9 + 11$ using the sigma notation.

Solution   Noting that the given sum is the sum of the first six odd positive integers and that an odd integer can be expressed as $2i + 1$ where i is an integer, we write

$$1 + 3 + 5 + 7 + 9 + 11 = \sum_{i=0}^{5} (2i + 1)$$

which can also be written

$$1 + 3 + 5 + 7 + 9 + 11 = \sum_{i=1}^{6} (2i - 1)$$

Do Exercise 25.   ■

Mathematical induction* may be used to prove the following formulas:

$$\sum_{i=1}^{n} i = \frac{n(n + 1)}{2} \tag{1}$$

$$\sum_{i=1}^{n} i^2 = \frac{n(n + 1)(2n + 1)}{6} \tag{2}$$

$$\sum_{i=1}^{n} i^3 = \frac{n^2(n + 1)^2}{4} \tag{3}$$

We can use these formulas to evaluate certain sums, as illustrated in the following example.

EXAMPLE 5   Evaluate **a.** $\displaystyle\sum_{i=1}^{20} i$, **b.** $\displaystyle\sum_{i=5}^{25} i^2$, and **c.** $\displaystyle\sum_{i=1}^{7} (i^3 + 2i^2 + 3i - 5)$.

Solution   **a.** Using Formula (1),

$$\sum_{i=1}^{20} i = \frac{20(20 + 1)}{2} = 10(21) = 210$$

b. To use Formula (2), first note that

$$\sum_{i=5}^{25} i^2 = \sum_{i=1}^{25} i^2 - \sum_{i=1}^{4} i^2$$

---

*You may find a discussion of Mathematical Induction in Appendix A.

Thus,

$$\sum_{i=5}^{25} i^2 = \frac{25(25 + 1)[2(25) + 1]}{6} - \frac{4(4 + 1)[2(4) + 1]}{6}$$

$$= \frac{25(26)(51)}{6} - \frac{4(5)(9)}{6}$$

$$= 5525 - 30 = 5495$$

c. Using the basic properties of the real numbers and Formulas (1), (2), and (3) we obtain

$$\sum_{i=1}^{7} (i^3 + 2i^2 + 3i - 5) = \sum_{i=1}^{7} i^3 + 2\sum_{i=1}^{7} i^2 + 3\sum_{i=1}^{7} i - \sum_{i=1}^{7} 5$$

$$= \frac{7^2(7 + 1)^2}{4} + 2\frac{7(7 + 1)[2(7) + 1]}{6} + 3\frac{7(7 + 1)}{2} - 7(5)$$

$$= 784 + 280 + 84 - 35 = 1113$$

Do Exercises 17 and 21.

■

Exercise Set 1.8

In Exercises 1–8, a set A is given and a function f is defined on A. In each case, evaluate $\sum_{x \in A} f(x)$.

1. $A = \{1, 3, 7, 10\}; f(x) = 3x + 1$

2. $A = \{0, 2, 5, 8, 11\}; f(x) = 2x - 4$

3. $A = \{-2, -1, 0, 3, 6\}; f(x) = x^2 + 1$

4. $A = \{-3, 1, 4, 5, 8\}; f(x) = -x^2 + 2$

5. $A = \{-4, 0, 2, 4, 5, 9\}; f(x) = x^3$

6. $A = \{-5, -1, 0, 3, 6, 10\}; f(x) = -x^3 + 3$

7. $A = \{-4, -3, -2, -1, 0, 1, 2, 3, 4\}; f(x) = x^3$

8. $A = \{-5, -2, 0, 1, 3, 6\}; f(x) = 5$

In Exercises 9–24, evaluate the given sum.

9. $\sum_{i=2}^{6} (3i + 5)$

10. $\sum_{i=3}^{10} (2i - 7)$

11. $\sum_{i=5}^{8} (i^2 + 3i - 2)$

12. $\sum_{i=6}^{10} (-i^2 + 3)$

13. $\sum_{i=-2}^{2} (i^2 + 1)$

14. $\sum_{i=1}^{150} i$

15. $\sum_{i=1}^{150} i^2$

16. $\sum_{i=1}^{150} i^3$

17. $\sum_{i=41}^{160} i$

18. $\sum_{i=41}^{120} i^2$

19. $\displaystyle\sum_{i=61}^{170} i^3$

20. $\displaystyle\sum_{i=1}^{100} (i^3 + 5i^2 - 7i + 6)$

21. $\displaystyle\sum_{i=1}^{120} (2i^3 - 5i^2 + 6i - 7)$

22. $\displaystyle\sum_{i=1}^{150} 6$

23. $\displaystyle\sum_{i=0}^{185} -9$

24. $\displaystyle\sum_{i=1}^{120} (-1)^i$

In Exercises 25–35, express the given sums using the Σ notation.

25. $1 + 3 + 5 + 7 + 9 + 11 + 13 + 15 + 17$

26. $3 + 6 + 9 + 12 + 15 + 18$

27. $5 + 8 + 11 + 14 + 17 + 20 + 23$

28. $5 + 10 + 15 + 20 + 25 + 30 + 35 + 40$

29. $1 + 4 + 9 + 16 + 25 + 36 + 49 + 64$

30. $16 + 25 + 36 + 49 + 64 + 81$

31. $8 + 27 + 64 + 125 + 216$

32. $2 + 2 + 2 + 2 + 2 + 2 + 2 + 2 + 2$

33. $1 - 1 + 1 - 1 + 1 - 1 + 1 - 1 + 1 - 1$

34. $-8 - 1 + 1 + 8 + 27 + 64 + 125$

35. $1 - 4 + 9 - 16 + 25 - 36 + 49 - 64 + 81$

1.9 Chapter Review

IMPORTANT SYMBOLS AND TERMS

$\subset$ [1.1]
$\cup$ [1.1]
$\cap$ [1.1]
$\in$ [1.1]
$\subseteq$ [1.1]
$\emptyset$ (or { }) [1.1]
$<$ [1.4]
$>$ [1.4]
$\leq$ [1.4]
$\geq$ [1.4]
$(-\infty, a)$ [1.4]
$(-\infty, a]$ [1.4]
(a, ∞) [1.4]
$(a, \infty]$ [1.4]
$[a, b]$ [1.4]
(a, b) [1.4]
$(a, b]$ [1.4]
$[a, b)$ [1.4]
b^x [1.6]
e^x [1.6]
f [1.5]
$f(x)$ [1.5]
log [1.7]
$\log_b$ [1.7]
ln [1.7]
A' [1.1]
Abscissa [1.5]
Base [1.4]
Bounded interval [1.4]

Closed interval [1.4]
Common logarithmic function [1.7]
Complement [1.1]
Completing squares [1.3]
Compound interest [1.6]
Compounding continuously [1.6]
Concave down [1.5]
Concave up [1.5]
Conditional equation [1.3]
Continuous function [1.3]
Coordinates [1.5]
Cube root [1.6]
Decreasing function [1.6]
Defective [1.3]
De Morgan's laws [1.2]
Dependent variable [1.5]
Disjoint [1.1]
Domain [1.5]
Effective rate of interest [1.6]
Element [1.1]
Empty set [1.1]
Equation in one variable [1.3]
Equilibrium price [1.3]
Equivalent [1.3]
Equivalent inequalities [1.4]

Exponent [1.6]
Exponential decay [1.6]
Exponential function with base b [1.6]
Function [1.5]
Graph [1.5]
Greater than [1.4]
Greater than or equal to [1.4]
Growing exponentially [1.6]
Half-life [1.6]
Half-open intervals [1.4]
Identity [1.3]
Image [1.5]
Increasing function [1.6]
Independent variable [1.5]
Inequalities [1.4]
Intersection [1.1]
Laws of Exponents [1.4]
Laws of logarithms [1.7]
Less than [1.4]
Less than or equal to [1.4]
Linear equation [1.3]
Linear function [1.5]
Logarithmic function with base b [1.7]
Member [1.5]
Natural exponential function [1.6]

Natural logarithmic function [1.7]
Nominal interest rate [1.6]
*n*th power of *a* [1.6]
*n*th root [1.6]
Open interval [1.4]
Open sentence [1.1]
Ordinate [1.5]
Ordered pair [1.5]
Order of the radical [1.6]
Parabola [1.5]
Point-slope form [1.5]
Population growth [1.6]
Principal *n*th root [1.6]
Proper subset [1.1]

Properties of inequalities [1.4]
Quadratic equation [1.3]
Quadratic formula [1.3]
Quadratic function [1.5]
Radical sign [1.6]
Radicand [1.6]
Range [1.5]
Redundant [1.3]
Replacement set [1.1]
Roster method [1.1]
Set [1.1]
Slope [1.5]
Slope-intercept form [1.5]

Solution [1.3]
Solution set [1.3]
Solving an equation [1.3]
Square root [1.6]
Subset [1.1]
Unbounded interval [1.4]
Union [1.1]
Universe [1.1]
Venn diagram [1.2]
Variable [1.1]
Vertex [1.5]
Word (Story) problems [1.3]
x-intercept [1.5]
y-intercept [1.5]

SUMMARY

A set is a collection of objects. If A and B are sets and $x \in A$ implies $x \in B$, we say that A is a subset of B and write $A \subseteq B$. In any discussion we have a universe U and each set under consideration is a subset of U. If A is a set, its complement is denoted A' and is the set of all elements in U but not in A. If A and B are sets, their union $A \cup B$ and intersection $A \cap B$ are defined as follows:

$$A \cup B = \{x \mid x \in A \text{ or } x \in B, \text{ or both}\}$$

and

$$A \cap B = \{x \mid x \in A \text{ and } x \in B\}$$

Venn diagrams may be used to represent sets pictorially.

A statement of equality containing one variable is called an equation. Two equations with the same replacement set are said to be equivalent if they have the same solution set. An equivalent equation is obtained if we do any of the following.

a. Add the same quantity to both sides of the equation.
b. Subtract the same quantity from both sides of the equation.
c. Multiply both sides by the same nonzero constant.
d. Divide both sides by the same nonzero constant.

If an equation involves a radical and we square both sides to eliminate the radical, we may obtain extra solutions. Hence, we must check which of the solutions of the new equation are solutions of the original one.

The solutions of $ax^2 + bx + c = 0$, where $a \neq 0$, may be found using the quadratic formula

$$x = \frac{-b \pm \sqrt{b^2 - 4ac}}{2a}$$

by factoring, or by completing squares. If a and b are real numbers and $b - a$ is positive, we say that a is less than b and write $a < b$. We also say that b is greater than a and write $b > a$. If $a < b$ or $a = b$, we write $a \leq b$ (read: "*a* is less than or equal to *b*"). Greater than or equal to is defined similarly. For real numbers a, b, c, and d,

1. Exactly one of the following is true: $a < b$, $a = b$, $a > b$.
2. $a < b$ if, and only if, $a + c < b + c$ (similarly for $\leq$).
3. If $0 < c$, then $a < b$ if, and only if, $ac < bc$ (similarly for $\leq$).
4. If $c < 0$, then $a < b$ if, and only if, $ac > bc$ (similarly for $\leq$).
5. If $a < b$ and $b < c$, then $a < c$ (similarly for $\leq$).
6. If $0 < a < b$, then $\frac{1}{b} < \frac{1}{a}$.
7. If $a < b$ and $c < d$, then $a + c < b + d$ (similarly for $\leq$). However, $a < b$ and $c < d$ does not imply $a - c < b - d$.

The process of solving linear inequalities in one variable is similar to that of solving linear equations. However, if we multiply or divide both sides of an inequality by a negative number, we must reverse the sense of the inequality.

A function is a rule that assigns to each element of a first set (the domain) a unique element of a second set. If f is the name of a function, and y corresponds to x, we write $y = f(x)$. The set $\{f(x) \mid x \in \text{domain of } f\}$ is the range of f. If the domain D_f and range of a function f are sets of real numbers, the set of points $(x, f(x))$ where $x \in D_f$ is called the graph of the function f. We have introduced the following functions.

1. The linear functions $y = f(x) = mx + b$, where m and b are real numbers. The graph of a linear function is a line. If (x_1, y_1) and (x_2, y_2) are two points on a line, the slope m of the line is

$$m = \frac{y_2 - y_1}{x_2 - x_1}, \text{ provided } x_2 - x_1 \neq 0$$

   If two lines have slopes m_1 and m_2, respectively, the lines are parallel if, and only if, $m_1 = m_2$; they are perpendicular if, and only if, $m_1 \cdot m_2 = -1$.
2. The quadratic function is $y = f(x) = ax^2 + bx + c$, where a, b, and c are constants and $a \neq 0$. Its graph is a parabola which is concave up if $a > 0$, and concave down if $a < 0$. The abscissa of the vertex is $\frac{-b}{2a}$
3. If b is a positive number and $b \neq 1$, the exponential function with base b is defined by $f(x) = b^x$ for all real numbers x. The exponential function serves as a mathematical model in many applications. The formula

   $$A = P(1 + i)^n$$

   is used to calculate the compound amount when money has been invested. It is also used to approximate the size of a population growing at a given rate.
4. If $b > 0$ and $b \neq 1$,

   $$y = \log_b x \text{ if, and only if, } x = b^y$$

The domain of the logarithmic function to the base b is the set of all positive real numbers and its range is the set of all real numbers. The graph passes through $(1, 0)$. Also,

$$\log_b(xy) = \log_b |x| + \log_b |y|, \text{ whenever } xy > 0$$

$$\log_b\left(\frac{x}{y}\right) = \log_b |x| - \log_b |y|, \text{ whenever } \frac{x}{y} > 0$$

$$\log_b(x^y) = y \log_b x, \text{ whenever } x > 0$$

$$\log_b b = 1$$

$$\log_b 1 = 0, \text{ and}$$

$$\log_b c = \frac{\log_a c}{\log_a b}, \text{ where } a, b \text{ are positive and not equal to 1.}$$

The bases most often used are 10 and e, where e is an irrational number approximately equal to 2.71828. An equation of the form $a^{f(x)} = b^{g(x)}$, where a and b are positive and not equal to 1, may be solved by first taking the logarithm of both sides.

SAMPLE EXAM QUESTIONS

1. Which of the following are true? Justify your answers.

   a. $2 \in (2, \infty)$ **b.** $\pi \in (3, 4)$ **c.** $\{2, 3, 5\} \in [2, 5]$
   d. $5 \in (-1, 5) \cap [5, 7]$ **e.** $(2, 3) \subseteq (-1, 4) \cap (1, 5)$
   f. $(4, 6) \cap [6, 9) \subseteq \{1, 3, 5\}$ **g.** $\{3, 6, 7\} \subseteq (-\infty, 4) \cup (5, 9)$
   h. $6 \subseteq \{5, 6, 7, 8\}$ **i.** $\emptyset \in [5, \infty)$

2. List all the subsets of the set $\{a, b, c\}$.

3. If the set A has five members, how many proper subsets does it have?

4. Use a Venn diagram to illustrate pictorially that if A and B are sets, then $A \cup (B \cap C) = (A \cup B) \cap (A \cup C)$.

5. A television station hires a man to poll 5000 viewers to find out how many people like the programs "Golden Girls," "Facts of Life," and "Family Ties" listing only those viewers who indicate a liking for at least one of the three programs. The director is troubled by the fact that the man gathered the results very quickly and doubts that he conducted the survey carefully. However, she is willing to pay him his fee provided the figures are consistent. These are the results:

Golden Girls	2093
Facts of Life	3534
Family Ties	3033
Facts of Life and Family Ties	2103
Facts of Life and Golden Girls	1523
Golden Girls and Family Ties	1323
All three	1013

   If the director asked you whether or not she should pay the man, what would you tell her? Justify your answer.

6. We want to enclose 4200 square meters in a rectangular region against an existing wall. The side against the wall does not require fencing. If we use 190 feet of fence, what are the dimensions of the rectangle?

7. Solve each equation.

   a. $\sqrt{x + 4} = 2x - 7$
   b. $\sqrt{7x + 22} = x + 4$

8. Solve by using the indicated method.
   a. $x^2 + 8x - 9 = 0$, (factoring)
   b. $x^2 - 11x + 18 = 0$, (factoring)
   c. $2x^2 + 16x - 66 = 0$, (completing squares)
   d. $3x^2 - 3x + 90 = 0$, (completing squares)

e. $-2x^2 + 5x - 2 = 0$, (quadratic formula)

f. $\frac{1}{3}x^2 + \frac{1}{6}x - 50 = 0$, (quadratic formula)

g. $2x^2 + 2x + 5 = 0$, (quadratic formula)

9. Solve each of the following inequalities and sketch the graphs of their solutions sets.

a. $3x - 2 < 5x - 10$

b. $\dfrac{x + 1}{3} + 1 \geq \dfrac{2x - 3}{7} + 2$

c. $|x - 2| < 5$

d. $|x + 3| \leq 2$

e. $|2 - x| \geq 5$

f. $|3 + x| + |5 - 2x| < 0$

g. $|2x - 3| < \dfrac{5}{4}$

h. $x^2 + 13x - 14 < 0$

i. $\dfrac{x^2 - 4}{x + 3} \geq 0$

j. $\dfrac{x^2 + x - 2}{15 + 2x - x^2} < 0$

k. $\dfrac{2x + 3}{x - 2} < 9$

l. $|2 - 3x| < 5x - 6$

10. Give an example to illustrate that $a < b$ and $c < d$ does not imply $a - c < b - d$.

In Exercises 11–12, an equation defines a function. In each case, find the domain of that function.

11. $g(x) = \sqrt{x - 2}$

12. $f(x) = \dfrac{x - 1}{x + 2}$

In Exercises 13–15, sketch the graph of the given function.

13. $f(x) = 2x + 3$

14. $g(x) = 2x^2 - 12x + 11$

15. $h(x) = -3x^2 + 24x - 33$

16. Julie has a job paying \$12 per hour. She is paid time and a half for the first 20 hours of overtime (between 40 and 60 hours in a given week), double time after that.
    a. If she works x hours during a given week, determine her income \$$I(x)$ in terms of x.
    b. Find $I(36)$.
    c. Find $I(52)$.
    d. Find $I(65)$.

17. Find the equation of each of the following lines in the requested form.
    a. Passing through $(-1, 2)$ and $(3, 4)$. Slope-intercept form.
    b. Passing through $(2, 5)$ and parallel to the graph of $2x + 3y = 5$. Point-slope form.
    c. Passing through $(-1, 3)$ and perpendicular to the graph of $x + 2y = 5$. Slope-intercept form.
    d. Passing through $(1, 3)$ and perpendicular to the x-axis.
    e. Passing through $(-1, 4)$ and perpendicular to the y-axis.

18. Sketch the graph of each function. (Use a calculator.)
    a. $y = f(x) = e^{-2x}$
    b. $y = g(x) = e^{x/2}$
    c. $y = h(x) = (2.5)^x$

19. Sketch the graph of the function f defined by $f(x) = \left(\frac{1}{3}\right)^x$.

20. The rabbit population on an island was estimated to be 15,000 at the beginning of 1987. Assuming that the population grows 12% per year, predict the approximate size of the rabbit population at the end of 1998.

21. If $700 is deposited at 6% interest compounded monthly, find the amount in the account (to the nearest cent) 3 years later. What is the interest earned?

22. Solve the given equations.
   a. $\log_2(5x - 8) = 5$ **b.** $\ln(3x - 5) = 2$
   c. $\log(x - 7) + \log(x + 8) = 2$ **d.** $2^{3x+1} = 3^{2x-3}$

23. Given $\log 2 \doteq .3010$ and $\log 3 \doteq .4771$, find:
   a. $\log 6$ **b.** $\log 8$ **c.** $\log \sqrt{3}$
   d. $\log \dfrac{4}{27}$ **e.** $\log 250$ **f.** $\log_2 10$

24. If $A = \{-1, 2, 3, 5, 9\}$ and $f(x) = x^2 + 1$, find $\sum_{x \in A} f(x)$.

25. Evaluate each of the following:
   a. $\sum_{i=1}^{20} i^2$ **b.** $\sum_{i=1}^{30} 5$

26. Evaluate: $\sum_{i=1}^{100} \left(\dfrac{1}{i} - \dfrac{1}{i+1} \right)$

27. If a bank pays 6% compounded monthly, how long will it take for a deposit to quadruple?

28. The Mason Company produces a gadget at a cost of $3.50 each. The fixed cost per month is $7000. The gadget sells for $5 each. Assuming that all gadgets produced can be sold, how many must be produced to realize a $9500 profit per month?

29. Mr. Greenthumb has a rectangular vegetable garden 60 feet by 40 feet. He will double the area of his garden by adding strips of equal width to one side and one end to get a new rectangular garden. Find the width of the strips.

30. The manager of a store wishes to mix two kinds of nuts to get 150 pounds of a mixture worth $4.10 a pound. Kind A is worth $3.50 a pound and kind B is worth $5 a pound. How many pounds of each kind must he use?

31. The manager of a store knows that 40 tennis racquets will be sold each week if the price is $90 per racquet, while 80 will be sold during the same period if the price is $70 per racquet.
   a. Derive an equation giving the price $p per racquet in terms of the number q of racquets sold each week assuming this equation is linear.
   b. Derive an equation giving the revenue $R per week in terms of q.
   c. If the cost to sell q racquets per week is $(60q + 250)$, how many racquets should be sold to maximize the profit? What is the price per racquet that will yield this maximum profit? What is the maximum profit?

32. A truck radiator with a 28-quart capacity has been prepared for fall driving with 2 quarts of antifreeze, but winter driving requires a solution which is 18.75% antifreeze. How many quarts of the fall solution should be withdrawn and replaced by pure antifreeze?

33. John and Mary invested part of $12,000 at 6% and the remainder at 8%. Their interest income in one year was the same as if they had invested the whole sum at 6.5%. How much did they invest at each rate?

34. A train travels 450 miles at a uniform speed. If its speed had been 10 miles per hour less, the trip would have taken 1 hour and 30 minutes longer. Find the actual speed of the train.

35. A truck radiator with a 28-quart capacity has been prepared for fall driving with 3 quarts of antifreeze, but winter driving requires a solution which is at least 25% but not more than 40% antifreeze. How many quarts of the fall solution should be withdrawn and replaced by pure antifreeze?

Systems of Equations and Inequalities

2

Many mathematical models that we encounter involve more than one variable. For example, a farmer may grow both apples and pears. In a mathematical model, x and y may represent the numbers of pounds of apples and pears, respectively, that he will produce. The manager of a grocery store may mix three kinds of nuts where x, y, and z, represent the number of pounds of each type A, B, and C, respectively. In this and subsequent chapters, we will consider open sentences with more than one variable.

2.1   Systems of Two Linear Equations in Two Variables

DEFINITION 2.1:   A *linear equation in two variables x and y* is an equation that can be written in the form $ax + by = c$, where a, b, and c are real numbers and $ab \neq 0$.

The replacement set for x and y is the set of all real numbers. An ordered pair of real numbers (x_0, y_0) is said to be a *solution of the equation* if, and only if, $ax_0 + by_0 = c$ is a true statement. For example, the ordered pair $(-2, 3)$ is a solution of the equation $2x + 3y = 5$ since $2(-2) + 3(3) = 5$ is true. On the other hand, the ordered pair $(3, 1)$ is not a solution of the same equation since $2(3) + 3(1) = 5$ is not true. When using other symbols for the two variables, consider one of them as the first variable and the other as the second, since a solution is an *ordered* pair of real numbers.

DEFINITION 2.2:   The set of all ordered pairs of real numbers that are solutions of the equation $ax + by = c$ is called the *solution set of the equation*. A linear equation in two variables has infinitely many solutions.

To see this, we need only replace one of the two variables by any real number and solve for the other.

EXAMPLE 1   Find three solutions of the equation $2x + 3y = 18$.

Solution   First, we replace x by 0 and get $0 + 3y = 18$ which yields $y = 6$. Thus, the ordered pair $(0, 6)$ is a solution. Next, we replace y by 0 and get $2x + 0 = 18$ which yields $x = 9$. Therefore, the ordered pair $(9, 0)$ is also a solution. We chose to replace x and y by 0 because this required only simple calculations. However, we may replace either variable by any real number we wish. If we replace x by 4.5, we get $2(4.5) + 3y = 18$ which yields $9 + 3y = 18$ and $y = 3$. Thus, the ordered pair $(4.5, 3)$ is a third solution.   ■

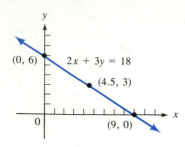

FIGURE 2.1

Graphing Systems of Two Linear Equations

We have seen that the graph of an equation of the form $ax + by = c$ is a straight line. For example, the graph of the equation of the previous example is the line passing through the points $(0, 6)$ and $(9, 0)$ (see Figure 2.1). Note that the point $(4.5, 3)$ is also on the graph. In fact, the coordinates of each point on the graph provide a solution of the equation.

If a_1, b_1, c_1, a_2, b_2, and c_2 are real numbers, the open sentence $a_1x + b_1y = c_1$ and $a_2x + b_2y = c_2$ is called a *system of two linear equations in two variables*. It is usually written as

$$\begin{cases} a_1x + b_1y = c_1 \\ a_2x + b_2y = c_2 \end{cases}$$

An ordered pair of real numbers (x_0, y_0) is a solution of the foregoing system if, and only if, it is a solution of both equations.

DEFINITION 2.3:   The *solution set* of the system is the intersection of the solution sets of both equations.

EXAMPLE 2   Solve the system

$$\begin{cases} 2x + 3y = 8 \\ 3x - 2y = -1 \end{cases}$$

graphically.

Solution   We know that the graph of each equation is a straight line. We sketch both graphs on the same Cartesian coodinate plane (see Figure 2.2). These two lines are graphical representations of the solution sets of the two equations of the system. Thus, the intersection of the two lines is a graphical representation of the solution of the system. It appears that this intersection is the single point $(1, 2)$. (See Figure 2.2.) We verify that $(1, 2)$ is indeed a solution of our system. We replace x by 1 and y by 2 in both equations. We get

$$2(1) + 3(2) = 8$$
$$3(1) - 2(2) = -1$$

Note that both of the equalities are true. Hence, $(1, 2)$ is a solution of the system. In fact, it is the only solution.   ■

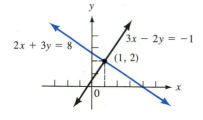

FIGURE 2.2

In general, it is not practical to solve a system of linear equations graphically. You would have to estimate the coordinates of the point of intersection of the two graphs, which may not be accurate. We shall, therefore, describe several algebraic methods to solve such systems. However, it is useful to consider the geometric approach because it sheds light on the nature of the solution set. Suppose, for example, that we have a system of two linear equations in the variables x and y. Let L_1 and L_2 be the graphs of the equations. Then, only one of three things can happen.

Characteristics of Graphs of Two Linear Equations

1. L_1 and L_2 intersect at exactly one point (see Figure 2.3a). In that case, the solution set of the system has exactly one member. We say that the system is *consistent and independent*.
2. L_1 and L_2 are parallel (see Figure 2.3b). In that case, we have no point of intersection, hence we have no solution. The solution set is the empty set and we say that the system is *inconsistent*.
3. L_1 and L_2 coincide. Then the coordinates of each point of L_1 are solutions of the given system, and the solution set has infinitely many members (see Figure 2.3c). In that case, we say that the system is *consistent and dependent*.

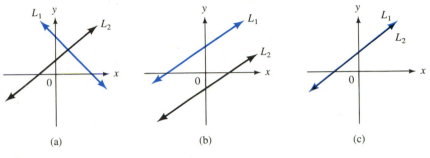

(a)   (b)   (c)

FIGURE 2.3

Solving by Elimination of Variables

You can solve a system of two linear equations in two variables using a number of algebraic methods. You can *eliminate a variable by addition or subtraction*. Given a system of two linear equations in two variables, an *equivalent system* is obtained if you multiply both sides of either equation by a nonzero constant or replace either equation by the sum, or the difference, of the two equations. Performing these two operations on a system results in an equivalent system where one of the two equations will have only one variable (the other variable having been eliminated). We solve the equation that has only one variable and replace this variable by its value in the other equation. We then solve this equation.

EXAMPLE 3 Solve the system

$$\begin{cases} 2x + 3y = -5 \\ x + 5y = -13 \end{cases}$$

Solution We first decide which variable to eliminate. Since the coefficient of x in the second equation is 1, we choose to eliminate x and multiply both sides of the second equation by -2 for a reason which will soon be clear. We obtain the equivalent system

$$\begin{cases} 2x + 3y = -5 \\ -2x - 10y = 26 \end{cases}$$

We now replace the second equation by the sum of the two equations to get the new equivalent system

$$\begin{cases} 2x + 3y = -5 \\ \quad\;\; -7y = 21 \end{cases}$$

Note that x has been eliminated from the second equation. We solve for y in the second equation to get

$$\begin{cases} 2x + 3y = -5 \\ \qquad y = -3 \end{cases}$$

Replacing y by -3 in the first equation, we obtain

$$\begin{cases} 2x - 9 = -5 \\ \qquad y = -3 \end{cases}$$

Adding 9 to both sides of the first equation then dividing both sides by 2 yields

$$\begin{cases} x = \;\;\; 2 \\ y = -3 \end{cases}$$

The solution of this last system is $(2, -3)$. Since each system is equivalent to the preceding one, the solution of the original system is also $(2, -3)$.
Do Exercise 3.   ■

EXAMPLE 4   Solve the system

$$\begin{cases} 7x + 2y = 31 \\ 3x - 5y = 25 \end{cases}$$

Solution   Decide which variable to eliminate. Since none of the coefficients is 1 and there is no advantage in choosing one variable over the other, eliminate y this time. We want an equivalent system where the y coefficients will have opposite signs but will have equal absolute values. Thus, we multiply both sides of the first equation by 5 and both sides of the second equation by 2. We obtain

$$\begin{cases} 35x + 10y = 155 \\ \;\; 6x - 10y = 50 \end{cases}$$

Replacing the second equation by the sum of the two equations, we get the equivalent system

$$\begin{cases} 35x + 10y = 155 \\ 41x \qquad\;\; = 205 \end{cases}$$

Solving the second equation for x we get

$$\begin{cases} 35x + 10y = 155 \\ \;\; x \qquad\quad = \;\;\; 5 \end{cases}$$

Replacing x by 5 in the first equation, we obtain

$$\begin{cases} 35(5) + 10y = 155 \\ \qquad\quad\;\; x = \;\;\; 5 \end{cases}$$

Subtracting 175 from both sides of the first equation, then dividing both sides by 10 yields

$$\begin{cases} y = -2 \\ x = 5 \end{cases}$$

The solution of the last system is $(5, -2)$. Since each system is equivalent to the preceding one, the solution of the given system is also $(5, -2)$.
Do Exercise 9. ∎

You can also solve a system of two linear equations in two variables by *eliminating a variable by substitution*. Solve for one of the variables in one of the two equations and substitute the expression obtained for that variable in the *other* equation.

EXAMPLE 5 Solve the system

$$\begin{cases} 3x + 2y = 6 \\ 5x + y = -4 \end{cases}$$

Solution Since the coefficient of y in the second equation is 1, it is easiest to solve for y in that equation. We get

$$y = -5x - 4$$

We now substitute $-5x - 4$ for y in the first equation and obtain the equivalent system

$$\begin{cases} 3x + 2(-5x - 4) = 6 \\ y = -5x - 4 \end{cases}$$

This system may be written

$$\begin{cases} -7x - 8 = 6 \\ y = -5x - 4 \end{cases}$$

Note that y has been eliminated from the first equation. We solve for x in that equation to get

$$\begin{cases} x = -2 \\ y = -5x - 4 \end{cases}$$

Replacing x by -2 in the second equation, we get

$$\begin{cases} x = -2 \\ y = -5(-2) - 4 \end{cases}$$

That is,

$$\begin{cases} x = -2 \\ y = 6 \end{cases}$$

Thus, the solution of the last and of the original system is $(-2, 6)$.
Do Exercise 19. ∎

EXAMPLE 6   Solve the system

$$\begin{cases} 5x + 3y = -3 \\ 3x - 7y = 29 \end{cases}$$

Solution   Since none of the coefficients is 1, there is no advantage in choosing one variable to eliminate over the other. We solve for x in the second equation and get

$$x = \frac{7y + 29}{3}$$

Substituting the foregoing expression for x in the first equation, we obtain the equivalent system

$$\begin{cases} 5 \cdot \dfrac{7y + 29}{3} + 3y = -3 \\ \qquad\quad 3x - 7y = 29 \end{cases}$$

We simplify the first equation by multiplying both sides by 3 to get

$$\begin{cases} 5(7y + 29) + 9y = -9 \\ \qquad\quad 3x - 7y = 29 \end{cases}$$

This system may be written

$$\begin{cases} \qquad 44y = -154 \\ 3x - 7y = 29 \end{cases}$$

Dividing both sides of the first equation by 44 we obtain

$$\begin{cases} \qquad y = \dfrac{-7}{2} \\ 3x - 7y = 29 \end{cases}$$

Replacing y by $\frac{-7}{2}$ in the second equation and solving for x we get

$$\begin{cases} y = \dfrac{-7}{2} \\ x = \dfrac{3}{2} \end{cases}$$

The solution of this system and consequently of the original system is $\left(\frac{3}{2}, \frac{-7}{2}\right)$.
Do Exercise 21.   ∎

Solving by Cramer's Rule

With the help of determinants, which we define below, we can solve a system of two linear equations in two variables efficiently using a method credited to the Swiss mathematician Gabriel Cramer (1704–1752) and called, appropriately, *Cramer's rule*. First, we define matrix and determinant.

DEFINITION 2.4: A square array of numbers of the form

$$\begin{bmatrix} a & b \\ c & d \end{bmatrix}$$

is called a *2 × 2* (read: "two-by-two") *matrix*. The *determinant* of this 2 × 2 matrix is the number $ad - bc$ and is usually denoted

$$\begin{vmatrix} a & b \\ c & d \end{vmatrix}$$

For example,

$$\begin{vmatrix} 3 & -5 \\ 2 & 4 \end{vmatrix} = 3(4) - 2(-5) = 12 + 10 = 22$$

Cramer's Rule

It can be shown (see Exercise 65) that if we let

$$\begin{vmatrix} a_1 & b_1 \\ a_2 & b_2 \end{vmatrix} = C, \quad \begin{vmatrix} c_1 & b_1 \\ c_2 & b_2 \end{vmatrix} = A_x, \quad \begin{vmatrix} a_1 & c_1 \\ a_2 & c_2 \end{vmatrix} = A_y,$$

then the systems

$$\begin{cases} a_1 x + b_1 y = c_1 \\ a_2 x + b_2 y = c_2 \end{cases}$$

and

$$\begin{cases} x = \dfrac{A_x}{C} \\ y = \dfrac{A_y}{C} \end{cases}$$

are equivalent, provided that $C \neq 0$.

Cramer's rule is easy to remember if we note the following:

1. C is the determinant of the matrix of coefficients of the variables in the original system.
2. A_x, which appears in the numerator of the fraction giving the value of x, is the determinant of the matrix obtained from the matrix of coefficients by replacing the column of coefficients of x by the column of constants on the right side of the original system.
3. A_y, which appears in the numerator of the fraction giving the value of y, is the determinant of the matrix obtained from the matrix of coefficients by replacing the column of coefficients of y by the column of constants on the right side of the original system.

EXAMPLE 7   Solve the system

$$\begin{cases} 2x + 3y = 6 \\ 5x + 4y = 1 \end{cases}$$

using Cramer's rule.

Solution   We find

$$C = \begin{vmatrix} 2 & 3 \\ 5 & 4 \end{vmatrix} = 2(4) - 5(3) = 8 - 15 = -7$$

$$A_x = \begin{vmatrix} 6 & 3 \\ 1 & 4 \end{vmatrix} = 6(4) - 1(3) = 24 - 3 = 21$$

$$A_y = \begin{vmatrix} 2 & 6 \\ 5 & 1 \end{vmatrix} = 2(1) - 5(6) = 2 - 30 = -28$$

Thus,

$$\begin{cases} x = \dfrac{A_x}{C} = \dfrac{21}{-7} = -3 \\ y = \dfrac{A_y}{C} = \dfrac{-28}{-7} = 4 \end{cases}$$

We see that the solution of the original system is $(-3, 4)$.
Do Exercise 35. ■

Inconsistent Systems and Consistent and Dependent Systems

The systems we have considered in the preceding examples were consistent and independent—that is, each system had exactly one solution. We now give an example of a system that is inconsistent (has no solution) and one that is consistent and dependent (its solution set is infinite).

EXAMPLE 8   Solve the system

$$\begin{cases} 2x + 3y = 13 \\ 6x + 9y = -2 \end{cases}$$

Solution   Multiplying the first equation by -3 and adding the result to the second equation to get a new second equation we obtain the equivalent system

$$\begin{cases} 2x + 3y = 13 \\ 0 = -41 \end{cases}$$

The second equality in this system is a compact form of the equation $0 \cdot x + 0 \cdot y = -41$. Clearly, this equation has no solution since any replacement of x and y by real numbers will lead to the false statement $0 = -41$. Thus, the solution set of this system and, therefore; of the original system, is the empty set. The system is inconsistent. Graph both linear equations of the system and note that the graphs are parallel lines.
Do Exercise 5. ■

EXAMPLE 9   Solve the system

$$\begin{cases} 3x - 5y = 7 \\ -6x + 10y = -14 \end{cases}$$

Solution   Multiplying the first equation by 2 and adding the result to the second equation to get a new second equation we have the equivalent system

$$\begin{cases} 3x - 5y = 7 \\ 0 = 0 \end{cases}$$

The second equality of this system is a compact form of the equation $0 \cdot x + 0 \cdot y = 0$. Any ordered pair of real numbers is a solution of this equation since replacing x and y by arbitrary real numbers always yields the true statement $0 = 0$. Thus, the solution set of the system is the solution set of the first equation $3x - 5y = 7$. If we let $x = c$, where c is an arbitrary real number, we obtain

$$3c - 5y = 7$$

Solving for y,

$$y = \frac{3c - 7}{5}$$

Hence, the solution set of the original system is

$$\left\{ \left(c, \frac{3c - 7}{5} \right) \,\middle|\, c \text{ is an arbitrary real number} \right\}$$

For example, if we replace c successively by $-6, 0, 4, 9$, and 14 we get the solutions $(-6, -5)$, $\left(0, \frac{-7}{5}\right)$, $(4, 1)$, $(9, 4)$, and $(14, 7)$. Of course, we could get as many solutions as we wish. Graph the two linear equations of the original system and note that the two graphs coincide.
Do Exercise 11.   ■

Nonlinear Systems

Sometimes a system of two nonlinear equations in two variables may be reduced to a system of linear equations by making an appropriate substitution. The next two examples involve such systems.

EXAMPLE 10   Solve the system

$$\begin{cases} 2x^2 - 3y^2 = -10 \\ 3x^2 + 2y^2 = 11 \end{cases}$$

Solution   The equations are nonlinear. However, if we let $u = x^2$ and $v = y^2$, we obtain the following system of two linear equations in u and v:

$$\begin{cases} 2u - 3v = -10 \\ 3u + 2v = 11 \end{cases}$$

Using any of the methods discussed, we easily find that the unique solution of this system is $(1, 4)$. However, this ordered pair is not a solution of the original system.

Recalling that $u = x^2$ and $v = y^2$, we write $1 = x^2$ and $4 = y^2$. Thus, the possible values of x are -1 and 1, while the possible values of y are -2 and 2. Consequently, the solution set of the original system is $\{(-1, -2), (-1, 2), (1, -2), (1, 2)\}$.
Do Exercise 53.   ■

EXAMPLE 11 Solve the system

$$\begin{cases} \dfrac{2}{x} - \dfrac{6}{y} = 3 \\ \dfrac{4}{x} - \dfrac{3}{y} = 3 \end{cases}$$

Solution We let $\frac{1}{x} = u$ and $\frac{1}{y} = v$, thus obtaining the system

$$\begin{cases} 2u - 6v = 3 \\ 4u - 3v = 3 \end{cases}$$

This system may be solved using any of the three methods we introduced earlier. Its solution is $\left(\frac{1}{2}, -\frac{1}{3}\right)$. Thus $\frac{1}{x} = \frac{1}{2}$ and $\frac{1}{y} = \frac{-1}{3}$. It follows that $x = 2$ and $y = -3$. The solution of the original system is $(2, -3)$.
Do Exercise 55.   ■

Applications of Systems of Linear Equations

The next two examples will illustrate some applications whose mathematical models are systems of linear equations.

EXAMPLE 12 Prior to a crucial NFL game between Denver and Seattle, a clever gambler met two enthusiastic football fans. The first fan was from Denver, and he was giving 2 to 1 odds on the Broncos; the second fan was from Seattle and he was giving 3 to 1 odds on the Seahawks. The clever gambler placed a bet with each fan so that she would win $100 as long as the game did not end up in a tie. How much did she bet with each fan?

Solution Suppose that the clever gambler bet $\$x$ with the Denver fan and $\$y$ with the Seattle fan. If Denver won the game, then the gambler would have to pay $\$x$ to the Denver fan but would receive $\$3y$ from the Seattle fan. Hence, her profit would be $\$(-x + 3y)$. Since she wanted to win $100, we obtain the first equation $-x + 3y = 100$. In exactly the same way, we see that if Seattle won the game, the gambler would win $\$2x$ from the Denver fan but would have to pay $\$y$ to the Seattle fan. Thus, $2x - y = 100$. We now have the system

$$\begin{cases} -x + 3y = 100 \\ 2x - y = 100 \end{cases}$$

This system can easily be solved. Its solution is $(80, 60)$. We conclude that the clever gambler bet $80 with the Denver fan and $60 with the Seattle fan. This example shows that sports fans should never bet on games, especially with strangers!
Do Exercise 59.   ■

EXAMPLE 13 A grocer has walnuts worth $4 a pound and Brazil nuts worth $5 a pound. How many pounds of each type must he add to get 150 pounds of a mixture worth $4.30 a pound?

Solution Let x be the number of pounds of walnuts and y be the number of pounds of Brazil nuts the grocer must mix. Since the mixture must weigh 150 pounds, the first equation is $x + y = 150$. The value of the walnuts is $4x$ and the value of the Brazil nuts is $5y$. Since the mixture must be worth $4.30 per pound, its value must be $645 since $(4.3)(150) = 645$. Clearly, the total value of the mixture is equal to the sum of the values of the components. Hence, the second equation is $4x + 5y = 645$. We obtain the system

$$\begin{cases} x + y = 150 \\ 4x + 5y = 645 \end{cases}$$

This system may be solved using any of the discussed methods. Its solution is $(105, 45)$. Thus, the grocer should mix 105 pounds of walnuts with 45 pounds of Brazil nuts to get 150 pounds of a mixture worth $4.30 a pound.
Do Exercise 61. ∎

Exercise Set 2.1

Solve each of the systems of linear equations in Exercises 1–17 using the method of elimination of a variable by addition or subtraction.

1. $\begin{cases} 2x - 3y = 8 \\ 3x + 5y = -7 \end{cases}$

2. $\begin{cases} 3x + y = -1 \\ 2x - 5y = -12 \end{cases}$

3. $\begin{cases} -2x + 3y = -9 \\ x - 4y = 7 \end{cases}$

4. $\begin{cases} 5x + 2y = -5 \\ 3x - 7y = -44 \end{cases}$

5. $\begin{cases} 2x - 7y = 17 \\ 6x - 21y = -2 \end{cases}$

6. $\begin{cases} 3x - 5y = 25 \\ 5x + 6y = 13 \end{cases}$

7. $\begin{cases} 7x - 3y = 37 \\ 2x + 6y = -10 \end{cases}$

8. $\begin{cases} 8x - 6y = -4 \\ 4x - 3y = -2 \end{cases}$

9. $\begin{cases} 5x + 8y = -34 \\ 3x - 5y = 9 \end{cases}$

10. $\begin{cases} 3x + 2y = 2 \\ 9x - 8y = -1 \end{cases}$

11. $\begin{cases} x + 3y = 5 \\ 4x + 12y = 20 \end{cases}$

12. $\begin{cases} .5x + .3y = 2.5 \\ 1.2x - 1.4y = -4.6 \end{cases}$

13. $\begin{cases} 5x - 3y = 38 \\ -4x + 7y = -35 \end{cases}$

14. $\begin{cases} 1.3x - 2.5y = 6 \\ 2.6x - 5y = -3 \end{cases}$

15. $\begin{cases} 5r + 8s = 46 \\ 2r - 5s = -37 \end{cases}$

16. $\begin{cases} 7u + 2v = 3 \\ -3u + 5v = 28 \end{cases}$

17. $\begin{cases} \dfrac{1}{4}t - \dfrac{1}{3}u = -\dfrac{1}{12} \\ \dfrac{1}{5}t + \dfrac{1}{6}u = \dfrac{3}{20} \end{cases}$

For Exercises 18–34, solve the systems of Exercises 1–17 using the method of elimination of a variable by substitution. For Exercise n solve the system of Exercise n − 17.

For Exercises 35–51, solve the systems of Exercises 1–17 using Cramer's rule. Whenever the determinant of the coefficient matrix is 0, use a different method. For Exercise n, solve the system of Exercise n ÷ 34.

In Exercises 52–57 solve the given system of nonlinear equations.

52. $\begin{cases} x^2 + y^2 = 13 \\ 2x^2 - 5y^2 = -37 \end{cases}$

53. $\begin{cases} 2x^2 - y^2 = -17 \\ 3x^2 + 2y^2 = 62 \end{cases}$

54. $\begin{cases} 2x^2 - 13y^2 = 59 \\ 3x^2 + 5y^2 = 113 \end{cases}$

55. $\begin{cases} \dfrac{5}{x} + \dfrac{8}{y} = -34 \\ \dfrac{3}{x} - \dfrac{5}{y} = 9 \end{cases}$

56. $\begin{cases} \dfrac{3}{x} + \dfrac{2}{y} = 2 \\ \dfrac{9}{x} - \dfrac{8}{y} = -1 \end{cases}$

57. $\begin{cases} -\dfrac{2}{x} + \dfrac{3}{y} = -9 \\ \dfrac{1}{x} - \dfrac{4}{y} = 7 \end{cases}$

58. Prior to a crucial game between the Minnesota Vikings and the Pittsburgh Steelers a clever gambler met two football fans. The first fan was giving 9 to 6 odds on the Vikings. The second fan was giving 7 to 5 odds on the Steelers. The clever gambler placed a bet with each fan so that he would win $110 as long as the game did not end up in a tie. How much did he bet with each fan?

59. Prior to an important NFL game between the Dallas Cowboys and the Los Angeles Rams an admirer of Jimmy the

Greek met two football fans. The first fan was giving 8 to 5 odds on the Cowboys, while the second fan was giving 7 to 4 odds on the Rams. She placed a bet with each fan so that she would win $360 as long as the game did not end up in a tie. How much did she bet with each fan?

60. A grocer has walnuts worth $3.50 a pound and Brazil nuts worth $4.50 a pound. How many pounds of each type must she mix in order to get 200 pounds of a mixture worth $4.15 a pound?

61. A grocer has two brands of coffee. Brand A is worth $4.25 a pound and brand B is worth $3.75 a pound. How many pounds of each brand must he mix in order to get 100 pounds of a mixture worth $3.90 a pound?

62. Julie invested part of the $10,000 she won at the lottery at a 6% annual rate of interest and the rest in a higher-risk bond paying an annual rate of interest of 9%. Her total annual income on the two investments is $690. How much did she invest at each rate?

63. John invested part of the $50,000 he inherited at a 6.5% annual rate of interest and the rest in a higher risk bond paying an annual rate of interest of 9.5%. His total annual income on the two investments is $3790. How much did he invest at each rate?

64. During last season, 6250 tickets were sold for a baseball game in the Seattle Kingdome. An adult pays $9 for a ticket, while a child under 16 pays only $3. If the total revenue from tickets for that game was $51,900, how many children and how many adults bought tickets?

*65. Prove Cramer's rule. (*Hint:* Use the method of elimination of a variable by addition to solve the system

$$\begin{cases} a_1x + b_1 y = c_1 \\ a_2x + b_2 y = c_2 \end{cases}$$

2.2 Systems of Three Linear Equations in Three Variables

In the present section we will consider systems of three linear equations in three variables. An example of such a system is

$$\begin{cases} 2x + 3y - 2z = 15 \\ 3x - y + 4z = -8 \\ -x + 4y + 2z = 7 \end{cases}$$

The ordered triple $(1, 3, -2)$ is a *solution* of this system since we get three true statements when we replace x, y, and z by 1, 3, and -2, respectively, in the three equations. The set of all solutions is called the *solution set* of the system.

Solving Linear Systems of Equations in Three Variables

The methods used to solve such systems are the same as those used in solving systems of two linear equations in two variables.*

---

*Although it is possible to interpret the solution set of a system of three linear equations in three variables as the intersection of three planes in a three-dimensional space, we shall not pursue this approach in this book. However, the student may find this geometric interpretation useful when considering the nature of the solution set of such systems. Clearly, the intersection of three planes will be either a single point, a straight line, a plane (if all three planes coincide), or the empty set. Thus, a system of three linear equations in three variables may have a single solution, an infinite number of solutions, or no solution.

As in the case of two systems of two equations, given a system of three linear equations in three variables, we obtain an *equivalent system* if we either multiply both sides of any equation by a nonzero constant or add to any equation one of the other two equations.

To shorten the explanations in the examples, we abbreviate the statement "both sides of the ith equation have been multiplied by the nonzero number k to yield the new ith equation" using the notation "$kE_i \rightarrow E_i$." We shall also abbreviate the statement "the members of the ith equation have been multiplied by the number k and the results have been added to the corresponding members of the jth equation to yield the new jth equation" using the notation "$kE_i + E_j \rightarrow E_j$." The meaning of "$k_1E_i + k_2E_j \rightarrow E_j$" should be easy to determine.

To find the solution to a given system of three linear equations in three variables, we eliminate one of the variables from two of the equations. We then eliminate one of the remaining variables from one of these two equations, and so on, until we obtain an equivalent system in the form

$$\begin{cases} x = x_0 \\ y = y_0 \\ z = z_0 \end{cases}$$

which gives the solution of the original system.

EXAMPLE 1 Solve the system

$$\begin{cases} 2x + 3y - 5z = -14 \\ 3x - 2y + 3z = 17 \\ 4x + 3y - 2z = -1 \end{cases}$$

Solution We wish to eliminate x from equations 2 and 3. We can eliminate x from equation 2 by multiplying equation 1 by 3, equation 2 by -2, and adding the results to yield a new equation 2. In the explanation following, we abbreviate this as $3E_1 + (-2)E_2 \rightarrow E_2$. We can also eliminate x from the third equation by multiplying the first equation by -2 and adding the result to equation 3 to yield the new equation 3. This is abbreviated as $(-2)E_1 + E_3 \rightarrow E_3$.

$$\begin{cases} 2x + 3y - 5z = -14 \\ 3x - 2y + 3z = 17 \\ 4x + 3y - 2z = -1 \end{cases} \qquad \begin{array}{l} \textit{Explanation} \\ 3E_1 + (-2)E_2 \rightarrow E_2 \\ (-2)E_1 + E_3 \rightarrow E_3 \end{array}$$

$$\begin{cases} 2x + 3y - 5z = -14 \\ 13y - 21z = -76 \\ {-3y} + 8z = 27 \end{cases}$$

Thus, in the new equivalent system x has been eliminated from equations 2 and 3. We now wish to eliminate y from equation 3. This can be accomplished by multiplying equation 2 by 3, equation 3 by 13, and adding the results to get the new equation 3.

We abbreviate this as $3E_2 + 13E_3 \to E_3$. The other necessary steps are described in abbreviated form in the explanation column.

Explanation

$$\begin{cases} 2x + 3y - 5z = -14 \\ 13y - 21z = -76 \\ -3y + 8z = 27 \end{cases}$$

$3E_2 + 13E_3 \to E_3$

$$\begin{cases} 2x + 3y - 5z = -14 \\ 13y - 21z = -76 \\ 41z = 123 \end{cases}$$

$\dfrac{1}{41}E_3 \to E_3$

$$\begin{cases} 2x + 3y - 5z = -14 \\ 13y - 21z = -76 \\ z = 3 \end{cases}$$

$21E_3 + E_2 \to E_2$

$$\begin{cases} 2x + 3y - 5z = -14 \\ 13y = -13 \\ z = 3 \end{cases}$$

$\dfrac{1}{13}E_2 \to E_2$

$$\begin{cases} 2x + 3y - 5z = -14 \\ y = -1 \\ z = 3 \end{cases}$$

$(-3)E_2 + 5E_3 + E_1 \to E_1$

$$\begin{cases} 2x = 4 \\ y = -1 \\ z = 3 \end{cases}$$

$\dfrac{1}{2}E_1 \to E_1$

$$\begin{cases} x = 2 \\ y = -1 \\ z = 3 \end{cases}$$

The solution of this last system is $(2, -1, 3)$. Since each system is equivalent to the preceding one, the solution of the original system is also $(2, -1, 3)$.
Do Exercise 1. ■

EXAMPLE 2   Solve the system

$$\begin{cases} x + 2y - 5z = -6 \\ -2x + 3y - z = -13 \\ 5x + 7y - 13z = -12 \end{cases}$$

Solution

Explanation

$$\begin{cases} x + 2y - 5z = -6 \\ -2x + 3y - z = -13 \\ 5x + 7y - 13z = -12 \end{cases}$$

$2E_1 + E_2 \to E_2$
$(-5)E_1 + E_3 \to E_3$

$$\begin{cases} x + 2y - 5z = -6 \\ 7y - 11z = -25 \\ -3y + 12z = 18 \end{cases} \qquad 3E_2 + 7E_3 \rightarrow E_3$$

$$\begin{cases} x + 2y - 5z = -6 \\ 7y - 11z = -25 \\ 51z = 51 \end{cases} \qquad \frac{1}{51}E_3 \rightarrow E_3$$

$$\begin{cases} x + 2y - 5z = -6 \\ 7y - 11z = -25 \\ z = 1 \end{cases} \qquad 11E_3 + E_2 \rightarrow E_2$$

$$\begin{cases} x + 2y - 5z = -6 \\ 7y = -14 \\ z = 1 \end{cases} \qquad \frac{1}{7}E_2 \rightarrow E_2$$

$$\begin{cases} x + 2y - 5z = -6 \\ y = -2 \\ z = 1 \end{cases} \qquad (-2)E_2 + 5E_3 + E_1 \rightarrow E_1$$

$$\begin{cases} x = 3 \\ y = -2 \\ z = 1 \end{cases}$$

The solution is $(3, -2, 1)$.
Do Exercise 3. ∎

It is also possible to eliminate a variable by substitution. We solve for one of the variables in terms of the other two variables using one of the three equations, then we substitute the expression obtained for that variable in the other two equations. We then have a system of two linear equations in two variables that we can solve by any of the methods discussed earlier.

EXAMPLE 3 Solve the system

$$\begin{cases} 2x + 4y - 5z = -8 \\ 3x - y + 2z = 14 \\ 5x + 3y - 7z = -2 \end{cases}$$

Solution Since the coefficient of y in the second equation is -1, it is best to solve for y in terms of x and z in that equation. We have

$$y = 3x + 2z - 14$$

Substituting this expression for y in equations 1 and 3, we get

$$\begin{cases} 2x + 4(3x + 2z - 14) - 5z = -8 \\ 5x + 3(3x + 2z - 14) - 7z = -2 \end{cases}$$

This system can be simplified to

$$\begin{cases} 14x + 3z = 48 \\ 14x - z = 40 \end{cases}$$

The solution of this system is easily found to be $(3, 2)$. Since $y = 3x + 2z - 14$, we get $y = 3(3) + 2(2) - 14 = -1$. Hence, the solution of the original system is $(3, -1, 2)$.

Do Exercise 17.   ∎

Consistent and Dependent Systems and Inconsistent Systems

The next two examples will involve a system whose solution set is infinite and another whose solution set is empty.

EXAMPLE 4   Solve the system

$$\begin{cases} x + y - 2z = 1 \\ x - y + 3z = 3 \\ 3x + y - z = 5 \end{cases}$$

Solution   We first eliminate x from equations 2 and 3.

$$\begin{array}{ll}
\begin{cases} x + y - 2z = 1 \\ x - y + 3z = 3 \\ 3x + y - z = 5 \end{cases} & \begin{array}{l} \textit{Explanation} \\[4pt] (-1)E_1 + E_2 \rightarrow E_2 \\ (-3)E_1 + E_3 \rightarrow E_3 \end{array}
\end{array}$$

$$\begin{cases} x + y - 2z = 1 \\ -2y + 5z = 2 \\ -2y + 5z = 2 \end{cases}$$

Notice that equations 2 and 3 of the system become identical. Thus, one may be left out, and we have a system of two linear equations in three variables.

$$\begin{cases} x + y - 2z = 1 \\ -2y + 5z = 2 \end{cases}$$

We may give z any value we wish and solve for the other two variables. Thus, let $z = c$ where c is an arbitrary real number. Then the system may be written

$$\begin{cases} x + y = 1 + 2c \\ -2y = 2 - 5c \end{cases}$$

This may be solved as follows:

$$\begin{array}{ll}
\begin{cases} x + y = 1 + 2c \\ -2y = 2 - 5c \end{cases} & \begin{array}{l} \textit{Explanation} \\[4pt] -\dfrac{1}{2}E_2 \rightarrow E_2 \end{array}
\end{array}$$

$$\begin{cases} x + y = 1 + 2c \\ \quad\quad y = -1 + \dfrac{5}{2}c \quad\quad (-1)E_2 + E_1 \rightarrow E_1 \end{cases}$$

$$\begin{cases} x = 2 - \dfrac{1}{2}c \\ y = -1 + \dfrac{5}{2}c \end{cases}$$

Thus, the solution set of the original system is $\left\{ \left(2 - \dfrac{c}{2}, -1 + \dfrac{5c}{2}, c\right) \mid \right.$ c is any real number$\left.\vphantom{\dfrac{c}{2}}\right\}$. We can produce as many solutions as we wish by giving c different values. This system is consistent and dependent. For example, if $c = 0$, we get the solution $(2, -1, 0)$; if $c = 2$, we get $(1, 4, 2)$; and if $c = -2$, we obtain $(3, -6, -2)$. You should verify that these three ordered triples are indeed solutions of the original system.

Do Exercise 21. ∎

EXAMPLE 5   Solve the system

$$\begin{cases} x + y - 2z = 3 \\ 2x - 3y + 3z = 2 \\ 5x - 10y + 11z = 7 \end{cases}$$

Solution   Eliminate x from equations 1 and 2, then y from equation 3 as follows:

Explanation

$$\begin{cases} x + y - 2z = 3 \\ 2x - 3y + 3z = 2 \quad\quad (-2)E_1 + E_2 \rightarrow E_2 \\ 5x - 10y + 11z = 7 \quad\quad (-5)E_1 + E_3 \rightarrow E_3 \end{cases}$$

$$\begin{cases} x + y - 2z = 3 \\ \quad -5y + 7z = -4 \\ \quad -15y + 21z = -8 \quad\quad (-3)E_2 + E_3 \rightarrow E_3 \end{cases}$$

$$\begin{cases} x + y - 2z = 3 \\ \quad -5y + 7z = -4 \\ \quad\quad\quad 0 = 4 \end{cases}$$

The last system has no solution since the third equality can be written $0 \cdot x + 0 \cdot y + 0 \cdot z = 4$, which yields a false statement each time x, y, and z are replaced by real numbers. Hence the solution set of our original system is empty. The system is inconsistent.

Do Exercise 15. ∎

Applications

We now give examples of word problems that make use of systems of three linear equations in three variables.

EXAMPLE 6   John, Mary, and Julie went to the same grocery store to buy sugar, coffee, and butter. John bought 5 pounds of sugar, 6 pounds of coffee, and 4 pounds of butter for $33; Mary bought 10 pounds of sugar, 2 pounds of coffee, and 12 pounds of butter for $40; while Julie paid $33 for 15 pounds of sugar, 3 pounds of coffee, and 6 pounds of butter. Find the cost per pound of each item.

Solution   Let x, y, and z be the prices in dollars per pound of sugar, coffee, and butter, respectively. Thus, John paid $\$5x$ for the sugar, $\$6y$ for the coffee, and $\$4z$ for the butter. Therefore, he spent $\$(5x + 6y + 4z)$. Since it was given that he spent $33, we have the first equation $5x + 6y + 4z = 33$. The other two equations are obtained in exactly the same way, giving us the system

$$\begin{cases} 5x + 6y + 4z = 33 \\ 10x + 2y + 12z = 40 \\ 15x + 3y + 6z = 33 \end{cases}$$

We now solve the system

Explanation

$$\begin{cases} 5x + 6y + 4z = 33 \\ 10x + 2y + 12z = 40 \\ 15x + 3y + 6z = 33 \end{cases} \qquad \begin{array}{l} (-2)E_1 + E_2 \to E_2 \\ (-3)E_1 + E_3 \to E_3 \end{array}$$

$$\begin{cases} 5x + 6y + 4z = 33 \\ -10y + 4z = -26 \\ -15y - 6z = -66 \end{cases} \qquad 3E_2 + (-2)E_3 \to E_3$$

$$\begin{cases} 5x + 6y + 4z = 33 \\ -10y + 4z = -26 \\ 24z = 54 \end{cases} \qquad \begin{array}{l} \dfrac{1}{2}E_2 \to E_2 \\[2mm] \dfrac{1}{24}E_3 \to E_3 \end{array}$$

$$\begin{cases} 5x + 6y + 4z = 33 \\ -5y + 2z = -13 \\ z = 2.25 \end{cases} \qquad (-2)E_3 + E_2 \to E_2$$

$$\begin{cases} 5x + 6y + 4z = 33 \\ -5y = -17.5 \\ z = 2.25 \end{cases} \qquad \dfrac{-1}{5}E_2 \to E_2$$

$$\begin{cases} 5x + 6y + 4z = 33 \\ y = 3.5 \\ z = 2.25 \end{cases} \qquad (-6)E_2 + (-4)E_3 + E_1 \to E_1$$

$$\begin{cases} 5x = 3 \\ y = 3.5 \\ z = 2.25 \end{cases} \qquad \dfrac{1}{5}E_1 \to E_1$$

$$\begin{cases} x = \quad .6 \\ y = 3.5 \\ z = 2.25 \end{cases}$$

We conclude that sugar, coffee, and butter cost $.60, $3.50, and $2.25 a pound, respectively.

Do Exercise 31. ■

A CLOSER LOOK

EXAMPLE 7 A refinery must provide one of its customers with 10,000 gallons of 88 octane gasoline. City regulations require that the vapor pressure index of the gasoline be 24. The refinery must mix three kinds of gasoline to obtain the appropriate mixture. (See the following table.)

	Octane rating	Vapor pressure index	Price per gallon
Regular unleaded	79	33	$.80
Premium unleaded	87	25	$.88
Super premium unleaded	98	14	$1.10

How many gallons of each gasoline should be mixed to fill the customer's order at a minimum cost?

Solution Let x, y, and z gallons be the quantities of regular unleaded, premium unleaded, and super premium unleaded, respectively, that must be used. Since 10,000 gallons must be provided, we have

$$x + y + z = 10,000$$

Since the octane rating of 10,000 gallons of mixture must be 88, and the octane rating of x gallons is 79, that of y gallons is 87, and that of z gallons is 98, we have

$$79x + 87y + 98z = 88(10,000)$$

Using the fact that the vapor pressure index of the 10,000 gallons of mixture must be 24, and that the vapor pressure indices of the x, y, and z gallons are 33, 25, and 14, respectively, we get

$$33x + 25y + 14z = 24(10,000)$$

To solve the system

$$\begin{cases} x + \quad y + \quad z = 10,000 \\ 79x + 87y + 98z = 88(10,000) \\ 33x + 25y + 14z = 24(10,000) \end{cases}$$

we proceed as follows:

$$
\begin{cases}
x + y + z = 10{,}000 \\
79x + 87y + 98z = 880{,}000 \\
33x + 25y + 14z = 240{,}000
\end{cases}
$$

Explanation
$(-79)E_1 + E_2 \rightarrow E_2$
$(-33)E_1 + E_3 \rightarrow E_3$

$$
\begin{cases}
x + y + z = 10{,}000 \\
8y + 19z = 90{,}000 \\
-8y - 19z = -90{,}000
\end{cases}
$$

The second and third equations are obviously equivalent. Thus, we delete the third equation to obtain

$$
\begin{cases}
x + y + z = 10{,}000 \\
8y + 19z = 90{,}000
\end{cases}
$$

Since z represents the number of gallons of super premium unleaded gasoline, it cannot be negative. Thus, we let $z = a$, where a is an arbitrary nonnegative number. From the second equation

$$
y = 11{,}250 - \frac{19a}{8}
$$

Replacing y by $11{,}250 - \frac{19a}{8}$ and z by a in the first equation, we get

$$
x = 10{,}000 - \left(11{,}250 - \frac{19a}{8}\right) - a = \frac{11a}{8} - 1250
$$

We must be sure that the values of x and y are nonnegative also. Thus, we require that

$$
11{,}250 - \frac{19a}{8} \geq 0 \quad \text{and} \quad \frac{11a}{8} - 1250 \geq 0
$$

This double inequality may be written

$$
\frac{10{,}000}{11} \leq a \leq \frac{90{,}000}{19}
$$

The solution set is

$$
\left\{ \left(\frac{11a}{8} - 1250,\ 11{,}250 - \frac{19a}{8},\ a \right) \ \middle|\ \frac{10{,}000}{11} \leq a \leq \frac{90{,}000}{19} \right\}
$$

Since the prices per gallon of the regular unleaded, premium unleaded, and super premium unleaded are \$.80, \$.88, and \$1.10, respectively, the total cost in terms of x, y, and z is \$$(.8x + .88y + 1.10z)$. We may express this cost in terms of a as follows:

$$
C = .8\left(\frac{11a}{8} - 1250\right) + .88\left(11{,}250 - \frac{19a}{8}\right) + 1.10a = 8900 + .11a
$$

Thus, the cost will be at a minimum when the value of a is minimized. That is, when $a = \frac{10,000}{11}$. The cost will be \$9000 since $8900 + .11\left(\frac{10,000}{11}\right) = 9000$. Also, $x = \frac{11}{8}\left(\frac{10,000}{11}\right) - 1250 = 0$, and $y = 11,250 - \frac{19}{8}\left(\frac{10,000}{11}\right) = \frac{100,000}{11}$. Thus, the refinery should use $\frac{100,000}{11}$ gallons of premium unleaded gasoline and $\frac{10,000}{11}$ gallons of super premium unleaded gasoline.

Do Exercise 39. ∎

Exercise Set 2.2

In Exercises 1–15, use the method of elimination of variables by addition or subtraction to solve the given systems.

1.
$\begin{cases} 2x + 3y - 5z = -1 \\ 3x - 2y + 4z = -3 \\ 6x + 4y - 3z = -1 \end{cases}$

2.
$\begin{cases} x + 3y + 7z = 5 \\ 2x - 4y + 2z = 8 \\ 5x - 2y + 3z = 10 \end{cases}$

3.
$\begin{cases} 2x + 3y + 2z = 8 \\ -x + 4y - 3z = 1 \\ 3x - 2y + 7z = 11 \end{cases}$

4.
$\begin{cases} 5x + 2y - 3z = 21 \\ 2x - 7y + 5z = -4 \\ 3x \qquad - 4z = 17 \end{cases}$

5.
$\begin{cases} x + y + 3z = 2 \\ 3x + 2y - z = -3 \\ 11x + 8y + 3z = 6 \end{cases}$

6.
$\begin{cases} 3x \qquad + 2z = -9 \\ \quad 2y + 5z = 4 \\ 3x + y + 3z = -7 \end{cases}$

7.
$\begin{cases} -2x + 3y + z = 4 \\ 3x - 2y + 2z = 14 \\ 4x + 3y - z = 6 \end{cases}$

8.
$\begin{cases} x + 2y - 5z = 3 \\ 2x - 3y + 3z = -2 \\ 5x + 3y - 12z = 7 \end{cases}$

9.
$\begin{cases} 3x + 2y - 5z = -2 \\ 2x + 3y - 8z = 5 \\ -x + 2y - 3z = 8 \end{cases}$

10.
$\begin{cases} 2x - 3y + 4z = 1 \\ 4x + 5y - 3z = 2 \\ 2x - 14y + 15z = 7 \end{cases}$

11.
$\begin{cases} x + y + z = 6 \\ 2x + y + 3z = 12 \\ x - y + 5z = 10 \end{cases}$

12.
$\begin{cases} 2x - 2y + 3z = 1 \\ x - 3y - 2z = -9 \\ x + y + z = 6 \end{cases}$

13.
$\begin{cases} x + y \qquad = -4 \\ x \qquad + z = 1 \\ 3x - y + 2z = 4 \end{cases}$

14.
$\begin{cases} 2x + 3y - z = -1 \\ 3x + y - 2z = -13 \\ x - 4y + 2z = 11 \end{cases}$

15.
$\begin{cases} 3x + y + z = 0 \\ x + y \qquad = 1 \\ 7x + 3y + 3z = 2 \end{cases}$

Solve each of the systems of linear equations in Exercises 16–27 by elimination of a variable by substitution.

16. The system of Exercise 2.

17. The system of Exercise 3.

18. The system of Exercise 5.

19. The system of Exercise 6.

20. The system of Exercise 7.

21. The system of Exercise 8.

22. The system of Exercise 9.

23. The system of Exercise 11.

24. The system of Exercise 12.

25. The system of Exercise 13.

26. The system of Exercise 14.

27. The system of Exercise 15.

28. A confectioner wishes to mix three kinds of candy to get 100 pounds of a mixture worth \$3.95 a pound. Kind A is worth \$3.50 a pound, while kinds B and C are worth \$4 and \$5 a pound, respectively. He wants to have the weight of kind A equal to the sum of the weights of kinds B and C. How many pounds of each kind of candy must he use?

29. A confectioner wishes to mix three kinds of candy to get 400 pounds of a mixture worth \$4.50 a pound. Kind A is worth \$4.20 a pound, while kinds B and C are worth \$5.25 and \$6 a pound, respectively. She wants the weight of kind A to be three times the sum of the weights of kinds B and C. How many pounds of each kind of candy must she use?

30. Mark won \$10,000 at the Washington State lottery. He invested his winnings in three bonds. Bond A was the safest investment paying 6.5% interest per year, while bonds B and C were not as safe but paid 8% and 12% interest per year, respectively. Mark wanted to be conservative so he invested four times as much in bond A than in the other two bonds combined. His yearly income from these investments is $700. How much did he invest in each bond?

31. Janet inherited $50,000. She invested that sum in three bonds. Bond A paid 6.2% interest per year, while bonds B and C paid 9% and 11% interest per year, respectively. She decided to invest as much in bond A as in the other two bonds combined. Her yearly income from these investments is $4000. How much did she invest in each bond?

32. Shirley, Norma, and Roy went to the same grocery store to buy flour, sugar, and butter. Shirley bought 20 pounds of flour, 40 pounds of sugar, and 5 pounds of butter for $39.25; Norma bought 10 pounds of flour, 20 pounds of sugar, and 10 pounds of butter for $36.50; while Roy paid $40 for 50 pounds of flour, 20 pounds of sugar, and 8 pounds of butter. Find the cost per pound of each item.

33. Ann, Greg, and George went to the same store to purchase items X, Y, and Z. Ann bought 2 units of X, 3 units of Y, and 1 unit of Z for a total cost of $57.75; while Greg bought 1 unit of X, 2 units of Y, and 3 units of Z for $67.50. George paid $85.25 for 2 units of X, 3 units of Y, and 3 units of Z. What is the price per unit of each item?

34. Three kinds of cargo were loaded on a ferry boat. Each unit of each kind requires a certain amount of space, weighs a certain number of pounds, and costs a certain amount, as given in the following table.

	Space in cubic feet	Weight in pounds	Cost in dollars
Kind A	5	50	32
Kind B	12	180	85
Kind C	7	85	96

If the total value of the cargo was $31,720, while the total weight and space were 51,200 pounds and 3740 cubic feets respectively, how many units of each kind were loaded on the ferryboat?

35. A construction company builds three types of houses. The number of units of roofing, concrete, and lumber required for each type of house are given in the following table.

	Roofing	Concrete	Lumber
Type A	3	5	4
Type B	2	6	7
Type C	5	4	6

If in a certain month the company used 51 units of roofing, 67 units of concrete, and 77 units of lumber, how many houses of each type were built that month?

36. An absentminded professor forgot how many students he has in his class. However, he remembers clearly that he has as many freshmen as he has sophomores and juniors together, that the number of freshmen exceeds the number of sophomores by 5 and of juniors by 15. How many students does he have?

37. The absentminded professor of Exercise 36 has a wife who does not like to reveal her age. One of her inconsiderate friends asked her what her age was and she replied: "My husband is 4 years older than I am. I was 24 when my son was born and in 20 years my age combined with my son's age will be only three halves of my husband's age." Find her age and those of her husband and son.

*38. A refinery must provide one of its customers with 25,200 gallons of 89 octane gasoline. State regulations require that the vapor pressure index of the gasoline be 21. The refinery must mix three kinds of gasoline to obtain the appropriate mixture. Information on the three kinds of gasoline is given in the table.

	Octane rating	Vapor pressure index	Price per gallon
Regular unleaded	76	34	$.85
Premium unleaded	88	22	$1.04
Super premium unleaded	97	13	$1.15

How many gallons of each gasoline should be mixed to fill the customer's order at a minimum cost?

*39. A hospital dietician wishes to mix three food items to get 200 pounds of a mixture containing 290 units of vitamins and 460 calories per pound. The relevant information on the three food items is shown in the table.

	Units of vitamins per pound	Number of calories per pound	Price per pound
Item A	150	600	$1.25
Item B	250	500	$1.35
Item C	420	330	$1.60

How many pounds of each item must she mix to minimize the cost?

*40. A company manufactures three products X, Y, and Z. The manufacturing of each unit of each product requires a certain number of hours from each of three departments A, B, and C, as is shown in the table.

Department	Hours required per unit of product		
	X	Y	Z
A	7	2	1
B	2.5	.7	.3
C	2.9	.85	.5

Each month the company has 230 hours available in department A, 81 hours in department B, and 97 hours in department C. Because of contractual obligations, all available hours must be used. If the profits per unit of products X, Y, and Z are, respectively, $239, $100, and $120, how many units of each product should be manufactured and sold per month to maximize the profit? What is the maximum profit?

2.3 Graphical Solutions of Systems of Equations

We have seen that an ordered pair (or triple) of real numbers is a solution of a system of equations in two (or three) variables if, and only if, it is a solution of each equation of the system. Consequently, the solution set of the system is the intersection of the solution sets of these equations. If the graphs of the equations of the system are sketched on the same set of coordinate axes, then we can determine geometrically the nature of the solution set of the system by considering the intersection of these graphs. In an earlier section (Example 2, Section 2.1) we gave a graphical solution of a system of two linear equations in two variables. In this section, we consider examples of systems of two equations in two variables where at least one of the two equations is nonlinear. The solutions of these systems will be found by sketching the graphs of the two equations and reading the coordinates of the points of intersection of these graphs. Since small errors in reading the coordinates of a point from a graph as well as small errors in drawing the graphs are unavoidable, solutions found graphically should always be checked by substitution in both equations.

EXAMPLE 1   Solve this system of equations graphically.

$$\begin{cases} y = x^2 - 2x - 2 \\ y = x + 2 \end{cases}$$

Solution   The graph of the first equation is a parabola that is concave up since the leading coefficient is positive. The abscissa of the vertex is 1 since in this case $\frac{-b}{2a} = \frac{-(-2)}{2(1)} = \frac{2}{2} = 1$. The ordinate of the vertex is found by replacing x by 1 in the first equation. It is -3. Hence, the vertex is the point $(1, -3)$. The parabola is sketched in Figure 2.4. The graph of the second equation is a straight line, which is also sketched in Figure 2.4.

From the graphs, it appears that there are two points of intersection with coordinates $(-1, 1)$ and $(4, 6)$. We substitute -1 for x and 1 for y in both equations of the system and obtain

$$\begin{cases} 1 = (-1)^2 - 2(-1) - 2 \\ 1 = -1 + 2 \end{cases}$$

which are both true. Similarly, we substitute 4 for x and 6 for y in both equations of the system and get

$$\begin{cases} 6 = 4^2 - 2(4) - 2 \\ 6 = 4 + 2 \end{cases}$$

which are also both true. Hence, the solutions of the system are $(-1, 1)$ and $(4, 6)$.
Do Exercise 1.

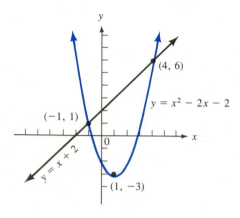

FIGURE 2.4

The system of the previous example could have easily been solved algebraically. For example, we could have substituted $x + 2$ for y in the first equation to obtain

$$x + 2 = x^2 - 2x - 2$$

which is equivalent to

$$x^2 - 3x - 4 = 0$$

and to

$$(x + 1)(x - 4) = 0$$

The solutions of these equations are -1 and 4. Replacing x successively by -1 and 4 in either of the two equations of the original system yields $y = 1$ and $y = 6$. Thus, the solutions are $(-1, 1)$ and $(4, 6)$.

In general, it is not advisable to solve a system of equations graphically if a simple algebraic solution is available. However, graphical solutions are useful in shedding

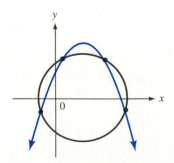

FIGURE 2.5

light on the nature of the solution set. For example, if the graph of one equation is a parabola and that of the other equation is a circle, we know that there will be at most four solutions. (See Figure 2.5.) Also, obtaining an approximate solution of a system graphically may be the best we can do when an algebraic solution is very difficult or not available as in the next two examples.

EXAMPLE 2 Solve this system of equations graphically.

$$\begin{cases} 15x - 8y + 19 = 0 \\ \qquad\qquad y = 2^x \end{cases}$$

Solution The graph of the first equation is a straight line. We easily find that the points $\left(0, \frac{19}{8}\right)$ and $(3, 8)$ are on the graph. Using this information we sketch the graph of the first equation in Figure 2.6. The graph of the second equation is that of an exponential function. We replace x successively by $-2, -1, 0, 1, 2, 3,$ and 4 to find the corresponding values of y. The results are tabulated below.

x	-2	-1	0	1	2	3	4
y	$\frac{1}{4}$	$\frac{1}{2}$	1	2	4	8	16

Hence, the points $\left(-2, \frac{1}{4}\right), \left(-1, \frac{1}{2}\right), (0, 1), (1, 2), (2, 4), (3, 8),$ and $(4, 16)$ are on the graph. We plot these points and sketch a smooth curve through them to get the graph of $y = 2^x$. Both graphs are shown in Figure 2.6. From the figure it appears that the graphs intersect at the points $\left(-1, \frac{1}{2}\right)$ and $(3, 8)$.

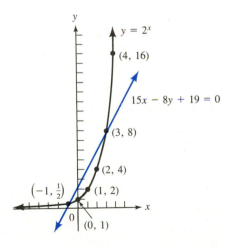

FIGURE 2.6

Replacing x by -1 and y by $\frac{1}{2}$ in both equations of the system, we obtain

$$\begin{cases} 15(-1) - 8\left(\dfrac{1}{2}\right) + 19 = 0 \\ \qquad\qquad\qquad \dfrac{1}{2} = 2^{-1} \end{cases}$$

which are both true. Thus, $\left(-1, \frac{1}{2}\right)$ is indeed a solution of the system. Replacing x by 3 and y by 8 in both equations, we get

$$\begin{cases} 15(3) - 8(8) + 19 = 0 \\ \qquad\qquad 8 = 2^3 \end{cases}$$

which are also both true. Hence $(3, 8)$ is the other solution of the system. The solutions are $\left(-1, \frac{1}{2}\right)$ and $(3, 8)$.

Do Exercise 11.

■

EXAMPLE 3   Solve this system graphically.

$$\begin{cases} y = \ln x \\ y = 2^{-x} \end{cases}$$

Solution   We replace x successively by .5, 1, 2, and 3 in the first equation and with the help of a calculator we find the approximate corresponding values of y. The results are tabulated below.

x	.5	1	2	3
y	-0.693147	0	0.693147	1.098612

Using the entries in this table, we sketch the graph of the first equation in Figure 2.7. Now we replace x successively by -1, 0, .5, 1, 2, and 3 in the second equation and find the corresponding values of y. The results are tabulated below.

x	-1	0	.5	1	2	3
y	2	1	0.7071	0.5	0.25	0.125

Using these results we sketch the graph of the second equation. Both graphs are shown in Figure 2.7.

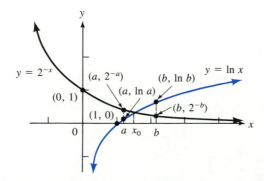

FIGURE 2.7

Notice that the graph of $y = 2^{-x}$ is falling while the graph of $y = \ln x$ is rising. If the graphs intersect at (x_0, y_0), then $a < x_0$ whenever $\ln a < 2^{-a}$ and $x_0 < b$ whenever $2^{-b} < \ln b$. Using a calculator*, find $2^{-x} - \ln x$ for some value of x. If the answer is positive, then $2^{-x} > \ln x$ and $x_0 > x$. If the answer is negative, then $2^{-x} < \ln x$ and $x_0 < x$. From the graphs, it appears that x_0 is near 1.5. So we start with $x = 1.5$. We find

$$2^{-1.5} - \ln 1.5 \doteq 0.3535 - 0.4054 = -0.0519 < 0$$

So $x_0 < 1.5$ and we try 1.4. We find

$$2^{-1.4} - \ln 1.4 \doteq 0.3789 - 0.3364 = 0.0425 > 0$$

Therefore $1.4 < x_0$. We now know that $1.4 < x_0 < 1.5$. We try 1.45 and find

$$2^{-1.45} - \ln 1.45 \doteq 0.3660 - 0.37156 = -0.00556 < 0$$

Thus, $x_0 < 1.45$. We try 1.44 and find

$$2^{-1.44} - \ln 1.44 \doteq 0.36856 - 0.36464 = 0.00392 > 0 \qquad \text{(See Figure 2.8.)}$$

Therefore, $1.44 < x_0$. We try 1.445 and find

$$2^{-1.445} - \ln 1.445 \doteq 0.36729 - 0.36810 = -0.00081 < 0$$

Thus, $1.44 < x_0 < 1.445$. We conclude that $x_0 = 1.44$ is correct up to two decimal places. An approximate solution of the system is $(1.44, 0.37)$, rounded off to two decimal places.

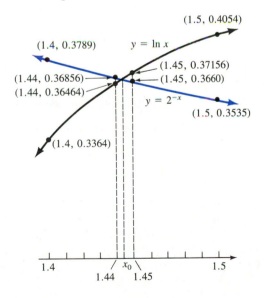

FIGURE 2.8

---

*Calculations such as those of Example 3 may be carried out most efficiently using a programmable calculator.

Exercise Set 2.3

In Exercises 1–15, solve the given system of equations graphically.

1. $\begin{cases} y = x^2 + 6x - 3 \\ y = 2x + 2 \end{cases}$

2. $\begin{cases} y = x^2 - 4x + 2 \\ y = -2x + 5 \end{cases}$

3. $\begin{cases} y = -x^2 + 2x + 5 \\ y = 2x + 1 \end{cases}$

4. $\begin{cases} y = -2x^2 + 8x + 7 \\ y = -2x + 15 \end{cases}$

5. $\begin{cases} y = x^2 + 3x + 1 \\ y = 2x^2 + 4x - 1 \end{cases}$

6. $\begin{cases} y = x^2 - 4x + 5 \\ y = 2x^2 - 5x + 3 \end{cases}$

7. $\begin{cases} y = -x^2 + 4x + 6 \\ y = x^2 + 6x + 2 \end{cases}$

8. $\begin{cases} y = \dfrac{7}{6}x^2 + \dfrac{1}{3} \\ y = 2^x + 1 \end{cases}$

9. $\begin{cases} y = 3^{-x} + 2 \\ y = -x^2 + \dfrac{1}{27}x + \dfrac{163}{27} \end{cases}$

10. $\begin{cases} y = 3^x \\ y = \dfrac{35}{4} + 2^{-x} \end{cases}$

11. $\begin{cases} y = 2^x \\ y = -x^2 + 7x - 4 \end{cases}$

12. $\begin{cases} y = -x^2 + 9x - 11 \\ y = 3^{x/2} \end{cases}$

13. $\begin{cases} y = 4^{-x} \\ y = \dfrac{1}{6}(3^{2x}) \end{cases}$

14. $\begin{cases} y = \ln(x + 1) \\ y = 3^{-2x} \end{cases}$

15. $\begin{cases} y = 2 \ln(x + 3) \\ y = 2^{-x/2} \end{cases}$

2.4 Systems of Linear Inequalities in Two Variables

Graphing Linear Inequalities in Two Variables

In this section, we introduce a method for finding the solution set of a system of linear inequalities in two variables graphically. Suppose we are asked to sketch the graph of the inequality $2x + 3y < 18$. By definition, a point P with coordinates (x_0, y_0) is on the graph of that inequality if, and only if, $2x_0 + 3y_0 < 18$ is a true statement. That is, if, and only if, (x_0, y_0) is a solution of the inequality $2x + 3y < 18$. Let L be the line with equation $2x + 3y = 18$. Note that this equation may be written $y = \left(-\frac{2}{3}\right)x + 6$. We have the following equivalent statements:

1. The point (x_0, y_0) is on the graph of the inequality $2x + 3y < 18$,
2. $2x_0 + 3y_0 < 18$,
3. $y_0 < \dfrac{-2x_0}{3} + 6$.

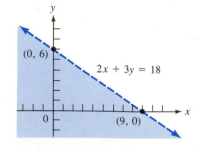

$2x + 3y = 18$

(0, 6)

(9, 0)

FIGURE 2.9

Since the point $\left(x_0, \left(\frac{-2}{3}\right)x_0 + 6\right)$ is on L, we conclude that P is on the graph of the given inequality if, and only if, P is below the line L. Thus, the graph of $2x + 3y < 18$ consists of all those points that are below the line L. (See Figure 2.9.) The graph is pictured as a shaded region in the plane. The line L is shown as a dashed line to indicate that it is not part of the graph of the inequality. If we were graphing $2x + 3y \le 18$, we would have drawn L as a solid line to indicate that it is included in the graph.

Notice that the line L divides the plane into three disjoint sets, namely:

1. The line L, which is the graph of $2x + 3y = 18$;
2. The region below L, which is the graph of $2x + 3y < 18$; and
3. The region above L, which is the graph of $2x + 3y > 18$.

The next two examples depict more complicated situations.

EXAMPLE 1 Sketch the graph of the system of inequalities.

$$\begin{cases} x < 6 \\ y > -4 \\ 3x + 2y \le 12 \end{cases}$$

Solution A point (x_0, y_0) is on the graph of the system if, and only if, it is on the graph of each inequality. Hence, the graph of the system of inequalities is the intersection of the three graphs. Clearly, the graph of the first inequality is the *half-plane* to the left of the line with equation $x = 6$. This is indicated in Figure 2.10 by drawing an arrow pointing to the left of that line. Similarly, the graph of the second inequality is the half-plane above the line with the equation $y = -4$. We indicate this in Figure 2.10 by drawing an arrow pointing upward.

Finally, we draw the line L with equation $3x + 2y = 12$. This line divides the plane into three disjoint subsets, one of which is a half-plane which is the graph of the third inequality of the system. To find which of the two half-planes is the graph of that inequality, we test an arbitrary point not on L. Although the choice of which point to use as a check is arbitrary, the origin is the most convenient because the calculations are easiest. The coordinates of the origin are $(0, 0)$. We replace x and y by 0 in the third inequality of the system and get $3 \cdot 0 + 2 \cdot 0 < 12$, which is certainly true. Therefore, the origin is on the graph of the third inequality. Consequently, the half-plane that contains the origin is the graph of the third inequality. We indicate this by placing an arrow by the line L pointing toward the origin. The graph of the given system is the intersection of the three half-planes, which is the region shaded in Figure 2.10. Note that the graphs of $x = 6$ and $y = -4$ have been drawn as dashed lines because they are not included in the graphs of $x < 6$ and $y > -4$, respectively. On the other hand, the graph of $3x + 2y = 12$ has been drawn as a solid line to show that it is part of the graph of $3x + 2y \le 12$.

Do Exercise 5. ■

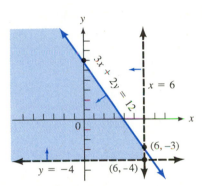

FIGURE 2.10

EXAMPLE 2 Sketch the graph of the system of inequalities

$$\begin{cases} 7x + y \ge 15 \\ x + y < 9 \\ x - 2y < 0 \end{cases}$$

Solution Let L_1, L_2, and L_3 be the graphs of the equations $7x + y = 15$, $x + y = 9$, and $x - 2y = 0$, respectively. To find the intersection of L_1 and L_2, we solve the system

$$\begin{cases} 7x + y = 15 \\ x + y = 9 \end{cases}$$

The solution of the system is $(1, 8)$. Thus, the point $(1, 8)$ is the intersection of L_1 and L_2. Similarly, to find the intersection of L_1 and L_3 we solve the system

$$\begin{cases} 7x + y = 15 \\ x - 2y = 0 \end{cases}$$

We find that the solution is (2, 1). Thus, the point (2, 1) is the intersection of L_1 and L_3. Finally, to find the intersection of L_2 and L_3 we solve the system

$$\begin{cases} x + y = 9 \\ x - 2y = 0 \end{cases}$$

The solution is (6, 3). Thus, the point (6, 3) is the intersection of L_2 and L_3. The usefulness of finding the three points of intersection is now evident when we sketch the three lines since we have obtained two points on each of the three lines. (See Figure 2.11.)

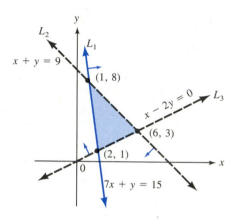

FIGURE 2.11

The line L_1 divides the plane into three disjoint regions, one of which is the graph of the inequality $7x + y > 15$. Using the origin, we replace x and y by 0 in the given inequality and obtain the false statement $7 \cdot 0 + 0 > 15$. The origin is not, therefore, on the graph of the first inequality and the half-plane that does not contain the origin is the graph of the first inequality. We indicate this by placing an arrow next to the line L_1 pointing away from the origin.

Similarly, we find that the origin does belong to the graph of the second inequality. Thus, we place an arrow next to the line L_2 pointing toward the origin. Finally, we see that the line L_3 divides the plane into three disjoint regions, one of which is the graph of the third inequality. We may not use the origin to determine which half-plane is that graph since the origin is on the line L_3. Since the third inequality is equivalent to $y > \left(\frac{1}{2}\right)x$, we conclude that the half-plane above the line L_3 is its graph. We indicate this by drawing an arrow pointing upward by the line L_3.

The graph of the given system is the intersection of the three half-planes we have obtained. It is the triangle shaded in Figure 2.11. Note that the side of the triangle that is part of the line L_1 is included in the graph and consequently has been drawn as a solid line, while the other two sides are not included in the graph and have been drawn as dashed lines.
Do Exercise 11.   ■

An Application

EXAMPLE 3   A company manufactures two types of gadgets, A and B. A requires 5 hours to assemble and 3 hours to paint, while B requires 2 hours to assemble and 1 hour to paint. The company employs 63 assemblers and 33 painters each working 8 hours a day. Let x and y denote the respective daily production of gadgets A and B. Graph the set of possible ordered pairs (x, y).

Solution   Obviously,

$$x \geq 0 \quad \text{and} \quad y \geq 0$$

since we cannot have negative production. Since each employee works 8 hours per day, the daily number of hours available for assembly is 504 because $63 \cdot 8 = 504$, and the daily number of hours available for painting is 264 because $33 \cdot 8 = 264$. The production of x gadgets A requires $5x$ assembly hours and $3x$ painting hours, while the production of y gadgets B requires $2y$ assembly hours and y painting hours. Thus, the total number of assembly hours needed is $5x + 2y$, while the total number of painting hours required is $3x + y$. Since these totals cannot exceed 504 and 264, respectively, we have the following system of inequalities:

$$\begin{cases} 5x + 2y \leq 504 \\ 3x + y \leq 264 \\ x \geq 0 \\ y \geq 0 \end{cases}$$

Using the same procedure as we did in Examples 1 and 2, we represent the solution set of this system as the shaded region in Figure 2.12. A question that naturally arises is this: Subject to the given restrictions and knowing the cost of production and the revenue for each of the gadgets, how should x and y be chosen to maximize the daily profit? We shall answer this question in Chapter 4.

Do Exercise 21.   ■

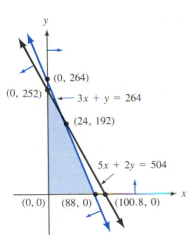

FIGURE 2.12

Exercise Set 2.4

In Exercises 1–17, sketch the graph of the given system of linear inequalities.

1. $\begin{cases} x + y > 1 \\ x < 2 \end{cases}$

2. $\begin{cases} x + 2y < 4 \\ y > -2 \end{cases}$

3. $\begin{cases} 2x + y < 6 \\ x - y > -3 \end{cases}$

4. $\begin{cases} 3x - 5y \geq 30 \\ x + y < 2 \end{cases}$

5. $\begin{cases} x + 2y \leq 4 \\ x > -1 \\ y > -2 \end{cases}$

6. $\begin{cases} 2x + y > 4 \\ y < 5 \\ x < 3 \end{cases}$

7. $\begin{cases} x - 3y < -3 \\ x < 1 \\ y > -1 \end{cases}$

8. $\begin{cases} 2x + 3y \geq 12 \\ x > 1 \\ y > 2 \end{cases}$

9. $\begin{cases} x + y \leq 1 \\ 2x - y \geq -4 \\ y \geq -2 \end{cases}$

10. $\begin{cases} x + 2y \geq 2 \\ x - 2y \geq 0 \\ x \leq 3 \end{cases}$

11. $\begin{cases} 2x + y \leq 6 \\ x - y > -3 \\ 2x + 3y > -6 \end{cases}$

12. $\begin{cases} 3x - 5y \leq 30 \\ x + y \geq 2 \\ 3x + 7y \leq 30 \end{cases}$

13. $\begin{cases} x + y \leq 1 \\ 2x - y \geq -4 \\ x - 2y \leq 4 \end{cases}$

14. $\begin{cases} x - y \geq -1 \\ x + y < 5 \\ 3x - 5y \leq -1 \end{cases}$

15. $\begin{cases} x + y \geq 1 \\ 2x - y \geq -4 \\ 7x - 8y \leq 22 \end{cases}$

16. $\begin{cases} 2x - y \geq -4 \\ x - 2y \leq 4 \\ x + y \geq 1 \\ x + y \leq 7 \end{cases}$

17. $\begin{cases} 3x - 4y > -19 \\ 5x + 3y > 7 \\ 2x - y < 5 \\ 4x + y < 19 \end{cases}$

18. A farmer has 200 acres on which he raises peanuts and corn. He has accepted orders requiring 10 acres of peanuts and 25 acres of corn. Moreover, he must follow regulations that the acreage for corn must be at least twice the acreage of peanuts. Let x and y denote the number of acres allocated to peanuts and corn, respectively. Graph the set of possible ordered pairs (x, y).

19. A farmer has 800 acres on which she will raise wheat and barley. She has accepted orders requiring 150 acres of wheat and 100 acres of barley. However, she must follow regulations that the acreage for wheat cannot exceed three times the acreage of barley. Let x and y denote the number of acres allocated to wheat and barley, respectively. Graph the set of possible ordered pairs (x, y).

20. An electronics company makes two models of VCRs. Model A takes 2 hours to assemble and 15 minutes to test, while model B takes 3 hours to assemble and 30 minutes to test. The plant is designed in such a way that a maximum of 60,000 hours per month are available for assembly, while a maximum of 8000 hours a month are available for testing. The company has accepted orders totaling 10,000 units of the A model and 2000 units of the B model. Let x and y denote the numbers of A and B models, respectively, the company can make per month. Graph the set of possible ordered pairs (x, y).

21. A camera company makes two models of movie cameras. The deluxe model takes 3 hours to assemble and 10 minutes to test, while the standard model takes 2 hours to assemble and 6 minutes to test. The plant is designed in such a way that a maximum of 38,000 hours per month are available for assembly, while a maximum of 2000 hours a month are available for testing. The company has accepted orders totaling 2500 of the deluxe model and 3600 of the standard model. Let x and y denote the numbers of deluxe and standard models, respectively, the company can make per month. Graph the set of possible ordered pairs (x, y).

22. An owner of a gasoline station can have at most 16,000 gallons of gasoline per day in his tanks. On a certain day, he estimates that his regular customers will purchase at least 5000 gallons of regular gasoline and at least 2000 gallons of premium gasoline. He also knows from experience that he cannot sell more than three times as much regular gasoline as premium gasoline. Let x and y denote, respectively, the number of gallons of regular and premium gasoline he can sell per day. Graph the set of possible ordered pairs (x, y).

23. An investor wishes to buy a maximum of 300 shares of common stock. Her broker recommends stocks A and B and suggests that to reduce the risk of a fluctuating market at least 50 shares of stock A and not more than 200 shares of stock B should be bought. He also recommends that the number of shares of stock A should not exceed twice that of stock B. Let x and y represent the numbers of shares of stocks A and B, respectively, that the investor will buy. Graph the set of all possible ordered pairs (x, y).

24. An investor wishes to buy a maximum of 500 shares of common stock. His broker recommends stocks A and B and suggests that to reduce the risk of a fluctuating market at least 50 shares of stock A and 100 shares of stock B should be bought. She also recommends that the number of shares of stock B should not exceed four times that of stock A. Let x and y represent the numbers of shares of stocks A and B, respectively, that the investor will buy. Graph the set of all possible ordered pairs (x, y).

25. Shirley makes dolls and blankets to be sold at her church bazaar. She can make a doll in 3 hours, while it takes her 5 hours to make a blanket. Her costs to make each doll and each blanket are $4 and $7, respectively. If she can spend at most 120 hours a month on this church work and if the church provides at most $166 a month to cover the costs of making the dolls and blankets, sketch the graph of the set of ordered pairs (x, y) where x and y are the numbers of dolls and blankets that Shirley can make each month.

26. A handyman makes baskets and vases to be sold at the local flea market. He can make a basket in 4 hours, while it takes him 3 hours to make a vase. His costs to make each basket and each vase are $2 and $5, respectively. If he can spend at most 36 hours a week on this project and if his cost on this work is not to exceed $32 per week, sketch the graph of the set of ordered pairs (x, y) where x and y are the numbers of baskets and vases that the handyman can make each week.

27. Same as Exercise 26 with the additional requirement that the handyman has accepted orders of three baskets and two vases per week.

28. Andre's motel has 120 rooms and a restaurant that can seat 45 people. Some of the rooms may be furnished to accommodate travelers who like to go first class, while the remaining rooms will be furnished to satisfy the average guest. The manager of the motel knows from experience that 60% of the first-class guests will eat in the motel restaurant, while only 30% of the other guests will eat there. If the manager furnishes x rooms for first-class guests and y rooms for ordinary guests, sketch the graph of the set of possible ordered pairs (x, y).

29. Same as Exercise 28 with the additional requirements that the number of rooms for first-class guests must be at least 10 and the number of rooms for ordinary guests must not exceed 100.

30. A nursing home dietician is planning the menus for the patients. She wants to provide a diet that has a minimum of 120 units of fat, 60 units of carbohydrates, and 90 units of proteins per patient per week. These goals can be met using two types of food. Type I contains 3 units of fats, 2 units of carbohydrates, and 6 units of proteins per pound; each pound of type II contains 12 units of fats, 4 units of carbohydrates, and 3 units of proteins. If the dietician plans to use x pounds of type I and y pounds of type II food per patient per week, graph the set of possible ordered pairs (x, y).

2.5 Applications to Business and Economics

Break-Even Analysis

If the cost to a company of producing q units of a commodity is $C(q)$ dollars and the revenue obtained from the sales of q units of this commodity is $R(q)$ dollars, then the company makes a profit whenever $R(q) > C(q)$. However, the company operates at a loss whenever $R(q) < C(q)$ and breaks even when $R(q) = C(q)$. The number of units produced and sold in the latter case is called the *break-even point*. In general, at the start of production of a commodity, a company operates at a loss because the revenue is not large enough to cover the *fixed costs* (the costs incurred by the company at a zero production level, such as rent and depreciation). To find the break-even point, we simply solve the system

$$\begin{cases} y = C(q) \\ y = R(q) \end{cases}$$

EXAMPLE 1   The cost $\$C(q)$ of producing q units of a certain commodity in a month is given by the equation $y = C(q) = 5q + 12{,}000$. If each unit of that commodity sells for $\$8$, find the following:

a. The fixed costs per month.
b. The number of units of the commodity that should be produced and sold per month to make sure that the business breaks even.

Solution   **a.** By definition, the fixed costs may be obtained by replacing q by 0 in the formula giving the cost function. Thus, the fixed costs are $\$12{,}000$ per month, since

$$C(0) = 5(0) + 12{,}000 = 12{,}000$$

b. Since each unit of the commodity sells for $\$8$, the revenue from the sales of q units is given by

$$y = R(q) = 8q$$

To find the break-even point, we solve the system

$$\begin{cases} y = 5q + 12{,}000 \\ y = 8q \end{cases}$$

Substituting $8q$ for y in the first equation we obtain

$$8q = 5q + 12{,}000$$

Thus,

$$3q = 12,000$$

and

$$q = 4000$$

Therefore, 4000 units of the commodity should be manufactured and sold each month to guarantee that the company breaks even. In Figure 2.13 we show that the company operates at a loss when production is between 0 and 4000 units and makes a profit when production is greater than 4000 units.

Do Exercise 1.

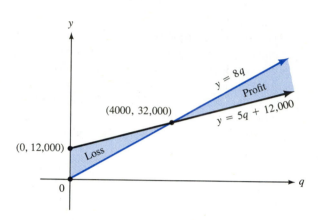

FIGURE 2.13

EXAMPLE 2   A company making personal computers found that the cost $C(q)$ of producing q PCs per month is given by

$$y = C(q) = .5q^2 + 1000q + 1,125,000$$

provided that q is between 0 and 2000. It also found that the revenue $R(q)$ from the sales of q PCs per month is given by $y = R(q) = -q^2 + 4000q$ provided that q is between 0 and 2000.

a. Find the level of production at which the company will break even.
b. At what level of production will the company make a profit?

Solution   **a.** To find the break-even point, it is sufficient to solve the system

$$\begin{cases} y = .5q^2 + 1000q + 1,125,000 \\ y = -q^2 + 4000q \end{cases}$$

Substituting $-q^2 + 4000q$ for y in the first equation, we obtain

$$-q^2 + 4000q = .5q^2 + 1000q + 1,125,000$$

This equation may be written

$$1.5q^2 - 3000q + 1,125,000 = 0$$

Factoring the left side of this equation, we get

$$1.5(q - 500)(q - 1500) = 0$$

Thus, $q = 500$ or $q = 1500$. We conclude that there are two break-even points. The company may manufacture 500 PCs per month with costs and revenues equal to \$1,750,000, or it may produce 1500 PCs per month with costs and revenues equal to \$3,750,000.

b. The company is making a profit whenever the revenues are larger than the costs. That is, whenever

$$R(q) > C(q)$$

Thus, we have

$$-q^2 + 4000q > .5q^2 + 1000q + 1,125,000$$

The previous inequality is equivalent to

$$1.5(q - 500)(q - 1500) < 0$$

Using the method which we presented in Section 0.8, we find that the solution set of the inequality is

$$\{q \mid 500 < q < 1500\}$$

Thus, the company will operate at a profit if it produces more than 500 but fewer than 1500 PCs per month. (See Figure 2.14.)

Do Exercise 3.

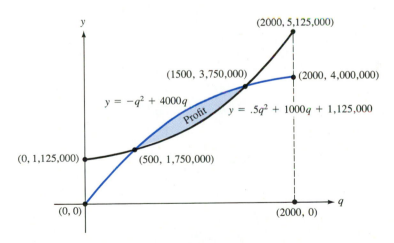

FIGURE 2.14

Commodity Demand and Supply

One of the most basic principles of economic theory is that the market price of a commodity is determined by the demand and supply. In general, the quantity q units of a product that consumers are willing to purchase depends on the price p dollars per unit at which the product is made available. The relationship between q and p is called the *demand function D* and the graph of the equation $q = D(p)$ is called the *demand curve*. Similarly, the quantity q units of a product that its supplier is willing to make available depends on the price p dollars per unit at which they are able to sell it. The relationship between q and p is called the *supply function S* and the graph of the equation $q = S(p)$ is called the *supply curve*. Increasing the price per unit tends to decrease the number of units consumers are willing to buy, so in general the demand function is decreasing while the supply function is increasing.

In the usual economy, the price will fluctuate and tend to adjust so that the quantity suppliers are willing to provide is equal to the quantity consumers are willing to purchase. When this occurs, we say that the *market is in equilibrium*. The corresponding price and quantity are called the *equilibrium price* and the *equilibrium quantity*, respectively. In the simplest case where the demand and supply functions are linear, the equilibrium price and quantity may be found by solving a system of linear equations as in the following example.

EXAMPLE 3   The demand for a commodity is described by the equation $q = D(p) = 15{,}000 - 150p$, where q is the quantity demanded (measured in thousands of units) and p is the price (measured in dollars per unit). The supply function for this commodity is described by the equation $q = S(p) = -200 + 100p$, where q is the quantity supplied (also measured in thousands of units) and p is the price (measured in dollars per unit). Find the equilibrium price and quantity.

Solution   The supply function indicates that the price cannot be less than $2 per unit and the demand function shows that the price cannot be greater than $100 per unit since quantities cannot be negative; therefore we have $-200 + 100p \geq 0$ and $15{,}000 - 150p \geq 0$. Also, both functions are linear and their graphs are straight lines in the *pq*-plane. The demand curve is a line with slope -150. Thus, as expected, the demand function is decreasing. The supply curve is a line with slope 100. Hence, the supply function is increasing. Since the two lines have different slopes, they are not parallel and will have a unique point of intersection. To find the coordinates of this point, we solve the system

$$\begin{cases} q = 15{,}000 - 150p \\ q = -200 + 100p \end{cases}$$

Substituting $-200 + 100p$ for q in the first equation we get

$$-200 + 100p = 15{,}000 - 150p$$

That is,

$$250p = 15{,}200$$

and

$$p = \frac{15{,}200}{250} = 60.80$$

Replacing p by 60.80 in the first equation of the system we get

$$q = 15{,}000 - 150(60.80) = 5880$$

Thus, we conclude that the equilibrium quantity of 5880 thousand units is demanded and supplied at the equilibrium price of $60.80 per unit. (See Figure 2.15.)
Do Exercise 7.

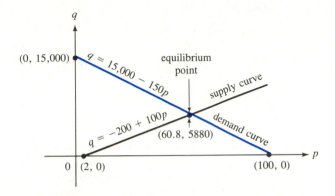

FIGURE 2.15

Sometimes the quantity demanded or supplied depends not only on the price of the commodity, but also on some other variable, such as the consumer's average income. In such cases, the equilibrium price and quantity will vary according to the value of the variable as in the following example.

EXAMPLE 4   A demand function for a commodity is described by $q = D(p) = -100p + 1500 + 10I$, where q is the quantity demanded (measured in thousands of units), p is the price (measured in dollars per unit), and I is the consumer's average income. The supply function for the same commodity is defined by $q = S(p) = 300\sqrt{8p - 60}$, where q is the quantity supplied (measured in thousands of units) and p is the price (measured in dollars per unit). Note that $p \geq 7.5$ since $8p - 60$ must be nonnegative.

a. Find the equilibrium price and quantity when the consumer's average income is given by $I = 350$.
b. Same as part **a** with $I = 590$.
c. Sketch the demand and supply curves on the same coordinate plane.

Solution   **a.** Since $q = D(p) = -100p + 1500 + 10I$, when $I = 350$, we have

$$q = D(p) = -100p + 5000$$

To find the equilibrium quantity and price, we solve the system

$$\begin{cases} q = -100p + 5000 \\ q = 300\sqrt{8p - 60} \end{cases}$$

Substituting $300\sqrt{8p - 60}$ for q in the first equation we get

$$300\sqrt{8p - 60} = -100p + 5000 \qquad (*)$$

Note that the foregoing equation shows that supply equals demand when p is the equilibrium price. Dividing both sides of equation (*) by 100 and then squaring both sides yields

$$9(8p - 60) = (-p + 50)^2$$

and

$$72p - 540 = p^2 - 100p + 2500$$

That is,

$$p^2 - 172p + 3040 = 0$$

Factoring the left side of this equation yields

$$(p - 20)(p - 152) = 0$$

Thus, $p = 20$ or $p = 152$.

Since we squared both sides of an equation, we may have introduced some extraneous solutions. We must check the validity of the solutions we obtained by replacing p by 20 in equation (*) to get

$$300\sqrt{8(20) - 60} = -100(20) + 5000$$

which yields the true statement

$$3000 = 3000$$

Thus, 20 is a solution.

Replacing p by 152 in equation (*) we get

$$300\sqrt{8(152) - 60} = -100(152) + 5000$$

which yields the false statement

$$10,200 = -10,200$$

Thus, we reject 152 and conclude that 20 is the only solution. Therefore, the equilibrium price is $20 per unit. We find the equilibrium quantity by replacing p by 20 in either the demand or supply equation. We get

$$q = D(20) = -100(20) + 5000$$
$$= 3000$$

Hence, the equilibrium quantity is 3,000,000 units (since q is in thousands).

b. When $I = 590$, we get

$$q = D(p) = -100p + 1500 + 10(590)$$
$$= -100p + 7400$$

To find the new equilibrium price and quantity, we solve the system

$$\begin{cases} q = -100p + 7400 \\ q = 300\sqrt{8p - 60} \end{cases}$$

We solve it in exactly the same way as we solved the system of part **a** and find that the only solution is $p = 32$. Replacing p by 32 in the first equation we get

$$q = -100(32) + 7400 = 4200$$

The equilibrium price is $32 per unit and the equilibrium quantity is 4200 thousand units.

c. The supply curve and the two demand curves are sketched in Figure 2.16. Note that the second demand curve was obtained by shifting the first demand curve to the right.

Do Exercise 9.

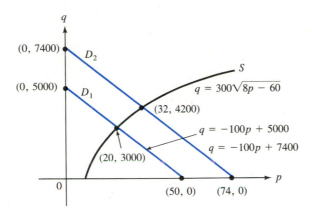

FIGURE 2.16

Taxes and Subsidies

Often, the government imposes taxes on certain commodities and in other cases the government pays subsidies to producers to encourage them to supply essential commodities to consumers at reasonable prices. Taxes and subsidies are variables which affect the equilibrium price and equilibrium quantity of a commodity. The next two examples illustrate this.

EXAMPLE 5   The demand and supply functions for a commodity are defined by

$$q = D(p) = -3p + 139$$
$$q = S(p) = 2p - 11$$

where q is the number of units (measured in thousands) and p is the price (measured in dollars per unit).

a. Find the equilibrium price and quantity.
b. Suppose that the government imposes a tax of $7.50 per unit on the commodity. Find the new equilibrium price and quantity.

Solution   **a.** To find the equilibrium point, we solve the system

$$\begin{cases} q = -3p + 139 \\ q = 2p - 11 \end{cases}$$

Substituting $2p - 11$ for q in the first equation we obtain

$$2p - 11 = -3p + 139$$

from which we get

$$p = 30$$

Replacing p by 30 in the first equation gives

$$q = -3(30) + 139$$
$$= 49$$

The equilibrium price is $30 per unit and the equilibrium quantity is 49,000 units.

b. We assume that the tax imposed is included in the price p paid by the consumer.* Thus, the demand equation does not change. However, when the consumers pay $p per unit, only $(p - 7.5)$ goes to the suppliers and $7.50 goes to the government. The supply equation becomes

$$q = 2(p - 7.5) - 11$$

or

$$q = 2p - 26$$

To find the new equilibrium point, we solve the system

$$\begin{cases} q = -3p + 139 \\ q = 2p - 26 \end{cases}$$

We find that $p = 33$ and $q = 40$. The new equilibrium price is $33 per unit and the new equilibrium quantity is 40,000 units. Note that the price per unit increased by $3 and the quantity demanded and supplied decreased by 9000 units. (See Figure 2.17.)

Do Exercise 13.

---

*If the tax is not included in the price, but is added after the purchase, it will not affect the supply equation. However, p must be replaced by $p + T$ in the demand equation. The new equilibrium price is obtained by solving

$$\begin{cases} q = D(p + T) \\ q = S(p) \end{cases}$$

See Exercise 15.

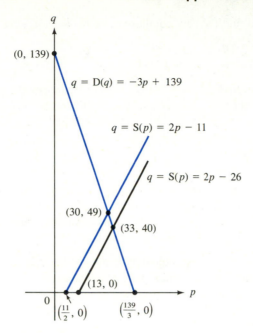

q

$(0, 139)$

$q = D(q) = -3p + 139$

$q = S(p) = 2p - 11$

$q = S(p) = 2p - 26$

$(30, 49)$

$(33, 40)$

$(13, 0)$

0

$\left(\frac{11}{2}, 0\right)$ $\left(\frac{139}{3}, 0\right)$

p

FIGURE 2.17

Observe that the graph of the supply equation has been shifted 7.5 units to the right when the tax was added.

EXAMPLE 6   The demand and supply functions for a commodity are defined by $q = D(p) = -2p + 89$ and $q = S(p) = 6p - 7$, where p is the price (measured in dollars per unit) and q is the quantity (measured in thousands of units).

a. Find the equilibrium price and quantity.
b. Suppose that the government pays the suppliers a subsidy of \$4 per unit. Find the new equilibrium price and quantity.

Solution   **a.** We must solve the system

$$\begin{cases} q = -2p + 89 \\ q = 6p - 7 \end{cases}$$

We find that $p = 12$ and $q = 65$ and conclude that the equilibrium price is \$12 per unit and the equilibrium quantity is 65,000.

b. The subsidy paid to the supplier does not affect the demand equation. However, when the price is \$$p$ per unit, the producers receive \$$(p + 4)$ since the consumers pay \$$p$ and the government pays \$4 for each unit produced and sold. The supply equation becomes

$$q = 6(p + 4) - 7$$

which may be written

$$q = 6p + 17$$

To find the new equilibrium point, we solve the following system

$$\begin{cases} q = -2p + 89 \\ q = 6p + 17 \end{cases}$$

We find that $p = 9$ and $q = 71$. We conclude that the new equilibrium price and quantity are $9 per unit and 71,000 units, respectively. Note that the equilibrium price has decreased by $3 per unit and the quantity demanded and supplied has increased by 6000 units.

Do Exercise 17. ■

EXAMPLE 7 The demand equation for a commodity is $q = -p^2 - p + 210$, where q is the number of units demanded (in millions) when $$p$ is the price per unit. The suppliers will produce and sell 48,000,000 units when the price is $8 per unit, but only 18,000,000 units when the price is $5 per unit.

a. Find the supply equation assuming that it is linear.
b. Find the equilibrium price and quantity.
c. What should the government subsidy be per unit to increase the demand by 42,000,000?

Solution **a.** It is given that $q = 48$ when $p = 8$ and that $q = 18$ when $p = 5$. Therefore, the points $(8, 48)$ and $(5, 18)$ are on the supply curve. Noting that the supply curve is a straight line, we find that its slope is 10 since $(48 - 18)/(8 - 5) = 10$. Therefore, the point-slope equation of the line is

$$q - 18 = 10(p - 5)$$

This equation may be written

$$q = 10p - 32$$

b. We find the equilibrium price and quantity solving the system

$$\begin{cases} q = -p^2 - p + 210 \\ q = 10p - 32 \end{cases}$$

Subtracting the first equation from the second, we obtain

$$0 = p^2 + 11p - 242$$

This equation may be factored as

$$0 = (p - 11)(p + 22)$$

Thus, $p = 11$ or $p = -22$. We consider only the positive value of p and conclude that the equilibrium price is $11. Replacing p by 11 in the supply equation, we obtain $q = 10 \cdot 11 - 32 = 78$. Hence, the equilibrium quantity is 78,000,000.

c. The demand is increased by 42,000,000 if the value of q is increased by 42. Therefore, the new value of q will be 120 since $78 + 42 = 120$. The subsidy paid by the government does not affect the demand equation so we can find the price paid by the consumers when $q = 120$. We find that

$$120 = -p^2 - p + 210$$

The foregoing equation may be written

$$p^2 + p - 90 = 0$$

The only positive solution of this equation is 9. Thus, the new equilibrium price (paid by the consumers) is $9 per unit. Suppose now that the government pays a subsidy of $$S$ per unit, so that the suppliers receive $$(9 + S)$ per unit. Recalling that the value of q is now 120, we replace p and q by $9 + S$ and 120, respectively, in the supply equation and obtain

$$120 = 10(9 + S) - 32$$

Thus,

$$120 = 10S + 58$$

and

$$S = 6.2$$

We conclude that the government should pay a subsidy of $6.20 per unit to increase the demand by 42,000,000 units.

Do Exercise 19. ■

Multiple Commodities

In the previous examples, we have assumed that the quantities of a commodity demanded and supplied are affected only by the price of that commodity, and in some cases by other variables such as consumer's average income, taxes, and subsidies. In many instances, the quantities of a commodity demanded and supplied depend not only on the price of that commodity but also on the price of related commodities. For example, an increase in the price of beef may create an increase in the demand for poultry. We shall conclude this section with an example involving more than one commodity. In all cases, the underlying principle is that the market equilibrium is reached when the supply of each commodity equals the demand for that commodity.*

EXAMPLE 8 Let D_x and S_x denote the quantities (measured in thousands of units) demanded and supplied, respectively, of commodity X. Let D_y and S_y denote the quantities (also measure in thousnds of units) demanded and supplied, respectively, of commodity Y. Suppose that p_x and p_y represent the prices (in dollars per unit) of commodities X and

---

*In principle, if we knew the demand and supply functions for every commodity in the economy, we could then compute the so-called *Walrasian general equilibrium*. In practice, if we consider n commodities and have n demand functions and n supply functions, we can get a system of n equations in n unknowns by setting $D_i = S_i$ for each $i = 1, 2, 3, \ldots, n$. We can then solve this system to get the equilibrium price of each of the n commodities. Finally, we obtain the equilibrium quantities traded for each of the n commodities. Cases where $n = 3$ are considered in the exercises.

Y, respectively. It is known that the demand and supply functions for these two commodities are interrelated as is shown by

$$D_x = 168 - 8p_x - 15p_y$$

$$S_x = -13 + 11p_x + 20p_y$$

$$D_y = 119 - 13p_x - 14p_y$$

$$S_y = -5 + 10p_y.$$

a. Find the equilibrium prices of these commodities.
b. Find the equilibrium quantities of these commodities.

Solution   **a.** The market will be in equilibrium if $D_x = S_x$ and $D_y = S_y$. Thus, we set

$$168 - 8p_x - 15p_y = -13 + 11p_x + 20p_y$$

and

$$119 - 13p_x - 14p_y = -5 + 10p_y$$

These two equations may be simplified to obtain

$$\begin{cases} 19p_x + 35p_y = 181 \\ 13p_x + 24p_y = 124 \end{cases}$$

The solution is

$$\begin{cases} p_x = 4 \\ p_y = 3 \end{cases}$$

Thus, the equilibrium price of commodity X is $4 per unit and that of commodity Y is $3 per unit.

b. Replacing each p_x by 4 and each p_y by 3 in the original equations giving the demand for each commodity, we obtain

$$D_x = 168 - 8 \cdot 4 - 15 \cdot 3 = 91$$

and

$$D_y = 119 - 13 \cdot 4 - 14 \cdot 3 = 25$$

Note that we could have just as well used the supply equations to get these results since at equilibrium the demand and supply are equal. We conclude that the equilibrium quantity of commodity X is 91,000 and that of commodity Y is 25,000.

Do Exercise 23.   ■

Exercise Set 2.5

1. The cost $\$C(q)$ of producing q units of a commodity in a week is given by

$$y = C(q) = 3q + 25,000.$$

If each unit of that commodity sells for $5, find the following:

a. The fixed cost per week.
b. The number of units of the commodity that should be produced and sold per week to make sure that the business breaks even.

2. The Triplenight Company found that the cost $\$C(q)$ of making q racquetball racquets per month is given by

$$y = 45q + 30,000.$$

Each racquet may be sold for $60. Find the following:

a. The fixed cost per month.
b. The number of racquets that should be made and sold each month so that the company breaks even.

3. A company manufacturing motorcycles found that the cost $C(q)$ of making q motorcycles per month is given by

$$y = C(q) = .25q^2 + 2000q + 525,000$$

provided that q is between 0 and 1500. It also found that the revenue $R(q)$ from the sales of q motorcycles per month is given by

$$y = R(q) = -2.25q^2 + 4500q$$

provided that q is between 0 and 1500.

a. Find the level of production at which the company will break even.
b. At what level of production will the company make a profit?

4. A company manufacturing television sets found that the cost $C(q)$ of making q deluxe models per week is given by

$$y = C(q) = .2q^2 + 560q + 120,000$$

provided that q is between 0 and 700. It also found that the revenue $R(q)$ from the sales of q deluxe sets per week is given by

$$y = R(q) = -q^2 + 1400q$$

provided that q is between 0 and 700.

a. Find the level of production at which the company will break even.
b. At what level of production will the company make a profit?

5. A company manufacturing VCRs found that the cost $C(q)$ of making q recorders per day is given by

$$y = C(q) = .5q^2 + 215q + 18,000$$

provided that q is between 0 and 100. It also found that the revenue $R(q)$ from the sales of q VCRs per day is given by

$$y = R(q) = -4q^2 + 800q$$

provided that q is between 0 and 100.

a. Find the level of production at which the company will break even.
b. At what level of production will the company make a profit?

6. The demand for a commodity can be described by

$$q = D(p) = 35,000 - 175p$$

where q is the quantity demanded (measured in thousands of units) and p is the price (measured in dollars per unit). The supply function for this commodity is described by

$$q = S(p) = -750 + 150p$$

where q is the quantity supplied (also measured in thousands of units) and p is the price (measured in dollars per unit). Find the equilibrium price and quantity.

7. The demand for a product is described by

$$q = D(p) = 75,000 - 120p$$

where q is the number of units demanded and p is the price (measured in dollars per unit). The supply function for this product has been found to be

$$q = S(p) = -3750 + 105p$$

where q is the number of units supplied and p is the price (measured in dollars per unit). Find the equilibrium price and quantity.

8. The demand for a commodity can be described by

$$q = D(p) = -p^2 - 10p + 11,000$$

where q is the quantity demanded (measured in thousands of units) and p is the price (measured in dollars per unit). The supply function for this commodity is

$$q = S(p) = -100 + 115p$$

where q is the quantity supplied (also measured in thousands of units) and p is the price (measured in dollars per unit). Find the equilibrium price and quantity.

9. The demand function for a commodity is

$$q = D(p) = -60p + 2640 + 12I$$

where q is the quantity demanded (measured in thousands of units), p is the price (measured in dollars per unit), and I is the consumer's average income measured in appropriately defined units. The supply function for the commodity is

$$q = S(p) = 42p - 60$$

where q is the quantity supplied measured in thousands of units and p is the price (measured in dollars per unit).

a. Find the equilibrium price and quantity when the consumer's average income is given by $I = 30$.
b. Same as part **a** with $I = 81$.

c. Sketch the demand and supply curves on the same coordinate plane.

10. The demand function for a product is

$$q = D(p) = -p^2 - 50p + 40{,}000 + 20I$$

where q is the number of units demanded when the price is $\$p$ per unit and I is the consumer's average income. The supply function for the product is

$$q = S(p) = 250p - 400$$

where q is the number of units supplied when the price is $\$p$ per unit.

a. Find the equilibrium price and quantity when the consumer's average income is given by $I = 500$.
b. Same as part **a** with $I = 775$.
c. Sketch the demand and supply curves on the same coordinate plane.

11. The demand function for a commodity is

$$q = D(p) = -p^2 - 20p + 4000 + 2I$$

where q is the quantity demanded (measured in thousands of units), p is the price (measured in dollars per unit), and I is the consumer's average income. The supply function for the commodity is

$$q = S(p) = p^2 + 28p - 510,$$

where q is the quantity supplied measured in thousands of units and p is the price (measured in dollars per unit).

a. Find the equilibrium price and quantity when the consumer's average income is given by $I = 850$.
b. Same as part **a** with $I = 1201$.
c. Sketch the demand and supply curves on the same coordinate plane.

12. The demand function for a commodity is

$$q = D(p) = -60p + 20{,}635 + 127I$$

where q is the number of units of the commodity demanded when the price is $\$p$ per unit and I is the consumer's average income. The supply function for the commodity is

$$q = S(p) = 250\sqrt{5p - 80}$$

where q is the number of units supplied when the price of the commodity is $\$p$ per unit. Note that $p \geq 16$ since $5p - 80 \geq 0$.

a. Find the equilibrium price and quantity when the consumer's average income is given by $I = 75$.
b. Same as part **a** with $I = 125$.

c. Sketch the demand and supply curves on the same coordinate plane.

13. The demand and supply functions for a commodity are

$$q = D(p) = -5p + 159$$
$$q = S(p) = 3p - 9$$

where q is the number of units (measured in thousands) and p is the price (measured in dollars per unit).

a. Find the equilibrium price and quantity.
b. Suppose that the government imposes a tax of $8 per unit on the commodity. Find the new equilibrium price and quantity.

14. A dealer can sell 650 units of a certain product when the price is $10 per unit but can sell only 130 units when the price is $50 per unit. The supply function for this product is

$$q = S(p) = 8p - 60$$

where q is the number of units supplied when the price is $\$p$ per unit.

a. Find the demand equation assuming that it is linear.
b. Find the equilibrium price and quantity.
c. Suppose that the government imposes a tax of $5.25 per unit on the commodity. Find the new equilibrium price and quantity.

15. When the price of a certain commodity is $12 per unit, the demand and supply are 1478 and 190 units, respectively; a price of $40 per unit changes the demand and supply to 582 and 862 units, respectively. Assuming that the demand and supply functions are linear, find the following:

a. The demand equation.
b. The supply equation.
c. The equilibrium price and quantity.
d. Suppose that the government imposes a tax of $7 per unit on the commodity. Find the new equilibrium price and quantity. (Assume the tax is not included in the price but is added after the purchase.)

16. The demand and supply functions for a commodity are

$$q = D(p) = -12p + 7000 \text{ and } q = S(p) = 18p - 2000$$

where p is the price per unit in dollars and q is the quantity demanded (in the first equation) and supplied (in the second equation).

a. Find the equilibrium price and quantity.
b. Suppose that the government pays the suppliers a subsidy of $50 per unit. Find the new equilibrium price and quantity.

17. The demand and supply functions for a commodity are known to be linear. If the price of the commodity is $5 per unit, the quantities demanded and supplied are 510 and 35, respectively. If the price is $20 per unit, the quantities demanded and supplied are 360 and 260, respectively.

    a. Find the demand equation.
    b. Find the supply equation.
    c. Find the equilibrium price and quantity.
    d. Suppose that the government pays the suppliers a subsidy of $5 per unit. Find the new equilibrium price and quantity.

18. The demand equation for a commodity is

    $$q = D(p) = -p^2 - 10p + 11,000$$

    where q is the number of units demanded when the price is p per unit. The supply equation is linear, and it is known that the suppliers will produce and sell 800 units when the price is $10 per unit, but when the price is $8 per unit they will produce and sell only 560 units.

    a. Find the supply equation.
    b. Find the equilibrium price and quantity.
    c. If the government provides a subsidy of $20 per unit, find the new equilibrium price and quantity.

19. The demand equation for milk in a French province is given by

    $$q = D(p) = -p^2 - p + 156$$

    where q is the number of liters of milk demanded (in millions) when the price is p francs per liter. Milk producers are willing to produce and sell 34 million liters of milk when the price is 4 francs per liter and 59 million liters when the price is 6 francs per liter.

    a. Find the supply equation assuming it is linear.
    b. Find the equilibrium price and quantity.
    c. The French government wishes to increase the demand for milk in that province to 114 million liters. What subsidy per liter must be provided to accomplish that goal?

20. The demand equation for potatoes in a state is given by

    $$q = D(p) = -p^2 - 3p + 270$$

    where q is the number of pounds (in millions) demanded when the price is p cents per pound. The farmers are willing to grow and sell 65,000,000 and 95,000,000 pounds of potatoes when the prices are $0.05 and $0.07 per pound, respectively.

    a. Find the supply equation assuming it is linear.
    b. Find the equilibrium price and quantity.

    c. The state government wishes to increase the demand by 42,000,000 pounds. What subsidy per pound must be provided to accomplish that goal?

21. The demand equation for a product is

    $$q = D(p) = -p^2 - 2p + 575$$

    where q is the number of units (in thousands) demanded when the price is p per unit. The supply equation is linear and it is known that the suppliers will produce and sell 231,000 units when the price is $10 per unit, but when the price is $5 per unit they will produce and sell only 81,000 units.

    a. Find the supply equation.
    b. Find the equilibrium price and quantity.
    c. If the government wants to increase the demand by 81,000, what subsidy per unit must it provide?

22. The demand equation for a commodity is

    $$q = D(p) = -p^2 - 2p + 323$$

    where q is the number of units demanded (in millions) when the price is p per unit. The supply equation is linear and it is known that the suppliers will produce and sell 140,000,000 units when the price is $7 per unit, but when the price is $4 per unit they will produce and sell only 77,000,000 units.

    a. Find the supply equation.
    b. Find the equilibrium price and quantity.
    c. If the government wants to increase the quantity demanded by 21,000,000 units, what subsidy per unit must it provide?

23. Let D_x and S_x denote the number of gallons of ice cream (measured in thousands) demanded and supplied, respectively. Let D_y and S_y denote the number of containers of ice cream topping (measured in hundreds) demanded and supplied, respectively. Suppose that p_x and p_y represent the prices (in dollars per unit) of ice cream and topping, respectively. It is known that the demand and supply functions for ice cream and topping are interrelated as shown by

    $$D_x = 16 - 2p_x - p_y$$
    $$S_x = -10 + 3p_x + 3p_y$$
    $$D_y = 30 - 2p_x - 3p_y$$
    $$S_y = -17 + 8p_x + 3p_y$$

    a. Find the equilibrium prices of ice cream and topping.
    b. Find the equilibrium quantities of ice cream and topping.

24. Let D_x and S_x denote the numbers of half-gallons of gourmet ice cream (measured in thousands) demanded and supplied,

respectively. Let D_y and S_y denote the number of half-gallons of regular ice cream (also measured in thousands) demanded and supplied, respectively. Suppose that p_x and p_y represent the prices (in dollars per half-gallon) of gourmet and regular ice cream, respectively. It is known that the demand and supply functions for gourmet and regular ice cream are inter-related as is shown by

$$D_x = 310 - 40p_x + 20p_y$$
$$S_x = -110 + 60p_x - 10p_y$$
$$D_y = 445 + 16p_x - 50p_y$$
$$S_y = -206 - 4p_x + 250p_y$$

a. Find the equilibrium prices of gourmet and regular ice cream.

b. Find the equilibrium quantities of gourmet and regular ice cream.

25. Suppose that when three commodities X, Y, and Z are priced at x, y, and z dollars per unit, respectively, the quantities demanded are D_x, D_y, and D_z, respectively (measured in thousands) and the quantities supplied are S_x, S_y, and S_z, respectively (also measured in thousands). The demand and supply functions for these commodities are interrelated as expressed by

$$D_x = 520 - 30x + 92y - 25z$$
$$S_x = -120 + 70x - 8y + 75z$$

$$D_y = 140 + 32x - 52y + 18z$$
$$S_y = -7 - 8x + 98y - 12z$$
$$D_z = 478 - 20x + 28y - 5z$$
$$S_z = -21 + 40x - 12y + 65z$$

Find the equilibrium prices and quantities for all three commodities.

26. Same as Exercise 25 with the following equations:

$$D_x = 150 - 10x + 85y - 17z$$
$$S_x = -8 + 90x - 15y + 83z$$
$$D_y = 280 + 42x - 20y + 81z$$
$$S_y = -17 - 8x + 120y - 19z$$
$$D_z = 310 - 17x + 51y - 12z$$
$$S_z = -44 + 63x - 9y + 88z$$

27. Same as Exercise 25 with the following equations:

$$D_x = 800 - 23x + 95y - 18z$$
$$S_x = -35 + 77x - 5y + 82z$$
$$D_y = 320 + 50x - 12y + 35z$$
$$S_y = -68 - 10x + 118y - 5z$$
$$D_z = 1310 - 10x + 42y - 20z$$
$$S_z = -112 + 80x - 8y + 100z$$

2.6 Chapter Review

IMPORTANT SYMBOLS AND TERMS

$kE_i \rightarrow E_i$ [2.2]
$kE_i + E_j \rightarrow E_j$ [2.2]
$k_1E_i + k_2E_j \rightarrow E_j$ [2.2]
Break-even point [2.5]
Consistent and dependent [2.1], [2.2]
Consistent and independent [2.1], [2.2]
Cramer's rule [2.1]
Demand curve [2.5]
Demand function [2.5]
Determinant [2.1]
Elimination of variables by addition [2.1]
Elimination of variables by substitution [2.1]

Equilibrium price [2.5]
Equilibrium quantity [2.5]
Equivalent systems [2.1], [2.2]
Fixed cost [2.5]
Graphical solution [2.1], [2.3]
Half-plane [2.4]
Inconsistent [2.1]
Linear equation in two variables [2.1]
Market equilibrium [2.5]
Matrix [2.1]
Nonlinear systems [2.1]
Ordered pair [2.1]
Solution [2.1]
Solution set [2.1], [2.2]

Subsidy [2.5]
Supply curve [2.5]
Supply function [2.5]
System of linear inequalities in two variables [2.4]
System of three linear equations in three variables [2.2]
System of two linear equations in two variables [2.1]
Tax [2.5]
Walrasian general equilibrium [2.5]

SUMMARY An equation that can be written in the form $ax + by = c$, where a, b, and c are real numbers and $ab \neq 0$ is called a linear equation in two variables x and y. An ordered pair of real numbers (x_0, y_0) is a solution of the equation if the statement $ax_0 + by_0 = c$ is true. The set of all solutions is called the solution set of the equation. If a_1, b_1, c_1, a_2, b_2, and c_2 are real numbers

$$\begin{cases} a_1 x + b_1 y = c_1 \\ a_2 x + b_2 y = c_2 \end{cases}$$

is called a system of two linear equations in two variables. The intersection of the solution sets of the two equations is the solution set of the system. We may solve the system by

a. finding the intersection of the graphs of the two equations,
b. eliminating one variable by addition or subtraction,
c. eliminating one variable by substitution, or
d. using Cramer's rule.

A square array of numbers of the form

$$\begin{bmatrix} a & b \\ c & d \end{bmatrix}$$

is called a 2×2 matrix. The determinant of this 2×2 matrix is the number $ad - bc$ and is denoted

$$\begin{vmatrix} a & b \\ c & d \end{vmatrix}$$

First make sure that the system is written as

$$\begin{cases} a_1 x + b_1 y = c_1 \\ a_2 x + b_2 y = c_2 \end{cases}$$

Then let

$$\begin{vmatrix} a_1 & b_1 \\ a_2 & b_2 \end{vmatrix} = C, \quad \begin{vmatrix} c_1 & b_1 \\ c_2 & b_2 \end{vmatrix} = A_x, \text{ and } \begin{vmatrix} a_1 & c_1 \\ a_2 & c_2 \end{vmatrix} = A_y$$

If $C \neq 0$, the solution of the system is $\left(\dfrac{A_x}{C}, \dfrac{A_y}{C} \right)$.

Sometimes the right substitutions in a nonlinear system of two equations can yield a linear system that can be solved. Then the values of the original variables can be obtained from the solutions of the linear system. We also consider systems of three linear equations in three variables.

$$\begin{cases} a_1 x + b_1 y + c_1 z = d_1 \\ a_2 x + b_2 y + c_2 z = d_2 \\ a_3 x + b_3 y + c_3 z = d_3 \end{cases}$$

We can get an equivalent system by multiplying both sides of an equation by a nonzero constant, by adding to any equation one of the other two equations, or by combining these two operations. Starting with a given system, we try to obtain an equivalent system in the form

$$\begin{cases} x = x_0 \\ y = y_0 \\ z = z_0 \end{cases}$$

whose solution is obviously (x_0, y_0, z_0). The necessary steps should be documented on the right side of each system using the following abbreviations: $kE_i \rightarrow E_i$ means "the ith equation is multiplied by the (nonzero) constant k to get the new ith equation," and $kE_i + E_j \rightarrow E_j$ means "the ith equation is multiplied by the constant k and the result is added to the jth equation to get the new jth equation." The meaning of $k_1 E_i + k_2 E_j \rightarrow E_j$ should be clear.

We can also eliminate a variable by substitution to obtain a linear system of two equations in two variables which then may be solved using any of the methods outlined earlier. A system of two equations in two variables, where at least one of the two equations is nonlinear, may be solved by graphing both equations and finding the coordinates of the points of intersection of the graphs.

An expression that can be written $ax + by < c$, $ax + by > c$, $ax + by \leq c$, or $ax + by \geq c$, where a, b, c are real numbers and $ab \neq 0$ is called a linear inequality in two variables. An ordered pair of real numbers (x_0, y_0) is a solution of one of these if a true statement results when x and y are replaced by x_0 and y_0, respectively. The graph of a linear inequality is a half-plane. If we have a system of n inequalities in the two variables x and y, we sketch the graph of each inequality on the same Cartesian coordinate plane. The intersection of the n half-planes is the graph of the system. We can get the coordinates of a vertex of the graph by choosing the appropriate two inequalities, changing the inequality signs to equal signs, and solving the resulting linear system of two equations in two variables.

In applications, if we know the cost (C) and revenue (R) functions, the break-even point is found by solving

$$\begin{cases} y = C(x) \\ y = R(x) \end{cases}$$

If $q = D(p)$ and $q = S(p)$ define the demand and supply functions, respectively, the market is in equilibrium when $D(p) = S(p)$. The equilibrium price and equilibrium quantity are found by solving

$$\begin{cases} q = D(p) \\ q = S(p) \end{cases}$$

If a tax of T dollars per unit is imposed by the government, the equilibrium price and quantity will be affected. If the tax is already included in the price the consumer pays, then the demand equation does not change, but we must replace p by $p - T$ in the supply equation since the supplier receives only $p - T$ dollars per unit, the other T dollars goes to the government. We get the new equilibrium point by solving the system

$$\begin{cases} q = D(p) \\ q = S(p - T) \end{cases}$$

If the tax is added after the purchase, the demand equation is affected since the consumer pays $p + T$ dollars per unit but the supply equation is unchanged. We get the new equilibrium point by solving the system

$$\begin{cases} q = D(p + T) \\ q = S(p) \end{cases}$$

If the government pays a subsidy of S dollars per unit, the demand equation is not affected. However, we replace p by $p + S$ in the supply equation since the suppliers receive p dollars (paid by the consumer) plus S dollars (paid by the government) for each unit sold. Suppose we have two commodities X and Y and when the prices of these commodities are p_x and p_y dollars per unit, respectively, the quantities demanded are D_x and D_y, respectively, and the quantities supplied are S_x and S_y, respectively. In general D_x, D_y, S_x, and S_y are given in terms of p_x and p_y. By letting $D_x = S_x$ and $D_y = S_y$, we obtain a system of two equations in the two variables p_x and p_y. The solution of this system gives us the equilibrium prices.

SAMPLE EXAM QUESTIONS   *In Exercises 1–8, solve the given system using the indicated method.*

1. $\begin{cases} 2x + 3y = -5 \\ 3x - 5y = 21 \end{cases}$ (graphically)

2. $\begin{cases} 3x - 2y = -9 \\ 2x + 4y = 10 \end{cases}$ (graphically)

3. $\begin{cases} 5x + 2y = 0 \\ 4x - 3y = 23 \end{cases}$ (substitution)

4. $\begin{cases} 3x + 7y = 10 \\ 5x - 2y = 3 \end{cases}$ (substitution)

5. $\begin{cases} 2x - 3y = 7 \\ 3x + 9y = -3 \end{cases}$ (addition)

6. $\begin{cases} 3x - 2y = -5 \\ 6x - 4y = 4 \end{cases}$ (addition)

7. $\begin{cases} 5x + y = -13 \\ 3x - 2y = -13 \end{cases}$ (Cramer's rule)

8. $\begin{cases} 2x + 3y = 11 \\ 3x - 7y = 5 \end{cases}$ (Cramer's rule)

In Exercises 9–12, solve the given system by first making an appropriate substitution.

9. $\begin{cases} 2x^2 + y^2 = 17 \\ 5x^2 - y^2 = 11 \end{cases}$

10. $\begin{cases} \frac{1}{3}x^2 + y^2 = 4 \\ \frac{1}{9}x^2 + 5y^2 = 6 \end{cases}$

11. $\begin{cases} \frac{5}{x} + \frac{1}{y} = -13 \\ \frac{3}{x} - \frac{2}{y} = -13 \end{cases}$

12. $\begin{cases} \frac{5}{x} + \frac{2}{y} = 8 \\ \frac{3}{x} - \frac{4}{y} = 10 \end{cases}$

Solve the systems of Exercises 13, 14, and 15 graphically.

13. $\begin{cases} y = -2x^2 + 8x + 13 \\ 6x - y = -9 \end{cases}$

14. $\begin{cases} 31x + 40y = 98 \\ y = 2^{-x} \end{cases}$

15. $\begin{cases} y = x^2 + 5x + 6 \\ y = -x^2 + 3x + 10 \end{cases}$

Solve the linear systems of Exercises 16–22 using the method of elimination of variables by addition.

16. $\begin{cases} 2x - 2y + 3z = 1 \\ x - 3y - 2z = -9 \\ x + y + z = 6 \end{cases}$

17. $\begin{cases} 12x - 12y - z = 0 \\ 6x + 6y + 2z = 90 \\ 2x + 3y + z = 36 \end{cases}$

18. $\begin{cases} 6x - 2y - z = 6 \\ x - 3y - 4z = 5 \\ 3x + y + 4z = 0 \end{cases}$

19. $\begin{cases} x + 2y - 7z = 75 \\ 6x - 2y + 3z = 75 \\ 2x + y + z = 0 \end{cases}$

20. $\begin{cases} 4x - y + z = 11 \\ 3x - 2y - z = 6 \\ 7x - 3y = 17 \end{cases}$

21. $\begin{cases} 2x + 3y - z = 2 \\ 5x - 2y + z = 3 \\ 9x + 4y - z = 9 \end{cases}$

22. $\begin{cases} 5x \quad\ - z = 16 \\ 3x - y - 2z = 9 \\ x + 2y + 3z = 4 \end{cases}$

Solve the linear systems of Exercises 23–26 by elimination of a variable by substitution.

23. The system of Exercise 16.

24. The system of Exercise 17.

25. The system of Exercise 18.

26. The system of Exercise 19.

27. The system of Exercise 20.

28. The system of Exercise 21.

29. The system of Exercise 22.

In Exercises 30–34, sketch the graph of the system of linear inequalities.

30. $\begin{cases} x > 5 \\ y > 2 \\ 5x + 3y \le 46 \end{cases}$

31. $\begin{cases} x > 2 \\ 2x - y \le 9 \\ x + y \le 12 \end{cases}$

32. $\begin{cases} 5x + 3y < 28 \\ x - y > -4 \\ x - 5y < 0 \end{cases}$

33. $\begin{cases} x \le 7 \\ y \le 8 \\ x \ge 0 \\ y \ge 0 \\ x + y \ge 5 \end{cases}$

34. $\begin{cases} x - y \ge -7 \\ x - 2y \le 4 \\ 3x + y \ge 19 \end{cases}$

35. Roy owns two stores. In 1987 the first store earned 10% of the investment in it, while the second store lost 5%, for a net gain of $47,500. In 1988 the stores earned 8% and 6%, respectively, for a net gain of $93,000. What is each store worth?

36. A mixture of 8 pounds of grade A coffee mixed with 10 pounds of grade B coffee is worth $3.50 a pound, and 14 pounds of grade A mixed with 4 pounds of grade B is worth $3.92 a pound. What is the price of each grade of coffee?

37. The wages per day of fifteen adults and eight teenagers amount to $1336. Two adults together earn $48 more per day than three teenagers. What are the wages of each adult and each teenager?

38. Six pounds of tea and fourteen pounds of coffee cost $77.40. If the price of coffee should decrease by 10% and that of tea should increase by 10%, the cost of the quantities would be $75.06. What is the price of each per pound?

39. If Kris works 2 days and Pete works 5 days on a certain job they can complete three eighths of the job. If Kris works 6 more days and Pete works 5 more days, they can finish the job. How long would it take either one to do the job alone?

40. A seamstress paid $99 for 3 yards of white, 5 yards of blue, and 2 yards of red material. Another seamstress paid $107 for 2 yards of white, 3 yards of blue, and 4 yards of red material. The first seamstress needed more material later and paid $13 for $\frac{1}{4}$ yard of white, $\frac{1}{3}$ yard of blue, and $\frac{1}{2}$ yard of red. Find the price per yard of each kind of material.

41. A company manufactures two types of rowing machines. Each deluxe model takes 4 hours to assemble and 2 hours to paint, while each standard model takes 3 hours to assemble and 1 hour and 12 minutes to paint. The company employs 78 assemblers and 36 painters each working 8 hours a day. Let x and y denote the respective daily production of deluxe and standard models, respectively. Sketch the graph of the set of possible ordered pairs (x, y).

42. A company makes two models of television sets. The deluxe model takes 6 hours to assemble and 20 minutes to test while the standard model takes 4 hours to assemble and 12 minutes to test. The plant is designed in such a way that a maximum of 19,000 hours are available each month for assembly and a maximum of 1000 hours are available per month for testing. The company has accepted orders totaling 1250 deluxe models and 1800 standard models per month. Let x and y denote the numbers of deluxe and standard models, respectively, the company manufactures per month. Sketch the graph of the set of possible ordered pairs (x, y).

43. Let D_x and S_x denote the numbers of pounds of gourmet chocolate demanded and supplied, respectively, when the price of gourmet chocolate is p_x dollars. Let D_y and S_y denote the numbers of pounds of regular chocolate demanded and supplied, respectively, when the price of regular chocolate is p_y dollars per pound. It is known that these quantities are interrelated as follows:

$$D_x = 387{,}500 - 50{,}000p_x + 25{,}000p_y$$

$$S_x = -137{,}500 + 75{,}000p_x - 12{,}500p_y$$

$$D_y = 556{,}250 + 20{,}000p_x - 62{,}500p_y$$

$$S_y = -257{,}500 - 5{,}000p_x + 312{,}500p_y$$

a. Find the equilibrium prices per pound of gourmet and regular chocolate.
b. Find the equilibrium quantities of gourmet and regular chocolate.

Matrices

3

It is frequently useful to arrange a collection of objects, or numbers, in a rectangular array. For example, an instructor may have a seating chart on which the rows and columns of the classroom are represented. If the student occupies a seat located in the fifth row from the front and the seventh column from the left wall, then the student's name would appear in row 5, column 7 of the chart. A used-car dealer may keep a record of his inventory in a similarly compact way. He may draw a chart on which the first column is for Chevrolets, the second column for Fords, the third column for Dodges, and so on. He reserves the first row for 1978 models, the second row for 1979 models, and the third row for 1980 models. If he enters a 5 in row 2, column 3 and a 0 in row 1, column 2, we know that he has five 1979 Dodges and no 1978 Fords.

> **DEFINITION 3.1:** In general, we define a *matrix* (plural *matrices*) to be a rectangular array of numbers, enclosed within brackets.

3.1 Matrix Notation and Matrix Arithmetic

Often matrices are denoted by boldface capital letters. For example we may use **A**, **B**, **S**, **T**, . . . to denote matrices. Since a matrix is a rectangular array of numbers, each of these numbers has a certain position in the array. If a number is in row i and column j of matrix **A**, it is called the *i-j entry* (element) of the matrix and it is denoted a_{ij} (read: "a sub i,j"). The first subscript always indicates the number of the (horizontal) row, while the second subscript always gives the (vertical) column number. We use a comma if either i or j has more than one digit. For example, if an entry of matrix **B** is in row 25, column 7, we denote it by $b_{25,7}$ (without the comma b_{257} would be ambiguous).

EXAMPLE 1 Let

$$\mathbf{B} = \begin{bmatrix} -1 & 5 & 8 & -7 \\ 2 & 3 & 5 & -4 \\ 6 & 7 & -3 & 13 \end{bmatrix}$$

Find b_{13} and b_{34}.

Solution Since b_{13} is the element in row 1, column 3, we have $b_{13} = 8$. Similarly, b_{34} is the entry in row 3, column 4. Thus, $b_{34} = 13$. ∎

Special Matrices

If a matrix has m rows and n columns, it is said to be of *dimension* $m \times n$ (read: "m by n").

DEFINITION 3.2: A $1 \times n$ matrix is called an *n-dimensional row vector*, while an $n \times 1$ matrix is called an *n-dimensional column vector*. A matrix of dimension $m \times m$ is called a *square matrix*. We often use the notation $[a_{ij}]_{m \times n}$ to denote an $m \times n$ matrix.

EXAMPLE 2   Find $\mathbf{B} = [b_{ij}]_{2 \times 3}$ if $b_{ij} = 3i - 2j + 1$.

Solution   The matrix $\mathbf{B}$ has two rows and three columns. Hence, there will be six entries that can be found using the formula $b_{ij} = 3i - 2j + 1$. To find b_{11} (the element in row 1, column 1), we replace both i and j by 1 and obtain $b_{11} = 2$ since $3(1) - 2(1) + 1 = 2$. Similarly, we find

$$b_{12} = 3(1) - 2(2) + 1 = 0$$
$$b_{13} = 3(1) - 2(3) + 1 = -2$$
$$b_{21} = 3(2) - 2(1) + 1 = 5$$
$$b_{22} = 3(2) - 2(2) + 1 = 3$$
$$b_{23} = 3(2) - 2(3) + 1 = 1$$

Thus,

$$\mathbf{B} = \begin{bmatrix} b_{11} & b_{12} & b_{13} \\ b_{21} & b_{22} & b_{23} \end{bmatrix} = \begin{bmatrix} 2 & 0 & -2 \\ 5 & 3 & 1 \end{bmatrix}$$

Do Exercise 3.

DEFINITION 3.3: The *diagonal elements* of a square matrix are those elements a_{ij} for which $i = j$. A square matrix for which all diagonal elements are equal and all other elements are 0 is called a *scalar matrix*.

For example,

$$\begin{bmatrix} 3 & 0 & 0 & 0 \\ 0 & 3 & 0 & 0 \\ 0 & 0 & 3 & 0 \\ 0 & 0 & 0 & 3 \end{bmatrix}$$

is a scalar matrix.

DEFINITION 3.4: If all diagonal elements of a scalar matrix are 1, the matrix is called a *multiplicative identity*.

For example,

$$\begin{bmatrix} 1 & 0 & 0 \\ 0 & 1 & 0 \\ 0 & 0 & 1 \end{bmatrix}$$

is the multiplicative identity matrix of order 3 and is often denoted I_3.

> **DEFINITION 3.5:**   If all entries of a matrix are 0, we say that the matrix is a *zero matrix* and denote it **0**. A zero matrix is frequently called an *additive identity*.

Finally, two matrices are said to be *equal* if, and only if, they have the same dimensions and the corresponding elements are equal. That is, $[a_{ij}]_{m \times n} = [b_{ij}]_{p \times q}$ if, and only if, $m = p$, $n = q$, and $a_{ij} = b_{ij}$ for all i and j where $1 \le i \le m$ and $1 \le j \le n$. When we make a statement about matrix elements a_{ij} and we add "for all i,j," we specifically mean for all integers i and j where $1 \le i \le m$ and $1 \le j \le n$, where m is the number of rows and n is the number of columns.

EXAMPLE 3   Find x, y, z, and w if

$$\begin{bmatrix} 2 & 3y \\ 3z & 5 \\ 6 & 4w \end{bmatrix} = \begin{bmatrix} 4x & 9 \\ 15 & 5 \\ 6 & 16 \end{bmatrix}$$

Solution   Both matrices have dimensions 3×2. We need only have equality between corresponding elements. Thus, we must have $2 = 4x$, $3y = 9$, $3z = 15$, $5 = 5$, $6 = 6$, and $4w = 16$. The fourth and fifth equalities are obviously true. The others will be true if $x = \frac{1}{2}$, $y = 3$, $z = 5$, and $w = 4$.
Do Exercise 23.   ■

Matrix Arithmetic

It is often useful to multiply each entry of a matrix by a fixed number as illustrated in the following example.

EXAMPLE 4   A gas station owner uses matrices to keep a record of the number of gallons of gasoline he sells each month. He owns two stations and records the number of gallons sold at station A in the first row and those sold at station B in the second row. He records the number of gallons of regular, unleaded, and premium unleaded each month in columns 1, 2, and 3, respectively. The following matrix is the record for June 1990.

Station	Regular	Unleaded	Premium unleaded
A	12,000	25,500	10,700
B	15,000	32,000	11,250

a. How many gallons of unleaded gasoline were sold at station B?

b. If his goal is to double the sales of all three kinds of gasoline at both stations, what will the goal matrix be for the next month?

Solution   a. Since we want to know how many gallons were sold at station B, we must look at the second row. The second column is used to record the sales of unleaded gasoline. The entry in row 2, column 2 gives us the number of gallons of unleaded gasoline sold at station B. That number is 32,000. Thus, 32,000 gallons of unleaded gasoline were sold at station B during the month of June 1990.

b. If the goal for the following month is to double the sales of all three kinds of gasoline at both stations, we must double every entry of the given matrix.

Station	Regular	Unleaded	Premium unleaded
A	24,000	51,000	21,400
B	30,000	64,000	22,500

DEFINITION 3.6:   Let $[a_{ij}]$ be an $m \times n$ matrix and let c be any scalar (number). Then the *scalar multiple* of $[a_{ij}]$ by c, $c[a_{ij}]$ is the $m \times n$ matrix $[b_{ij}]$ where $b_{ij} = ca_{ij}$ for all i,j.

EXAMPLE 5   Find $3A$ if

$$A = \begin{bmatrix} 2 & 5 & -3 \\ -1 & 0 & 7 \end{bmatrix}$$

Solution $3A = \begin{bmatrix} 3(2) & 3(5) & 3(-3) \\ 3(-1) & 3(0) & 3(7) \end{bmatrix} = \begin{bmatrix} 6 & 15 & -9 \\ -3 & 0 & 21 \end{bmatrix}$

Do Exercise 13.

We introduce the idea for adding and subtracting matrices in the following example.

EXAMPLE 6   A used-car dealer sells only Chevrolets, Fords, and Renaults. She owns a lot in Seattle and another in Portland. She has two matrices, one for the inventory at each lot. Columns 1, 2, and 3 are used for the Chevrolets, Fords, and Renaults, respectively. Rows 1, 2, 3, and 4 are used to record the 1984, 1985, 1986, and 1987 models, respectively. The matrix **S** is the record for the Seattle lot, while matrix **P** is the record for the Portland lot.

$$S = \begin{bmatrix} 2 & 6 & 4 \\ 0 & 3 & 2 \\ 3 & 2 & 0 \\ 6 & 0 & 2 \end{bmatrix} \quad \text{and} \quad P = \begin{bmatrix} 5 & 7 & 3 \\ 1 & 4 & 6 \\ 5 & 1 & 0 \\ 3 & 4 & 2 \end{bmatrix}$$

Find a single matrix **C** that will show the total number of cars of each kind.

Solution The dealer has two 1984 Chevrolets in Seattle and five in Portland. Thus, the total number of 1984 Chevrolets is seven. We shall enter a 7 as the 1-1 entry of the matrix **C**. We obtain the other entries similarly. We have

$$\mathbf{S} + \mathbf{P} = \begin{bmatrix} 2+5 & 6+7 & 4+3 \\ 0+1 & 3+4 & 2+6 \\ 3+5 & 2+1 & 0+0 \\ 6+3 & 0+4 & 2+2 \end{bmatrix} = \begin{bmatrix} 7 & 13 & 7 \\ 1 & 7 & 8 \\ 8 & 3 & 0 \\ 9 & 4 & 4 \end{bmatrix}$$ ∎

In general, matrices can be added and subtracted only if they have the same dimensions.

DEFINITION 3.7: If $\mathbf{A} = [a_{ij}]$ and $\mathbf{B} = [b_{ij}]$, where $\mathbf{A}$ and $\mathbf{B}$ are two $m \times n$ matrices, their *sum* and *difference* are the matrices denoted $\mathbf{A} + \mathbf{B}$ and $\mathbf{A} - \mathbf{B}$, respectively. If

$$\mathbf{A} + \mathbf{B} = [s_{ij}] \text{ and } \mathbf{A} - \mathbf{B} = [d_{ij}]$$

then

$$s_{ij} = a_{ij} + b_{ij} \text{ and } d_{ij} = a_{ij} - b_{ij} \text{ for all } i,j$$

It is easy to verify that addition of matrices is commutative and associative. That is, for any three $m \times n$ matrices $\mathbf{A}$, $\mathbf{B}$, and $\mathbf{C}$, it is true that

$$\mathbf{A} + \mathbf{B} = \mathbf{B} + \mathbf{A}$$

and

$$\mathbf{A} + (\mathbf{B} + \mathbf{C}) = (\mathbf{A} + \mathbf{B}) + \mathbf{C}$$

This follows from the commutative and associative properties of addition of numbers. These properties carry over to addition of matrices because to add matrices, we simply add corresponding elements.

EXAMPLE 7 Let the matrices $\mathbf{A}$, $\mathbf{B}$, $\mathbf{C}$, and $\mathbf{D}$ be defined as follows:

$$\mathbf{A} = \begin{bmatrix} 2 & 3 & -1 \\ 5 & -2 & 6 \end{bmatrix} \mathbf{B} = \begin{bmatrix} 4 & -2 & 6 \\ -7 & 0 & 3 \end{bmatrix}$$

$$\mathbf{C} = \begin{bmatrix} 3 & -4 \\ 1 & 0 \\ 3 & 4 \end{bmatrix} \mathbf{D} = \begin{bmatrix} 5 & 8 \\ 9 & 6 \\ -2 & 3 \end{bmatrix}$$

Find the following:

a. $\mathbf{A} + \mathbf{B}$ **b.** $\mathbf{C} - \mathbf{D}$

c. $2\mathbf{A} + (-3)\mathbf{B}$ **d.** $\mathbf{A} + \mathbf{C}$

Solution   a. $A + B = \begin{bmatrix} 2 & 3 & -1 \\ 5 & -2 & 6 \end{bmatrix} + \begin{bmatrix} 4 & -2 & 6 \\ -7 & 0 & 3 \end{bmatrix}$

$= \begin{bmatrix} 2+4 & 3+(-2) & -1+6 \\ 5+(-7) & -2+0 & 6+3 \end{bmatrix}$

$= \begin{bmatrix} 6 & 1 & 5 \\ -2 & -2 & 9 \end{bmatrix}$

b. $C - D = \begin{bmatrix} 3 & -4 \\ 1 & 0 \\ 3 & 4 \end{bmatrix} - \begin{bmatrix} 5 & 8 \\ 9 & 6 \\ -2 & 3 \end{bmatrix}$

$= \begin{bmatrix} 3-5 & -4-8 \\ 1-9 & 0-6 \\ 3-(-2) & 4-3 \end{bmatrix} = \begin{bmatrix} -2 & -12 \\ -8 & -6 \\ 5 & 1 \end{bmatrix}$

c. $2A + (-3)B = 2\begin{bmatrix} 2 & 3 & -1 \\ 5 & -2 & 6 \end{bmatrix} + (-3)\begin{bmatrix} 4 & -2 & 6 \\ -7 & 0 & 3 \end{bmatrix}$

$= \begin{bmatrix} 2(2) & 2(3) & 2(-1) \\ 2(5) & 2(-2) & 2(6) \end{bmatrix} + \begin{bmatrix} (-3)4 & (-3)(-2) & (-3)6 \\ (-3)(-7) & (-3)0 & (-3)3 \end{bmatrix}$

$= \begin{bmatrix} 4 & 6 & -2 \\ 10 & -4 & 12 \end{bmatrix} + \begin{bmatrix} -12 & 6 & -18 \\ 21 & 0 & -9 \end{bmatrix}$

$= \begin{bmatrix} 4+(-12) & 6+6 & -2+(-18) \\ 10+21 & -4+0 & 12+(-9) \end{bmatrix}$

$= \begin{bmatrix} -8 & 12 & -20 \\ 31 & -4 & 3 \end{bmatrix}$

d. $A + C = \begin{bmatrix} 2 & 3 & -1 \\ 5 & -2 & 6 \end{bmatrix} + \begin{bmatrix} 3 & -4 \\ 1 & 0 \\ 3 & 4 \end{bmatrix}$

The addition in part **d** cannot be performed since **A** is a 2×3 matrix, while **C** is a 3×2 matrix and matrices can be added, or subtracted, only if they have the same dimensions.

Do Exercise 19. ■

Exercise Set 3.1

1. Give the dimensions of each of the following matrices.

$A = \begin{bmatrix} 1 & 2 & -3 \\ 3 & 0 & 9 \end{bmatrix}$, $B = \begin{bmatrix} 2 & 7 \\ 4 & -6 \\ 3 & 5 \end{bmatrix}$,

$C = \begin{bmatrix} 5 & 2 & 4 & 0 \\ 4 & 0 & -3 & 7 \\ 2 & 5 & 9 & 0 \end{bmatrix}$

2. For the matrices of Exercise 1, find:

   a. a_{11}, a_{12}, a_{22}, and a_{23} if $A = [a_{ij}]$
   b. b_{12}, b_{21}, b_{22}, and b_{32} if $B = [b_{ij}]$
   c. c_{23}, c_{24}, c_{31}, and c_{34} if $C = [c_{ij}]$

3. Write the 2×3 matrix $[a_{ij}]$ if $a_{ij} = 3i + 2j$ for all i, j.

4. Write the 3×2 matrix $[b_{ij}]$ if $b_{ij} = 5i - 3j$ for all i, j.

5. Write the 3×3 matrix $[c_{ij}]$ if $c_{ij} = i^2 + 2j$ for all i, j.

6. Write the 4×3 matrix $[d_{ij}]$ if $d_{ij} = 3i + j^2$ for all i, j.

7. Write the 4×4 matrix $[a_{ij}]$ if

$$a_{ij} = \begin{cases} 0 & \text{whenever } i = j \\ 2i + j & \text{whenever } i \neq j. \end{cases}$$

8. Write the 4×4 matrix $[b_{ij}]$ if

$$b_{ij} = \begin{cases} 0 & \text{whenever } i = j \\ i + 2j & \text{whenever } i \neq j. \end{cases}$$

9. Write the 3×4 matrix $[c_{ij}]$ if

$$c_{ij} = \begin{cases} i^2 & \text{whenever } i = j \\ 3i + j & \text{whenever } i \neq j. \end{cases}$$

10. Write the 4×3 matrix $[d_{ij}]$ if

$$d_{ij} = \begin{cases} j^3 & \text{whenever } i = j \\ i + 5j & \text{whenever } i \neq j. \end{cases}$$

11. Write the 4×4 matrix $[a_{ij}]$ if

$$a_{ij} = \begin{cases} (-1)^j & \text{whenever } i = j \\ 2i & \text{whenever } i \neq j. \end{cases}$$

In Exercises 12–20, perform the indicated operations. If an operation is not possible, state why.

12. $4 \begin{bmatrix} 2 & 5 \\ 7 & 0 \end{bmatrix}$

13. $-3 \begin{bmatrix} 3 & 4 & -2 \\ 2 & -4 & 0 \\ 7 & 6 & 12 \end{bmatrix}$

14. $6 \begin{bmatrix} 3 & 7 & -2 & 8 \\ 5 & -3 & 9 & 0 \end{bmatrix}$

15. $\begin{bmatrix} 2 & -4 & 5 & -7 \\ 3 & 6 & -2 & 9 \end{bmatrix} + \begin{bmatrix} 0 & 0 & -1 \\ -3 & 5 & 9 \end{bmatrix}$

16. $\begin{bmatrix} 4 & 2 \\ 2 & -1 \end{bmatrix} - \begin{bmatrix} 3 & -5 \\ 1 & -3 \end{bmatrix}$

17. $2 \begin{bmatrix} 3 & 2 \\ 9 & 0 \\ 4 & 2 \\ 6 & 5 \end{bmatrix} + 3 \begin{bmatrix} 8 & 5 \\ -1 & 3 \\ 3 & -2 \\ -3 & 0 \end{bmatrix}$

18. $(-2) \begin{bmatrix} 5 & 7 & 3 \\ 4 & 1 & 6 \end{bmatrix} + 4 \begin{bmatrix} 4 & -9 \\ 3 & 7 \end{bmatrix}$

19. $5 \begin{bmatrix} 3 & 8 & 1 \\ 5 & 1 & -2 \\ 9 & 2 & 3 \end{bmatrix} - 2 \begin{bmatrix} 2 & -9 & 4 \\ 0 & 3 & 7 \\ 5 & 0 & 2 \end{bmatrix}$

20. $4 \begin{bmatrix} 5 & 4 \\ 2 & 9 \\ 1 & 0 \\ 5 & 3 \end{bmatrix} + 2 \begin{bmatrix} 2 & -3 \\ 3 & 6 \\ 4 & 2 \\ 4 & 2 \end{bmatrix} - 3 \begin{bmatrix} 6 & 1 \\ 5 & -2 \\ 0 & 3 \\ 6 & 1 \end{bmatrix}$

In Exercises 21–28, find the values of the variables that make the given matrix equations true statements.

21. $\begin{bmatrix} 3 & x \\ z & 6 \end{bmatrix} = \begin{bmatrix} y & 5 \\ 7 & 6 \end{bmatrix}$

22. $\begin{bmatrix} 5 + x & 6 + 2y \\ z - 3 & w + 10 \end{bmatrix} = \begin{bmatrix} 2x + 2 & 3y + 4 \\ 3z - 13 & 3w + 12 \end{bmatrix}$

23. $\begin{bmatrix} 2 - x & 3 + y \\ 5 & 2 + z \\ 4w & 7 \end{bmatrix} = \begin{bmatrix} 3x - 2 & 2y + 1 \\ 5 & 3z - 6 \\ 2w + 6 & 7 \end{bmatrix}$

24. $\begin{bmatrix} 5 & 2x \\ 2z & 5 \end{bmatrix} + \begin{bmatrix} 3y & 2 \\ 3 & -w \end{bmatrix} = \begin{bmatrix} 11 & 0 \\ 3z + 6 & 3 \end{bmatrix}$

25. $\begin{bmatrix} x + y & 2 \\ 5 & 2x + y \end{bmatrix} + \begin{bmatrix} x + 2y & 1 \\ 2 & x - 3y \end{bmatrix}$
$= \begin{bmatrix} 5 & 3 \\ 7 & -12 \end{bmatrix}$

26. $\begin{bmatrix} 4x + y & 3x \\ 4 & 3y + 4 \end{bmatrix} + \begin{bmatrix} 3x - 1 & 3x \\ 2 & 2y + 15 \end{bmatrix} =$
$\begin{bmatrix} 2x - 2y & 6x \\ 6 & 2x \end{bmatrix}$

27. $\begin{bmatrix} x + y & 2 & 1 \\ 3 & 2x - y & 2 \\ 5 & 3y & 2y - 2z \end{bmatrix}$
$+ \begin{bmatrix} x + z & 3 & -1 \\ 4 & x - y & 3 \\ 1 & y & y \end{bmatrix}$
$= \begin{bmatrix} 7 - 2z & 5 & 0 \\ 7 & -z & 5 \\ 6 & 4y & 3 + x \end{bmatrix}$

28. $\begin{bmatrix} 2x + y & 1 \\ 3y & x + y \\ x & 7 \end{bmatrix} + \begin{bmatrix} 3x + y & 2 \\ 2y & 2 + x \\ z - 6 & 1 \end{bmatrix}$
$= \begin{bmatrix} 4x - 3z + 5 & 3 \\ 5y & 1 + z \\ y - z & 8 \end{bmatrix}$

29. A used-car dealer uses matrices to keep track of his inventory. He uses columns 1, 2, 3, and 4 for Fords, Chevrolets, Dodges, and Buicks, respectively; and rows 1, 2, 3, 4, and 5 for 1983, 1984, 1985, 1986, and 1987 models, respectively. He owns two lots. The following matrices show his inventory on these lots.

$$\begin{bmatrix} 1 & 0 & 4 & 6 \\ 3 & 2 & 1 & 4 \\ 4 & 0 & 2 & 4 \\ 0 & 3 & 1 & 5 \\ 2 & 4 & 1 & 0 \end{bmatrix}$$

$$\begin{bmatrix} 0 & 2 & 5 & 3 \\ 5 & 0 & 4 & 2 \\ 2 & 5 & 0 & 7 \\ 0 & 3 & 5 & 2 \\ 5 & 3 & 1 & 0 \end{bmatrix}$$

Write a single matrix giving his total inventory.

30. Four racquetball players decide to hold a round-robin tournament during the summer. They each play the other three players three games in each of the two halves of the tournament and use matrices to keep track of the results. In the following two matrices, a_{ij} is the number of games player i won against player j in each half of the tournament, respectively.

$$\begin{bmatrix} 0 & 2 & 1 & 3 \\ 1 & 0 & 2 & 1 \\ 2 & 1 & 0 & 1 \\ 0 & 2 & 2 & 0 \end{bmatrix}$$

$$\begin{bmatrix} 0 & 1 & 3 & 1 \\ 2 & 0 & 3 & 1 \\ 0 & 0 & 0 & 2 \\ 2 & 2 & 1 & 0 \end{bmatrix}$$

a. Why is it that in both matrices, $a_{ij} + a_{ji} = 3$ for all i, j with $i \neq j$?
b. Why is it that in both matrices $a_{ii} = 0$ for each i?
c. Write a single matrix showing the record for the tournament.
d. Which player won the most games for the whole tournament?

31. Same as Exercise 30, this time with five players and the following two matrices.

$$\begin{bmatrix} 0 & 2 & 2 & 1 & 3 \\ 1 & 0 & 3 & 1 & 2 \\ 1 & 0 & 0 & 2 & 0 \\ 2 & 2 & 1 & 0 & 2 \\ 0 & 1 & 3 & 1 & 0 \end{bmatrix}$$

$$\begin{bmatrix} 0 & 3 & 1 & 2 & 1 \\ 0 & 0 & 2 & 1 & 3 \\ 2 & 1 & 0 & 3 & 0 \\ 1 & 2 & 0 & 0 & 1 \\ 2 & 0 & 3 & 2 & 0 \end{bmatrix}$$

32. A manufacturer has plants at four different locations and warehouses at three different locations. In the following matrix, a_{ij} gives the cost (in dollars) of transporting a unit of the product from plant i to warehouse j.

$$\begin{bmatrix} 5 & 4 & 6 \\ 2 & 6 & 4 \\ 4 & 2 & 8 \\ 6 & 5 & 2 \end{bmatrix}$$

If the cost is increased uniformly by 20%, what is the corresponding matrix?

33. A sporting goods company has a plant in Montreal and another in Chicago. At each plant it manufactures rowing machines and exercise cycles. It has three models A, B, and C of each. The following two matrices give the numbers (in thousands) of each machine of each type manufactured in 1990, where in each matrix rows 1 and 2 are used for the rowing machines and exercise cycles, respectively, and columns 1, 2, and 3 are used for models A, B, and C, respectively.

$$\mathbf{M} = \begin{bmatrix} 40 & 24 & 32 \\ 36 & 16 & 20 \end{bmatrix}$$

$$\mathbf{C} = \begin{bmatrix} 20 & 15 & 25 \\ 35 & 30 & 20 \end{bmatrix}$$

a. Write a single matrix showing the total production for both plants in 1990.
b. If production is increased by 25% in the Montreal plant (matrix $\mathbf{M}$) and by 20% in the Chicago plant (matrix $\mathbf{C}$),

what is the matrix for total production for the following year?

34. The following matrices give the enrollment figures at some university in the fall, winter, and spring of 1990. Columns 1, 2, 3, and 4 are used for the number (in hundreds) of freshmen, sophomores, juniors, and seniors, respectively. Rows 1, 2, 3, 4, and 5 are used to record the number of business, engineering, arts and sciences, education, and nursing majors, respectively.

$$\begin{bmatrix} 2 & 2.1 & 2.5 & 2.3 \\ 2 & 2.2 & 2.3 & 2.2 \\ 3 & 2.9 & 2.8 & 2.7 \\ 1 & 1.1 & 1 & 1.2 \\ 2 & 1.9 & 2.1 & 2.1 \end{bmatrix}$$

$$\begin{bmatrix} 2.3 & 2 & 2.2 & 2.4 \\ 1.9 & 2.1 & 2.1 & 1.9 \\ 2.9 & 2.7 & 2.9 & 2.8 \\ 1.1 & 1 & 1.2 & 1.1 \\ 2.1 & 2 & 2.2 & 2.1 \end{bmatrix}$$

$$\begin{bmatrix} 2 & 1.9 & 2.1 & 2.2 \\ 2.1 & 2.3 & 1.9 & 2.2 \\ 2.8 & 2.6 & 3 & 2.6 \\ 1.2 & 1.2 & 1.1 & 1 \\ 2.2 & 2.1 & 2.3 & 2.1 \end{bmatrix}$$

Write a matrix that gives the average enrollment per class and per major for the year 1990.

3.2 Matrix Multiplication

In the preceding section, you learned to add and subtract matrices that have the same dimensions. You also learned to multiply a matrix by a scalar. In this section, you will learn about the multiplication of matrices. We begin with the simpler case where both factors are vectors. (Recall from Definition 3.2 that a $1 \times n$ matrix is called an n-dimensional row vector and an $n \times 1$ matrix is called an n-dimensional column vector.) To help you understand the definition, we begin with the following example.

EXAMPLE 1 The Essers are considering having their daughter attend Seattle University in Fall 1992. They obtain some information in the following tabular form.

Cost (in dollars)		
Tuition per credit	Living expenses per month	Average price of a book
150	460	35

The number of items they will have to pay for are given in the following table.

Number of credits	Number of months	Number of books
45	9	15

a. Represent the item costs by a matrix.
b. Represent the numbers of each item needed by a matrix.
c. Find the estimated total cost for the academic year.

Solution   **a.** As in the table, the first, second, and third columns are used to represent the cost per credit hour, the average monthly living expenses, and average cost per book, respectively. We obtain the matrix

$$[\ 150 \quad 460 \quad 35 \ ]$$

The foregoing is a 3-dimensional row vector.

b. Again, if we use the first, second, and third columns to represent the number of credits needed, the number of months, and the number of books needed, we get the matrix

$$[\ 45 \quad 9 \quad 15 \ ]$$

This matrix is a 3-dimensional row vector also.

c. The total cost of tuition is $6750 since $(150)(45) = 6750$. The estimated cost of living expenses is $4140 since $(460)(9) = 4140$. The estimated cost of books is $525 since $(35)(15) = 525$. Hence, the estimated total cost is $11,415 because $6750 + 4140 + 525 = 11,415$.

$$(150)(45) + (460)(9) + (35)(15) = 11,415 \qquad ■$$

We see that the total cost is determined by multiplying the first entry of one vector by the first entry of the other vector, the second entry of one vector by the second entry of the other, and the third entry of one vector by the third entry of the other and then adding the three products. This is an example of the useful dot product of vectors, which we now define.

Dot Product of Vectors

> **DEFINITION 3.8:**   Suppose that **A** and **B** are two n-dimensional vectors. If we multiply the first entry of **A** by the first entry of **B**, then we multiply the second entry of **A** by the second entry of **B**, and if we continue this process until we have multiplied the nth entry of **A** by the nth entry of **B**, and if we finally add the n products, the final result is called the *dot product* (*inner product*) of the two vectors and is denoted **A** · **B**.

The two vectors *must* have the same dimensions. However, both can be row vectors ($1 \times n$ matrices), column vectors ($n \times 1$ matrices), or one of each.

EXAMPLE 2

Let $\mathbf{A} = [\ 2 \quad 3 \quad -4 \quad 0 \ ]$, $\mathbf{B} = [\ -2 \quad 3 \quad 2 \quad 4 \ ]$, and $\mathbf{C} = \begin{bmatrix} 3 \\ 5 \\ 2 \\ -1 \end{bmatrix}$

a. Find **A** · **B**.
b. Find **A** · **C**.

Solution   **a.** $\mathbf{A} \cdot \mathbf{B} = [\ 2 \quad 3 \quad -4 \quad 0\ ] \cdot [\ -2 \quad 3 \quad 2 \quad 4\ ]$

$$= 2(-2) + 3(3) + (-4)2 + 0(4) = -3$$

The dot product $\mathbf{A} \cdot \mathbf{B}$ is -3.

b. $\mathbf{A} \cdot \mathbf{C} = [2 \quad 3 \quad -4 \quad 0] \cdot \begin{bmatrix} 3 \\ 5 \\ 2 \\ -1 \end{bmatrix}$

$$= 2(3) + 3(5) + (-4)2 + 0(-1) = 13$$

The inner product $\mathbf{A} \cdot \mathbf{C}$ is 13. ∎

Matrix Multiplication

In the preceding section, we saw that if we know the dimensions of a matrix and if we can find each of its entries, then we can write the matrix. We begin our description of matrix multiplication by simply stating that if $\mathbf{A}$ and $\mathbf{B}$ are two matrices whose product is a matrix $\mathbf{C}$, then c_{ij} (the i-j entry of the product) is the dot product of the ith row of $\mathbf{A}$ by the jth column of $\mathbf{B}$. Therefore, we can find c_{ij} only if the ith row of $\mathbf{A}$ and the jth column of $\mathbf{B}$ are vectors with the same dimensions. That is, c_{ij} can be found only if the number of columns of $\mathbf{A}$ is the same as the number of rows of $\mathbf{B}$. In that case, $\mathbf{A}$ is said to be *conformable to* $\mathbf{B}$ for multiplication. Before stating the definition, we give the following example.

EXAMPLE 3   Let $\mathbf{A} = \begin{bmatrix} 2 & -1 & 3 \\ 4 & 0 & -5 \end{bmatrix}$ and $\mathbf{B} = \begin{bmatrix} 5 & -2 \\ 9 & 3 \\ 1 & 0 \end{bmatrix}$.

a. Verify that $\mathbf{A}$ is conformable to $\mathbf{B}$ for multiplication.
b. Find all possible c_{ij}'s where each c_{ij} is the dot product of the ith row of $\mathbf{A}$ by the jth column of $\mathbf{B}$.
c. Arrange the results in matrix form.

Solution   **a.** $\mathbf{A}$ has three columns and $\mathbf{B}$ has three rows. Therefore, $\mathbf{A}$ is conformable to $\mathbf{B}$ for multiplication.
b. To find c_{11}, we calculate the dot product of row 1 of $\mathbf{A}$ by column 1 of $\mathbf{B}$. We get

$$[\ 2 \quad -1 \quad 3\ ] \cdot \begin{bmatrix} 5 \\ 9 \\ 1 \end{bmatrix} = 2(5) + (-1)9 + 3(1) = 4$$

Similarly, we find c_{12} by calculating the dot product of row 1 of $\mathbf{A}$ by column 2 of $\mathbf{B}$. We obtain

$$[\ 2 \quad -1 \quad 3\ ] \cdot \begin{bmatrix} -2 \\ 3 \\ 0 \end{bmatrix} = 2(-2) + (-1)3 + 3(0) = -7$$

The entry c_{21} is found by calculating the dot product of row 2 of **A** by column 1 of **B**. We get

$$[\,4 \quad 0 \quad -5\,] \cdot \begin{bmatrix} 5 \\ 9 \\ 1 \end{bmatrix} = 4(5) + 0(9) + (-5)1 = 15$$

The dot product of row 2 of **A** by column 2 of **B** gives c_{22}.

$$[\,4 \quad 0 \quad -5\,] \cdot \begin{bmatrix} -2 \\ 3 \\ 0 \end{bmatrix} = 4(-2) + 0(3) + (-5)0 = -8$$

c. Arranging the c_{ij}'s we obtained in part **b** in matrix form we get

$$\mathbf{C} = \begin{bmatrix} c_{11} & c_{12} \\ c_{21} & c_{22} \end{bmatrix} = \begin{bmatrix} 4 & -7 \\ 15 & -8 \end{bmatrix}$$

As you will soon see, this matrix is the product **AB**. The matrix **C** found in Example 3 has 2 rows because **A** had 2 rows and we could find c_{ij} for $i = 1$ and for $i = 2$. Since **B** had 2 columns, we could find c_{ij} for $j = 1$ and for $j = 2$.

In general, if **A** is $m \times n$ and **B** is $n \times p$, **A** is conformable to **B** for multiplication (since the number of columns of **A** is the same as the number of rows of **B**) and the product **AB** can be found. Suppose that the product is $[c_{ij}]$. For each value of j, c_{ij} can be found for $i = 1, 2, \ldots, m$. (See Example 3.) Thus, the product has m rows. Also for each value of i, c_{ij} can be found for $j = 1, 2, \ldots, p$. Therefore, the product has p columns. Formally, we have the following definition.

DEFINITION 3.9:　Suppose that **A** is an $m \times n$ matrix and **B** is an $n \times p$ matrix. Then the *product* **AB** is the $m \times p$ matrix $[c_{ij}]$ where for all i ($i = 1, 2, \ldots, m$), and all j ($j = 1, 2, \ldots, p$), c_{ij} is the dot product of the ith row of **A** by the jth column of **B**.

To test whether a matrix **A** is conformable to a matrix **B** for multiplication and to determine the dimensions of the product **AB** in the case that it is, you may find the following procedure helpful. Let **A** be an $m \times n$ matrix and **B** be a $p \times q$ matrix. Write the dimensions in a staggered form as shown.

$$m \times n$$
$$p \times q$$
$$\uparrow$$

These must match (n and p).

When they do, **A** is conformable to **B** for multiplication. In that case, cross them out and get the dimensions of the product **AB**. For example, if **A** is 5 × 3 and **B** is 3 × 4, we write

$$5 \times 3$$
$$3 \times 4$$
$$\uparrow$$

These match (3 and 3).

Therefore **A** is conformable to **B** for multiplication. Cross out the 3s,

$$5 \times \cancel{3}$$
$$\cancel{3} \times 4$$

The product **AB** is 5 × 4.

EXAMPLE 4   Let **A** and **B** be the matrices of Example 3.

a. Verify that **B** is conformable to **A** for multiplication.
b. Find the product **BA**.

Solution   **a.** We write

$$3 \times 2$$
$$2 \times 3$$
$$\uparrow$$

These match.

Thus **B** is conformable to **A** for multiplication. Cross out the 2s,

$$3 \times \cancel{2}$$
$$\cancel{2} \times 3$$

The product **BA** is 3 × 3.
b. Let **BA** = $[d_{ij}]$. Then

$$d_{11} = [\,5 \quad -2\,] \cdot \begin{bmatrix} 2 \\ 4 \end{bmatrix} = 5(2) + (-2)4 = 2$$

$$d_{12} = [\,5 \quad -2\,] \cdot \begin{bmatrix} -1 \\ 0 \end{bmatrix} = 5(-1) + (-2)0 = -5$$

$$d_{13} = [\,5 \quad -2\,] \cdot \begin{bmatrix} 3 \\ -5 \end{bmatrix} = 5(3) + (-2)(-5) = 25$$

$$d_{21} = [\,9 \quad 3\,] \cdot \begin{bmatrix} 2 \\ 4 \end{bmatrix} = 9(2) + 3(4) = 30$$

$$d_{22} = [\,9 \quad 3\,] \cdot \begin{bmatrix} -1 \\ 0 \end{bmatrix} = 9(-1) + 3(0) = -9$$

$$d_{23} = [\,9 \quad 3\,] \cdot \begin{bmatrix} 3 \\ -5 \end{bmatrix} = 9(3) + 3(-5) = 12$$

$$d_{31} = [\ 1 \quad 0\ ] \cdot \begin{bmatrix} 2 \\ 4 \end{bmatrix} = 1(2) + 0(4) = 2$$

$$d_{32} = [\ 1 \quad 0\ ] \cdot \begin{bmatrix} -1 \\ 0 \end{bmatrix} = 1(-1) + 0(0) = -1$$

$$d_{33} = [\ 1 \quad 0\ ] \cdot \begin{bmatrix} 3 \\ -5 \end{bmatrix} = 1(3) + 0(-5) = 3$$

Thus,

$$\mathbf{BA} = \begin{bmatrix} d_{11} & d_{12} & d_{13} \\ d_{21} & d_{22} & d_{23} \\ d_{31} & d_{32} & d_{33} \end{bmatrix} = \begin{bmatrix} 2 & -5 & 25 \\ 30 & -9 & 12 \\ 2 & -1 & 3 \end{bmatrix}$$ ■

Matrix Multiplication Is Not Commutative

In Examples 3 and 4 we found the products **AB** and **BA**, respectively, for the same two matrices **A** and **B**. The products were not equal since **AB** was a 2×2 matrix, while **BA** was 3×3.

In general, if **A** is conformable to **B** for multiplication, **B** need not be conformable to **A** for multiplication. For example, if **A** is 3×5 and **B** is 5×2, then **A** is conformable to **B** but **B** is not conformable to **A**. Even in cases where both **AB** and **BA** can be found and have the same dimensions, the products are not necessarily equal. That is, multiplication of matrices is not commutative. We illustrate this with another example.

EXAMPLE 5   Find **AB** and **BA** if $\mathbf{A} = \begin{bmatrix} 2 & 5 \\ 3 & 1 \end{bmatrix}$ and $\mathbf{B} = \begin{bmatrix} 3 & 2 \\ 4 & 7 \end{bmatrix}$.

Solution   In exactly the same way as we performed multiplication in the preceding examples, we get

$$\mathbf{AB} = \begin{bmatrix} 2 & 5 \\ 3 & 1 \end{bmatrix} \cdot \begin{bmatrix} 3 & 2 \\ 4 & 7 \end{bmatrix} = \begin{bmatrix} 26 & 39 \\ 13 & 13 \end{bmatrix}$$

and

$$\mathbf{BA} = \begin{bmatrix} 3 & 2 \\ 4 & 7 \end{bmatrix} \cdot \begin{bmatrix} 2 & 5 \\ 3 & 1 \end{bmatrix} = \begin{bmatrix} 12 & 17 \\ 29 & 27 \end{bmatrix}$$ ■

For practice, you will find it valuable to carry out the details of the multiplication in this example. Note that $\mathbf{AB} \neq \mathbf{BA}$, even though both are 2×2 matrices. However, the statement $\mathbf{AB} \neq \mathbf{BA}$ does not mean that we can never have equality. Consider the following example.

EXAMPLE 6   A student and his girlfriend are discussing the fact that matrix multiplication is not commutative. The student says that he randomly wrote many 2×2 matrices and multiplied them in different order, but he never got the same answer. He concluded that the equality $\mathbf{AB} = \mathbf{BA}$ could never be obtained. His clever girlfriend bet him a lunch that she could get this equality as often as she pleased. She won the bet. Explain how she did it.

Solution She probably started with an arbitrary square matrix, say $\mathbf{A} = \begin{bmatrix} 2 & 5 \\ 4 & 1 \end{bmatrix}$. Then, let

$$\mathbf{B} = \mathbf{A}^2 = \begin{bmatrix} 2 & 5 \\ 4 & 1 \end{bmatrix} \cdot \begin{bmatrix} 2 & 5 \\ 4 & 1 \end{bmatrix} = \begin{bmatrix} 4+20 & 10+5 \\ 8+4 & 20+1 \end{bmatrix} = \begin{bmatrix} 24 & 15 \\ 12 & 21 \end{bmatrix}$$

Now,

$$\mathbf{AB} = \begin{bmatrix} 2 & 5 \\ 4 & 1 \end{bmatrix} \cdot \begin{bmatrix} 24 & 15 \\ 12 & 21 \end{bmatrix} = \begin{bmatrix} 48+60 & 30+105 \\ 96+12 & 60+21 \end{bmatrix} = \begin{bmatrix} 108 & 135 \\ 108 & 81 \end{bmatrix}$$

Also,

$$\mathbf{BA} = \begin{bmatrix} 24 & 15 \\ 12 & 21 \end{bmatrix} \cdot \begin{bmatrix} 2 & 5 \\ 4 & 1 \end{bmatrix} = \begin{bmatrix} 48+60 & 120+15 \\ 24+84 & 60+21 \end{bmatrix} = \begin{bmatrix} 108 & 135 \\ 108 & 81 \end{bmatrix}$$

We see that $\mathbf{AB} = \mathbf{BA}$. This is not surprising since both are equal to $\mathbf{A}^3$.*
Do Exercise 21. ■

Properties of Matrix Multiplication

Another fact about matrix multiplication which is different from multiplication of numbers is that it is possible to have $\mathbf{AB} = \mathbf{0}$, even though neither $\mathbf{A}$ nor $\mathbf{B}$ is the zero matrix.

EXAMPLE 7 Find $\mathbf{AB}$ if

$$\mathbf{A} = \begin{bmatrix} 1 & -1 \\ -1 & 1 \end{bmatrix} \text{ and } \mathbf{B} = \begin{bmatrix} 2 & 2 \\ 2 & 2 \end{bmatrix}$$

Solution We easily see that

$$\mathbf{AB} = \begin{bmatrix} 1 & -1 \\ -1 & 1 \end{bmatrix} \cdot \begin{bmatrix} 2 & 2 \\ 2 & 2 \end{bmatrix} = \begin{bmatrix} 1(2)+(-1)2 & 1(2)+(-1)2 \\ (-1)2+1(2) & (-1)2+1(2) \end{bmatrix}$$
$$= \begin{bmatrix} 0 & 0 \\ 0 & 0 \end{bmatrix}$$

Note that $\mathbf{AB} = \mathbf{0}$, although neither $\mathbf{A}$ nor $\mathbf{B}$ is a zero matrix.
Do Exercise 23. ■

There are, however, some properties of matrices which are like the familiar properties of numbers.

---

*We may have $\mathbf{AB} = \mathbf{BA}$ even if each factor is not a power of the same matrix $\mathbf{C}$. (See Exercise 43.)

Properties of Matrices

1. Multiplication of matrices is *associative*. That is, if **A**, **B**, and **C** are $m \times n$, $n \times p$, and $p \times q$ matrices, respectively, then

$$\mathbf{A(BC) = (AB)C}$$

2. Multiplication of matrices is *distributive over addition*. That is, if **A**, **B**, and **C** are matrices, **A** is $m \times n$, and both **B** and **C** are $n \times p$, then

$$\mathbf{A(B + C) = AB + AC}$$

In the preceding section, we defined the multiplicative identity of order n which we denoted $\mathbf{I}_n$. We now state a third property of matrix multiplication which justifies the name of the matrix $\mathbf{I}_n$.

3. If **A** is an $m \times n$ matrix, the following are true:

$$\mathbf{AI}_n = \mathbf{A} \quad \text{and} \quad \mathbf{I}_m\mathbf{A} = \mathbf{A}$$

When we multiply **A** on the right by the identity matrix, we must choose $\mathbf{I}_n$ so that **A** is conformable to $\mathbf{I}_n$. For a similar reason, we must choose $\mathbf{I}_m$ when we multiply **A** on the left.

EXAMPLE 8   Let $\mathbf{A} = \begin{bmatrix} 2 & 5 & -3 \\ 7 & -2 & 6 \end{bmatrix}$. Find $\mathbf{I}_2\mathbf{A}$ and $\mathbf{AI}_3$.

Solution   The identity matrix $\mathbf{I}_2$ is 2×2 and **A** is 2×3. Hence, $\mathbf{I}_2$ is conformable to **A** for multiplication and the product matrix $\mathbf{I}_2\mathbf{A}$ is 2×3. We get

$$\mathbf{I}_2\mathbf{A} = \begin{bmatrix} 1 & 0 \\ 0 & 1 \end{bmatrix} \cdot \begin{bmatrix} 2 & 5 & -3 \\ 7 & -2 & 6 \end{bmatrix}$$

$$= \begin{bmatrix} 1(2)+0(7) & 1(5)+0(-2) & 1(-3)+0(6) \\ 0(2)+1(7) & 0(5)+1(-2) & 0(-3)+1(6) \end{bmatrix}$$

$$= \begin{bmatrix} 2 & 5 & -3 \\ 7 & -2 & 6 \end{bmatrix} = \mathbf{A}$$

The matrix **A** is 2×3 and the identity matrix $\mathbf{I}_3$ is 3×3. Therefore, **A** is conformable to $\mathbf{I}_3$ for multiplication and the product matrix $\mathbf{AI}_3$ is 2×3. We obtain

$$\mathbf{AI}_3 = \begin{bmatrix} 2 & 5 & -3 \\ 7 & -2 & 6 \end{bmatrix} \cdot \begin{bmatrix} 1 & 0 & 0 \\ 0 & 1 & 0 \\ 0 & 0 & 1 \end{bmatrix}$$

$$= \begin{bmatrix} 2(1)+5(0)+(-3)0 & 2(0)+5(1)+(-3)0 & 2(0)+5(0)+(-3)1 \\ 7(1)+(-2)0+6(0) & 7(0)+(-2)1+6(0) & 7(0)+(-2)0+6(1) \end{bmatrix}$$

$$= \begin{bmatrix} 2 & 5 & -3 \\ 7 & -2 & 6 \end{bmatrix} = \mathbf{A}$$

Applications

EXAMPLE 9 A manufacturer builds regular, deluxe, and super sailboats. Each comes in two models, A and B. The following table gives the quantities the manufacturer plans to build in a certain month.

	Regular	Deluxe	Super
Model A	5	4	3
Model B	6	3	2

The material needed for each type is given by the following table in appropriately chosen units.

	Wood	Paint	Canvas
Regular	5	4	6
Deluxe	6	5	10
Super	8	7	15

Finally, the cost per unit (in thousands of dollars) of each of the materials used is given in tabular form.

	Cost per unit
Wood	2
Paint	.3
Canvas	3

a. Represent the given tables by matrices **Q**, **M**, and **C**, respectively.
b. Find the product **QM** and interpret the result.
c. Find the matrix **(QM)C** and interpret the result.

Solution **a.** Using the values from the tables, we have

$$\mathbf{Q} = \begin{bmatrix} 5 & 4 & 3 \\ 6 & 3 & 2 \end{bmatrix}, \mathbf{M} = \begin{bmatrix} 5 & 4 & 6 \\ 6 & 5 & 10 \\ 8 & 7 & 15 \end{bmatrix}, \text{ and } \mathbf{C} = \begin{bmatrix} 2 \\ .3 \\ 3 \end{bmatrix}$$

b.

$$\mathbf{QM} = \begin{bmatrix} 5 & 4 & 3 \\ 6 & 3 & 2 \end{bmatrix} \cdot \begin{bmatrix} 5 & 4 & 6 \\ 6 & 5 & 10 \\ 8 & 7 & 15 \end{bmatrix}$$

$$= \begin{bmatrix} 25+24+24 & 20+20+21 & 30+40+45 \\ 30+18+16 & 24+15+14 & 36+30+30 \end{bmatrix}$$

$$= \begin{bmatrix} 73 & 61 & 115 \\ 64 & 53 & 96 \end{bmatrix}$$

Note that 73 is the dot product of row 1 of matrix **Q** by column 1 of matrix **M**. Since row 1 of matrix **Q** represents the A models and column 1 of matrix **M** represents the number of units of wood, 73 represents the total number of units of wood needed to build all the A models. Similarly, 61 units of paint and 115 units of canvas are needed to build all the A models. Also, to build all the B models, 64 units of wood, 53 units of paint, and 96 units of canvas are needed.

c.

$$(\mathbf{QM})\mathbf{C} = \begin{bmatrix} 73 & 61 & 115 \\ 64 & 53 & 96 \end{bmatrix} \cdot \begin{bmatrix} 2 \\ .3 \\ 3 \end{bmatrix}$$

$$= \begin{bmatrix} 146+18.3+345 \\ 128+15.9+288 \end{bmatrix} = \begin{bmatrix} 509.3 \\ 431.9 \end{bmatrix}$$

Since the cost per unit is given in thousands of dollars, 509.3 indicates that the total cost to build all the A models is $509,300 and the total cost to build all the B models is $431,900.

Do Exercise 37.   ■

To illustrate another application of matrix multiplication, we introduce the *basic principle of counting*, which is discussed in greater detail in Chapter 6. Suppose that an event (such as a flight from one city to another) can occur in p different ways and once it has occurred in any of these, a second event (such as a connecting flight) can occur in q different ways, then the two events can occur, in that order, in $p \cdot q$ different ways.

EXAMPLE 10   The following two tables give the number of daily nonstop flights between certain cities in the United States.

From To → ↓	Chicago	Kansas City	Dallas
Seattle	5	3	2
San Francisco	6	4	7
Los Angeles	7	3	9

From To → ↓	Boston	New York
Chicago	4	4
Kansas City	5	6
Dallas	3	5

Use matrix multiplication to find the total number of flights from each of the West Coast cities to each of the East Coast cities with exactly one overnight stop in either Chicago, Kansas City, or Dallas.

Solution Let $\mathbf{W}$ and $\mathbf{E}$ be matrices representing the first and second tables, respectively. Then,

$$\mathbf{W} = \begin{bmatrix} 5 & 3 & 2 \\ 6 & 4 & 7 \\ 7 & 3 & 9 \end{bmatrix} \text{ and } \mathbf{E} = \begin{bmatrix} 4 & 4 \\ 5 & 6 \\ 3 & 5 \end{bmatrix}$$

The product $\mathbf{WE}$ is a 3×2 matrix $[c_{ij}]$. We first calculate c_{11} to get

$$c_{11} = 5(4) + 3(5) + 2(3) = 20 + 15 + 6 = 41$$

On any day there are five flights from Seattle to Chicago and, on the next day, four flights from Chicago to Boston. Therefore, there are 20 flights from Seattle to Boston with one overnight stop in Chicago. Similarly, we see that there are 15 flights from Seattle to Boston with one overnight stop in Kansas City and 6 flights from Seattle to Boston with one overnight stop in Dallas. It is now clear that there are 41 flights from Seattle to Boston with exactly one overnight stop in Chicago, Kansas City, or Dallas and $c_{11} = 41$.

We complete the multiplication of the two matrices as shown.

$$\mathbf{WE} = \begin{bmatrix} 5 & 3 & 2 \\ 6 & 4 & 7 \\ 7 & 3 & 9 \end{bmatrix} \cdot \begin{bmatrix} 4 & 4 \\ 5 & 6 \\ 3 & 5 \end{bmatrix}$$

$$= \begin{bmatrix} 20+15+6 & 20+18+10 \\ 24+20+21 & 24+24+35 \\ 28+15+27 & 28+18+45 \end{bmatrix} = \begin{bmatrix} 41 & 48 \\ 65 & 83 \\ 70 & 91 \end{bmatrix}$$

Using the product matrix, we can write a table that gives the number of flights between the three western cities and the two eastern cities with exactly one overnight stop in Chicago, Kansas City, or Dallas.

From To → ↓	Boston	New York
Seattle	41	48
San Francisco	65	83
Los Angeles	70	91

Do Exercise 39. ■

We now introduce a procedure that we shall use in greater detail later.

EXAMPLE 11 Consider the system $\begin{cases} 2x + y = -3 \\ 5x + 3y = 4 \end{cases}$.

a. Write the system as one matrix equation.

b. Write the left side of this equation as a product of a 2×2 matrix and the column vector $\begin{bmatrix} x \\ y \end{bmatrix}$.

Solution   **a.** We first write $\begin{bmatrix} 2x & + & y \\ 5x & + & 3y \end{bmatrix} = \begin{bmatrix} -3 \\ 4 \end{bmatrix}$.*

Since two matrices having the same dimensions are equal if, and only if, the corresponding elements are equal, the foregoing matrix equation is clearly equivalent to the original system of two equations.

b. $2x + y$ is the dot product of the vectors $[\ 2 \quad 1\ ]$ and $\begin{bmatrix} x \\ y \end{bmatrix}$, while $5x + 3y$ is the dot product of the vectors $[\ 5 \quad 3\ ]$ and $\begin{bmatrix} x \\ y \end{bmatrix}$. We can factor the matrix on the left side of the matrix equation of part **a** to get

$$\begin{bmatrix} 2 & 1 \\ 5 & 3 \end{bmatrix} \cdot \begin{bmatrix} x \\ y \end{bmatrix} = \begin{bmatrix} -3 \\ 4 \end{bmatrix}$$

Do Exercise 31.   ■

In the preceding example, we wrote the linear system

$$\begin{cases} 2x + y = -3 \\ 5x + 3y = 4 \end{cases}$$

as a single matrix equation $\mathbf{AX} = \mathbf{B}$, where

$$\mathbf{A} = \begin{bmatrix} 2 & 1 \\ 5 & 3 \end{bmatrix}, \mathbf{X} = \begin{bmatrix} x \\ y \end{bmatrix}, \text{ and } \mathbf{B} = \begin{bmatrix} -3 \\ 4 \end{bmatrix}$$

The matrices $\mathbf{A}$, $\mathbf{X}$, and $\mathbf{B}$ are called the *coefficient matrix*, the *variable vector*, and the *value vector*, respectively. (The matrices $\mathbf{X}$ and $\mathbf{B}$ are 2×1, or 2-dimensional column vectors.)

In general, a system of m linear equations in n variables can be written as a single matrix equation

$$\mathbf{AX} = \mathbf{B}$$

Where $\mathbf{A}$ is an $m \times n$ coefficient matrix, $\mathbf{X}$ is an n-dimensional column vector whose entries are the variables of the system, and $\mathbf{B}$ is an m-dimensional value vector whose entries are the constants appearing on the right sides of the equations of the system.

EXAMPLE 12   Write the system as a single matrix equation.

$$\begin{cases} 3w & + 2y - 3z = 5 \\ & 5x + 3y + 5z = 2 \\ 2w + 4x & - 6y + 2z = 7 \end{cases}$$

*Students often mistakenly think of the matrix on the left as a 2×2 matrix. However, each of the expressions $2x + y$ and $5x + 3y$ represents a single quantity and therefore that matrix has only one column. It is a 2×1 matrix.

Solution The coefficient matrix is **A** where

$$A = \begin{bmatrix} 3 & 0 & 2 & -3 \\ 0 & 5 & 3 & 5 \\ 2 & 4 & -6 & 2 \end{bmatrix}$$

while the variable vector **X** and value vector **B** are defined by

$$X = \begin{bmatrix} w \\ x \\ y \\ z \end{bmatrix} \text{ and } B = \begin{bmatrix} 5 \\ 2 \\ 7 \end{bmatrix}$$

You should verify that **AX** = **B**.* ■

Exercise Set 3.2

In Exercises 1–5, find the products **AB**, **BA**, **AC**, **CA**, **BC**, *and* **CB**. *If any of these cannot be found, state why.*

1. $A = \begin{bmatrix} 1 & 3 & -2 \\ 0 & 2 & 5 \\ 2 & 4 & 0 \end{bmatrix}$, $B = \begin{bmatrix} 2 & 7 \\ 3 & 5 \\ 7 & 2 \end{bmatrix}$,

$C = \begin{bmatrix} -1 & 2 & 1 \\ 9 & 0 & 3 \end{bmatrix}$

2. $A = \begin{bmatrix} 0 & -1 \\ 3 & 4 \\ 9 & 3 \\ 4 & 1 \end{bmatrix}$, $B = \begin{bmatrix} 3 & 5 \\ 8 & 0 \\ 2 & 5 \\ 3 & 7 \end{bmatrix}$,

$C = \begin{bmatrix} 4 & 6 & 3 \\ 3 & 0 & -1 \end{bmatrix}$

3. $A = \begin{bmatrix} 2 & 5 \\ 1 & 0 \end{bmatrix}$, $B = \begin{bmatrix} 3 & 5 & -2 \\ 9 & 3 & 0 \\ 0 & -1 & 3 \end{bmatrix}$,

$C = \begin{bmatrix} 2 & 5 & 1 \end{bmatrix}$

4. $A = \begin{bmatrix} 3 & 2 & 5 \\ 0 & 1 & 0 \\ 2 & 5 & 1 \\ 3 & -1 & 0 \end{bmatrix}$, $B = \begin{bmatrix} 3 & 2 & 1 \\ 5 & 3 & 0 \end{bmatrix}$,

$C = \begin{bmatrix} 3 & 1 & -8 \\ 0 & 2 & 5 \\ 3 & 2 & 0 \end{bmatrix}$

5. $A = \begin{bmatrix} 5 & 0 & 8 & 1 \\ 5 & 1 & 0 & 2 \\ 0 & 2 & 5 & 7 \end{bmatrix}$, $B = \begin{bmatrix} 9 & 5 & -2 \\ 0 & 3 & 2 \end{bmatrix}$,

$C = \begin{bmatrix} 5 \\ 3 \\ 6 \end{bmatrix}$

In Exercises 6–10, perform the indicated operations.

6. $\begin{bmatrix} 2 & 3 & -4 \\ 3 & 0 & 2 \end{bmatrix} \cdot \left(\begin{bmatrix} 3 & 2 \\ 4 & 0 \\ 2 & -2 \end{bmatrix} + \begin{bmatrix} 5 & -2 \\ 3 & 1 \\ 7 & 5 \end{bmatrix} \right)$

*When writing a system of m linear equations in n variables in the form **AX** = **B**, make sure that the system has been written in such a way that the variables are in the same order in all the equations and that blank spaces appear in place of missing variables.

7. $4\begin{bmatrix} 3 & 1 & 7 \\ 0 & 4 & -2 \end{bmatrix}$

$\cdot \left(2\begin{bmatrix} 1 & 2 & -3 \\ 0 & -1 & 5 \\ 7 & 0 & 2 \end{bmatrix} + 5\begin{bmatrix} 3 & 6 & 0 \\ 3 & 0 & 1 \\ 5 & 4 & 9 \end{bmatrix} \right)$

8. $-4\begin{bmatrix} 5 & 3 \\ 3 & 6 \\ 4 & -3 \end{bmatrix} \cdot \left(3\begin{bmatrix} 5 & 3 & -1 \\ 6 & 0 & 9 \end{bmatrix} \right.$

$\left. + (-4)\begin{bmatrix} 3 & 4 & -2 \\ 7 & -4 & 6 \end{bmatrix} \right)$

9. $\left(\begin{bmatrix} 2 & 4 \\ 4 & 6 \\ 5 & -4 \end{bmatrix} + \begin{bmatrix} 3 & 5 \\ 6 & 7 \\ 0 & 3 \end{bmatrix} \right) \cdot \begin{bmatrix} 2 & 5 & 7 \\ 4 & 2 & 3 \end{bmatrix}$

10. $\left(2\begin{bmatrix} 5 & 2 & 9 \\ 7 & 4 & 3 \\ 0 & 3 & 0 \end{bmatrix} + (-2)\begin{bmatrix} 3 & 1 & 3 \\ 2 & 4 & 0 \\ 3 & 1 & 5 \end{bmatrix} \right)$

$\cdot \begin{bmatrix} 2 & 4 & 7 & 1 \\ 3 & 8 & 3 & -2 \\ 2 & 7 & 9 & -5 \end{bmatrix} \Big)$

In Exercises 11–20, find both products **AB** *and* **BA**.

11. $\mathbf{A} = \begin{bmatrix} 2 & 5 & -1 \\ 4 & 0 & 7 \end{bmatrix}$ and $\mathbf{B} = \begin{bmatrix} 3 & -2 \\ 0 & 4 \\ 5 & 1 \end{bmatrix}$

12. $\mathbf{A} = \begin{bmatrix} 5 & -2 & 4 \\ 2 & 9 & 0 \end{bmatrix}$ and $\mathbf{B} = \begin{bmatrix} 0 & 2 \\ 4 & 1 \\ 6 & -1 \end{bmatrix}$

13. $\mathbf{A} = \begin{bmatrix} 3 & 1 & 7 & -9 \\ 0 & 4 & 2 & 3 \end{bmatrix}$ and $\mathbf{B} = \begin{bmatrix} 3 & 5 \\ 0 & -3 \\ 3 & 7 \\ 6 & 1 \end{bmatrix}$

14. $\mathbf{A} = \begin{bmatrix} 2 & 4 \\ 3 & 0 \end{bmatrix}$ and $\mathbf{B} = \begin{bmatrix} 3 & 1 \\ 1 & 5 \end{bmatrix}$

15. $\mathbf{A} = \begin{bmatrix} 5 & 2 \\ 7 & 3 \end{bmatrix}$ and $\mathbf{B} = \begin{bmatrix} 3 & -2 \\ -7 & 5 \end{bmatrix}$

16. $\mathbf{A} = \begin{bmatrix} 4 & 3 \\ 7 & 1 \end{bmatrix}$ and $\mathbf{B} = \begin{bmatrix} 2 & 7 \\ 8 & 4 \end{bmatrix}$

17. $\mathbf{A} = \begin{bmatrix} 1 & 2 & 3 \\ 4 & 5 & 6 \\ 7 & 8 & 9 \end{bmatrix}$ and $\mathbf{B} = \begin{bmatrix} 9 & 8 & 7 \\ 6 & 5 & 4 \\ 3 & 2 & 1 \end{bmatrix}$

18. $\mathbf{A} = \begin{bmatrix} 5 & -5 & 4 \\ 0 & 3 & 1 \\ 2 & -1 & 2 \end{bmatrix}$ and

$\mathbf{B} = \begin{bmatrix} 7 & 6 & -17 \\ 2 & 2 & -5 \\ -6 & -5 & 15 \end{bmatrix}$

19. $\mathbf{A} = \begin{bmatrix} 3 & 1 & 2 \\ 0 & 7 & -2 \\ 3 & 1 & 3 \end{bmatrix}$ and $\mathbf{B} = \begin{bmatrix} 0 & 2 & 0 \\ 3 & 6 & 1 \\ 2 & 1 & -6 \end{bmatrix}$

20. $\mathbf{A} = \begin{bmatrix} 1 & -4 & 0 & 1 \\ 5 & 0 & -5 & 4 \\ 0 & 1 & 3 & 1 \\ 2 & 2 & -1 & 2 \end{bmatrix}$ and

$\mathbf{B} = \begin{bmatrix} -28 & 35 & 34 & -73 \\ -1 & 1 & 1 & -2 \\ -8 & 10 & 10 & -21 \\ 25 & -31 & -30 & 65 \end{bmatrix}$

21. Find two 2×2 matrices **A** and **B** such that $\mathbf{AB} = \mathbf{BA}$.

22. Find two 3×3 matrices **A** and **B** such that $\mathbf{AB} = \mathbf{BA}$.

23. Find a 2×2 nonzero matrix **A** such that $\mathbf{AB} = 0$, where

$$\mathbf{B} = \begin{bmatrix} 6 & 10 \\ 3 & 5 \end{bmatrix}$$

24. Same as Exercise 23 with $\mathbf{B} = \begin{bmatrix} 8 & 18 \\ 4 & 9 \end{bmatrix}$.

25. Let $\mathbf{A} = \begin{bmatrix} 3 & 5 \\ 2 & -1 \\ 4 & 7 \end{bmatrix}$. Verify that $\mathbf{I}_3\mathbf{A} = \mathbf{AI}_2 = \mathbf{A}$.

26. Let $\mathbf{A} = \begin{bmatrix} 3 & 2 & 6 & 5 \\ 2 & 4 & 1 & 0 \end{bmatrix}$. Verify that $\mathbf{I}_2\mathbf{A} = \mathbf{AI}_4 = \mathbf{A}$.

27. Find the products **AB** and **BA** if $\mathbf{A} = \begin{bmatrix} 7 & 5 \\ 4 & 3 \end{bmatrix}$ and $\mathbf{B} = \begin{bmatrix} 3 & -5 \\ -4 & 7 \end{bmatrix}$.

28. Same as Exercise 27 with $\mathbf{A} = \begin{bmatrix} 11 & 6 \\ 9 & 5 \end{bmatrix}$ and $\mathbf{B} = \begin{bmatrix} 5 & -6 \\ -9 & 11 \end{bmatrix}$.

29. Same as Exercise 27 with $\mathbf{A} = \begin{bmatrix} 3 & -5 & 8 \\ -4 & 7 & 13 \end{bmatrix}$ and $\mathbf{B} = \begin{bmatrix} 7 & 5 \\ 4 & 3 \\ 0 & 0 \end{bmatrix}$.

30. Same as Exercise 27 with

$$A = \begin{bmatrix} 5 & 17 & 3 \\ 2 & 7 & 5 \end{bmatrix} \text{ and } B = \begin{bmatrix} 7 & -17 \\ -2 & 5 \\ 0 & 0 \end{bmatrix}.$$

In Exercises 31–35, write the linear system as a single matrix equation $AX = B$, where A is the coefficient matrix, X is the variable vector, and B is the value vector.

31. $\begin{cases} 5x + 2y = -3 \\ 7x + 3y = 5 \end{cases}$
32. $\begin{cases} 7x + 5y = -2 \\ 4x + 3y = 8 \end{cases}$

33. $\begin{cases} 11x + 6y - 7 = 0 \\ 9x + 5y + 5 = 0 \end{cases}$
34. $\begin{cases} 5x - 5y + 4z = 3 \\ z + 3y = 1 \\ 2x - y + 2z = -2 \end{cases}$

35. $\begin{cases} x + w - 4y = -2 \\ 4w + 5x - 5z = 3 \\ y + 3z + w = 5 \\ 2x + 2y - z + 2w = -1 \end{cases}$

36. A manufacturer builds regular, deluxe, and super sofas. Each comes in two models, A and B. The following table gives the quantities he plans to build in a certain month.

	Regular	Deluxe	Super
Model A	7	8	5
Model B	4	9	3

The material needed for each type is given by the following table in appropriately chosen units.

	Wood	Upholstery	Foam
Regular	3	6	9
Deluxe	4	8	10
Super	5	11	12

Finally, the cost per unit (in dollars) of each of the materials used is given in tabular form.

	Cost per unit
Wood	30
Upholstery	80
Foam	15

  a. Represent the tables by matrices **Q**, **M**, and **C**, respectively.
  b. Find the product **QM** and interpret the result.
  c. Find the matrix **(QM)C** and interpret the result.

37. A builder can purchase materials from three different suppliers, but because of transportation costs she will buy all materials from the same supplier. The prices per unit (in

dollars) charged by each of the suppliers are given in the following table.

	Bricks	Concrete	Lumber	Glass
Supplier A	8	5	9	3
Supplier B	7	6	8	2
Supplier C	9	5	7	2

The builder has contracted four different projects. The amount of materials required for each of the projects is given in another table.

	Project I	Project II	Project III	Project IV
Bricks	1000	1500	1300	1400
Concrete	700	950	875	910
Lumber	2200	1900	2100	1875
Glass	800	950	875	650

Represent the first table by a matrix **P** and the second table by a matrix **R**.

  a. Find the product **PR** and interpret the result.
  b. Which supplier should the contractor use for each of the four projects?

38. The following two tables give the number of daily nonstop flights between certain cities in the United States.

From To → ↓	Minneapolis	Denver	Houston
Philadelphia	3	4	5
Washington	6	5	8
Miami	4	3	6

From To → ↓	Portland	San Diego
Minneapolis	3	5
Denver	5	7
Houston	4	6

Use matrix multiplication to find the total number of flights from each of the East Coast cities to each of the West Coast cities with exactly one overnight stop in either Minneapolis, Denver, or Houston.

39. The following two tables give the number of daily nonstop flights between certain cities in Canada.

From To → ↓	Edmonton	Winnipeg	Calgary
Quebec	2	4	3
Montreal	3	2	4
Toronto	4	3	2

From To → ↓	Vancouver	Victoria
Edmonton	3	2
Winnipeg	4	3
Calgary	5	1

Use matrix multiplication to find the total number of flights from each of the eastern cities to each of the western cities with exactly one overnight stop in either Edmonton, Winnipeg, or Calgary.

40. Suppose four exams are given during a semester and a professor assigns weights of .2, .2, .2, and .4 to exams 1, 2, 3, and the final, respectively. Then, if a student receives scores of 80, 75, 82, and 90 on the four exams, that student's average is 83.4 since .2(80) + .2(75) + .2(82) + .4(90) = 16 + 15 + 16.4 + 36 = 83.4. Suppose that four mathematics professors teaching the same course assign different weights to the four exams during a semester according to the following table.

Weight assignment	Exam I	Exam II	Exam III	Final Exam
Professor W	.1	.1	.1	.7
Professor X	.2	.2	.2	.4
Professor Y	.1	.2	.3	.4
Professor Z	.15	.15	.15	.55

Five students who took the course received the scores displayed in the following table.

Grades	Jeremy	Shirley	Kathy	Andre	Ted
Exam I	80	85	90	65	70
Exam II	92	88	92	72	67
Exam III	87	89	88	81	72
Final Exam	92	90	85	83	99

Let **W** and **S** be matrices representing tables 1 and 2, respectively. Find the product **WS** and interpret the result.

41. Same as Exercise 40 with the following tables.

Weight assignment	Exam I	Exam II	Exam III	Final Exam
Professor X	.15	.15	.2	.5
Professor Y	.3	.3	.3	.1
Professor Z	.25	.2	.25	.3

Grades	Anay	Janet	Alan	Carl	Dave	Burnett
Exam I	87	92	90	78	98	25
Exam II	90	95	85	89	79	45
Exam III	88	96	95	94	97	32
Final Exam	79	98	92	93	100	51

*42. Recall from Section 2.1 that if $\mathbf{A} = \begin{bmatrix} a & b \\ c & d \end{bmatrix}$, then the determinant of **A** is denoted det**A** and is the number $ad - bc$. Show that if **A** and **B** are two 2×2 matrices, then $\det(\mathbf{AB}) = \det\mathbf{A} \cdot \det\mathbf{B}$.

*43. Let $\mathbf{A} = \begin{bmatrix} 2 & 1 \\ 6 & 3 \end{bmatrix}$ and $\mathbf{B} = \begin{bmatrix} 0 & 2 \\ 12 & 2 \end{bmatrix}$.

   a. Show that $\mathbf{AB} = \mathbf{BA}$.
   b. Use the result of Exercise 42 to show that **A** and **B** are not powers of the same 2×2 matrix **C**.

3.3 Solution of Linear Systems by Row Reduction

Much of the work that was done in Chapter 2 to solve a system of n linear equations in n variables can be simplified using matrix notation. The ideas are essentially the same, but the equations are written in a more compact form. We need only remember that the important data in a system of linear equations are the set of coefficients on the left sides of the equations and the set of constants on the right. In a matrix, the variables and the equal signs may be omitted if we are careful to place each coefficient in the position that indicates the variable it would be attached to.

Augmented Matrix of a Linear System

EXAMPLE 1 Write a matrix that represents the following system of linear equations:

$$\begin{cases} 2x + 3z = -5 \\ 3y - 4z + 6x = 2 \\ x + 3z - 5y = 7 \end{cases}$$

Solution We first write the system in such a way that the variables appear in the same order in all three equations. We make sure that all three variables appear in all three equations, using a zero coefficient where a variable is missing. We get

$$\begin{cases} 2x + 0y + 3z = -5 \\ 6x + 3y - 4z = 2 \\ x - 5y + 3z = 7 \end{cases}$$

We can now write the matrix

$$\begin{bmatrix} 2 & 0 & 3 & -5 \\ 6 & 3 & -4 & 2 \\ 1 & -5 & 3 & 7 \end{bmatrix}$$

where it is understood that the elements of the first column are the x-coefficients, the elements of the second column are the y-coefficients, those of the third column are the z-coefficients, and those of the fourth column are the constant terms.
Do Exercise 3. ■

A matrix such as this is usually called the *augmented matrix of the system*. It is customary to draw a vertical line where the equal signs would appear in the system of equations. Thus, if the matrix equation representing a linear system is $\mathbf{AX} = \mathbf{B}$, the augmented matrix of the system is $\mathbf{A} \mid \mathbf{B}$.

EXAMPLE 2 The following is the augmented matrix of a system of linear equations in w, x, y, and z. Write the system of equations.

$$\begin{bmatrix} 2 & -3 & 5 & 1 & -2 \\ 0 & 1 & 0 & 3 & 4 \\ 3 & 0 & 2 & -4 & 5 \\ 2 & 5 & 1 & 3 & 0 \end{bmatrix}$$

Solution The first, second, third, and fourth columns give the coefficients of w, x, y, and z, respectively, while the fifth column gives the constant terms on the right sides of the equations. The system is

$$\begin{cases} 2w - 3x + 5y + z = -2 \\ x + 3z = 4 \\ 3w + 2y - 4z = 5 \\ 2w + 5x + y + 3z = 0 \end{cases}$$

Do Exercise 7. ■

Recall from Chapter 2 that given a system of linear equations, we can get an equivalent linear system if we multiply both sides of an equation by a nonzero constant or add to any equation one of the other equations. We can also interchange any two equations. In the examples in Chapter 2, we used these methods, sometimes in combination, to arrive at an equivalent linear system in the form

$$\begin{cases} x & = a \\ y & = b \\ z = c \end{cases}$$

We found that the solution of the last system, and therefore of the original system, was the ordered triple (a, b, c). In the same spirit, we shall perform "row operations" on the augmented matrix until we obtain an augmented matrix in the form

$$\begin{bmatrix} 1 & 0 & 0 & a \\ 0 & 1 & 0 & b \\ 0 & 0 & 1 & c \end{bmatrix}$$

Elementary Row Operations

The following example will help clarify the technique.

EXAMPLE 3 Solve the following system as we did in Chapter 2, but this time simultaneously work the same steps on the augmented matrix of the system

$$\begin{cases} 2x + 9y = 3 \\ x + 5y = -2 \end{cases}$$

Solution

Linear System   *Augmented Matrix*

$$\begin{cases} 2x + 9y = 3 \\ x + 5y = -2 \end{cases}$$
Interchange the two equations (rows).
$$\begin{bmatrix} 2 & 9 & 3 \\ 1 & 5 & -2 \end{bmatrix}$$

$$\begin{cases} x + 5y = -2 \\ 2x + 9y = 3 \end{cases}$$
Multiply the first equation (row) by -2 and add to the second equation (row).
$$\begin{bmatrix} 1 & 5 & -2 \\ 2 & 9 & 3 \end{bmatrix}$$

$$\begin{cases} x + 5y = -2 \\ -y = 7 \end{cases}$$
Multiply the second equation (row) by -1.
$$\begin{bmatrix} 1 & 5 & -2 \\ 0 & -1 & 7 \end{bmatrix}$$

$$\begin{cases} x + 5y = -2 \\ y = -7 \end{cases}$$
Multiply the second equation (row) by -5 and add to the first equation (row).
$$\begin{bmatrix} 1 & 5 & -2 \\ 0 & 1 & -7 \end{bmatrix}$$

$$\begin{cases} x = 33 \\ y = -7 \end{cases}$$
$$\begin{bmatrix} 1 & 0 & 33 \\ 0 & 1 & -7 \end{bmatrix}$$

The solution of the system is the ordered pair $(33, -7)$.
Do Exercise 9.

**Procedure to Solve a Linear System of *n*
Equations in *n* Variables by Row Reduction**

Step 1. Write the augmented matrix for the given system. If we perform any of the following row operations on the augmented matrix, we obtain an augmented matrix for an equivalent system.

a. Interchange any two rows. The symbol $R_i \leftrightarrow R_j$ means "interchange the *i*th and *j*th rows."

b. Multiply any row by a nonzero constant. The symbol $kR_i \rightarrow R_i$ means "the *i*th row is multiplied by the nonzero constant *k* to obtain the new *i*th row."

c. Add to any row a constant multiple of another row. The symbol $kR_i + R_j \rightarrow R_j$ means "multiply the *i*th row by the constant *k* and add this product to the *j*th row to obtain the new *j*th row."

d. We may combine some of these steps. The meaning of the symbol $mR_i + nR_j \rightarrow R_j$ should now be clear.

Step 2. Perform a sequence of elementary row operations on the augmented matrix for the given system until we get an $n \times (n + 1)$ augmented matrix whose first *n* columns form the identity matrix $\mathbf{I}_n$. The solution of the given system can then be read from the $(n + 1)$st column of the last augmented matrix.

 We may perform more than one row operation in each step. However, we must clearly indicate to the right of the augmented matrix which operations we are performing. This is an essential part of the solution and serves two purposes:

1. In complicated cases it is virtually impossible for anyone to follow what was done unless the proper documentation is written along side the augmented matrices.
2. If you discover a wrong solution, you will have a much easier time in locating the error if each operation has been clearly described.

The format is such that the documentation written on the right side describes what was done to the augmented matrix directly to the left to get the next augmented matrix. The next examples will familiarize you with the form.

EXAMPLE 4 Use row reduction to solve the system $\begin{cases} 2x + 6y = 12 \\ 6x + 21y = -3 \end{cases}$.

Solution The augmented matrix for the system is

$$\begin{bmatrix} 2 & 6 & | & 12 \\ 6 & 21 & | & -3 \end{bmatrix}$$

We now perform a sequence of elementary row operations on this matrix.

$$\begin{bmatrix} 2 & 6 & | & 12 \\ 6 & 21 & | & -3 \end{bmatrix} \qquad \begin{array}{l} \left(\tfrac{1}{2}\right)R_1 \rightarrow R_1 \\ \left(\tfrac{1}{3}\right)R_2 \rightarrow R_2 \end{array}$$

$$\begin{bmatrix} 1 & 3 & | & 6 \\ 2 & 7 & | & -1 \end{bmatrix} \qquad (-2)R_1 + R_2 \to R_2$$

$$\begin{bmatrix} 1 & 3 & | & 6 \\ 0 & 1 & | & -13 \end{bmatrix} \qquad (-3)R_2 + R_1 \to R_1$$

$$\begin{bmatrix} 1 & 0 & | & 45 \\ 0 & 1 & | & -13 \end{bmatrix}$$

The first two columns of the last augmented matrix form the identity matrix I_2. Hence, we are done. The solution of the original system is the ordered pair $(45, -13)$ since the last augmented matrix represents the system

$$\begin{cases} x & = & 45 \\ & y = & -13 \end{cases}$$

Do Exercise 13.   ■

EXAMPLE 5   Solve the following system by row reduction:

$$\begin{cases} 5x - 5y + 4z = 17 \\ 3y + z = 6 \\ 2x - y + 2z = 9 \end{cases}$$

Solution   The augmented matrix for the system is

$$\begin{bmatrix} 5 & -5 & 4 & | & 17 \\ 0 & 3 & 1 & | & 6 \\ 2 & -1 & 2 & | & 9 \end{bmatrix}$$

We perform elementary row operations on this augmented matrix until the part of the resulting matrix to the left of the vertical bar is the identity matrix I_3. That is, until the last augmented matrix is in the form $I_3 \mid C$ where C is a 3-dimensional column vector. The basic idea is to get a 1 in the 1-1 position and 0's elsewhere in the first column, a 1 in the 2-2 position and 0's elsewhere in the second column, and a 1 in the 3-3 position and 0's elsewhere in the third column. There are many ways of doing this. We could start by multiplying row 1 by $\frac{1}{5}$ to get a 1 in the 1-1 position. However, this would result in fractions leading to awkward calculations. Another way to get a 1 in the 1-1 position is to multiply row 3 by -2 and add the result to row 1. This is the way we begin.

$$\begin{bmatrix} 5 & -5 & 4 & | & 17 \\ 0 & 3 & 1 & | & 6 \\ 2 & -1 & 2 & | & 9 \end{bmatrix} \qquad (-2)R_3 + R_1 \to R_1$$

$$\begin{bmatrix} 1 & -3 & 0 & | & -1 \\ 0 & 3 & 1 & | & 6 \\ 2 & -1 & 2 & | & 9 \end{bmatrix} \qquad (-2)R_1 + R_3 \to R_3$$

$$\begin{bmatrix} 1 & -3 & 0 & | & -1 \\ 0 & 3 & 1 & | & 6 \\ 0 & 5 & 2 & | & 11 \end{bmatrix} \qquad R_2 \leftrightarrow R_3$$

(Since we already have a 1 in column 3, we will work on that column.)

$$\begin{bmatrix} 1 & -3 & 0 & | & -1 \\ 0 & 5 & 2 & | & 11 \\ 0 & 3 & 1 & | & 6 \end{bmatrix} \qquad (-2)R_3 + R_2 \to R_2$$

$$\begin{bmatrix} 1 & -3 & 0 & | & -1 \\ 0 & -1 & 0 & | & -1 \\ 0 & 3 & 1 & | & 6 \end{bmatrix} \qquad \begin{array}{l} (-3)R_2 + R_1 \to R_1 \\ (3)R_2 + R_3 \to R_3 \\ (-1)R_2 \to R_2 \end{array}$$

$$\begin{bmatrix} 1 & 0 & 0 & | & 2 \\ 0 & 1 & 0 & | & 1 \\ 0 & 0 & 1 & | & 3 \end{bmatrix}$$

The first three columns of the last matrix form the identity matrix $\mathbf{I}_3$. Hence, we are finished. The solution of the original system of equations is the ordered triple $(2, 1, 3)$. Verify that if we replace x, y, and z in the original system by 2, 1, and 3, respectively, we indeed get three true statements.
Do Exercise 15. ■

A CLOSER LOOK In the previous example, we worked on the first column, then on the third column to take advantage of the fact that a 1 already appeared in the third column. Basically, it does not matter in what order we work as long as we remember to avoid fractions as much as possible, and once we have a column, say the jth one, in the desired form ($a_{jj} = 1$ and all other elements $= 0$), to keep it in that form. That is, we should not add a nonzero multiple of the jth row to another row, because this operation would replace one of the 0's in the jth column by a nonzero number. On occasion, if the integers m and n appear in a column, it is advantageous to look for integers p and q so that $pm + qn = 1$. For example, if 5 and 7 appear in a column, $3(5) + (-2)7 = 1$. We illustrate this idea in the next example.

EXAMPLE 6 Use the method of row reduction to solve the system

$$\begin{cases} 13x + 7y + 17z = 0 \\ 15x + 5y + 14z = 2 \\ 37x \quad\quad + 12z = 13 \end{cases}$$

Solution The augmented matrix of the system is

$$\begin{bmatrix} 13 & 7 & 17 & | & 0 \\ 15 & 5 & 14 & | & 2 \\ 37 & 0 & 12 & | & 13 \end{bmatrix}$$

Since none of the entries in the first three columns is a 1 and since we wish to avoid fractions, we begin with column 2. The reason for this choice is the fact that one of the entries in the second column is a 0 and also that $3(5) + (-2)7 = 1$. Hence, we multiply row 1 by -2, row 2 by 3, and add the results to get the new row 2. We proceed with this first step. The explanation for each of the subsequent steps is given to the right of each augmented matrix.

$$\begin{bmatrix} 13 & 7 & 17 & 0 \\ 15 & 5 & 14 & 2 \\ 37 & 0 & 12 & 13 \end{bmatrix} \qquad (-2)R_1 + 3R_2 \rightarrow R_2$$

$$\begin{bmatrix} 13 & 7 & 17 & 0 \\ 19 & 1 & 8 & 6 \\ 37 & 0 & 12 & 13 \end{bmatrix} \qquad (-7)R_2 + R_1 \rightarrow R_1$$

$$\begin{bmatrix} -120 & 0 & -39 & -42 \\ 19 & 1 & 8 & 6 \\ 37 & 0 & 12 & 13 \end{bmatrix} \qquad \left(\tfrac{1}{3}\right)R_1 \rightarrow R_1$$

$$\begin{bmatrix} -40 & 0 & -13 & -14 \\ 19 & 1 & 8 & 6 \\ 37 & 0 & 12 & 13 \end{bmatrix} \qquad R_1 + R_3 \rightarrow R_3$$

$$\begin{bmatrix} -40 & 0 & -13 & -14 \\ 19 & 1 & 8 & 6 \\ -3 & 0 & -1 & -1 \end{bmatrix} \qquad \begin{aligned} (-13)R_3 + R_1 &\rightarrow R_1 \\ 8R_3 + R_2 &\rightarrow R_2 \\ (-1)R_3 &\rightarrow R_3 \end{aligned}$$

$$\begin{bmatrix} -1 & 0 & 0 & -1 \\ -5 & 1 & 0 & -2 \\ 3 & 0 & 1 & 1 \end{bmatrix} \qquad \begin{aligned} (-5)R_1 + R_2 &\rightarrow R_2 \\ 3R_1 + R_3 &\rightarrow R_3 \\ (-1)R_1 &\rightarrow R_1 \end{aligned}$$

$$\begin{bmatrix} 1 & 0 & 0 & 1 \\ 0 & 1 & 0 & 3 \\ 0 & 0 & 1 & -2 \end{bmatrix}$$

Since the first three columns of the last augmented matrix form the identity matrix I_3, we are finished. The solution of the given system is obtained from the last column of the last augmented matrix. We find that the solution is the ordered triple $(1, 3, -2)$. Verify that if we replace x, y, and z in the given system by 1, 3, and -2, respectively, then we obtain three true statements.
Do Exercise 19. ■

More General Linear Systems

Thus far we have only considered linear systems where the number of equations and variables are the same. We will now discuss briefly the more general case of a linear system of m equations in n variables where m and n need not be equal. We shall see that a linear system has either a unique solution, no solution, or an infinite number of solutions. In general, we proceed as we did in the previous examples until we get an augmented matrix which is in row-reduced echelon form.

> **DEFINITION 3.10:** An augmented matrix is said to be a *row-reduced echelon matrix* if all of the following conditions are met.
>
> 1. All rows consisting entirely of 0's are below all rows with nonzero entries;
> 2. In each row with nonzero entries, the first nonzero entry is a 1;
> 3. If a 1 is the first nonzero entry of a row and appears in column j, then all other entries in that column are 0's; and
> 4. The first nonzero entry in any row, except for the first row, is to the right of the first nonzero entry in each preceding row.
>
> The variables corresponding to columns that contain a leading 1 are called *leading variables*.

The nature of the solution set of the linear system can be deduced from an augmented matrix which is in row-reduced echelon form. For example, if after performing a sequence of elementary row operations on a linear system we obtain a row-reduced echelon matrix that contains a row in which all entries except the last one are 0's, then the system has no solution; it is inconsistent. If this does *not* happen, then the system will have *at least one solution*. The reasons for this will be apparent in the following examples.

EXAMPLE 7 Solve the linear system

$$\begin{cases} x + 3y - z = 2 \\ 3x + 5y + 2z = 0 \\ 4x + 12y - 4z = 5 \end{cases}$$

Solution The augmented matrix of the system is

$$\left[\begin{array}{ccc|c} 1 & 3 & -1 & 2 \\ 3 & 5 & 2 & 0 \\ 4 & 12 & -4 & 5 \end{array}\right]$$

Since we have a 1 in the 1-1 position, we begin working with column 1.

$$\left[\begin{array}{ccc|c} 1 & 3 & -1 & 2 \\ 3 & 5 & 2 & 0 \\ 4 & 12 & -4 & 5 \end{array}\right] \qquad \begin{array}{l} (-3)R_1 + R_2 \rightarrow R_2 \\ (-4)R_1 + R_3 \rightarrow R_3 \end{array}$$

$$\left[\begin{array}{ccc|c} 1 & 3 & -1 & 2 \\ 0 & -4 & 5 & -6 \\ 0 & 0 & 0 & -3 \end{array}\right]$$

We need not go any further. All entries in the third row are 0's except for the last entry. We can conclude that the given system has no solution—it is inconsistent—

because the last row represents the equation

$$0x + 0y + 0z = -3$$

which clearly has no solution since any replacement of x, y, and z by real numbers will yield 0 on the left side and $0 \neq -3$.
Do Exercise 17. ■

EXAMPLE 8 Solve the system

$$\begin{cases} x - y = 3 \\ x + y = 7 \\ x - 2y = 1 \end{cases}$$

Solution The augmented matrix of the system is

$$\begin{bmatrix} 1 & -1 & | & 3 \\ 1 & 1 & | & 7 \\ 1 & -2 & | & 1 \end{bmatrix}$$

We perform the following elementary row operations.

$$\begin{bmatrix} 1 & -1 & | & 3 \\ 1 & 1 & | & 7 \\ 1 & -2 & | & 1 \end{bmatrix} \quad \begin{array}{l} (-1)R_1 + R_2 \to R_2 \\ (-1)R_1 + R_3 \to R_3 \end{array}$$

$$\begin{bmatrix} 1 & -1 & | & 3 \\ 0 & 2 & | & 4 \\ 0 & -1 & | & -2 \end{bmatrix} \quad \left(\tfrac{1}{2}\right)R_2 \to R_2$$

$$\begin{bmatrix} 1 & -1 & | & 3 \\ 0 & 1 & | & 2 \\ 0 & -1 & | & -2 \end{bmatrix} \quad \begin{array}{l} R_2 + R_1 \to R_1 \\ R_2 + R_3 \to R_3 \end{array}$$

$$\begin{bmatrix} 1 & 0 & | & 5 \\ 0 & 1 & | & 2 \\ 0 & 0 & | & 0 \end{bmatrix}$$

The last matrix is in row-reduced echelon form. All entries in the third row are 0's. This row represents the equation $0x + 0y = 0$, which is satisfied for all values of x and y. This last row can thus be ignored and from the first two rows we see that $x = 5$ and $y = 2$. Hence, the system has a unique solution, the ordered pair $(5, 2)$.
Do Exercise 33. ■

It may be useful to think of this example geometrically. Each equation is linear in x and y; hence, the graph of each equation is a straight line. Since the system has a unique solution, the three straight lines intersect at a single point.

EXAMPLE 9   Solve the following system of three linear equations in four variables w, x, y, and z.

$$\begin{cases} 3w + 6x + y + 18z = 9 \\ w + 2x + 5z = 2 \\ 2w + 4x + 3y + 19z = 13 \end{cases}$$

Solution   The augmented matrix of the system is

$$\left[\begin{array}{cccc|c} 3 & 6 & 1 & 18 & 9 \\ 1 & 2 & 0 & 5 & 2 \\ 2 & 4 & 3 & 19 & 13 \end{array}\right]$$

We then perform a sequence of elementary row operations, giving the explanations on the right sides of the matrices.

$$\left[\begin{array}{cccc|c} 3 & 6 & 1 & 18 & 9 \\ 1 & 2 & 0 & 5 & 2 \\ 2 & 4 & 3 & 19 & 13 \end{array}\right] \quad R_1 \leftrightarrow R_2$$

$$\left[\begin{array}{cccc|c} 1 & 2 & 0 & 5 & 2 \\ 3 & 6 & 1 & 18 & 9 \\ 2 & 4 & 3 & 19 & 13 \end{array}\right] \quad \begin{array}{l}(-3)R_1 + R_2 \to R_2 \\ (-2)R_1 + R_3 \to R_3\end{array}$$

$$\left[\begin{array}{cccc|c} 1 & 2 & 0 & 5 & 2 \\ 0 & 0 & 1 & 3 & 3 \\ 0 & 0 & 3 & 9 & 9 \end{array}\right] \quad (-3)R_2 + R_3 \to R_3$$

$$\left[\begin{array}{cccc|c} 1 & 2 & 0 & 5 & 2 \\ 0 & 0 & 1 & 3 & 3 \\ 0 & 0 & 0 & 0 & 0 \end{array}\right]$$

The last augmented matrix is in row-reduced echelon form. All entries in the third row are 0's. Thus the system is consistent.

Observe that w and y are leading variables. The nonleading variables, x and z in this case, end up being parameters* and the leading variables are expressed in terms of these parameters. The first two rows correspond to the equations

$$w + 2x + 5z = 2$$
$$y + 3z = 3$$

These can be written

$$w = 2 - 2x - 5z$$
$$y = 3 - 3z$$

---

*A parameter is an arbitrary constant in a mathematical expression that distinguishes specific cases. For example, in $y = mx + b$, m and b are parameters, and giving these certain values specifies which line is represented by the equation.

We can give the parameters x and z any values we wish, say a and b, respectively, and obtain

$$w = 2 - 2a - 5b$$
$$y = 3 - 3b$$

Thus, the solution set of the given system is the set of all ordered quadruples $(2 - 2a - 5b, a, 3 - 3b, b)$, where a and b are arbitrary real numbers. We therefore have infinitely many solutions. For example, for $a = b = 0$ we get the solution $(2, 0, 3, 0)$ and for $a = 1$, $b = 2$ we get another solution $(-10, 1, -3, 2)$. You should check that these are indeed solutions of the original system.
Do Exercise 35.   ■

In general, if the number of variables in a linear system is greater than the number of equations, the system will have infinitely many solutions provided that it is consistent. Later we shall see that the graph of a linear equation in three variables is a plane. You might find it helpful to consider a special case which illustrates the remark we just made. For example, if we have a linear system of two equations in three variables, then the graphs of these equations are two planes. Only one of the following three things can happen:

1. The two planes are parallel; they have no points in common, hence there are no solutions—the system is inconsistent.
2. The two planes coincide; they have infinitely many points in common, hence there are infinitely many solutions.
3. The intersection of the two planes is a straight line. Again, there are infinitely many common points and thus infinitely many solutions.

EXAMPLE 10   Solve the system

$$\begin{cases} 2x - 4y + 6z = 3 \\ 3x - 6y + 9z = 5 \end{cases}$$

Solution   The augmented matrix of the system is

$$\begin{bmatrix} 2 & -4 & 6 & \bigm| & 3 \\ 3 & -6 & 9 & \bigm| & 5 \end{bmatrix}$$

We proceed as we did in the previous examples.

$$\begin{bmatrix} 2 & -4 & 6 & \bigm| & 3 \\ 3 & -6 & 9 & \bigm| & 5 \end{bmatrix} \qquad (-3)R_1 + 2R_2 \rightarrow R_2$$

$$\begin{bmatrix} 2 & -4 & 6 & \bigm| & 3 \\ 0 & 0 & 0 & \bigm| & 1 \end{bmatrix}$$

All entries, except the last one, of the second row of this augmented matrix are 0's. Thus, the system has no solutions; it is inconsistent.
Do Exercise 31.   ■

Solving Several Linear Systems Simultaneously

In certain applications, it is necessary to solve several linear systems which differ only in the constants on the right sides of the equations. In such cases, the method of row reduction can be used to solve all systems simultaneously as in the following example.

EXAMPLE 11 Consider the following four linear equations:

$$x + 3y = 15$$
$$x + 3y = -6$$
$$2x - y = -19$$
$$2x - y = 23$$

with graphs L_1, L_2, L_3, and L_4, respectively.

a. Show that L_1 is parallel to L_2 and that L_3 is parallel to L_4.
b. Find the vertices of the parallelogram formed by these four lines.

Solution **a.** We write the first two equations in the form

$$y = \left(-\tfrac{1}{3}\right)x + 5 \quad \text{and} \quad y = \left(-\tfrac{1}{3}\right)x - 2$$

The slopes of L_1 and of L_2 are equal, therefore L_1 is parallel to L_2. Similarly, we can show that the slopes of L_3 and L_4 are equal.

b. To find the four vertices of the parallelogram formed by these four lines, we must find the points of intersection of L_1 and L_3, L_2 and L_4, L_1 and L_4, and, L_2 and L_3. We must, therefore, solve the systems

$$\begin{cases} x + 3y = 15 \\ 2x - y = -19 \end{cases} \qquad \begin{cases} x + 3y = -6 \\ 2x - y = 23 \end{cases}$$

$$\begin{cases} x + 3y = 15 \\ 2x - y = 23 \end{cases} \qquad \begin{cases} x + 3y = -6 \\ 2x - y = -19 \end{cases}$$

Observe that the left sides of all four systems are identical. We can combine the four augmented matrices of these systems in a single matrix and apply the row operations to all four simultaneously. We have

$$\begin{bmatrix} 1 & 3 & | & 15 & | & -6 & | & 15 & | & -6 \\ 2 & -1 & | & -19 & | & 23 & | & 23 & | & -19 \end{bmatrix} \qquad (-2)R_1 + R_2 \rightarrow R_2$$

$$\begin{bmatrix} 1 & 3 & | & 15 & | & -6 & | & 15 & | & -6 \\ 0 & -7 & | & -49 & | & 35 & | & -7 & | & -7 \end{bmatrix} \qquad \left(-\tfrac{1}{7}\right)R_2 \rightarrow R_2$$

$$\begin{bmatrix} 1 & 3 & | & 15 & | & -6 & | & 15 & | & -6 \\ 0 & 1 & | & 7 & | & -5 & | & 1 & | & 1 \end{bmatrix} \qquad (-3)R_2 + R_1 \rightarrow R_1$$

$$\begin{bmatrix} 1 & 0 & | & -6 & | & 9 & | & 12 & | & -9 \\ 0 & 1 & | & 7 & | & -5 & | & 1 & | & 1 \end{bmatrix} \qquad (-3)R_2 + R_1 \rightarrow R_1$$

The first two columns of the last matrix form the identity $\mathbf{I}_2$. Hence we are finished. The first three columns form the augmented matrix

$$\begin{bmatrix} 1 & 0 & | & -6 \\ 0 & 1 & | & 7 \end{bmatrix}$$

which indicates that the solution of the first system is $x = -6$ and $y = 7$. Therefore the point of intersection of L_1 and L_3 is the point $(-6, 7)$. Similarly, we find that the other three vertices of the parallelogram are $(9, -5)$, $(12, 1)$, and $(-9, 1)$. (See Figure 3.1.)

Do Exercise 47.

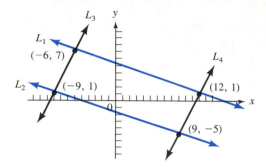

FIGURE 3.1

Applications

EXAMPLE 12 (Optional)

A firm produces three commodities at four different plants. In the following matrix, a_{ij} is the number of units of the ith commodity produced at the jth plant per day.

$$\begin{bmatrix} 3 & 4 & 5 & 2 \\ 2 & 3 & 2 & 1 \\ 0 & 2 & 2 & 3 \end{bmatrix}$$

The firm must fill orders of 202, 121, and 98 units of commodities 1, 2, and 3, respectively. How many days will it take each plant to produce enough commodities to fill the orders?

Solution

Let w, x, y, and z be the number of days that plants 1, 2, 3, and 4, respectively, must produce a commodity. Then, the numbers of units of commodity 1 produced at plant 1 is $3w$, while plants 2, 3, and 4 produce $4x$, $5y$, and $2z$ units of commodity 1. Thus the total number of units of commodity 1 produced is

$$3w + 4x + 5y + 2z$$

Since 202 units of commodity 1 have been ordered, the first equation is

$$3w + 4x + 5y + 2z = 202$$

Similarly, the number of units of commodity 2 produced at plants 1, 2, 3, and 4 are $2w$, $3x$, $2y$, and z, respectively. Since 121 units of commodity 2 have been ordered, we get the second equation

$$2w + 3x + 2y + z = 121$$

The third equation is

$$2x + 2y + 3z = 98$$

and the linear system is

$$\begin{cases} 3w + 4x + 5y + 2z = 202 \\ 2w + 3x + 2y + z = 121 \\ \phantom{2w + {}} 2x + 2y + 3z = 98 \end{cases}$$

The augmented matrix of this system is

$$\left[\begin{array}{cccc|c} 3 & 4 & 5 & 2 & 202 \\ 2 & 3 & 2 & 1 & 121 \\ 0 & 2 & 2 & 3 & 98 \end{array}\right]$$

We now can perform the following elementary row operations.

$$\left[\begin{array}{cccc|c} 3 & 4 & 5 & 2 & 202 \\ 2 & 3 & 2 & 1 & 121 \\ 0 & 2 & 2 & 3 & 98 \end{array}\right] \qquad (-1)R_2 + R_1 \rightarrow R_1$$

$$\left[\begin{array}{cccc|c} 1 & 1 & 3 & 1 & 81 \\ 2 & 3 & 2 & 1 & 121 \\ 0 & 2 & 2 & 3 & 98 \end{array}\right] \qquad (-2)R_1 + R_2 \rightarrow R_2$$

$$\left[\begin{array}{cccc|c} 1 & 1 & 3 & 1 & 81 \\ 0 & 1 & -4 & -1 & -41 \\ 0 & 2 & 2 & 3 & 98 \end{array}\right] \qquad \begin{array}{l}(-1)R_2 + R_1 \rightarrow R_1 \\ (-2)R_2 + R_3 \rightarrow R_3\end{array}$$

$$\left[\begin{array}{cccc|c} 1 & 0 & 7 & 2 & 122 \\ 0 & 1 & -4 & -1 & -41 \\ 0 & 0 & 10 & 5 & 180 \end{array}\right] \qquad \left(\tfrac{1}{10}\right)R_3 \rightarrow R_3$$

$$\left[\begin{array}{cccc|c} 1 & 0 & 7 & 2 & 122 \\ 0 & 1 & -4 & -1 & -41 \\ 0 & 0 & 1 & .5 & 18 \end{array}\right] \qquad \begin{array}{l}(-7)R_3 + R_1 \rightarrow R_1 \\ 4R_3 + R_2 \rightarrow R_2\end{array}$$

$$\left[\begin{array}{cccc|c} 1 & 0 & 0 & -1.5 & -4 \\ 0 & 1 & 0 & 1 & 31 \\ 0 & 0 & 1 & .5 & 18 \end{array}\right]$$

The last matrix is in row-reduced echelon form. The leading variables are w, x, and y, and z is a nonleading variable. If we let $z = a$, the last augmented matrix corresponds to the linear system

$$\begin{cases} w -1.5a = -4 \\ x + a = 31 \\ y + .5a = 18 \end{cases}$$

Thus, we have

$$\begin{cases} w = 1.5a - 4 \\ x = 31 - a \\ y = 18 - .5a \\ z = a \end{cases}$$

There are infinitely many solutions. All ordered quadruples $(1.5a - 4, 31 - a, 18 - .5a, a)$, where a is an arbitrary real number, are solutions of the linear system. However, we must consider only nonnegative values of $w, x, y,$ and z since those variables denote the number of days that plants 1, 2, 3, and 4 have to produce the three commodities to fill the orders. Therefore, $1.5a - 4 \geq 0$, $31 - a \geq 0$, $18 - .5a \geq 0$ and $a \geq 0$. The first, second, and third of the four inequalities are equivalent to $a \geq \frac{8}{3}$, $a \leq 31$, and $a \leq 36$, respectively. Since $0 < \frac{8}{3} < 31 < 36$, the four inequalities will be satisfied simultaneously if $\frac{8}{3} \leq a \leq 31$.

Although it is not necessary to do so, we may want our solutions to be integers. Since $1.5a$ and $.5a$ appear in two of the solutions, we choose an even integer between $\frac{8}{3}$ and 31 for the value of a. For example, if we let $a = 4$, then a feasible solution is $(2, 27, 16, 4)$, which means that the first, second, third, and fourth plants should produce the three commodities for 2, 27, 16, and 4 days, respectively. On the other hand, if we let $a = 20$, then we obtain another feasible solution, namely $(26, 11, 8, 20)$, and the production of plants 1, 2, 3, and 4 should last 26, 11, 8, and 20 days, respectively. You may find other feasible solutions to this problem.
Do Exercise 57.   ■

EXAMPLE 13   Suppose that when three commodities X, Y, and Z are priced at x, y, and z dollars per unit, respectively, the quantities demanded are D_x, D_y, and D_z, respectively, (measured in thousands) and the quantities supplied are S_x, S_y, and S_z, respectively, (also measured in thousands). The demand and supply functions for these commodities are interrelated as expressed by

$$D_x = 1050 - 25x + 94y - 30z$$
$$S_x = -126 + 80x - 10y + 75z$$
$$D_y = 813 + 30x - 54y + 15z$$
$$S_y = -15 - 10x + 94y - 10z$$
$$D_z = 920 - 15x + 36y - 10z$$
$$S_z = -26 + 50x - 12y + 60z$$

Find the equilibrium prices and quantities for all three commodities.

Solution   Since at equilibrium supply equals demand, we obtain a system of three equations by setting $D_x = S_x$, $D_y = S_y$, and $D_z = S_z$. We obtain

$$\begin{cases} 1050 - 25x + 94y - 30z = -126 + 80x - 10y + 75z \\ 813 + 30x - 54y + 15z = -15 - 10x + 94y - 10z \\ 920 - 15x + 36y - 10z = -26 + 50x - 12y + 60z \end{cases}$$

The foregoing system may be written

$$\begin{cases} -105x + 104y - 105z = -1176 \\ 40x - 148y + 25z = -828 \\ -65x + 48y - 70z = -946 \end{cases}$$

We first write the augmented matrix of the system, then solve the system using row operations.

$$\begin{bmatrix} -105 & 104 & -105 & -1176 \\ 40 & -148 & 25 & -828 \\ -65 & 48 & -70 & -946 \end{bmatrix} \qquad \left(\tfrac{1}{40}\right)R_2 \to R_2 \\ R_2 \leftrightarrow R_1$$

$$\begin{bmatrix} 1 & -3.7 & 0.625 & -20.7 \\ -105 & 104 & -105 & -1176 \\ -65 & 48 & -70 & -946 \end{bmatrix} \qquad 105R_1 + R_2 \to R_2 \\ 65R_1 + R_3 \to R_3$$

$$\begin{bmatrix} 1 & -3.7 & 0.625 & -20.7 \\ 0 & -284.5 & -39.375 & -3349.5 \\ 0 & -192.5 & -29.375 & -2291.5 \end{bmatrix} \qquad \left(-\tfrac{1}{284.5}\right)R_2 \to R_2$$

$$\begin{bmatrix} 1 & -3.7 & 0.625 & -20.7 \\ 0 & 1 & \frac{315}{2276} & \frac{6699}{569} \\ 0 & -192.5 & -29.375 & -2291.5 \end{bmatrix} \qquad (3.7)R_2 + R_1 \to R_1 \\ (192.5)R_2 + R_3 \to R_3$$

$$\begin{bmatrix} 1 & 0 & \frac{2588}{2276} & \frac{13008}{569} \\ 0 & 1 & \frac{315}{2276} & \frac{6699}{569} \\ 0 & 0 & \frac{-6220}{2276} & \frac{-14306}{569} \end{bmatrix} \qquad \left(\tfrac{-2276}{6220}\right)R_3 \to R_3$$

$$\begin{bmatrix} 1 & 0 & \frac{2588}{2276} & \frac{13008}{569} \\ 0 & 1 & \frac{315}{2276} & \frac{6699}{569} \\ 0 & 0 & 1 & \frac{92}{10} \end{bmatrix} \qquad \left(\tfrac{-2588}{2276}\right)R_3 + R_1 \to R_1 \\ \left(\tfrac{-315}{2276}\right)R_3 + R_2 \to R_2$$

$$\begin{bmatrix} 1 & 0 & 0 & 12.4 \\ 0 & 1 & 0 & 10.5 \\ 0 & 0 & 1 & 9.2 \end{bmatrix}$$

This matrix is in reduced-row echelon form. The equilibrium prices for products X, Y, and Z are \$12.40, \$10.50, and \$9.20 per unit, respectively. The equilibrium quantity for product X is obtained by replacing x, y, and z by 12.4, 10.5, and 9.2, respectively, in the equation giving D_x. (We could also use the equation giving S_x since $D_x = S_x$.) We get $1050 - 25(12.4) + 94(10.5) - 30(9.2) = 1451$. The equilibrium quantity for product X is 1,451,000. Similarly, we find that the equilibrium quantities for products Y and Z are 756,000 and 1,020,000, respectively.
Do Exercise 63. ■

Homogeneous Linear Systems

We conclude this section with a brief discussion of a special type of system of equations.

DEFINITION 3.11: A linear system, in which all constant terms are 0, is called a *homogeneous* system. The solution in which all variables are given the value 0 is called the *trivial solution*. Any other type of solution is called a *nontrivial solution*.

For example,

$$\begin{cases} 2x + 3y - 4z = 0 \\ 5x - 3y + 6z = 0 \\ 7x + 5y - 3z = 0 \end{cases}$$

is a homogeneous system and $(0, 0, 0)$ is the trivial solution for that system. In general, if a homogeneous linear system has more unknowns than equations, then nontrivial solutions can also be found as illustrated in the following example.

EXAMPLE 14   Solve the following homogeneous linear system:

$$\begin{cases} 3x + 2y - 5z = 0 \\ 2x + 3y + 5z = 0 \end{cases}$$

Solution   The augmented matrix of the system is

$$\left[\begin{array}{ccc|c} 3 & 2 & -5 & 0 \\ 2 & 3 & 5 & 0 \end{array} \right]$$

We then perform a sequence of elementary row operations, giving the explanations on the right sides of the matrices.

$$\left[\begin{array}{ccc|c} 3 & 2 & -5 & 0 \\ 2 & 3 & 5 & 0 \end{array} \right] \qquad (-1)R_2 + R_1 \rightarrow R_1$$

$$\left[\begin{array}{ccc|c} 1 & -1 & -10 & 0 \\ 2 & 3 & 5 & 0 \end{array} \right] \qquad (-2)R_1 + R_2 \rightarrow R_2$$

$$\left[\begin{array}{ccc|c} 1 & -1 & -10 & 0 \\ 0 & 5 & 25 & 0 \end{array} \right] \qquad \left(\tfrac{1}{5}\right)R_2 \rightarrow R_2$$

$$\left[\begin{array}{ccc|c} 1 & -1 & -10 & 0 \\ 0 & 1 & 5 & 0 \end{array} \right] \qquad R_2 + R_1 \rightarrow R_1$$

$$\left[\begin{array}{ccc|c} 1 & 0 & -5 & 0 \\ 0 & 1 & 5 & 0 \end{array} \right]$$

The last augmented matrix is in reduced-row echelon form. As expected, because we had more unknowns than equations, one column (the third) has no entry that can be the first nonzero entry of its row. So, in the corresponding system of equations, z does not appear as the first variable in either equation. We give z an arbitrary value, say t, and solve the system

$$\begin{cases} x - 5t = 0 \\ y + 5t = 0 \end{cases}$$

Thus, $x = 5t$ and $y = -5t$. The solution set of the system is $\{(5t, -5t, t) \mid t$ is any real number$\}$. The trivial solution is obtained by letting $t = 0$, and $(5, -5, 1)$ is obtained by letting $t = 1$. Other solutions may be found.
Do Exercise 41.   ■

A procedure for solving a linear system by row reduction follows.

Solving a Linear System by Row Reduction

Step 1. Write the augmented matrix for the given system.

Step 2. Perform a sequence of elementary row operations on the augmented matrix until you obtain an augmented matrix that is in row-reduced echelon form.

You now have exactly one of three possibilities.

1. There is no solution in the case where there is a row in which all entries except the last one are 0's.
2. There is a unique solution in the case where every variable is a leading variable.
3. There are infinitely many solutions in the case where there are no rows in which all entries except the last one are 0's, and at least one variable is a nonleading variable so that the leading variables are expressed in terms of at least one parameter.

Exercise Set 3.3

In Exercises 1–4, write the augmented matrix of the given linear system.

1. $\begin{cases} 2x + 3y = -5 \\ 3x - 5y = 17 \end{cases}$

2. $\begin{cases} 3y + 12 = 6x \\ 5x - 15 = 2y \end{cases}$

3. $\begin{cases} 2y + 5z - 7x = 4 \\ 3x + 7z = 0 \\ 2z - 5y = 7 \end{cases}$

4. $\begin{cases} x + 2y - 5w = 3 \\ z + 2x - 3y = 7 \\ 5y + 2w = 3 \end{cases}$

In Exercises 5–8, write a linear system corresponding to the given augmented matrix.

5. $\begin{bmatrix} 2 & 0 & | & 3 \\ 5 & 3 & | & -2 \end{bmatrix}$

6. $\begin{bmatrix} 0 & 5 & 3 & | & 2 \\ 2 & 4 & 9 & | & 0 \\ -3 & 0 & 2 & | & 8 \end{bmatrix}$

7. $\begin{bmatrix} 2 & 0 & 3 & 4 & | & 7 \\ -3 & 2 & 5 & 0 & | & 2 \\ 2 & -1 & 0 & 4 & | & -3 \end{bmatrix}$

8. $\begin{bmatrix} \frac{1}{2} & 3 & \frac{5}{4} & 2 & 4 & | & -1 \\ 7 & 3 & 0 & 0 & 2 & | & \frac{1}{3} \\ 0 & 2 & \frac{1}{4} & 3 & \frac{3}{2} & | & 4 \\ -3 & 0 & 1 & \frac{1}{5} & 3 & | & 0 \end{bmatrix}$

In Exercises 9–40, solve the given linear system. If the system is inconsistent, give the reason.

9. $\begin{cases} 2x + y = 5 \\ 6x + 3y = 1 \end{cases}$

10. $\begin{cases} 6x + 17y = -1 \\ x + 3y = 0 \end{cases}$

11. $\begin{cases} 8x + 12y = 16 \\ 6x + 9y = 12 \end{cases}$

12. $\begin{cases} 3x + 7y = -1 \\ 2x + 5y = -1 \end{cases}$

13. $\begin{cases} 6x + 15y = 2 \\ 4x + 10y = 1 \end{cases}$

14. $\begin{cases} 15x + 21y = 6 \\ 2x + 6y = 4 \end{cases}$

15. $\begin{cases} 5x - 5y + 4z = 8 \\ 3y + z = 5 \\ 2x - y + 2z = 5 \end{cases}$

16. $\begin{cases} x - 3y = -7 \\ 2x + 2y + 3z = 5 \\ 2x - y + 2z = -2 \end{cases}$

17. $\begin{cases} x + 2y + 3z = 2 \\ 3x + 3y = 4 \\ 2x + y - 3z = 3 \end{cases}$

18. $\begin{cases} 7x + 6y - 17z = 49 \\ 2x + 2y - 5z = 15 \\ -6x - 5y + 15z = -42 \end{cases}$

19. $\begin{cases} 3x + 2y - 7z = 3 \\ 2x + 3y + 5z = 15 \\ 2x + 2y - 5z = 3 \end{cases}$

20. $\begin{cases} x - 2z = 2 \\ 3x + 2y - 7z = 11 \\ 4x + 5y - 10z = 20 \end{cases}$

21. $\begin{cases} 2x + y - 3z = 2 \\ 3x + 2y + 4z = 1 \\ 8x + 5y + 5z = 4 \end{cases}$ **22.** $\begin{cases} -x - 2y - 2z = -10 \\ 4x + y + 5z = 2 \\ 2x - y + 2z = -7 \end{cases}$

23. $\begin{cases} 3x - y + 3z = -1 \\ 2x + 2y + 3z = 8 \\ -3y - z = -10 \end{cases}$ **24.** $\begin{cases} 2x + 3y + z = 4 \\ 9x + 11y + 4z = 11 \\ 6x + 4y + 2z = -2 \end{cases}$

25. $\begin{cases} 2x - y + 2z = 14 \\ 4x + y + 5z = 37 \\ 11x - 8y + 10z = 69 \end{cases}$ **26.** $\begin{cases} 2x + 5y + z = 1 \\ 3x + 4y + 5z = 0 \\ 8x + 13y + 11z = 2 \end{cases}$

27. $\begin{cases} 7w - 8x - 5y + 6z = 27 \\ w - 4x + z = 7 \\ 3w - 11x + 3y + 4z = 22 \\ 4w - 6x - y + 4z = 18 \end{cases}$

28. $\begin{cases} -w + x + y - 2z = -4 \\ -8w + 15x + 14y - 33z = -53 \\ w + x + y - 3z = -2 \\ -5w - x + 5z = -1 \end{cases}$

29. $\begin{cases} x + 3y + z = 1 \\ 5w + x - 2y + 5z = 3 \\ w - 2x + 6y + 3z = 3 \\ 2w + 3x + 2y + 3z = 2 \end{cases}$

30. $\begin{cases} 2w + 2x - y + 2z = 22 \\ 4w + 5x + y + 5z = 38 \\ 7w + 2x - 6y + 6z = 61 \\ 3w - 2x - y + 3z = 6 \end{cases}$

31. $\begin{cases} 2x + 3y - z = 2 \\ 4x + 6y - 2z = 3 \end{cases}$ **32.** $\begin{cases} 3x + y + 5z = 1 \\ 2x + 3y - 2z = 6 \end{cases}$

33. $\begin{cases} x + 3y = -7 \\ 3x + y = 3 \\ 2x - 5y = 19 \end{cases}$ **34.** $\begin{cases} x + 5y = 6 \\ 2x - y = 4 \\ 3x + 2y = 7 \end{cases}$

35. $\begin{cases} 4w + 5x + 2y - 5z = -2 \\ 2w + y + 6z = 22 \\ 2w + 2x + y - z = 3 \end{cases}$

36. $\begin{cases} -w - 2x - 2y + 3z = -20 \\ 4w + 5x + y - 2z = 16 \\ 2w + 2x - y + z = -5 \end{cases}$

37. $\begin{cases} w + x - 3y = 0 \\ 2w + 2y + 3z = 13 \\ 8w + 3x - 4y + 8z = 34 \end{cases}$

38. $\begin{cases} 3w - x + 3y + 2z = 2 \\ 2w + 2x + 3y - z = -2 \\ -3x - y + 3z = 5 \end{cases}$

39. $\begin{cases} 3w + 7x + 2y - 7z = 21 \\ 2w + 4x + 2y - 5z = 17 \\ 2w + 3x + 3y + 5z = -10 \end{cases}$

40. $\begin{cases} x - 3y = -7 \\ 2x + 2y + 3z = 13 \\ 2x - y + 2z = 3 \\ x + y - z = 4 \end{cases}$

In Exercises 41–45, find the solution set of the given homogeneous system.

41. $\begin{cases} 4w + 5x + 2y - 5z = 0 \\ 2w + y + 6z = 0 \\ 2w + 2x + y - z = 0 \end{cases}$

42. $\begin{cases} -w - 2x - 2y + 3z = 0 \\ 4w + 5x + y - 2z = 0 \\ 2w + 2x - y + z = 0 \end{cases}$

43. $\begin{cases} w + x - 3y = 0 \\ 2w + 2y + 3z = 0 \\ 8w + 3x - 4y + 8z = 0 \end{cases}$

44. $\begin{cases} 3w - x + 3y + 2z = 0 \\ 2w + 2x + 3y - z = 0 \\ -3x - y + 3z = 0 \end{cases}$

45. $\begin{cases} 3w + 7x + 2y - 7z = 0 \\ 2w + 4x + 2y - 5z = 0 \\ 2w + 3x + 3y + 5z = 0 \end{cases}$

In Exercises 46–51, solve the linear systems of parts a, b, and c simultaneously.

46. $\begin{cases} 2x - 3y = d_1 \\ -3x + 5y = d_2, \text{ where} \end{cases}$
   a. $d_1 = 3, d_2 = 0$; **b.** $d_1 = -2, d_2 = 5$; **c.** $d_1 = 4, d_2 = 7$.

47. $\begin{cases} 3x - 7y = d_1 \\ -2x + 5y = d_2, \text{ where} \end{cases}$
   a. $d_1 = 2, d_2 = 5$; **b.** $d_1 = -4, d_2 = 7$; **c.** $d_1 = 3, d_2 = 6$.

48. $\begin{cases} 3x + 5y = d_1 \\ 4x + 7y = d_2, \text{ where} \end{cases}$
   a. $d_1 = 0, d_2 = 4$; **b.** $d_1 = 11, d_2 = 6$; **c.** $d_1 = 3, d_2 = 2$.

49.
$$\begin{cases} x \quad - 2z = d_1 \\ 2x + 2y - 5z = d_2 \\ 4x + 5y - 10z = d_3, \text{ where} \end{cases}$$
a. $d_1 = 2$, $d_2 = 0$, $d_3 = 7$; **b.** $d_1 = 0$, $d_2 = 3$, $d_3 = 5$;
c. $d_1 = 7$, $d_2 = -2$, $d_3 = 4$.

50.
$$\begin{cases} 3x + 2y - 7z = d_1 \\ 2x + 2y - 5z = d_2 \\ 2x + 3y + 5z = d_3, \text{ where} \end{cases}$$
a. $d_1 = 1$, $d_2 = 3$, $d_3 = 0$; **b.** $d_1 = 2$, $d_2 = 7$, $d_3 = 4$;
c. $d_1 = 4$, $d_2 = -5$, $d_3 = 6$.

51.
$$\begin{cases} 5x - 5y + 4z = d_1 \\ \quad\quad 3y + z = d_2 \\ 2x - y + 2z = d_3, \text{ where} \end{cases}$$
a. $d_1 = 9$, $d_2 = 5$, $d_3 = 1$; **b.** $d_1 = 6$, $d_2 = 1$, $d_3 = 7$;
c. $d_1 = 2$, $d_2 = 13$, $d_3 = -5$.

52. The Trebon Electronic Company manufactures VCRs. Their deluxe model requires 5 hours of assembly time and 2 hours of testing time, while their regular model requires 3 hours of assembly time and 1 hour of testing time. Assembly time and testing time available for production of VCRs vary from day to day according to the following schedule.

Time available (in hours)	Monday	Tuesday	Wednesday	Thursday	Friday
For assembly	68	66	70	71	69
For testing	26	25	27	27	26

For each day of the week, let x and y be the numbers of deluxe and regular models the company manufactures.

a. Write five linear systems of equations knowing that all available assembly and testing time is used each day.
b. Solve the five systems simultaneously.

53. A trucking company has two types of trucks, large and medium, that transport two kinds of cargo. The number of units that each type of truck can carry is given in the following table.

Number of units of cargo transported per day	Truck type	
	Large	Medium
Kind A	120	50
Kind B	80	30

The number of units of each kind of cargo the company must move each day of the week is given in the following table.

	Monday	Tuesday	Wednesday	Thursday	Friday
Kind A	750	820	730	1020	1040
Kind B	490	540	470	660	680

For each day of the week, let x and y be the numbers of large and medium trucks, respectively, the company schedules to transport the two commodities.

a. Write five linear systems of equations knowing that all units of cargo will have been transported each day.
b. Solve the five systems simultaneously.

54. The Companie de Bateaux de Quebec makes aluminum boats. Type A requires 7 hours of assembly time and 2 hours of painting time, while type B requires 4 hours of assembly time and 1 hour of painting time. Assembly and painting time varies from week to week each month. The times available in each department (in hours) is given in the table below.

	Week 1	Week 2	Week 3	Week 4
Assembly time	200	218	227	247
Painting time	55	60	62	68

Let x and y denote the numbers of type A and B boats, respectively, that the company will manufacture each week.

a. Write four linear systems of equations in x and y assuming that all assembly and painting time is used each week.
b. Solve the four systems simultaneously.

55. A television manufacturing company makes three models of television sets. The super model requires 5 hours of wiring time, 6 hours of assembly time, and 2 hours of testing time. The deluxe model requires 3 hours of wiring time, 4 hours of assembly time, and 1 hour of testing time. The regular model requires 2, 3, and 1 hours of wiring, assembly, and testing time, respectively. The number of hours available in each of the three departments varies from month to month. The number of hours available in the first three months of the year is given in the table below.

	January	February	March
Wiring time	690	675	785
Assembly time	910	900	930
Testing time	280	275	320

Let x, y, and z denote the numbers of super, deluxe, and regular models, respectively, the company manufactures the first three months of the year.

a. Write three linear systems of equations assuming that all available time in all three departments is used.
b. Solve all three systems simultaneously.

56. An airline company flies three types of cargo planes: jumbo, large, and medium. On Mondays, Wednesdays, and Fridays these planes transport three kinds of freight: kind A, kind B, and kind C. The number of units of freight that each plane is able to transport is given in the table below.

	Jumbo	Large	Medium
Kind A	50	35	20
Kind B	60	40	35
Kind C	55	45	30

The number of units of each kind of cargo the company is committed to transport is given in the following table.

	Monday	Wednesday	Friday
Kind A	405	470	505
Kind B	520	585	655
Kind C	510	580	615

Let x, y, and z be the number of jumbo, large, and medium planes, respectively, the company must schedule to each of the days.

a. Write three systems of linear equations in x, y, and z, assuming that on each flight, each plane transports the maximum number of units of each kind of freight.

b. Solve the three systems simultaneously.

57. A manufacturer can produce three types of refrigerators called models A, B, and C. Each model takes a certain number of hours to assemble and a certain number of hours to test as shown in the following table.

	Model A	Model B	Model C
Assembly time (in hours)	7	5	4
Testing time	2	1.5	1

Each day, the assembly and testing departments have a maximum of 285 and 78 hours available, respectively. How many refrigerators of each type can the company produce if it operates at full capacity? If the profit for each model is $50, $60, and $40, respectively, how many of each model should be produced and sold to maximize profit? What is the maximum profit?

58. Repeat Exercise 57 if each day the assembly and testing departments have a maximum of 304 and 85 hours available, respectively.

59. A company manufactures three models of vacuum cleaners: standard, regular, and deluxe. The production of these requires the use of two machines I and II. The time (in minutes) needed on each machine for each model is given in the following table.

	Model		
	Standard	Regular	Deluxe
Time on machine I	20	30	35
Time on machine II	15	20	30

In a given week, machine I is available for 35 hours and machine II is available for 28 hours. How many of each model should be manufactured each week to have all the available time on each machine utilized? If the profit on each of the standard, regular, and deluxe models is $45, $56, and $60, respectively, how many of each model should be produced and sold in order to maximize profit? What is the maximum profit?

60. Repeat Exercise 59 if in a given week machine I is available for 33 hours and machine II is available for 27 hours.

61. A manufacturer can produce four types of inflatable rubber boats: small, medium, large, and super large. Each boat requires a certain number of hours from the cutting, assembly, and packaging departments as shown in the table.

	Small	Medium	Large	Super large
Cutting time (in hours)	1	1.5	2	2.5
Assembly time (in hours)	1.5	2	2.5	3
Packaging time (in hours)	.3	.4	.7	.6

The maximum number of hours available each day in the cutting, assembly, and packaging departments are 295, 385, and 85, respectively. How many boats of each type can be produced each day if the plant operates at full capacity? If the profit on each model is $40, $50, $75, and $90, respectively, how many of each model should be produced and sold in order to maximize profit? What is the maximum profit?

62. Repeat Exercise 61 if the maximum number of hours available each day in the cutting, assembly, and packaging departments is 275, 385, and 93, respectively.

63. Suppose that when three commodities X, Y, and Z are priced at x, y, and z dollars per unit, respectively, the quantities demanded are D_x, D_y, and D_z, respectively (measured in thousands), and the quantities supplied are S_x, S_y, and S_z, respectively (also measured in thousands). The demand and supply functions for these commodities are interrelated as expressed by

$D_x = 520 - 30x + 92y - 25z$

$S_x = -120 + 70x - 8y + 75z$

$D_y = 140 + 32x - 52y + 18z$

$S_y = -7 - 8x + 98y - 12z$

$D_z = 478 - 20x + 28y - 5z$

$S_z = -21 + 40x - 12y + 65z$

Find the equilibrium prices and quantities for all three commodities.

64. Same as Exercise 63 with the following equations:

$D_x = 150 - 10x + 85y - 17z$

$S_x = -8 + 90x - 15y + 83z$

$D_y = 280 + 42x - 20y + 81z$

$S_y = -17 - 8x + 120y - 19z$

$D_z = 310 - 17x + 51y - 12z$

$S_z = -44 + 63x - 9y + 88z$

65. Same as Exercise 63 with the following equations:

$D_x = 800 - 23x + 95y - 18z$

$S_x = -35 + 77x - 5y + 82z$

$D_y = 320 + 50x - 12y + 35z$

$S_y = -68 - 10x + 118y - 5z$

$D_z = 1310 - 10x + 42y - 20z$

$S_z = -112 + 80x - 8y + 100z$

3.4 Multiplicative Inverse

The following example illustrates the concept of the multiplicative inverse of matrices.

EXAMPLE 1 Find the products **AB** and **BA** if

$$\mathbf{A} = \begin{bmatrix} 2 & 1 \\ 5 & 3 \end{bmatrix} \text{ and } \mathbf{B} = \begin{bmatrix} 3 & -1 \\ -5 & 2 \end{bmatrix}$$

Solution Since both matrices are 2 × 2, each is conformable to the other for multiplication. Both products are defined and are 2 × 2 matrices. We get

$$\mathbf{AB} = \begin{bmatrix} 2 & 1 \\ 5 & 3 \end{bmatrix} \cdot \begin{bmatrix} 3 & -1 \\ -5 & 2 \end{bmatrix} = \begin{bmatrix} 2(3)+1(-5) & 2(-1)+1(2) \\ 5(3)+3(-5) & 5(-1)+3(2) \end{bmatrix}$$

$$= \begin{bmatrix} 1 & 0 \\ 0 & 1 \end{bmatrix} = \mathbf{I}_2$$

$$\mathbf{BA} = \begin{bmatrix} 3 & -1 \\ -5 & 2 \end{bmatrix} \cdot \begin{bmatrix} 2 & 1 \\ 5 & 3 \end{bmatrix} = \begin{bmatrix} 3(2)+(-1)5 & 3(1)+(-1)3 \\ (-5)2+2(5) & (-5)1+2(3) \end{bmatrix}$$

$$= \begin{bmatrix} 1 & 0 \\ 0 & 1 \end{bmatrix} = \mathbf{I}_2$$

We have obtained the identity matrix $\mathbf{I}_2$ for both products **AB** and **BA**. ∎

Recall that if a and b are numbers such that the product ab is the number 1, then b is said to be the multiplicative inverse of a and we write $b = a^{-1}$. For example, .2 is the multiplicative inverse of 5 since $5(.2) = 1$. The number 0 has no multiplicative inverse since $0(b) = 0 \neq 1$ for all numbers b. This holds true for matrices as well.

> **DEFINITION 3.12:** If **A** and **B** are $n \times n$ matrices and $\mathbf{AB} = \mathbf{BA} = \mathbf{I}_n$, then we say that **B** is the *multiplicative inverse* of **A** and we write $\mathbf{B} = \mathbf{A}^{-1}$. Note that **A** is also the multiplicative inverse of **B** and $\mathbf{A} = \mathbf{B}^{-1}$.

Just as the number 0 has no multiplicative inverse, not all square matrices have multiplicative inverses, as you can see in the following example.

EXAMPLE 2   Show that the matrix $\begin{bmatrix} 2 & -3 \\ 6 & -9 \end{bmatrix}$ does not have an inverse.

Solution   Suppose that the given matrix had an inverse

$$\begin{bmatrix} a & b \\ c & d \end{bmatrix}$$

Then we would have

$$\begin{bmatrix} 2 & -3 \\ 6 & -9 \end{bmatrix} \cdot \begin{bmatrix} a & b \\ c & d \end{bmatrix} = \begin{bmatrix} 1 & 0 \\ 0 & 1 \end{bmatrix}$$

But

$$\begin{bmatrix} 2 & -3 \\ 6 & -9 \end{bmatrix} \cdot \begin{bmatrix} a & b \\ c & d \end{bmatrix} = \begin{bmatrix} 2a-3c & 2b-3d \\ 6a-9c & 6b-9d \end{bmatrix}$$

It follows that

$$\begin{bmatrix} 2a-3c & 2b-3d \\ 6a-9c & 6b-9d \end{bmatrix} = \begin{bmatrix} 1 & 0 \\ 0 & 1 \end{bmatrix}$$

The truth of the foregoing equality would imply that the system

$$\begin{cases} 2a - 3c = 1 \\ 6a - 9c = 0 \end{cases}$$

has a solution. But $6a - 9c = 0$ is equivalent to $2a - 3c = 0$. Since $2a - 3c$ cannot equal both 0 and 1, the given matrix does not have an inverse.
Do Exercise 3.   ∎

We shall discuss this further when we learn how to find the multiplicative inverse of a matrix.

> **DEFINITION 3.13:**   A square matrix that has an inverse is said to be *invertible* (or *nonsingular*). A square matrix that does not have an inverse is said to be a *singular* matrix.

If a square matrix has a multiplicative inverse, it is unique. To demonstrate that this is the case, suppose that **B** and **C** are both inverses of **A**. Then by definition we have

$$\mathbf{AB} = \mathbf{BA} = \mathbf{I}_n \quad \text{and} \quad \mathbf{AC} = \mathbf{CA} = \mathbf{I}_n$$

We write

$$\mathbf{C} = \mathbf{CI}_n = \mathbf{C(AB)} = \mathbf{(CA)B} = \mathbf{I}_n\mathbf{B} = \mathbf{B}$$

Thus, **B** and **C** are the same matrix.

EXAMPLE 3   Find $\mathbf{AB}$ and $\mathbf{BA}$ if $\mathbf{A} = \begin{bmatrix} 7 & 2 & 5 \\ 3 & 1 & 2 \end{bmatrix}$ and $\mathbf{B} = \begin{bmatrix} 1 & -2 \\ -3 & 7 \\ 0 & 0 \end{bmatrix}$.

Solution   Clearly, $\mathbf{A}$ is conformable to $\mathbf{B}$ for multiplication and the product $\mathbf{AB}$ is a 2×2 matrix. We get

$$\mathbf{AB} = \begin{bmatrix} 7 & 2 & 5 \\ 3 & 1 & 2 \end{bmatrix} \cdot \begin{bmatrix} 1 & -2 \\ -3 & 7 \\ 0 & 0 \end{bmatrix}$$

$$= \begin{bmatrix} 7(1)+2(-3)+5(0) & 7(-2)+2(7)+5(0) \\ 3(1)+1(-3)+2(0) & 3(-2)+1(7)+2(0) \end{bmatrix} = \begin{bmatrix} 1 & 0 \\ 0 & 1 \end{bmatrix} = \mathbf{I}_2$$

Since $\mathbf{B}$ is 3×2 and $\mathbf{A}$ is 2×3, $\mathbf{B}$ is conformable to $\mathbf{A}$ for multiplication and the product $\mathbf{BA}$ is a 3×3 matrix. We find that

$$\mathbf{BA} = \begin{bmatrix} 1 & -2 \\ -3 & 7 \\ 0 & 0 \end{bmatrix} \cdot \begin{bmatrix} 7 & 2 & 5 \\ 3 & 1 & 2 \end{bmatrix} = \begin{bmatrix} 1 & 0 & 1 \\ 0 & 1 & -1 \\ 0 & 0 & 0 \end{bmatrix}$$

Although $\mathbf{AB}$ is the identity matrix of order 2, $\mathbf{A}$ and $\mathbf{B}$ are not inverses of each other because $\mathbf{BA}$ is not an identity matrix.   ■

    The previous example illustrates the fact that two matrices $\mathbf{A}$ and $\mathbf{B}$ are inverses of each other if, and only if, both matrices are *square* and $\mathbf{AB} = \mathbf{BA} = \mathbf{I}_n$.

Procedure to Find the Inverse

There are several procedures to find the inverse of an invertible matrix. We shall introduce the first of these in the present section. Another will be discussed in Section 3.6.

EXAMPLE 4   Find the inverse of the matrix $\begin{bmatrix} 7 & 5 \\ 4 & 3 \end{bmatrix}$ and check your result.

Solution   Let the inverse be the matrix

$$\begin{bmatrix} x_1 & y_1 \\ x_2 & y_2 \end{bmatrix}$$

Then,

$$\begin{bmatrix} 7 & 5 \\ 4 & 3 \end{bmatrix} \cdot \begin{bmatrix} x_1 & y_1 \\ x_2 & y_2 \end{bmatrix} = \begin{bmatrix} 1 & 0 \\ 0 & 1 \end{bmatrix}$$

This can be written

$$\begin{bmatrix} 7x_1 + 5x_2 & 7y_1 + 5y_2 \\ 4x_1 + 3x_2 & 4y_1 + 3y_2 \end{bmatrix} = \begin{bmatrix} 1 & 0 \\ 0 & 1 \end{bmatrix}$$

This matrix equation is equivalent to the system

$$\begin{cases} 7x_1 + 5x_2 = 1 \\ 4x_1 + 3x_2 = 0 \\ 7y_1 + 5y_2 = 0 \\ 4y_1 + 3y_2 = 1 \end{cases}$$

We may consider the first two equations as a linear system in x_1 and x_2, and the last two equations as a linear system in y_1 and y_2. In so doing, the two systems have the same coefficient matrix. Thus, we may solve them simultaneously as we did in Section 3.3. (See Example 11 and Exercises 46–51 of that section.) The augmented matrix for both systems is

$$\left[\begin{array}{cc|c|c} 7 & 5 & 1 & 0 \\ 4 & 3 & 0 & 1 \end{array}\right]$$

We perform elementary row operations on this augmented matrix to solve the two systems simultaneously.

$$\left[\begin{array}{cc|c|c} 7 & 5 & 1 & 0 \\ 4 & 3 & 0 & 1 \end{array}\right] \qquad (-2)R_2 + R_1 \rightarrow R_1$$

$$\left[\begin{array}{cc|c|c} -1 & -1 & 1 & -2 \\ 4 & 3 & 0 & 1 \end{array}\right] \qquad \begin{array}{l} 4R_1 + R_2 \rightarrow R_2 \\ (-1)R_1 \rightarrow R_1 \end{array}$$

$$\left[\begin{array}{cc|c|c} 1 & 1 & -1 & 2 \\ 0 & -1 & 4 & -7 \end{array}\right] \qquad \begin{array}{l} R_2 + R_1 \rightarrow R_1 \\ (-1)R_2 \rightarrow R_2 \end{array}$$

$$\left[\begin{array}{cc|c|c} 1 & 0 & 3 & -5 \\ 0 & 1 & -4 & 7 \end{array}\right]$$

Thus, $x_1 = 3$, $x_2 = -4$, $y_1 = -5$, and $y_2 = 7$. The inverse of the given matrix is

$$\begin{bmatrix} 3 & -5 \\ -4 & 7 \end{bmatrix}$$

We can check our result by multiplication.

$$\begin{bmatrix} 3 & -5 \\ -4 & 7 \end{bmatrix} \cdot \begin{bmatrix} 7 & 5 \\ 4 & 3 \end{bmatrix} = \begin{bmatrix} 3(7)+(-5)4 & 3(5)+(-5)3 \\ (-4)7+7(4) & (-4)5+7(3) \end{bmatrix}$$

$$= \begin{bmatrix} 1 & 0 \\ 0 & 1 \end{bmatrix}$$

We leave it to you to check that

$$\begin{bmatrix} 7 & 5 \\ 4 & 3 \end{bmatrix} \cdot \begin{bmatrix} 3 & -5 \\ -4 & 7 \end{bmatrix} = \begin{bmatrix} 1 & 0 \\ 0 & 1 \end{bmatrix}$$

We can conclude that our result is correct.
Do Exercise 5. ∎

EXAMPLE 5 Find the inverse of the matrix $\mathbf{A}$ if $\mathbf{A} = \begin{bmatrix} 5 & 2 & -6 \\ 7 & -3 & 4 \\ 3 & 5 & -12 \end{bmatrix}$.

Solution   Suppose that the inverse of the given matrix is

$$\begin{bmatrix} x_1 & y_1 & z_1 \\ x_2 & y_2 & z_2 \\ x_3 & y_3 & z_3 \end{bmatrix}$$

Then,

$$\begin{bmatrix} 5 & 2 & -6 \\ 7 & -3 & 4 \\ 3 & 5 & -12 \end{bmatrix} \cdot \begin{bmatrix} x_1 & y_1 & z_1 \\ x_2 & y_2 & z_2 \\ x_3 & y_3 & z_3 \end{bmatrix} = \begin{bmatrix} 1 & 0 & 0 \\ 0 & 1 & 0 \\ 0 & 0 & 1 \end{bmatrix}$$

where the x_i's, y_i's, and z_i's need to be determined.

Performing the multiplication on the left, and equating each entry of the product to the corresponding entry of the matrix $\mathbf{I}_3$ yields the following three linear systems.

$$\begin{cases} 5x_1 + 2x_2 - 6x_3 = 1 \\ 7x_1 - 3x_2 + 4x_3 = 0 \\ 3x_1 + 5x_2 - 12x_3 = 0 \end{cases} \qquad \begin{cases} 5y_1 + 2y_2 - 6y_3 = 0 \\ 7y_1 - 3y_2 + 4y_3 = 1 \\ 3y_1 + 5y_2 - 12y_3 = 0 \end{cases}$$

and

$$\begin{cases} 5z_1 + 2z_2 - 6z_3 = 0 \\ 7z_1 - 3z_2 + 4z_3 = 0 \\ 3z_1 + 5z_2 - 12z_3 = 1 \end{cases}$$

The three linear systems have the same coefficient matrix; therefore, they can be solved simultaneously as follows.

$$\left[\begin{array}{ccc|c|c|c} 5 & 2 & -6 & 1 & 0 & 0 \\ 7 & -3 & 4 & 0 & 1 & 0 \\ 3 & 5 & -12 & 0 & 0 & 1 \end{array}\right] \qquad (-2)R_3 + R_1 \rightarrow R_1$$

$$\left[\begin{array}{ccc|c|c|c} -1 & -8 & 18 & 1 & 0 & -2 \\ 7 & -3 & 4 & 0 & 1 & 0 \\ 3 & 5 & -12 & 0 & 0 & 1 \end{array}\right] \qquad \begin{array}{l} 7R_1 + R_2 \rightarrow R_2 \\ 3R_1 + R_3 \rightarrow R_3 \\ (-1)R_1 \rightarrow R_1 \end{array}$$

$$\left[\begin{array}{ccc|c|c|c} 1 & 8 & -18 & -1 & 0 & 2 \\ 0 & -59 & 130 & 7 & 1 & -14 \\ 0 & -19 & 42 & 3 & 0 & -5 \end{array}\right] \qquad (-3)R_3 + R_2 \rightarrow R_2$$

Because we could not easily get a 1 in the 2-2 or 3-3 position without introducing fractions, we multiplied row 3 by -3 and added the result to row 2 to get smaller integers (still hoping to avoid fractions!).

$$\left[\begin{array}{ccc|c|c|c} 1 & 8 & -18 & -1 & 0 & 2 \\ 0 & -2 & 4 & -2 & 1 & 1 \\ 0 & -19 & 42 & 3 & 0 & -5 \end{array}\right] \qquad \left(\tfrac{-1}{2}\right)R_2 \rightarrow R_2$$

$$\left[\begin{array}{ccc|c|c|c} 1 & 8 & -18 & -1 & 0 & 2 \\ 0 & 1 & -2 & 1 & \tfrac{-1}{2} & \tfrac{-1}{2} \\ 0 & -19 & 42 & 3 & 0 & -5 \end{array}\right] \qquad \begin{array}{l} (-8)R_2 + R_1 \rightarrow R_1 \\ (19)R_2 + R_3 \rightarrow R_3 \end{array}$$

$$\begin{bmatrix} 1 & 0 & -2 & | & -9 & | & 4 & | & 6 \\ 0 & 1 & -2 & | & 1 & | & \frac{-1}{2} & | & \frac{-1}{2} \\ 0 & 0 & 4 & | & 22 & | & \frac{-19}{2} & | & \frac{-29}{2} \end{bmatrix} \qquad \left(\tfrac{1}{4}\right)R_3 \to R_3$$

$$\begin{bmatrix} 1 & 0 & -2 & | & -9 & | & 4 & | & 6 \\ 0 & 1 & -2 & | & 1 & | & \frac{-1}{2} & | & \frac{-1}{2} \\ 0 & 0 & 1 & | & \frac{11}{2} & | & \frac{-19}{8} & | & \frac{-29}{8} \end{bmatrix} \qquad \begin{array}{l} 2R_3 + R_1 \to R_1 \\ 2R_3 + R_2 \to R_2 \end{array}$$

$$\begin{bmatrix} 1 & 0 & 0 & | & 2 & | & \frac{-3}{4} & | & \frac{-5}{4} \\ 0 & 1 & 0 & | & 12 & | & \frac{-21}{4} & | & \frac{-31}{4} \\ 0 & 0 & 1 & | & \frac{11}{2} & | & \frac{-19}{8} & | & \frac{-29}{8} \end{bmatrix}$$

Since the first three columns of the last augmented matrix form the identity matrix $\mathbf{I}_3$, we are done. We see that

$$x_1 = 2, \quad y_1 = \tfrac{-3}{4}, \quad z_1 = \tfrac{-5}{4}$$

$$x_2 = 12, \quad y_2 = \tfrac{-21}{4}, \quad z_2 = \tfrac{-31}{4}$$

$$x_3 = \tfrac{11}{2}, \quad y_3 = \tfrac{-19}{8}, \quad z_3 = \tfrac{-29}{8}$$

Hence,

$$\mathbf{A}^{-1} = \begin{bmatrix} 2 & \frac{-3}{4} & \frac{-5}{4} \\ 12 & \frac{-21}{4} & \frac{-31}{4} \\ \frac{11}{2} & \frac{-19}{8} & \frac{-29}{8} \end{bmatrix}$$

For convenience, we write

$$\mathbf{A}^{-1} = \left(\tfrac{1}{8}\right)\begin{bmatrix} 16 & -6 & -10 \\ 96 & -42 & -62 \\ 44 & -19 & -29 \end{bmatrix}$$

To check our result, we perform the following multiplication.

$$\left(\tfrac{1}{8}\right)\begin{bmatrix} 16 & -6 & -10 \\ 96 & -42 & -62 \\ 44 & -19 & -29 \end{bmatrix} \cdot \begin{bmatrix} 5 & 2 & -6 \\ 7 & -3 & 4 \\ 3 & 5 & -12 \end{bmatrix}$$

$$= \left(\tfrac{1}{8}\right)\begin{bmatrix} 8 & 0 & 0 \\ 0 & 8 & 0 \\ 0 & 0 & 8 \end{bmatrix} = \begin{bmatrix} 1 & 0 & 0 \\ 0 & 1 & 0 \\ 0 & 0 & 1 \end{bmatrix}$$

Similarly, we find that

$$\begin{bmatrix} 5 & 2 & -6 \\ 7 & -3 & 4 \\ 3 & 5 & -12 \end{bmatrix}\left(\tfrac{1}{8}\right)\begin{bmatrix} 16 & -6 & -10 \\ 96 & -42 & -62 \\ 44 & -19 & -29 \end{bmatrix}$$

$$= \left(\tfrac{1}{8}\right)\begin{bmatrix} 8 & 0 & 0 \\ 0 & 8 & 0 \\ 0 & 0 & 8 \end{bmatrix} = \begin{bmatrix} 1 & 0 & 0 \\ 0 & 1 & 0 \\ 0 & 0 & 1 \end{bmatrix}$$

Since in both cases the multiplications yielded the identity matrix I_3, we conclude that our result is correct.

You should carry out the details of the two matrix multiplications.
Do Exercise 11. ■

A CLOSER LOOK In the third step of the sequence of elementary row operations, we could have obtained a 1 in the 2-2 position by multiplying row 2 by $\frac{-1}{59}$. However, this operation would have introduced cumbersome fractions. We could also have obtained a 1 in the 2-2 position by multiplying row 2 by 9, row 3 by -28, and adding the results to get the new row 2 since $9(-59) + (-28)(-19) = -531 + 532 = 1$. We did not use this approach, although it would have delayed the introduction of fractions, because we wished to illustrate the fact that at each step we have several options. We should always try to determine quickly which option leads to the simplest solution. In this case, it would be difficult to readily find integers m and n such that $m(-59) + n(-19) = 1$; although, as we have shown above, such integers exist.

 In general, we use the following procedure to find the inverse of an $n \times n$ matrix **A**. We start with the augmented matrix $A \mid I_n$. If a sequence of elementary row operations transforms this matrix into $I_n \mid B$, then the matrix **B** is A^{-1}. However, if at any step we obtain all 0's in any row to the left of the vertical line, the matrix **A** does not have an inverse; it is singular.

EXAMPLE 6 Find the inverse, if it exists, of the matrix $\begin{bmatrix} 3 & 5 & 2 \\ 5 & 9 & -4 \\ 2 & 3 & 5 \end{bmatrix}$.

Solution

$$\left[\begin{array}{ccc|ccc} 3 & 5 & 2 & 1 & 0 & 0 \\ 5 & 9 & -4 & 0 & 1 & 0 \\ 2 & 3 & 5 & 0 & 0 & 1 \end{array}\right] \quad (-1)R_3 + R_1 \rightarrow R_1$$

$$\left[\begin{array}{ccc|ccc} 1 & 2 & -3 & 1 & 0 & -1 \\ 5 & 9 & -4 & 0 & 1 & 0 \\ 2 & 3 & 5 & 0 & 0 & 1 \end{array}\right] \quad \begin{array}{l} (-5)R_1 + R_2 \rightarrow R_2 \\ (-2)R_1 + R_3 \rightarrow R_3 \end{array}$$

$$\left[\begin{array}{ccc|ccc} 1 & 2 & -3 & 1 & 0 & -1 \\ 0 & -1 & 11 & -5 & 1 & 5 \\ 0 & -1 & 11 & -2 & 0 & 3 \end{array}\right] \quad \begin{array}{l} (-1)R_2 + R_3 \rightarrow R_3 \\ (-1)R_2 \rightarrow R_2 \end{array}$$

$$
\left[
\begin{array}{ccc|ccc}
1 & 2 & -3 & 1 & 0 & -1 \\
0 & 1 & -11 & 5 & -1 & -5 \\
0 & 0 & 0 & 3 & -1 & -2
\end{array}
\right]
$$

We may stop. We have obtained all 0's to the left of the vertical line in the third row. We conclude that the given matrix does not have an inverse. It is singular.
Do Exercise 15.   ■

EXAMPLE 7   Use the result of Example 5 to solve the system

$$
\begin{cases}
5x + 2y - 6z = 8 \\
7x - 3y + 4z = -24 \\
3x + 5y - 12z = 16 *
\end{cases}
$$
                                                                                                  ■

Solution   We first write the system as a matrix equation.

$$
\begin{bmatrix}
5 & 2 & -6 \\
7 & -3 & 4 \\
3 & 5 & -12
\end{bmatrix}
\cdot
\begin{bmatrix}
x \\
y \\
z
\end{bmatrix}
=
\begin{bmatrix}
8 \\
-24 \\
16
\end{bmatrix}
$$

We then multiply both sides of the equation (on the *left* of each side) by the inverse of the coefficient matrix found in Example 5.

$$
\frac{1}{8}
\begin{bmatrix}
16 & -6 & -10 \\
96 & -42 & -62 \\
44 & -19 & -29
\end{bmatrix}
\cdot
\begin{bmatrix}
5 & 2 & -6 \\
7 & -3 & 4 \\
3 & 5 & -12
\end{bmatrix}
\cdot
\begin{bmatrix}
x \\
y \\
z
\end{bmatrix}
$$

$$
= \frac{1}{8}
\begin{bmatrix}
16 & -6 & -10 \\
96 & -42 & -62 \\
44 & -19 & -29
\end{bmatrix}
\cdot
\begin{bmatrix}
8 \\
-24 \\
16
\end{bmatrix}
$$

After performing the multiplications we get

$$
\begin{bmatrix}
1 & 0 & 0 \\
0 & 1 & 0 \\
0 & 0 & 1
\end{bmatrix}
\cdot
\begin{bmatrix}
x \\
y \\
z
\end{bmatrix}
=
\frac{1}{8}
\begin{bmatrix}
112 \\
784 \\
344
\end{bmatrix}
$$

That is,

$$
\begin{bmatrix}
x \\
y \\
z
\end{bmatrix}
=
\begin{bmatrix}
14 \\
98 \\
43
\end{bmatrix}
$$

---

*The procedure illustrated in this example is not as efficient as it may appear because it is usually as tedious to find the inverse of the coefficient matrix as it is to solve the linear system by the method described in Section 3.3.

The solution of the original linear system is the ordered triple (14, 98, 43).
Do Exercise 31. ∎

EXAMPLE 8 Find the inverse, if it exists, of the following matrix

$$\begin{bmatrix} 2 & 2 & -1 & 2 \\ 4 & 5 & 1 & 5 \\ 7 & 2 & -6 & 6 \\ 3 & -2 & -1 & 3 \end{bmatrix}$$

Solution We proceed as follows.

$$\left[\begin{array}{cccc|cccc} 2 & 2 & -1 & 2 & 1 & 0 & 0 & 0 \\ 4 & 5 & 1 & 5 & 0 & 1 & 0 & 0 \\ 7 & 2 & -6 & 6 & 0 & 0 & 1 & 0 \\ 3 & -2 & -1 & 3 & 0 & 0 & 0 & 1 \end{array}\right]$$
$R_2 + R_1 \rightarrow R_1$
$6R_2 + R_3 \rightarrow R_3$
$R_2 + R_4 \rightarrow R_4$

$$\left[\begin{array}{cccc|cccc} 6 & 7 & 0 & 7 & 1 & 1 & 0 & 0 \\ 4 & 5 & 1 & 5 & 0 & 1 & 0 & 0 \\ 31 & 32 & 0 & 36 & 0 & 6 & 1 & 0 \\ 7 & 3 & 0 & 8 & 0 & 1 & 0 & 1 \end{array}\right]$$
$R_2 \leftrightarrow R_3$

$$\left[\begin{array}{cccc|cccc} 6 & 7 & 0 & 7 & 1 & 1 & 0 & 0 \\ 31 & 32 & 0 & 36 & 0 & 6 & 1 & 0 \\ 4 & 5 & 1 & 5 & 0 & 1 & 0 & 0 \\ 7 & 3 & 0 & 8 & 0 & 1 & 0 & 1 \end{array}\right]$$
$(-1)R_1 + R_4 \rightarrow R_4$

$$\left[\begin{array}{cccc|cccc} 6 & 7 & 0 & 7 & 1 & 1 & 0 & 0 \\ 31 & 32 & 0 & 36 & 0 & 6 & 1 & 0 \\ 4 & 5 & 1 & 5 & 0 & 1 & 0 & 0 \\ 1 & -4 & 0 & 1 & -1 & 0 & 0 & 1 \end{array}\right]$$
$(-7)R_4 + R_1 \rightarrow R_1$
$(-36)R_4 + R_2 \rightarrow R_2$
$(-5)R_4 + R_3 \rightarrow R_3$

$$\left[\begin{array}{cccc|cccc} -1 & 35 & 0 & 0 & 8 & 1 & 0 & -7 \\ -5 & 176 & 0 & 0 & 36 & 6 & 1 & -36 \\ -1 & 25 & 1 & 0 & 5 & 1 & 0 & -5 \\ 1 & -4 & 0 & 1 & -1 & 0 & 0 & 1 \end{array}\right]$$
$(-5)R_1 + R_2 \rightarrow R_2$
$(-1)R_1 + R_3 \rightarrow R_3$
$R_1 + R_4 \rightarrow R_4$
$(-1)R_1 \rightarrow R_1$

$$\left[\begin{array}{cccc|cccc} 1 & -35 & 0 & 0 & -8 & -1 & 0 & 7 \\ 0 & 1 & 0 & 0 & -4 & 1 & 1 & -1 \\ 0 & -10 & 1 & 0 & -3 & 0 & 0 & 2 \\ 0 & 31 & 0 & 1 & 7 & 1 & 0 & -6 \end{array}\right]$$
$35R_2 + R_1 \rightarrow R_1$
$10R_2 + R_3 \rightarrow R_3$
$(-31)R_2 + R_4 \rightarrow R_4$

$$\left[\begin{array}{cccc|cccc} 1 & 0 & 0 & 0 & -148 & 34 & 35 & -28 \\ 0 & 1 & 0 & 0 & -4 & 1 & 1 & -1 \\ 0 & 0 & 1 & 0 & -43 & 10 & 10 & -8 \\ 0 & 0 & 0 & 1 & 131 & -30 & -31 & 25 \end{array}\right]$$

Since the first four columns of the last augmented matrix form the identity matrix $\mathbf{I}_4$, we are done. The inverse of the given matrix is

$$\begin{bmatrix} -148 & 34 & 35 & -28 \\ -4 & 1 & 1 & -1 \\ -43 & 10 & 10 & -8 \\ 131 & -30 & -31 & 25 \end{bmatrix}$$

You should check the correctness of the result by multiplying the given matrix by the matrix we just obtained to verify that the product is indeed the identity matrix $\mathbf{I}_4$.*
Do Exercise 21.   ■

Application

EXAMPLE 9   A Seattle merchant owns three grocery stores and wishes to mix two kinds of nuts. Kind A is worth \$4 a pound and kind B is worth \$6 a pound. He would like to get 100 pounds of mixture to sell at \$5.20 per pound for his central store, 120 pounds of mixture to sell at \$4.80 per pound for his northern store, and 150 pounds of mixture to sell at \$5.40 per pound for his southern store. How many pounds of each kind of nut must he mix for each of his three stores?

Solution   Let q pounds be the quantity of mixture he wishes to obtain for a store and c dollars per pound be the selling price of the mixture at that store. Thus, for the central store $q = 100$ and $c = \$5.20$, for the northern store $q = 120$ and $c = \$4.80$, and for the southern store $q = 150$ and $c = \$5.40$. Let x pounds of kind A nuts and y pounds of kind B nuts be the quantities he must mix in each store to get the desired result. Then, $x + y = q$. Since kind A costs \$4 a pound, the cost of x pounds is \$4x. Similarly, we find that the cost of y pounds of kind B is \$6y. It follows that the cost of the mixture at any store is \$$(4x + 6y)$. But q pounds of mixture at c dollars a pound is worth qc dollars. Thus, we have the following linear system in x and y:

$$\begin{cases} x + y = q \\ 4x + 6y = qc \end{cases}$$

The matrix equation for the system is

$$\begin{bmatrix} 1 & 1 \\ 4 & 6 \end{bmatrix} \cdot \begin{bmatrix} x \\ y \end{bmatrix} = \begin{bmatrix} q \\ qc \end{bmatrix}$$

We find the inverse of the coefficient matrix using row reduction.

$$\left[\begin{array}{cc|cc} 1 & 1 & 1 & 0 \\ 4 & 6 & 0 & 1 \end{array}\right] \qquad (-4)R_1 + R_2 \rightarrow R_2$$

_____

*In advanced texts, it is proved that if $\mathbf{A}$ and $\mathbf{B}$ are $n \times n$ matrices and $\mathbf{AB} = \mathbf{I}_n$, $\mathbf{BA} = \mathbf{I}_n$ also. Hence, when you have found the inverse $\mathbf{B}$ of some $n \times n$ matrix $\mathbf{A}$ and you wish to check the result, it is sufficient to verify through multiplication that $\mathbf{AB} = \mathbf{I}_n$. It is not necessary to check that $\mathbf{BA} = \mathbf{I}_n$ also.

$$\left[\begin{array}{cc|cc} 1 & 1 & 1 & 0 \\ 0 & 2 & -4 & 1 \end{array}\right] \qquad (\tfrac{1}{2})R_2 \rightarrow R_2$$

$$\left[\begin{array}{cc|cc} 1 & 1 & 1 & 0 \\ 0 & 1 & -2 & \tfrac{1}{2} \end{array}\right] \qquad (-1)R_2 + R_1 \rightarrow R_1$$

$$\left[\begin{array}{cc|cc} 1 & 0 & 3 & \tfrac{-1}{2} \\ 0 & 1 & -2 & \tfrac{1}{2} \end{array}\right]$$

The inverse of the coefficient matrix is

$$\left[\begin{array}{cc} 3 & \tfrac{-1}{2} \\ -2 & \tfrac{1}{2} \end{array}\right]$$

We multiply both sides of the matrix equation (on the left of each side) by this inverse. We get

$$\left[\begin{array}{cc} 3 & \tfrac{-1}{2} \\ -2 & \tfrac{1}{2} \end{array}\right] \cdot \left[\begin{array}{cc} 1 & 1 \\ 4 & 6 \end{array}\right] \cdot \left[\begin{array}{c} x \\ y \end{array}\right] = \left[\begin{array}{cc} 3 & \tfrac{-1}{2} \\ -2 & \tfrac{1}{2} \end{array}\right] \cdot \left[\begin{array}{c} q \\ qc \end{array}\right]$$

Thus,

$$\left[\begin{array}{cc} 1 & 0 \\ 0 & 1 \end{array}\right] \cdot \left[\begin{array}{c} x \\ y \end{array}\right] = \left[\begin{array}{c} 3q - (\tfrac{1}{2})qc \\ -2q + (\tfrac{1}{2})qc \end{array}\right]$$

That is,

$$\left[\begin{array}{c} x \\ y \end{array}\right] = q\left[\begin{array}{c} 3 - (\tfrac{1}{2})c \\ -2 + (\tfrac{1}{2})c \end{array}\right]$$

Since $q = 100$ and $c = 5.2$ in the central store, we have

$$\left[\begin{array}{c} x \\ y \end{array}\right] = 100\left[\begin{array}{c} 3 - (\tfrac{1}{2})5.2 \\ -2 + (\tfrac{1}{2})5.2 \end{array}\right] = 100\left[\begin{array}{c} 3 - 2.60 \\ -2 + 2.60 \end{array}\right] = \left[\begin{array}{c} 40 \\ 60 \end{array}\right]$$

The merchant should mix 40 pounds of kind A nuts with 60 pounds of kind B nuts for his central store. Similarly, $q = 120$ and $c = \$4.80$ in the northern store. Therefore,

$$\left[\begin{array}{c} x \\ y \end{array}\right] = 120\left[\begin{array}{c} 3 - (\tfrac{1}{2})4.8 \\ -2 + (\tfrac{1}{2})4.8 \end{array}\right] = 120\left[\begin{array}{c} 3 - 2.4 \\ -2 + 2.4 \end{array}\right] = \left[\begin{array}{c} 72 \\ 48 \end{array}\right]$$

The merchant should mix 72 pounds of kind A nuts with 48 pounds of kind B nuts for his northern store.

Finally, for the southern store we have $q = 150$ and $c = 5.4$. Thus,

$$\left[\begin{array}{c} x \\ y \end{array}\right] = 150\left[\begin{array}{c} 3 - (\tfrac{1}{2})5.4 \\ -2 + (\tfrac{1}{2})5.4 \end{array}\right] = 150\left[\begin{array}{c} 3 - 2.7 \\ -2 + 2.7 \end{array}\right] = \left[\begin{array}{c} 45 \\ 105 \end{array}\right].$$

The merchant should mix 45 pounds of kind A nuts with 105 pounds of kind B nuts for his southern store.
Do Exercise 45.

Exercise Set 3.4

In Exercises 1–4, show that the given matrix does not have an inverse.

1. $\begin{bmatrix} 1 & 0 \\ 1 & 0 \end{bmatrix}$

2. $\begin{bmatrix} 2 & 1 \\ 6 & 3 \end{bmatrix}$

3. $\begin{bmatrix} 3 & 2 \\ 9 & 6 \end{bmatrix}$

4. $\begin{bmatrix} 5 & 7 \\ 20 & 28 \end{bmatrix}$

In Exercises 5–24, find the inverse, if it exists, of the given matrix.

5. $\begin{bmatrix} 5 & 7 \\ 2 & 3 \end{bmatrix}$

6. $\begin{bmatrix} 6 & 17 \\ 1 & 3 \end{bmatrix}$

7. $\begin{bmatrix} 9 & 13 \\ 2 & 3 \end{bmatrix}$

8. $\begin{bmatrix} -4 & 9 \\ 3 & -7 \end{bmatrix}$

9. $\begin{bmatrix} 3 & 8 \\ 5 & 12 \end{bmatrix}$

10. $\begin{bmatrix} 9 & 3 \\ 8 & 5 \end{bmatrix}$

11. $\begin{bmatrix} 5 & -5 & 4 \\ 0 & 3 & 1 \\ 2 & -1 & 2 \end{bmatrix}$

12. $\begin{bmatrix} 1 & -3 & 0 \\ 2 & 2 & 3 \\ 2 & -1 & 2 \end{bmatrix}$

13. $\begin{bmatrix} 7 & 6 & -17 \\ 2 & 2 & -5 \\ -6 & -5 & 15 \end{bmatrix}$

14. $\begin{bmatrix} 3 & 2 & -7 \\ 2 & 2 & -5 \\ 2 & 3 & -5 \end{bmatrix}$

15. $\begin{bmatrix} 1 & 0 & -2 \\ 2 & 5 & -6 \\ 4 & 5 & -10 \end{bmatrix}$

16. $\begin{bmatrix} -1 & -2 & -2 \\ 4 & 1 & 5 \\ 2 & -1 & 2 \end{bmatrix}$

17. $\begin{bmatrix} -7 & 1 & 1 \\ 3 & 2 & 3 \\ 5 & 1 & 2 \end{bmatrix}$

18. $\begin{bmatrix} 3 & -1 & 3 \\ 2 & 2 & 3 \\ 0 & -3 & -1 \end{bmatrix}$

19. $\begin{bmatrix} 2 & -1 & 7 \\ 3 & 9 & 0 \\ 5 & 0 & 2 \end{bmatrix}$

20. $\begin{bmatrix} 3 & 1 & 4 \\ -3 & -4 & 2 \\ -3 & 2 & -10 \end{bmatrix}$

21. $\begin{bmatrix} 7 & -8 & -5 & 6 \\ 1 & -4 & 0 & 1 \\ 3 & -11 & 3 & 4 \\ 4 & -6 & -1 & 4 \end{bmatrix}$

22. $\begin{bmatrix} -1 & 1 & 1 & -2 \\ -8 & 15 & 14 & -33 \\ 1 & 1 & 1 & -3 \\ -5 & -1 & 0 & 5 \end{bmatrix}$

23. $\begin{bmatrix} 0 & 1 & 3 & 1 \\ 5 & 1 & -2 & 5 \\ 1 & -2 & 6 & 3 \\ 2 & 3 & 2 & 3 \end{bmatrix}$

24. $\begin{bmatrix} 2 & 2 & -1 & 2 \\ 4 & 5 & 1 & 5 \\ 7 & 2 & -6 & 6 \\ 3 & -2 & -1 & 3 \end{bmatrix}$

In Exercises 25–44, write the given linear system as a matrix equation and solve that equation by using the inverses found in Exercises 5–24.

25. $\begin{cases} 5x + 7y = 2 \\ 2x + 3y = 5 \end{cases}$

26. $\begin{cases} 6x + 17y = -1 \\ x + 3y = 3 \end{cases}$

27. $\begin{cases} 9x + 13y = -3 \\ 2x + 3y = 5 \end{cases}$

28. $\begin{cases} -4x + 9y = 13 \\ 3x - 7y = -4 \end{cases}$

29. $\begin{cases} 3x + 8y = -1 \\ 5x + 12y = 17 \end{cases}$

30. $\begin{cases} 9x + 3y = 6 \\ 8x + 5y = 15 \end{cases}$

31. $\begin{cases} 5x - 5y + 4z = 3 \\ 3y + z = 1 \\ 2x - y + 2z = 5 \end{cases}$

32. $\begin{cases} x - 3y = -5 \\ 2x + 2y + 3z = -7 \\ 2x - y + 2z = 9 \end{cases}$

33. $\begin{cases} 7x + 6y - 17z = 0 \\ 2x + 2y - 5z = 3 \\ -6x - 5y + 15z = 7 \end{cases}$

34. $\begin{cases} 3x + 2y - 7z = 15 \\ 2x + 2y - 5z = -4 \\ 2x + 3y - 5z = 7 \end{cases}$

35. $\begin{cases} x - 2z = 3 \\ 2x + 5y - 6z = 0 \\ 4x + 5y - 10z = 5 \end{cases}$

36. $\begin{cases} -x - 2y - 2z = -4 \\ 4x + y + 5z = 11 \\ 2x - y + 2z = 5 \end{cases}$

37. $\begin{cases} -7x + y + z = 3 \\ 3x + 2y + 3z = 1 \\ 5x + y + 2z = 5 \end{cases}$

38. $\begin{cases} 3x - y + 3z = 12 \\ 2x + 2y + 3z = 0 \\ -3y - z = 17 \end{cases}$

39. $\begin{cases} 2x - y + 7z = 8 \\ 3x + 9y = 7 \\ 5x + 2z = 1 \end{cases}$

40. $\begin{cases} 3x + y + 4z = 9 \\ -3x - 4y + 2z = 19 \\ -3x + 2y - 10z = 3 \end{cases}$

41. $\begin{cases} 7w - 8x - 5y + 6z = 1 \\ w - 4x + z = 3 \\ 3w - 11x + 3y + 4z = 1 \\ 4w - 6x - y + 4z = 7 \end{cases}$

42.
$$\begin{cases} w + x + y - 2z = 0 \\ -8w + 15x + 14y - 33z = 2 \\ w + x + y - 3z = 3 \\ -5w - x + 5z = 0 \end{cases}$$

43.
$$\begin{cases} x + 3y + z = 8 \\ 5w + x - 2y + 5z = 3 \\ w - 2x + 6y + 3z = 5 \\ 2w + 3x + 2y + 3z = 7 \end{cases}$$

44.
$$\begin{cases} 2w + 2x - y + 2z = 6 \\ 4w + 5x + y + 5z = 2 \\ 7w + 2x - 6y + 6z = -1 \\ 3w - 2x - y + 3z = 1 \end{cases}$$

45. A Chicago merchant owns three grocery stores and wishes to mix two kinds of nuts. Kind A is worth $3.60 a pound and kind B is worth $4.60 a pound. She would like to get 200 pounds of mixture to sell at $3.80 per pound for her central store, 300 pounds of mixture to sell at $4 a pound for her northern store, and 250 pounds of mixture to sell at $4.20 per pound for her southern store. How many pounds of each kind of nut must she mix for each of her three stores?

46. A Montreal confectioner owns three stores and wishes to mix candy worth $4.25 per pound with candy worth $5.25 per pound for each of the three stores. He wants to get 200 pounds of an assortment worth $4.45 a pound for store A, 230 pounds of an assortment worth $4.45 a pound for store B, and 250 pounds of an assortment worth $5.05 a pound for store C. How much of each kind must he mix for each of the three stores?

47. A concert hall has 6000 seats. Each evening the manager divides the seats into two sections. The tickets are worth $7 in section A and $10 in section B. Assuming that all tickets can be sold, how many seats must he assign to each section on weekdays, on Saturdays, and on Sundays to get the following revenues: $48,000 on each of the weekdays, $57,000 on Saturdays, an $51,000 on Sundays.

48. A football stadium has 60,000 seats. The manager divides the stadium into two sections for the exhibition games, the games the home team plays outside its division, and the games it plays within its division. The tickets are worth $19 in section A and $14 in section B. Assuming that all tickets can be sold, how many seats must he assign to each section on days of an exhibition game, a game outside the division, and a game within the division to bring in the following revenues: $970,000 for each exhibition game, $990,000 for each game outside the division, and $1,120,000 for each game within the division.

49. Use the inverse of the coefficient matrix to solve the five systems obtained in Exercise 52, Section 3.3.

50. Use the inverse of the coefficient matrix to solve the five systems obtained in Exercise 53, Section 3.3.

51. Use the inverse of the coefficient matrix to solve the four systems obtained in Exercise 54, Section 3.3.

52. Use the inverse of the coefficient matrix to solve the three systems obtained in Exercise 55, Section 3.3.

53. Use the inverse of the coefficient matrix to solve the three systems obtained in Exercise 56, Section 3.3.

54. In the preceding section we have shown that a homogeneous system of linear equations has a nontrivial solution whenever the number of unknowns is larger than the number of equations. Show that if the number of equations is the same as the number of unknowns, the only solution of the system is the trivial solution whenever the coefficient matrix has an inverse.

55. Show that the only solution of the following homogeneous system is the trivial solution.
$$\begin{cases} 2x + 5y = 0 \\ 3x + 7y = 0 \end{cases}$$

56. Same as Exercise 55 for the system
$$\begin{cases} 3x + 2y - 3z = 0 \\ x + 2y - 3z = 0 \\ 4x + 3y + 3z = 0 \end{cases}$$

3.5 Determinants

In Section 2.1 we defined the determinant of a 2×2 matrix and showed that we could solve a linear system of two equations in two variables using a rule called Cramer's rule. In this section we wish to apply this concept to any square matrix. We shall describe how we assign to each $n \times n$ matrix a number called the determinant of that matrix. We shall do this inductively.*

Cofactors

We begin by defining the determinant of the 1×1 matrix $[c]$ as the number c. We then define the cofactor of an entry of a square matrix.

> **DEFINITION 3.14:**   The *cofactor* C_{ij} of the *i-j* entry of a square matrix **A** is the number $(-1)^{i+j}$ times the determinant of the square matrix obtained from **A** by deleting its *i*th row and *j*th column.

(As of now we can obtain only the cofactors of the entries of a 2×2 matrix since we have not yet defined the determinant of square matrices other than 1×1 matrices.)

If we consider the 2×2 matrix

$$\mathbf{A} = \begin{bmatrix} a_{11} & a_{12} \\ a_{21} & a_{22} \end{bmatrix}$$

and let C_{ij} denote the cofactor of each a_{ij}, we have

$$C_{11} = (-1)^{1+1}\det \begin{bmatrix} a_{11} & a_{12} \\ a_{21} & a_{22} \end{bmatrix} = \det[a_{22}] = a_{22}$$

$$C_{12} = (-1)^{1+2}\det \begin{bmatrix} a_{11} & a_{12} \\ a_{21} & a_{22} \end{bmatrix} = (-1)\det[a_{21}] = -a_{21}$$

$$C_{21} = (-1)^{2+1}\det \begin{bmatrix} a_{11} & a_{12} \\ a_{21} & a_{22} \end{bmatrix} = (-1)\det[a_{12}] = -a_{12}$$

$$C_{22} = (-1)^{2+2}\det \begin{bmatrix} a_{11} & a_{12} \\ a_{21} & a_{22} \end{bmatrix} = \det[a_{11}] = a_{11}$$

We now calculate the dot product of each row of the matrix **A** by the corresponding row of cofactors. We get

$$\begin{aligned} [\, a_{11} \quad a_{12} \,] \cdot [\, C_{11} \quad C_{12} \,] &= [\, a_{11} \quad a_{12} \,] \cdot [\, a_{22} \quad -a_{21} \,] \\ &= a_{11}(a_{22}) + a_{12}(-a_{21}) \\ &= a_{11}a_{22} - a_{12}a_{21} \end{aligned}$$

Also,

$$\begin{aligned} [\, a_{21} \quad a_{22} \,] \cdot [\, C_{21} \quad C_{22} \,] &= [\, a_{21} \quad a_{22} \,] \cdot [\, -a_{12} \quad a_{11} \,] \\ &= a_{21}(-a_{12}) + a_{22}(a_{11}) \\ &= -a_{21}a_{12} + a_{22}a_{11} = a_{11}a_{22} - a_{12}a_{21} \end{aligned}$$

---

*The process known as definition by induction consists of two parts: **a.** we define the determinant of a 1×1 matrix and **b.** assuming we have defined the determinant of an $n \times n$ matrix for fixed $n \geq 1$, we further define the determinant of a $(n + 1) \times (n + 1)$ matrix.

In both cases we get the same result. This common value of the two dot products is defined as the determinant of the 2×2 matrix **A**. This definition agrees with that given in Section 2.1. Thus,

$$\det \begin{bmatrix} a_{11} & a_{12} \\ a_{21} & a_{22} \end{bmatrix} = a_{11}a_{22} - a_{12}a_{21}$$

EXAMPLE 1   Calculate the determinant of the 2×2 matrix

$$\begin{bmatrix} 3 & -2 \\ 5 & 7 \end{bmatrix}$$

Solution $\det \begin{bmatrix} 3 & -2 \\ 5 & 7 \end{bmatrix} = 3(7) - (-2)5$

$$= 21 + 10 = 31$$

Do Exercise 1.   ■

Now that we know how to calculate the determinant of a 2×2 matrix, we can find each of the cofactors of the nine entries of a 3×3 matrix.

EXAMPLE 2   Calculate the cofactors of the nine entries of the 3×3 matrix

$$\mathbf{A} = \begin{bmatrix} 3 & 5 & -2 \\ 2 & 1 & 4 \\ 6 & 3 & 8 \end{bmatrix}$$

Solution   The cofactor C_{11} is found by multiplying $(-1)^{1+1}$ by the determinant of the 2×2 matrix obtained from **A** by deleting row 1 and column 1. Thus,

$$C_{11} = (-1)^{1+1} \det \begin{bmatrix} 3 & 5 & -2 \\ 2 & 1 & 4 \\ 6 & 3 & 8 \end{bmatrix}$$

$$= (-1)^{1+1} \det \begin{bmatrix} 1 & 4 \\ 3 & 8 \end{bmatrix}$$

$$= 1(8) - 4(3) = 8 - 12 = -4$$

Similarly, C_{12} is obtained by multiplying $(-1)^{1+2}$ by the determinant of the 2×2 matrix obtained from **A** by deleting row 1 and column 2. Thus,

$$C_{12} = (-1)^{1+2} \det \begin{bmatrix} 3 & 5 & -2 \\ 2 & 1 & 4 \\ 6 & 3 & 8 \end{bmatrix}$$

$$= (-1)^{1+2} \det \begin{bmatrix} 2 & 4 \\ 6 & 8 \end{bmatrix}$$

$$= (-1)[2(8) - 4(6)] = (-1)[16 - 24]$$

$$= (-1)(-8) = 8$$

Similarly,

$$C_{13} = (-1)^{1+3} \det \begin{bmatrix} 3 & 5 & -2 \\ 2 & 1 & 4 \\ 6 & 3 & 8 \end{bmatrix}$$

$$= (-1)^{1+3} \det \begin{bmatrix} 2 & 1 \\ 6 & 3 \end{bmatrix}$$

$$= 2(3) - 1(6) = 6 - 6 = 0$$

$$C_{21} = (-1)^{2+1} \det \begin{bmatrix} 3 & 5 & -2 \\ 2 & 1 & 4 \\ 6 & 3 & 8 \end{bmatrix}$$

$$= (-1)^{2+1} \det \begin{bmatrix} 5 & -2 \\ 3 & 8 \end{bmatrix}$$

$$= (-1)[5(8) - (-2)3] = (-1)[40 + 6]$$

$$= (-1)46 = -46$$

$$C_{22} = (-1)^{2+2} \det \begin{bmatrix} 3 & 5 & -2 \\ 2 & 1 & 4 \\ 6 & 3 & 8 \end{bmatrix}$$

$$= (-1)^{2+2} \det \begin{bmatrix} 3 & -2 \\ 6 & 8 \end{bmatrix}$$

$$= 3(8) - (-2)6 = 24 + 12 = 36$$

$$C_{23} = (-1)^{2+3} \det \begin{bmatrix} 3 & 5 & -2 \\ 2 & 1 & 4 \\ 6 & 3 & 8 \end{bmatrix}$$

$$= (-1)^{2+3} \det \begin{bmatrix} 3 & 5 \\ 6 & 3 \end{bmatrix}$$

$$= (-1)[3(3) - 5(6)] = (-1)[9 - 30]$$

$$= (-1)(-21) = 21$$

$$C_{31} = (-1)^{3+1} \det \begin{bmatrix} 3 & 5 & -2 \\ 2 & 1 & 4 \\ 6 & 3 & 8 \end{bmatrix}$$

$$= (-1)^{3+1} \det \begin{bmatrix} 5 & -2 \\ 1 & 4 \end{bmatrix}$$

$$= 5(4) - (-2)1 = 20 + 2 = 22$$

$$C_{32} = (-1)^{3+2} \det \begin{bmatrix} 3 & 5 & -2 \\ 2 & 1 & 4 \\ 6 & 3 & 8 \end{bmatrix}$$

$$= (-1)^{3+2} \det \begin{bmatrix} 3 & -2 \\ 2 & 4 \end{bmatrix}$$

$$= (-1)[3(4) - (-2)2] = (-1)[12 + 4]$$
$$= (-1)16 = -16$$

$$C_{33} = (-1)^{3+3} \det \begin{bmatrix} 3 & 5 & -2 \\ 2 & 1 & 4 \\ 6 & 3 & 8 \end{bmatrix}$$

$$= (-1)^{3+3} \det \begin{bmatrix} 3 & 5 \\ 2 & 1 \end{bmatrix}$$

$$= 3(1) - 5(2) = 3 - 10 = -7$$

EXAMPLE 3   Calculate the dot product of each row of the matrix of the previous example by the corresponding row of cofactors and compare the three results.

Solution   We have

$$[\ a_{11} \quad a_{12} \quad a_{13}\ ] \cdot [\ C_{11} \quad C_{12} \quad C_{13}\ ] =$$
$$[\ 3 \quad 5 \quad -2\ ] \cdot [\ -4 \quad 8 \quad 0\ ] = 3(-4) + 5(8) + (-2)0$$
$$= -12 + 40 + 0 = 28$$

$$[\ a_{21} \quad a_{22} \quad a_{23}\ ] \cdot [\ C_{21} \quad C_{22} \quad C_{23}\ ] =$$
$$[\ 2 \quad 1 \quad 4\ ] \cdot [\ -46 \quad 36 \quad 21\ ] = 2(-46) + 1(36) + 4(21)$$
$$= -92 + 36 + 84 = 28$$

$$[\ a_{31} \quad a_{32} \quad a_{33}\ ] \cdot [\ C_{31} \quad C_{32} \quad C_{33}\ ] =$$
$$[\ 6 \quad 3 \quad 8\ ] \cdot [\ 22 \quad -16 \quad -7\ ] = 6(22) + 3(-16) + 8(-7)$$
$$= 132 - 48 - 56 = 28$$

Note that 28 was obtained all three times.

    The fact that the same number was obtained for the three dot products is not a coincidence. It can be proved, but we shall not do it here, that if **A** is any 3×3 matrix, then the dot product of any row of **A** by the corresponding row of cofactors is equal to the dot product of any other row of **A** by the corresponding row of cofactors. This common value is, by definition, the determinant of the matrix **A**. Thus, from the results of Example 3, we can write

$$\det \begin{bmatrix} 3 & 5 & -2 \\ 2 & 1 & 4 \\ 6 & 3 & 8 \end{bmatrix} = 28$$

Definition of Determinant

Now that we know how to find the determinant of a 3×3 matrix, we may define the determinant of a 4×4 matrix as the dot product of any row of that matrix by the corresponding row of cofactors. It can be proved that the result is independent of the choice of the row.

DEFINITION 3.15:   In general, once we have defined the determinant of an $n \times n$ matrix, then the *determinant of an $(n + 1) \times (n + 1)$ matrix* is defined as the dot product of any row of that matrix by the corresponding row of cofactors.*

Again, it can be proved that this dot product is independent of the choice of the row.

It is not practical to use the definition of the determinant of a square matrix for computing its value. We already have seen that

$$\det \begin{bmatrix} a_{11} & a_{12} \\ a_{21} & a_{22} \end{bmatrix} = a_{11}a_{22} - a_{12}a_{21}$$

We now derive a formula for the determinant of a 3×3 matrix.

EXAMPLE 4   Use the definition of determinant to derive a formula for the determinant of a 3×3 matrix.

Solution   Consider an arbitrary 3×3 matrix $\mathbf{A}$, where

$$\mathbf{A} = \begin{bmatrix} a_{11} & a_{12} & a_{13} \\ a_{21} & a_{22} & a_{23} \\ a_{31} & a_{32} & a_{33} \end{bmatrix}$$

We first calculate the cofactors of each of the entries of row 1. We have

$$C_{11} = (-1)^{1+1} \det \begin{bmatrix} a_{11} & a_{12} & a_{13} \\ a_{21} & a_{22} & a_{23} \\ a_{31} & a_{32} & a_{33} \end{bmatrix}$$

$$= (-1)^{1+1} \det \begin{bmatrix} a_{22} & a_{23} \\ a_{32} & a_{33} \end{bmatrix} = a_{22}a_{33} - a_{23}a_{32}$$

---

* If $\mathbf{A} = \begin{bmatrix} a_{11} & a_{12} & a_{13} & \cdots & a_{1n} \\ a_{21} & a_{22} & a_{23} & \cdots & a_{2n} \\ \vdots & \vdots & \vdots & & \vdots \\ a_{n1} & a_{n2} & a_{n3} & \cdots & a_{nn} \end{bmatrix}$ is an $n \times n$ matrix, then the symbol []

$\begin{vmatrix} a_{11} & a_{12} & a_{13} & \cdots & a_{1n} \\ a_{21} & a_{22} & a_{23} & \cdots & a_{2n} \\ \vdots & \vdots & \vdots & & \vdots \\ a_{n1} & a_{n2} & a_{n3} & \cdots & a_{nn} \end{vmatrix}$ is often used instead of det$\mathbf{A}$ to denote the determinant of the

matrix $\mathbf{A}$.

$$C_{12} = (-1)^{1+2} \det \begin{bmatrix} a_{11} & a_{12} & a_{13} \\ a_{21} & a_{22} & a_{23} \\ a_{31} & a_{32} & a_{33} \end{bmatrix}$$

$$= (-1)^{1+2} \det \begin{bmatrix} a_{21} & a_{23} \\ a_{31} & a_{33} \end{bmatrix} = (-1)^{1+2}(a_{21}a_{33} - a_{23}a_{31})$$

$$= -a_{21}a_{33} + a_{23}a_{31}$$

$$C_{13} = (-1)^{1+3} \det \begin{bmatrix} a_{11} & a_{12} & a_{13} \\ a_{21} & a_{22} & a_{23} \\ a_{31} & a_{32} & a_{33} \end{bmatrix}$$

$$= (-1)^{1+3} \det \begin{bmatrix} a_{21} & a_{22} \\ a_{31} & a_{32} \end{bmatrix} = a_{21}a_{32} - a_{22}a_{31}$$

Thus,

$$\det A = [\, a_{11} \quad a_{12} \quad a_{13} \,] \cdot [\, C_{11} \quad C_{12} \quad C_{13} \,]$$
$$= a_{11}C_{11} + a_{12}C_{12} + a_{13}C_{13}$$
$$= a_{11}(a_{22}a_{33} - a_{23}a_{32}) + a_{12}$$
$$(-a_{21}a_{33} + a_{23}a_{31}) + a_{13}(a_{21}a_{32} - a_{22}a_{31})$$
$$= a_{11}a_{22}a_{33} - a_{11}a_{23}a_{32} - a_{12}a_{21}a_{33} + a_{12}a_{23}a_{31}$$
$$+ a_{13}a_{21}a_{32} - a_{13}a_{22}a_{31}.$$

There are six terms in the last line of the sum, three with a plus sign and three with a minus sign. Each term is a product of three factors, one from each row and each column. There are two mnemonic devices commonly used to remember the formula giving the determinant of the 3 × 3 matrix

$$A = \begin{bmatrix} a_{11} & a_{12} & a_{13} \\ a_{21} & a_{22} & a_{23} \\ a_{31} & a_{32} & a_{33} \end{bmatrix}$$

These are illustrated schematically by the following diagrams. In each case, we multiply the three entries that are strung together by an arrow, assigning to each term the sign that is indicated at the tail of the corresponding arrow. In the second diagram, we duplicated columns 1 and 2 to the right of the matrix **A** to be able to extend the arrows.

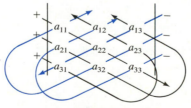

First device

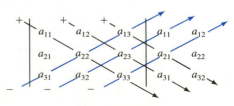

Second device

EXAMPLE 5   Find the determinant of the 3×3 matrix

$$\begin{bmatrix} 3 & -2 & 5 \\ 4 & 1 & 2 \\ 7 & 3 & -6 \end{bmatrix}$$

Solution   We use the first device to find which entries are multiplied together.

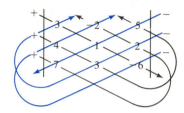

Thus,

$$\det \begin{bmatrix} 3 & -2 & 5 \\ 4 & 1 & 2 \\ 7 & 3 & -6 \end{bmatrix} = 3(1)(-6) + 4(3)(5) + 7(2)(-2)$$
$$-5(1)(7) - 2(3)(3) - (-6)(4)(-2)$$
$$= -18 + 60 - 28 - 35 - 18 - 48$$
$$= -87 \qquad \blacksquare$$

Unfortunately, there is no simple device to calculate the determinant of an $n \times n$ matrix when n is larger than 3. (See Exercise 56.) Thus, we usually depend on the definition, as is illustrated in the following example.

EXAMPLE 6   Find the determinant of the matrix

$$\mathbf{A} = \begin{bmatrix} 3 & 5 & -1 & 6 \\ 2 & 0 & 3 & 0 \\ 5 & 2 & 5 & 2 \\ 7 & 1 & 2 & 7 \end{bmatrix}$$

Solution   Since two of the entries in row 2 are 0's, it is easiest to compute the dot product of row 2 by the corresponding row of cofactors. With the usual notation for cofactors, we get

$$\det \mathbf{A} = [\, 2 \quad 0 \quad 3 \quad 0 \,] \cdot [\, C_{21} \quad C_{22} \quad C_{23} \quad C_{24} \,]$$
$$= 2C_{21} + 0C_{22} + 3C_{23} + 0C_{24} = 2C_{21} + 3C_{23}$$

We do not need to calculate the cofactors C_{22} and C_{24}. We find

$$C_{21} = (-1)^{2+1} \det \begin{bmatrix} 5 & -1 & 6 \\ 2 & 5 & 2 \\ 1 & 2 & 7 \end{bmatrix}$$

$$= (-1)[5 \cdot 5 \cdot 7 + (-1)2 \cdot 1 + 6 \cdot 2 \cdot 2$$
$$- 1 \cdot 5 \cdot 6 - 2 \cdot 2 \cdot 5 - 7 \cdot (-1)2]$$
$$= (-1)[175 - 2 + 24 - 30 - 20 + 14] = (-1)(161) = -161$$

Also,

$$C_{23} = (-1)^{2+3} \det \begin{bmatrix} 3 & 5 & 6 \\ 5 & 2 & 2 \\ 7 & 1 & 7 \end{bmatrix}$$
$$= (-1)[3 \cdot 2 \cdot 7 + 5 \cdot 2 \cdot 7 + 6 \cdot 1 \cdot 5$$
$$- 7 \cdot 2 \cdot 6 - 1 \cdot 2 \cdot 3 - 7 \cdot 5 \cdot 5]$$
$$= (-1)[42 + 70 + 30 - 84 - 6 - 175] = (-1)(-123) = 123$$

Thus,

$$\det \mathbf{A} = 2(-161) + 3(123) = -322 + 369$$
$$= 47$$

Do Exercise 15. ∎

Properties of Determinants

It is clear from the previous example that if we wish to calculate the determinant of a matrix using the definition, it is best to choose the row which has the most 0's. In fact, we can use elementary row operations to get a new matrix that has a row with all but one entry equal to 0 and whose determinant is related to the determinant of the original matrix. We will state the necessary properties of determinants to justify this process after the following definition.

DEFINITION 3.16: If $\mathbf{A}$ is an $m \times n$ matrix, its *transpose* is the $n \times m$ matrix whose i-j entry is the j-i entry of the matrix $\mathbf{A}$. The rows of matrix $\mathbf{A}$ become the columns of the transpose and the columns of matrix $\mathbf{A}$ become the rows of the transpose, while preserving their order. The transpose of a matrix $\mathbf{A}$ is denoted $\mathbf{A}^T$.

EXAMPLE 7   Find the transpose of the matrix $\mathbf{A}$ where

$$\mathbf{A} = \begin{bmatrix} 2 & 3 & 7 \\ 9 & 0 & 4 \end{bmatrix}$$

Solution   Since $\mathbf{A}$ is a 2×3 matrix, its transpose is a 3×2 matrix. Furthermore, row 1 of $\mathbf{A}$ is column 1 of $\mathbf{A}^T$ and row 2 of $\mathbf{A}$ is column 2 of $\mathbf{A}^T$. We write

$$\mathbf{A}^T = \begin{bmatrix} 2 & 9 \\ 3 & 0 \\ 7 & 4 \end{bmatrix}$$
∎

We now state (but do not prove) some basic properties of determinants which are useful for calculating the determinant of a square matrix.

Basic Properties of Determinants

1. If $\mathbf{A}$ is a square matrix, then $\det\mathbf{A} = \det\mathbf{A}^T$.
2. If $\mathbf{A}$ is a square matrix, then $\det\mathbf{A}$ is equal to the dot product of any column of $\mathbf{A}$ by the corresponding column of cofactors.
3. If $\mathbf{A}$ is a square matrix and $\mathbf{B}$ is obtained from $\mathbf{A}$ by interchanging two rows (columns), then $\det\mathbf{B} = -\det\mathbf{A}$.
4. If $\mathbf{A}$ is a square matrix with two identical rows (columns), then $\det\mathbf{A} = 0$.
5. If $\mathbf{A}$ is a square matrix and $\mathbf{B}$ is obtained from $\mathbf{A}$ by multiplying one of the rows (columns) of $\mathbf{A}$ by a number k, then $\det\mathbf{B} = k \cdot \det\mathbf{A}$.
6. If $\mathbf{A}$ is a square matrix and $\mathbf{B}$ is obtained from $\mathbf{A}$ by replacing a row (column) of $\mathbf{A}$ by the sum of that row (column) and a constant multiple of another row (column) of $\mathbf{A}$, then $\det\mathbf{B} = \det\mathbf{A}$.

Evaluation of Determinants

Using these properties repeatedly, we can get a new matrix with many entries which are 0, and whose determinant is a known multiple of that of the original matrix. We illustrate with the following example.

EXAMPLE 8   Evaluate the determinant of the matrix

$$\mathbf{A} = \begin{bmatrix} 3 & 5 & 7 & 8 \\ 7 & 10 & 3 & 2 \\ 2 & -5 & 8 & 3 \\ 2 & 15 & 9 & 5 \end{bmatrix}$$

Solution   Using Property 5, we can factor 5 out of column 2. Thus, we get

$$\det\mathbf{A} = \det\begin{bmatrix} 3 & 5 & 7 & 8 \\ 7 & 10 & 3 & 2 \\ 2 & -5 & 8 & 3 \\ 2 & 15 & 9 & 5 \end{bmatrix} = 5 \cdot \det\begin{bmatrix} 3 & 1 & 7 & 8 \\ 7 & 2 & 3 & 2 \\ 2 & -1 & 8 & 3 \\ 2 & 3 & 9 & 5 \end{bmatrix}$$

$$= 5 \cdot \det\begin{bmatrix} 3 & 1 & 7 & 8 \\ 1 & 0 & -11 & -14 \\ 5 & 0 & 15 & 11 \\ -7 & 0 & -12 & -19 \end{bmatrix}$$

where the last matrix was obtained by using Property 6 three times ($(-2)R_1+R_2 \rightarrow R_2$, $R_1+R_3 \rightarrow R_3$, $(-3)R_1+R_4 \rightarrow R_4$). Since column 2 of the last matrix has three 0's, we will find the determinant by calculating the dot product of column 2 by the corresponding column of cofactors. Since the 2-2, 3-2, and 3-4 entries of the last matrix are all 0, we need not calculate their corresponding cofactors. Therefore, we need to find C_{12}. We get

$$C_{12} = (-1)^{1+2} \det \begin{bmatrix} 1 & -11 & -14 \\ 5 & 15 & 11 \\ -7 & -12 & -19 \end{bmatrix}$$

$$= (-1)^{1+2} \det \begin{bmatrix} 1 & -11 & -14 \\ 0 & 70 & 81 \\ 0 & -89 & -117 \end{bmatrix}$$

$$= (-1)^{1+2}(-1)^{1+1} \det \begin{bmatrix} 70 & 81 \\ -89 & -117 \end{bmatrix}$$

$$= (-1)[70(-117) - 81(-89)] = (-1)(-981) = 981$$

Thus,

$$\det A = \det \begin{bmatrix} 3 & 5 & 7 & 8 \\ 7 & 10 & 3 & 2 \\ 2 & -5 & 8 & 3 \\ 2 & 15 & 9 & 5 \end{bmatrix} = 5 \cdot \det \begin{bmatrix} 3 & 1 & 7 & 8 \\ 7 & 2 & 3 & 2 \\ 2 & -1 & 8 & 3 \\ 2 & 3 & 9 & 5 \end{bmatrix}$$

$$= 5 \cdot \det \begin{bmatrix} 3 & 1 & 7 & 8 \\ 1 & 0 & -11 & -14 \\ 5 & 0 & 15 & 11 \\ -7 & 0 & -12 & -19 \end{bmatrix}$$

$$= 5(981) = 4905 \qquad \blacksquare$$

We now give another example to illustrate how the evaluation of a determinant can be done using the basic properties of determinants, as we did in the previous example, but writing the solution in a more compact way. We give the explanations on the right side using our earlier convention that $kR_i + R_j \to R_j$ means k times row i added to row j gives the new row j, and $kC_i + C_j \to C_j$ means k times column i added to column j gives the new column j.

EXAMPLE 9 Evaluate the determinant of the matrix

$$\begin{bmatrix} 2 & 3 & 4 & 6 \\ 5 & 2 & 7 & -3 \\ 4 & 8 & -4 & 12 \\ 8 & 4 & -2 & 3 \end{bmatrix}$$

Solution Using the properties of determinants, we get

$$\det \begin{bmatrix} 2 & 3 & 4 & 6 \\ 5 & 2 & 7 & -3 \\ 4 & 8 & -4 & 12 \\ 8 & 4 & -2 & 3 \end{bmatrix} \qquad \text{Factoring } -4 \text{ out of row 3}$$

$$= (-4)\det \begin{bmatrix} 2 & 3 & 4 & 6 \\ 5 & 2 & 7 & -3 \\ -1 & -2 & 1 & -3 \\ 8 & 4 & -2 & 3 \end{bmatrix}$$

$(-4)R_3 + R_1 \rightarrow R_1$
$(-7)R_3 + R_2 \rightarrow R_2$
$2R_3 + R_4 \rightarrow R_4$

$$= (-4)\det \begin{bmatrix} 6 & 11 & 0 & 18 \\ 12 & 16 & 0 & 18 \\ -1 & -2 & 1 & -3 \\ 6 & 0 & 0 & -3 \end{bmatrix}$$

Calculating the dot product of column 1 by the corresponding column of cofactors and using the fact that three entries in that column are 0's.

$$= (-4)(1)C_{33}$$

$$= (-4)(1)(-1)^{3+3} \det \begin{bmatrix} 6 & 11 & 18 \\ 12 & 16 & 18 \\ 6 & 0 & -3 \end{bmatrix}$$

Factoring 6 out of column 1 and 3 out of column 3.

$$= (-4)(6)(3)\det \begin{bmatrix} 1 & 11 & 6 \\ 2 & 16 & 6 \\ 1 & 0 & -1 \end{bmatrix}$$

$C_1 + C_3 \rightarrow C_3$

$$= (-72)\det \begin{bmatrix} 1 & 11 & 7 \\ 2 & 16 & 8 \\ 1 & 0 & 0 \end{bmatrix}$$

Calculating the dot product of row 3 by the corresponding row of cofactors.

$$= (-72)(1)C_{31}$$

$$= (-72)\det \begin{bmatrix} 11 & 7 \\ 16 & 8 \end{bmatrix}$$

$$= (-72)[11 \cdot 8 - 16 \cdot 7]$$

$$= (-72)(88 - 112)$$

$$= (-72)(-24)$$

$$= 1728$$

Do Exercise 15.

■

Cramer's Rule

We now expand Cramer's rule to solve linear systems of n equations in n variables.

Cramer's Rule

Suppose a linear system of n equations in n variables is written as a matrix equation as follows:

AX = B

where

$$\mathbf{A} = [a_{ij}]$$

$$X = \begin{bmatrix} x_1 \\ x_2 \\ \vdots \\ x_n \end{bmatrix} \quad \text{and} \quad B = \begin{bmatrix} b_1 \\ b_2 \\ \vdots \\ b_n \end{bmatrix}$$

For each i, let A_i be the matrix obtained from the coefficient matrix A by replacing the ith column by the n-dimensional column vector B. The linear system has a unique solution $(c_1, c_2, \ldots, c_n)$ if, and only if, $\det A$ is nonzero. In that case,

$$c_i = \frac{\det A_i}{\det A}, \text{ for each } i = 1, 2, \ldots, n$$

EXAMPLE 10 Use Cramer's rule to solve the system

$$\begin{cases} 2x + 3y - 2z + 2 = 0 \\ z + 3x = 0 \\ 2y + 5x + 1 = 0 \end{cases}$$

Solution We rearrange the terms of each equation so that all three variables are in the same order and eliminate any constant terms on the left sides of the equations.

$$\begin{cases} 2x + 3y - 2z = -2 \\ 3x \quad\quad + z = 0 \\ 5x + 2y \quad\quad = -1 \end{cases}$$

Thus, the coefficient matrix is

$$A = \begin{bmatrix} 2 & 3 & -2 \\ 3 & 0 & 1 \\ 5 & 2 & 0 \end{bmatrix}$$

The matrix A_x is obtained from A by replacing the first column (the x-coefficients) by the three-dimensional column vector B, where

$$B = \begin{bmatrix} -2 \\ 0 \\ -1 \end{bmatrix}$$

Thus,

$$A_x = \begin{bmatrix} -2 & 3 & -2 \\ 0 & 0 & 1 \\ -1 & 2 & 0 \end{bmatrix}$$

Similarly, we get

$$A_y = \begin{bmatrix} 2 & -2 & -2 \\ 3 & 0 & 1 \\ 5 & -1 & 0 \end{bmatrix}$$

and

$$A_z = \begin{bmatrix} 2 & 3 & -2 \\ 3 & 0 & 0 \\ 5 & 2 & -1 \end{bmatrix}$$

Thus,

$$\det A = \det \begin{bmatrix} 2 & 3 & -2 \\ 3 & 0 & 1 \\ 5 & 2 & 0 \end{bmatrix} = -1$$

$$\det A_x = \det \begin{bmatrix} -2 & 3 & -2 \\ 0 & 0 & 1 \\ -1 & 2 & 0 \end{bmatrix} = 1$$

$$\det A_y = \det \begin{bmatrix} 2 & -2 & -2 \\ 3 & 0 & 1 \\ 5 & -1 & 0 \end{bmatrix} = -2$$

$$\det A_z = \det \begin{bmatrix} 2 & 3 & -2 \\ 3 & 0 & 0 \\ 5 & 2 & -1 \end{bmatrix} = -3$$

It follows that,

$$x = \frac{\det A_x}{\det A} = \frac{1}{-1} = -1; \quad y = \frac{\det A_y}{\det A} = \frac{-2}{-1} = 2; \quad z = \frac{\det A_z}{\det A} = \frac{-3}{-1} = 3$$

The solution of the given linear system is the ordered triple $(-1, 2, 3)$.*
Do Exercise 31.

■

Equation of a Line

We conclude this section with an interesting device to find the equation of a straight line through two given points.

EXAMPLE 11   **a.** Show that an equation of a straight line passing through the points (a_1, b_1) and (a_2, b_2) is

$$\det \begin{bmatrix} x & y & 1 \\ a_1 & b_1 & 1 \\ a_2 & b_2 & 1 \end{bmatrix} = 0$$

---

*In general, Cramer's rule is not the most efficient method of solving linear systems, except possibly in the case of two linear equations in two variables. However, Cramer's rule is important theoretically. For example, it allows us to determine that a linear system of n equations in n variables has a unique solution whenever the coefficient matrix has a nonzero determinant. In particular, the trivial solution is the only solution of a homogeneous system of n linear equations in n variables whenever the coefficient matrix has a nonzero determinant.

b. Use this result to find an equation of the line passing through the points (7, 3) and (5, 2).

Solution **a.** If we replace x by a_1 and y by b_1 on the left side, rows 1 and 2 are identical. So, by Property 4, the determinant on the left is 0 and we conclude that (a_1, b_1) is a solution of the equation. Thus, the point (a_1, b_1) is on the graph of the equation. Similarly, we can show that the point (a_2, b_2) is also on the graph of the equation. If we expand the left side of the equation, we get

$$A + Bx + Cy = 0$$

where A, B, and C are constants obtained as follows:

$$A = \det \begin{bmatrix} a_1 & b_1 \\ a_2 & b_2 \end{bmatrix}, \ B = -\det \begin{bmatrix} 1 & b_1 \\ 1 & b_2 \end{bmatrix}, \ C = \det \begin{bmatrix} 1 & a_1 \\ 1 & a_2 \end{bmatrix}$$

Thus, we have verified that the given equation is linear, and that the two given points are on its graph.

b. An equation of the line through the points (7, 3) and (5, 2) is

$$\det \begin{bmatrix} x & y & 1 \\ 7 & 3 & 1 \\ 5 & 2 & 1 \end{bmatrix} = 0$$

Expanding the left side, we get

$$3x + 14 + 5y - 15 - 7y - 2x = 0$$

which simplifies to

$$x - 2y - 1 = 0$$

Do Exercise 57. ■

Exercise Set 3.5

In Exercises 1–16, evaluate the determinant of the given matrix.

1. $\begin{bmatrix} 2 & 3 \\ 5 & 4 \end{bmatrix}$

2. $\begin{bmatrix} 3 & 4 \\ -2 & 5 \end{bmatrix}$

3. $\begin{bmatrix} 5 & 7 \\ 4 & 6 \end{bmatrix}$

4. $\begin{bmatrix} 5 & -1 \\ 2 & 6 \end{bmatrix}$

5. $\begin{bmatrix} 5 & 6 & 3 \\ 4 & 5 & -1 \\ 1 & 0 & 0 \end{bmatrix}$

6. $\begin{bmatrix} 3 & 7 & 5 \\ 2 & 5 & 1 \\ 1 & 3 & 3 \end{bmatrix}$

7. $\begin{bmatrix} 4 & -1 & 3 \\ 1 & 5 & -1 \\ 1 & 3 & -2 \end{bmatrix}$

8. $\begin{bmatrix} 1 & 4 & 1 \\ 2 & -1 & -2 \\ 1 & 1 & -1 \end{bmatrix}$

9. $\begin{bmatrix} 1 & -2 & 1 \\ 2 & -1 & -1 \\ 1 & 1 & 4 \end{bmatrix}$

10. $\begin{bmatrix} -1 & -2 & -3 \\ 2 & 4 & 1 \\ -1 & 2 & 0 \end{bmatrix}$

11. $\begin{bmatrix} 4 & -2 & 0 \\ 1 & 2 & 2 \\ 1 & 2 & 3 \end{bmatrix}$

12. $\begin{bmatrix} 2 & 1 & 0 \\ 1 & 0 & -1 \\ 0 & 1 & -1 \end{bmatrix}$

13. $\begin{bmatrix} 2 & -1 & 2 \\ 4 & 1 & 5 \\ 11 & -8 & 10 \end{bmatrix}$

14. $\begin{bmatrix} 6 & 4 & -14 \\ 2 & 2 & -5 \\ 2 & 3 & -5 \end{bmatrix}$

15. $\begin{bmatrix} 7 & -8 & -5 & 6 \\ 3 & -12 & 0 & 3 \\ 3 & -11 & 3 & 4 \\ 4 & -6 & -1 & 4 \end{bmatrix}$

16. $\begin{bmatrix} 2 & 2 & -5 & 2 \\ 4 & 5 & 5 & 5 \\ 7 & 2 & -30 & 6 \\ 3 & -2 & -5 & 3 \end{bmatrix}$

In Exercises 17–20, solve the given equation for x and check your result.

17. $\det \begin{bmatrix} 1 & x & -2 \\ 2 & 3 & 2 \\ 4 & 9 & -2 \end{bmatrix} = 0$

18. $\det \begin{bmatrix} 6x & 3 & 2 \\ 2 & x & 2 \\ -2 & -1 & 1 \end{bmatrix} = 0$

19. $\det \begin{bmatrix} x & 2x+1 & x \\ x^2 & x+1 & 3 \\ 1 & 0 & 0 \end{bmatrix} = -3$

20. $\det \begin{bmatrix} 2x+1 & 2x & -17 \\ 2 & x-1 & -5 \\ -2x & -5 & 15 \end{bmatrix} = 1$

In Exercises 21–30, solve the given system using Cramer's rule.

21. $\begin{cases} 3x + 2y = -2 \\ 7x + 5y = 3 \end{cases}$

22. $\begin{cases} 3x - 4y = -2 \\ -7x + 9y = 3 \end{cases}$

23. $\begin{cases} 5x - 3y = 7 \\ 2x + 7y = 11 \end{cases}$

24. $\begin{cases} 5x + 2y = -4 \\ 3x - 11y = -39 \end{cases}$

25. $\begin{cases} x + 13y = 34 \\ 2x - 3y = -19 \end{cases}$

26. $\begin{cases} 5x + 2y = 3 \\ 10x + 4y = -2 \end{cases}$

27. $\begin{cases} -4x + 3y = 15 \\ 3x - 7y = -16 \end{cases}$

28. $\begin{cases} 3x - 13y = 6 \\ 2x + 4y = 4 \end{cases}$

29. $\begin{cases} -5x + 7y = 64 \\ 3x + 5y = 26 \end{cases}$

30. $\begin{cases} 3x - 17y = -97 \\ 2x + 15y = 67 \end{cases}$

Exercises 31–45: In each Exercise n ($31 \le n \le 45$), use Cramer's rule to solve the linear system of Exercise n − 30, Section 2.2.

Exercises 46–55: In each Exercise n ($46 \le n \le 55$), use Cramer's rule to solve the linear system of Exercise n − 18, Section 2.2.

56. Two mnemonic devices used to calculate the determinant of a 3 × 3 matrix are given on page 253. Apply one of the devices to the matrix of Exercise 16 and show that it does not yield the correct value.

In Exercises 57–62, write in determinant form an equation of the line passing through the two points whose coordinates are given. Then expand and simplify.

57. (2, 3), (5, 6)

58. (2, −3), (−3, −4)

59. (1, 7), (3, −6)

60. (5, 1), (7, 8)

61. (4, −2), (−5, −3)

62. (2, 4), (5, −2)

In Exercises 63–67, use the properties of determinants to find detB.

63. **A** is a 2 × 2 matrix, detA = 3 and **B** = 5**A**.
64. **A** is a 3 × 3 matrix, detA = 7 and **B** = 3**A**.
65. **A** is a 4 × 4 matrix, detA = −2 and **B** = −3**A**.
66. detA = −8 and **B** is obtained by interchanging rows 2 and 3 of **A**.
67. detA = −6 and **B** = **A**^{T}.

It can be shown that if the vertices of a triangle have coordinates (a_1, b_1), (a_2, b_2) and (a_3, b_3), then the area of the triangle is given by the absolute value of

$$\left(\frac{1}{2}\right)\det \begin{bmatrix} a_1 & b_1 & 1 \\ a_2 & b_2 & 1 \\ a_3 & b_3 & 1 \end{bmatrix}$$

In Exercises 68–71, use a determinant to find the area of the triangle whose vertices are the points with the given coordinates.

68. (1, 2), (2, −3), (−2, −3)

69. (3, 1), (5, −2), (−3, −2)

70. (5, 3), (−2, 4), (2, −3)

71. (3, 7), (3, −1), (−4, −6)

3.6 Inverse of a Square Matrix Using Adjoints

In Section 3.4, we used elementary row operations to find the inverse of a nonsingular square matrix. In this section, we describe a different way of finding inverses. This way is often more efficient for small matrices (2 × 2 or 3 × 3), but is not very effective for larger matrices.

Adjoint of a Square Matrix

In Example 2 of the preceding section we found the nine cofactors of the matrix

$$\mathbf{A} = \begin{bmatrix} 3 & 5 & -2 \\ 2 & 1 & 4 \\ 6 & 3 & 8 \end{bmatrix}$$

We have tabulated the results. In this table the column headed a_{ij} gives the entries of matrix $\mathbf{A}$, while the column headed C_{ij} gives the corresponding cofactors.

i	j	a_{ij}	C_{ij}
1	1	3	-4
1	2	5	8
1	3	-2	0
2	1	2	-46
2	2	1	36
2	3	4	21
3	1	6	22
3	2	3	-16
3	3	8	-7

We have already observed that if we calculate the dot product of any row of matrix $\mathbf{A}$ by the corresponding row of cofactors, we obtain the determinant of matrix $\mathbf{A}$.

$$[\, 3 \quad 5 \quad -2\,] \cdot [\, -4 \quad 8 \quad 0\,] = 28$$
$$[\, 2 \quad 1 \quad 4\,] \cdot [\, -46 \quad 36 \quad 21\,] = 28$$
$$[\, 6 \quad 3 \quad 8\,] \cdot [\, 22 \quad -16 \quad -7\,] = 28$$

Suppose now we calculate the dot product of any row by a row of cofactors corresponding to some different row. For example,

$$[\, a_{11} \quad a_{12} \quad a_{13}\,] \cdot [\, C_{21} \quad C_{22} \quad C_{23}\,]$$
$$= [\, 3 \quad 5 \quad -2\,] \cdot [\, -46 \quad 36 \quad 21\,]$$
$$= 3(-46) + 5(36) + (-2)21 = -138 + 180 + -42 = 0$$

You can verify that the following are also true.

$$[\, a_{11} \quad a_{12} \quad a_{13}\,] \cdot [\, C_{31} \quad C_{32} \quad C_{33}\,]$$
$$= [\, 3 \quad 5 \quad -2\,] \cdot [\, 22 \quad -16 \quad -7\,] = 0$$
$$[\, a_{21} \quad a_{22} \quad a_{23}\,] \cdot [\, C_{11} \quad C_{12} \quad C_{13}\,]$$
$$= [\, 2 \quad 1 \quad 4\,] \cdot [\, -4 \quad 8 \quad 0\,] = 0$$
$$[\, a_{21} \quad a_{22} \quad a_{23}\,] \cdot [\, C_{31} \quad C_{32} \quad C_{33}\,]$$
$$= [\, 2 \quad 1 \quad 4\,] \cdot [\, 22 \quad -16 \quad -7\,] = 0$$

$$[\ a_{31} \quad a_{32} \quad a_{33}\ ] \cdot [\ C_{11} \quad C_{12} \quad C_{13}\ ]$$
$$= [\ 6 \quad 3 \quad 8\ ] \cdot [\ -4 \quad 8 \quad 0\ ] = 0$$

$$[\ a_{31} \quad a_{32} \quad a_{33}\ ] \cdot [\ C_{21} \quad C_{22} \quad C_{23}\ ]$$
$$= [\ 6 \quad 3 \quad 8\ ] \cdot [\ -46 \quad 36 \quad 21\ ] = 0$$

Consider the matrix $\mathbf{C}$ whose entries are the cofactors of $\mathbf{A}$.

$$\mathbf{C} = \begin{bmatrix} C_{11} & C_{12} & C_{13} \\ C_{21} & C_{22} & C_{23} \\ C_{31} & C_{32} & C_{33} \end{bmatrix} = \begin{bmatrix} -4 & 8 & 0 \\ -46 & 36 & 21 \\ 22 & -16 & -7 \end{bmatrix}$$

Then,

$$\mathbf{C}^T = \begin{bmatrix} -4 & -46 & 22 \\ 8 & 36 & -16 \\ 0 & 21 & -7 \end{bmatrix}$$

The first row of matrix $\mathbf{C}$ is the first column of $\mathbf{C}^T$, the second row of matrix $\mathbf{C}$ is the second column of $\mathbf{C}^T$, and the third row of matrix $\mathbf{C}$ is the third column of $\mathbf{C}^T$. Recalling that in matrix multiplication the i-j entry of the product is the dot product of the ith row of the first factor by the jth column of the second factor, we see that the i-j entry of the product $\mathbf{A}\mathbf{C}^T$ is the dot product of the ith row of $\mathbf{A}$ by the jth column of $\mathbf{C}^T$. That is to say, it is the dot product of the ith row of $\mathbf{A}$ by the jth row of $\mathbf{C}$. It follows from the preceding results that the i-j entry of the product $\mathbf{A}\mathbf{C}^T$ is 28 when $i = j$ and 0 when $i \neq j$. Therefore,

$$\mathbf{A}\mathbf{C}^T = \begin{bmatrix} 28 & 0 & 0 \\ 0 & 28 & 0 \\ 0 & 0 & 28 \end{bmatrix} = 28\begin{bmatrix} 1 & 0 & 0 \\ 0 & 1 & 0 \\ 0 & 0 & 1 \end{bmatrix}$$

DEFINITION 3.17: The *adjoint* of an $n \times n$ matrix $\mathbf{A}$ is denoted adj$\mathbf{A}$ and is the transpose of the $n \times n$ matrix $\mathbf{C}$ whose i-j entry is the cofactor of the i-j entry of the matrix $\mathbf{A}$ for each $i = 1, 2, \ldots, n$ and each $j = 1, 2, \ldots, n$.

EXAMPLE 1   Find the adjoint of the matrix

$$\mathbf{A} = \begin{bmatrix} 1 & 3 & -5 \\ 6 & 1 & 2 \\ 4 & -3 & 7 \end{bmatrix}$$

Solution   The calculations are given in the table.

i	j	a_{ij}	**Calculations**	**Cofactor** C_{ij}
1	1	1	$(-1)^{1+1} \det \begin{bmatrix} 1 & 2 \\ -3 & 7 \end{bmatrix} = 7 + 6 =$	13
1	2	3	$(-1)^{1+2} \det \begin{bmatrix} 6 & 2 \\ 4 & 7 \end{bmatrix} = -(42 - 8) =$	-34
1	3	-5	$(-1)^{1+3} \det \begin{bmatrix} 6 & 1 \\ 4 & -3 \end{bmatrix} = -18 - 4 =$	-22
2	1	6	$(-1)^{2+1} \det \begin{bmatrix} 3 & -5 \\ -3 & 7 \end{bmatrix} = -(21 - 15) =$	-6
2	2	1	$(-1)^{2+2} \det \begin{bmatrix} 1 & -5 \\ 4 & 7 \end{bmatrix} = 7 + 20 =$	27
2	3	2	$(-1)^{2+3} \det \begin{bmatrix} 1 & 3 \\ 4 & -3 \end{bmatrix} = -(-3 - 12) =$	15
3	1	4	$(-1)^{3+1} \det \begin{bmatrix} 3 & -5 \\ 1 & 2 \end{bmatrix} = 6 + 5 =$	11
3	2	-3	$(-1)^{3+2} \det \begin{bmatrix} 1 & -5 \\ 6 & 2 \end{bmatrix} = -(2 + 30) =$	-32
3	3	7	$(-1)^{3+3} \det \begin{bmatrix} 1 & 3 \\ 6 & 1 \end{bmatrix} = 1 - 18 =$	-17

Thus, the cofactor matrix $\mathbf{C}$ is

$$\begin{bmatrix} C_{11} & C_{12} & C_{13} \\ C_{21} & C_{22} & C_{23} \\ C_{31} & C_{32} & C_{33} \end{bmatrix} = \begin{bmatrix} 13 & -34 & -22 \\ -6 & 27 & 15 \\ 11 & -32 & -17 \end{bmatrix}$$

By definition, $\text{adj}\mathbf{A} = \mathbf{C}^T$. Thus,

$$\text{adj}\mathbf{A} = \begin{bmatrix} 13 & -6 & 11 \\ -34 & 27 & -32 \\ -22 & 15 & -17 \end{bmatrix}$$

EXAMPLE 2   **a.** Calculate the product $\mathbf{A}(\text{adj}\mathbf{A})$ for the matrix $\mathbf{A}$ of Example 1.
b. Find $\det\mathbf{A}$ for the same matrix $\mathbf{A}$.

Solution   **a.** $\mathbf{A}\,\text{adj}\mathbf{A} = \begin{bmatrix} 1 & 3 & -5 \\ 6 & 1 & 2 \\ 4 & -3 & 7 \end{bmatrix} \begin{bmatrix} 13 & -6 & 11 \\ -34 & 27 & -32 \\ -22 & 15 & -17 \end{bmatrix}$

$$= \begin{bmatrix} 13 - 102 + 110 & -6 + 81 - 75 & 11 - 96 + 85 \\ 78 - 34 - 44 & -36 + 27 + 30 & 66 - 32 - 34 \\ 52 + 102 - 154 & -24 - 81 + 105 & 44 + 96 - 119 \end{bmatrix}$$

$$= \begin{bmatrix} 21 & 0 & 0 \\ 0 & 21 & 0 \\ 0 & 0 & 21 \end{bmatrix}$$

b. Note that 21 was obtained by multiplying each row of matrix **A** by the corresponding row of cofactors. Hence, $\det \mathbf{A} = 21$. ∎

Finding the Inverse Using Adjoints

We now give the following theorems.

THEOREM 3.1:*   If **A** is any $n \times n$ matrix, then

$$\mathbf{A}(\text{adj}\mathbf{A}) = (\text{adj}\mathbf{A})\mathbf{A} = (\det\mathbf{A})\mathbf{I}_n$$

THEOREM 3.2:   If **A** and **B** are $n \times n$ matrices, then

$$\det(\mathbf{AB}) = (\det\mathbf{A})(\det\mathbf{B})$$

THEOREM 3.3:*   An $n \times n$ matrix **A** is invertible if, and only if, $\det\mathbf{A} \neq 0$. In that case,

$$\mathbf{A}^{-1} = \frac{1}{\det\mathbf{A}} \, \text{adj}\mathbf{A}$$

EXAMPLE 3   Find $\mathbf{A}^{-1}$ if $\mathbf{A} = \begin{bmatrix} 5 & -8 \\ -4 & 7 \end{bmatrix}$.

Solution   We see that **A** is invertible since

$$\det\mathbf{A} = 5(7) - (-4)(-8) = 35 - 32 = 3$$

---

*See Exercises 25–28 for the proofs of Theorems 3.1 and 3.3. The proof of Theorem 3.2 is too technical at this level and may be found in more advanced books on matrix algebra.

Also, the cofactors are

$$C_{11} = (-1)^{1+1}(7) = 7$$
$$C_{12} = (-1)^{1+2}(-4) = 4$$
$$C_{21} = (-1)^{2+1}(-8) = 8$$
$$C_{22} = (-1)^{2+2}(5) = 5$$

Therefore, the cofactor matrix is

$$\begin{bmatrix} 7 & 4 \\ 8 & 5 \end{bmatrix}$$

and the adjoint is

$$\begin{bmatrix} 7 & 8 \\ 4 & 5 \end{bmatrix}$$

It follows that

$$\mathbf{A}^{-1} = \left(\frac{1}{3}\right)\begin{bmatrix} 7 & 8 \\ 4 & 5 \end{bmatrix} = \begin{bmatrix} \frac{7}{3} & \frac{8}{3} \\ \frac{4}{3} & \frac{5}{3} \end{bmatrix}$$

Do Exercise 9.   ■

EXAMPLE 4   Find $\mathbf{A}^{-1}$ for the matrix $\mathbf{A}$ of Example 1.

Solution   We found in Example 2 that $\det\mathbf{A} = 21$; hence, it is nonzero. It follows that the matrix $\mathbf{A}$ is invertible. We also found in Example 1 that

$$\text{adj}\mathbf{A} = \begin{bmatrix} 13 & -6 & 11 \\ -34 & 27 & -32 \\ -22 & 15 & -17 \end{bmatrix}$$

Thus,

$$\mathbf{A}^{-1} = \left(\frac{1}{21}\right)\begin{bmatrix} 13 & -6 & 11 \\ -34 & 27 & -32 \\ -22 & 15 & -17 \end{bmatrix}$$

Do Exercise 13.   ■

Procedure to Find the Inverse of an $n \times n$ Matrix A Using the Adjoint

Step 1.   Calculate $\det\mathbf{A}$ by finding the dot product of any row of $\mathbf{A}$ by the corresponding row of cofactors. Recall that a cofactor C_{ij} is the product $(-1)^{i+j}$ times the determinant of the matrix obtained from $\mathbf{A}$ by deleting row i and column j. If $\det\mathbf{A} = 0$, the matrix does not have an inverse. Otherwise we proceed to Step 2.

Step 2.   For each $i = 1, 2, \ldots, n$ and $j = 1, 2, \ldots, n$ find the cofactor C_{ij}, if not already found in Step 1.

Step 3.   Find the matrix adjA. Recall that this matrix is the transpose of the cofactor matrix which may be written from the results of Steps 1 and 2.

Step 4.   The inverse A^{-1} is obtained by dividing each entry of adjA found in Step 3 by detA found in Step 1.

Exercise Set 3.6

Exercises 1–24: Do Exercises 1–24 of Section 3.4 using the procedure summarized at the end of this section.

*25. Prove that if two rows of a square matrix A are identical, then detA = 0. (*Hint:* Suppose that row i and row j are identical. Let B be the matrix obtained from A by interchanging rows i and j. Then, A = B and detB = −detA.)

*26. Prove that if A is an $n \times n$ matrix, then the dot product of any row by the row of cofactors corresponding to a different row is 0. (*Hint:* Suppose $i \neq j$. We want to show that

$(a_{i1}\ a_{i2}\ .\ .\ .\ a_{in}) \cdot (C_{j1}\ C_{j2}\ .\ .\ .\ C_{jn}) = 0$. Obtain a matrix B from A by replacing row j by row i. Then matrix B has two identical rows and therefore its determinant is 0 by Exercise 25. Now find detB by calculating the dot product of row j of matrix B by the corresponding row of cofactors.)

*27. Prove Theorem 3.1 (*Hint:* Use the result of Exercise 26 and the definition of determinant.)

*28. Prove Theorem 3.3 (*Hint:* Use Theorems 3.1 and 3.2.)

3.7 Applications in Economics and Cryptography: Input-Output Analysis and Coded Messages

Input-Output Models: The Closed System

Matrix multiplication and the multiplicative inverse of a matrix provide tools to solve some important problems in economics. We begin with a discussion of *input-output models*, which were first introduced in the late 1940s by the American economist Wassily Leontief. (In 1973, Leontief received the Nobel Prize in Economics.) The purpose of the models is to allow economists to predict the production levels needed to meet anticipated demands.

Leontief made the assumption that the model representing the production of all industries in an economy is a linear system. The simplest model is the *closed* system where the output of each industry serves only as input to all other industries within the system and no other inputs are needed. An *open* system consists of a productive segment made up of a certain number of industries, and a nonproductive (consumer) segment creating a demand for commodities over and above their uses for production. It is assumed that whatever is produced is consumed. We begin with a simple example of a closed system.

EXAMPLE 1   Suppose that in a simple society there are three types of individuals: the carpenters who build all housing, the farmers who grow all the food, and the tailors who make all the clothing. Each type of individual will consume goods produced not only by the other types but also by his own type. Carpenters must have shelters, farmers must eat, and tailors must wear clothes. Assume that all commodities produced are consumed and that the fraction of each commodity consumed by each is given in the following table.

	Housing	Food	Clothing
Carpenters	20%	30%	30%
Farmers	45%	40%	25%
Tailors	35%	30%	45%

Is it possible, under these assumptions, to have *interior equilibrium*? That is, is it possible for consumption and production to be equal?

Solution   Let us measure consumption and production in dollars and suppose that the carpenters, farmers, and tailors have total incomes of x, y, and z dollars per year, respectively. From the table, it is clear that the carpenters spend $\$.20x$ on housing, $\$.30y$ on food, and $\$.30z$ on clothing. Since we assume that they spend exactly the amount they earn, we have the equation

$$.2x + .3y + .3z = x$$

Similarly, using the fact that the farmers spend $\$.45x$, $\$.40y$, and $\$.25z$ on housing, food, and clothing, respectively, we get a second equation

$$.45x + .4y + .25z = y$$

The third equation is obtained similarly, giving the system

$$\begin{cases} .2x + .3y + .3z = x \\ .45x + .4y + .25z = y \\ .35x + .3y + .45z = z \end{cases}$$

This system is equivalent to

$$\begin{cases} -.8x + .3y + .3z = 0 \\ .45x - .6y + .25z = 0 \\ .35x + .3y - .55z = 0 \end{cases}$$

We see that the ordered triple $(0, 0, 0)$ is a solution of this system. However, the determinant of the coefficient matrix is also zero. Hence, this solution is not unique. We seek some solution (a, b, c) where a, b, and c are positive. We use the method of row reduction discussed in Section 3.3.

$$\begin{bmatrix} -.8 & .3 & .3 & | & 0 \\ .45 & -.6 & .25 & | & 0 \\ .35 & .3 & -.55 & | & 0 \end{bmatrix} \quad R_1 + R_2 \rightarrow R_2$$

$$\begin{bmatrix} -.8 & .3 & .3 & | & 0 \\ -.35 & -.3 & .55 & | & 0 \\ .35 & .3 & -.55 & | & 0 \end{bmatrix} \quad \begin{aligned} R_2 + R_3 &\rightarrow R_3 \\ R_2 + R_1 &\rightarrow R_1 \end{aligned}$$

$$\begin{bmatrix} -1.15 & 0 & .85 & | & 0 \\ -.35 & -.3 & .55 & | & 0 \\ 0 & 0 & 0 & | & 0 \end{bmatrix} \quad \left(\frac{-1}{1.15}\right)R_1 \rightarrow R_1$$

$$\begin{bmatrix} 1 & 0 & \frac{-17}{23} & | & 0 \\ -.35 & -.3 & .55 & | & 0 \\ 0 & 0 & 0 & | & 0 \end{bmatrix} \quad .35R_1 + R_2 \rightarrow R_2$$

$$\begin{bmatrix} 1 & 0 & \frac{-17}{23} & | & 0 \\ 0 & -.3 & \frac{67}{230} & | & 0 \\ 0 & 0 & 0 & | & 0 \end{bmatrix} \quad \left(\frac{-1}{.3}\right)R_2 \rightarrow R_2$$

$$\begin{bmatrix} 1 & 0 & \frac{-17}{23} & | & 0 \\ 0 & 1 & \frac{-67}{69} & | & 0 \\ 0 & 0 & 0 & | & 0 \end{bmatrix}$$

The last augmented matrix represents the system

$$\begin{cases} x = \left(\frac{17}{23}\right)z \\ y = \left(\frac{67}{69}\right)z \end{cases}$$

We have an infinite number of solutions. If we let $z = 69c$, where c is a positive constant, then $x = \left(\frac{17}{23}\right)69c = 51c$ and $y = \left(\frac{67}{69}\right)69c = 67c$. Thus, the set of solutions is the set of ordered triples $(51c, 67c, 69c)$, where c is an arbitrary positive number. Therefore, the carpenters', farmers', and tailors' incomes are in the ratios 51:67:69.

∎

Input-Output Models: The Open System

We now discuss the more interesting open Leontief system.

EXAMPLE 2   Assume that we have a simple economy with only two industries I and II. To produce 1 unit of its own product, industry I needs inputs of .52 units of its own product and .44 units of the product of industry II. Industry II needs inputs of .25 units of the product of industry I and .25 units of its own product to produce 1 unit of its own product. The commodities that are not used internally by the industries themselves are called *final demand*. These goods are produced for the consumer. Find the production necessary to provide for final demands of

a. 32 units of the product of industry I and 54 units of the product of industry II.
b. 60 units of the product of industry I and 48 units of the product of industry II.

Solution   **a.** The inputs are x and y units of the products of industries I and II, respectively. Of the x units produced by industry I, $.52x$ units are used by industry I, $.25y$ units are used by industry II, and 32 units are provided for final demand. These three quantities added must give the total output of industry I, which is x units. Thus,

$$.52x + .25y + 32 = x$$

Similarly, of the y units produced by industry II, $.44x$ units are used by industry I, $.25y$ units are used by industry II, and 54 units are provided for final demand. These three quantities added must give the total output of industry II, which is y units. Therefore,

$$.44x + .25y + 54 = y$$

Writing these two equations together gives the linear system

$$\begin{cases} .52x + .25y + 32 = x \\ .44x + .25y + 54 = y \end{cases}$$

Although this system can easily be solved, we proceed as follows to illustrate the general model that we wish to discuss. We first write the system as a matrix equation

$$\mathbf{BX} + \mathbf{D} = \mathbf{X}$$

where

$$\mathbf{B} = \begin{bmatrix} .52 & .25 \\ .44 & .25 \end{bmatrix}$$

$$\mathbf{X} = \begin{bmatrix} x \\ y \end{bmatrix} \quad \text{and} \quad \mathbf{D} = \begin{bmatrix} 32 \\ 54 \end{bmatrix}$$

We get the following sequence of equivalent matrix equations:

$$\mathbf{BX} + \mathbf{D} = \mathbf{X}$$
$$\mathbf{X} - \mathbf{BX} = \mathbf{D}$$
$$(\mathbf{I}_2 - \mathbf{B})\mathbf{X} = \mathbf{D}$$
$$(\mathbf{I}_2 - \mathbf{B})^{-1}(\mathbf{I}_2 - \mathbf{B})\mathbf{X} = (\mathbf{I}_2 - \mathbf{B})^{-1}\mathbf{D}$$
$$\mathbf{I}_2\mathbf{X} = (\mathbf{I}_2 - \mathbf{B})^{-1}\mathbf{D}$$
$$\mathbf{X} = (\mathbf{I}_2 - \mathbf{B})^{-1}\mathbf{D} \qquad (*)$$

Now,

$$\mathbf{I}_2 - \mathbf{B} = \begin{bmatrix} 1 & 0 \\ 0 & 1 \end{bmatrix} - \begin{bmatrix} .52 & .25 \\ .44 & .25 \end{bmatrix} = \begin{bmatrix} .48 & -.25 \\ -.44 & .75 \end{bmatrix}$$

We easily find that

$$(\mathbf{I}_2 - \mathbf{B})^{-1} = \begin{bmatrix} 3 & 1 \\ 1.76 & 1.92 \end{bmatrix}$$

Thus,

$$\mathbf{X} = \begin{bmatrix} 3 & 1 \\ 1.76 & 1.92 \end{bmatrix}\begin{bmatrix} 32 \\ 54 \end{bmatrix}$$

$$= \begin{bmatrix} 3(32)+1(54) \\ 1.76(32)+1.92(54) \end{bmatrix} = \begin{bmatrix} 150 \\ 160 \end{bmatrix}$$

We conclude that industry I should produce 150 units while industry II should produce 160 units.

b. We need only use the demand matrix $\mathbf{D} = \begin{bmatrix} 60 \\ 48 \end{bmatrix}$ in equation (*) to get

$$\mathbf{X} = \begin{bmatrix} 3 & 1 \\ 1.76 & 1.92 \end{bmatrix} \begin{bmatrix} 60 \\ 48 \end{bmatrix}$$

$$= \begin{bmatrix} 3(6)+1(48) \\ 1.76(60)+1.92(48) \end{bmatrix} = \begin{bmatrix} 228 \\ 197.76 \end{bmatrix}$$

We conclude that industry I should produce 228 units and industry II should produce 197.76 units.

Do Exercise 1.   ∎

Input-Output Models Applied

We now generalize the ideas presented in this last example. Assume that we have an economy consisting of n industries, each one of which produces a single commodity. Assume further that these are interdependent; that is, each uses some of the others' output in order to operate. In addition, each industry produces some goods for the consumer (nonproductive) segment of the economy that are called *final demands*. We assume that production of one unit of the jth commodity (output) requires a quantity b_{ij} of the ith commodity (input). These provide the entries of the $n \times n$ matrix

$$\mathbf{B} = \begin{bmatrix} b_{11} & b_{12} & \cdots & b_{1n} \\ b_{21} & b_{22} & \cdots & b_{2n} \\ \vdots & & & \\ b_{n1} & b_{n2} & \cdots & b_{nn} \end{bmatrix}$$

where the entries of the jth column give the input requirements for production of one unit of the jth commodity. That is to say, if the unit is the dollar, then b_{ij} gives the dollar value that industry j must purchase from industry i to produce one dollar worth of its own commodity. The matrix $\mathbf{B}$ is often called the *technological coefficient matrix* or the *input-output matrix*. (The matrix

$$\begin{bmatrix} .52 & .25 \\ .44 & .25 \end{bmatrix}$$

in the previous example is a technological coefficient matrix).*

Now, for $i = 1, 2, \ldots, n$, let x_i dollars be the total output of the ith industry and let d_i dollars be the demand for the product of the ith industry by the nonproductive segment (final demand). The x_i dollars output of the ith industry is divided up as the n inputs used by the n industries and the final demand. The first industry requires $b_{i1}x_1$

---

*Each entry in a technological coefficient matrix is between 0 and 1, and the sum of the entries in each column is never greater than 1. In fact, in the preceding example, the sum of the entries in columns 1 and 2 were .96 and .5, respectively. The "missing" .04 and .5 in the two columns is attributed to external sources of supply. We will illustrate this in Example 3 where primary inputs will be involved.

dollars of the output of the ith industry since it requires b_{i1} dollar of that industry for each dollar it produces and its total output is x_1. Similarly, the second industry requires $b_{i2}x_2$ dollars of the output of the ith industry, and so on. Since the interindustry demand for the product of the ith industry added to the final (consumer) demand for that product must equal the total output of that industry, we have

$$\underbrace{b_{i1}x_1 + b_{i2}x_2 + b_{i3}x_3 + \cdots + b_{in}x_n +}_{\substack{\text{Output of the } i\text{th industry used by} \\ \text{the } n \text{ industries.}}} \underbrace{d_i}_{\text{Final demand.}} = x_i$$

The foregoing is true for $i = 1, 2, 3, \ldots, n$. Thus, we have the linear system

$$\begin{cases} b_{11}x_1 + b_{12}x_2 + b_{13}x_3 + \cdots + b_{1n}x_n + d_1 = x_1 \\ b_{21}x_1 + b_{22}x_2 + b_{23}x_3 + \cdots + b_{2n}x_n + d_2 = x_2 \\ \quad \vdots \\ b_{n1}x_1 + b_{n2}x_2 + b_{n3}x_3 + \cdots + b_{nn}x_n + d_n = x_n \end{cases}$$

Recalling that

$$\mathbf{B} = \begin{bmatrix} b_{11} & b_{12} & \cdots & b_{1n} \\ b_{21} & b_{22} & \cdots & b_{2n} \\ \vdots & & & \\ b_{n1} & b_{n2} & \cdots & b_{nn} \end{bmatrix} \quad \text{and letting } \mathbf{X} = \begin{bmatrix} x_1 \\ x_2 \\ \vdots \\ x_n \end{bmatrix}, \mathbf{D} = \begin{bmatrix} d_1 \\ d_2 \\ \vdots \\ d_n \end{bmatrix}$$

we may write the system as a single matrix equation

$$\mathbf{BX} + \mathbf{D} = \mathbf{X}$$

where $\mathbf{B}$ is the technological coefficient matrix, $\mathbf{X}$ is the *total output vector*, and $\mathbf{D}$ is the *demand vector* of the consumer segment (final demand).

We can solve for the vector $\mathbf{X}$ using the methods of matrix algebra as follows:

$$\mathbf{D} = \mathbf{X} - \mathbf{BX}$$
$$= (\mathbf{I}_n - \mathbf{B})\mathbf{X}$$

Thus, if the matrix $\mathbf{I}_n - \mathbf{B}$ is invertible, we get

$$\mathbf{X} = (\mathbf{I}_n - \mathbf{B})^{-1}\mathbf{D} \quad *$$

The following example will illustrate a situation in which we are given the industry inputs and the final demands in tabular form. In this example, the industries will be using *primary inputs* which include such inputs as labor, raw material, and land. The primary inputs used by each industry are given in the last row of the table.

---

*In more advanced books, it is proved that if all entries of the matrix $\mathbf{B}$ are nonnegative and if for each $j = 1, 2, \ldots, n$ the sum of all the entries in the jth column is less than 1, then the matrix $\mathbf{I}_n - \mathbf{B}$ is invertible and all the entries of its inverse are nonnegative.

EXAMPLE 3   Suppose we have a simple economy consisting of three industries: agriculture, housing, and manufacturing. The interaction between these industries is given in the following table, where the units are in millions of dollars per year.

	Agriculture	Housing	Manufacturing	Final demands	Total output
Agriculture	3000	4050	2840	2110	12,000
Housing	3600	2700	5680	1520	13,500
Manufacturing	1800	4050	2840	5510	14,200
Primary inputs	3600	2700	2840		

Assuming that the technological coefficient matrix remains constant over the periods of time under consideration, find the outputs needed from each industry if the final demands change to the following, where the units are in millions of dollars per year.

a. 774 for agriculture, 1290 for housing, and 1032 for manufacturing.
b. 516 for agriculture, 774 for housing, and 1548 for manufacturing.
c. 1806 for agriculture, 2322 for housing, and 2838 for manufacturing.

Solution   **a.** We must first find the technological coefficient matrix. If we add the entries in each of the first three columns of the table we obtain 12,000, 13,500, and 14,200, which are the total outputs of agriculture, housing, and manufacturing, respectively. This agrees with the fact that for each industry, the total value of the output is equal to the sum of the values of all inputs, and all outputs are used by either the producing segment or the nonproducing segment (final demand) of the economy.

Agriculture uses 3000, 3600, and 1800 millions of dollars per year of agriculture, housing, and manufacturing, respectively, to produce 12,000 millions of dollars per year. Therefore, to produce 1 dollar per year, it will use .25, .3, and .15 dollars per year of agriculture, housing, and manufacturing, respectively, since $\frac{3000}{12,000} = .25$, $\frac{3600}{12,000} = .3$ and $\frac{1800}{12,000} = .15$.

Similarly, we find that to produce 1 dollar per year, housing uses .3, .2, and .3 dollars of agriculture, housing, and manufacturing, respectively, since $\frac{4050}{13,500} = .3$, $\frac{2700}{13,500} = .2$, and $\frac{4050}{13,500} = .3$.

Finally, we find that to produce 1 dollar per year, manufacturing uses .2, .4, and .2 dollars of agriculture, housing, and manufacturing, respectively, since $\frac{2840}{14,200} = .2$, $\frac{5680}{14,200} = .4$, and $\frac{2840}{14,200} = .2$.

Hence, the technological coefficient matrix is

$$\mathbf{B} = \begin{bmatrix} .25 & .3 & .2 \\ .3 & .2 & .4 \\ .15 & .3 & .2 \end{bmatrix}$$

As in the preceding example, the output vector is given by

$$\mathbf{X} = (\mathbf{I}_3 - \mathbf{B})^{-1}\mathbf{D}$$

where

$$D = \begin{bmatrix} 774 \\ 1290 \\ 1032 \end{bmatrix}$$

$$I_3 - B = \begin{bmatrix} 1 & 0 & 0 \\ 0 & 1 & 0 \\ 0 & 0 & 1 \end{bmatrix} - \begin{bmatrix} .25 & .3 & .2 \\ .3 & .2 & .4 \\ .15 & .3 & .2 \end{bmatrix}$$

$$= \begin{bmatrix} .75 & -.3 & -.2 \\ -.3 & .8 & -.4 \\ -.15 & -.3 & .8 \end{bmatrix}$$

Since each entry of matrix B is nonnegative and the sum of the three elements of each column of that matrix is less than 1, we know that $I_3 - B$ is invertible and that the entries of its inverse are nonnegative. We find the nine cofactors as follows:

i	j	C_{ij}
1	1	$(-1)^{1+1}[.8(.8) - (-.3)(-.4)] = .52$
1	2	$(-1)^{1+2}[(-.3)(.8) - (-.15)(-.4)] = .3$
1	3	$(-1)^{1+3}[(-.3)(-.3) - (-.15)(.8)] = .21$
2	1	$(-1)^{2+1}[(-.3)(.8) - (-.3)(-.2)] = .3$
2	2	$(-1)^{2+2}[.75(.8) - (-.15)(-.2)] = .57$
2	3	$(-1)^{2+3}[.75(-.3) - (-.15)(-.3)] = .27$
3	1	$(-1)^{3+1}[(-.3)(-.4) - (.8)(-.2)] = .28$
3	2	$(-1)^{3+2}[.75(-.4) - (-.3)(-.2)] = .36$
3	3	$(-1)^{3+3}[.75(.8) - (-.3)(-.3)] = .51$

To find the determinant of $I_3 - B$, we calculate the dot product of its first row by the corresponding row of cofactors. We get

$$(.75 \quad -.3 \quad -.2) \cdot (.52 \quad .3 \quad .21) =$$
$$.75(.52) + (-.3)(.3) + (-.2)(.21) = 0.258$$

Therefore,

$$\det(I_3 - B) = 0.258$$

Also,

$$\text{adj}(I_3 - B) = \begin{bmatrix} .52 & .3 & .21 \\ .3 & .57 & .27 \\ .28 & .36 & .51 \end{bmatrix}^T$$

$$= \begin{bmatrix} .52 & .3 & .28 \\ .3 & .57 & .36 \\ .21 & .27 & .51 \end{bmatrix}$$

Thus,

$$(\mathbf{I}_3 - \mathbf{B})^{-1} = (1/0.258) \begin{bmatrix} .52 & .3 & .28 \\ .3 & .57 & .36 \\ .21 & .27 & .51 \end{bmatrix}$$

$$= \begin{bmatrix} \frac{520}{258} & \frac{300}{258} & \frac{280}{258} \\ \frac{300}{258} & \frac{570}{258} & \frac{360}{258} \\ \frac{210}{258} & \frac{270}{258} & \frac{510}{258} \end{bmatrix}$$

It follows that

$$\mathbf{X} = \begin{bmatrix} \frac{520}{258} & \frac{300}{258} & \frac{280}{258} \\ \frac{300}{258} & \frac{570}{258} & \frac{360}{258} \\ \frac{210}{258} & \frac{270}{258} & \frac{510}{258} \end{bmatrix} \cdot \begin{bmatrix} 774 \\ 1290 \\ 1032 \end{bmatrix}$$

$$= \begin{bmatrix} 4180 \\ 5190 \\ 4020 \end{bmatrix}$$

Agriculture, housing, and manufacturing must produce 4180, 5190, and 4020 millions of dollars per year, respectively.

b. When the final demand vector is

$$\mathbf{D} = \begin{bmatrix} 516 \\ 774 \\ 1548 \end{bmatrix}$$

the output vector is obtained as follows:

$$\mathbf{X} = \begin{bmatrix} \frac{520}{258} & \frac{300}{258} & \frac{280}{258} \\ \frac{300}{258} & \frac{570}{258} & \frac{360}{258} \\ \frac{210}{258} & \frac{270}{258} & \frac{510}{258} \end{bmatrix} \cdot \begin{bmatrix} 516 \\ 774 \\ 1548 \end{bmatrix}$$

$$= \begin{bmatrix} 3620 \\ 4470 \\ 4290 \end{bmatrix}$$

Hence, the production of agriculture, housing, and manufacturing must be 3620, 4470, and 4290 millions of dollars per year, respectively.

c. Finally, when the final demand vector is

$$\mathbf{D} = \begin{bmatrix} 1806 \\ 2322 \\ 2838 \end{bmatrix}$$

the output vector is obtained as follows:

$$
X = \begin{bmatrix} \frac{520}{258} & \frac{300}{258} & \frac{280}{258} \\ \frac{300}{258} & \frac{570}{258} & \frac{360}{258} \\ \frac{210}{258} & \frac{270}{258} & \frac{510}{258} \end{bmatrix} \cdot \begin{bmatrix} 1806 \\ 2322 \\ 2838 \end{bmatrix}
$$

$$
= \begin{bmatrix} 9420 \\ 11{,}190 \\ 9510 \end{bmatrix}
$$

We conclude that the production of agriculture, housing, and manufacturing must be 9420, 11,190, and 9510 millions of dollars per year, respectively.

Do Exercise 17.   ■

If an economist wishes to calculate the output vectors X for a number of different final demand vectors D, then it is more convenient to first calculate $(I_n - B)^{-1}$ and to use the formula $X = (I_n - B)^{-1}D$ than it is to use the row reduction method to solve for X for each D.

Encoding and Decoding (Optional)

Another interesting application of the inverse of a matrix is encoding and decoding secret messages. The art of putting messages into code or cipher is called *cryptography* and dates back to the ancient Greeks. Since World War II, matrices and their inverses have been used effectively to encode and decode messages. The following example illustrates the technique.

EXAMPLE 4   Use the matrix A to encode the message "Mathematics can be fun!" where

$$
A = \begin{bmatrix} 5 & 2 & -6 \\ 7 & -3 & 4 \\ 3 & 5 & -12 \end{bmatrix}
$$

Solution   We first assign a number to each letter of the message. Let $a = 1$, $b = 2$, $c = 3$, . . . , $z = 26$. Let the number 27 correspond to a blank space and the number 28 correspond to the exclamation mark. Thus,

M A T H E M A T I C S – C A N – B E – F U N !
13 1 20 8 5 13 1 20 9 3 19 27 3 1 14 27 2 5 27 6 21 14 28

Since we are using a 3×3 matrix to encode the message, we separate the numbers into groups of three and write these as column vectors

$$
\begin{bmatrix} 13 \\ 1 \\ 20 \end{bmatrix} \begin{bmatrix} 8 \\ 5 \\ 13 \end{bmatrix} \begin{bmatrix} 1 \\ 20 \\ 9 \end{bmatrix} \begin{bmatrix} 3 \\ 19 \\ 27 \end{bmatrix} \begin{bmatrix} 3 \\ 1 \\ 14 \end{bmatrix} \begin{bmatrix} 27 \\ 2 \\ 5 \end{bmatrix} \begin{bmatrix} 27 \\ 6 \\ 21 \end{bmatrix} \begin{bmatrix} 14 \\ 28 \\ 27 \end{bmatrix}
$$

Note that we entered a 27 as the third entry of the last vector since there are 23 letters, which is not a multiple of 3. Because 27 corresponds to a blank space, the message is not altered by this entry. We now multiply each of these vectors, on the left, by the given matrix **A**. We get

$$\begin{bmatrix} 5 & 2 & -6 \\ 7 & -3 & 4 \\ 3 & 5 & -12 \end{bmatrix} \begin{bmatrix} 13 \\ 1 \\ 20 \end{bmatrix} = \begin{bmatrix} -53 \\ 168 \\ -196 \end{bmatrix}$$

$$\begin{bmatrix} 5 & 2 & -6 \\ 7 & -3 & 4 \\ 3 & 5 & -12 \end{bmatrix} \begin{bmatrix} 8 \\ 5 \\ 13 \end{bmatrix} = \begin{bmatrix} -28 \\ 93 \\ -107 \end{bmatrix}$$

$$\begin{bmatrix} 5 & 2 & -6 \\ 7 & -3 & 4 \\ 3 & 5 & -12 \end{bmatrix} \begin{bmatrix} 1 \\ 20 \\ 9 \end{bmatrix} = \begin{bmatrix} -9 \\ -17 \\ -5 \end{bmatrix}$$

$$\begin{bmatrix} 5 & 2 & -6 \\ 7 & -3 & 4 \\ 3 & 5 & -12 \end{bmatrix} \begin{bmatrix} 3 \\ 19 \\ 27 \end{bmatrix} = \begin{bmatrix} -109 \\ 72 \\ -220 \end{bmatrix}$$

$$\begin{bmatrix} 5 & 2 & -6 \\ 7 & -3 & 4 \\ 3 & 5 & -12 \end{bmatrix} \begin{bmatrix} 3 \\ 1 \\ 14 \end{bmatrix} = \begin{bmatrix} -67 \\ 74 \\ -154 \end{bmatrix}$$

$$\begin{bmatrix} 5 & 2 & -6 \\ 7 & -3 & 4 \\ 3 & 5 & -12 \end{bmatrix} \begin{bmatrix} 27 \\ 2 \\ 5 \end{bmatrix} = \begin{bmatrix} 109 \\ 203 \\ 31 \end{bmatrix}$$

$$\begin{bmatrix} 5 & 2 & -6 \\ 7 & -3 & 4 \\ 3 & 5 & -12 \end{bmatrix} \begin{bmatrix} 27 \\ 6 \\ 21 \end{bmatrix} = \begin{bmatrix} 21 \\ 255 \\ -141 \end{bmatrix}$$

$$\begin{bmatrix} 5 & 2 & -6 \\ 7 & -3 & 4 \\ 3 & 5 & -12 \end{bmatrix} \begin{bmatrix} 14 \\ 28 \\ 27 \end{bmatrix} = \begin{bmatrix} -36 \\ 122 \\ -142 \end{bmatrix}$$

The message is transmitted as "−53, 168, −196, −28, 93, −107, −9, −17, −5, −109, 72, −220, −67, 74, −154, 109, 203, 31, 21, 255, −141, −36, 122, −142." The person who receives the message knows the matrix **A** that was used to encode the message and proceeds to separate the numbers into groups of three, writes these groups as column vectors, and multiplies each of these vectors—on the left—by $\mathbf{A}^{-1}$, thus getting back the original set of vectors. Then the message is deciphered by replacing each 1 by "a," each 2 by "b," and so on.
Do Exercise 25. ■

EXAMPLE 5   The matrix of the preceding example was used to encode the message: "−90, 141, −235, 73, −42, 138, 109, 183, 43, −137, 114, −290, −128, 125, −283." Decode the message.

Solution   Since the matrix is a 3×3 matrix, we separate these numbers into groups of three and write these groups as column vectors as follows:

$$\begin{bmatrix} -90 \\ 141 \\ -235 \end{bmatrix} \quad \begin{bmatrix} 73 \\ -42 \\ 138 \end{bmatrix} \quad \begin{bmatrix} 109 \\ 183 \\ 43 \end{bmatrix} \quad \begin{bmatrix} -137 \\ 114 \\ -290 \end{bmatrix} \quad \begin{bmatrix} -128 \\ 125 \\ -283 \end{bmatrix}$$

The inverse of the matrix that was used to encode the message was found in Example 5, Section 3.4. It is

$$\mathbf{A}^{-1} = \left(\frac{1}{8}\right) \begin{bmatrix} 16 & -6 & -10 \\ 96 & -42 & -62 \\ 44 & -19 & -29 \end{bmatrix}$$

We now multiply each of the vectors we obtained by this inverse. We get

$$\left(\frac{1}{8}\right) \begin{bmatrix} 16 & -6 & -10 \\ 96 & -42 & -62 \\ 44 & -19 & -29 \end{bmatrix} \begin{bmatrix} -90 \\ 141 \\ -235 \end{bmatrix} = \begin{bmatrix} 8 \\ 1 \\ 22 \end{bmatrix} = \begin{bmatrix} H \\ A \\ V \end{bmatrix}$$

$$\left(\frac{1}{8}\right) \begin{bmatrix} 16 & -6 & -10 \\ 96 & -42 & -62 \\ 44 & -19 & -29 \end{bmatrix} \begin{bmatrix} 73 \\ -42 \\ 138 \end{bmatrix} = \begin{bmatrix} 5 \\ 27 \\ 1 \end{bmatrix} = \begin{bmatrix} E \\ [] \\ A \end{bmatrix}$$

$$\left(\frac{1}{8}\right) \begin{bmatrix} 16 & -6 & -10 \\ 96 & -42 & -62 \\ 44 & -19 & -29 \end{bmatrix} \begin{bmatrix} 109 \\ 183 \\ 43 \end{bmatrix} = \begin{bmatrix} 27 \\ 14 \\ 9 \end{bmatrix} = \begin{bmatrix} [] \\ N \\ I \end{bmatrix}$$

$$\left(\frac{1}{8}\right) \begin{bmatrix} 16 & -6 & -10 \\ 96 & -42 & -62 \\ 44 & -19 & -29 \end{bmatrix} \begin{bmatrix} -137 \\ 114 \\ -290 \end{bmatrix} = \begin{bmatrix} 3 \\ 5 \\ 27 \end{bmatrix} = \begin{bmatrix} C \\ E \\ [] \end{bmatrix}$$

$$\left(\frac{1}{8}\right) \begin{bmatrix} 16 & -6 & -10 \\ 96 & -42 & -62 \\ 44 & -19 & -29 \end{bmatrix} \begin{bmatrix} -128 \\ 125 \\ -283 \end{bmatrix} = \begin{bmatrix} 4 \\ 1 \\ 25 \end{bmatrix} = \begin{bmatrix} D \\ A \\ Y \end{bmatrix}$$

Thus, the message is HAVE A NICE DAY.
Do Exercise 21.   ■

In spite of the fact that this type of encoding and decoding is simple to do, it is nevertheless difficult to break. In practice, long messages are divided into large groups which are written as n-dimensional column vectors where n may be as large as 30. Then each vector is multiplied, on the left, by an $n \times n$ invertible matrix $\mathbf{A}$, to encode the message. The message can be decoded by writing it as a group of n-dimensional column vectors and multiplying each of these, on the left, by $\mathbf{A}^{-1}$. Finding the inverse of a large matrix requires a huge amount of time if done longhand. However, government agencies involved in encoding and decoding have access to computers, which greatly facilitates this process.

Exercise Set 3.7

In Exercises 1–10, a technological coefficient matrix for a simple economy is given. In each case, find the necessary output by each industry to meet the projected final demands.

1. $\begin{bmatrix} .25 & .50 \\ .40 & .40 \end{bmatrix}$

   a. 240 units from industry I and 360 units from industry II.
   b. 320 units from industry I and 450 units from industry II.
   c. 380 units from industry I and 290 units from industry II.

2. $\begin{bmatrix} .40 & .68 \\ .25 & .30 \end{bmatrix}$

   a. 550 units from industry I and 460 units from industry II.
   b. 480 units from industry I and 560 units from industry II.
   c. 395 units from industry I and 470 units from industry II.

3. $\begin{bmatrix} .60 & .40 \\ .35 & .40 \end{bmatrix}$

   a. 580 units from industry I and 490 units from industry II.
   b. 620 units from industry I and 600 units from industry II.
   c. 710 units from industry I and 850 units from industry II.

4. $\begin{bmatrix} .30 & .40 \\ .625 & .50 \end{bmatrix}$

   a. 890 units from industry I and 750 units from industry II.
   b. 750 units from industry I and 645 units from industry II.
   c. 900 units from industry I and 865 units from industry II.

5. $\begin{bmatrix} .20 & .46 \\ .50 & .40 \end{bmatrix}$

   a. 780 units from industry I and 820 units from industry II.
   b. 690 units from industry I and 850 units from industry II.
   c. 950 units from industry I and 670 units from industry II.

6. $\begin{bmatrix} .20 & .30 & .40 \\ .30 & .20 & .30 \\ .20 & .40 & .20 \end{bmatrix}$

   a. 680, 780, and 850 units from industries I, II, and III, respectively.
   b. 590, 705, and 574 units from industries I, II, and III, respectively.
   c. 470, 680, and 725 units from industries I, II, and III, respectively.

7. $\begin{bmatrix} .30 & .25 & .30 \\ .20 & .40 & .20 \\ .40 & .30 & .20 \end{bmatrix}$

   a. 550, 800, and 470 units from industries I, II, and III, respectively.
   b. 650, 745, and 475 units from industries I, II, and III, respectively.
   c. 470, 750, and 840 units from industries I, II, and III, respectively.

8. $\begin{bmatrix} .25 & .30 & .40 \\ .30 & .20 & .10 \\ .40 & .40 & .40 \end{bmatrix}$

   a. 450, 540, and 630 units from industries I, II, and III, respectively.
   b. 650, 805, and 580 units from industries I, II, and III, respectively.
   c. 870, 970, and 850 units from industries I, II, and III, respectively.

9. $\begin{bmatrix} .10 & .40 & .30 \\ .50 & .25 & .20 \\ .20 & .30 & .40 \end{bmatrix}$

   a. 380, 540, and 730 units from industries I, II, and III, respectively.
   b. 670, 890, and 500 units from industries I, II, and III, respectively.
   c. 600, 785, and 435 units from industries I, II, and III, respectively.

10. $\begin{bmatrix} .25 & .10 & .20 \\ .40 & .20 & .30 \\ .30 & .40 & .40 \end{bmatrix}$

    a. 860, 870, and 580 units from industries I, II, and III, respectively.
    b. 490, 825, and 745 units from industries I, II, and III, respectively.

c. 970, 880, and 925 units from industries I, II, and III, respectively.

In Exercises 11–20, a table describes the interaction between sectors of a simple hypothetical economy. In each of these, find the output needed by each sector to meet the anticipated final demands. The units are in millions of dollars per year.

11.

	Industry I	Industry II	Final demands	Total output
Industry I	455	1050	315	1820
Industry II	728	840	532	2100
Primary inputs	637	210		

a. The final demands change to 450 for industry I and 640 for industry II. What are the new primary input requirements?

b. The final demands change to 650 for industry I and 780 for industry II. What are the new primary input requirements?

c. The final demands change to 760 for industry I and 860 for industry II. What are the new primary input requirements?

12.

	Industry I	Industry II	Final demands	Total output
Industry I	1280	1904	16	3200
Industry II	800	840	1160	2800
Primary inputs	1120	56		

a. The final demands change to 50 for industry I and 940 for industry II. What are the new primary input requirements?

b. The final demands change to 120 for industry I and 1780 for industry II. What are the new primary input requirements?

c. The final demands change to 180 for industry I and 1260 for industry II. What are the new primary input requirements?

13.

	Industry I	Industry II	Final demands	Total output
Industry I	1512	720	288	2520
Industry II	882	720	198	1800
Primary inputs	126	360		

a. The final demands change to 250 for industry I and 340 for industry II. What are the new primary input requirements?

b. The final demands change to 380 for industry I and 580 for industry II. What are the new primary input requirements?

c. The final demands change to 460 for industry I and 420 for industry II. What are the new primary input requirements?

14.

	Industry I	Industry II	Final demands	Total output
Industry I	750	1640	110	2500
Industry II	1562.5	2050	487.5	4100
Primary inputs	187.5	410		

a. The final demands change to 100 for industry I and 540 for industry II. What are the new primary input requirements?

b. The final demands change to 150 for industry I and 870 for industry II. What are the new primary input requirements?

c. The final demands change to 60 for industry I and 950 for industry II. What are the new primary input requirements?

15.

	Industry I	Industry II	Final demands	Total output
Industry I	652	1886	722	3260
Industry II	1630	1640	830	4100
Primary inputs		574		

a. The final demands change to 850 for industry I and 740 for industry II. What are the new primary input requirements?

b. The final demands change to 920 for industry I and 710 for industry II. What are the new primary input requirements?

c. The final demands change to 670 for industry I and 840 for industry II. What are the new primary input requirements?

16.

	Industry I	Industry II	Industry III	Final demands	Total output
Industry I	720	1230	1280	370	3600
Industry II	1080	820	960	1240	4100
Industry III	720	1640	640	200	3200
Primary inputs	1080	410	320		

a. The final demand for industry I products increases by 20%; for industry II products, it decreases by 10%; and for industry III products, it increases by 50%.

b. The final demands change to 650 for industry I, 1100 for industry II, and 510 for industry III.

17.

	Industry I	Industry II	Industry III	Final demands	Total output
Industry I	1560	1025	1380	1235	5200
Industry II	1040	1640	920	500	4100
Industry III	2080	1230	920	370	4600
Primary inputs	520	205	1380		

a. The final demand for industry I products increases by 10%; for industry II products, it decreases by 15%; and for industry III products, it increases by 30%.

b. The final demands change to 950 for industry I, 850 for industry II, and 720 for industry III.

18.

	Industry I	Industry II	Industry III	Final demands	Total output
Industry I	1675	1200	2920	905	6700
Industry II	2010	800	73	460	4000
Industry III	2680	1600	2920	100	7300
Primary inputs	335	400	730		

a. The final demand for industry I products decreases by 20%; for industry II products, it increases by 10%; and for industry III products, it increases by 70%.

b. The final demands change to 980 for industry I, 870 for industry II, and 320 for industry III.

19.

	Industry I	Industry II	Industry III	Final demands	Total output
Industry I	420	1760	1140	880	4200
Industry II	2100	1100	760	440	4400
Industry III	840	1320	1520	120	3800
Primary inputs	840	220	380		

a. The final demand for industry I products increases by 15%; for industry II products, it decreases by 20%; and for industry III products, it increases by 80%.

b. The final demands change to 750 for industry I, 1250 for industry II, and 480 for industry III.

20.

	Industry I	Industry II	Industry III	Final demands	Total output
Industry I	1630	623	1490	2777	6520
Industry II	2608	1246	2235	141	6230
Industry III	1956	2492	2980	22	7450
Primary inputs	326	1869	745		

a. The final demand for industry I products decreases by 30%; for industry II products, it increases by 70%; and for industry III products, it increases by 90%.

b. The final demands change to 3250 for industry I, 380 for industry II, and 95 for industry III.

21. The matrix

$$\begin{bmatrix} 1 & 0 & 2 \\ 2 & 1 & 4 \\ 3 & 5 & 5 \end{bmatrix}$$

was used to encode a message. Suppose that you received the message: "46, 110, 207, 61, 141, 251, 35, 79, 138, 66, 159, 306." Decode that message.

22. The matrix

$$\begin{bmatrix} 1 & 0 & 0 \\ 3 & 1 & 2 \\ 2 & 0 & 1 \end{bmatrix}$$

was used to encode a message. Suppose that you received the message: "5, 49, 20, 15, 124, 57, 12, 57, 30, 5, 60, 19, 20, 105, 49, 19, 122, 57, 8, 75, 34, 20, 101, 58, 27, 117, 62, 1, 71, 29, 25, 132, 71, 27, 117, 62, 9, 63, 29." Decode that message.

23. The matrix

$$\begin{bmatrix} 1 & 0 & 1 \\ 0 & 1 & 1 \\ 2 & 0 & 3 \end{bmatrix}$$

was used to encode a message. Suppose that you received the message: "40, 41, 101, 31, 52, 89, 10, 27, 25, 45, 52, 117, 29, 26, 83." Decode that message.

24. Use the matrix of Exercise 21 to encode the following message:

DO NOT DRINK AND DRIVE

25. Use the matrix of Exercise 22 to encode the following message:

EDUCATION IS A PRIVILEGE

26. Use the matrix of Exercise 23 to encode the following message:

RACQUETBALL IS FUN

3.8 Markov Chains (Optional)

In this section we begin the study of sequences of events, where at any stage an element of chance is present. Such a sequence is called a *stochastic process*. This sequence may be such that at any stage an outcome is independent of all previous outcomes. In this case, it is a sequence of *Bernoulli trials*. For example, a sequence of coin tosses or a sequence of rolls of a die are sequences of Bernoulli trials. However, in general, each outcome in a sequential process depends on what has happened previously. For example, the price of a stock on a certain day depends on the behavior of that stock on preceding days. Similarly, the results of an election may be affected by the results of previous elections. The simplest case occurs when at any stage the outcome depends only on the outcome at the immediately preceding stage and is not affected by earlier outcomes. Such a process is called a *Markov Chain*. Andrei Markov (1856–1922) was a Russian mathematician who has been given much of the credit for introducing the systematic study of stochastic processes.

Introductory Terminology

We begin with a simple discussion and we assume that you know some of the basic terminology of games of chance.

Suppose that private educators have found that 15% of high school students attend private schools, while the remaining 85% attend public school. They wish to increase the percentage of students attending private schools, so they hire a team of trained recruiters. These recruiters carry out a campaign and collect the following information. Among all students who are attending private schools this year, 82% will attend private schools again the following year, while 18% will switch to public schools. On the other hand, among all students who are attending public schools this year, 6% will attend private schools the following year, while 94% will remain in public schools. Let us calculate what percentages of students will attend private schools and public schools the following year, starting with 15% attending private schools and 85% attending public schools.

If there are only 100 students attending school this year, 15 would be in private schools, while 85 would be in public schools. Since 82% of the students attending private schools remain in private schools, of the 15 who are attending private schools, 12.3 would still attend private schools. (Do not be be concerned with the .3. Remember that we are dealing with percentages!) However, of the 85 who are now attending public schools, 5.1 would switch to private schools, since 85(.06) = 5.1. Thus, of the 100 students who are in school this year, 17.4 will be in private schools the following year since 12.3 + 5.1 = 17.4. We conclude that 17.4% of all students will be attending private schools the following year. Note that

$$(.15)(.82) + (.85)(.06) = .174$$

is the dot product of the vectors

$$[\ .15 \quad .85 \ ] \text{ and } \begin{bmatrix} .82 \\ .06 \end{bmatrix}$$

Similarly, of these 100 students, 82.6 will be in public schools the following year, since $15(.18) + 85(.94) = 82.6$. We conclude that 82.6% of all students will be attending public schools the following year. Again, we note that

$$(.15)(.18) + (.85)(.94) = .826$$

is the dot product of the vectors

$$[.15 \quad .85] \text{ and } \begin{bmatrix} .18 \\ .94 \end{bmatrix}$$

Combining these results, we write

$$[.15 \quad .85] \begin{bmatrix} .82 & .18 \\ .06 & .94 \end{bmatrix} = [.174 \quad .826]$$

The vectors $[.15 \quad .85]$ and $[.174 \quad .826]$ are called *state vectors*, while the matrix

$$\begin{bmatrix} .82 & .18 \\ .06 & .94 \end{bmatrix}$$

is called the *transition matrix* of the system. Note that the entries of the first row add up to 1, which is to be expected since these entries give the percentages of students from private schools who will be attending private schools and public schools, respectively, the following year. Of course, we are assuming that no student drops out of school.

For the same reason, the entries of the second row of the transition matrix, as well as the entries of both state vectors add up to 1. (See Exercises 37 and 38.) Also, all entries of the transition matrix and of any state vector must be nonnegative. It should now be clear that if the transition matrix remains the same, we can calculate how the percentages change from year to year simply by multiplying on the right the present state vector by the transition matrix to obtain the new state vector.

Thus, the state vectors after 1, 2, 3, . . . , 13 years are obtained as follows:

$$\mathbf{V}_1 = \mathbf{V}_0\mathbf{T} = [.1500 \quad .8500] \begin{bmatrix} .82 & .18 \\ .06 & .94 \end{bmatrix} = [.1740 \quad .8260]$$

$$\mathbf{V}_2 = \mathbf{V}_1\mathbf{T} = [.1740 \quad .8260] \begin{bmatrix} .82 & .18 \\ .06 & .94 \end{bmatrix} = [.1922 \quad .8078]$$

$$\mathbf{V}_3 = \mathbf{V}_2\mathbf{T} = [.1922 \quad .8078] \begin{bmatrix} .82 & .18 \\ .06 & .94 \end{bmatrix} = [.2061 \quad .7939]$$

$$\mathbf{V}_4 = \mathbf{V}_3\mathbf{T} = [.2061 \quad .7939] \begin{bmatrix} .82 & .18 \\ .06 & .94 \end{bmatrix} = [.2166 \quad .7834]$$

$$\mathbf{V}_5 = \mathbf{V}_4\mathbf{T} = [.2166 \quad .7834] \begin{bmatrix} .82 & .18 \\ .06 & .94 \end{bmatrix} = [.2246 \quad .7754]$$

$$\mathbf{V}_6 = \mathbf{V}_5\mathbf{T} = [.2246 \quad .7754] \begin{bmatrix} .82 & .18 \\ .06 & .94 \end{bmatrix} = [.2307 \quad .7693]$$

$$\mathbf{V}_7 = \mathbf{V}_6\mathbf{T} = [.2307 \quad .7693] \begin{bmatrix} .82 & .18 \\ .06 & .94 \end{bmatrix} = [.2353 \quad .7647]$$

$$V_8 = V_7T = [\ .2353 \quad .7647\ ] \begin{bmatrix} .82 & .18 \\ .06 & .94 \end{bmatrix} = [\ .2388 \quad .7612\ ]$$

$$V_9 = V_8T = [\ .2388 \quad .7612\ ] \begin{bmatrix} .82 & .18 \\ .06 & .94 \end{bmatrix} = [\ .2415 \quad .7585\ ]$$

$$V_{10} = V_9T = [\ .2415 \quad .7585\ ] \begin{bmatrix} .82 & .18 \\ .06 & .94 \end{bmatrix} = [\ .2435 \quad .7565\ ]$$

$$V_{11} = V_{10}T = [\ .2435 \quad .7565\ ] \begin{bmatrix} .82 & .18 \\ .06 & .94 \end{bmatrix} = [\ .2451 \quad .7549\ ]$$

$$V_{12} = V_{11}T = [\ .2451 \quad .7549\ ] \begin{bmatrix} .82 & .18 \\ .06 & .94 \end{bmatrix} = [\ .2463 \quad .7537\ ]$$

$$V_{13} = V_{12}T = [\ .2463 \quad .7537\ ] \begin{bmatrix} .82 & .18 \\ .06 & .94 \end{bmatrix} = [\ .2472 \quad .7528\ ]$$

Steady-State Vector

Note that we have rounded off the results to four decimal places. We can see that there was not much change in the last five state vectors. In fact, it appears that these vectors seem to be approaching the vector [.25   .75]. If we use this vector as a state vector, we get

$$[\ .2500 \quad .7500\ ] \begin{bmatrix} .82 & .18 \\ .06 & .94 \end{bmatrix} = [\ .2500 \quad .7500\ ]$$

We see that the transition of the vector [.25   .75] leads to the same state vector. We call this vector the *steady-state vector*.

The following example illustrates an interesting characteristic of Markov chains.

EXAMPLE 1 Using the same transition matrix as in the prior discussion, find the first seven state vectors if initially 21% of the students attended private schools and 79% attended public schools.

Solution The initial state vector is $V_0 = [\ .21 \quad .79\ ]$. Thus,

$$V_1 = V_0T = [\ .21 \quad .79\ ] \begin{bmatrix} .82 & .18 \\ .06 & .94 \end{bmatrix} = [\ .2196 \quad .7804\ ]$$

Continuing, we get the following results:

$$V_2 = V_1T = [\ .2269 \quad .7731\ ]$$
$$V_3 = V_2T = [\ .2324 \quad .7676\ ]$$
$$V_4 = V_3T = [\ .2366 \quad .7634\ ]$$
$$V_5 = V_4T = [\ .2398 \quad .7602\ ]$$
$$V_6 = V_5T = [\ .2422 \quad .7578\ ]$$
$$V_7 = V_6T = [\ .2441 \quad .7559\ ]$$

Again, we see that the state vectors seem to stabilize at [.25   .75]. This is not a coincidence. It follows from Theorem 3.4.

Do Exercise 25.

DEFINITION 3.18: A transition matrix **T** is said to be *regular* if for some positive integer n, all entries of the nth power of **T** are positive.

THEOREM 3.4: If **T** is a regular transition matrix, then the successive state vectors approach some fixed state vector **V** regardless of the initial state vector. Furthermore, this vector **V** has the property that **VT** = **V**. For this reason, it is called the steady-state vector for the system.

It is useful to be able to find the steady-state vector without calculating a succession of state vectors and guessing the steady state from the resulting sequence. We illustrate how this can be done in the next two examples.

EXAMPLE 2   Suppose that the team of recruiters hired by the private educators want to predict the percentages of students attending private and public schools, respectively, after a number of years, without calculating the different state vectors. How would they proceed?

Solution   The transition matrix is

$$\begin{bmatrix} .82 & .18 \\ .06 & .94 \end{bmatrix}$$

Assume that the steady-state vector is [x y]. Then, we have

$$[x \quad y] \begin{bmatrix} .82 & .18 \\ .06 & .94 \end{bmatrix} = [x \quad y]$$

This matrix equation is equivalent to the system

$$\begin{cases} .82x + 0.6y = x \\ .18x + .94y = y \end{cases}$$

This system can be written

$$\begin{cases} -.18x + .06y = 0 \\ .18x - .06y = 0 \end{cases}$$

These two equations are obviously equivalent. Hence, we keep only one of them and write it with the equation $x + y = 1$ (which must be satisfied since the vector [x y] is a state vector), to obtain the system

$$\begin{cases} x + \quad y = 1 \\ -.18x + .06y = 0 \end{cases}$$

Using row reduction to solve the system, we get

$$\begin{bmatrix} 1 & 1 & | & 1 \\ -.18 & .06 & | & 0 \end{bmatrix} \qquad .18R_1 + R_2 \rightarrow R_2$$

$$\left[\begin{array}{cc|c} 1 & 1 & 1 \\ 0 & .24 & .18 \end{array}\right] \quad \left(\tfrac{1}{.24}\right)R_2 \to R_2$$

$$\left[\begin{array}{cc|c} 1 & 1 & 1 \\ 0 & 1 & .75 \end{array}\right] \quad (-1)R_2 + R_1 \to R_1$$

$$\left[\begin{array}{cc|c} 1 & 0 & .25 \\ 0 & 1 & .75 \end{array}\right]$$

Thus, $x = .25$ and $y = .75$. We conclude that the steady-state vector of the system is $[\ .25 \quad .75\ ]$. The recruiters could have predicted that, after a number of years, enrollments would stabilize to 25% of the student population attending private schools and 75% attending public schools.

Do Exercise 21. ■

EXAMPLE 3 The price of a certain stock varies from day to day. On any given day, the price may increase, remain unchanged, or decrease. Furthermore, the change on any given day affects the change on the following day. The transition matrix is described by the following table.

Tomorrow's change

		Increase	Unchanged	Decrease
Today's change	Increase	35%	40%	25%
	Unchanged	25%	60%	15%
	Decrease	35%	40%	25%

The interpretation of the first row of the table is that if the stock increased today, there is a 35% chance that it will increase tomorrow, a 40% chance that its value will remain unchanged tomorrow, and a 25% chance that its value will decrease tomorrow. The second and third rows have similar interpretations.

a. If the value of the stock went up today, what are the chances that its value will increase, remain the same, or decrease in four days?

b. Find the steady-state vector and interpret the result.

Solution **a.** The transition matrix is $\mathbf{T} = \begin{bmatrix} .35 & .4 & .25 \\ .25 & .6 & .15 \\ .35 & .4 & .25 \end{bmatrix}$.

Since the value of the stock went up today, the chance of its value going up is 100%, while the chances of its value remaining the same or decreasing are 0%. Thus, the initial state vector is $\mathbf{V}_0 = [\ 1 \quad 0 \quad 0\ ]$. If we let $\mathbf{V}_1$, $\mathbf{V}_2$, $\mathbf{V}_3$, and $\mathbf{V}_4$ be the state vectors after 1, 2, 3, and 4 days, respectively, we have

$$\mathbf{V}_1 = \mathbf{V}_0\mathbf{T} = [\ 1 \quad 0 \quad 0\ ]\begin{bmatrix} .35 & .4 & .25 \\ .25 & .6 & .15 \\ .35 & .4 & .25 \end{bmatrix} = [\ .35 \quad .4 \quad .25\ ]$$

Similarly, we get

$$\mathbf{V}_2 = \mathbf{V}_1\mathbf{T} = [\ .31 \qquad .48 \qquad .21\ ]$$

$$\mathbf{V}_3 = \mathbf{V}_2\mathbf{T} = [\ .302 \qquad .496 \qquad .202\ ]$$

$$\mathbf{V}_4 = \mathbf{V}_3\mathbf{T} = [\ .3004 \qquad .4992 \qquad .2004\ ]$$

Considering the three entries of the state vector $\mathbf{V}_4$, we conclude that after 4 days, the value of the stock has a 30.04% chance of increasing, a 49.92% chance of remaining the same, and a 20.04% chance of decreasing.

b. Let $[\ x \quad y \quad z\ ]$ be the steady-state vector. We know such a vector exists because the transition matrix $\mathbf{T}$ is regular. Then we have $x + y + z = 1$, with x, y, and z nonnegative.

$$[\ x \quad y \quad z\ ]\begin{bmatrix} .35 & .4 & .25 \\ .25 & .6 & .15 \\ .35 & .4 & .25 \end{bmatrix} = [\ x \quad y \quad z\ ]$$

This matrix equation is equivalent to the linear system

$$\begin{cases} .35x + .25y + .35z = x \\ .40x + .60y + .40z = y \\ .25x + .15y + .25z = z \end{cases}$$

This system, with the added equation $x + y + z = 1$, may be written

$$\begin{cases} x + \quad y + .\quad z = 1 \\ -.65x + .25y + .35z = 0 \\ .40x - .40y + .40z = 0 \\ .25x + .15y - .75z = 0 \end{cases}$$

The solution is $x = .3$, $y = .5$, and $z = .2$. The steady-state vector is $[\ .3 \quad .5 \quad .2\ ]$. This means that after a certain length of time, the behavior of the stock will stabilize so that each day, the chances for its value to increase, remain the same, or decrease will be 30%, 50%, 20%, respectively.

Do Exercise 31.   ■

Nonregular Transition Matrix

We conclude this section with an example to illustrate that if the transition matrix fails to be regular, then a steady-state vector may not exist.

EXAMPLE 4   Consider the nonregular transition matrix $\mathbf{T}$ where $\mathbf{T} = \begin{bmatrix} 0 & 1 \\ 1 & 0 \end{bmatrix}$.

Show that if we start with a state vector $[\ a \quad b\ ]$ with $a \neq b$, then $[\ a \quad b\ ]\mathbf{T}^n$ does not approach a limiting vector.

Solution   Note that

$$[\ a \quad b\ ]\begin{bmatrix} 0 & 1 \\ 1 & 0 \end{bmatrix} = [\ b \quad a\ ] \quad \text{and} \quad [\ b \quad a\ ]\begin{bmatrix} 0 & 1 \\ 1 & 0 \end{bmatrix} = [\ a \quad b\ ]$$

Thus, $[a \quad b]T = [b \quad a]$, $[a \quad b]T^2 = [a \quad b]$, $[a \quad b]T^3 = [b \quad a]$, $[a \quad b]T^4 = [a \quad b]$, $\cdots$. In general, $[a \quad b]T^n = [a \quad b]$, when n is even and $[a \quad b]T^n = [b \quad a]$, when n is odd. Since $a \neq b$, $[a \quad b]T^n$ does not approach a limiting steady-state vector. ∎

Exercise Set 3.8

In Exercises 1–6, determine whether the given vector is a state vector. If a vector is not a state vector, explain why.

1. $\left[\frac{1}{3} \quad \frac{1}{6} \quad \frac{1}{6} \right]$

2. $\left[\frac{1}{4} \quad \frac{1}{2} \quad \frac{1}{4} \right]$

3. $[.7 \quad .6 \quad -.3]$

4. $[1 \quad 2 \quad 5 \quad 6]$

5. $[.15 \quad .25 \quad .6]$

6. $[.1 \quad .23 \quad .42 \quad .25]$

In Exercises 7–12, determine whether the given matrix is a transition matrix of some system. If a matrix is not a transition matrix, explain why.

7. $\begin{bmatrix} .3 & .4 \\ .7 & .6 \end{bmatrix}$

8. $\begin{bmatrix} .2 & .5 & .3 \\ .4 & .1 & .5 \end{bmatrix}$

9. $\begin{bmatrix} .4 & .8 & -.2 \\ .3 & .4 & .3 \\ .5 & .1 & .4 \end{bmatrix}$

10. $\begin{bmatrix} \frac{1}{3} & \frac{1}{6} & \frac{1}{2} \\ \frac{1}{5} & \frac{11}{20} & \frac{1}{4} \\ 0 & 1 & 0 \end{bmatrix}$

11. $\begin{bmatrix} \frac{1}{5} & \frac{1}{3} & \frac{7}{15} \\ \frac{2}{7} & \frac{1}{4} & \frac{13}{28} \\ \frac{2}{3} & \frac{1}{9} & \frac{2}{9} \end{bmatrix}$

12. $\begin{bmatrix} .2 & .3 & .5 \\ .5 & .1 & .4 \\ .9 & .05 & .05 \\ .8 & .15 & .05 \end{bmatrix}$

In Exercises 13–18, determine whether the given transition matrix is regular. If a transition matrix is not regular, explain why.

13. $\begin{bmatrix} .2 & .8 \\ 1 & 0 \end{bmatrix}$

14. $\begin{bmatrix} 1 & 0 \\ 0 & 1 \end{bmatrix}$

15. $\begin{bmatrix} 0 & 1 \\ 1 & 0 \end{bmatrix}$

16. $\begin{bmatrix} 0 & 1 \\ .3 & .7 \end{bmatrix}$

17. $\begin{bmatrix} .2 & .4 & .4 \\ .1 & .7 & .2 \\ .6 & 0 & .4 \end{bmatrix}$

18. $\begin{bmatrix} 0 & 0 & 1 \\ 1 & 0 & 0 \\ 0 & 1 & 0 \end{bmatrix}$

In Exercises 19–24, a transition matrix of a system is given. The matrices are regular since all entries are positive. In each case find the steady-state vector.

19. $\begin{bmatrix} .3 & .7 \\ .3 & .7 \end{bmatrix}$

20. $\begin{bmatrix} .675 & .325 \\ .175 & .825 \end{bmatrix}$

21. $\begin{bmatrix} .78 & .22 \\ .18 & .82 \end{bmatrix}$

22. $\begin{bmatrix} .595 & .405 \\ .095 & .905 \end{bmatrix}$

23. $\begin{bmatrix} .15 & .1 & .75 \\ .25 & .3 & .45 \\ .3125 & .55 & .1375 \end{bmatrix}$

24. $\begin{bmatrix} .3 & .4 & .3 \\ .16 & .78 & .06 \\ .2 & .1 & .7 \end{bmatrix}$

In Exercises 25–30, a transition matrix T of a system is given along with an initial state vector V_0. These transition matrices are regular since they have positive entries. In each case find a sequence V_0, V_1, V_2, V_3, . . . , V_n of state vectors long enough so that you can guess the steady-state vector V. Then check your answer by showing that $VT = V$.

25. $T = \begin{bmatrix} .91 & .09 \\ .03 & .97 \end{bmatrix}$; $V_0 = [.15 \quad .85]$

26. $T = \begin{bmatrix} .4 & .6 \\ .15 & .85 \end{bmatrix}$; $V_0 = [.3 \quad .7]$

27. $T = \begin{bmatrix} .585 & .415 \\ .085 & .915 \end{bmatrix}$; $V_0 = [.25 \quad .75]$

28. $T = \begin{bmatrix} .715 & .285 \\ .215 & .785 \end{bmatrix}$; $V_0 = [.35 \quad .65]$

29. $T = \begin{bmatrix} .3 & .5 & .2 \\ .2 & .7 & .1 \\ .1 & .4 & .5 \end{bmatrix}$; $V_0 = [.35 \quad .52 \quad .13]$

30. $T = \begin{bmatrix} .575 & .3 & .125 \\ .2 & .3 & .5 \\ .4 & .5 & .1 \end{bmatrix}$; $V_0 = [.35 \quad .52 \quad .13]$

31. The price of a certain stock varies from day to day. On any given day, the price may increase, remain unchanged, or decrease. Furthermore, the change on any given day affects the change on the following day. The transition matrix is described by the following table.

Tomorrow's change

	Increase	Unchanged	Decrease
Increase	25%	30%	45%
Today's change Unchanged	30%	40%	30%
Decrease	55%	20%	25%

a. If the value of the stock went up on January 5, what are the chances that its value will increase, remain the same, or decrease on January 9 of the same year?

b. If the value of the stock was unchanged on March 23, what are the chances that its value will increase, remain the same, or decrease on March 28 of the same year?

c. If the value of the stock went down on September 23, what are the chances that its value will increase, remain the same, or decrease on September 29 of the same year?

32. Same as Exercise 31 with the following table.

Tomorrow's change

	Increase	Unchanged	Decrease
Increase	15%	20%	65%
Today's change Unchanged	25%	50%	25%
Decrease	45%	15%	40%

a. If the value of the stock went up on April 27, what are the chances that its value will increase, remain the same, or decrease on April 30 of the same year?

b. If the value of the stock was unchanged on November 24, what are the chances that its value will increase, remain the same, or decrease on November 29 of the same year?

c. If the value of the stock went down on June 4, what are the chances that its value will increase, remain the same, or decrease on June 10 of the same year?

33. Assume that a nation is controlled by three main political parties, A, B, and C. The result of each election depends somewhat on which party is in control at the time of the election. The following transition matrix describes the chances that the nation will be controlled by one of the parties after an election, given the possible different parties in control at the time of the election.

In control after the election

	A	B	C
A	65%	20%	15%
In control now B	25%	60%	15%
C	35%	25%	40%

Given that the elections take place every three years and that party A won in 1987, what are the chances that

a. party B will take control in 1993?
b. party A will take control in 1996?
c. party C will take control in 1999?

34. Same as Exercise 33 if the transition matrix is described by the following table.

In control after the election

	A	B	C
A	63%	17%	20%
In control now B	30%	55%	15%
C	40%	20%	40%

Given that the elections take place every four years and that party A won in 1988, what are the chances that

a. party B will take control in 1996?
b. party A will take control in 2000?
c. party C will take control in 2004?

35. The spring weather conditions in Seattle are influenced from day to day by the weather conditions on the previous day. The transition matrix is described by the following table.

Tomorrow

	Rainy	Cloudy	Sunny
Rainy	60%	30%	10%
Today Cloudy	30%	55%	15%
Sunny	20%	20%	60%

Given that it is sunny on a Monday, what are the chances that

a. it will be sunny on Thursday of the same week?
b. it will be cloudy on Friday of the same week?
c. it will be rainy on Saturday of the same week?

36. The summer weather conditions in Quebec are influenced from day to day by the weather conditions on the previous day. The transition matrix is described by the following table.

Tomorrow

	Rainy	Cloudy	Sunny
Rainy	40%	30%	30%
Today Cloudy	30%	45%	25%
Sunny	10%	15%	75%

Given that it is rainy on a Monday, what are the chances that

a. it will be sunny on Thursday of the same week?
b. it will be cloudy on Friday of the same week?
c. it will be rainy on Saturday of the same week?

37. Given that $\mathbf{V} = [\ x \quad y\ ]$ is a state vector and that

$$\mathbf{T} = \begin{bmatrix} a & b \\ c & d \end{bmatrix}$$

is a transition matrix of a system, show that $\mathbf{VT}$ is a state vector.

38. Given that $\mathbf{V} = [\ x \quad y \quad z\ ]$ is a state vector and that

$$\mathbf{T} = \begin{bmatrix} a & b & c \\ d & e & f \\ g & h & i \end{bmatrix}$$

is a transition matrix of a system, show that $\mathbf{VT}$ is a state vector.

39. Show that if a transition matrix has a 1 on its main diagonal, it cannot be regular.

3.9 Chapter Review

IMPORTANT SYMBOLS AND TERMS

$\mathbf{A}_i$ [3.5]
a_{ij} [3.1]
$[a_{ij}]m \times n$ [3.1]
$\mathbf{A}^T$ [3.5]
DetA [3.5]
Additive identity [3.1]
Adjoint [3.6]
Associative [3.2]
Augmented matrix [3.3]
Basic principle of counting [3.2]
Bernoulli trials [3.8]
Closed system [3.7]
Coefficient matrix [3.2]
Cofactor [3.5]
Conformable [3.2]
Cryptography [3.7]
Decoding [3.7]
Demand vector [3.7]
Determinant [3.5]

Diagonal element [3.1]
Difference [3.1]
Dimensions [3.1]
Distributive [3.2]
Dot product (inner product) [3.2]
Elementary row operation [3.3]
Encoding [3.7]
Final demands [3.7]
i-j entry [3.1]
Input-Output model [3.7]
Interior equilibrium [3.7]
Invertible (nonsingular) [3.4]
Leading variable [3.3]
Matrix [3.1]
Multiplicative identity [3.1]
Multiplicative inverse [3.2]
n-dimensional column vector [3.1]

n-dimensional row vector [3.1]
Open Leontief system [3.7]
Product [3.2]
Row-reduced echelon matrix [3.3]
Scalar matrix [3.1]
Singular [3.4]
Square matrix [3.1]
Stochastic process [3.8]
Technological coefficient matrix (input-output matrix) [3.7]
Total output vector [3.7]
Transpose [3.5]
Zero matrix [3.1]
Scalar multiplication [3.1]
Sum [3.1]
Value vector [3.2]
Variable vector [3.2]

SUMMARY

A rectangular array of numbers enclosed within brackets is called a matrix. We use the notation $[\mathbf{a}_{ij}]_{m \times n}$ to denote a matrix with m rows, n columns, and the number $\mathbf{a}_{ij}$ in its ith row and jth column. A $1 \times n$ matrix is called an n-dimensional row vector, an $n \times 1$ matrix is called an n-dimensional column vector. A zero matrix is a matrix all of whose entries are zero, and is called an additive identity. If α is a scalar (number), and $\mathbf{A}$ is an $m \times n$ matrix, $\alpha\mathbf{A}$ is the $m \times n$ matrix $\mathbf{B}$ such that $b_{ij} = \alpha a_{ij}$ for all i, j. Two matrices may be added or subtracted, if and only if they have the same dimensions. Addition of matrices is commutative, and associative. The dot product of two n-dimensional vectors is the sum of the n products obtained by multiplying corresponding entries of the two vectors. If $\mathbf{A}$ is an $m \times n$ matrix and $\mathbf{B}$ is an $n \times p$ matrix, the product $\mathbf{AB}$ is the $m \times p$ matrix $\mathbf{C}$ whose i-j entry is the dot product of the ith row of $\mathbf{A}$ by the jth column of $\mathbf{B}$, for all i, j. Matrix multiplication is associative and distributive over addition. However, multiplication is not commutative. Also it is possible to have

$\mathbf{AB} = 0$ where neither $\mathbf{A}$ nor $\mathbf{B}$ is the zero matrix. Because of this, we may have $\mathbf{AB} = \mathbf{AC}$ even though $\mathbf{B} \neq \mathbf{C}$. The $n \times n$ matrix $[\mathbf{e}_{ij}]$ where $\mathbf{e}_{ij} = 1$ if $i = j$, and 0 if $i \neq j$ is called the multiplicative identity of order n and is denoted $\mathbf{I}_n$. If $\mathbf{A}$ and $\mathbf{B}$ are $n \times n$ matrices and $\mathbf{AB} = \mathbf{BA} = \mathbf{I}_n$, then we say that $\mathbf{B}$ is the multiplicative inverse of $\mathbf{A}$ and write $\mathbf{B} = \mathbf{A}^{-1}$. Furthermore, we say that $\mathbf{A}$ and $\mathbf{B}$ are invertible (nonsingular). The system

$$\begin{cases} a_{11}x_1 + a_{12}x_2 + \cdots + a_{1n}x_n = b_1 \\ a_{21}x_1 + a_{22}x_2 + \cdots + a_{2n}x_n = b_2 \\ \qquad \vdots \\ a_{m1}x_1 + a_{m2}x_2 + \cdots + a_{mn}x_n = b_m \end{cases}$$

may be written $\mathbf{AX} = \mathbf{B}$ where

$$\mathbf{A} = \begin{bmatrix} a_{11} & a_{12} & \cdots & a_{1n} \\ a_{21} & a_{22} & \cdots & a_{2n} \\ & & \vdots & \\ a_{m1} & a_{m2} & \cdots & a_{mn} \end{bmatrix} \quad \mathbf{X} = \begin{bmatrix} x_1 \\ x_2 \\ \vdots \\ x_n \end{bmatrix} \text{ and } \mathbf{B} = \begin{bmatrix} b_1 \\ b_2 \\ \vdots \\ b_m \end{bmatrix}$$

The matrices $\mathbf{A}$, $\mathbf{X}$, and $\mathbf{B}$ are called the coefficient matrix, the variable vector, and the value vector, respectively. If $m = n$ and $\mathbf{A}$ is invertible, we may find $\mathbf{A}^{-1}$ and multiply both sides of $\mathbf{AX} = \mathbf{B}$ (*on the left*) by $\mathbf{A}^{-1}$ to get $\mathbf{X} = \mathbf{A}^{-1}\mathbf{B}$, from which the solution of the original system may be read. The augmented matrix of the system of m linear equations in n variables

$$\begin{cases} a_{11}x_1 + a_{12}x_2 + \cdots + a_{1n}x_n = b_1 \\ a_{21}x_1 + a_{22}x_2 + \cdots + a_{2n}x_n = b_2 \\ \qquad \vdots \\ a_{m1}x_1 + a_{m2}x_2 + \cdots + a_{mn}x_n = b_m \end{cases}$$

is the $m \times (n + 1)$ matrix

$$\begin{bmatrix} a_{11} & a_{12} & \cdots & a_{1n} & | & b_1 \\ a_{21} & a_{22} & \cdots & a_{2n} & | & b_2 \\ & & \vdots & & | & \\ a_{m1} & a_{m2} & \cdots & a_{mn} & | & b_m \end{bmatrix}$$

and is written in compact form $\mathbf{A}|\mathbf{B}$. Given the augmented matrix of a linear system, we get the augmented matrix of an equivalent linear system if we perform any of the following row operations: interchange row i and row $j(R_i \leftrightarrow R_j)$; multiply row i by the nonzero constant k to obtain the new row $i(kR_i \rightarrow R_i)$; multiply row i by the constant k and add the result to row j to obtain the new row $j(kR_i + R_j \rightarrow R_j)$; and multiply row i by the constant k_1 and add the result to k_2 times row j, where k_2 is a nonzero constant, to obtain the new row $j(k_1R_i + k_2R_j \rightarrow R_j)$. We perform these operations until we obtain a matrix in row-reduced echelon form, from which we can determine the solutions of the system.

To find the multiplicative inverse of an $n \times n$ matrix $\mathbf{A}$ we start with the $n \times 2n$ matrix $\mathbf{A} \mid \mathbf{I}_n$. If a sequence of elementary row operations transforms this matrix into $\mathbf{I}_n \mid \mathbf{B}$, then the matrix **B** is the inverse of **A**. However, if at any step we obtain all 0's in any row to the left of the vertical line, the matrix **A** does not have an inverse, it is singular.

The determinant of a square matrix is a number that can easily be calculated for 2×2 and 3×3 matrices but, in general, not for larger matrices.

To solve a system of n linear equations in n variables using Cramer's rule we proceed as follows.

Step 1. Arrange all variables so that corresponding variables are underneath each other and make sure all constants are on the right of the equal signs in all n equations.

Step 2. Write the system as a single matrix equation $\mathbf{AX} = \mathbf{B}$, where **A** is the coefficient matrix

$$\mathbf{X} = \begin{bmatrix} x_1 \\ x_2 \\ \vdots \\ x_n \end{bmatrix} \text{ and } \mathbf{B} = \begin{bmatrix} b_1 \\ b_2 \\ \vdots \\ b_n \end{bmatrix}$$

Step 3. For each $i \in \{1, 2, \ldots, n\}$, let $\mathbf{A}_i$ be the matrix obtained by replacing the ith column of **A** by the n-dimensional vector **B**.

Step 4. Calculate $\det \mathbf{A}$. If $\det \mathbf{A} = 0$, either the system has no solution, or infinitely many solutions, which we can determine by using the row reduction method. If $\det \mathbf{A} \neq 0$, calculate $\det \mathbf{A}_i$ for each $i \in \{1, 2, \ldots, n\}$. The system has a unique solution $(c_1, c_2, \ldots, c_n)$, where for each i, $c_i = \det \mathbf{A}_i / \det \mathbf{A}$. We can also find the multiplicative inverse of a nonsingular square matrix **A** using determinants.

In an open Leontief system, we have an economy consisting of n industries, each of which produces a single commodity. Each uses some of the others' output in order to operate. In addition, each industry produces some goods for the consumer (nonproductive) segment of the economy. We assume that production of one unit of the jth commodity (output) requires a quantity b_{ij} of the ith commodity (input), the $n \times n$ matrix $\mathbf{B} = [b_{ij}]$ is called the technological coefficient matrix (input-output matrix). We let x_i dollars be the total output of the ith industry and d_i dollars be the demand for the product of the ith industry by the nonproductive segment. Then, we have

$$\mathbf{BX} + \mathbf{D} = \mathbf{X}$$

where,

$$\mathbf{X} = \begin{bmatrix} x_1 \\ x_2 \\ \vdots \\ x_n \end{bmatrix} \text{ and } \mathbf{D} = \begin{bmatrix} d_1 \\ d_2 \\ \vdots \\ d_n \end{bmatrix}$$

The n-dimensional vectors **X** and **D** are called the total output vector and the demand vector, respectively. Given the technological matrix **B** and the demand vector **D**, we can find the production necessary to meet the final demands by solving the matrix equation $\mathbf{BX} + \mathbf{D} = \mathbf{X}$. We get

$$\mathbf{X} = (\mathbf{I}_n - \mathbf{B})^{-1} \mathbf{D}$$

Matrix multiplication can also be used to encode and decode messages.

A stochastic process is a sequence of events where at each stage an element of chance is present. If at any stage an outcome is independent of all previous outcomes, we have a sequence of Bernoulli trials. If at any stage the outcome depends only on the outcome of the immediately preceding stage and is not affected by earlier outcomes, the process is called a Markov chain. Given that outcome O_i has occurred at some trial, the chance that outcome O_j will occur on the next trial is p_{ij}, where $0 \le p_{ij} \le 1$, the $n \times n$ matrix $[p_{ij}]$ is called the transition matrix. For each i, we have $\sum_{j=1}^{n} p_{ij} = 1$, since at any stage one of the n possible outcomes must occur.

If it is given that at the initial trial, the chance of outcome O_i occurring is p_i, the vector $[\,p_1 \quad p_2 \quad \cdots \quad p_n\,]$ is called the initial state vector. We must have $0 \le p_i \le 1$ for each i and $p_1 + p_2 + \cdots + p_n = 1$. If we let $\mathbf{V}_0$ denote the initial state vector and $\mathbf{T}$ denote the transition matrix, then $\mathbf{V}_1 = \mathbf{V}_0\mathbf{T}$ is the state vector for the first trial, $\mathbf{V}_2 = \mathbf{V}_1\mathbf{T}$ is the state vector for the second trial, $\mathbf{V}_3 = \mathbf{V}_2\mathbf{T}$ is the state vector for the third trial, and so on. For the kth trial, the jth entry of the vector $\mathbf{V}_k$ gives the chance that the jth outcome will occur. A transition matrix $\mathbf{T}$ is said to be regular if for some positive integer n, all entries of the nth power of $\mathbf{T}$ are positive. If $\mathbf{T}$ is a regular transition matrix, then the successive state vectors approach some fixed state vector $\mathbf{V}$ regardless of the initial vector. Furthermore, this vector has the property that $\mathbf{VT} = \mathbf{V}$. If $\mathbf{T}$ is a regular $n \times n$ transition matrix, we can find the steady state vector as follows. We let $X = [\,x_1, x_2, \ldots, x_n\,]$ so that $X\mathbf{T} = \mathbf{X}$. Also, we have $x_1 + x_2 + \cdots + x_n = 1$. The last two equalities yield a system of linear equations which can be solved for $x_1, x_2, \ldots, x_n$.

SAMPLE EXAM QUESTIONS

1. Write the 2×3 matrix $[a_{ij}]$ if $a_{ij} = 2i - 3j$ for all i, j.

2. Write the 3×2 matrix $[b_{ij}]$ if $b_{ij} = (-1)^{i+j}(2i + j^2)$ for all i, j.

3. Write the 4×4 matrix $[c_{ij}]$ if

$$c_{ij} = \begin{cases} 0 & \text{if } i = j \\ -i^2 + 3j & \text{whenever } i \ne j. \end{cases}$$

4. Perform the indicated operations.

a. $4\begin{bmatrix} 2 & 3 \\ -1 & 5 \end{bmatrix}$

b. $-3\begin{bmatrix} 4 & 5 & 0 \\ -3 & 5 & 7 \\ -2 & 1 & 9 \end{bmatrix} + 5\begin{bmatrix} 2 & 7 & 1 \\ -5 & 3 & 9 \\ -3 & 0 & 4 \end{bmatrix}$

5. For each equality, find the value of the variables which make the matrix equation a true statement.

a. $\begin{bmatrix} x & 4 \\ -1 & 7 \end{bmatrix} = \begin{bmatrix} 3 & y \\ -1 & z \end{bmatrix}$

b. $\begin{bmatrix} x & 2 & 3 \\ -5 & y & 3 \\ -2 & 6 & z \end{bmatrix} = \begin{bmatrix} 6 & 2 & 3 \\ -5 & 2 & 3 \\ -2 & 6 & 3 \end{bmatrix}$

c. $\begin{bmatrix} x+2 & 3 & 5 \\ -9 & y-3 & 3 \\ 4 & 7 & z+4 \end{bmatrix} = \begin{bmatrix} 2x-1 & 3 & 5 \\ -9 & 1+2y & 3 \\ 4 & 7 & 5z \end{bmatrix}$

6. Let $A = \begin{bmatrix} 2 & 4 & 0 \\ 3 & 1 & 2 \\ 4 & 9 & 3 \end{bmatrix}$, $B = \begin{bmatrix} 3 & 1 \\ 4 & 7 \\ 2 & 5 \end{bmatrix}$, $C = \begin{bmatrix} 1 & 0 & 5 \\ 4 & 2 & 8 \end{bmatrix}$.

Find the following if possible. If not possible, state why.

a. The product AB. b. The product CA.
c. The product AC. d. The product CB.
e. The product BC. f. The inverse A^{-1}.
g. The product BA. h. The inverse B^{-1}.

7. a. Find two 2×2 matrices A and B such that $AB \neq BA$.
   b. Find two 2×2 matrices A and B such that $AB = BA$, where neither A nor B is the identity or the zero matrix.

8. Write the following system of linear equations as a single matrix equation:
$$\begin{cases} 5w + 2x + 7y - 3z = 2 \\ -3w + 6x + 3y - 4z = 7 \\ 2w + 5x + 7y - 5z = 2 \end{cases}$$

9. Solve each of the following linear systems by row reduction.

a. $\begin{cases} 3x + 4y - 5z = 4 \\ -4x - 2y + 5z = 6 \end{cases}$

b. $\begin{cases} 5x + 3y - 4z = -5 \\ -6x + 4y - 3z = -25 \\ x + 4y - 6z = -20 \end{cases}$

10. Solve the linear systems of parts a, b, and c simultaneously.
$$\begin{cases} 2x + 5y = d_1 \\ 7x + 17y = d_2 \end{cases}$$
a. $d_1 = 5, d_2 = -3$
b. $d_1 = 3, d_2 = -7$
c. $d_1 = -2, d_2 = 6$

11. Solve the linear systems of parts a, b, and c simultaneously.
$$\begin{cases} 2x + 6y + 3z = d_1 \\ 2x + 7y + 5z = d_2 \\ x + 3y + 2z = d_3 \end{cases}$$
a. $d_1 = 3, d_2 = -4, d_3 = 7$
b. $d_1 = 0, d_2 = -3, d_3 = 5$
c. $d_1 = -5, d_2 = 1, d_3 = -3$

12. Find the inverse of each of the following matrices, if it exists.

a. $\begin{bmatrix} 9 & 11 \\ 4 & 5 \end{bmatrix}$

b. $\begin{bmatrix} 9 & 3 \\ 12 & 4 \end{bmatrix}$

$$\textbf{c.} \begin{bmatrix} 2 & 4 & 3 \\ 6 & 9 & 8 \\ 1 & 2 & 2 \end{bmatrix} \quad (\textit{Hint: } \text{Use the adjoint.})$$

$$\textbf{d.} \begin{bmatrix} 2 & 4 & 1 \\ 3 & 7 & 3 \\ 5 & 10 & 3 \end{bmatrix}$$

$$\textbf{e.} \begin{bmatrix} 3 & 2 & 2 \\ 9 & 3 & 5 \\ 6 & 1 & 3 \end{bmatrix}$$

13. Solve the following systems. First write the given system as a matrix equation and then multiply both sides by the inverse of the coefficient matrix.

   a. $\begin{cases} 9x + 11y = 3 \\ 4x + 5y = -2 \end{cases}$

   b. $\begin{cases} 2x + 4y + 3z = -3 \\ 6x + 9y + 8z = 6 \\ x + 2y + 2z = 18 \end{cases}$ (Use the result of Exercise 12**c**.)

   c. $\begin{cases} 2x + 4y + z = -3 \\ 3x + 7y + 3z = 2 \\ 5x + 10y + 3z = 4 \end{cases}$ (Use the result of Exercise 12**d**.)

14. Evaluate each of the following determinants:

   a. $\det \begin{bmatrix} 5 & 2 \\ 3 & 4 \end{bmatrix}$

   b. $\det \begin{bmatrix} 1 & 2 & 3 \\ 3 & 1 & 6 \\ 5 & 2 & 7 \end{bmatrix}$

   c. $\det \begin{bmatrix} 3 & 0 & 6 & 9 \\ 5 & -1 & 9 & 14 \\ 4 & 2 & 6 & -2 \\ 7 & 5 & 10 & -1 \end{bmatrix}$ (*Hint:* Use the basic properties of determinants.)

15. Solve each of the following systems using Cramer's rule.

   a. $\begin{cases} 2x + 5y = -11 \\ 3x - 2y = 12 \end{cases}$

   b. $\begin{cases} 2x + 3y - 4z = -12 \\ 3x - 2y + 5z = 13 \\ 5x + y - 3z = -15 \end{cases}$

16. Assume that we have a simple economy with only two industries I and II. Assume that the technological coefficient matrix for that economy is

$$\begin{bmatrix} .5 & .5 \\ .3 & .2 \end{bmatrix}$$

where the entries of the first column are the amounts industry I uses from its own industry and from industry II to produce one unit of its own commodity, while the second column gives the amounts industry II uses from industry I and from its own industry to produce one unit of its own commodity. Find the production necessary to provide for final demands of

a. 250 units of the product of industry I and 300 units of the product of industry II.
b. 400 units of the product of industry I and 500 units of the product of industry II.

17. Suppose we have a simple economy consisting of three industries: manufacturing, housing, and agriculture. The interaction between these industries is shown in the following table, where the units are in millions of dollars per year.

	Agriculture	Housing	Manufacturing	Final demands	Total output
Agriculture	3300	3100	3960	640	11,000
Housing	2200	4960	2640	2600	12,400
Manufacturing	4400	3720	2640	2440	13,200
Primary inputs	1100	620	3960		

Assuming that the technological coefficient matrix remains constant over the period under consideration, find the outputs needed from each industry if the final demands change to the following, where the units are in millions of dollars per year.

a. 660 for agriculture, 960 for housing, and 564 for manufacturing.
b. 910 for agriculture, 1043 for housing, and 665 for manufacturing.
c. 940 for agriculture, 1500 for housing, and 1680 for manufacturing.

18. Use the matrix $\begin{bmatrix} 3 & 1 \\ 5 & 2 \end{bmatrix}$ to encode the message: THIS EXAM IS HARD.

19. The matrix of Exercise 18 was used to encode a message. Then the message was sent to you to decode during your exam. This is what you received: "36, 65, 49, 83, 93, 159, 66, 111, 60, 109." What did the message say?

20. Which of the following is a state vector? If a vector is not a state vector, explain why.

a. [.5   .7   −.2]   **b.** [.4   0   .6]
c. [.13   .27   .6]   **d.** [.15   .25   −.2   .8]
e. [.3   .2   1.5   .6]

21. Which of the following is a transition matrix of a system? If a matrix is not a transition matrix, explain why.

a. $\begin{bmatrix} .2 & .6 \\ .8 & .4 \end{bmatrix}$ **b.** $\begin{bmatrix} .3 & .7 \\ .6 & .4 \end{bmatrix}$

c. $\begin{bmatrix} .1 & .3 & .6 \\ .2 & .5 & .3 \end{bmatrix}$ **d.** $\begin{bmatrix} .2 & .1 & .7 \\ .4 & .3 & .3 \\ .1 & .1 & .8 \end{bmatrix}$

22. Which of the following transition matrices are regular? If a transition matrix is not regular, explain why.

a. $\begin{bmatrix} .1 & .9 \\ 0 & 1 \end{bmatrix}$ 　　　　　　　　　**b.** $\begin{bmatrix} .4 & .6 \\ 1 & 0 \end{bmatrix}$

c. $\begin{bmatrix} .1 & .3 & .6 \\ .2 & 0 & .8 \\ .4 & .5 & .1 \end{bmatrix}$

23. The matrix $\begin{bmatrix} .74 & .26 \\ .24 & .76 \end{bmatrix}$ is the transition matrix of a system. Find the steady-state vector of that system.

24. The matrix **T** is the transition matrix of a system and the vector $\mathbf{v}_0$ is the initial state vector. Find the state vectors $\mathbf{v}_1$ and $\mathbf{v}_2$.

$$\begin{bmatrix} .87 & .13 \\ .21 & .79 \end{bmatrix}; \mathbf{v}_0 = [\ .32 \quad .68\ ]$$

25. The fall weather conditions in Montreal are influenced from day to day by the weather conditions on the previous day. The transition matrix is described by the following table.

		Tomorrow		
		Rainy	**Cloudy**	**Sunny**
	Rainy	55%	35%	10%
Today	Cloudy	50%	45%	5%
	Sunny	15%	25%	60%

Given that it is cloudy on a Monday, what are the chances that

a. it will be sunny on Wednesday of the same week?
b. it will be cloudy on Thursday of the same week?
c. it will be rainy on Friday of the same week?

26. A manufacturer builds regular, deluxe, and super sailboats. Each comes in two models, A and B. The following table gives the quantities which are scheduled to be built in a certain month.

	Regular	**Deluxe**	**Super**
Model A	6	6	3
Model B	5	2	4

The material needed for each type is given by the following table in appropriate units:

	Wood	**Paint**	**Canvas**
Regular	7	3	5
Deluxe	8	4	10
Super	9	6	17

Finally, the cost per unit (in thousands of dollars) of each of the materials used is given in tabular form.

	Cost per unit
Wood	3
Paint	.2
Canvas	3

a. Represent the given tables by matrices **Q**, **M**, and **C**, respectively.
b. Calculate the product **QM** and interpret the result.
c. Calculate the product (**QM**)**C** and interpret the result.

27. The Bonnechance Company manufactures slot machines. Each deluxe model requires 8 hours of assembly time and 3 hours of testing time, while each regular model requires 5 hours of assembly time and 2 hours of testing time. The time available for assembly and testing varies according to the following schedule:

Time available (in hours)	Monday	Tuesday	Wednesday	Thursday	Friday
For assembly	70	75	64	75	62
For testing	20	30	28	33	25

For each day of the week, let x and y denote the numbers of deluxe and regular models the company manufactures.

a. Write five linear systems of equations knowing that all available time for assembly and testing is used each day.
b. Solve the five systems simultaneously.

Linear Programming

4

Of the many applications of mathematics in business and economics, linear programming is perhaps the most useful. We are often faced with the task of maximizing or minimizing a function subject to constraints. For example, a farmer may have a number of acres on which to grow corn and peanuts. He may already have accepted orders that will require him to use a given number of those acres for corn and a given number for peanuts. He may also be required to follow regulations concerning the number of acres he may use to grow corn or peanuts. All these conditions represent constraints or limitations on how the farmer may grow his crops. If x and y are the number of acres he will use for corn and peanuts, respectively, then his profit may also be expressed in terms of x and y. The problem then is to maximize profit and yet have x and y satisfy all the constraints. The profit, expressed in terms of x and y, is often called the *objective function*, while x and y are called the *decision variables*. It is useful to introduce the following definition.

> **DEFINITION 4.1:**   The set of points (x, y) for which x and y satisfy all the constraints is called the *region of feasible solutions*.

If the constraints and the objective function are expressed as linear inequalities and a linear equation, respectively, then we have a *linear programming problem*. In this chapter, we shall first introduce a *graphical method* for solving linear programming problems involving only two variables. In practice, however, most linear programming problems involve many variables and many constraints. Consequently, we shall also discuss the *simplex method*, which provides, with the help of computers, an efficient way of solving these problems.

4.1 The Geometric Approach

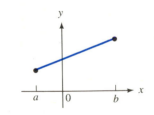

FIGURE 4.1

A linear function of one variable defined on a closed bounded interval $[a, b]$ will assume its maximum and minimum values at the endpoints of that interval. It is easy to see this geometrically since the graph of a linear equation is a straight line. When we draw the portion of that line over the interval $[a, b]$, the highest and lowest points of the graph will be endpoints. (See Figure 4.1.)

Geometrically there are two ways of looking at a linear programming problem with two variables. We shall discuss both ways briefly before giving specific examples. As we shall see, just as the graph of a function of one variable is often a curve in the coordinate plane, the graph of a function of two variables is often a surface in three-dimensional space. And just as the graph of a linear function of one variable is a straight line, the graph of a linear function of two variables is a plane.

Bounded Regions

To help you understand the concept, visualize a house with a polygonal base and flat but slanted roof. The highest point of the roof will be at one corner of the house and the lowest point will be at another corner. (See Figure 4.2.) For the same reason,

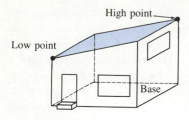

FIGURE 4.2

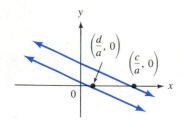

FIGURE 4.3

when the set of points (x, y), which satisfy all the constraints of a given problem (region of feasible solutions), happens to be a polygon and its interior and the objective function is linear (defined by $z = ax + by + c$ where a, b, c are constants), then the maximum and minimum values of z over that polygon and its interior will occur at the corners of the polygon.

Thus, to find the maximum and minimum values of z when the point (x, y) is on the polygon, or in its interior, we need only evaluate the expression $ax + by + c$ at each corner of the polygon. Of all the numbers that are obtained, the largest is the maximum value of the function and the smallest is its minimum value.

For a slightly different way of looking at a linear programming problem geometrically, we will recall a few facts about linear equations in two variables. First, the equation $ax + by = c$, where a, b, and c are real numbers, is called a linear equation in two variables. If at least a or b is not 0, its graph is a straight line intersecting the x-axis at $\left(\frac{c}{a}, 0\right)$ if $a \neq 0$ and the y-axis at $\left(0, \frac{c}{b}\right)$ if $b \neq 0$. The slope of the graph of that equation is $\frac{-a}{b}$ if $b \neq 0$. Since two lines with slopes m_1 and m_2 are parallel if, and only if, $m_1 = m_2$, the graphs of $ax + by = c$ and $ax + by = d$ are parallel since both have slope $\frac{-a}{b}$, when $b \neq 0$, and both are parallel to the y-axis when $b = 0$. It is also true that the graph of $ax + by = c$ is farther away from the origin than the graph of $ax + by = d$ if $|d| < |c|$, since these lines intersect the x-axis at $\left(\frac{c}{a}, 0\right)$ and $\left(\frac{d}{a}, 0\right)$. (See Figure 4.3.)

EXAMPLE 1   Sketch the graphs of $2x + 3y = 6$ and $2x + 3y = 12$. The two lines are parallel. Which graph is farther away from the origin and why?

Solution   Let L_1 and L_2 be the graphs of the first and second equations, respectively. The slope of each line is $\frac{-2}{3}$. Hence, the two lines are parallel. (See Figure 4.4.)

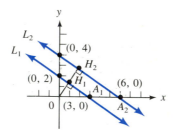

FIGURE 4.4

The x-intercept of L_1 is $(3, 0)$ and that of L_2 is $(6, 0)$. Thus, L_2 is farther away from the origin than L_1. You can see in Figure 4.4 that the right triangles OH_1A_1 and OH_2A_2 are similar. The hypotenuse OA_2 is longer than the hypotenuse OA_1, so OH_2 must be longer than OH_1. ◼

At this point, you should review Section 2.4 on systems of linear inequalities. Recall that the graph of a single linear inequality in two variables is a half-plane and therefore, the graph of a system of linear inequalities is the intersection of half-planes. If this intersection is bounded (that is, enclosed in a circle centered at the origin), then it is called a *bounded convex polygonal region*.

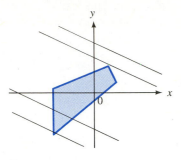

FIGURE 4.5

Suppose we have $z = ax + by + c$ where a, b, and c are real numbers and we consider only values of x and y so that (x, y) is in a bounded convex polygonal region. Then, as claimed earlier, the largest and smallest values of z will occur at corners of that region. To see this, we need only replace z by several constants c_1, c_2, c_3, ..., c_n and observe that the graphs of all equations $ax + by + c = c_i$ are parallel (see Figure 4.5) and the graphs with large values for $|c_i|$ will be the farthest away from the origin. Some of these lines will have points in common with the convex polygonal region. Some will not because $|c_i|$ is either too large or too small. As you can see from Figure 4.5, the lines corresponding to the largest and smallest values of $|c_i|$ which still have points in common with the convex polygonal region are those that touch the region at a corner.

EXAMPLE 2 Find the maximum and minimum values of z if $z = 5x + 2y + 1$ and x and y are subject to the following constraints:

$$3x + 2y \leq 210$$
$$x - y \leq 20$$
$$x - 3y \geq -150$$
$$x \geq 0 \text{ and } y \geq 0$$

Solution We sketch the system

$$\begin{cases} 3x + 2y \leq 210 \\ x - y \leq 20 \\ x - 3y \geq -150 \\ x \geq 0 \text{ and } y \geq 0 \end{cases}$$

as in Section 2.4. (See Figure 4.6.) The graph is a bounded convex polygonal region with vertices (corners) $(0, 0)$, $(20, 0)$, $(50, 30)$, $(30, 60)$, and $(0, 50)$.

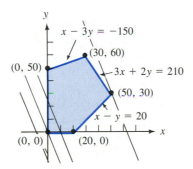

FIGURE 4.6

We evaluate $5x + 2y + 1$ at each vertex:

at $(0, 0)$, $z = 5(0) + 2(0) + 1 = 0 + 0 + 1 = 1$,
at $(20, 0)$, $z = 5(20) + 2(0) + 1 = 100 + 0 + 1 = 101$,
at $(50, 30)$, $z = 5(50) + 2(30) + 1 = 250 + 60 + 1 = 311$,
at $(30, 60)$, $z = 5(30) + 2(60) + 1 = 150 + 120 + 1 = 271$, and
at $(0, 50)$, $z = 5(0) + 2(50) + 1 = 0 + 100 + 1 = 101$.

Of the five numbers obtained, 311 is the largest and 1 is the smallest. Thus, the maximum value of z is 311; it occurs at (50, 30). The minimum value of z is 1; it occurs at (0, 0).

Do Exercise 9.   ∎

EXAMPLE 3 Find the maximum value of z if $z = 3x + 2y + 6$, where x and y are subject to the constraints of Example 2.

Solution The set of feasible solutions is the same as in Example 2 and is drawn in Figure 4.7.

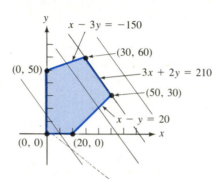

FIGURE 4.7

We evaluate z at each of the six corners:

at (0, 0), $z = 3(0) + 2(0) + 6 = 0 + 0 + 6 = 6$,

at (20, 0), $z = 3(20) + 2(0) + 6 = 60 + 0 + 6 = 66$,

at (50, 30), $z = 3(50) + 2(30) + 6 = 150 + 60 + 6 = 216$,

at (30, 60), $z = 3(30) + 2(60) + 6 = 90 + 120 + 6 = 216$, and

at (0, 50), $z = 3(0) + 2(50) + 6 = 0 + 100 + 6 = 106$.

The largest number obtained is 216. Thus, the maximum value of z is 216. However, it occurs at both corners (50, 30) and (30, 60). In that case, the maximum value of z is attained not only at these two points, but at every point of the segment joining these two corners. For example, the point (40, 45) is on that segment and $3(40) + 2(45) + 6 = 120 + 90 + 6 = 216$. The lines whose equations are $c = 3x + 2y + 6$, for different values of c, are parallel to the edge joining the endpoints (50, 30) and (30, 60). Therefore the largest value of c for which such a line intersects the region of feasible solutions is 216. In this case, the intersection of the line and the polygonal region is one of the edges of the region. (See Figure 4.7.)

Do Exercise 11.   ∎

EXAMPLE 4 A farmer has 150 acres of land on which he may grow corn and wheat. His costs to plant the wheat and the corn are $20 per acre and $10 per acre, respectively. He has at most $2000 to spend for planting his crops. He already has accepted orders which require that he plant at least 10 acres of wheat and 20 acres of corn. His profit per acre is $150 for wheat and $120 for corn. How many acres of each should he plant to maximize his profit?

Solution   Let x denote the number of acres of wheat and y the number of acres of corn that he should plant. Since he has only 150 acres of land available,

$$x + y \leq 150$$

His total cost (measured in dollars) of planting the wheat will be $20x$ and that of planting the corn will be $10y$. Hence, his total cost will be $(20x + 10y)$. Since he has at most \$2000 available, we have

$$20x + 10y \leq 2000$$

which is equivalent to

$$2x + y \leq 200$$

Because of the orders he has accepted, he must grow at least 10 acres of wheat and 20 acres of corn. Therefore, it is also necessary that $x \geq 10$ and $y \geq 20$. His total profit will be \$150x from the wheat and \$120y from the corn. If z dollars is his total profit,

$$z = 150x + 120y$$

Our problem is to maximize z subject to the constraints

$$\begin{cases} x + y \leq 150 \\ 2x + y \leq 200 \\ \quad\;\; x \geq \;\; 10 \\ \quad\;\; y \geq \;\; 20 \end{cases}$$

The graph of this system is shown in Figure 4.8. The vertex $(50, 100)$ is obtained by solving the system

$$\begin{cases} x + y = 150 \\ 2x + y = 200 \end{cases}$$

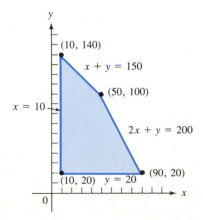

FIGURE 4.8

The other vertices are obtained similarly. We must now evaluate z at each of the four corners of the region of feasible solutions shaded in Figure 4.8.

At $(10, 20)$, $z = 150(10) + 120(20) = 1500 + 2400 = 3900$,

at $(90, 20)$, $z = 150(90) + 120(20) = 13,500 + 2400 = 15,900$,

at $(50, 100)$, $z = 150(50) + 120(100) = 7500 + 12,000 = 19,500$,

and at $(10, 140)$, $z = 150(10) + 120(140) = 1500 + 16,800 = 18,300$.

The largest of these four values of z is 19,500, which occurs when $x = 50$ and $y = 100$. We conclude that the farmer should grow 50 acres of wheat and 100 acres of corn to realize a maximum profit of \$19,500.
Do Exercise 19. ■

Unbounded Regions

It should be clear that if the region of feasible solutions is a *bounded* convex polygonal region, then a linear function will always assume both a maximum and a minimum value and each will always occur at a vertex. But what happens when the region of feasible solutions is unbounded? It may help to first look at the case of a linear function of one variable defined on the unbounded interval $[a, \infty)$. If the graph has positive slope, then the linear function has a minimum at the left endpoint of the interval and it has no maximum. (See Figure 4.9a.) If the graph has negative slope, then the function has a maximum at the left endpoint of the interval and it has no minimum. (See Figure 4.9b.)

FIGURE 4.9

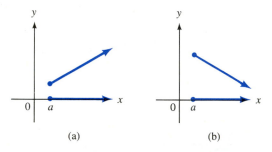

(a)   (b)

 Similarly, when we consider $z = ax + by + c$ and the region of feasible solutions is unbounded, then z may, or may not, have a maximum or a minimum value (see Exercise 17). However, if a maximum or a minimum exists, then it will occur at one of the corners of the region of feasible solutions.

EXAMPLE 5   Find an optimum (maximum or minimum) value of z, if it exists, where

$$z = 2x + 3y + 13$$

and

$$4x - y + 10 \geq 0$$
$$x - 10y + 50 \leq 0$$
$$x + y \geq 60$$

Solution   The region of feasible solutions is shown in Figure 4.10. This region is unbounded and it has two corners, (10, 50), and (50, 10).

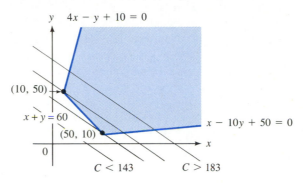

FIGURE 4.10

We wish to find a maximum or minimum value of z, if it exists. First we evaluate z at the two corners. At (10, 50), we get $z = 183$ and at (50, 10) we get $z = 143$. We may be tempted to state that 143 is the minimum value of z and 183 is the maximum value of z. But let us not be too hasty. Consider the graphs of the equations

$$2x + 3y + 13 = 183 \quad \text{and} \quad 2x + 3y + 13 = 143$$

The two lines are parallel and they intersect the region of feasible solutions at the points (10, 50) and (50, 10), respectively. It is clear from Figure 4.10 that if we draw the graph of $2x + 3y + 13 = c$, the graph will not intersect the region of feasible solutions if $c < 143$, but will always intersect the region if $c > 183$. Therefore z has no maximum value. It has a minimum value, however, which is 143. This occurs at (50, 10).

Do Exercise 13.   ■

In summary, when we wish to solve geometrically a linear programming problem with two variables, we proceed as follows:

Step 1.   Graph the system of linear inequalities that defines all constraints.

Step 2.   Find the coordinates of each corner of the region by solving the appropriate pairs of linear equations.

Step 3.   If the region of feasible solutions is bounded and the objective function is nonconstant, then both a maximum and a minimum value of the objective function will exist. In this case, evaluate the objective function at each corner of the region. The largest number obtained is the maximum value of the objective function and the smallest is its minimum. If an optimum value occurs at two corners, then it occurs at every point on the line segment joining these two corners.

Step 4. If the region of feasible solutions is unbounded, evaluate the objective function $z = ax + by + c$ at each corner of the region. If c_1 and c_2 are the largest and smallest numbers obtained, respectively, then if a maximum exists, it will be c_1; if a minimum exists, it will be c_2. Replace z by c_1 and by c_2 and draw the graphs of the two equations. The graphs will intersect the region of feasible solutions at certain corners. Then draw the graph of an equation $c_3 = ax + by + c$ where c_3 is larger than c_1. If this graph does not intersect the region of feasible solutions, then c_1 is the maximum value of z. If this graph does intersect the region, then z has no maximum. Similarly, draw the graph of an equation $c_4 = ax + by + c$ where c_4 is less than c_2. If this graph does not intersect the region of feasible solutions, then the minimum value of z exists and is c_2. If this graph does intersect the region, then z has no minimum.

Exercise Set 4.1

In Exercises 1–6, a region of feasible solutions is shown and an equation defining an objective function is given. In each case, find the maximum and minimum value of z.

5. $z = .25x + .45y - .75$

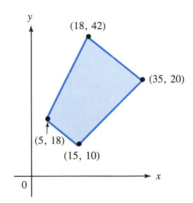

1. $z = 2x + 5y + 3$

2. $z = 3x + 2y - 5$

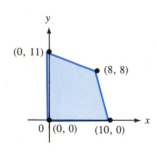

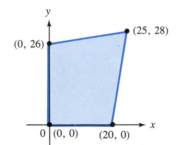

6. $z = -4x + 7y - 9$

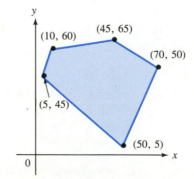

3. $z = 5x + 3y - 7$

4. $z = 4x + 7y + 5$

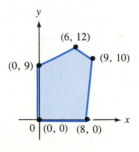

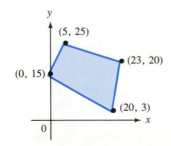

7. Find the maximum and minimum values of z if $z = 4x + 7y$ and

$$3x + 5y \leq 180$$
$$2x - 3y + 70 \geq 0$$
$$x \geq 0 \text{ and } y \geq 0$$

8. Find the maximum and minimum values of z if $z = 3x + 5y$ and

$$2x + 7y \leq 285$$
$$3x - 5y + 115 \geq 0$$
$$x \geq 0 \text{ and } y \geq 0$$

9. Find the maximum and minimum values of z if $z = 5x - 13y$ and

$$5x + 13y \leq 359$$
$$4x - 5y + 67 \geq 0$$
$$x \geq 5 \text{ and } y \geq 12$$

10. Find the maximum and minimum values of z if $z = -3x + 5y$ and

$$5x + 9y \leq 363$$
$$3x - 7y + 179 \geq 0$$
$$x \geq 7 \text{ and } y \geq 20$$

11. Find the maximum and minimum values of z if $z = 2x + 2y - 13$ and

$$2x - y \leq 20$$
$$x + y \leq 220$$
$$-2x + y \leq 70$$
$$x \geq 20 \text{ and } y \geq 80$$

12. Find the maximum and minimum values of z if $z = 7x - 3y + 25$ and

$$8x - 5y + 340 \geq 0$$
$$x + y \leq 120$$
$$2x + y + 40 \geq 0$$
$$x \leq 100 \text{ and } y \geq -20$$

13. If it exists, find the maximum or minimum value of z if $z = 2x + 5y - 7$ and

$$x - 2y \leq 10$$
$$-2x + y \leq 10$$
$$x \geq 10 \text{ and } y \geq 5$$

14. If it exists, find the maximum or minimum value of z if $z = -5x + 3y + 12$ and

$$x - y \leq 50$$
$$x + y \leq 150$$
$$x - y + 50 \geq 0$$

15. If possible, find the maximum or minimum value of z if $z = 4x - 5y + 150$ and

$$x - 7y + 50 \leq 0$$
$$3x + 4y \geq 100$$
$$4x - 3y + 200 \geq 0$$

16. If possible, find the maximum or minimum value of z if $z = 2x + y + 10$ and

$$x - y + 50 \geq 0$$
$$3x - y + 50 \geq 0$$
$$2x + 3y \geq 40$$
$$x - 2y + 15 \leq 0$$

17. Let $z = x - y$ and

$$x \geq 0$$
$$y \geq 0$$

Show that z has neither a maximum nor a minimum value.

Exercises 18–30 were given in Exercise Set 2.4 without an objective function to optimize. Refer to the corresponding exercises to obtain the regions of feasible solutions.

18. If the profit is $350 per acre of peanuts and $280 per acre of corn, write the profit z as a function of x and y. How many acres of each should the farmer plant to maximize his profit?

19. Suppose that the profit is $370 per acre of wheat and $290 per acre of barley. Write the profit z as a function of x and y. How many acres of each should the farmer grow to maximize her profit?

20. Suppose that the company makes a $120 profit on each model A and a $95 profit on each model B. Express the total profit per month z as a function of x and y. How many of each model should be produced each month to maximize the profit?

21. Assume that the profit per deluxe model is $350 and that per standard model is $275. Express the total profit per month z as a function of x and y. How many of each model should be produced per month to maximize the profit?

22. Suppose that regular and premium gasoline sell for $1.05 per gallon and $1.15 per gallon, respectively. Express the revenue per day z as a function of x and y. How many gallons of each type should be sold per day in order to maximize the revenue?

23. Suppose that stock A sells at $80 a share and stock B sells at $70 a share. Write the total cost z as a function of x and y. How many shares of each kind should the investor buy to minimize her total cost?

24. Suppose that stock A sells at $120 a share and stock B sells at $95 a share. Write the cost z as a function of x and y. How many shares of each kind should the investor buy to minimize his total cost?

25. If the church makes a profit of $10 per doll and $25 per blanket, how many dolls and blankets should Shirley make per month to maximize the profit?

26. If the handyman makes a profit of $3 per basket and $6 per vase, how many baskets and vases should he make per week to maximize his profit?

27. If the handyman makes a profit of $5 per basket and $4 per vase, how many baskets and vases should he make per week to maximize his profit?

28. If the charge per night is $90 per first-class room and $55 per ordinary room, how many rooms of each type should there be to maximize the revenue?

29. If the charge per night is $95 per first-class room and $50 per ordinary room, how many rooms of each type should there be to maximize the revenue?

30. If type I food costs $3 per pound and type II costs $3.75 per pound, how many pounds of each kind should be used per week to minimize cost?

4.2 The Three-Dimensional Coordinate System

In Section 4.1, we discussed how a linear programming problem with two variables can be solved geometrically. However, in the applications, we usually have many variables and many constraints. It is necessary in such cases to use a method other than the geometric one. To describe this method effectively, we will restrict ourselves, at the start, to problems in which there are two or three decision variables. In doing so, we will be able to justify geometrically some of the steps we must take in an algebraic method. Then we shall be able to expand the method to cases with many decision variables and many constraints. First, however, we need to briefly introduce the three-dimensional Cartesian coordinate system. A more detailed presentation will be given in Section 11.1.

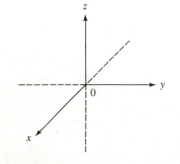

FIGURE 4.11

The System of Coordinates

A three-dimensional coordinate system is obtained by adding a third axis to the familiar xy-Cartesian plane. This third axis is usually called the z-axis and is perpendicular to the xy-plane through the origin. The xyz-coordinate system is pictured in Figure 4.11. This system is often called a right-handed system because, if a right hand is placed so that the index finger points in the positive direction of the x-axis and the middle finger points in the direction of the positive y-axis, then the thumb points in the positive direction of the z-axis. (See Figure 4.12.)

Given an ordered (a, b, c) of real numbers, the *graph* of that ordered triple in the coordinate system can be described as follows. We find the points on the x and y axes

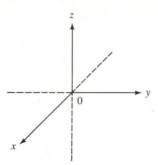

FIGURE 4.12

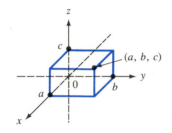

FIGURE 4.13

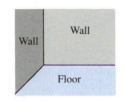

FIGURE 4.14

at directed distances a and b units from the origin, respectively, as we do when we locate (a, b) in the xy-plane. Then we find the point on the z-axis at a directed distance c from the origin. Now through each of the three points we construct a plane perpendicular to the axis on which the point lies. (See Figure 4.13.)

These three planes intersect at a unique point P, which is the graph of the ordered triple (a, b, c). We also say that a, b, c, are the *coordinates* of the point P: a is the x-coordinate or first coordinate, b is the y-coordinate or second coordinate, and c is the z-coordinate or third coordinate. Note that any point on the x-axis has coordinates $(x, 0, 0)$, any point on the y-axis has coordinates $(0, y, 0)$, and any point on the z-axis has coordinates $(0, 0, z)$. Similarly, points on the xy-plane have coordinates $(x, y, 0)$, points on the xz-plane have coordinates $(x, 0, z)$, and points on the yz-plane have coordinates $(0, y, z)$. The three coordinate planes divide the three-dimensional space into eight unbounded regions called *octants*. The region that consists of all points with coordinates (a, b, c), where a, b, and c are positive, is called the *first octant*. The other seven octants are not given special names. It is easy to visualize a three-dimensional coordinate system using two walls and the floor of a room. (See Figure 4.14.)

EXAMPLE 1   Plot the points $(3, 4, 2)$ and $(3, 2, -4)$.

Solution   The point $(3, 4, 2)$ is shown in Figure 4.15a and the graph of $(3, 2, -4)$ is shown in Figure 4.15b.
Do Exercise 1.   ■

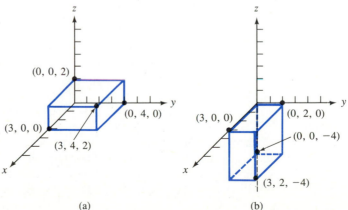

FIGURE 4.15   (a)   (b)   ■

Graphs of Linear Equations

The graph of an equation $ax + by = c$, where a, b, and c are real numbers with at least one of the numbers a and $b \neq 0$, is a straight line in the xy-plane. Similarly, if a, b, c, and d are real numbers, with at least one of the numbers a, b, and $c \neq 0$, the graph of the equation $ax + by + cz = d$ is a plane in the three-dimensional coordinate system. We find the intercepts of that plane with the three axes by letting two of the variables x, y, and z equal 0 and solve for the third variable. Since three noncollinear points determine a plane, it is usually sufficient to find the three intercepts of the plane with the three axes in order to sketch the plane whose equation is given.

EXAMPLE 2 Sketch the graph of $2x + 3y + 5z = 60$.

Solution To find the x-intercept, replace y and z by 0 and solve for x. We get

$2x = 60$, so that $x = 30$

The x-intercept is the point $(30, 0, 0)$.

To find the y-intercept, replace x and z by 0 and solve for y. We obtain

$3y = 60$, so that $y = 20$

The y-intercept is the point $(0, 20, 0)$.

The z-intercept is found by replacing x and y by 0 and solving for z. We get

$5z = 60$, so that $z = 12$

The z-intercept is $(0, 0, 12)$.

We now plot $(30, 0, 0)$, $(0, 20, 0)$, and $(0, 0, 12)$. These points are the three vertices of a first octant triangle. (See Figure 4.16.) This triangle is part of the plane which is the graph of $2x + 3y + 5z = 60$.

Do Exercise 5.

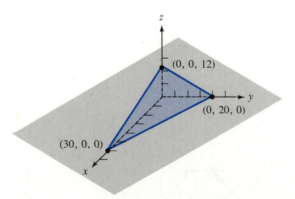

FIGURE 4.16

The plane whose equation is $ax + by + cz = d$ divides the three-dimensional space into three subsets, the plane itself and two half-spaces. One of these half-spaces is the graph of the inequality $ax + by + cz < d$, while the other is the graph of $ax + by + cz > d$. For example, in Figure 4.16, the half-space bounded by the plane and which contains the origin is the graph of $2x + 3y + 5z < 60$, since $2(0) + 3(0) + 5(0) < 60$ is true. The other half-space is the graph of $2x + 3y + 5z > 60$.

In general, two planes coincide, are parallel, or intersect along a straight line. In the latter case, a third plane may pass through that line, be parallel to it, or intersect it at a single point. To find the intersection of three planes, we need only solve the system of three linear equations whose graphs are the given planes.

EXAMPLE 3   **a.** Find the intersection of the planes P_1, P_2, and P_3, whose equations are,

$$x - \ y + 2z = \ 6$$
$$-x + 3y + 4z = 12$$
$$4x + 6y - 7z = 24$$

respectively.

b. Sketch the region in the first octant bounded by these three planes and the three coordinate planes.

Solution   **a.** We must solve the linear system

$$\begin{cases} x - \ y + 2z = \ 6 \\ -x + 3y + 4z = 12 \\ 4x + 6y - 7z = 24 \end{cases}$$

We use any of the techniques of Chapters 2 and 3 to obtain $x = 5$, $y = 3$, and $z = 2$. The point $(5, 3, 2)$ is, therefore, the point of intersection of the three planes.

b. It is easy to verify that $(6, 0, 0)$ is the x-intercept of both P_1 and P_3, $(0, 4, 0)$ is the y-intercept of both P_2 and P_3, and $(0, 0, 3)$ is the z-intercept of both P_1 and P_2. Hence, $(6, 0, 0)$, $(0, 0, 3)$, and $(5, 3, 2)$ are the vertices of a triangle that is part of the plane P_1. The points $(0, 4, 0)$, $(0, 0, 3)$, and $(5, 3, 2)$ are the vertices of a triangle that is part of the plane P_2. The points $(6, 0, 0)$, $(0, 4, 0)$, and $(5, 3, 2)$ are the vertices of a triangle that is part of the plane P_3. These triangles are sketched in Figure 4.17. They form part of the boundary of the region in the first octant bounded by the three given planes and by the three coordinate planes.

Do Exercise 9.

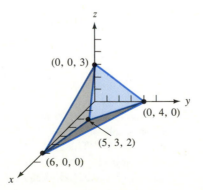

FIGURE 4.17

As we shall see in the next section, this region is the graph of the system

$$\begin{cases} x - y + 2z \le 6 \\ -x + 3y + 4z \le 12 \\ 4x + 6y - 7z \le 24 \\ x \ge 0,\, y \ge 0,\, z \ge 0 \end{cases}$$

If we have a set of n linear equations where each is in the two variables x and y and is the graph of a straight line in the xy-plane it may be desirable to find all possible points of intersection of pairs of these lines. Since each of these points is the intersection of two lines, we must solve all systems of two equations that can be obtained from the given set of n equations. The number of systems is the same as the number of subsets, each with two members, that we obtain from a set with n members.

Similarly, if we have a set of n linear equations where each is in the three variables x, y, and z and is the graph of a plane in the xyz-coordinate system, it may be desirable to find all possible points of intersection of triples of these planes. Since each of these points is the intersection of three planes, we must solve all systems of three equations that can be obtained from the given set of n equations. The number of systems is the same as the number of subsets, each with three members, that we obtain from a set with n members.

We shall show in Chapter 6 that the number of subsets, each with k elements, that can be obtained from a set of n elements is $\frac{n!}{k!(n-k)!}$, where the symbol $n!$ (read "n factorial") represents the product of the first n positive integers. For example, $5! = 5 \cdot 4 \cdot 3 \cdot 2 \cdot 1 = 120$.

EXAMPLE 4   We have a set of six planes. Assume that the maximum number of points of intersection we can get is finite. What is that number?

Solution   We have six linear equations. The coordinates of each point of intersection of three planes are obtained by solving a system of three linear equations. Each system is obtained by choosing three equations from the given set of six equations. Hence, the number of systems obtained is the same as the number of subsets, each with three members, that can be obtained from a set with six members. This number is 20 since

$$\frac{6!}{3!(6-3)!} = \frac{6!}{3!\,3!}$$
$$= \frac{6(5)(4)}{3!}$$
$$= \frac{120}{6} = 20$$

Note that 20 is the maximum number of points of intersection that we can have. Several of the systems may have identical solutions. In that case, the number of points of intersection will be less than 20. This will happen if more than three planes pass through the same point. Similarly, some of the systems may have no solution; for example, if two of the planes are parallel. Again, this will result in less than 20 points

of intersection. On the other hand, if we had not assumed that the number of points of intersection is finite, it is possible for all six planes to intersect along the same line. In that case, the number of points of intersection is infinite.
Do Exercise 21.   ∎

Exercise Set 4.2

1. Plot the points **a.** $(2, 4, 7)$, **b.** $(-3, 4, 5)$, and **c.** $(2, -3, 4)$.

2. Plot the points **a.** $(3, 5, 1)$, **b.** $(-2, 5, 7)$, and **c.** $(3, -4, 6)$.

3. Plot the points **a.** $(-2, 3, 6)$, **b.** $(2, -4, -5)$, and **c.** $(7, -5, 8)$.

4. Sketch the graph of the equation $2x + 3y + 4z = 24$.

5. Sketch the graph of the equation $5x + 2y + 3z = 60$.

6. Sketch the graph of the equation $x + 2y - 3z = 18$.

7. Sketch the graph of the equation $-x + 5y + 3z = 30$.

8. Sketch the graph of the equation $-3x + 4y - 3z = 36$.

In Exercises 9–13, the equations of three planes P_1, P_2, and P_3 are given. In each case:

a. Find the intersection of the three planes.
b. Sketch the region in the first octant bounded by the three planes and the three coordinate planes.

9. $\begin{aligned} 9x - 7y + 3z &= 9 \\ -11x + 6y + 4z &= 12 \\ 8x + 4y - 5z &= 8 \end{aligned}$

10. $\begin{aligned} 6x - 7y + 12z &= 12 \\ -14x + 3y + 12z &= 12 \\ 6x + 3y - 8z &= 12 \end{aligned}$

11. $\begin{aligned} 20x - 38y + 25z &= 100 \\ -x + y + z &= 4 \\ 24x + 30y - 33z &= 120 \end{aligned}$

12. $\begin{aligned} 21x - 37y + 9z &= 63 \\ -6x + 7y + z &= 7 \\ x + 3y - z &= 3 \end{aligned}$

13. $\begin{aligned} 7x - 10y + 14z &= 42 \\ -13x + 12y + 20z &= 60 \\ 5x + 6y - 13z &= 30 \end{aligned}$

In Exercises 14–21, assume that the maximum number of points of intersection is finite.

14. Five lines have been drawn in a plane. What is the maximum number of points of intersection of pairs of these lines?

15. Same as Exercise 14 with six lines.

16. Same as Exercise 14 with seven lines.

17. Same as Exercise 14 with eight lines.

18. Four planes have been drawn in the three-dimensional space. What is the maximum finite number of points of intersection of triples of these planes?

19. Same as Exercise 18 with five planes.

20. Same as Exercise 18 with seven planes.

21. Same as Exercise 18 with nine planes.

4.3 Introduction to the Simplex Method

In Section 4.1, the linear programming problem was discussed geometrically. However, this approach is cumbersome unless there are only two decision variables and relatively few constraints. In practice, there are usually many decision variables and constraints. In such cases, we use the *simplex method*. The name of this method comes from a geometric object. This object is analogous to the convex polygonal region introduced earlier and, in more advanced work, is called a simplex. We need not concern ourselves with this object now, but we should understand that many of its properties are similar to those of a convex polygonal region so that the techniques used to solve problems with only two decision variables may be extended to problems with many variables.

Interest in linear programming developed during World War II when military forces wanted to find efficient ways to bring supplies to the front. It was then that the simplex method was developed. Today, it has far-reaching applications in the fields of business and economics.

As stated, there are often many decision variables; we use symbols such as x_1, x_2, x_3, x_4, . . . , x_n to represent them. The variable names are convenient when computers are used to solve linear programming problems.

> We will at first restrict our discussion to the so-called *standard maximum linear programming problems*. In these problems, all constraints are in the form
>
> $$a_{i1}x_1 + a_{i2}x_2 + a_{i3}x_3 + \cdots + a_{in}x_n \leq b_i$$
>
> where x_1, x_2, x_3, . . . , x_n are decision variables and a_{i1}, a_{i2}, a_{i3}, . . . , a_{in}, and b_i are constants and
>
> 1. All variables are nonnegative.
> 2. All constant terms b_i are nonnegative.
> 3. The objective function is expressed in the form
>
> $$z = c_1x_1 + c_2x_2 + c_3x_3 + \cdots + c_nx_n$$
>
> where c_1, c_2, c_3, . . . , c_n are constants.
> 4. We are trying to maximize the objective function.

We will discuss minimization briefly at the end of this chapter.

When we considered problems with only two decision variables, we noted that the graph of $z = ax + by + c$ is a plane and asked you to visualize a house with a slanted flat roof over a convex polygonal floor. We argued that such a house would clearly have its highest point at the corner. To find this point, we could measure the height of the house at each of its corners; the largest of the numbers would be the highest point. Similarly, when we wanted to find the maximum value of the objective function, we evaluated that function at each corner of the region of feasible solutions and concluded that the largest number obtained was the maximum value of the function. A disadvantage of this method is that it is necessary to find the coordinates of each corner of the region of feasible solutions.

To understand the idea behind the simplex method, let us go back to visualizing the house with a slanted flat roof. Imagine that a squirrel is running along the edges of the roof and he has decided to reach the highest point. He will surely succeed if he adopts the following strategy. Starting at a corner, he runs along the edge of the roof from that corner provided he is going uphill by doing so. If both edges starting at that corner are going up, he chooses the steeper edge. He keeps doing this, from corner to corner, until he reaches a corner where no edge is going up. He clearly has reached the highest point. The advantage of this strategy is that he will likely reach the highest point without having tested each corner. The idea behind the simplex method is similar to the squirrel's strategy.

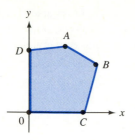

FIGURE 4.18

We will illustrate the technique geometrically and first restrict the discussion to cases where the number of decision variables is two or three. Therefore, the set of feasible solutions is a convex polygonal region in the case of two variables and a convex polyhedral region in the case of three variables. We shall say that two corners of such a region are *adjacent* if they are connected by an edge of the region. In Figure 4.18, A and B are adjacent, while A and C are not. In Figure 4.19, A and C are adjacent while A and B are not.

The simplex method is a procedure that allows us to move from one corner of the region of feasible solutions to an adjacent corner in such a way that the value of the objective function is increased at each step, until we either find the maximum value of the function or see that no maximum exists. The following theorem, stated without proof, justifies this procedure.

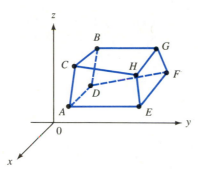

FIGURE 4.19

THEOREM 4.1: Let A be a corner of the region of feasible solutions of a linear programming problem. If the value of the objective function is not maximum at A, then there is an edge of the region of feasible solutions, starting at A, along which the values of the objective function increase as we move away from A.

Our goal is to develop an algebraic way to identify corners of the region of feasible solutions and also to find a way to move from one corner to an adjacent corner.

Slack Variables

The first step is to replace the inequalities by equations through the introduction of new variables called *slack variables*. For example, suppose that a farmer has 200 acres of land on which he wishes to grow wheat and barley. Let x and y represent the numbers of acres he uses for wheat and barley, respectively. Then

$$x + y \le 200$$

Since he may not use all of his land for these two grains, let s represent the number of acres on which he plants neither wheat nor barley. Then

$$x + y + s = 200$$

since each of the 200 acres is used for wheat, barley, or neither. Of course, s is 0 if the farmer uses all of his land for wheat and barley, but s will always be nonnegative. The variable s represents the "slack," the acreage not used for the two grains.

DEFINITION 4.2:   If each constraint which is described by an inequality

$$a_{i1}x_1 + a_{i2}x_2 + a_{i3}x_3 + \cdots + a_{in}x_n \leq b_i$$

is converted to an equation

$$a_{i1}x_1 + a_{i2}x_2 + a_{i3}x_3 + \cdots + a_{in}x_n + s_i = b_i,$$

where $s_i \geq 0$, then each s_i is called a *slack variable*.

EXAMPLE 1   Sketch the graph of the inequality $2x + 3y \leq 12$. Then introduce a variable s to change the inequality to an equation. Finally, discuss geometrically the situation when **a.** $s > 0$, **b.** $s = 0$, and **c.** $s < 0$.

Solution   The graph is shown in Figure 4.20.

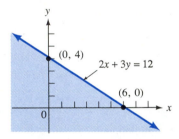

FIGURE 4.20

We introduce the variable s by writing

$$2x + 3y + s = 12$$

As we know, the graph of $2x + 3y = 12$ is a line L which divides the plane into three subsets: the open half-plane containing the origin (the shaded region in Figure 4.20), the line L, and the other open half-plane (not shaded in Figure 4.20). The coordinates of each point in the first half-plane satisfy the strict inequality $2x + 3y < 12$, those of each point on the line satisfy the equation $2x + 3y = 12$, and those of each point in the nonshaded half-plane satisfy the strict inequality $2x + 3y > 12$. Consequently, suppose that (a, b, c) satisfies

$$2x + 3y + s = 12$$

That is, suppose that

$$2a + 3b + c = 12$$

Then,

a. If $c > 0$, $2a + 3b < 12$ and the point (a, b) is in the shaded half-plane in Figure 4.20.

b. If $c = 0$, $2a + 3b = 12$ and the point (a, b) is on the line L.
c. If $c < 0$, $2a + 3b > 12$ and the point (a, b) is in the nonshaded half-plane.
Do Exercise 1. ■

 In Example 1, the variable s was not called a slack variable since we allowed it to assume negative values. However, this example and the next two examples illustrate that a slack variable must be nonnegative for the corresponding point to be in the region of feasible solutions.

EXAMPLE 2 Give a discussion similar to that of Example 1 for $3x + 5y + 2z \leq 30$.

Solution The graph of $3x + 5y + 2z = 30$ is a plane P, part of which is shown in Figure 4.21.

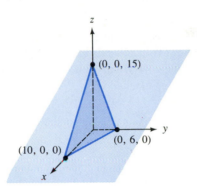

FIGURE 4.21

This plane divides the three-dimensional space into three subsets: the open half-space containing the origin, the plane P, and the open half-space not containing the origin. The coordinates of each point in the first half-space satisfy the strict inequality $3x + 5y + 2z < 30$, those of each point of the plane P satisfy the equation $2x + 5y + 2z = 30$, and those of each point in the other half-space satisfy the strict inequality $3x + 5y + 2z > 30$. Converting the inequality to an equation by introducing the variable s, we get

$$3x + 5y + 2z + s = 30$$

Suppose now that (a, b, c, d) is a solution of the foregoing equation. That is, suppose that $3a + 5b + 2c + d = 30$ is true. Then,

a. If $d > 0$, $3a + 5b + 2c < 30$ and the point (a, b, c) is in the half-space containing the origin.
b. If $d = 0$, $3a + 5b + 2c = 30$ and the point (a, b, c) is on the plane P.
c. If $d < 0$, $3a + 5b + 2c > 30$ and the point (a, b, c) is in the half-space not containing the origin.
Do Exercise 7. ■

EXAMPLE 3 **a.** Sketch the graph of the system

$$x + 2y \leq 8$$
$$x + y \leq 6$$
$$x \geq 0, y \geq 0$$

b. Introduce variables to convert the first two inequalities to equations.

c. Discuss the relationship between the values of the variables and the position of points in the coordinate plane.

Solution **a.** Let L_1 be the graph of $x + 2y = 8$ and L_2 be the graph of $x + y = 6$. The ordered pair $(4, 2)$ is the solution of the system

$$\begin{cases} x + 2y = 8 \\ x + y = 6 \end{cases}$$

Hence, the point $(4, 2)$ is the intersection of the lines L_1 and L_2. Each of these lines divides the plane into half-planes. The two half-planes containing the origin are the graphs of the inequalities

$$x + 2y < 8 \quad \text{and} \quad x + y < 6.$$

The half-plane to the right of the y-axis is the graph of the inequality $x > 0$ and the half-plane above the x-axis is the graph of the inequality $y > 0$. Thus, the intersection of these four half-planes, together with its boundary, is the graph of the given system of inequalities shown as the shaded region in Figure 4.22.

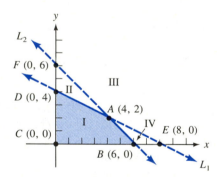

FIGURE 4.22

The corners (vertices) A, B, C, and D of the convex polygonal region have coordinates $(4, 2)$, $(6, 0)$, $(0, 0)$, and $(0, 4)$, respectively.

b. We introduce the variables s and t so that the first two inequalities are converted to equations.

$$x + 2y + s = 8$$
$$x + y + t = 6.$$

c. For convenience, let $E(8, 0)$ be the x-intercept of L_1 and $F(0, 6)$ be the y-intercept of L_2. (See Figure 4.22.) The lines L_1 and L_2 divide the first quadrant into four regions labeled I, II, III, and IV. Region I is the interior of the convex polygonal region $ABCD$; region II is the interior of triangle ADF; region III is the unbounded set to the right of the y-axis, the lines L_1 and L_2, and above the x-axis; and region IV is the interior of triangle ABE. Now suppose that (a, b, c, d) is a solution of the system

$$\begin{cases} x + 2y + s = 8 \\ x + y + t = 6 \end{cases}$$

That is, suppose that

$$a + 2b + c = 8 \quad \text{and} \quad a + b + d = 6.$$

Then, using the same reasoning as in Example 1, we see that the point (a, b) will be in different regions depending on the following conditions:

1. In region I if a, b, c, and d are positive.
2. In region II if a, b, and d are positive while c is negative.
3. In region III if a and b are positive while c and d are negative.
4. In region IV if a, b, and c are positive while d is negative.
5. On the boundary of region I if one or two of a, b, c, and d are 0 while the others are positive. For example, if both c and d are 0, the point will be on both L_1 and L_2, so it must be point A.

You should consider other cases. However, for the purpose of what will follow, we wish to stress that given the linear system

$$\begin{cases} x + 2y + s = 8 \\ x + \ y + t = 6 \end{cases}$$

and a solution (a, b, c, d) of that system, we can associate a point (a, b) in the coordinate plane and locate it vis-a-vis the lines L_1, L_2, the x-axis, and the y-axis, depending on the signs of a, b, c, and d. We emphasize that (a, b) is *inside* the convex polygonal region $ABCD$ if a, b, c, and d are all positive, and that (a, b) is at a corner (vertex) of the region if two of the four numbers a, b, c, and d are 0. Specifically, A is associated with the solution $(4, 2, 0, 0)$, B with the solution $(6, 0, 2, 0)$, C with the solution $(0, 0, 8, 6)$, and D with the solution $(0, 4, 0, 2)$. Also E is associated with the solution $(8, 0, 0, -2)$ and F is associated with the solution $(0, 6, -4, 0)$.

Do Exercise 13. ■

Vertices of the Region of Feasible Solutions

The next example is lengthy and detailed; we will omit all but the essential steps. It is important that you understand the role of slack variables in analyzing a system of linear inequalities. You are urged to become familiar with the steps taken.

EXAMPLE 4

Consider the system $\begin{cases} \qquad\ y + \ z \leq \ 5 \\ 5x + 7y - \ 3z \leq \ 35 \\ 15x + \ y + 21z \leq 105 \\ x \geq 0, y \geq 0, z \geq 0 \end{cases}$

Introduce slack variables to convert the first three inequalities to equations. Use the values of these slack variables to describe geometrically the graph of the given system of inequalities and sketch the graph.

Solution We introduce three slack variables r, s, and t and write

$$\begin{aligned}
y + \quad z + r &= \quad 5 \\
5x + 7y - \quad 3z + s &= \quad 35 \\
15x + \quad y + 21z + t &= 105 \\
x \geq 0,\ y \geq 0,\ z \geq 0,\ r \geq 0,\ s \geq 0,\ t \geq 0
\end{aligned}$$

Now consider the equations

$$y + \quad z = \quad 5 \tag{1}$$
$$5x + 7y - \quad 3z = \quad 35 \tag{2}$$
$$15x + \quad y + 21z = 105 \tag{3}$$
$$x = \quad 0 \tag{4}$$
$$y = \quad 0 \tag{5}$$
$$z = \quad 0 \tag{6}$$

The graph of each equation is a plane in three dimensions. Let P_i denote the graph of equation (i) for $i = 1$, 2, 3, 4, 5, and 6. Note that P_4 is the yz-plane, P_5 is the xz-plane, and P_6 is the xy-plane. To find the points of intersection of these planes, we solve linear systems of three equations taken from the six equations. For example, $(4, 3, 2)$ is the solution of the system

$$\begin{cases}
\quad\quad y + \quad z = \quad 5 & (1) \\
5x + 7y - \quad 3z = \quad 35 & (2) \\
15x + \quad y + 21z = 105 & (3)
\end{cases}$$

Hence, the point $(4, 3, 2)$ is the intersection of the planes P_1, P_2, and P_3. Since we form each system by taking three equations from a set of six equations, the number of different systems we can obtain is 20 since $\frac{6!}{3!(6-3)!} = \frac{6!}{3!3!} = 20$. The 20 subsets of the set $\{1, 2, 3, 4, 5, 6\}$, each with three elements, are $\{1, 2, 3\}$, $\{1, 2, 4\}$, $\{1, 2, 5\}$, $\{1, 2, 6\}$, $\{1, 3, 4\}$, $\{1, 3, 5\}$, $\{1, 3, 6\}$, $\{1, 4, 5\}$, $\{1, 4, 6\}$, $\{1, 5, 6\}$, $\{2, 3, 4\}$, $\{2, 3, 5\}$, $\{2, 3, 6\}$, $\{2, 4, 5\}$, $\{2, 4, 6\}$, $\{2, 5, 6\}$, $\{3, 4, 5\}$, $\{3, 4, 6\}$, $\{3, 5, 6\}$, $\{4, 5, 6\}$. With each of these subsets, we associate a linear system of three equations taken from the original set of six equations. For example, the system

$$\begin{cases}
\quad\quad y + \quad z = \quad 5 & (1) \\
5x + 7y - \quad 3z = \quad 35 & (2) \\
15x + \quad y + 21z = 105 & (3)
\end{cases}$$

is associated with the subset $\{1, 2, 3\}$, since it is made up of equations (1), (2), and (3); and the system

$$\begin{cases}
\quad\quad y + \quad z = \ 5 & (1) \\
5x + 7y - \quad 3z = 35 & (2) \\
\quad\quad\quad\quad x = \ 0 & (4)
\end{cases}$$

is associated with the subset $\{1, 2, 4\}$, since it is made up of equations (1), (2), and (4). The table gives the solutions of the 20 linear systems of three equations.

System	Solution
{1, 2, 3}	(4, 3, 2)
{1, 2, 4}	(0, 5, 0)
{1, 2, 5}	(10, 0, 5)
{1, 2, 6}	(0, 5, 0)
{1, 3, 4}	(0, 0, 5)
{1, 3, 5}	(0, 0, 5)
{1, 3, 6}	$\left(\frac{20}{3}, 5, 0\right)$
{1, 4, 5}	(0, 0, 5)
{1, 4, 6}	(0, 5, 0)
{1, 5, 6}	no solution
{2, 3, 4}	$\left(0, 7, \frac{14}{3}\right)$
{2, 3, 5}	(7, 0, 0)
{2, 3, 6}	(7, 0, 0)
{2, 4, 5}	$\left(0, 0, \frac{-35}{3}\right)$
{2, 4, 6}	(0, 5, 0)
{2, 5, 6}	(7, 0, 0)
{3, 4, 5}	(0, 0, 5)
{3, 4, 6}	(0, 105, 0)
{3, 5, 6}	(7, 0, 0)
{4, 5, 6}	(0, 0, 0)

Notice that there are duplications. For example, (0, 5, 0) is the solution of the systems represented by {1, 2, 4}, {1, 2, 6}, {1, 4, 6}, and {2, 4, 6}. This is because the planes P_1, P_2, P_4, and P_6 intersect at (0, 5, 0). Although the steps used to solve each of the 20 systems were left out, it is clear that this process is time consuming. In the future we will not need to find the solutions of all 20 systems. Remember that we are interested only in graphing the system of inequalities

$$\begin{cases} y + z \le 5 \\ 5x + 7y - 3z \le 35 \\ 15x + y + 21z \le 105 \\ x \ge 0, \ y \ge 0, \ z \ge 0 \end{cases}$$

Now for each of the 20 solutions in the table to the left, we calculate the values of the variables r, s, and t, so that (x, y, z, r, s, t) is a solution of the system

$$\begin{cases} y + z + r = 5 \\ 5x + 7y - 3z + s = 35 \\ 15x + y + 21z + t = 105 \end{cases}$$

The results are given in the following table.

System	Solution
{1, 2, 3}	(4, 3, 2, 0, 0, 0)
{1, 2, 4}	(0, 5, 0, 0, 0, 100)
{1, 2, 5}	(10, 0, 5, 0, 0, −150)
{1, 2, 6}	(0, 5, 0, 0, 0, 100)
{1, 3, 4}	(0, 0, 5, 0, 50, 0)
{1, 3, 5}	(0, 0, 5, 0, 50, 0)
{1, 3, 6}	$\left(\frac{20}{3}, 5, 0, 0, \frac{-100}{3}, 0\right)$
{1, 4, 5}	(0, 0, 5, 0, 50, 0)
{1, 4, 6}	(0, 5, 0, 0, 0, 100)
{1, 5, 6}	no solution
{2, 3, 4}	$\left(0, 7, \frac{14}{3}, \frac{-20}{3}, 0, 0\right)$
{2, 3, 5}	(7, 0, 0, 5, 0, 0)
{2, 3, 6}	(7, 0, 0, 5, 0, 0)
{2, 4, 5}	$\left(0, 0, \frac{-35}{3}, \frac{50}{3}, 0, 350\right)$
{2, 4, 6}	(0, 5, 0, 0, 0, 100)
{2, 5, 6}	(7, 0, 0, 5, 0, 0)
{3, 4, 5}	(0, 0, 5, 0, 50, 0)
{3, 4, 6}	(0, 105, 0, −100, 0, 0)
{3, 5, 6}	(7, 0, 0, 5, 0, 0)
{4, 5, 6}	(0, 0, 0, 5, 35, 105)

Recall that x, y, z, r, s, and t are required to be nonnegative. Thus, among the ordered sextuples listed, we can reject all those with a negative entry. Each of the ordered sextuples listed has at least three 0's. Recall the significance of each 0. Consider, for example, the sextuple (4, 3, 2, 0, 0, 0). This means that $x = 4$, $y = 3$, $z = 2$, $r = 0$, $s = 0$, and $t = 0$. The fact that $r = 0$ indicates that the point (4, 3, 2) is on the plane P_1. That point is also on the plane P_2 since $s = 0$, and on the plane P_3 since $t = 0$. Since $x = 4$, $y = 3$, and $z = 2$ satisfy the system of linear inequalities, we conclude that the point is on the graph of that system. In fact, it is a corner of the graph since it is the intersection of three of the planes which bound the graph. We now list all five of the ordered sextuples that have no negative entry. They are $(4, 3, 2, 0, 0, 0)$, $(0, 5, 0, 0, 0, 100)$, $(0, 0, 5, 0, 50, 0)$, $(7, 0, 0, 5, 0, 0)$, and $(0, 0, 0, 5, 35, 105)$. Thus, the points $(4, 3, 2)$, $(0, 5, 0)$, $(0, 0, 5)$, $(7, 0, 0)$, and $(0, 0, 0)$ are the corners of the graph of the given system of inequalities. Let us label these points A, B, C, D, and E, respectively. The graph is the polyhedron whose faces are the triangles ABD, ABC, ADC, BCE, BDE, and CDE. (See Figure 4.23.) This polyhedron has nine edges.

Do Exercise 27.

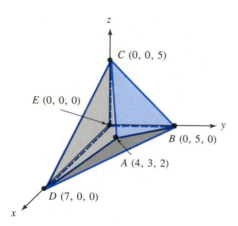

FIGURE 4.23

In the next section, we shall introduce a good technique to identify the corners of the graph of a system of linear inequalities. We shall also show how to move from any corner to an adjacent corner.

Exercise Set 4.3

In Exercises 1–12, sketch the graph of the given inequality. Then introduce a variable s to convert the inequality to an equation. Finally, discuss geometrically the situation when **a.** *s > 0,* **b.** *s = 0, and* **c.** *s < 0.*

1. $3x + y \leq 15$

2. $x + 6y \leq 18$

3. $2x + 3y \leq 12$

4. $2x + 5y \leq 30$

5. $2x + y \leq 14$

6. $4x + 3y \leq 24$

7. $4x + 5y + 4z \leq 40$

8. $12x + 8y + 9z \leq 72$

9. $4x + y + 4z \leq 16$

10. $10x + 6y + 5z \leq 90$

11. $2x + 2y + 3z \leq 12$

12. $2x + 4y + 5z \leq 30$

In Exercises 13–28, a system of linear inequalities is given.

a. *Sketch the graph of the system.*

b. *Introduce variables to convert the inequalities that have more than one variable to equations.*

c. *Discuss the relationship between the values of the variables and the position of points in the appropriate coordinate systems.*

13. $x + y \leq 7$
$3x + y \leq 15$
$x \geq 0, y \geq 0$

14. $x + y \leq 8$
$x + 6y \leq 18$
$x > 0, y \geq 0$

15. $x + 2y \leq 7$
 $2x + 3y \leq 12$
 $x \geq 0, y \geq 0$

16. $2x + 5y \leq 25$
 $3x + 2y \leq 21$
 $x \geq 0, y \geq 0$

17. $2x + 5y \leq 30$
 $2x + y \leq 14$
 $x \geq 0, y \geq 0$

18. $7x + 2y \leq 105$
 $x + y \leq 20$
 $x \geq 0, y \geq 0$

19. $x + y \leq 10$
 $x + 2y \leq 17$
 $3x + 2y \leq 27$
 $x \geq 0, y \geq 0$

20. $x + y \leq 17$
 $2x + y \leq 26$
 $4x + 3y \leq 58$
 $x \geq 0, y \geq 0$

21. $2x + y \leq 18$
 $x + 2y \leq 21$
 $4x + 5y \leq 72$
 $x \geq 0, y \geq 0$

22. $8x + 9y \leq 260$
 $5x + 3y \leq 131$
 $x + 2y \leq 50$
 $x \geq 0, y \geq 0$

23. $4x + 5y + 4z \leq 40$
 $12x + 9y + 8z \leq 96$
 $x \geq 0, y \geq 0, z \geq 0$

24. $12x + 8y + 9z \leq 72$
 $x + y + z \leq 7$
 $x \geq 0, y \geq 0, z \geq 0$

25. $4x + y + 4z \leq 16$
 $2x + 3y + 7z \leq 18$
 $x \geq 0, y \geq 0, z \geq 0$

26. $10x + 6y + 5z \leq 90$
 $8x + 12y + 13z \leq 108$
 $x \geq 0, y \geq 0, z \geq 0$

27. $2x + y + z \leq 10$
 $2x + 2y + 3z \leq 14$
 $x \geq 0, y \geq 0, z \geq 0$

28. $2x + 4y + 5z \leq 30$
 $4x + 4y + 5z \leq 40$
 $x \geq 0, y \geq 0, z \geq 0$

4.4 Moving from Corner to Corner

Now we shall discuss algebraically how to identify corners of the region of feasible solutions of a linear programming problem and how to go from any corner to an adjacent corner. This is the final step before describing the simplex method in Section 4.5. Again, because we wish our presentation to have a geometric flavor, we shall restrict our examples to cases with two or three variables. However, the procedure can be generalized to any number of variables.

Basic Feasible Solutions

In the preceding section, we showed that by introducing slack variables we could convert all constraints, except the nonnegative ones, to equations. In doing so, we always obtain a system of linear equations with more variables than equations. In general, as we saw in Section 3.3, such a system has infinitely many solutions. In the preceding section, we also observed that given a solution of the linear system obtained by introducing slack variables, we could locate a point in the appropriate coordinate system in relation to the boundary of the region of feasible solutions. In particular, we could tell whether a point was outside the region, on the boundary, a vertex, or inside the region. Since we are primarily interested in the vertices of the region of feasible solutions, recall that in each example given so far, if an ordered n-tuple was a solution of the linear system obtained by introducing slack variables and it represented a vertex of the region of feasible solutions, then the number of 0's in the n-tuple was at least as large as the number of decision variables in the problem. This is not a coincidence. It derives from the following theorem, which we state without proof.

THEOREM 4.2:   Suppose that in a linear programming problem we have m decision variables $x_1, x_2, \ldots, x_m$, subject to n constraints, and suppose that we introduce n slack variables $s_1, s_2, \ldots, s_n$, thus obtaining a system of n equations in $m + n$ variables. Suppose further that an optimum solution exists. Then, there exists a solution $(a_1, a_2, \ldots, a_m, b_1, b_2, \ldots, b_n)$ of the linear system where at least m of the numbers $a_1, a_2, \ldots, a_m, b_1, b_2, \ldots, b_n$ are 0 with the objective function attaining its optimum value at $(a_1, a_2, \ldots, a_m)$.

You need not memorize the foregoing theorem, but you should remember the following.

In trying to locate where the objective function in a linear programming problem attains its maximum, look for solutions of the linear system of equations obtained by introducing slack variables, where the number of 0's is at least as large as the number of decision variables and all other values are nonnegative.

At this point, you may wish to review Section 3.3, where you learned to solve a system of linear equations by row reduction working on the augmented matrix of the system. The following example will illustrate the concepts we need to remember.

EXAMPLE 1   You are given a system of three equations in the five variables x_1, x_2, x_3, x_4, and x_5. After performing a sequence of elementary row operations on the augmented matrix of the system, you obtain

$$\begin{bmatrix} 2 & 0 & 1 & 3 & 0 & \vline & 4 \\ -7 & 1 & 0 & 8 & 0 & \vline & -3 \\ 4 & 0 & 0 & 9 & 1 & \vline & 9 \end{bmatrix}$$

Find three solutions of the system.

Solution   This augmented matrix represents

$$\begin{cases} 2x_1 \quad\quad + x_3 + 3x_4 \quad\quad = \quad 4 \\ -7x_1 + x_2 \quad\quad + 8x_4 \quad\quad = -3 \\ 4x_1 \quad\quad\quad\quad + 9x_4 + x_5 = \quad 9 \end{cases}$$

Since x_3, x_2, and x_5 appear only in the first, second, and third equations, respectively, we can solve for these variables in terms of x_1 and x_4 as follows:

$$\begin{cases} x_3 = \quad 4 - 2x_1 - 3x_4 \\ x_2 = -3 + 7x_1 - 8x_4 \\ x_5 = \quad 9 - 4x_1 - 9x_4 \end{cases}$$

We may now give any values to x_1 and x_4 and obtain corresponding values for x_3, x_2, and x_5.

We begin with the easiest and let $x_1 = x_4 = 0$. We get

$$x_3 = 4 - 2(0) - 3(0) = 4$$
$$x_2 = -3 + 7(0) - 8(0) = -3$$
$$x_5 = 9 - 4(0) - 9(0) = 9$$

The corresponding solution is $(0, -3, 4, 0, 9)$.
   We let $x_1 = 0$ and $x_4 = 1$ and get

$$x_3 = 4 - 2(0) - 3(1) = 1$$
$$x_2 = -3 + 7(0) - 8(1) = -11$$
$$x_5 = 9 - 4(0) - 9(1) = 0$$

The corresponding solution is $(0, -11, 1, 1, 0)$.
   Finally, we let $x_1 = 1$ and $x_4 = 0$ and obtain

$$x_3 = 4 - 2(1) - 3(0) = 2$$
$$x_2 = -3 + 7(1) - 8(0) = 4$$
$$x_5 = 9 - 4(1) - 9(0) = 5$$

A third solution is $(1, 4, 2, 0, 5)$.
Do Exercise 3.   ■

EXAMPLE 2   You are given a system of three equations in the six variables x_1, x_2, x_3, x_4, x_5, and x_6. After performing a sequence of elementary row operations on the augmented matrix of the system, you obtain

$$\begin{bmatrix} 3 & 0 & 2 & 1 & 6 & 0 & | & 7 \\ 2 & 0 & 7 & 0 & 9 & 1 & | & 8 \\ -1 & 1 & 3 & 0 & 7 & 0 & | & -3 \end{bmatrix}$$

Find an obvious solution of the system.

Solution   Note that columns 4, 6, and 2, in that order, form the 3×3 identity matrix. Thus, the entries of the last column give the values of x_4, x_6, and x_2 provided that the other variables are given the value 0. An obvious solution is $(0, -3, 0, 7, 0, 8)$.
Do Exercise 5.   ■

> **DEFINITION 4.3:**   In general, if we have a linear system of m equations in n variables where $n > m$, and if we have performed a sequence of elementary row operations on the augmented matrix of the system to obtain a matrix in which m of the columns, written in some order, form the identity matrix $\mathbf{I}_m$, then the variables corresponding to those columns are called *basic variables*. The other $(n - m)$ variables are called the *nonbasic variables*.

   If we give the value 0 to all nonbasic variables, the values of the basic variables may be read directly from the last column of the last augmented matrix. We now show

how a matrix can be used to identify the corners of a region of feasible solutions of a linear programming problem and how to move from any corner to an adjacent corner. For the time being, we consider only the constraints of the problem, rather than the objective function, focusing instead on the region of feasible solutions.

Entering and Leaving Variables

EXAMPLE 3   Consider the system

$$x + 2y \leq 10$$
$$x + y \leq 6$$
$$2x + y \leq 11$$
$$x \geq 0, y \geq 0$$

a. Sketch the graph of the system.
b. Introduce slack variables to convert the first three inequalities to equations.
c. Use the augmented matrix of the linear system obtained in part **b** to identify one corner of the graph of part **a** and use elementary row operations on that matrix to move from corner to corner.

Solution   a. The graph is shown in Figure 4.24.

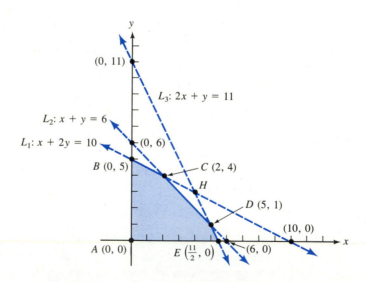

FIGURE 4.24

b. We introduce slack variables r, s, and t by writing

$$\begin{cases} x + 2y + r = 10 \\ x + y + s = 6 \\ 2x + y + t = 11 \end{cases}$$
$$x \geq 0, y \geq 0, r \geq 0, s \geq 0, t \geq 0$$

c. The augmented matrix of the system is

$$
\left[\begin{array}{ccccc|c}
1 & 2 & 1 & 0 & 0 & 10 \\
1 & 1 & 0 & 1 & 0 & 6 \\
2 & 1 & 0 & 0 & 1 & 11
\end{array}\right]
$$

Note that columns 3, 4, and 5 that give the r, s, and t coefficients, respectively, form the identity matrix I_3. Thus, we begin by selecting r, s, and t as basic variables and x, y as nonbasic variables. We say that $\{r, s, t\}$ is the *basis*. Then we let $x = y = 0$, and from the last column of the matrix we read $r = 10$, $s = 6$, and $t = 11$. The first solution is $(0, 0, 10, 6, 11)$. This is a feasible solution since all entries are nonnegative. We call it a *basic feasible solution*. It corresponds to the corner of the graph of part **a** that has coordinates $(0, 0)$. This corner has been labeled A in Figure 4.24.

We would like to move to corner B. We see that A is the intersection of the x and y axes. Now we wish to move along the y-axis. Thus, x must remain 0 and y must become positive. That is, x will still be a nonbasic variable, while y will become a basic variable. We say that y is the *entering variable* and we indicate this on the matrix by drawing an arrow below the column of y coefficients as indicated in the next matrix. That column is called the *pivot column*.

	Basis	x	y	r	s	t	Values of basic variables
←	r	1	②	1	0	0	10
	s	1	1	0	1	0	6
	t	2	1	0	0	1	11
			↑				

Since y is entering the basis, one present element of the basis must leave. Should it be r, s, or t? We go to the point B and we see that B is not only on the y-axis, but also on the line L_1, the graph of $x + 2y = 10$. Thus, r must be 0 since the first equation of our system was $x + 2y + r = 0$. That is, r must become nonbasic. We say that r is the *leaving variable* (or *departing variable*) and indicate this with an arrow pointing out to the left of the matrix by the row representing r. That row is called the *pivot row*. The number 2 which is in the pivot column and in the pivot row is called the *pivot element* and is circled. Now we perform elementary row operations on the augmented matrix of the system to convert the pivot element to the number 1 and all other elements in the pivot column to 0. We proceed as follows:

Basis	x	y	r	s	t	Values of basic variables
r	1	2	1	0	0	10
s	1	1	0	1	0	6
t	2	1	0	0	1	11

$\left(\tfrac{1}{2}\right)R_1 \to R_1$

Values of

Basis	x	y	r	s	t	basic variables	
r	$\frac{1}{2}$	1	$\frac{1}{2}$	0	0	5	
s	1	1	0	1	0	6	$-R_1 + R_2 \rightarrow R_2$
t	2	1	0	0	1	11	$-R_1 + R_3 \rightarrow R_3$

Values of

Basis	x	y	r	s	t	basic variables
y	$\frac{1}{2}$	1	$\frac{1}{2}$	0	0	5
s	$\frac{1}{2}$	0	$\frac{-1}{2}$	1	0	1
t	$\frac{3}{2}$	0	$\frac{-1}{2}$	0	1	6

The basis is now $\{y, s, t\}$ and the values of the basic variables can be read from the last column: $y = 5$, $s = 1$, and $t = 6$. The nonbasic variables are x and r and are both 0. The solution of the linear system is $(0, 5, 0, 1, 6)$. The first two entries give us $(0, 5)$, which are the coordinates of the corner B. Now we wish to move along the line L_1 and go to adjacent corner C. Since we are staying on L_1, the value of r must remain 0; that is, r will remain nonbasic. Since we are leaving the y-axis, x will be positive. That is, x will enter the basis. We indicate this with an arrow below the matrix under the column of x coefficients. Which of the variables y, s, or t must leave the basis? Since the point C is on the line L_2 which is the graph of the equation $x + y = 6$, we must have $s = 0$ because the second equation of our system was $x + y + s = 6$. Thus, s must become nonbasic and will be the leaving variable. We indicate this with an arrow to the left of the augmented matrix by the s row. (See the following matrix.)

Values of

	Basis	x	y	r	s	t	basic variables
	y	$\frac{1}{2}$	1	$\frac{1}{2}$	0	0	5
$\leftarrow$	s	$\left(\frac{1}{2}\right)$	0	$\frac{-1}{2}$	1	0	1
	t	$\frac{3}{2}$	0	$\frac{-1}{2}$	0	1	6

$\uparrow$

Thus, the pivot element is in column 1, row 2 and is circled. Now we must perform elementary row operations to convert the pivot element to 1 and all other entries in the pivot column to 0.

Values of

Basis	x	y	r	s	t	basic variables	
y	$\frac{1}{2}$	1	$\frac{1}{2}$	0	0	5	
s	$\frac{1}{2}$	0	$\frac{-1}{2}$	1	0	1	$2R_2 \rightarrow R_2$
t	$\frac{3}{2}$	0	$\frac{-1}{2}$	0	1	6	

Basis	x	y	r	s	t	Values of basic variables	
y	$\frac{1}{2}$	1	$\frac{1}{2}$	0	0	5	$\left(\frac{-1}{2}\right)R_2 + R_1 \to R_1$
s	1	0	-1	2	0	2	
t,	$\frac{3}{2}$	0	$\frac{-1}{2}$	0	1	6	$\left(\frac{-3}{2}\right)R_2 + R_3 \to R_3$

Basis	x	y	r	s	t	Values of basic variables
y	0	1	1	-1	0	4
x	1	0	-1	2	0	2
t	0	0	1	-3	1	3

Note that r and s are the nonbasic variables and we set them equal to 0. From the last column, we read $y = 4$, $x = 2$, and $t = 3$. Thus, the solution of the linear system is $(2, 4, 0, 0, 3)$. The first two entries give us the coordinates $(2, 4)$ of C. Since the other entries are nonnegative, we have a feasible solution to the given system of inequalities. Since two of the entries in $(2, 4, 0, 0, 3)$ are 0, we have a basic feasible solution.

We leave it as an exercise for you to proceed from C to D and then from D to E, and from E back to A.

Do Exercise 13. ■

Selecting the Leaving Variable

In the preceding example, we depended on the graph of the system of inequalities to select the entering and leaving variables each time we moved from one corner to an adjacent corner. However, we are striving for a way to do this without a diagram. The way the entering variable is selected will be discussed in the next section. We shall now show how the leaving variable should be selected once the entering variable has been chosen.

Let us go back for a moment to the previous example where we wanted to go from corner B to corner C. Suppose that we decided that t should be the leaving variable. Then t would have become nonbasic and its value would have been 0. We would have obtained the point H on L_3. However, the point H is on the side of the line L_2 which does not contain the origin, so the value of s would have been negative and the solution would not have been a feasible solution. The underlying principle is as follows: once an entering variable has been chosen, the leaving variable should be chosen so that after performing elementary row operations to convert the pivot element to a 1 and all other elements in the pivot column to 0, *all entries in the last column of the resulting matrix should be nonnegative*. We use this principle to arrive at a systematic way of choosing the leaving variable.

Suppose that we have an $m \times n$ augmented matrix, with $m < (n - 1)$ such that m of the first $(n - 1)$ columns, arranged in the proper order, form the identity matrix $\mathbf{I}_m$ and suppose further that all entries in the last column are nonnegative. If we have selected the jth variable to be the entering variable, the jth column is the pivot column.

In the following matrix we write both the pivot column and the last column, which at this stage gives the values of the basic variables, and we omit all other columns.

$$\begin{bmatrix} & & a_{1j} & & & b_1 \\ & & a_{2j} & & & b_2 \\ \cdots & & a_{3j} & \cdots & & b_3 \\ & & \vdots & & & \vdots \\ & & a_{mj} & & & b_m \end{bmatrix}$$

We wonder whether a_{kj} is a good candidate for the pivot element. If it were, we would perform elementary row operations on the matrix to convert a_{kj} to a 1 and all other entries in the pivot column to 0. Our first step might be to multiply row k by $1/a_{kj}$. By doing this, the corresponding entry in the last column becomes b_k/a_{kj}. This already puts some restrictions on a_{kj}. It cannot be 0. Furthermore, it should not be negative since the quotient b_k/a_{kj} gives the value of the new basic variable and that value must be nonnegative (remember that b_k is nonnegative). Thus, as a first step, we note that the pivot element should be positive. But what do we do if we have several positive entries in the pivot column? Suppose that both a_{pj} and a_{kj} are positive. We choose a_{kj} over a_{pj} for the pivot element provided that by converting a_{kj} to 1 and a_{pj} to 0 through elementary row operations, the corresponding entries in the last column will remain nonnegative. Clearly, a_{pj} will be converted to a 0 by multiplying row k by $-a_{pj}/a_{kj}$ and adding the result to row p to obtain the new row p. The new entry in row p, last column will be $b_p - (a_{pj}/a_{kj})b_k$. This last element must be nonnegative. Thus, we have

$$b_p - (a_{pj}/a_{kj})b_k \geq 0$$

Multiplying both sides by a_{kj} (which is legal since we know a_{kj} is positive), we get

$$b_p a_{kj} - b_k a_{pj} \geq 0$$

Equivalently,

$$b_p a_{kj} \geq b_k a_{pj}$$

or

$$b_p/a_{pj} \geq b_k/a_{kj}$$

Procedure for Finding the Pivot

Having chosen the entering variable, from which we get the pivot column, we determine the pivot as follows. We divide each positive entry of the pivot column into the corresponding entry in the last column. If b_k/a_{kj} is the smallest quotient obtained, a_{kj} is chosen as the pivot element. Note that b_k/a_{kj} is necessarily nonnegative since all b_i's are nonnegative. In case of a tie between b_k/a_{kj} and b_q/a_{qj}, either a_{kj} or a_{qj} may be chosen.

EXAMPLE 4   Consider the system of inequalities of Example 4, Section 4.3.

$$y + \quad z \leq \quad 5$$
$$5x + 7y - \quad 3z \leq \quad 35$$
$$15x + \quad y + 21z \leq 105$$
$$x \geq 0,\ y \geq 0,\ z \geq 0$$

Find three vertices of the graph of this system using the method just described.

Solution   We first introduce slack variables r, s, and t and write

$$\begin{cases} \quad\quad y + \quad z + r \quad\quad\quad\quad = \quad 5 \\ 5x + 7y - \quad 3z \quad\quad + s \quad\quad = \quad 35 \\ 15x + \quad y + 21z \quad\quad\quad\quad + t = 105 \end{cases}$$

The augmented matrix is

Basis	x	y	z	r	s	t	Values of basic variables
r	0	1	1	1	0	0	5
s	5	7	−3	0	1	0	35
t	15	1	21	0	0	1	105

We choose r, s, and t as our initial basic variables since the r, s, and t columns form the identity matrix $\mathbf{I}_3$. Thus, x, y, and z are nonbasic variables and we set them equal to 0. The first basic feasible solution is $(0, 0, 0, 5, 35, 105)$. The first three entries of this sextuple give us the coordinates $(0, 0, 0)$ of the origin, which is a vertex of the graph of the original system of inequalities. Suppose that we wish to have y as the entering variable. All three entries in the y column are positive. We divide each entry into the corresponding element in the last column. We get $\frac{5}{1} = 5$, $\frac{35}{7} = 5$, and $\frac{105}{1} = 105$. It is a tie between the first two quotients. We may choose either 1 or 7 to be the pivot element. We choose 1, since that choice will lead to simpler calculations. This means that r will be the leaving variable. We perform elementary row operations as follows to convert the other entries in the y column to 0.

Basis	x	y	z	r	s	t	Values of basic variables	
← r	⓪	1	1	1	0	0	5	$(-7)R_1 + R_2 \rightarrow R_2$
s	5	7	−3	0	1	0	35	$(-1)R_1 + R_3 \rightarrow R_3$
t	15	1	21	0	0	1	105	

Basis	x	y	z	r	s	t	Values of basic variables
y	0	1	1	1	0	0	5
s	5	0	−10	−7	1	0	0
t	15	0	20	−1	0	1	100

We see from the last column that $y = 5$, $s = 0$, and $t = 100$. The other variables are nonbasic and we set them equal to 0. We get the basic feasible solution $(0, 5, 0, 0, 0, 100)$. The first three entries give us the coordinates $(0, 5, 0)$ of a second vertex of the graph of the original system.

Next we choose z to be the entering variable. Then the pivot column is the z column. Only two of its entries are positive. We divide them into the corresponding entries of the last column. We get $\frac{5}{1} = 5$ and $\frac{100}{20} = 5$. We have a tie again and choose the 1 to be the pivot, which means that y will be the leaving variable, and we proceed.

Basis	x	y	z	r	s	t	Values of basic variables	
$\leftarrow$ y	0	①	1	1	0	0	5	$10R_1 + R_2 \rightarrow R_2$
s	5	0	-10	-7	1	0	0	$(-20)R_1 + R_3 \rightarrow R_3$
t	15	0	20	-1	0	1	100	

Basis	x	y	z	r	s	t	Values of basic variables
z	0	1	1	1	0	0	5
s	5	10	0	3	1	0	50
t	15	-20	0	-21	0	1	0

From this, we read the values of the basic variables $z = 5$, $s = 50$, and $t = 0$. The other variables are nonbasic and have the value 0. The solution is $(0, 0, 5, 0, 50, 0)$. The first three entries give the coordinates $(0, 0, 5)$ of another vertex of the graph of the original system.

Do Exercise 17.   ■

Unbounded Region of Feasible Solutions

In the preceding two examples, we saw that a pivot column was determined by the choice of an entering variable. We then considered only the positive entries of that column as candidates for the pivot element. The following example should help shed light on the significance of having a pivot column whose entries are all negative.

EXAMPLE 5   Consider the system

$$-2x + y \leq 1$$
$$x - y \leq 10$$
$$5x - 8y \leq 25$$
$$x \geq 0, y \geq 0$$

a. Sketch the graph of the system.
b. Introduce slack variables to convert the first three inequalities to equations, then move from corner to corner of the graph using the method introduced in this section.

Solution   **a.** The graph is shown in Figure 4.25. The graphs of the equations $-2x + y = 1$, $x - y = 10$, and $5x - 8y = 25$ have been labeled L_1, L_2, and L_3, respectively. The four corners of the graph have been denoted A, B, C, and D.

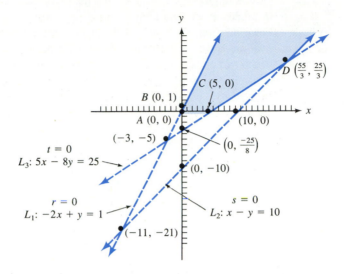

FIGURE 4.25

b. We introduce the slack variables r, s, and t and write

$$\begin{cases} -2x + y + r \qquad\quad = 1 \\ \quad\; x - y \quad\;\; + s \quad\; = 10 \\ \quad 5x - 8y \qquad\quad + t = 25 \end{cases}$$

The augmented matrix is

$$\begin{bmatrix} -2 & 1 & 1 & 0 & 0 & | & 1 \\ 1 & -1 & 0 & 1 & 0 & | & 10 \\ 5 & -8 & 0 & 0 & 1 & | & 25 \end{bmatrix}$$

Since the third, fourth, and fifth columns form $\mathbf{I}_3$, we select r, s, and t as our initial basis. The values of the basic variables are read from the last column. They are $r = 1$, $s = 10$, and $t = 25$. The nonbasic variables are x and y and are given the value 0. Thus, a basic feasible solution is $(0, 0, 1, 10, 25)$. The first two entries give the coordinates $(0, 0)$ of the origin, which is a corner of the region of feasible solutions. It has been labeled A in Figure 4.25. Suppose that we select y as the entering variable as indicated by the arrow below the y column in the augmented matrix below.

	Basis	x	y	r	s	t	Values of basic variables
←	r	-2	①	1	0	0	1
	s	1	-1	0	1	0	10
	t	5	-8	0	0	1	25

The only positive entry in the pivot column is the 1 in the r row. Hence, r will be the leaving variable as indicated by the arrow on the left of the matrix. Therefore 1 is the pivot element and it is circled. We must perform elementary row operations on this augmented matrix so that the pivot element is converted to a 1 (in this case it already is 1), and then all other entries in the pivot column are converted to 0. From here on, we shall not document on the right side the type of elementary row operations we are performing as we have done earlier, since we shall always proceed in the same way. That is, once a_{kj} has been selected as the pivot element, we shall multiply row k by $1/a_{kj}$ to get a new row k. This will convert a_{kj} to 1. Also for each $p \neq k$, we shall multiply row k by $-a_{pj}/a_{kj}$ and add the result to row p to get a new row p. This will convert a_{pj} to 0 as required, since $a_{pj} + (-a_{pj}/a_{kj})a_{kj} = a_{pj} - a_{pj} = 0$.

	Basis	x	y	r	s	t	Values of basic variables
←	r	-2	①	1	0	0	1
	s	1	-1	0	1	0	10
	t	5	-8	0	0	1	25

Basis	x	y	r	s	t	Values of basic variables
y	-2	1	1	0	0	1
s	-1	0	1	1	0	11
t	-11	0	8	0	1	33

The new basis is $\{y, s, t\}$. The values of the basic variables are $y = 1$, $s = 11$, and $t = 33$. The nonbasic variables are x and r and we give them the value 0. The new feasible solution is $(0, 1, 0, 11, 33)$. The first two entries give the coordinates $(0, 1)$ of corner B of the graph of the given system. The nonbasic variables are x and r. Therefore, one of these two must be the next entering variable. If we choose r, the basis will be $\{r, s, t\}$, which was the initial basis, and we will be back at the initial corner A. So, we choose x instead. But that choice makes the first column the pivot column and all entries in that column are negative. Why is this so? Notice in Figure 4.25 that point B is the only corner of the region of feasible solutions that is on line L_1 and the region of feasible solutions is unbounded. Thus, if we wish to keep $r = 0$, that is to keep r as a nonbasic variable and enter x as the new basic variable, we cannot choose an entering variable in a way which will keep us within the region of feasible solutions. The following will illustrate this fact.

Suppose we choose s as the leaving variable, then -1 becomes the pivot element and we proceed as shown.

	Basis	x	y	r	s	t	Values of basic variables
	y	-2	1	1	0	0	1
←	s	-1	0	1	1	0	11
	t	-11	0	8	0	1	33

Basis	x	y	r	s	t	Values of basic variables
y	0	1	-1	-2	0	-21
x	1	0	-1	-1	0	-11
t	0	0	-3	-11	1	-88

The corresponding solution is $(-11, -21, 0, 0, -88)$. This is not a feasible solution since several of the values of the variables are negative. The first two entries give the coordinates $(-11, -21)$ of the point of intersection of the lines L_1 and L_2. You will find it instructive to rework part **b** of this example choosing, at the beginning, x instead of y as the entering variable. This should give the point C with coordinates $(5, 0)$. The next time, choose y as the entering variable, thus obtaining the point D with coordinates $\left(\frac{55}{3}, \frac{25}{3}\right)$. Since point D is the only corner of the region of feasible solutions which is on line L_2, we will have the same situation as we had when we reached point B.

Do Exercise 25. ■

Procedure for Moving from Corner to Corner

Given a set of "$\leq$" constraints for a linear programming problem, we identify the basic feasible solutions as follows:

Step 1. Convert each "$\leq$" constraint to an equation by introducing a slack variable for each constraint.

Step 2. Write the augmented matrix of the system of equations obtained in Step 1. The columns representing the slack variables introduced in Step 1 form an identity matrix. Thus, we choose the slack variables as our initial basis and their values are read from the last column. All other variables are given the value 0. This gives us our initial basic feasible solution.

Step 3. Having chosen one of the nonbasic variables to become a basic variable (the entering variable), which at this point we do arbitrarily, the column representing that variable becomes the pivot column. Divide all positive entries of the pivot column into the corresponding entry of the last column. If b_k/a_{kj} is the smallest quotient obtained, a_{kj} is chosen as the pivot element. In case of a tie between b_k/a_{kj} and b_p/a_{pj}, either a_{kj} or a_{pj} may be chosen.

Step 4. Perform elementary row operations on the augmented matrix to convert the pivot element to a 1 and all other entries in the pivot column to 0. From now on we shall refer to this process as *pivoting*. The new basis consists of the original basis with one of its members replaced by the entering variable. The values of the basic variables are read from the last column and the nonbasic variables are given the value 0 to get the next basic feasible solution.

Step 5. Choose another entering variable and return to Step 3.

Exercise Set 4.4

In Exercises 1–6, the given matrix has been obtained by performing elementary row operations on the augmented matrix of a linear system of equations. Give the solution that can be read directly from this matrix.

1. $\begin{bmatrix} 2 & 0 & -3 & 1 & 7 & | & 5 \\ 3 & 1 & 9 & 0 & 6 & | & -3 \end{bmatrix}$

2. $\begin{bmatrix} 3 & 1 & 7 & 2 & 0 & | & 9 \\ 5 & 0 & 8 & 1 & 1 & | & 7 \end{bmatrix}$

3. $\begin{bmatrix} 0 & 3 & 1 & 5 & 0 & 8 & | & -2 \\ 1 & 7 & 0 & 9 & 0 & 2 & | & 13 \\ 0 & -3 & 0 & 4 & 1 & 9 & | & 17 \end{bmatrix}$

4. $\begin{bmatrix} 2 & 3 & 0 & 0 & 5 & 1 & | & 4 \\ 9 & -2 & 1 & 0 & 3 & 0 & | & -5 \\ -8 & 5 & 0 & 1 & 6 & 0 & | & 7 \end{bmatrix}$

5. $\begin{bmatrix} 0 & 0 & 3 & 1 & 7 & 0 & | & 5 \\ 1 & 0 & 8 & 0 & 6 & 0 & | & -2 \\ 0 & 1 & 9 & 0 & 8 & 0 & | & 3 \\ 0 & 0 & 1 & 0 & 6 & 1 & | & 13 \end{bmatrix}$

6. $\begin{bmatrix} 6 & 1 & 7 & 0 & 3 & 0 & 0 & | & 8 \\ 4 & 0 & -9 & 0 & 6 & 1 & 0 & | & 7 \\ 5 & 0 & 3 & 1 & 9 & 0 & 0 & | & 6 \\ -4 & 0 & 6 & 0 & -5 & 0 & 1 & | & 8 \end{bmatrix}$

In Exercises 7–24, a system of linear inequalities is given. In each case, find the coordinates of the indicated number of corners of the graph of the system using the method described in this section. Do not sketch the graphs. For Exercise n, where $7 \le n \le 22$, use the system of linear inequalities of Exercise $n + 6$, Exercise Set 4.3.

7. Three corners.

8. Three corners.

9. Three corners.

10. Three corners.

11. Three corners.

12. Three corners.

13. Four corners.

14. Four corners.

15. Four corners.

16. Four corners.

17. Four corners.

18. Four corners.

19. Four corners.

20. Four corners.

21. Four corners.

22. Four corners.

23.
$$\begin{aligned} x + y + z &\le 15 \\ x + y &\le 10 \\ 4x - 3y &\le 12 \\ -3x + 4y &\le 12 \\ x \ge 0, y \ge 0, z &\ge 0 \end{aligned}$$
Five corners.

24.
$$\begin{aligned} 2x + 3y + 5z &\le 60 \\ -y + z &\le 4 \\ 3y - z &\le 24 \\ 2y + 3z &\le 27 \\ x \ge 0, y \ge 0, z &\ge 0 \end{aligned}$$
Five corners.

In Exercises 25–28, show, without sketching the graphs, that the region of feasible solutions of the given set of constraints is unbounded.

25.
$$\begin{aligned} -3x + y &\le 3 \\ x - y &\le 5 \\ 2x - y &\le 15 \\ x \ge 0, y &\ge 0 \end{aligned}$$

26.
$$\begin{aligned} -x + y &\le 10 \\ x - 3y &\le 9 \\ x - 2y &\le 10 \\ x \ge 0, y &\ge 0 \end{aligned}$$

27.
$$\begin{aligned} -4x + y &\le 6 \\ -3x + y &\le 7 \\ x - 4y &\le 5 \\ x - 3y &\le 6 \\ x \ge 0, y &\ge 0 \end{aligned}$$

28.
$$\begin{aligned} -10x + 3y &\le 30 \\ -7x + 3y &\le 39 \\ 3x - 5y &\le 21 \\ 7x - 4y &\le 72 \\ x \ge 0, y &\ge 0 \end{aligned}$$

4.5 The Simplex Method

We are ready for the simplex method. In Section 4.3, we learned how to convert the set of "$\le$" constraints to a linear system of equations by using slack variables. In Section 4.4, we used the augmented matrix of the system to identify corners of the region of feasible solutions and to move from any corner to an adjacent corner. It is a simple matter now to incorporate the objective function into the system.

Description of the Simplex Method

Since the objective function is defined by

$$z = c_1x_1 + c_2x_2 + c_3x_3 + \cdots + c_nx_n$$

and since the linear equations obtained from the constraints have all their variables on the left side, we first write

$$z - c_1x_1 - c_2x_2 - c_3x_3 - \cdots - c_nx_n = 0$$

We then add this equation to the linear system obtained from the "$\leq$" constraints by introducing slack variables. We then proceed exactly as in the preceding section treating z as a basic variable throughout. In this manner, we can read the value of the objective function in the right column as we move from corner to corner. It is useful to have the following definition.

> **DEFINITION 4.4:** Performing the necessary row operations to convert the pivot element to 1 and the other entries in the pivot column to 0 is called *pivoting*.

EXAMPLE 1 Find the maximum value of the objective function $z = 5x_1 + 4x_2$ subject to the constraints

$$\begin{aligned} x_1 + x_2 &\leq 5 \\ 3x_1 + 2x_2 &\leq 12 \\ x_1 \geq 0, \, x_2 &\geq 0 \end{aligned}$$

Solution We introduce slack variables s_1 and s_2 and write the constraints as

$$\begin{aligned} x_1 + x_2 + s_1 \qquad &= 5 \\ 3x_1 + 2x_2 + \qquad + s_2 &= 12 \\ x_1 \geq 0, \, x_2 \geq 0, \, s_1 \geq 0, \, s_2 &\geq 0 \end{aligned}$$

We now write the objective function as

$$z - 5x_1 - 4x_2 = 0$$

Adding this to these two equations, we get

$$\begin{cases} \quad x_1 + x_2 + s_1 \qquad = 5 \\ \quad 3x_1 + 2x_2 + \qquad + s_2 = 12 \\ z - 5x_1 - 4x_2 \qquad \qquad = 0 \end{cases}$$

The augmented matrix is

$$\begin{bmatrix} 0 & 1 & 1 & 1 & 0 & 5 \\ 0 & 3 & 2 & 0 & 1 & 12 \\ 1 & -5 & -4 & 0 & 0 & 0 \end{bmatrix}$$

It is clear that columns 4, 5, and 1, in that order, form the identity I_3. Thus, we let $\{s_1, s_2, z\}$ be our initial basis. From this point on, we will display the augmented matrix and variables in a slightly different format. This is often called a *tableau*. The initial tableau of the problem is

Basis	z	x_1	x_2	s_1	s_2	Values of basic variables
s_1	0	1	1	1	0	5
s_2	0	3	2	0	1	12
z	1	-5	-4	0	0	0

We use a horizontal line to separate the coefficients of the objective function from those of the constraints. We give x_1 and x_2 the value 0 since they are nonbasic variables. Consequently, the initial value of the objective function is 0, which can be read in the lower right corner of the tableau. From the tableau, we see that the initial basic feasible solution is (0, 0, 5, 12). The first two entries give (0, 0), the origin, as a corner of the region of feasible solutions. Since the objective function is $z = 5x_1 + 4x_2$, its value will increase if either x_1 or x_2 becomes positive. Since the coefficient of x_1 is larger than that of x_2, the value of z will increase more per unit increase of x_1; thus, it seems reasonable to use x_1 as the entering variable. (See Example 2.) Because the largest coefficient in the system is 5, that of x_1, the bottom entry in the x_1 column is -5, which is the *most negative* entry in the last row of the tableau. Since $\frac{12}{3}$ is less than $\frac{5}{1}$, we select s_2 as the leaving variable. Thus, 3 is the pivot element and is circled in the initial tableau.

Tableau I

Basis	z	x_1	x_2	s_1	s_2	Values of basic variables
s_1	0	1	1	1	0	5
s_2	0	③	2	0	1	12
z	1	-5	-4	0	0	0

The numbers in the bottom row, except for the last one, are called *indicators* because they are used to determine the entering variable. Pivoting, we get the next tableau.

Tableau II

Basis	z	x_1	x_2	s_1	s_2	Values of basic variables
s_1	0	0	$\frac{1}{3}$	1	$\frac{-1}{3}$	1
x_1	0	1	$\frac{2}{3}$	0	$\frac{1}{3}$	4
z	1	0	$\frac{-2}{3}$	0	$\frac{5}{3}$	20

We see from the tableau that $x_1 = 4$, $x_2 = 0$ (since it is a nonbasic variable), $s_1 = 1$, $s_2 = 0$ (since it is the other nonbasic variable), and $z = 20$. This means that our second basic feasible solution is (4, 0, 1, 0). The first two entries give the coordinates (4, 0) of another corner of the region of feasible solutions, and the value of the objective function at that corner is 20. The last row of the tableau indicates that

$$z - \left(\tfrac{2}{3}\right)x_2 + \left(\tfrac{5}{3}\right)s_2 = 20$$

which we can write

$$z = \left(\tfrac{2}{3}\right)x_2 - \left(\tfrac{5}{3}\right)s_2 + 20$$

The values of x_2 and s_2 are 0, so $z = 20$. Clearly the value of z will increase if s_2 remains 0 and x_2 becomes positive. Thus, we choose x_2 as our entering variable. The bottom entry in the x_2 column is $\tfrac{-2}{3}$ which is the most negative indicator. Since $1/(1/3)$ is less than $4/(2/3)$, we choose s_1 to be the leaving variable and $\tfrac{1}{3}$ becomes the pivot element as indicated in the next tableau.

Tableau III

Basis	z	x_1	x_2	s_1	s_2	Values of basic variables
s_1	0	0	$\tfrac{1}{3}$	1	$\tfrac{-1}{3}$	1
x_1	0	1	$\tfrac{2}{3}$	0	$\tfrac{1}{3}$	4
z	1	0	$\tfrac{-2}{3}$	0	$\tfrac{5}{3}$	20

Pivoting, we obtain the next tableau.

Tableau IV

Basis	z	x_1	x_2	s_1	s_2	Values of basic variables
x_2	0	0	1	3	-1	3
x_1	0	1	0	-2	1	2
z	1	0	0	2	1	22

All indicators in the bottom row are nonnegative. Thus, the value of z cannot be increased. To see this, note that the last row of the tableau indicates that

$$z + 2s_1 + s_2 = 22$$

Since s_1 and s_2 are 0 (they are nonbasic variables), the value of z is 22, which could have been read from the tableau. However, if we write

$$z = -2s_1 - s_2 + 22$$

it is clear that the value of z will decrease if either s_1 or s_2 is made positive while the other remains 0. Thus, neither s_1 nor s_2 can be the entering variable. But these were the only two choices. Thus, we are through, and 22 is the maximum value of the objective function. It occurs at (2, 3).

Do Exercise 3.

The z column was not changed throughout the problem. It never does since we wish z to remain a basic variable so that its value may be read from the tableau at each step. Consequently, that column will be omitted in future tableaus.

Procedure of the Simplex Method

Step 1. Convert the "$\leq$" constraints to equations by introducing slack variables. Write the objective function in the form

$$z - c_1x_1 - c_2x_2 - c_3x_3 - \cdots - c_nx_n = 0$$

Step 2. For the system obtained in Step 1 (with the objective equation written last), write the associated tableau, leaving out the z-column. This is the initial tableau. By choosing the slack variables to be the basic variables, we obtain the initial basic feasible solution at the origin. The value of the objective function appears in the lower right corner of the tableau.

Step 3. Determine whether the maximum value of the objective function has been attained by examining the bottom row of the tableau for negative indicators. If there are no negative indicators, the maximum has been attained. Its value is the entry in the lower right corner of the tableau. Otherwise, choose as the entering variable the one corresponding to the most negative indicator in the bottom row.

Step 4. The entering variable determines the pivot column. Divide each positive entry in the pivot column into its corresponding element in the last column (the current values of the basic variables). Choose as the leaving variable the one corresponding to the smallest nonnegative quotient. In case of a tie, choose either.

Step 5. The leaving (departing) variable determines the pivot row. The element in the pivot row and the pivot column is the pivot element. If the pivot element is a_{kj}, multiply row k by $1/a_{kj}$ to obtain the new row k, and for each $p \neq k$, add to row p row k multiplied by $(-a_{pj}/a_{kj})$ to obtain the new row p. In this manner, the pivot element has been converted to 1 and all other entries of the pivot column have been converted to 0. Performing Step 5 is referred to as pivoting. Return to Step 3.

A Remark on the Choice of the Entering Variable

The manner in which we choose the entering variable does not necessarily give the best choice. It gives the best choice for each increase of *one unit* of the entering variable. Recall the example in Section 4.3 where we asked you to visualize a house with a slanted flat roof and a squirrel trying to get to the highest point by running along the edges of the roof from corner to corner. His strategy was to go along the steeper edge when he was at a corner from which both edges were going uphill. This does not guarantee that the next corner he will reach will be higher than the one he would have reached if he had made the other choice because the steeper edge could be much shorter than the other edges. However, since our squirrel is nearsighted, he cannot see how long the edges are and he must make a choice. Hence, it is reasonable for him to choose the steeper edge although it may not be the best choice. Our situation is similar. In Step 3, we make the most reasonable choice of the entering variable because this is the best we can do. The next example should illustrate this point.

EXAMPLE 2   Consider the objective function $z = 5x + 10y$ subject to the constraints

$$-x + y \leq 5$$
$$-x + 2y \leq 15$$
$$x + 7y \leq 120$$
$$x \geq 0, y \geq 0$$

a. Find the maximum value of z using the simplex method.
b. Repeat part **a** except that on the first choice of the entering variable make the "wrong" choice and select x as the entering variable instead of y.
c. Explain geometrically why the solution in **b** is simpler than the solution in **a**.

Solution   **a.** We first introduce slack variables r, s, and t and we write the objective function with all its variables on the left.

$$\begin{cases} -x + y + r \phantom{{}+s+t} = 5 \\ -x + 2y \phantom{{}+y} + s \phantom{{}+t} = 15 \\ x + 7y \phantom{{}+s} + t = 120 \\ z - 5x - 10y \phantom{{}+s+t} = 0 \end{cases}$$

We then set up the associated tableau which represents the initial solution of the problem, leaving out the z-column.

Tableau I

Basis	x	y	r	s	t	Values of basic variables
r ←	-1	①	1	0	0	5
s	-1	2	0	1	0	15
t	1	7	0	0	1	120
z	-5	-10	0	0	0	0

The most negative indicator in the bottom row is -10. Thus, we select y as the entering variable, as indicated by the arrow at the bottom of the y column. Since $\frac{5}{1}$ is less than both $\frac{15}{2}$ and $\frac{120}{7}$, we select r as the leaving variable. The pivot element is circled. It is already 1. Pivoting, we obtain the next tableau.

Tableau II

Basis	x	y	r	s	t	Values of basic variables
y	-1	1	1	0	0	5
s ←	①	0	-2	1	0	5
t	8	0	-7	0	1	85
z	-15	0	10	0	0	50

Thus, $y = 5$, $s = 5$, $t = 85$, $z = 50$, and $x = r = 0$. The second basic feasible solution is $(0, 5, 0, 5, 85)$ and the value of z is 50 at that point. Since there is a negative indicator in the bottom row, the value of z can be improved. We choose x as the entering variable. The x column becomes the pivot column. Since $\frac{5}{1}$ is less than $\frac{85}{8}$, we choose s as the leaving variable. The pivot element is circled. It is already 1. We need only convert the other entries in the pivot column to 0.

Tableau III

Basis	x	y	r	s	t	Values of basic variables
y	0	1	-1	1	0	10
x	1	0	-2	1	0	5
t	0	0	⑨	-8	1	45
z	0	0	-20	15	0	125

The next basic feasible solution is $(5, 10, 0, 0, 45)$. The value of the objective function is 125. Since there is a negative indicator in the last row, the value of z can be improved. The most negative indicator is -20; hence, we select r as our next entering variable. Note that 9 is the only positive entry in the pivot column. Thus, we choose t as the leaving variable. The pivot element is circled. Pivoting again, we get the next tableau.

Tableau IV

Basis	x	y	r	s	t	Values of basic variables
y	0	1	0	$\frac{1}{9}$	$\frac{1}{9}$	15
x	1	0	0	$\frac{-7}{9}$	$\frac{2}{9}$	15
r	0	0	1	$\frac{-8}{9}$	$\frac{1}{9}$	5
z	0	0	0	$\frac{-25}{9}$	$\frac{20}{9}$	225

The corresponding basic feasible solution is $(15, 15, 5, 0, 0)$, and the value of z is 225. There is still a negative indicator in the bottom row but it is the only one. Consequently, it is the most negative and we choose s as the entering variable. Since $\frac{1}{9}$ is the only positive entry in the pivot column, it must be the pivot. Pivoting, we get

Tableau V

Basis	x	y	r	s	t	Values of basic variables
s	0	9	0	1	1	135
x	1	7	0	0	1	120
r	0	8	1	0	1	125
z	0	25	0	0	5	600

There are no negative indicators in the bottom row. Thus, the value of the objective function cannot be improved. The maximum value is 600. It occurs at (120, 0).

b. Let us consider the initial tableau again.

Tableau I

Basis	x	y	r	s	t	Values of basic variables
r	−1	1	1	0	0	5
s	−1	2	0	1	0	15
t	①	7	0	,0	1	120
z	−5	−10	0	0	0	0

The most negative indicator in the bottom row is −10. For this reason, we chose y as our original entering variable. Suppose we had decided not to follow the usual procedure and had chosen x as our first entering variable. Since 1 is the only positive entry in the x column, it would have to be the pivot element. Therefore, t would have to be the leaving variable. Converting the other entries in the pivot column to 0, we would have obtained

Tableau II

Basis	x	y	r	s	t	Values of basic variables
r	0	8	1	0	1	125
s	0	9	0	1	1	135
x	1	7	0	0	1	120
z	0	25	0	0	5	600

Since there are no negative indicators in the bottom row, we have reached the maximum value of the objective function. This value is 600 as obtained in part **a**.

c. Note that we obtained the maximum value in part **b** much faster than in part **a**. Why did this happen? It will be easier to understand if we solve the same problem using the geometric approach, as we did in Section 4.1. The graph of the region of feasible solutions is shown in Figure 4.26.

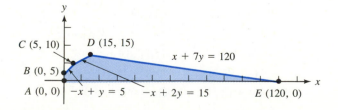

FIGURE 4.26

The vertices of the region have been found to be the points A, B, C, D, and E with coordinates $(0, 0)$, $(0, 5)$, $(5, 10)$, $(15, 15)$, and $(120, 0)$, respectively. The value of the objective function at each of these points is found as follows:

at $(0, 0)$, $z = 0$; at $(0, 5)$, $z = 50$; at $(5, 10)$, $z = 125$;
at $(15, 15)$, $z = 225$; and at $(120, 0)$, $z = 600$.

We again see that the maximum value is 600 and is attained at E with coordinates $(120, 0)$. Using the initial tableau, the initial basic solution was the origin. The maximum value of the objective function occurs at the adjacent corner E. By choosing x as the entering variable, we get to the corner E in one step. However, by choosing y as the first entering variable, we go through the corners B, C, and D before getting to E. Note that the values of z which we calculated at these corners are 50, 125, 225, and 600, respectively, and these are the values which appeared in the lower right corners of the second, third, fourth, and fifth tableaus. By going from A to B, we increased y by only 5 units and the objective function by 50 even though we increased the objective function by 10 for each unit increase in y. However, by going from A to E, although we increased the objective function by only 5 for each unit increase in x, we increased x by 120 units and the objective function by 600. ∎

In spite of what happened in the preceding example, in problems with many decision variables and many constraints, the choice of the entering variable will be based on the most negative indicator in the bottom row of the tableau. We must do this because in such cases we do not have the benefit of a graph or the simplicity of the situation in the foregoing example to decide otherwise. Clearly, the number of steps necessary to solve a linear programming problem using the simplex method is much less than the number of steps necessary when we use the method of Section 4.3. Using that method, we had to solve many systems of linear equations and many of the solutions obtained were not feasible. Therefore, the simplex method is much more efficient.

EXAMPLE 3   Maximize $z = 2x_1 + 3x_2 + 5x_3 + 2x_4$ subject to

$$x_1 - x_2 + x_3 + 2x_4 \leq 3$$
$$x_1 + 2x_2 + x_3 + x_4 \leq 11$$
$$2x_1 + x_2 + 4x_4 \leq 10$$
$$x_1 \geq 0, x_2 \geq 0, x_3 \geq 0, x_4 \geq 0$$

Solution   We introduce slack variables s_1, s_2, and s_3 and write

$$x_1 - x_2 + x_3 + 2x_4 + s_1 = 3$$
$$x_1 + 2x_2 + x_3 + x_4 + s_2 = 11$$
$$2x_1 + x_2 + 4x_4 + s_3 = 10$$
$$x_1 \geq 0, x_2 \geq 0, x_3 \geq 0, x_4 \geq 0, s_1 \geq 0, s_2 \geq 0, s_3 \geq 0$$

We now write the objective function as

$$z - 2x_1 - 3x_2 - 5x_3 - 2x_4 = 0$$

The initial tableau is

Tableau I

Basis	x_1	x_2	x_3	x_4	s_1	s_2	s_3	Values of basic variables
$\leftarrow$ s_1	1	−1	①	2	1	0	0	3
s_2	1	2	1	1	0	1	0	11
s_3	2	1	0	4	0	0	1	10
z	−2	−3	−5	−2	0	0	0	0

$\uparrow$

The most negative indicator in the bottom row is -5. Thus, we select x_3 as the entering variable. Since $\frac{3}{1}$ is less than $\frac{11}{1}$, s_1 is the leaving variable and the circled 1 is the pivot element. Pivoting, we get

Tableau II

Basis	x_1	x_2	x_3	x_4	s_1	s_2	s_3	Values of basic variables
x_3	1	−1	1	2	1	0	0	3
$\leftarrow$ s_2	0	③	0	−1	−1	1	0	8
s_3	2	1	0	4	0	0	1	10
z	3	−8	0	8	5	0	0	15

$\uparrow$

Since there is a negative indicator in the bottom row, we choose x_2 as our next entering variable. Since $\frac{8}{3}$ is less than $\frac{10}{1}$, s_2 is selected as the leaving variable. The circled 3 is the pivot element. Pivoting, we get

Tableau III

Basis	x_1	x_2	x_3	x_4	s_1	s_2	s_3	Values of basic variables
x_3	1	0	1	$\frac{5}{3}$	$\frac{2}{3}$	$\frac{1}{3}$	0	$\frac{17}{3}$
x_2	0	1	0	$\frac{-1}{3}$	$\frac{-1}{3}$	$\frac{1}{3}$	0	$\frac{8}{3}$
s_3	2	0	0	$\frac{13}{3}$	$\frac{1}{3}$	$\frac{-1}{3}$	1	$\frac{22}{3}$
z	3	0	0	$\frac{16}{3}$	$\frac{7}{3}$	$\frac{8}{3}$	0	$\frac{109}{3}$

There are no negative indicators in the bottom row. Thus, the maximum value of the objective function has been attained. It is $\frac{109}{3}$ and occurs at $\left(0, \frac{8}{3}, \frac{17}{3}, 0\right)$ since $x_2 = \frac{8}{3}$, $x_3 = \frac{17}{3}$, and $x_1 = x_4 = 0$.
Do Exercise 13. ■

To appreciate the effectiveness of the simplex method, you are urged to compare this example to Example 4, Section 4.3. If we had to find all vertices of the region in that example and evaluate the objective function at each vertex, the work would have been overwhelming.

Checking Your Calculations

A CLOSER LOOK   Suppose that in a maximization problem in standard form, there are m "$\leq$" constraints and therefore m slack variables are introduced. In any tableau used in the problem, the m entries in the bottom row and in the columns corresponding to the slack variables form an m-dimensional row vector. If we calculate the dot product of that vector in the *kth tableau* with the m-dimensional column vector formed by the values of these m variables read in the last column of the *initial tableau*, the result should always give the value of the objective function in the *kth tableau*. For instance, in the first tableau of the previous example, the three-dimensional column vector formed by the initial values of the slack variables (read in the last column)

was $\begin{bmatrix} 3 \\ 11 \\ 10 \end{bmatrix}$. The three-dimensional row vectors formed by the numbers read in the

bottom row and in the columns corresponding to the slack variables s_1, s_2, and s_3 in the second and third tableaus are $\begin{bmatrix} 5 & 0 & 0 \end{bmatrix}$ and $\begin{bmatrix} \frac{7}{3} & \frac{8}{3} & 0 \end{bmatrix}$, respectively. Note that

$$\begin{bmatrix} 5 & 0 & 0 \end{bmatrix} \cdot \begin{bmatrix} 3 \\ 11 \\ 10 \end{bmatrix} = 3(5) + 11(0) + 10(0) = 15$$

and 15 is in the lower right corner of the second tableau. Similarly,

$$\begin{bmatrix} \frac{7}{3} & \frac{8}{3} & 0 \end{bmatrix} \cdot \begin{bmatrix} 3 \\ 11 \\ 10 \end{bmatrix} = 3\left(\frac{7}{3}\right) + 11\left(\frac{8}{3}\right) + 10(0) = \frac{109}{3}$$

and $\frac{109}{3}$ is in the lower right corner of the third tableau. (You may find this useful in checking your steps when solving a maximization problem in standard form by the simplex method.) If the solution requires several tableaus and the arithmetic is troublesome, then for each tableau, calculate the dot product of the row vector (which is formed by the entries in the bottom row and the columns corresponding to the slack variables) and the column vector (which is formed using the initial values of the slack variables in the initial tableau). The result should be equal to the value of the objective function that appears in the lower right corner. If it is equal, there is no guarantee that the work is correct, but if it is not, then you know there is an error in the solution.

An Application

EXAMPLE 4   Jack bought 7 acres of land on which he wishes to grow wheat, barley, and corn. The cost of planting one acre is $120 for wheat, $80 for barley, and $90 for corn. Jack has only $720 available for planting. The anticipated profit per acre is $1000 for wheat,

$800 for barley, and $400 for corn. How many acres of each should he plant to maximize his profit?

Solution Let x, y, and z denote the numbers of acres of wheat, barley, and corn, respectively, that Jack will plant. Then the total number of acres planted will be $x + y + z$. We have

$$x + y + z \leq 7$$

The planting costs (measured in dollars) will be $120x$ for wheat, $80y$ for barley, and $90z$ for corn. Thus, the total cost of planting is $(120x + 80y + 90z)$. Since Jack has only $720 available, we have

$$120x + 80y + 90z \leq 720$$

or

$$12x + 8y + 9z \leq 72$$

Jack will make $1000x$ profit on the wheat, $800y$ profit on the barley, and $400z$ profit on the corn. Thus, his total profit will be $\$(1000x + 800y + 400z)$. The problem is how to maximize w where

$$w = 1000x + 800y + 400z$$

subject to

$$\begin{aligned} x + y + z &\leq 7 \\ 12x + 8y + 9z &\leq 72 \\ x \geq 0, y \geq 0, z &\geq 0 \end{aligned}$$

We begin as usual and write

$$\begin{aligned} x + y + z + r &= 7 \\ 12x + 8y + 9z \quad\quad + s &= 72 \end{aligned}$$

We then write the objective function with all variables on the left.

$$w - 1000x - 800y - 400z = 0$$

The initial tableau is

Tableau I

Basis	x	y	z	r	s	Values of basic variables
r	1	1	1	1	0	7
← s	⑫	8	9	0	1	72
w	-1000	-800	-400	0	0	0

↑

Since -1000 is the most negative indicator in the bottom row, we choose x as the entering variable. We then choose s as the leaving variable, since $\frac{72}{12}$ is less than $\frac{7}{1}$. Thus, 12 is the circled pivot element. We pivot to get the second tableau.

Tableau II

Basis	x	y	z	r	s	Values of basic variables
r	0	$\frac{1}{3}$	$\frac{1}{4}$	1	$\frac{-1}{12}$	1
x	1	$\frac{2}{3}$	$\frac{3}{4}$	0	$\frac{1}{12}$	6
w	0	$\frac{-400}{3}$	350	0	$\frac{250}{3}$	6000

Since we have a negative indicator in the bottom row, the value of the objective function, which is now 6000, may be improved. We choose y as the entering variable. Comparing $1/(1/3) = 3$ to $6/(2/3) = 9$, we choose r to be the leaving variable. Thus, $\frac{1}{3}$ is the new pivot element. Again, we pivot to obtain the third tableau.

Tableau III

Basis	x	y	z	r	s	Values of basic variables
y	0	1	$\frac{3}{4}$	3	$\frac{-1}{4}$	3
x	1	0	$\frac{1}{4}$	-2	$\frac{1}{4}$	4
w	0	0	450	400	50	6400

Since there are no negative indicators in the bottom row, the value of the objective function cannot be improved. It is 6400 and occurs when $x = 4$, $y = 3$, and $z = 0$. Thus, Jack should plant 4 acres of wheat, 3 acres of barley, and no corn.
Do Exercise 29.   ■

Unbounded Feasible Solution Set in the Optimizing Direction

We conclude this section with a brief discussion of cases where we have an unbounded feasible solution set in the optimizing direction and where we have multiple optimum solutions. Suppose, as in the next example, we have a linear programming problem that gives rise to a tableau in which an entering variable has been selected and all entries above the indicator are negative or 0. Then, the linear programming problem has an unbounded feasible solution set in the optimizing direction.

EXAMPLE 5   Maximize $z = 3x + 4y$ subject to

$$-x + y \leq 5$$
$$-3x + 5y \leq 35$$
$$x \geq 0, y \geq 0$$

Solution   As usual we get

$$-x + y + s_1 = 5$$
$$-3x + 5y + s_2 = 35$$
$$z - 3x - 4y = 0$$

The initial tableau is

Tableau I

Basis	x	y	s_1	s_2	Values of basic variables
s_1	-1	①	1	0	5
s_2	-3	5	0	1	35
z	-3	-4	0	0	0

The most negative indicator is -4. Thus, y is the entering variable. Comparing ratios, we see that s_1 is the leaving variable. Pivoting, we get

Tableau II

Basis	x	y	s_1	s_2	Values of basic variables
y	-1	1	1	0	5
s_2	②	0	-5	1	10
z	-7	0	4	0	20

Since -7 is the only negative indicator, we select x as the next entering variable. The only positive entry in the pivot column is 2. Thus, the leaving variable is s_2. Pivoting, we get

Tableau III

Basis	x	y	s_1	s_2	Values of basic variables
y	0	1	$\frac{-3}{2}$	$\frac{1}{2}$	10
x	1	0	$\frac{-5}{2}$	$\frac{1}{2}$	5
z	0	0	$\frac{-27}{2}$	$\frac{7}{2}$	55

The only negative indicator is $\frac{-27}{2}$. Thus, s_1 is the entering variable and the s_1 column is the pivot column. Note that all entries in the pivot column are negative. Thus, the region of feasible solutions is unbounded in the optimizing direction and z has no maximum value. You would find it instructive to solve this example geometrically.
Do Exercise 3.   ■

Multiple Optimum Solutions

We suggested earlier that you visualize a house with a convex polygonal base and a slanted flat roof to see that such a house would have its highest point at a corner. If two adjacent corners of the roof are the same altitude, then every point on the edge connecting these two corners are at that altitude. Similarly, if in a linear programming problem the objective function attains its maximum value at two adjacent vertices of the region of feasible solutions, then it attains its maximum value at each point on the edge connecting these two vertices. In the case where we have n decision variables, if the function attains its maximum value at $(x_1, x_2, x_3, \ldots, x_n)$ and also at $(y_1, y_2, y_3, \ldots, y_n)$ then, if for each $i \in \{1, 2, \ldots, n\}$ we let $z_i = (1 - t)x_i + ty_i$, where $0 \le t \le 1$, the function attains its maximum value at $(z_1, z_2, z_3, \ldots, z_n)$ also. (See Exercises 38, 39, and 40.)

Suppose now that we have a linear programming problem that creates a tableau in which there are no negative indicators. In this case, an optimum solution has been reached. But what if one or more of the indicators corresponding to nonbasic variables are 0? Then, by using one of these nonbasic variables as an entering variable, we may get a different basic feasible solution. This would give the same value of the objective function, since in pivoting, the bottom row would remain unchanged. In such a case, both basic feasible solutions, together with every point on the edge joining them, would yield the maximum value of the objective function as is illustrated in the next example.

EXAMPLE 6   Maximize $z = 10x + 6y$ subject to

$$x + y \le 10$$
$$5x + 3y \le 36$$
$$2x + y \le 14$$
$$x \ge 0, y \ge 0$$

Solution   As usual, we write

$$x + y + s_1 = 10$$
$$5x + 3y + s_2 = 36$$
$$2x + y + s_3 = 14$$
$$z - 10x - 6y = 0$$
$$x \ge 0, y \ge 0$$

The initial tableau is

Tableau I

Basis	x	y	s_1	s_2	s_3	Values of basic variables
s_1	1	1	1	0	0	10
s_2	5	3	0	1	0	36
s_3	②	1	0	0	1	14
z	-10	-6	0	0	0	0

Since -10 is the most negative indicator, x is the entering variable. Comparing ratios, we see that s_3 is the leaving variable. Pivoting, we get

Tableau II

Basis	x	y	s_1	s_2	s_3	Values of basic variables
s_1	0	$\frac{1}{2}$	1	0	$\frac{-1}{2}$	3
s_2	0	$\left(\frac{1}{2}\right)$	0	1	$\frac{-5}{2}$	1
x	1	$\frac{1}{2}$	0	0	$\frac{1}{2}$	7
z	0	-1	0	0	5	70

Since -1 is the only negative indicator, y is the next entering variable. Comparing ratios, we see that s_2 is the leaving variable. Pivoting, we get

Tableau III

Basis	x	y	s_1	s_2	s_3	Values of basic variables
s_1	0	0	1	-1	$\left(2\right)$	2
y	0	1	0	2	-5	2
x	1	0	0	-1	3	6
z	0	0	0	2	0	72

There are no negative indicators in the bottom row, so the optimum solution has been reached. The maximum value of z is 72 and it occurs when $x = 6$, $y = 2$, $s_1 = 2$, and $s_2 = s_3 = 0$. That is, it occurs at $(6, 2)$. Since s_3 is a nonbasic variable and the indicator in the s_3 column is 0, we shall check for multiple solutions. We treat s_3 as an entering variable. Comparing ratios, we see that s_1 is the leaving variable. Pivoting, we get

Tableau IV

Basis	x	y	s_1	s_2	s_3	Values of basic variables
s_3	0	0	$\frac{1}{2}$	$\frac{-1}{2}$	1	1
y	0	1	$\frac{5}{2}$	$\frac{-1}{2}$	0	7
x	1	0	$\frac{-3}{2}$	$\frac{1}{2}$	0	3
z	0	0	0	2	0	72

The bottom row is the same as in the preceding tableau. We have, therefore, no negative indicators and we still have an optimum solution. The maximum value of z is 72 and occurs when $x = 3$, $y = 7$, $s_1 = s_2 = 0$, and $s_3 = 1$. That is, it occurs at $(3, 7)$. Note that $(3, 7)$ and $(6, 2)$ are two adjacent vertices of the region of feasible solutions. Since z has the value 72 at both points, it will have the value 72 at every point on the edge connecting these two vertices. That is, z will have the value 72 at each (x, y) where

$$x = (1 - t)3 + 6t, \text{ and } y = (1 - t)7 + 2t, 0 \leq t \leq 1$$

When $t = 0$, we get the endpoint $(3, 7)$ and when $t = 1$, we obtain the other endpoint $(6, 2)$. When $t = \frac{1}{2}$ we get the point $\left(\frac{9}{2}, \frac{9}{2}\right)$. We verify that, as expected, at that point the value of z is 72 by calculating

$$z = 10\left(\frac{9}{2}\right) + 6\left(\frac{9}{2}\right) = 45 + 27 = 72$$

Do Exercise 17.

Exercise Set 4.5

In Exercises 1–37, use the simplex method to solve the given problems.

1. Maximize $z = 2x + 4y$
subject to

$$x + 5y \leq 75$$
$$2x + 5y \leq 90$$
$$x + y \leq 33$$
$$x \geq 0, y \geq 0$$

2. Maximize $z = 3x + 3y$
subject to

$$x + 5y \leq 22$$
$$3x + 7y \leq 42$$
$$x \geq 0, y \geq 0$$

3. Maximize $z = 3x + 2y$
subject to

$$-2x + 3y \leq 3$$
$$-x + 4y \leq 8$$
$$x \geq 0, y \geq 0$$

4. Maximize $z = 20x + 5y$
subject to

$$x - y \leq 2$$
$$x + 2y \leq 16$$
$$x + y \leq 10$$
$$x \geq 0, y \geq 0$$

5. Maximize $z = 5x + 5y$
subject to

$$x - y \leq 60$$
$$8x + 5y \leq 600$$
$$2x + y \leq 90$$
$$-x + y \leq 60$$
$$x \geq 0, y \geq 0$$

6. Maximize $z = 3x + 5y$
subject to

$$x - y \leq 5$$
$$7x - 5y \leq 49$$
$$x \geq 0, y \geq 0$$

7. Maximize $z = 14x + 24y$
subject to

$$7x + 3y \leq 210$$
$$x + 7y \leq 140$$
$$2x + 5y \leq 118$$
$$x \geq 0, y \geq 0$$

8. Maximize $z = 12x + 15y$
subject to

$$x + 3y \leq 30$$
$$4x + 5y \leq 64$$
$$2x + y \leq 26$$
$$x \geq 0, y \geq 0$$

9. Maximize $w = 8x + 9y + 6z$ subject to

$$4x + 5z \leq 210$$
$$x + y \leq 48$$
$$x \leq 45$$
$$x \geq 0, y \geq 0, z \geq 0$$

10. Maximize $w = 2x - 24y + 8z$ subject to

$$4x + 3y - z \leq 7$$
$$x - y - z \leq 7$$
$$-x - y + z \leq 14$$
$$x \geq 0, y \geq 0, z \geq 0$$

11. Maximize $w = 2x + 4y + 12z$ subject to

$$x + y + 2z \leq 14$$
$$y + 4z \leq 12$$
$$x \geq 0, y \geq 0, z \geq 0$$

12. Maximize $w = 5x + y + z$ subject to

$$x + y + z \leq 66$$
$$-x + 3y + 2z \leq 72$$
$$2x + y + 3z \leq 108$$
$$x \geq 0, y \geq 0, z \geq 0$$

13. Maximize $w = x + y + z$ subject to

$$x + 2y + z \leq 168$$
$$2x + y + z \leq 280$$
$$x + y + 3z \leq 210$$
$$x \geq 0, y \geq 0, z \geq 0$$

14. Maximize $w = 12x + 10y + 2z$ subject to

$$3x + 9y - 2z \leq 10$$
$$2x + 4y - z \leq 4$$
$$-4x + y + z \leq 6$$
$$x \geq 0, y \geq 0, z \geq 0$$

15. Maximize $w = 10x + 25y + 15z$ subject to

$$x + 4y + 2z \leq 200$$
$$4x + 6y + 5z \leq 750$$
$$x \geq 0, y \geq 0, z \geq 0$$

16. Maximize $w = 3x + 6y + 3z$ subject to

$$2x + y + 3z \leq 24$$
$$x + 2y \leq 12$$
$$2x + z \leq 8$$
$$x \geq 0, y \geq 0, z \geq 0$$

17. Maximize $w = 8x + 2y + 3z$ subject to

$$8x + 10y - 9z \leq 80$$
$$30x - 25y + 60z \leq 300$$
$$-22x + 35y + 56z \leq 280$$
$$x \geq 0, y \geq 0, z \geq 0$$

18. Maximize $w = 10x + 20y + 15z$ subject to

$$x + 5y + 2z \leq 80$$
$$x + 3y + 4z \leq 60$$
$$x \geq 0, y \geq 0, z \geq 0$$

19. Maximize $w = 3x + 6y - 3z$ subject to

$$x + y + z \leq 40$$
$$2x + y - 3z \leq 20$$
$$x - y + 2z \leq 30$$
$$x \geq 0, y \geq 0, z \geq 0$$

20. Maximize $w = 2x + 24y + 4z$ subject to

$$x + 5y + 2z \leq 420$$
$$2x + y + 3z \leq 980$$
$$x + 7y \leq 560$$
$$x \geq 0, y \geq 0, z \geq 0$$

21. Maximize $w = 3x + 4y + 5z$ subject to

$$3x + 2y \leq 60$$
$$3x + 2y + 3z \leq 72$$
$$2y + 3z \leq 48$$
$$x \geq 0, y \geq 0, z \geq 0$$

22. Maximize $z = 3x_1 + 2x_3$ subject to

$$x_1 + x_2 \leq 16$$
$$x_1 - 2x_2 \leq 28$$
$$x_3 - 2x_4 \leq 8$$
$$x_3 + x_4 \leq 20$$
$$x_1 \geq 0, x_2 \geq 0, x_3 \geq 0, x_4 \geq 0$$

23. Maximize $z = 5x_1 + 10x_2 + 5x_3 + 25x_4$ subject to

$$3x_1 + x_2 + 2x_3 + x_4 \leq 10$$
$$x_1 + 2x_2 + x_3 + x_4 \leq 5$$
$$x_1 \geq 0, x_2 \geq 0, x_3 \geq 0, x_4 \geq 0$$

24. Maximize $z = 2x_1 + 3x_2 - x_3 + 5x_4$ subject to

$$x_1 + x_2 - x_3 + 2x_4 \leq 8$$
$$2x_1 - x_2 + x_3 + x_4 \leq 7$$
$$3x_1 - x_2 + 2x_3 + x_4 \leq 9$$
$$x_1 \geq 0, x_2 \geq 0, x_3 \geq 0, x_4 \geq 0$$

25. A farmer has 100 acres of land on which to grow tomatoes and peas. The cost of planting each acre of tomatoes is $50. Each acre of peas will cost $25 to plant. The farmer's budget for planting is $6250. If the anticipated profits are $100 per acre of peas and $150 per acre of tomatoes, how many acres of each should be planted to maximize profit? What is the maximum profit?

26. A farmer has 25 acres on which to grow apples and pears. Because of an agreement, he cannot grow more than 20 acres of apples and 15 acres of pears. If the profit per acre is $600 for apples and $400 for pears, how many acres of each should be planted to maximize profits? What is the maximum profit?

27. A farmer has 100 acres on which he wishes to grow oranges and lemons. He has been advised not to grow more than 50 acres of lemons. The cost of harvesting is $150 per acre of oranges and $300 per acre of lemons. He has only $18,000 budgeted for harvesting. If the profit per acre is $500 for oranges and $250 for lemons, how many acres of each should be planted to maximize profit? What is the maximum profit?

28. A company makes two gadgets, A and B. Each gadget A requires 1 hour on machine I, 1 hour on machine II, 2 hours on machine III, and 1 hour on machine IV. Each gadget B requires 2 hours on machine I, 1 hour on machine II, 1 hour on machine III, and 10 hours on machine IV. The time available per week for the production of these two gadgets is 48 hours on machine I, 40 hours on machine II, 50 hours on machine III, and 200 hours on machine IV. If the profit on each gadget A is $300 and on each gadget B it is $200, and if the company can sell as many gadgets as it produces, how many of each gadget should be produced per week to maximize profit? What is the maximum profit?

29. A company makes products A, B, and C. The times required for each unit of these products on each of the three machines are given in the following table.

	Product A	Product B	Product C
Machine I	3 hours	5 hours	3 hours
Machine II	3 hours	2 hours	0 hour
Machine III	1 hour	2 hours	0 hour

The amount of time available per day on each type of machine is 450 hours on machine I, 60 hours on machine II, and 120 hours on machine III. If the profits per unit of products A, B, and C are $70, $100, and $40, respectively, how many units of each product should be manufactured per day to maximize profit? What is the maximum profit?

30. A camera company makes two models of movie cameras. The deluxe model takes 3 hours to assemble, 6 minutes to test, and 16 minutes to pack. The regular model takes 4 hours to assemble, 3 minutes to test, and 4 minutes to pack. The number of work hours available per day for assembling, testing, and packing are 300, 5, and $\frac{32}{3}$, respectively. If the profit per deluxe model is $400 and that per regular model is $150, and if the company can sell as many cameras as it can produce, how many deluxe and how many regular models should be produced per day to maximize profit? What is the maximum profit?

31. A farmer has 40 acres of land on which he wishes to plant peanuts, wheat, and barley. The cost of planting each acre is $100 for peanuts, $50 for wheat, and $100 for barley. The cost of harvesting each acre is $300 for peanuts, $200 for wheat, and $100 for barley. The farmer has $3000 available for planting and $8000 available for harvesting. If the profits per acre are $100 for peanuts, $300 for wheat, and $400 for barley, how many acres of each should be planted to maximize profit? What is the maximum profit?

32. A company manufactures products A, B, and C, each of which is made from raw materials I, II, and III. The unit requirements for each of the three products are tabulated.

	Each Unit of Product Requires		
	Raw material I	Raw material II	Raw material III
Product A	1 unit	1 unit	1 unit
Product B	3 units	6 units	9 units
Product C	6 units	3 units	10 units

The maximum quantities available each day are 960 units of raw material I, 1800 units of raw material II, and 2740 units of raw material III. The cost to manufacture each unit of products A, B, and C are $5, $37, and $120, respectively. The company can sell as many units as it can produce at the following prices per unit: $13 for A, $69 for B, and $168 for C. How many units per day of each product should be manufactured to maximize profit? What is the maximum profit?

33. A company has 1000 pounds of peanuts, 1200 pounds of cashew nuts, and 800 pounds of Brazil nuts. It wishes to package three varieties of mixed nuts. The quantities (in pounds) of each kind of nut required to make each pound of each variety of mixed nuts are given in the following table.

	Variety I	Variety II	Variety III
Peanuts	$\frac{1}{2}$	$\frac{1}{3}$	0
Cashew nuts	$\frac{1}{2}$	$\frac{1}{3}$	$\frac{1}{2}$
Brazil nuts	0	$\frac{1}{3}$	$\frac{1}{2}$

Variety I sells for $3 per pound, variety II sells for $4 per pound, and variety III sells for $5 per pound. How many pounds of each variety should be mixed to maximize revenue? What is the maximum revenue?

34. A candy manufacturer has 320 pounds of chocolate, 180 pounds of fruit, and 240 pounds of nuts. She produces three types of candy. The quantities (in pounds) required to make each box of each type are given in the following table.

	Type A	Type B	Type C
Chocolate	$\frac{1}{4}$	$\frac{1}{2}$	1
Fruit	$\frac{1}{2}$	$\frac{1}{4}$	$\frac{1}{4}$
Nuts	$\frac{1}{4}$	$\frac{1}{2}$	$\frac{3}{4}$

If she can sell each box of type A for $5, each box of type B for $6, and each box of type C for $10, how many boxes of each should she make to maximize revenue? What is the maximum revenue?

35. A company manufactures products A, B, and C. The numbers of hours of machine time and of finishing time necessary to produce each unit of each product are given in the following table.

	Machine Time	Finishing Time
Product A	1	1
Product B	2	3
Product C	3	5

The number of work hours of machine time and of finishing time available each day are 50 and 60, respectively. The profits per unit realized from A, B, and C are $10, $20, and $40, respectively. How many units of each should be manufactured each day to maximize profit? What is the maximum profit?

36. A company manufactures small boats. Each type requires materials as indicated in the following table.

	Type A	Type B	Type C
Aluminum	2 units	3 units	6 units
Wood	3 units	4 units	1 unit
Paint	3 units	5 units	2 units

Each week, the company has available 600 units of aluminum, 480 units of wood, and 240 units of paint. If the profit per boat is $300 for A, $500 for B, and $300 for C, and if the company can sell all the boats it can produce, how many boats of each type should be produced each week to maximize profit? What is the maximum profit?

37. A contractor has 75 acres of land, $30,000,000 in capital, and 800,000 labor hours available to build houses. The amounts of resources required for each type of house are given in the following table.

	Type A	Type B	Type C
Land	.25 acre	.25 acre	1 acre
Capital	$50,000	$150,000	$100,000
Labor hours	6000	4000	2000

If the contractor's profit is $20,000 on each A house, $40,000 on each B house, and $60,000 on each C house, how many houses of each type should he build to maximize profit? What is the maximum profit?

38. Suppose that the function defined by

$$z = f(x, y) = mx + ny,$$

where m and n are constants, assumes the same value at (x_1, y_1) and (x_2, y_2). That is,

$$mx_1 + ny_1 = mx_2 + ny_2 = A$$

Show that if $x = (1 - t)x_1 + tx_2$ and $y = (1 - t)y_1 + ty_2$, where $0 \le t \le 1$, then $f(x, y) = A$ also. The points (x, y) are the points on the line segment joining (x_1, y_1) and (x_2, y_2). (*Hint:* Replace x and y by $(1 - t)x_1 + tx_2$ and $(1 - t)y_1 + ty_2$, respectively, in the right side of the equation $f(x, y) = mx + ny$, and simplify.)

39. Suppose that the function defined by

$$w = f(x, y, z) = mx + ny + pz$$

where m, n, and p are constants, assumes the same value at (x_1, y_1, z_1) and (x_2, y_2, z_2). That is,

$$mx_1 + ny_1 + pz_1 = mx_2 + ny_2 + pz_2 = A$$

Show that if $x = (1 - t)x_1 + tx_2$, $y = (1 - t)y_1 + ty_2$, and $z = (1 - t)z_1 + tz_2$ where $0 \le t \le 1$, then $f(x, y, z) = A$ also. The points (x, y, z) are the points on the line segment joining (x_1, y_1, z_1) and (x_2, y_2, z_2). (See the hint in Exercise 38.)

40. Let $Z = f(x_1, x_2, \ldots, x_n) = c_1x_1 + c_2x_2 + \cdots + c_nx_n$, where $c_1, c_2, \ldots, c_n$ are constants and suppose that Z has the same value at $(a_1, a_2, \ldots, a_n)$ and at $(b_1, b_2, \ldots, b_n)$. That is, suppose that

$$c_1a_1 + c_2a_2 + \cdots + c_na_n$$
$$= c_1b_1 + c_2b_2 + \cdots + c_nb_n = A$$

Let

$$z_1 = (1 - t)a_1 + tb_1$$
$$z_2 = (1 - t)a_2 + tb_2$$
$$\vdots \qquad \vdots \qquad \vdots$$
$$z_n = (1 - t)a_n + tb_n$$

where $0 \le t \le 1$. Show that $f(z_1, z_2, \ldots, z_n) = A$ also. (See the hint in Exercise 38.)

4.6 Minimization and Duality (Optional)

In the preceding section we studied only maximization problems in standard form. In this section, we discuss minimization problems in standard form. This form differs from that of maximums only in the fact that all "$\le$" inequalities are reversed and the coefficients in the equation defining the objective function must be nonnegative. We shall solve a minimization problem in standard form by applying the simplex method to a corresponding maximization problem in standard form. We shall first show how to obtain the corresponding maximization problem, then illustrate the technique with examples. We conclude with two examples that will help show why the technique works.

The Dual Problem

Recall the definition of the transpose of a matrix from Section 3.5. If $\mathbf{A}$ is an $m \times n$ matrix, its transpose $\mathbf{A}^T$ is the $n \times m$ matrix whose i-j entry is the j-i entry of $\mathbf{A}$. That is, the ith row of $\mathbf{A}^T$ is the ith column of $\mathbf{A}$. For example, if

$$\mathbf{A} = \begin{bmatrix} 1 & 5 & 9 \\ 0 & 3 & 4 \end{bmatrix}$$

then

$$\mathbf{A}^T = \begin{bmatrix} 1 & 0 \\ 5 & 3 \\ 9 & 4 \end{bmatrix}$$

Consider the following minimization problem. Minimize $z = c_{11}x_1 + c_{12}x_2 + \cdots + c_{1n}x_n$ where all the coefficients c_{ij} are nonnegative and the variables x_i are subject to

$$
\begin{aligned}
a_{11}x_1 + a_{12}x_2 + \cdots + a_{1n}x_n &\geq b_1 \\
a_{21}x_1 + a_{22}x_2 + \cdots + a_{2n}x_n &\geq b_2 \\
&\;\;\vdots \\
a_{m1}x_1 + a_{m2}x_2 + \cdots + a_{mn}x_n &\geq b_m \\
x_1 \geq 0, \; x_2, \geq 0, \; \ldots, \; x_n &\geq 0
\end{aligned}
$$

All b_i's need not be nonnegative. We write the first m inequalities as equalities, *without introducing additional variables*. The equality defining the objective function may be written on the last line with the z on the right side of the equal sign as follows:

$$
\begin{aligned}
a_{11}x_1 + a_{12}x_2 + \cdots + a_{1n}x_n &= b_1 \\
a_{21}x_1 + a_{22}x_2 + \cdots + a_{2n}x_n &= b_2 \\
&\;\;\vdots \\
a_{m1}x_1 + a_{m2}x_2 + \cdots + a_{mn}x_n &= b_m \\
c_{11}x_1 + c_{12}x_2 + \cdots + c_{1n}x_n &= z
\end{aligned}
$$

The augmented matrix of the system is

$$\begin{bmatrix} a_{11} & a_{12} & \cdots & a_{1n} & b_1 \\ a_{21} & a_{22} & \cdots & a_{2n} & b_2 \\ \cdot & \cdot & & \cdot & \cdot \\ \cdot & \cdot & & \cdot & \cdot \\ \cdot & \cdot & & \cdot & \cdot \\ a_{m1} & a_{m2} & \cdots & a_{mn} & b_m \\ c_{11} & c_{12} & \cdots & c_{1n} & z \end{bmatrix}$$

Finally, we write the transpose of the matrix.

$$\begin{bmatrix} a_{11} & a_{21} & \cdots & a_{m1} & c_{11} \\ a_{12} & a_{22} & \cdots & a_{m2} & c_{12} \\ \cdot & \cdot & & \cdot & \cdot \\ \cdot & \cdot & & \cdot & \cdot \\ \cdot & \cdot & & \cdot & \cdot \\ a_{1n} & a_{2n} & \cdots & a_{mn} & c_{1n} \\ b_1 & b_2 & \cdots & b_m & z \end{bmatrix}$$

Using this matrix, we write the following maximization problem, which is called the *dual* of our original minimization problem.

Maximize $Z = b_1 y_1 + b_2 y_2 + \cdots + b_m y_m$, where the y_j's are subject to

$$a_{11} y_1 + a_{21} y_2 + \cdots + a_{m1} y_m \leq c_{11}$$
$$a_{12} y_1 + a_{22} y_2 + \cdots + a_{m2} y_m \leq c_{12}$$
$$\begin{matrix} \cdot & \cdot & & \cdot & \cdot \\ \cdot & \cdot & & \cdot & \cdot \\ \cdot & \cdot & & \cdot & \cdot \end{matrix}$$
$$a_{1n} y_1 + a_{2n} y_2 + \cdots + a_{mn} y_m \leq c_{1n}$$
$$y_1 \geq 0, \, y_2 \geq 0, \, \ldots, \, y_m \geq 0$$

The original minimization problem is called the *primal* problem. We used different names for the variables in the dual to emphasize that it is a different problem.

EXAMPLE 1   Write the dual of the following primal minimization problem.
Minimize $z = 3x + 5y$, where x and y are subject to

$$2x + 5y \geq 4$$
$$3x + 7y \geq -2$$
$$x - 6y \geq 3$$
$$x \geq 0, \, y \geq 0$$

Solution   We change the first three constraint inequalities to equations and write

$$2x + 5y = 4$$
$$3x + 7y = -2$$
$$x - 6y = 3$$
$$3x + 5y = z$$

The augmented matrix of this system is

$$\begin{bmatrix} 2 & 5 & 4 \\ 3 & 7 & -2 \\ 1 & -6 & 3 \\ 3 & 5 & z \end{bmatrix}$$

The transpose is

$$\begin{bmatrix} 2 & 3 & 1 & 3 \\ 5 & 7 & -6 & 5 \\ 4 & -2 & 3 & z \end{bmatrix}$$

The last row of the transpose gives the coefficients of the equation defining the new objective function, while the first two rows give the coefficients of the new constraint inequalities. The dual problem is the following.

Maximize $Z = 4u - 2v + 3w$, where u, v, and w are subject to

$$2u + 3v + w \leq 3$$
$$5u + 7v - 6w \leq 5$$
$$u \geq 0, v \geq 0, w \geq 0$$

As required, the constants 3 and 5 of the first two inequalities are nonnegative. This results from the fact that these quantities were the coefficients of the equation defining the original objective function, which in a minimization problem in standard form are nonnegative.   ■

Solving Minimization Problems in Standard Form

The following theorem, which is proved in more advanced texts, allows us to solve a minimization problem in standard form by solving the dual problem.

> **THEOREM 4.3:**   The objective function z of a minimization problem in standard form assumes a minimum value if, and only if, the objective function Z of the dual problem assumes a maximum value. Furthermore, the minimum value of z is equal to the maximum value of Z. The coordinates of the point where the objective function z attains its minimum value are read in the final tableau of the dual problem: in the bottom row and in the columns corresponding to the slack variables used to solve the dual problem.

EXAMPLE 2   Minimize $z = 54x + 22y$ subject to

$$9x + y \geq 144$$
$$x + y \geq 120$$
$$x \geq 0, y \geq 0$$

Solution The matrix of the primal problem is

$$\begin{bmatrix} 9 & 1 & 144 \\ 1 & 1 & 120 \\ 54 & 22 & z \end{bmatrix}$$

The transpose is

$$\begin{bmatrix} 9 & 1 & 54 \\ 1 & 1 & 22 \\ 144 & 120 & z \end{bmatrix}$$

This is the matrix of the dual problem. Thus, we can formulate the dual problem as follows. Maximize $Z = 144u + 120v$ subject to

$$9u + v \le 54$$
$$u + v \le 22$$
$$u \ge 0, v \ge 0$$

Using slack variables in the usual way, we obtain the system

$$\begin{cases} 9u + v + r & = 54 \\ u + v & + s = 22 \\ Z - 144u - 120v & = 0 \end{cases}$$

The initial tableau for this system is given with the entering and leaving variables and the pivot element as indicated.

Tableau I

Basis	u	v	r	s	Values of basic variables
r	⑨	1	1	0	54
s	1	1	0	1	22
Z	-144	-120	0	0	0

Pivoting, we get

Tableau II

Basis	u	v	r	s	Values of basic variables
u	1	$\frac{1}{9}$	$\frac{1}{9}$	0	6
s	0	$\frac{8}{9}$	$\frac{-1}{9}$	1	16
Z	0	-104	16	0	864

The entering and leaving variables and the pivot element are indicated above. Pivoting, we obtain

Tableau III

Basis	u	v	r	s	Values of basic variables
u	1	0	$\frac{1}{8}$	$\frac{-1}{8}$	4
v	0	1	$\frac{-1}{8}$	$\frac{9}{8}$	18
Z	0	0	3	117	2736

All indicators in the bottom row are nonnegative, so an optimum solution has been reached. The maximum value of Z is 2736. It occurs when $u = 4$ and $v = 18$. Therefore, the minimum value of z in the primal problem is 2736. It occurs at (3, 117) since 3 and 117 appear in the bottom row and in the columns corresponding to the slack variables r and s of the final tableau. Hence, the minimum value of z occurs when $x = 3$ and $y = 117$.
Do Exercise 1.   ■

The next example illustrates the fact that in a minimization problem in standard form, the constants on the right of the constraint inequalities may be negative, with the exception of those expressing the nonnegativity of the decision variables.

EXAMPLE 3   Minimize $z = 4x + 7y$ subject to

$$x + y \geq 14$$
$$x - 3y \leq 6$$
$$x \geq 0, y \geq 0$$

Solution   We must first multiply the second inequality by -1 to change the "$\leq$" to "$\geq$."
Minimize z if $z = 4x + 7y$ subject to

$$x + y \geq 14$$
$$-x + 3y \geq -6$$
$$x \geq 0, y \geq 0$$

The problem is now in standard form. The matrix of the primal problem is

$$\begin{bmatrix} 1 & 1 & 14 \\ -1 & 3 & -6 \\ 4 & 7 & z \end{bmatrix}$$

The transpose is

$$\begin{bmatrix} 1 & -1 & 4 \\ 1 & 3 & 7 \\ 14 & -6 & z \end{bmatrix}$$

This is the matrix of the dual problem. Thus, we can formulate the dual problem as follows. Maximize $Z = 14u - 6v$ subject to

$$u - v \leq 4$$
$$u + 3v \leq 7$$
$$u \geq 0, v \geq 0$$

Using slack variables, we obtain

$$\begin{cases} u - v + r = 4 \\ u + 3v + s = 7 \\ Z - 14u + 6v = 0 \end{cases}$$

The initial tableau is now given with the entering and leaving variables and the pivot element indicated.

Tableau I

Basis	u	v	r	s	Values of basic variables
r	①	-1	1	0	4
s	1	3	0	1	7
Z	-14	6	0	0	0

Pivoting, we get

Tableau II

Basis	u	v	r	s	Values of basic variables
u	1	-1	1	0	4
s	0	④	-1	1	3
Z	0	-8	14	0	56

The entering and leaving variables and the pivot element are indicated. Pivoting, we get

Tableau III

Basis	u	v	r	s	Values of basic variables
u	1	0	$\frac{3}{4}$	$\frac{1}{4}$	$\frac{19}{4}$
v	0	1	$\frac{-1}{4}$	$\frac{1}{4}$	$\frac{3}{4}$
Z	0	0	12	2	62

All indicators in the bottom row are nonnegative. Hence, the maximum of Z has been reached. It is 62 and occurs when $u = \frac{19}{4}$ and $v = \frac{3}{4}$. Therefore, the minimum value of z in the primal problem is 62. It occurs at $(12, 2)$ since 12 and 2 appear in the bottom row and in the columns corresponding to the slack variables r and s of the final tableau. Hence, the minimum value of z occurs when $x = 12$ and $y = 2$.
Do Exercise 21. ■

EXAMPLE 4 John must supplement his daily intake of vitamins A, B, and C. He should add at least 24 units of vitamin A, 7 units of vitamin B, and 18 units of vitamin C. John can buy vitamin pills at store S and at store P. The store S pills cost $15 per bottle of 100 pills, while the store P pills cost $18 per bottle of 200 pills. The vitamin content of each type of pill is given in the following table.

	Vitamin A	Vitamin B	Vitamin C
Store S pill	6 units	2 units	3 units
Store P pill	4 units	1 unit	6 units

How many pills from each store must he take each day to minimize his cost?

Solution Let x and y denote the numbers of pills he takes daily from store S and store P, respectively. The cost per day of the store S pills is $.15x$ since the cost per pill is given by $\frac{15}{100} = .15$. Also, the cost per day of the store P pills is $.09y$ since the cost per pill is given by $\frac{18}{200} = .09$. Hence, his total daily cost is $(.15x + .09y)$. This is the quantity we wish to minimize. There are some constraints. John will get $6x$ units of vitamin A from the x pills from store S and $4y$ units of vitamin A from the y pills from store P. Therefore, his total intake of vitamin A from these pills is $(6x + 4y)$ units. Since this intake must be at least 24 units, we write

$$6x + 4y \geq 24$$

Similarly, we get

$$2x + y \geq 7 \quad \text{and} \quad 3x + 6y \geq 18$$

Simplifying these inequalities, we can state the minimization problem in standard form as follows:

Minimize $z = .15x + .09y$ subject to

$$3x + 2y \geq 12$$
$$2x + y \geq 7$$
$$x + 2y \geq 6$$
$$x \geq 0, y \geq 0$$

The matrix of the primal problem is

$$\begin{bmatrix} 3 & 2 & 12 \\ 2 & 1 & 7 \\ 1 & 2 & 6 \\ .15 & .09 & z \end{bmatrix}$$

The transpose is

$$\begin{bmatrix} 3 & 2 & 1 & .15 \\ 2 & 1 & 2 & .09 \\ 12 & 7 & 6 & z \end{bmatrix}$$

This is the matrix of the dual problem. Thus, we can formulate the dual problem as follows.

Maximize $Z = 12u + 7v + 6w$ subject to

$$3u + 2v + w \le .15$$
$$2u + v + 2w \le .09$$
$$u \ge 0, v \ge 0, w \ge 0$$

Using slack variables, we get

$$\begin{cases} 3u + 2v + w + r & = .15 \\ 2u + v + 2w & + s = .09 \\ Z - 12u - 7v - 6w & = 0 \end{cases}$$

The initial tableau is given below with the entering and leaving variables and the pivot element as indicated.

Tableau I

Basis	u	v	w	r	s	Values of basic variables
r	3	2	1	1	0	.15
s	②	1	2	0	1	.09
Z	-12	-7	-6	0	0	0

Pivoting, we obtain

Tableau II

Basis	u	v	w	r	s	Values of basic variables
r	0	$\frac{1}{2}$	-2	1	$\frac{-3}{2}$	.015
u	1	$\frac{1}{2}$	1	0	$\frac{1}{2}$	.045
Z	0	-1	6	0	6	.54

The entering and leaving variables and the pivot element are indicated. Pivoting, we get

Tableau III

Basis	u	v	w	r	s	Values of basic variables
v	0	1	-4	2	-3	.03
u	1	0	3	-1	-1	.03
Z	0	0	2	2	3	.57

All indicators in the bottom row are nonnegative. The maximum value of Z has been reached. It is .57 and occurs at $u = .03$ and $v = .03$. Therefore, the minimum value of z in the primal problem is .57. It occurs at the point $(2, 3)$ since 2 and 3 appear in the bottom row and in the columns corresponding to the slack variables r and s of the final tableau. Thus, the minimum value of z occurs when $x = 2$ and $y = 3$. We conclude that John should take two of the store S pills and three of the store P pills per day to minimize his cost. His daily cost is $.57.
Do Exercise 23.   ■

Warning About Simplifications in the Dual Problem

Students are usually encouraged to simplify equations and inequalities before solving them. This is good practice because in general simplication does make subsequent calculations easier. When we use the method described in this section, it is correct to simplify a constraint inequality in the primal problem by multiplying it by a positive number. However, *a constraint inequality should not be multiplied by a positive number in a dual maximization problem if that maximization problem is used to solve a minimization problem*. The next example is given to illustrate this difficulty.

EXAMPLE 5   Minimize $z = 12x + 16y$ subject to

$$x + 2y \geq 40$$
$$3x + 2y \geq 72$$
$$x \geq 0, y \geq 0$$

Solution   The matrix of the primal problem is

$$\begin{bmatrix} 1 & 2 & 40 \\ 3 & 2 & 72 \\ 12 & 16 & z \end{bmatrix}$$

The transpose is

$$\begin{bmatrix} 1 & 3 & 12 \\ 2 & 2 & 16 \\ 40 & 72 & z \end{bmatrix}$$

This is the matrix of the dual problem. Thus, we can formulate the dual problem as follows:

Maximize $Z = 40u + 72v$ subject to

$$u + 3v \leq 12$$
$$2u + 2v \leq 16$$
$$u \geq 0, v \geq 0$$

At this point, we might be tempted to multiply both sides of the second inequality by $\frac{1}{2}$. However, we should not do so. Instead, we continue by using slack variables to obtain the system

$$\begin{cases} u + 3v + r & = 12 \\ 2u + 2v & + s = 16 \\ Z - 40u - 72v & = 0 \end{cases}$$

The initial tableau for this system showing the entering and leaving variables and the pivot element is

Tableau I

Basis	u	v	r	s	Values of basic variables
r	1	③	1	0	12
s	2	2	0	1	16
Z	-40	-72	0	0	0

Pivoting, we obtain

Tableau II

Basis	u	v	r	s	Values of basic variables
v	$\frac{1}{3}$	1	$\frac{1}{3}$	0	4
s	$\frac{4}{3}$	0	$\frac{-2}{3}$	1	8
Z	-16	0	24	0	288

The entering and leaving variables and the pivot element are indicated. Pivoting, we obtain

Tableau III

Basis	u	v	r	s	Values of basic variables
v	0	1	$\frac{1}{2}$	$\frac{-1}{4}$	2
u	1	0	$\frac{-1}{2}$	$\frac{3}{4}$	6
Z	0	0	16	12	384

All indicators in the bottom row are nonnegative. Hence, the maximum value of Z has been reached. It is 384 and occurs when $u = 6$ and $v = 2$. Therefore, the minimum value of z in the primal problem is also 384. It occurs at (16, 12). Hence, the minimum value of z occurs when $x = 16$ and $y = 12$. Suppose that we had simplified the constraint inequality in the dual maximization problem as follows:

$$u + 3v \leq 12$$
$$2u + 2v \leq 16$$
$$u \geq 0, v \geq 0$$

Multiplying both sides of the second inequality by $\frac{1}{2}$ we would obtain

$$u + 3v \leq 12$$
$$u + v \leq 8$$
$$u \geq 0, v \geq 0$$

We use slack variables as usual to get the system

$$\begin{cases} u + 3v + r = 12 \\ u + v + s = 8 \\ Z - 40u - 72v = 0 \end{cases}$$

The initial tableau showing the entering and leaving variables and the pivot element is

Tableau I

Basis	u	v	r	s	Values of basic variables
r	1	③	1	0	12
s	1	1	0	1	8
Z	-40	-72	0	0	0

Pivoting, we obtain

Tableau II

Basis	u	v	r	s	Values of basic variables
v	$\frac{1}{3}$	1	$\frac{1}{3}$	0	4
s	②⁄₃	0	$\frac{-1}{3}$	1	4
Z	-16	0	24	0	288

The entering and leaving variables and the pivot element are indicated. Pivoting, we obtain

Tableau III

Basis	u	v	r	s	Values of basic variables
v	0	1	$\frac{1}{2}$	$\frac{-1}{4}$	2
u	1	0	$\frac{-1}{2}$	$\frac{3}{2}$	6
Z	0	0	16	24	384

All indicators in the bottom row are nonnegative. Hence, the maximum value of Z is 384. It occurs when $u = 6$ and $v = 2$. This is the same answer that we obtained earlier for the *dual maximization problem*. Reading the bottom line of the final tableau, we conclude that the minimum value of z in the primal problem is 384. It occurs at (16, 24). Hence, the minimum value of z occurs when $x = 16$ and $y = 24$. But this answer is *wrong*. The minimum value of z is 384. But it occurs when $x = 16$ and $y = 12$. You are urged to give some thought to the reason why simplifying a constraint inequality does not alter the answer to the dual problem, but does alter the answer to the primal problem. ∎

Relationship Between a Primal Minimization Problem and its Dual

The last two examples are given to help you see why the solution of a primal minimization problem in standard form and that of its dual are equal.

A CLOSER LOOK

EXAMPLE 6   The Rouquier brothers own a company that manufactures products A and B. The profit from product A is $5280 per unit, while the profit from product B is $1650 per unit. The manufacturing of these products requires the use of two machines. The time required on each machine for each product is given in the following table.

	Number of Hours Required to Produce One Unit of	
	Product A	Product B
Machine I	11	2
Machine II	3	1

The time available on each machine per year is 2310 hours on machine I and 990 hours on machine II. Assuming that every unit of each product made can be sold, use the simplex method to determine how many units of each product should be manufactured each year to maximize the yearly profit. What is the maximum yearly profit?

Solution Let x and y denote the numbers of units per year of products A and B, respectively. The profit realized from x units of A is $5280x$ and that from y units of B is $1650y$. Hence, the total profit is $(5280x + 1650y)$. We wish to maximize this quantity. The number of hours used on machine I is $11x$ in the manufacturing of A and $2y$ hours in the manufacturing of B. Hence, the total number of hours machine I is used is $11x + 2y$. Since the total number of hours available on machine I is 2310, we have

$$11x + 2y \leq 2310$$

Similarly,

$$3x + y \leq 990$$

Our problem is to maximize $z = 5280x + 1650y$ subject to

$$11x + 2y \leq 2310$$
$$3x + \ y \leq \ \ 990$$
$$x \geq 0, y \geq 0$$

The initial tableau is

Tableau I

Basis	x	y	r	s	Values of basic variables
r	⑪	2	1	0	2310
s	3	1	0	1	990
z	-5280	-1650	0	0	0

Pivoting, we get

Tableau II

Basis	x	y	r	s	Values of basic variables
x	1	$\frac{2}{11}$	$\frac{1}{11}$	0	210
s	0	$\frac{5}{11}$	$\frac{-3}{11}$	1	360
z	0	-690	480	0	1,108,800

Pivoting, we get

Tableau III

Basis	x	y	r	s	Values of basic variables
x	1	0	$\frac{1}{5}$	$\frac{-2}{5}$	66
y	0	1	$\frac{-3}{5}$	$\frac{11}{5}$	792
z	0	0	66	1518	1,655,280

There are no negative indicators in the bottom row, so an optimum solution has been reached. The maximum profit per year is $1,655,280 reached at $x = 66$ and $y = 792$. The Rouquier company should manufacture 66 units of A and 792 units of B per year.

■

EXAMPLE 7 One of the Rouquier brothers of Example 6 decides that instead of making the two products, they could rent the machine time to another company. Suppose that machine I is rented for u dollars per hour and machine II is rented for v dollars per hour.

a. Express the amount of rent the company would receive per year in terms of u and v.
b. Write the inequalities that must be satisfied for this alternative to be at least as profitable as making the two products.
c. Since the Rouquier company is competing with other companies, they must try to minimize the amount of rent they charge while still making a profit. Give a geometric solution to find the minimum total amount of rent they receive each year and how much per hour is charged for each machine.

Solution **a.** Since machine I is rented for 2310 hours at u dollars per hour, the yearly rental for machine I is $2310u$. Similarly, machine II is rented for 990 hours at v dollars per hour, therefore, the yearly rental for machine II is $990v$. The total yearly income from the rental is $(2310u + 990v)$.

b. In producing one unit of product A, machine I is used 11 hours and machine II is used 3 hours. The value of the rental for 11 hours of machine I is $11u$. For the 3 hours of machine II, it is $3v$. The value of the rental to produce one unit of product A is $(11u + 3v)$. This amount must be at least as large as the profit the Rouquier brothers would make by producing one unit of product A; otherwise, they should make the product rather than rent the machines. We have

$$11u + 3v \geq 5280$$

Similarly, in producing one unit of product B, machine I is used 2 hours and machine II is used 1 hour. The value of the rental for these 2 hours of machine I is $2u$. For the 1 hour of machine II it is v. The value of the rental to produce one unit of product B is $(2u + v)$. This amount must be at least as large as the profit the Rouquier brothers would make by producing one unit of product

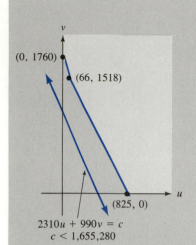

FIGURE 4.27

B; otherwise, they should make the product rather than rent the machines. We have

$$2u + v \geq 1650$$

c. The sum received for the rental per year is z dollars, where $z = 2310u + 990v$. Thus, to be competitive, the Rouquier brothers must minimize z subject to

$$11u + 3v \geq 5280$$
$$2u + v \geq 1650$$
$$u \geq 0, v \geq 0$$

The geometric solution to this minimization problem appears in Figure 4.27. The region of feasible solutions is shaded. The region of feasible solutions is unbounded. It has three vertices: $(0, 1760)$, $(66, 1518)$, and $(825, 0)$. We evaluate z at each vertex.

At $(0, 1760)$, $z = 2310(0) + 990(1760) = 1,742,400$,

at $(66, 1518)$, $z = 2310(66) + 990(1518) = 1,655,280$,

and at $(825, 0)$, $z = 2310(825) + 990(0) = 1,905,750$.

The smallest of these three values is 1,655,280. If $c < 1,655,280$, the graph of the equation $2310u + 990v = c$ does not intersect the region of feasible solutions. Therefore, the minimum value of z is 1,655,280 and it occurs when $u = 66$ and $v = 1518$. Therefore, the Rouquier company should charge \$66 per hour for machine I and \$1518 per hour for machine II. The minimum total amount they can receive for the rent and be as well off as if they made the two products is \$1,655,280. Note that this amount is equal to the maximum profit they can make if they do, in fact, make the two products. Observe that the maximization problem of Example 6 is the dual of the minimization problem of Example 7.

Do Exercise 34.   ■

Procedure to Solve a Minimization Problem in Standard Form

Step 1.   Be sure the problem is in standard form.

Step 2.   Find the dual maximization problem by

    a. Writing the given "$\geq$" inequalities as equations together with the equation defining the objective function, which has the dependent variable on the right side and appears last.

    b. Writing the augmented matrix of the system obtained in **a**.

    c. Writing the transpose of the matrix obtained in **b**.

    d. Writing the system of equations corresponding to the matrix obtained in **c**.

    e. Replacing all equal signs by "$\leq$" in all but the last equation.

    f. *Not* simplifying any of the constraint inequalities in the dual maximization problem just obtained.

> **Step 3.** Use the simplex method to get the final tableau of the maximization problem obtained in Step 2.
>
> **Step 4.** Read the solution to the original minimization problem as follows. The minimum value of the objective function is found in the lower right corner of the final tableau obtained in Step 3. This minimum value occurs at a point whose coordinates are read in the bottom row and in the columns corresponding to the slack variables of the same final tableau.

Exercise Set 4.6

In Exercises 1–33, use the method described in the procedure at the end of this section to solve a minimization problem.

1. Minimize $z = 45x + 27y$ subject to

$$x + y \geq 45$$
$$3x + y \geq 81$$
$$x \geq 0, y \geq 0$$

2. Minimize $z = 12x + 16y$ subject to

$$x + 2y \geq 40$$
$$3x + 2y \geq 72$$
$$x \geq 0, y \geq 0$$

3. Minimize $z = 57x + 133y$ subject to

$$2x + 5y \geq 210$$
$$5x + 3y \geq 240$$
$$x \geq 0, y \geq 0$$

4. Minimize $z = 10x + 4y$ subject to

$$-x + y \geq 2$$
$$11x - 4y \geq 55$$
$$x \geq 0, y \geq 0$$

5. Minimize $z = 27x + 20y$ subject to

$$4x + 5y \geq 260$$
$$3x + y \geq 96$$
$$x \geq 0, y \geq 0$$

6. Minimize $z = 10x + 18y$ subject to

$$x + y \geq 15$$
$$x + 3y \geq 21$$
$$x \geq 0, y \geq 0$$

7. Minimize $z = 16x + 25y$ subject to

$$2x + y \geq 7$$
$$x + y \geq 5$$
$$2x + 5y \geq 16$$
$$x \geq 0, y \geq 0$$

8. Minimize $z = 1974x + 1470y$ subject to

$$7x + 3y \geq 210$$
$$x + 7y \geq 140$$
$$2x + 5y \geq 118$$
$$x \geq 0, y \geq 0$$

9. Minimize $w = 210x + 48y + 45z$ subject to

$$4x + y + z \geq 8$$
$$y \geq 9$$
$$5x \geq 6$$
$$x \geq 0, y \geq 0, z \geq 0$$

10. Minimize $w = 7x + 7y + 14z$ subject to

$$4x + y - z \geq 2$$
$$3x - y - z \geq -24$$
$$-x - y + z \geq 8$$
$$x \geq 0, y \geq 0, z \geq 0$$

11. Minimize $w = 66x + 72y + 108z$ subject to

$$x - y + 2z \geq 5$$
$$x + 3y + z \geq 1$$
$$x + 2y + 3z \geq 1$$
$$x \geq 0, y \geq 0, z \geq 0$$

12. Minimize $w = 168x + 280y + 210z$ subject to

$$x + 2y + z \geq 14$$
$$2x + y + z \geq 14$$
$$x + y + 3z \geq 14$$
$$x \geq 0, y \geq 0, z \geq 0$$

13. Minimize $z = 200x + 750y$ subject to

$$x + 4y \geq 20$$
$$4x + 6y \geq 50$$
$$2x + 5y \geq 30$$
$$x \geq 0, y \geq 0$$

14. Minimize $w = 24x + 12y$ subject to

$$2x + y + 2z \geq 3$$
$$x + 2y \geq 6$$
$$3x + z \geq 3$$
$$x \geq 0, y \geq 0, z \geq 0$$

15. Minimize $z = 80x + 60y$ subject to

$$x + y \geq 10$$
$$5x + 3y \geq 20$$
$$2x + 4y \geq 15$$
$$x \geq 0, y \geq 0$$

16. Minimize $w = 40x + 20y + 30z$ subject to

$$x + 2y + z \geq 3$$
$$x + y - z \geq 6$$
$$-x + 3y - 2z \leq 3$$
$$x \geq 0, y \geq 0, z \geq 0$$

17. Minimize $w = 420x + 910y + 560z$ subject to

$$x + 2y + z \geq 2$$
$$5x + y + 7z \geq 24$$
$$2x + 3y \geq 4$$
$$x \geq 0, y \geq 0, z \geq 0$$

18. Minimize $w = 60x + 72y + 48z$ subject to

$$x + y \geq 3$$
$$x + y + z \geq 6$$
$$3y + 3z \geq 15$$
$$x \geq 0, y \geq 0, z \geq 0$$

19. Minimize $z = 16x_1 + 28x_2 + 8x_3 + 20x_4$ subject to

$$x_1 + x_2 \geq 3$$
$$x_1 - 2x_2 \geq 0$$
$$x_3 + x_4 \geq 2$$
$$-2x_3 + x_4 \geq 0$$
$$x_1 \geq 0, x_2 \geq 0,$$
$$x_3 \geq 0, x_4 \geq 0$$

20. Minimize $z = 10x + 5y$ subject to

$$3x + y \geq 5$$
$$x + 2y \geq 10$$
$$2x + y \geq 5$$
$$x + y \geq 25$$
$$x \geq 0, y \geq 0$$

21. Minimize $w = 8x + 7y + 9z$ subject to

$$x + 2y + 3z \geq 2$$
$$x - y - z \geq 3$$
$$x - y - 2z \leq 1$$
$$2x + y + z \geq 5$$
$$x \geq 0, y \geq 0, z \geq 0$$

22. A publishing company is organizing a fishing trip for its 200 employees. Small and large boats are available at costs of $300 and $1200 each, respectively. A small boat can take six employees and requires one skipper, while a large boat can take 20 employees but requires one skipper and two baiters. The charter company has, at most, 31 people available who can be skippers or baiters. How many small boats and large boats should the publishing company charter to minimize the cost? What is the minimum cost?

23. A politician hired a consultant who advised him that for each dollar spent on radio advertising, 10 Republicans and 30 Democrats can be reached; while for each dollar spent on television advertising, 20 Republicans and 20 Democrats can be reached. The politician wishes to reach at least 700,000 Republicans and 1,200,000 Democrats. Find the amounts the politician should spend on television and radio advertising to reach his goal at a minimum cost. What is the minimum cost?

24. A Jesuit university found that for each dollar spent advertising in a nonreligious publication, 40 Catholic and 80 non-Catholic families can be reached. For each dollar spent on advertising in a Catholic publication, 100 Catholic and 30 non-Catholic families can be reached. The university wants to reach at least 330,000 Catholic and 235,000 non-Catholic

families. How much should be spent on advertising in each publication to reach the goal at a minimum cost? What is the minimum cost?

25. A corporation produces chemicals X, Y, and Z at sites A, B, and C. The daily production of each chemical at each of the sites is given in the following table, where the units are tons.

	Site A	Site B	Site C
Chemical X	6	3	3
Chemical Y	5	5	5
Chemical Z	4	8	4

The corporation has received orders of 270 tons of X, 300 tons of Y, and 288 tons of Z. The daily cost of operation for each site is $30,000, $36,000, and $27,000, respectively. How many days should the corporation operate each site to be able to fill the orders at a minimum operating cost? What is the minimum operating cost?

26. A dietitian has to plan a special diet for athletes using two types of food, A and B. The numbers of units of iron, calcium, vitamin A, and cholesterol contained in each ounce of each type of food are given in the following table.

	Food A	Food B
Iron	18	9
Calcium	24	20
Vitamin A	7	35
Cholesterol	9	5

The daily minimum requirements for an athlete are 180 units of iron, 320 units of calcium, and 210 units of vitamin A. How many ounces of each type of food should be used daily to minimize the intake of cholesterol? What is the minimum intake of cholesterol per day?

27. A dietitian in a school cafeteria has to plan a diet for school children using foods X, Y, and Z. The numbers of units of calcium, vitamin B, iron, and cholesterol in each ounce of each type of food are given in the following table.

	Food X	Food Y	Food Z
Calcium	15	10	10
Vitamin B	8	6	2
Iron	10	6	2
Cholesterol	16	8	4

The minimum daily requirements for each child are 160 units of calcium, 56 units of vitamin B, and 48 units of iron. How many ounces of each of the three types of food should be mixed for each child to minimize the daily intake of cholesterol? What is the minimum daily intake of cholesterol?

28. A dietician in the local hospital is planning meals for patients using foods I, II, and III. The amounts of protein and calcium, and the costs per unit of each type of food are given in the following table.

	Type I	Type II	Type III
Protein	8g	4g	20g
Calcium	80mg	20mg	20mg
Cost	40c	12c	40c

The minimum requirements for each patient are 80 grams of protein and 600 milligrams of calcium. How many units of each type of food should be used for each patient to minimize the total cost of food per patient per day? What is that minimum cost?

29. A dairy company pasteurizes milk at three different sites: Arlington, Bellingham, and Mount Vernon. The daily production of 2% and 4% milk (in thousands of gallons) as well as the daily cost of operation are given for each site in the table.

	2%	4%	Daily cost in dollars
Arlington	10	20	20,000
Bellingham	10	10	14,000
Mount Vernon	20	10	24,000

Each month, the company must fill orders of 630,000 gallons of 2% and 360,000 gallons of 4% milk. How many days per month should each site operate to minimize the total operating cost per month? What is the minimum total monthly operating cost?

30. The dean of the business school at a local university is faced with the following problem. University policies dictate that during an academic year, each assistant professor teaches 12 credits, each associate professor teaches 10 credits, and each full professor teaches 8 credits, and that the number of assistant professors be at least twice the number of full professors. On the average, each assistant professor publishes one research paper per year, each associate professor publishes two research papers per year, and each full professor publishes three research papers per year. The enrollment in the school creates a need of at least 470 credits taught per academic year. Because of accreditation, the school's faculty is expected to publish a total of at least 80 research papers per academic year. The average salaries per academic year for assistant professors, associate professors, and full professors are $20,000, $33,000, and $45,000, respectively. How many professors should be at each rank to minimize the teaching budget of the school? What is the minimum teaching budget?

31. An electronics company makes its VCRs at sites A and B and has two distributors I and II. Site A can manufacture at most 750 VCRs per week, while site B can manufacture at

most 600 VCRs a week. Distributors I and II must receive at least 450 and 750 VCRs each week, respectively. The transportation costs per VCR from each site to each distributor are given in the following table.

From	To	
	Distributor I	Distributor II
Site A	$6.00	$12.00
Site B	$15.00	$9.00

How many VCRs should be shipped each week from each site to each distributor to minimize the cost of transportation? What is the minimum cost?

32. A company manufactures its refrigerators at plants A and B and has two distributors I and II. Plant A can manufacture at most 48 refrigerators per week, while plant B can manufacture at most 100 refrigerators a week. Distributors I and II must receive at least 60 and 80 refrigerators each week, respectively. The transportation costs per refrigerator from each plant to each distributor are given in the following table.

From	To	
	Distributor I	Distributor II
Plant A	$20.00	$25.00
Plant B	$30.00	$25.00

How many refrigerators should be shipped each week from each plant to each distributor to minimize the total cost of transportation? What is the minimum cost?

33. An RV manufacturer builds its motor homes at factories A and B and has two distributors I and II. Factory A can manufacture at most 25 motor homes a month, while factory B can manufacture 25 motor homes a month at most. Distributors I and II must receive at least 20 and 30 motor homes each month, respectively. The transportation costs per motor home from each factory to each distributor are given in the following table.

From	To	
	Distributor I	Distributor II
Factory A	$200.00	$250.00
Factory B	$225.00	$175.00

How many motor homes should be shipped each week from each factory to each distributor to minimize the cost of transportation? What is that minimum cost?

34. Verify that the maximization problem of Example 6 of this section is the dual of the minimization problem of Example 7. Also check that the coordinates of the point where the function of Example 7 attains its minimum appear in the bottom row of the final tableau of Example 6.

4.7 The Big *M* Method

In the preceding sections, we presented methods for solving maximization and minimization problems in standard form. For a linear programming problem in standard form, we always begin with a basic feasible solution in which all decision variables are 0. However, for a maximization problem which is not in standard form, such a basic feasible solution may not exist. In this section, we show how the simplex method may be used in such cases. We begin with a simple example which we first solve geometrically. We shall use this example to illustrate the idea behind the new method.

EXAMPLE 1   Maximize $P = 3x + 2y$ subject to

$$x - y \leq 1$$
$$x + y \leq 9$$
$$2x + y \geq 5$$
$$x \geq 0,\ y \geq 0$$

Solution   We first graph the equations

$$x - y = 1$$
$$x + y = 9$$
$$2x + y = 5$$

(See Figure 4.28.) We then shade the region that represents the set of feasible solutions. The vertices of this region are (2, 1), (5, 4), (0, 9), and (0, 5). We find that

$P = 8$ at (2, 1) since $3(2) + 2(1) = 8$
$P = 23$ at (5, 4) since $3(5) + 2(4) = 23$
$P = 18$ at (0, 9) since $3(0) + 2(9) = 18$
$P = 10$ at (0, 5) since $3(0) + 2(5) = 10$

Hence, the maximum value of P is 23 and occurs at (5, 4).   ■

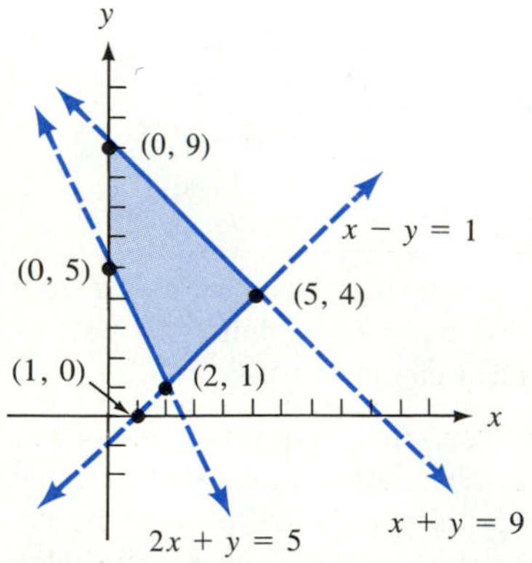

FIGURE 4.28

The Modified Problem

We cannot start the simplex method as we did in the standard problem by letting the decision variables be 0, since, as we see in Figure 4.28, the point (0, 0) is not in the region of feasible solutions. We must find some way to get to one of the vertices of the region of feasible solutions. As we often do in mathematics, we modify our problem so that the techniques already available may apply. We first introduce slack variables to change the "≤" constraints to equalities. We obtain

$$x - y + s_1 = 1, \text{ where } s_1 \geq 0$$

and

$$x + y + s_2 = 9, \text{ where } s_2 \geq 0$$

We now introduce a *surplus variable* to change the "≥" constraint to an equality. At this time, it serves our purpose best to name the surplus variable z, and we write

$$2x + y - z = 5, \text{ where } z \geq 0$$

If we modify our problem, thinking of z as a new decision variable, so that the constraints are

$$x - y \leq 1$$
$$x + y \leq 9$$
$$2x + y - z \leq 5$$
$$x \geq 0, \ y \geq 0, \ z \geq 0$$

we have a problem in standard form. The region of feasible solutions of the modified problem is sketched in Figure 4.29, where in the xy-plane we also show the region of feasible solutions of the original problem. Note that the edge connecting the points A and B belongs to both regions. Adding slack variables to change the "≤" to equalities, we obtain

$$x - y + s_1 = 1$$
$$x + y + s_2 = 9$$
$$2x + y - z + s_3 = 5$$
$$x \geq 0, \ y \geq 0, \ z \geq 0, \ s_1 \geq 0, \ s_2 \geq 0, \ s_3 \geq 0$$

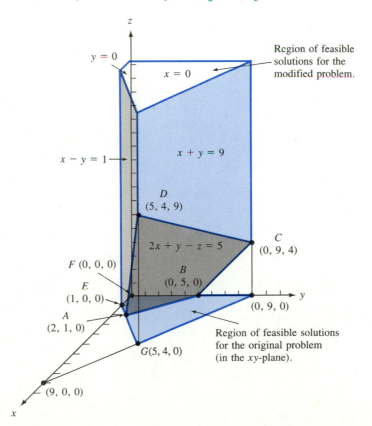

FIGURE 4.29

The slack variable s_3 introduced here is called an *artificial variable* for the original problem. In Figure 4.29, the region of feasible solutions for the modified problem is an unbounded region and its vertices are the points A, B, C, D, E, F, with coordinates $(2, 1, 0)$, $(0, 5, 0)$, $(0, 9, 4)$, $(5, 4, 9)$, $(1, 0, 0)$, and $(0, 0, 0)$, respectively. The points A and B and the projections of the points C and D onto the xy-plane are vertices of the region of feasible solutions of the original problem. For this reason, it is desirable for the optimum solution of the modified problem to occur at one of the four points A, B, C, or D. That is, to occur on the plane whose equation is $2x + y - z = 5$. Hence, *it is desirable for the value of the artificial variable s_3 to be 0*, since the constraint $2x + y - z + s_3 = 5$ must be satisfied.

At this point, we cannot expect the original objective function to attain its maximum value at one of the points A, B, C, or D. We must, therefore, modify the objective function also. Since we wish to have the new objective function attain its maximum value at a point where the value of the artificial variable is 0, we impose a very large "penalty," M, for each unit value of the artificial variable. That is, we subtract $s_3 M$ from the value of the objective function where M is a very large positive number. The modified objective function is defined by

$$P_M = 3x + 2y - s_3 M$$

It is clear that if P_M has a maximum value, it must be at a point where $s_3 = 0$.

We now solve the modified problem using the simplex method. We shall then show how this solution is related to the solution of the original problem. Finally, we shall describe the general method to solve linear programming problems that are not in standard form.

EXAMPLE 2 Use the simplex method to solve the modified problem. That is, maximize P_M if $P_M = 3x + 2y - s_3 M$, where

$$x - y + s_1 = 1$$
$$x + y + s_2 = 9$$
$$2x + y - z + s_3 = 5$$
$$x \geq 0, \ y \geq 0, \ z \geq 0, \ s_1 \geq 0, \ s_2 \geq 0, \ s_3 \geq 0$$

Solution We proceed as we did in Section 4.5. We write the equation defining the objective function as

$$P_M - 3x - 2y + s_3 M = 0$$

The first tableau is

Tableau I

Basis	x	y	z	s_1	s_2	s_3	Values of basic variables
s_1	1	−1	0	1	0	0	1
s_2	1	1	0	0	1	0	9
s_3	2	1	−1	0	0	1	5
P_M	−3	−2	0	0	0	M	0

Here $x = y = z = 0$, $s_1 = 1$, $s_2 = 9$, and $s_3 = 5$. Note first that $(0, 0, 0)$ is a feasible solution for the modified problem (See Figure 4.29), but the corresponding point $(0, 0)$ is not a feasible solution for the original problem. We observe that the value of the new objective function is $-5M$ since $3(0) + 2(0) - 5(M) = -5M$. This is not the value that is in the lower right corner of the tableau. We modify this by changing the M at the bottom of the s_3 column to 0. $((-M)R_3 + R_4 \to R_4)$. We obtain

Tableau II

Basis	x	y	z	s_1	s_2	s_3	Values of basic variables
s_1	①	-1	0	1	0	0	1
s_2	1	1	0	0	1	0	9
s_3	2	1	1	0	0	1	5
P_M	$-2M - 3$	$-M - 2$	M	0	0	0	$-5M$

Since M is a very large positive number, $-2M - 3$ is the most negative indicator in the bottom row of the tableau. Thus, we choose x as the entering variable. The quotient $\frac{1}{1}$ is smaller than both quotients $\frac{9}{1}$ and $\frac{5}{2}$. So we choose s_1 as the leaving variable. Pivoting, we obtain

Tableau III

Basis	x	y	z	s_1	s_2	s_3	Values of basic variables
x	1	-1	0	1	0	0	1
s_2	0	2	0	-1	1	0	8
s_3	0	③	-1	-2	0	1	3
P_M	0	$-3M - 5$	M	$2M + 3$	0	0	$-3M + 3$

Here $x = 1$, $y = z = s_1 = 0$, $s_2 = 8$, and $s_3 = 3$. The point E with coordinates $(1, 0, 0)$ is a corner of the region of feasible solutions for the modified problem but is not in the region of feasible solutions of the original problem. (See Figures 4.28 and 4.29, where in Figure 4.28 the point E has coordinates $(1, 0)$.) The most negative indicator in the bottom row is $-3M - 5$. We choose y as the entering variable and s_3 as the leaving variable since $\frac{3}{3}$ is less than $\frac{8}{2}$. Pivoting, we obtain

Tableau IV

Basis	x	y	z	s_1	s_2	s_3	Values of basic variables
x	1	0	$\frac{-1}{3}$	$\frac{1}{3}$	0	$\frac{1}{3}$	2
s_2	0	0	$\frac{2}{3}$	$\frac{1}{3}$	1	$\frac{-2}{3}$	6
y	0	1	$\frac{-1}{3}$	$\frac{-2}{3}$	0	$\frac{1}{3}$	1
P_M	0	0	$\frac{-5}{3}$	$\frac{-1}{3}$	0	$M + \frac{5}{3}$	8

So, $x = 2$, $y = 1$, $z = 0$, $s_1 = 0$, $s_2 = 6$, and $s_3 = 0$. The point A with coordinates $(2, 1, 0)$ is a corner of the region of feasible solutions of the modified problem. (See Figure 4.29.) It is also a corner of the region of feasible solutions of the original problem. In the xy-plane the point A has coordinates $(2, 1)$. We have reached a point which is on the common edge of both regions. The value of the artificial variable s_3 is 0. Therefore, we no longer need to carry the s_3 column in succeeding tableaus. We delete that column and obtain the simplified tableau.

Tableau V

Basis	x	y	z	s_1	s_2	Values of basic variables
x	1	0	$\frac{-1}{3}$	$\frac{1}{3}$	0	2
s_2	0	0	$\frac{2}{3}$	$\frac{1}{3}$	1	6
y	0	1	$\frac{-1}{3}$	$\frac{-2}{3}$	0	1
P_M	0	0	$\frac{-5}{3}$	$\frac{-1}{3}$	0	8

Clearly, z and s_2 should be entering and leaving variables, respectively. Pivoting, we obtain

Tableau VI

Basis	x	y	z	s_1	s_2	Values of basic variables
x	1	0	0	$\frac{1}{2}$	$\frac{1}{2}$	5
z	0	0	1	$\frac{1}{2}$	$\frac{3}{2}$	9
y	0	1	0	$\frac{-1}{2}$	$\frac{1}{2}$	4
P_M	0	0	0	$\frac{1}{2}$	$\frac{5}{2}$	23

Here $x = 5$, $y = 4$, $z = 9$, and $s_1 = s_2 = 0$. The point D with coordinates $(5, 4, 9)$ is a corner of the region of feasible solutions of the modified problem. Also its projection onto the xy-plane is the point G, it has coordinates $(5, 4)$ and is a corner of the region of feasible solutions of the original problem. There are no negative indicators in the bottom row. Thus, the maximum value of the objective function has been attained. It is read in the lower right corner of the final tableau. The maximum value of P_M is 23. It turns out that 23 is also the maximum value of the objective function of the original problem. To see this, suppose that we could find feasible values of x, y, z, s_1, and s_2, which would be solutions of the original system and would yield a value of P larger than 23. Using these values along with $s_3 = 0$, we would get a value of P_M larger than 23, contradicting the fact that we found the maximum value of P_M to be 23.

The previous example serves as an illustration of the steps that we follow to solve a linear programming problem that is not in standard form using the Big *M* Method.

Procedure: The Big *M* Method

Step 1.   Make sure that all constraints have nonnegative constants on the right side. If necessary, multiply both sides of a constraint by -1, remembering to reverse the sense of the inequality sign if the constraint is an inequality.

Step 2.   Add a slack variable s_i on the left side of each "$\leq$" constraint to obtain an equality.

Step 3.   Subtract a surplus variable s_j and add an artificial variable a_i on the left side of each "$\geq$" constraint to obtain an equality.

Step 4.   Add an artificial variable a_j on the left side of each equality constraint.

Step 5.   For each artificial variable a_i which was added, subtract Ma_i from the right side of the equation defining the objective function P. Use the same constant M for all artificial variables. Rename the objective function P_M to obtain the modified problem. *All constraint, slack, surplus, and artificial variables must be nonnegative.*

Step 6.   Form the preliminary simplex tableau for the modified problem obtained in the first five steps.

Step 7.   Eliminate the M's in the bottom row of the preliminary simplex tableau from all columns which correspond to artificial variables by *adding all the rows corresponding to artificial variables, multiplying the result by* $-M$, *and adding the product to the bottom row*. At this stage, all artificial variables are basic variables. In this way, we obtain the initial simplex tableau.

Step 8.   Solve the modified problem using the simplex method on the tableau obtained in Step 7. A tableau may be simplified as soon as all artificial variables have the value 0. We can then delete all columns corresponding to the artificial variables and continue until the solution of the modified problem is complete.

Step 9.   The solution of the original problem is related to that of the modified problem as follows:

   a. If the modified problem has no solution, neither does the original problem.
   b. If we obtain a solution of the modified problem where one of the artificial variables has a positive value, the original problem has no solution.
   c. If we obtain a solution of the modified problem where all artificial variables are 0, then the solution of the original problem is obtained by deleting all artificial variables in the solution of the modified problem.

We now give several examples to illustrate the Big *M* Method.

EXAMPLE 3 Formulate the modified problem for the following, but do not solve the modified problem. Maximize $P = 2x + 3y - 4z$ subject to

$$2x + 5y - 7z \leq 2$$
$$x - 2y - 3z \leq -4$$
$$x + 4y + 5z = 20$$
$$2x + 3y - z = -3$$
$$x \geq 0, y \geq 0, z \geq 0$$

Solution In order to have positive constants on the right sides of the first four constraints, multiply both sides of the second inequality by -1, reversing the sense of the inequality. Also multiply both sides of the second equality by -1. The constraints are written

$$2x + 5y - 7z \leq 2$$
$$-x + 2y + 3z \geq 4$$
$$x + 4y + 5z = 20$$
$$-2x - 3y + z = 3$$
$$x \geq 0, y \geq 0, z \geq 0$$

Now add a slack variable s_1 to the left side of the first inequality. Subtract a surplus variable s_2 from, and add an artificial variable a_1 to, the left side of the second inequality. Also add artificial variables a_2 and a_3 to the left sides of the first and second equalities, respectively. Write the first four constraints as equalities. We obtain

$$2x + 5y - 7z + s_1 = 2$$
$$-x + 2y + 3z - s_2 + a_1 = 4$$
$$x + 4y + 5z + a_2 = 20$$
$$-2x - 3y + z + a_3 = 3$$
$$x \geq 0, y \geq 0, z \geq 0, s_1 \geq 0, s_2 \geq 0, a_1 \geq 0, a_2 \geq 0, a_3 \geq 0$$

We subtract $(a_1 + a_2 + a_3)M$ from the right side of the equation defining the objective function and rename the objective function P_M. We get

$$P_M = 2x + 3y - 4z - a_1M - a_2M - a_3M$$

The modified problem is to maximize P_M subject to

$$2x + 5y - 7z + s_1 = 2$$
$$-x + 2y + 3z - s_2 + a_1 = 4$$
$$x + 4y + 5z + a_2 = 20$$
$$-2x - 3y + z + a_3 = 3$$
$$x \geq 0, y \geq 0, z \geq 0, s_1 \geq 0, s_2 \geq 0, a_1 \geq 0, a_2 \geq 0, a_3 \geq 0$$

■

EXAMPLE 4 Maximize $P = 5x + 8y + z$ subject to

$$2x + 3y + z \le 14$$
$$4x + 3y + z \ge 18$$
$$x + 2y + z = 9$$
$$x \ge 0,\ y \ge 0,\ z \ge 0$$

Solution First we introduce a slack variable s_1 in the first constraint, a surplus variable s_2 and an artificial variable a_1 in the second constraint, and an artificial variable a_2 in the third constraint. Then we subtract $M(a_1 + a_2)$ from the right side of the equality defining the objective function P. We rename the objective function P_M and obtain the modified problem. Maximize $P_M = 5x + 8y + z - a_1M - a_2M$ subject to

$$2x + 3y + z + s_1 \qquad = 14$$
$$4x + 3y + z - s_2 + a_1 = 18$$
$$x + 2y + z + a_2 \qquad = 9$$
$$x \ge 0,\ y \ge 0,\ z \ge 0,\ s_1 \ge 0,\ s_2 \ge 0,\ a_1 \ge 0,\ \text{and } a_2 \ge 0$$

The equation defining the objective function is written

$$P_M - 5x - 8y - z + a_1M + a_2M = 0$$

The preliminary simplex tableau for the modified problem is

Tableau I

Basis	x	y	z	s_1	s_2	a_1	a_2	Values of basic variables
s_1	2	3	1	1	0	0	0	14
a_1	4	3	1	0	-1	1	0	18
a_2	1	2	1	0	0	0	1	9
P_M	-5	-8	-1	0	0	M	M	0

Now add rows 2 and 3, corresponding to the artificial variables and multiply the result by $-M$. Add the product to row 4 to obtain the initial simplex tableau for the modified problem.

Tableau II

Basis	x	y	z	s_1	s_2	a_1	a_2	Values of basic variables
s_1	2	3	1	1	0	0	0	14
a_1	4	3	1	0	-1	1	0	18
← a_2	1	②	1	0	0	0	1	9
P_M	$-5M - 5$	$-5M - 8$	$-2M - 1$	0	M	0	0	$-27M$

↑

The most negative indicator is $-5M - 8$. Therefore, y is the entering variable. Since $\frac{9}{2}$ is less than $\frac{18}{3}$ and $\frac{14}{3}$, a_2 is the leaving variable. Pivoting, we obtain

Tableau III

Basis	x	y	z	s_1	s_2	a_1	a_2	Values of basic variables
s_1	$\frac{1}{2}$	0	$\frac{-1}{2}$	1	0	0	$\frac{-3}{2}$	$\frac{1}{2}$
a_1	$\frac{5}{2}$	0	$\frac{-1}{2}$	0	-1	1	$\frac{-3}{2}$	$\frac{9}{2}$
y	$\frac{1}{2}$	1	$\frac{1}{2}$	0	0	0	$\frac{1}{2}$	$\frac{9}{2}$
P_M	$\frac{-5M}{2} - 1$	0	$\frac{M}{2} + 3$	0	M	0	$\frac{5M}{2} + 4$	$\frac{-9M}{2} + 36$

The only negative indicator is $\frac{-5M}{2} - 1$. Thus, x is the entering variable. Since $\frac{1}{2}/\frac{1}{2}$ is less than $\frac{9}{2}/\frac{5}{2}$ and $\frac{9}{2}/\frac{1}{2}$, s_1 is the leaving variable. Pivoting, we obtain

Tableau IV

Basis	x	y	z	s_1	s_2	a_1	a_2	Values of basic variables
x	1	0	-1	2	0	0	-3	1
a_1	0	0	2	-5	-1	1	6	2
y	0	1	1	-1	0	0	2	4
P_M	0	0	$-2M + 2$	$5M + 2$	M	0	$-5M + 1$	$-2M + 37$

The most negative indicator is $-5M + 1$. Thus, a_2 is the entering variable. Since $\frac{2}{6}$ is less than $\frac{4}{2}$, a_1 is the leaving variable. Pivoting, we obtain

Tableau V

Basis	x	y	z	s_1	s_2	a_1	a_2	Values of basic variables
x	1	0	0	$\frac{-1}{2}$	$\frac{-1}{2}$	$\frac{1}{2}$	0	2
a_2	0	0	$\frac{1}{3}$	$\frac{-5}{6}$	$\frac{-1}{6}$	$\frac{1}{6}$	1	$\frac{1}{3}$
y	0	1	$\frac{1}{3}$	$\frac{2}{3}$	$\frac{1}{3}$	$\frac{-1}{3}$	0	$\frac{10}{3}$
P_M	0	0	$\frac{-M + 5}{3}$	$\frac{5M + 17}{6}$	$\frac{M + 1}{6}$	$\frac{5M - 1}{6}$	0	$\frac{-M + 110}{3}$

Since $\frac{-M + 5}{3}$ is the only negative indicator, z is the entering variable. Comparing ratios, we see that a_2 should be the leaving variable. Pivoting, we get

Tableau VI

Basis	x	y	z	s_1	s_2	a_1	a_2	Values of basic variables
x	1	0	0	$\frac{-1}{2}$	$\frac{-1}{2}$	$\frac{1}{2}$	0	2
z	0	0	1	$\frac{-5}{2}$	$\frac{-1}{2}$	$\frac{1}{2}$	3	1
y	0	1	0	$\frac{3}{2}$	$\frac{1}{2}$	$\frac{-1}{2}$	-1	3
P_M	0	0	0	7	1	$M - 1$	$M - 5$	35

All indicators are positive. Thus, 35 is the maximum value of P_M. It occurs where $x = 2$, $y = 3$, $z = 1$, and $s_1 = s_2 = a_1 = a_2 = 0$. Since both artificial variables are 0, 35 is also the maximum value of P. It occurs at the point $(2, 3, 1)$.
Do Exercise 5. ■

EXAMPLE 5   Maximize $P = 5x + 6y$ subject to

$$2x + 3y \le 6$$
$$4x + 5y \ge 20$$
$$x \ge 0, y \ge 0$$

Solution   Add a slack variable s_1 to the left side of the first constraint, subtract a surplus variable s_2 and add an artificial variable a_1 to the left side of the second constraint to change these constraints to equalities. Next, subtract $a_1 M$ from the right side of the equation defining the objective function and rename the objective function P_M to get the modified problem. Maximize $P_M = 5x + 6y - a_1 M$ subject to

$$2x + 3y + s_1 \quad\quad = 6$$
$$4x + 5y - s_2 + a_1 = 20$$
$$x \ge 0, y \ge 0, s_1 \ge 0, s_2 \ge 0, a_1 \ge 0$$

We write the equation defining the objective function as

$$P_M - 5x - 6y + a_1 M = 0$$

The preliminary tableau for the modified problem is

Tableau I

Basis	x	y	s_1	s_2	a_1	Values of basic variables
s_1	2	3	1	0	0	6
a_1	4	5	0	-1	1	20
P_M	-5	-6	0	0	M	0

Multiply row 2 by $-M$ and add the result to row 3 to change the bottom entry of the column corresponding to a_1 to 0. We obtain the initial simplex tableau for the modified problem.

Tableau II

Basis	x	y	s_1	s_2	a_1	Values of basic variables
s_1	2	③	1	0	0	6
a_1	4	5	0	-1	1	20
P_M	$-4M - 5$	$-5M - 6$	0	M	0	$-20M$

Clearly $-5M - 6$ is the most negative indicator, so y is the entering variable. Since $\frac{6}{3}$ is less than $\frac{20}{5}$ we choose s_1 to be the leaving variable. Hence 3 is the pivot. Pivoting, we obtain

Tableau III

Basis	x	y	s_1	s_2	a_1	Values of basic variables
y	$\frac{2}{3}$	1	$\frac{1}{3}$	0	0	2
a_1	$\frac{2}{3}$	0	$\frac{-5}{3}$	-1	1	10
P_M	$\frac{-2M}{3} - 1$	0	$\frac{5M}{3} + 2$	M	0	$-10M + 12$

Since $(-2M/3) - 1$ is the only negative indicator, x is the entering variable and, comparing ratios, y should be the leaving variable. Pivoting, we obtain

Tableau IV

Basis	x	y	s_1	s_2	a_1	Values of basic variables
x	1	$\frac{3}{2}$	$\frac{1}{2}$	0	0	3
a_1	0	-1	-2	-1	1	8
P_M	0	$M + \frac{3}{2}$	$2M + \frac{5}{2}$	M	0	$-8M + 15$

All indicators in the bottom row are nonnegative. Thus, the simplex procedure terminates with $x = 3$, $y = s_1 = s_2 = 0$, $a_1 = 8$, and $P_M = -8M + 15$. Since the artificial variable a_1 has a positive value, the original problem has no solution. **Do Exercise 3.** ■

Solving a Minimization Problem

In order to locate the point where a function attains its minimum value, it suffices to locate the point where the negative of the function attains its maximum value. For example, if a function f attains its minimum value at (x_0, y_0, z_0), then $f(x_0, y_0, z_0) \leq f(x, y, z)$ for all (x, y, z) in the domain of f. Consequently, $-f(x_0, y_0, z_0) \geq -f(x, y, z)$ for all (x, y, z) in the domain of $-f$. That is, $-f$ attains its maximum at (x_0, y_0, z_0). Furthermore, $\min f = f(x_0, y_0, z_0) = -[-f(x_0, y_0, z_0)] = -\max(-f)$.

EXAMPLE 6   Minimize $C = 6x + 4y + 4z$ subject to

$$5x + 7y + 4z \leq 65$$
$$5x + 5y + 4z \geq 57$$
$$x + 2y + 4z = 25$$
$$x \geq 0,\ y \geq 0,\ z \geq 0$$

Solution   Let $P = -C = -6x - 4y - 4z$ and try to find the maximum value of P subject to the same constraints. First, add a slack variable s_1 to the left side of the first constraint changing it to an equality. Change the second constraint to an equality by subtracting a surplus variable s_2 and adding an artificial variable a_1 to its left side. Then, add an artificial variable a_2 to the left side of the equality constraint. Subtract $(a_1 + a_2)M$ from the right side of the equality defining P and rename the objective function to get the modified problem. Maximize $P_M = -6x - 4y - 4z - a_1M - a_2M$ subject to

$$5x + 7y + 4z + s_1 \qquad\qquad = 65$$
$$5x + 5y + 4z - s_2 + a_1 \quad = 57$$
$$x + 2y + 4z \qquad\quad + a_2 = 25$$
$$x \geq 0,\ y \geq 0,\ z \geq 0,\ s_1 \geq 0,\ s_2 \geq 0,\ a_1 \geq 0,\ a_2 \geq 0$$

Write the equation defining the objective function in the following form

$$P_M + 6x + 4y + 4z + a_1M + a_2M = 0$$

The preliminary tableau for the modified problem is

Tableau I

Basis	x	y	z	s_1	s_2	a_1	a_2	Values of basic variables
s_1	5	7	4	1	0	0	0	65
a_1	5	5	4	0	-1	1	0	57
a_2	1	2	4	0	0	0	1	25
P_M	6	4	4	0	0	M	M	0

Now add rows 2 and 3, corresponding to the artificial variables. Then, multiply the result by $-M$ and add the product to row 4 to obtain the initial simplex tableau for the modified problem.

Tableau II

Basis	x	y	z	s_1	s_2	a_1	a_2	Values of basic variables
s_1	5	7	4	1	0	0	0	65
a_1	5	5	4	0	-1	1	0	57
a_2	1	2	④	0	0	0	1	25
P_M	$-6M + 6$	$-7M + 4$	$-8M + 4$	0	M	0	0	$-82M$

Since the most negative indicator is $-8M + 4$, z is the entering variable. Comparing $\frac{25}{4}$, $\frac{57}{4}$, and $\frac{65}{4}$, $\frac{25}{4}$ is the smallest. Thus, a_2 is the leaving variable. Pivoting, we obtain

Tableau III

Basis	x	y	z	s_1	s_2	a_1	a_2	Values of basic variables
s_1	4	5	0	1	0	0	-1	40
a_1	④	3	0	0	-1	1	-1	32
z	$\frac{1}{4}$	$\frac{1}{2}$	1	0	0	0	$\frac{1}{4}$	$\frac{25}{4}$
P_M	$-4M + 5$	$-3M + 2$	0	0	M	0	$2M - 1$	$-32M - 25$

Since the most negative indicator is $-4M + 5$, x is the entering variable. Comparing ratios, we see that a_1 should be the leaving variable. Pivoting, we get

Tableau IV

Basis	x	y	z	s_1	s_2	a_1	a_2	Values of basic variables
s_1	0	2	0	1	1	-1	0	8
x	1	$\frac{3}{4}$	0	0	$\frac{-1}{4}$	$\frac{1}{4}$	$\frac{-1}{4}$	8
z	0	$\frac{5}{16}$	1	0	$\frac{1}{16}$	$\frac{-1}{16}$	$\frac{5}{16}$	$\frac{17}{4}$
P_M	0	$\frac{-7}{4}$	0	0	$\frac{5}{4}$	$M - \frac{5}{4}$	$M + \frac{1}{4}$	-65

Here, $x = 8$, $y = 0$, $z = \frac{17}{4}$, $s_1 = 8$, and $s_2 = a_1 = a_2 = 0$. Since both artificial variables are equal to 0, delete the two columns corresponding to these variables to obtain the simplified tableau.

Tableau V

Basis	x	y	z	s_1	s_2	Values of basic variables
s_1	0	②	0	1	1	8
x	1	$\frac{3}{4}$	0	0	$\frac{-1}{4}$	8
z	0	$\frac{5}{16}$	1	0	$\frac{1}{16}$	$\frac{17}{4}$
P_M	0	$\frac{-7}{4}$	0	0	$\frac{5}{4}$	-65

Since $\frac{-7}{4}$ is the only negative indicator, y is the entering variable. Comparing ratios, we find that s_1 is the leaving variable. Pivoting, we obtain

Tableau VI

Basis	x	y	z	s_1	s_2	Values of basic variables
y	0	1	0	$\frac{1}{2}$	$\frac{1}{2}$	4
x	1	0	0	$\frac{-3}{8}$	$\frac{-5}{8}$	5
z	0	0	1	$\frac{-5}{32}$	$\frac{-3}{32}$	3
P_M	0	0	0	$\frac{7}{8}$	$\frac{17}{8}$	-58

Since all indicators in the bottom row are nonnegative, an optimum solution has been reached. The maximum value of P_M is -58 and occurs when $x = 5$, $y = 4$, $z = 3$, and $s_1 = s_2 = a_1 = a_2 = 0$. Thus, the maximum value of P is also -58. Recalling that $P = -C$, the minimum value of C is 58, since $-(-58) = 58$. This minimum value occurs at the point $(5, 4, 3)$.

Do Exercise 11.   ■

An Application

EXAMPLE 7   An oil refinery's facilities allow the blending of, at most, 9 million gallons per week. Some antiknock ingredient must be used. Each million gallons of grade A gasoline requires 4 pounds of the ingredient, each million gallons of grade B gasoline requires 5 pounds of the ingredient, while each million gallons of grade C gasoline requircs only 2 pounds. Only 36 pounds of the ingredient is available each week. Because of market conditions, the sum of twice the output of grade A and half the output of grade B must exceed the output of grade C by at least 6 million gallons. Suppose that the refinery makes profits of 11 cents per gallon on grade A gasoline, 14 cents per gallon on grade B gasoline, and 6 cents per gallon on grade C gasoline. How many gallons of each grade of gasoline should be blended per week to maximize the weekly profit? What is the maximum weekly profit?

Solution   Let x, y, and z be the numbers of millions of gallons of grade A, grade B, and grade C gasoline, respectively, that the refinery will blend each week. Since the facilities allow the blending of at most 9 million gallons per week, we have

$$x + y + z \leq 9$$

The amounts of antiknock ingredient required is $4x$ pounds for the x million gallons of grade A, $5y$ pounds for the y million gallons of grade B, and $2z$ pounds for the z million gallons of grade C. Since only 36 pounds of the ingredient are available per week we have

$$4x + 5y + 2z \leq 36$$

Because of market conditions, $2x + \frac{1}{2}y$ must exceed z by at least 6. Thus,

$$2x + \frac{1}{2}y - z \geq 6$$

or

$$4x + y - 2z \geq 12$$

The profits on grade A gasoline, grade B gasoline, and grade C gasoline, in tens of thousands of dollars, will be $11x$, $14y$, and $6z$, respectively. Thus, the total profit P, also in tens of thousands of dollars, is given by

$$P = 11x + 14y + 6z$$

The problem is to maximize P, subject to

$$\begin{aligned}
x + y + z &\leq 9 \\
4x + 5y + 2z &\leq 36 \\
4x + y - 2z &\geq 12 \\
x \geq 0, \ y \geq 0, \ z &\geq 0
\end{aligned}$$

We change the first three inequalities to equalities by adding a slack variable s_1 to the left side of the first inequality, adding a slack variable s_2 to the left side of the second inequality, and subtracting a surplus variable s_3 and adding an artificial variable a_1 to the left side of the third inequality. We then subtract $a_1 M$ from the right side of the equality defining the objective function, where M is a very large positive number, and rename the objective function P_M to get the modified problem.

Maximize $P_M = 11x + 14y + 6z - a_1 M$ subject to

$$\begin{aligned}
x + y + z + s_1 &= 9 \\
4x + 5y + 2z + s_2 &= 36 \\
4x + y - 2z - s_3 + a_1 &= 12 \\
x \geq 0, \ y \geq 0, \ z \geq 0, \ s_1 \geq 0, \ s_2 \geq 0, \ s_3 \geq 0, \ a_1 &\geq 0
\end{aligned}$$

Write the equation defining the objective function as

$$P_M - 11x - 14y - 6z + a_1 M = 0$$

Then write the preliminary simplex tableau for the modified problem.

Tableau I

Basis	x	y	z	s_1	s_2	s_3	a_1	Values of basic variables
s_1	1	1	1	1	0	0	0	9
s_2	4	5	2	0	1	0	0	36
a_1	4	1	-2	0	0	-1	1	12
P_M	-11	-14	-6	0	0	0	M	0

Now multiply row 3, corresponding to the artificial variable a_1, by $-M$ and add the result to row 4 to obtain the initial simplex tableau for the modified problem.

Tableau II

Basis	x	y	z	s_1	s_2	s_3	a_1	Values of basic variables
s_1	1	1	1	1	0	0	0	9
s_2	4	5	2	0	1	0	0	36
a_1	④	1	-2	0	0	-1	1	12
P_M	$-4M - 11$	$-M - 14$	$2M - 6$	0	0	M	0	$-12M$

Since $-4M - 11$ is the most negative indicator, x is the entering variable. Comparing ratios, we see that a_1 is the leaving variable. Pivoting, we obtain

Tableau III

Basis	x	y	z	s_1	s_2	s_3	a_1	Values of basic variables
s_1	0	$\frac{3}{4}$	$\frac{3}{2}$	1	0	$\frac{1}{4}$	$\frac{-1}{4}$	6
s_2	0	4	4	0	1	1	-1	24
x	1	$\frac{1}{4}$	$\frac{-1}{2}$	0	0	$\frac{-1}{4}$	$\frac{1}{4}$	3
P_M	0	$\frac{-45}{4}$	$\frac{-23}{2}$	0	0	$\frac{-11}{4}$	$M + \frac{11}{4}$	33

Here, $x = 3$, $y = 0$, $z = 0$, $s_1 = 6$, $s_2 = 24$, and $s_3 = a_1 = 0$. Since the value of the artificial variable is 0, delete the column corresponding to the artificial variable to obtain the simplified tableau.

Tableau IV

Basis	x	y	z	s_1	s_2	s_3	Values of basic variables
s_1	0	$\frac{3}{4}$	$\left(\frac{3}{2}\right)$	1	0	$\frac{1}{4}$	6
s_2	0	4	4	0	1	1	24
x	1	$\frac{1}{4}$	$\frac{-1}{2}$	0	0	$\frac{-1}{4}$	3
P_M	0	$\frac{-45}{4}$	$\frac{-23}{2}$	0	0	$\frac{-11}{4}$	33

Since $\frac{-23}{2}$ is the most negative indicator in the bottom row, z is the entering variable. Comparing ratios, we see that s_1 is the leaving variable. Pivoting, we obtain

Tableau V

Basis	x	y	z	s_1	s_2	s_3	Values of basic variables
z	0	$\frac{1}{2}$	1	$\frac{2}{3}$	0	$\frac{1}{6}$	4
s_2	0	$\left(2\right)$	0	$\frac{-8}{3}$	1	$\frac{1}{3}$	8
x	1	$\frac{1}{2}$	0	$\frac{1}{3}$	0	$\frac{-1}{6}$	5
P_M	0	$\frac{-11}{2}$	0	$\frac{23}{3}$	0	$\frac{-5}{6}$	79

Since $\frac{-11}{2}$ is the most negative indicator in the bottom row, y is the entering variable. Comparing ratios, we see that s_2 is the leaving variable. Pivoting, we obtain

Tableau VI

Basis	x	y	z	s_1	s_2	s_3	Values of basic variables
z	0	0	1	$\frac{4}{3}$	$\frac{-1}{4}$	$\frac{1}{12}$	2
y	0	1	0	$\frac{-4}{3}$	$\frac{1}{2}$	$\frac{1}{6}$	4
x	1	0	0	1	$\frac{-1}{4}$	$\frac{-1}{4}$	3
P_M	0	0	0	$\frac{1}{3}$	$\frac{11}{4}$	$\frac{1}{12}$	101

All indicators in the bottom row are nonnegative. Thus, an optimum solution has been reached. The maximum value of P_M is 101 and occurs when $x = 3$, $y = 4$, $z = 2$, and $s_1 = s_2 = s_3 = a_1 = 0$. Since the value of the artificial variable is 0, 101 is also the maximum value of the original objective function P and occurs when $x = 3$, $y = 4$, and $z = 2$. Thus, the refinery should blend 3 million gallons of type A gasoline, 4 million gallons of type B gasoline, and 2 million gallons of type C gasoline per week to maximize the weekly profit. The weekly profit will be $1,010,000, since $101(10,000) = 1,010,000$.

Do Exercise 19.

Exercise Set 4.7

Use the Big M Method to solve all problems. If for some problem a simpler technique is available, use that technique only to check the answer you obtained with the Big M Method.

1. Maximize $P = 3x + 2y$ subject to

$$2x + 3y \leq 36$$
$$-x + y \geq 2$$
$$x \geq 0, y \geq 0$$

2. Maximize $P = 2x + y$ subject to

$$x - y + 50 \geq 0$$
$$3x - y + 50 \geq 0$$
$$2x + 3y \geq 40$$
$$x - 2y + 15 \leq 0$$
$$x \geq 0, y \geq 0$$

3. Maximize $P = 3x + 2y$ subject to

$$2x + 3y \geq 36$$
$$x + 4y \leq 8$$
$$x \geq 0, y \geq 0$$

4. Maximize $P = 4x + 3y$ subject to

$$2x - y \leq -3$$
$$-x + 2y \geq 12$$
$$x + y \leq 18$$
$$x \geq 0, y \geq 0$$

5. Maximize $P = 4x + 4y + 3z$ subject to

$$2x + 4y + 3z \geq 48$$
$$2x + y + 3z \leq 45$$
$$2x + 6y + z \leq 56$$
$$x + 2y + 3z = 30$$
$$x \geq 0, y \geq 0, z \geq 0$$

6. Minimize P where P is the function of Exercise 5 subject to the same constraints.

7. Maximize $P = 3x + 2y + 5z$ subject to

$$-4x + y - 3z \leq -29$$
$$6x + 3y \leq 39$$
$$5x + y + 6z \leq 61$$
$$x + y + z = 12$$
$$x \geq 0, y \geq 0, z \geq 0$$

8. Minimize $C = x + 5y + 3z$ subject to the constraints of Exercise 7.

9. Minimize $C = 2x + 3y + 5z$ subject to the constraints of Exercise 7.

10. Maximize $P = 6x + 11y + 12z$ subject to

$$2x + 3y + 4z \leq 34$$
$$5x + 5y + 8z \geq 64$$
$$x + 2y + 4z = 26$$
$$x \geq 0, y \geq 0, z \geq 0$$

11. Minimize $C = 9x + 4y + 4z$ subject to the constraints of Exercise 10.

12. Maximize $P = 5x + 6y + 3z$ subject to

$$3x + 2y + 3z \leq 33$$
$$7x + 3y + 6z \geq 62$$
$$2x + y + 3z = 25$$
$$x \geq 0, y \geq 0, z \geq 0$$

13. Minimize $C = 9x + 2y + 3z$ subject to the constraints of Exercise 12.

14. Maximize $P = 4x + 3y + 5z$ subject to

$$-13x + 33y + 4z \leq 183$$
$$16x + 24y - 13z \leq 219$$
$$9x - 4y + 8z \leq 121$$
$$2x + 3y + z \geq 30$$
$$x \geq 0, y \geq 0, z \geq 0$$

15. Minimize $C = 4x + 3y + 5z$ subject to the constraints of Exercise 14.

16. Maximize $P = 11x + 5y + z$ subject to

$$3x + 4y + z \leq 35$$
$$3x + 2y + z \leq 29$$
$$5x + 4y + z \geq 49$$
$$2x + 3y + z = 25$$
$$x \geq 0, y \geq 0, z \geq 0$$

17. A farmer has 18 acres on which he may grow peanuts and corn. Because of market conditions the acreage for corn must exceed twice the acreage for peanuts by at least 3 acres and twice the acreage for corn must exceed the acreage for peanuts by at least 12 acres. Let x and y denote the numbers of acres allocated to peanuts and corn, respectively. If his profit is \$400 per acre of peanuts and \$300 per acre of corn, write his profit P as a function of x and y. How many acres of each should he plant to maximize his profit?

18. A farmer has 800 acres on which she may grow wheat and barley. She has accepted orders for 150 acres of wheat and 100 acres of barley. Moreover, she must follow regulations that require that four times the acreage for barley must exceed the acreage for wheat by at least 200 acres. Let x and y denote the numbers of acres allocated to wheat and barley, respectively. Suppose that her profit is $370 per acre of wheat and $290 per acre of barley. Write her total profit P as a function of x and y. How many acres of each should she grow to maximize profit?

19. An electronics company makes two models of VCRs. Model A takes 2 hours to assemble and 15 minutes to test, while model B takes 3 hours to assemble and 30 minutes to test. A maximum of 60,000 hours per month are available for assembly, while a maximum of 8000 hours a month are available for testing. Market studies indicate that 15 times the monthly production of model B should exceed the monthly production of the A model by at least 3000. Let x and y denote the numbers of A and B models, respectively, the company can make per month. Suppose that the company makes a $120 profit on each model A and a $95 profit on each model B. Express the total profit per month P as a function of x and y. How many of each model should be produced each month to maximize the profit per month?

20. A camera company makes two models of movie cameras. The deluxe model takes 3 hours to assemble and 10 minutes to test, while the standard model takes 2 hours to assemble and 6 minutes to test. A maximum of 38,000 hours per month are available for assembly, while a maximum of 2000 hours a month are available for testing. Because of market conditions, the management has been advised that three times the monthly production of the standard model should exceed the monthly production of deluxe models by at least 6000. Let x and y denote the numbers of deluxe and standard models, respectively, the company can make per month. Assume that the profit per deluxe model is $450 and that per standard model is $175. Express the total profit per month P as a function of x and y. How many of each model should be produced per month to maximize the profit per month?

21. An owner of a gasoline station can buy at most 1600 gallons of gasoline per day from his supplier. On a certain day he estimates that his regular customers will purchase at least 500 gallons of regular gasoline and at least 200 gallons of premium gasoline. He also knows from experience that each day five times the number of gallons of premium sold exceeds the number of gallons of regular sold by at least 2000. Let x and y denote, respectively, the numbers of gallons of regular and premium gasoline he can sell per day. Suppose that regular and premium gasoline sell for $1.05 per gallon and $1.15 per gallon, respectively. Express the revenue per day R dollars as a function of x and y. How many gallons of each type should be sold per day to maximize his revenue per day?

22. An investor wishes to buy 300 shares of common stock. Her broker recommends stocks A, B, and C and suggests that to reduce the risk of a fluctuating market, twice the number of shares of stock A plus the number of shares of stock B should not exceed 275. He also recommends that twice the number of shares of stock A plus the number of shares of stock C should not exceed 325 and that seven times the number of shares of stock A should exceed the number of shares of stock C by at least 125. Let x, y, and z represent the numbers of shares of stocks A, B, and C, respectively, that the investor will buy. Suppose that stock A sells at $80 a share, stock B sells at $50 a share, and stock C sells at $70 a share. Write the total cost C as a function of x, y, and z. How many shares of each kind should the investor buy to minimize her total cost?

23. An investor wishes to buy 500 shares of common stock. His broker recommends stocks A, B, and C and suggests that to reduce the risk of a fluctuating market, the number of shares of stock A plus the number of shares of stock B should not exceed 400. She also recommends that the number of shares of stock A should exceed that of stock B by less than 100 and that twice the number of shares of stock A should exceed the number of shares of stock C by at least 100. Let x, y, and z represent the numbers of shares of stocks A, B, and C, respectively, that the investor will buy. Suppose that stock A sells at $20 a share, stock B sells at $35 a share, and stock C sells at $25 a share. Write the total cost C as a function of x, y, and z. How many shares of each kind should the investor buy to minimize his total cost?

24. Henri's motel has 120 rooms and a restaurant that can seat 45 people. During the tourist season, the motel will be full each day. Some of the rooms may be furnished to accommodate travelers who like to go first class, while some of the remaining rooms will be furnished to satisfy the second-class guests. The remaining rooms will be furnished for third-class guests. The manager of the motel knows from experience that 60% of the first-class guests, 30% of the second-class guests, and 10% of the third-class guests will eat in the motel restaurant. Because of market conditions, not more than 90 rooms should be furnished for first- and second-class guests. If the manager furnishes x rooms for first-class guests, y rooms for second-class guests, and z rooms for third-class guests, and if the charge per night is $90 per first-class room, $55 per second-class room, and $30 per third-class room, write the revenue per night R as a function of x, y, and z. How many rooms of each type should there be to maximize the revenue per night?

25. A nursing home dietician is planning menus for patients. She wishes to provide a diet which has a minimum of 120 units of fat, 60 units of carbohydrates, and 90 units of proteins, and a maximum of 70 units of cholesterol per patient per month. These goals can be met using two types of food. Type I contains 3 units of fats, 2 units of carbohydrates, 6 units of proteins, and 2 units of cholesterol per pound, while each pound of type II contains 12 units of fats, 4 units of carbohydrates, 3 units of proteins, and 3 units of cholesterol. If the dietician plans to use x pounds of type I and y pounds of type II food per patient per month, and if type I food costs $3 per pound and type II costs $3.75 per pound, write the cost C per patient per month as a function of x and y. How many pounds of each kind of food should be used per month to minimize the cost?

26. An oil refinery's facilities allow the blending of at most 11 million gallons per week. Some antiknock ingredient must be used. Each million gallons of grade A gasoline requires 6 pounds of the ingredient, each million gallons of grade B gasoline requires 4 pounds of the ingredient, while each million gallons of grade C gasoline requires only 2 pounds. Only 38 pounds of the ingredient are available each week. Because of market conditions, the sum of the output of grade A and twice the output of grade B must exceed the output of grade C by at least 1 million gallons. Suppose that the refinery makes profits of $.15 per gallon on grade A gasoline, $.12 per gallon on grade B gasoline, and $.08 per gallon on grade C gasoline. How many gallons of each grade of gasoline should be blended per week to maximize the weekly profit? What is this profit?

27. The Tallgrass Company makes motors for its lawn mowers at two different sites, A and B. Site A can make at most 900 motors a week while site B can make at most 500 motors a week. The motors are shipped to two assembly plants I and II. The plants have already accepted orders so that plant I must assemble at least 600 lawn mowers a week and plant II must assemble at least 700 lawn mowers a week. The shipping costs per motor are as follows:

From site A to plant I: $8

From site A to plant II: $6

From site B to plant I: $7

From site B to plant II: $5

Let x_1, x_2, x_3, and x_4 denote the numbers of motors shipped from site A to plant I, from site A to plant II, from site B to plant I, and from site B to plant II, respectively. Write the total shipping cost $\$C$ as a function of x_1, x_2, x_3, and x_4. What weekly shipping schedule will minimize the total shipping cost?

28. The chairperson of the mathematics department must decide how many professors, associate professors, and assistant professors the department will have. The administration requires that the number of assistant professors must exceed the sum of the numbers of professors and associate professors by at least nine. Exactly 202 sections of mathematics courses must be taught each academic year and the instructional budget available is $1,015,000. On the average, professors earn $50,000, associate professors earn $45,000, and assistant professors earn $25,000. The department must have a good research publication record to satisfy accreditation requirements. On the average, professors publish five research papers and teach five sections a year, associate professors publish four research papers and teach six sections a year, and assistant professors publish .5 research paper and teach seven sections a year. How many professors, associate professors, and assistant professors should be in the department to maximize the number of research publications per year?

4.8 Chapter Review

IMPORTANT SYMBOLS AND TERMS

$C(n,k) = n!/[k!(n − k)!]$ [4.2]
$n!$
Adjacent corners [4.3]
Artificial variable [4.7]
Basic feasible solution [4.4]
Basic variables [4.4]
Basis [4.4]
Big M Method [4.7]

Bounded [4.1]
Convex polygonal region [4.1]
Coordinates [4.2]
Decision variable [4.1]
Dual maximization problem [4.6]
Entering variable [4.4]
First octant [4.2]
Graphical method [4.1]

Graph of an ordered triple of real numbers (a, b, c) [4.2]
Indicators [4.5]
Intersection of planes [4.2]
kth tableau [4.5]
Leaving variable [4.4]
Linear programming [4.1]
Modified problem [4.7]
Multiple optimum solutions [4.5]

n factorial
Nonbasic variables [4.4]
Objective function [4.1],
[4.5]
Octant [4.2]
Pivot column [4.4]
Pivot element [4.4]
Pivoting [4.4]

Pivot row [4.4]
Preliminary simplex tableau
[4.7]
Primal problem [4.6]
Right-handed system [4.2]
Set of feasible solutions
[4.1]
Simplex method [4.3]

Slack variables [4.3]
Surplus variable [4.7]
Tableau [4.4]
Three-dimensional Cartesian
coordinate system [4.2]
Unbounded [4.1]
Unbounded sets of feasible
solutions [4.5]

SUMMARY   In a linear programming problem, we either maximize or minimize a linear function, called the objective function, subject to linear constraints which are expressed as inequalities. It is possible to use the graphical method to solve a linear programming problem involving only two variables, but we use the simplex method for solving problems with more variables. Given a problem, we first form a mathematical model by introducing some decision variables and writing a linear objective function. We translate the constraints of the problem as linear inequalities and/or equations and write inequalities which show that all constraints are nonnegative. If the objective function is $z = ax + by + d$, and we wish to solve the problem graphically, we proceed as follows:

Step 1. We sketch the graph of the system of linear constraints to get the region of feasible solutions. If this region is empty, the problem has no solution. Otherwise, we find the corners of the region by solving the appropriate systems of equations.

Step 2. If the region of feasible solutions is bounded, then both a maximum and minimum value of the objective function exist. In that case, we evaluate the objective function at each corner of the region. The largest of the numbers obtained is the maximum value of the objective function, and the smallest is its minimum. If an optimum value occurs at two corners, then it occurs at every point on the segment joining the two corners.

Step 3. If the region of feasible solutions is unbounded, evaluate the objective function at each corner of the region. If c_1 and c_2 are the largest and smallest numbers obtained, respectively, then if a maximum exists, it will be c_1, and if a minimum exists, it will be c_2. We sketch the graph of $d = ax + by + c$ for some $d > c_1$. If this graph intersects the region of feasible solutions, the objective function has no maximum. Otherwise c_1 is the maximum. We proceed similarly with some $d < c_2$ to check for a minimum.

The simplex method is used if there are more than two decision variables.
If the problem is a maximization problem in standard form, use the procedure on page 342.
If the problem is a minimization problem in standard form, use the procedure on page 372.
If the problem is not in standard form, use the Big *M* Method procedure on page 381.

SAMPLE EXAM QUESTIONS

1. Suppose the vertices of the set of feasible solutions for a linear programming problem are $(1, 5)$, $(3, 8)$, $(5, 4)$, and $(2, 1)$ and the objective function is $z = 3x + 2y$. Find the maximum and minimum value of z.

2. Same as Exercise 1 if the vertices are $(2, 4)$, $(4, 7)$, $(10, 5)$, $(8, 3)$, and $(3, 1)$ and the objective function is $z = 5x - 2y$.

3. Same as Exercise 1 if the vertices are $(1, 1, 1)$, $(4, 3, 5)$, $(5, 2, 3)$, and $(7, 4, 6)$ and the objective function is $z = 2x + 5y + 3z$.

Solve Exercises 4, 5, and 6 geometrically.

4. Find the maximum and minimum value of $z = 3x + 11y$ subject to

$$2x + y \leq 24$$
$$2x + 5y \leq 40$$
$$x \geq 0 \text{ and } y \geq 0$$

5. Find the maximum and minimum value of $z = 3x + 4y$ subject to

$$x + 7y \geq 17$$
$$x + y \geq 5$$
$$-4x + 5y \leq 8$$
$$11x + 5y \leq 115$$
$$x \geq 0 \text{ and } y \geq 0$$

6. Find an optimum value of $z = 2x + 3y$ subject to

$$x + 7y \geq 17$$
$$x + y \geq 5$$
$$-4x + 5y \leq 8$$
$$x \geq 0 \text{ and } y \geq 0$$

7. Plot the points **a.** $(-1, 2, 3)$, **b.** $(2, 4, 5)$, and **c.** $(-2, 3, 5)$.

8. Sketch the graph of the equation $3x + 5y + 2z = 120$.

9. Suppose that five lines have been drawn in a plane and that the number of points of intersection is finite. What is the maximum number of points of intersection?

10. Suppose that five planes have been drawn in a three-dimensional Cartesian coordinate system and that the number of points of intersection is finite. What is the maximum number of points of intersection?

11. The equations

$$-10x + 12y + 9z = 72$$
$$18x + 15y - 4z = 90$$
$$16x - 7y + 10z = 80$$

are the equations of planes P_1, P_2, and P_3, respectively.

a. Find the intersection of the three planes.
b. Sketch the region in the first octant bounded by the three planes and the three coordinate planes.
c. Using the information of **a** and **b**, find the maximum value of $w = 2x + 5y + 3z$ subject to

$$-10x + 12y + 9z \leq 72$$
$$18x + 15y - 4z \leq 90$$
$$16x - 7y + 10z \leq 80$$
$$x \geq 0, y \geq 0, \text{ and } z \geq 0$$

12. The following has been obtained by performing elementary row operations on the augmented matrix of a linear system of equations in the variables $x_1, x_2, \ldots, x_7$. This system has infinitely many solutions. Give the solution which can be read directly from the given matrix.

$$\begin{bmatrix} 3 & -5 & 0 & 1 & 4 & 0 & -6 & 7 \\ -1 & 4 & 1 & 0 & 7 & 0 & 9 & -5 \\ 4 & 3 & 0 & 0 & -2 & 1 & 7 & 3 \end{bmatrix}$$

13. For each of the following systems of linear inequalities, find the coordinates of the indicated number of corners of the graph of the system by introducing slack variables and obtaining a sequence of tableaux. Start with the origin. Do not sketch the graphs.

a. $2x + y \le 24$

$2x + 5y \le 40$

$x \ge 0$ and $y \ge 0$, (3 corners)

b. $x + y \le 15$

$5x + 3y \le 55$

$5x + 2y \le 50$

$x \ge 0$ and $y \ge 0$, (4 corners)

c. $-10x + 12y + 9z \le 72$

$18x + 15y - 4z \le 90$

$16x - 7y + 10z \le 80$

$x \ge 0$, $y \ge 0$, and $z \ge 0$, (4 corners)

Solve each of the following using the simplex method.

14. Find the maximum value of $z = 2x + 7y$ subject to

$2x + y \le 24$

$2x + 5y \le 40$

$x \ge 0$ and $y \ge 0$

15. Find the maximum value of $z = 3x + 8y$ subject to

$-2x + y \le 2$

$-2x + 3y \le 14$

$x \ge 0$ and $y \ge 0$

16. Find the maximum value of $z = 2x + y$ subject to

$x + y \le 15$

$5x + 3y \le 55$

$5x + 2y \le 50$

$x \ge 0$ and $y \ge 0$

17. Maximize $w = 2x + 5y + 3z$ subject to

$-10x + 12y + 9z \le 72$

$18x + 15y - 4z \le 90$

$16x - 7y + 10z \le 80$

$x \ge 0$, $y \ge 0$, and $z \ge 0$

18. Maximize $z = 6x + 9y$ subject to

$3x + 7y \le 49$

$2x + 3y \le 26$

$x + y \le 12$

$x \ge 0$ and $y \ge 0$

19. Maximize $w = 15x + 3y + 3z$
subject to

$$-x + 3y + 2z \leq 216$$
$$x + y + z \leq 198$$
$$2x + y + 3z \leq 324$$
$$x \geq 0, y \geq 0, z \geq 0$$

20. Minimize $z = 9x + 7y$ subject to

$$3x + 4y \geq 120$$
$$9x + 2y \geq 180$$
$$x \geq 0, y \geq 0$$

21. Minimize $z = 5x + 2y$ subject to

$$3x + y \geq 50$$
$$x + y \geq 30$$
$$x + 4y \geq 60$$
$$x \geq 0, y \geq 0$$

22. Minimize $w = 33x + 54y + 36z$
subject to

$$x + 2y - z \geq 10$$
$$x + y + 3z \geq 2$$
$$x + 3y + 2z \geq 2$$
$$x \geq 0, y \geq 0, z \geq 0$$

23. A company makes gadgets A and B. The production of each requires time on machines I, II, and III. Gadget A requires 2 hours on I, 1 hour on II, and 1 hour on III; gadget B requires 1 hour on I, 1 hour on II, and 2 hours on III. The time available on these machines per week is 200 hours on I, 140 hours on II, and 240 hours on III. The profit on each A is $90, and the profit on each B is $120. The company can sell all it produces. How many of each should it make per week to maximize profit? What is that profit?

24. A company builds three types of sailboats. The deluxe model requires 3 units of wood, 2 units of canvas, and 1 unit of paint; the super model requires 5 units of wood, 1 unit of canvas, and 2 units of paint; and the regular model requires 2 units of wood, 1 unit of canvas, and 1 unit of paint. Each month, the company has, at most, 69 units of wood, 24 units of canvas, and 21 units of paint available. If the profits are $4000 per deluxe model, $2000 per super model, and $3000 per regular model, and if the company can sell all the boats it can build each month, how many of each model should be built each month to maximize profit? What is this profit?

25. A dairy company processes milk at two locations. Its Arlington plant produces 50 gallons of 4% and 100 gallons of 2% milk per hour and costs $168 per hour to operate, while the Marysville plant produces 100 gallons of 4% and 150 gallons of 2% milk per hour and costs $288 per hour to operate. The company needs a total daily production of at least 1200 gallons of 4% and 2250 gallons of 2% milk. How many hours per day should each plant operate to minimize cost?

26. Use the Big M Method to maximize $z = 5x + 4y$ subject to

$$3x + 8y \leq 120$$
$$3x + 2y \leq 46$$
$$3x - 5y \leq -10$$
$$x \geq 0, y \geq 0$$

27. Use the Big M Method to maximize $w = 2x + 5y + 3z$ subject to

$$4x - y + 3z \geq 29$$
$$x + y + z = 12$$
$$6x + 3y \leq 39$$
$$5x + y + 6z \leq 61$$
$$x \geq 0, y \geq 0, z \geq 0$$

28. Use the Big M Method to minimize $w = 3x - 2y + 7z$ subject to the constraints of Exercise 27.

29. Use the Big M Method to maximize $w = 7x + 4y + z$ subject to the constraints of Exercise 27.

Sequences and Mathematics of Finance

5

Often, events occur sequentially. That is, there will be a first event, then a second, then a third, and so on. In this chapter we describe a type of function that serves as a mathematical model for solving problems involving sequential outcomes or events. Special business applications will also be discussed.

5.1 Sequences, Arithmetic and Geometric Progressions

> **DEFINITION 5.1:** Suppose n is a positive integer and for each i in $\{1, 2, 3, \ldots, n\}$ we associate a real number a_i. Then, the list $a_1, a_2, a_3, \ldots, a_n$ is called a *finite sequence*. If for each j in $\{1, 2, 3, \ldots, n, \ldots\}$ we associate a real number b_j, then the list $b_1, b_2, b_3, \ldots, b_n, \ldots$ is called an *infinite sequence*.

In other words, a finite sequence is a function with domain $\{1, 2, \ldots, n\}$ and an infinite sequence is a function with domain $\{1, 2, 3, \ldots, n, \ldots\}$. If s is the name of a sequence, we write s_i instead of the usual function notation $s(i)$. The function values are called the *terms* of the sequence. Thus, s_1 is called the first term of the sequence, s_2 is the second term, and so on. The methods used to describe sequences are the same as those used to describe functions.

EXAMPLE 1   The sequence s is defined by $s_n = \frac{2}{n+1}$. Find the first four terms of that sequence.

Solution   The first term is 1, since $s_1 = \frac{2}{1+1} = 1$. The second term is $s_2 = \frac{2}{2+1} = \frac{2}{3}$; the third term is $\frac{1}{2}$; and the fourth term is $\frac{2}{5}$.
Do Exercise 3.   ■

EXAMPLE 2   The sequence t is defined by $t_n = (-1)^{[n/2]}n^2$. Find the first five terms of the sequence.

Solution   Recall that if x is a real number, $[x]$ denotes the largest integer less than or equal to x. Thus,

$$t_1 = (-1)^{[1/2]}1^2 = (-1)^0 \cdot 1 = 1$$
$$t_2 = (-1)^{[2/2]}2^2 = (-1)^1 \cdot 4 = -4$$
$$t_3 = (-1)^{[3/2]}3^2 = (-1)^1 \cdot 9 = -9$$
$$t_4 = (-1)^{[4/2]}4^2 = (-1)^2 \cdot 16 = 16$$
$$t_5 = (-1)^{[5/2]}5^2 = (-1)^2 \cdot 25 = 25$$

Do Exercise 7.   ■

EXAMPLE 3   In Section 1.4, it was shown that if P dollars in an account earns interest at an annual rate r (written as a decimal) compounded k times per year, then the amount A_n dollars in the account at the end of the nth conversion period is

$$A_n = P(1 + i)^n$$

where $i = \frac{r}{k}$. If \$1000 is deposited in an account at 8% interest compounded quarterly, find the amount in the account at the end of each of the first six conversion periods.

Solution The interest per conversion period is .02 since $\frac{.08}{4} = .02$. We get the following amounts in the account:

$$A_1 = 1000(1 + .02)^1 = 1020$$

$$A_2 = 1000(1 + .02)^2 = 1040.40$$

$$A_3 = 1000(1 + .02)^3 \doteq 1061.21$$

$$A_4 = 1000(1 + .02)^4 \doteq 1082.43$$

$$A_5 = 1000(1 + .02)^5 \doteq 1104.08$$

$$A_6 = 1000(1 + .02)^6 \doteq 1126.16$$

These results were obtained with a calculator and were rounded to two decimal places. **Do Exercise 31.** ∎

We will briefly discuss two types of sequences which occur so frequently that they have been given special names. They are called *arithmetic progressions* and *geometric progressions*.

Arithmetic Progressions

An arithmetic progression is a sequence that starts with an arbitrary number as the first term and then the same constant is added to obtain successive terms. For example, 2, 5, 8, 11, 14, and 17 are the first six terms of an arithmetic progression. We started with 2 as our first term and kept adding 3 to obtain successive terms.

DEFINITION 5.2: A sequence in the form a, $a+d$, $a+2d$, . . . is called an *arithmetic progression* (abbreviated A.P.) with *first term a* and *common difference d*.

EXAMPLE 4 The first term of an arithmetic progression is 6 and its common difference is 2. Find the first six terms of this A.P.

Solution It is given that $a_1 = 6$. Thus, $a_2 = 6 + 2 = 8$, $a_3 = 8 + 2 = 10$, $a_4 = 10 + 2 = 12$, $a_5 = 12 + 2 = 14$, and $a_6 = 14 + 2 = 16$.
Do Exercise 9. ∎

If an A.P. has first term a and common difference d, then $a_1 = a$, $a_2 = a + d$, $a_3 = a_2 + d = (a + d) + d = a + 2d$, $a_4 = a_3 + d = (a + 2d) + d = a + 3d$ and, in general,

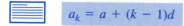

$$a_k = a + (k - 1)d$$

EXAMPLE 5   The 21st and 35th terms of an A.P. are 64 and 106, respectively. Find the first three terms.

Solution   We have

$$a_{21} = a + (21 - 1)d = a + 20d = 64$$

and

$$a_{35} = a + (35 - 1)d = a + 34d = 106$$

Subtracting these two equalities, we obtain

$$(a + 34d) - (a + 20d) = 106 - 64$$

That is,

$$14d = 42$$

and

$$d = \frac{42}{14} = 3$$

Replacing d by 3 in the expression for a_{21}, we obtain

$$a + 20 \cdot 3 = 64$$

so that $a = 64 - 60 = 4$. It follows that $a_1 = 4$, $a_2 = 4 + 3 = 7$, and $a_3 = 7 + 3 = 10$.

Do Exercise 13.   ■

Now we wish to derive a formula for the sum S_n of the first n terms of an A.P. with first term a and common difference d. We first write

$$S_n = a + (a + d) + (a + 2d) + \cdots + (a + (n - 2)d) + (a + (n - 1)d)$$

Writing this sum in reverse order, we get

$$S_n = (a + (n - 1)d) + (a + (n - 2)d) + \cdots + (a + 2d) + (a + d) + a$$

Each of the sums on the right of the equalities has n terms. When we add these equalities, we have a sum of n identical terms.

$$2S_n = [2a + (n - 1)d] + [2a + (n - 1)d] + \cdots + [2a + (n - 1)d]$$
$$= n[2a + (n - 1)d]$$

Thus,

$$S_n = \frac{n[2a + (n - 1)d]}{2}$$

If we note that $2a + (n - 1)d = a + [a + (n - 1)d] = a_1 + a_n$, we obtain a simpler formula

$$S_n = \frac{n(a_1 + a_n)}{2}$$

Since $(a_1 + a_n)/2$ is the average of the first and nth terms and n is the number of terms in the sum, we may state the following:

> To find the sum of the first n terms of an A.P., we need only calculate the average of the first and nth terms and multiply this average by n, the number of terms.

EXAMPLE 6 Find the sum of the first 50 terms of the A.P. of Example 5.

Solution We found that the first term a_1 is 4 and the common difference d is 3. Thus, the 50th term is 151 since $4 + (50 - 1)3 = 4 + 49 \cdot 3 = 4 + 147 = 151$. The average of the 1st and 50th terms is $\frac{4 + 151}{2} = 77.5$. Hence, the sum of the first 50 terms is $50 \cdot 77.5 = 3875$.

Do Exercise 19. ■

Geometric Progressions

A geometric progression is a sequence where each successive term is obtained by multiplying the preceding term by the same constant. For example, 3, 6, 12, 24, 48, 96 are the first six terms of a geometric progression. We start with 3 as the first term and keep multiplying by 2 to obtain successive terms.

> **DEFINITION 5.3:** A sequence in the form $a, ar, ar^2, \ldots$ is called a *geometric progression* (abbreviated G.P.) with *first term a* and *common ratio r*.

EXAMPLE 7 The first term of a G.P. is 5 and its common ratio is -2. Find the second, third, fourth, and fifth terms of this G.P.

Solution The second term is

$$a_2 = a_1 \cdot r = 5 \cdot (-2) = -10$$

Similarly

$$a_3 = a_2 \cdot r = (-10) \cdot (-2) = 20$$
$$a_4 = a_3 \cdot r = (20) \cdot (-2) = -40$$
$$a_5 = a_4 \cdot r = (-40) \cdot (-2) = 80$$

Do Exercise 37. ■

If a G.P. has first term a and common ratio r, then $a_1 = a$, $a_2 = ar$, $a_3 = a_2 r = (ar)r = ar^2$, $a_4 = a_3 r = (ar^2)r = ar^3$ and, in general,

$$a_k = ar^{k-1}$$

EXAMPLE 8 Suppose that the 10th and 13th terms of a G.P. are 39,366 and $-1{,}062{,}882$, respectively. Find its first three terms.

Solution We have

$$a_{10} = ar^{10-1} = ar^9 = 39{,}366 \tag{1}$$

and

$$a_{13} = ar^{13-1} = ar^{12} = -1{,}062{,}882 \tag{2}$$

Dividing the second equality by the first, we obtain

$$r^3 = -27$$

Thus, $r = -3$. Replacing r by -3 in (1), we get $a(-3)^9 = 39{,}366$. Thus,

$$-19{,}683a = 39{,}366$$

and

$$a = -2$$

The first three terms are -2, 6, and -18 since $(-2)(-3) = 6$ and $6(-3) = -18$.
Do Exercise 41. ■

In applications, it is often necessary to add the first n terms of a G.P. If a G.P. has first term a and common ratio r, then the sum of the first n terms S_n is

$$S_n = a + ar + ar^2 + ar^3 + \cdots + ar^{n-2} + ar^{n-1} \tag{1}$$

Multiplying both sides by r, we obtain

$$S_n r = ar + ar^2 + ar^3 + ar^4 + \cdots + ar^{n-1} + ar^n \tag{2}$$

Subtracting the second equality from the first, we get

$$S_n - S_n r = a - ar^n$$

Thus,

$$S_n(1 - r) = a(1 - r^n)$$

and if $1 - r \neq 0$, we have

$$S_n = \frac{a(1 - r^n)}{1 - r}$$

In the case that $1 - r = 0$, $r = 1$, and the G.P. is the constant sequence a, a, a, . . . , the sum of the first n terms is na.

EXAMPLE 9 The first term of a G.P. is 5 and the common ratio is 2. Find the sum of the first 20 terms.

Solution We use the formula

$$S_n = \frac{a(1 - r^n)}{1 - r}$$

with $a = 5$, $r = 2$, and $n = 20$. We obtain

$$S_{20} = \frac{5(1 - 2^{20})}{1 - 2}$$

$$= \frac{5(1 - 1,048,576)}{-1}$$

$$= 5,242,875$$

The sum of the first 20 terms of the G.P. is 5,242,875.
Do Exercise 47.   ■

The next three examples illustrate how arithmetic and geometric progressions occur in applications.

Applications

EXAMPLE 10   Office equipment was purchased for $30,000 and is assumed to have a scrap value of $12,000 after 6 years. Determine the value of the equipment V_k after k years, $0 < k \leq 6$, assuming that the value is depreciated linearly.

Solution   Since the depreciation is linear, each year the value of the equipment will be $3000 less than the preceding year since $\frac{30,000 - 12,000}{6} = 3000$. Thus the value $\$V_1$ at the end of the first year is given by

$$V_1 = 30,000 - 3000 = 27,000$$

The values $\$V_2$, $\$V_3$, $\$V_4$, $\$V_5$, and $\$V_6$ are obtained as follows:

$$V_2 = 27,000 - 3000 = 24,000$$
$$V_3 = 24,000 - 3000 = 21,000$$
$$V_4 = 21,000 - 3000 = 18,000$$
$$V_5 = 18,000 - 3000 = 15,000$$
$$V_6 = 15,000 - 3000 = 12,000$$

Therefore, 30,000, 27,000, 24,000, 21,000, 18,000, 15,000, and 12,000 are the first seven terms of an A.P. whose first term is 30,000 and common difference is 3000.
Do Exercise 27.   ■

EXAMPLE 11   The value of a building which cost its owner $200,000 is declining by 10% per year. What will the building be worth at the end of the kth year for $k = 1, 2, 3, 4$, and 5?

Solution   Let V_k be the value of the building at the end of the kth year. The value declined by 10% the first year. Thus, at the end of the first year, the value of the building is 90% of the original value.

$$V_1 = 200,000(.9) = 180,000$$

At the end of the first year, the building is worth $180,000. Similarly, we get

$$V_2 = 180,000(.9) = 162,000$$
$$V_3 = 162,000(.9) = 145,800$$
$$V_4 = 145,800(.9) = 131,220$$
$$V_5 = 131,220(.9) = 118,098$$

At the end of the second, third, fourth, and fifth years, the value of the building will be $162,000, $145,800, $131,220, and $118,098, respectively. Therefore, 200,000, 180,000, 162,000, 145,800, 131,220, and 118,098 are the first six terms of a G.P. whose first term is 200,000 and common ratio is .9.
Do Exercise 29.

■

EXAMPLE 12 Mary's bank pays a nominal rate of interest of 6% compounded monthly. She deposits $200 on the first of each month. How much is in Mary's account immediately after the sixth deposit?

Solution We shall use the formula $A_n = P(1 + i)^n$ to determine the value of each deposit immediately after she makes the sixth deposit. We have $i = \frac{.06}{12} = .005$. Immediately after the sixth deposit, the value of the sixth deposit is still $200. The fifth deposit has been in the bank 1 month; thus, its value is $200(1 + .005) = $200(1.005)$. The fourth deposit has been in the bank 2 months; thus, its value is $200(1 + .005)^2 = $200(1.005)^2$. The values of the third, second, and first deposits are $200(1.005)^3$, $200(1.005)^4$, and $200(1.005)^5$, respectively. (See Figure 5.1.) Therefore, the balance in Mary's account is the sum

$$200 + 200(1.005) + 200(1.005)^2 + 200(1.005)^3 + 200(1.005)^4 + 200(1.005)^5$$
$$= \frac{200(1 - 1.005^6)}{1 - 1.005} \doteq 1215.10$$

This is the sum of the first six terms of a G.P. with first term 200 and common ratio 1.005. This result was obtained with a calculator and was rounded to two decimal places. The balance in Mary's account is $1215.10.
Do Exercise 57.

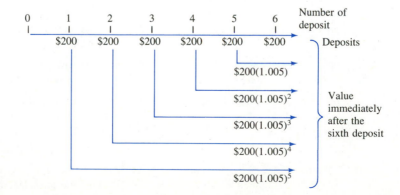

FIGURE 5.1

■

Exercise Set 5.1

In Exercises 1–8, find the first five terms of each of the sequences.

1. $s_n = \dfrac{2}{2n - 1}$ **2.** $t_n = \dfrac{3}{2n + 1}$

3. $a_n = \dfrac{-2}{n^2 + 1}$ **4.** $s_n = \dfrac{(-1)^n}{3n - 2}$

5. $t_n = \dfrac{(-1)^{n+1}n}{n + 1}$ **6.** $a_n = \dfrac{2n - 1}{2n + 1}$

7. $s_n = \dfrac{(-1)^{[n/3]}\,(n + 3)}{3n - 1}$ **8.** $t_n = \dfrac{n^2 + n + 1}{n^2 + n - 1}$

9. The first term of an A.P. is -4 and its common difference is 3. Find the first ten terms.

10. The first term of an A.P. is 12 and its common difference is 5. Find the first five terms.

11. The first term of an A.P. is -5 and its common difference is -2. Find the first eight terms.

12. The second term of an A.P. is 4 and its common difference is 4. Find the first ten terms.

13. The 20th and 30th terms of an A.P. are 73 and 113, respectively. Find the first five terms.

14. The 32nd and 43rd terms of an A.P. are 95 and 128, respectively. Find the first seven terms.

15. The 16th and 36th terms of an A.P. are -20 and -60, respectively. Find the first five terms.

16. The 17th and 51st terms of an A.P. are 76 and 246, respectively. Find the first six terms.

17. The 19th and 63rd terms of an A.P. are 46 and 178, respectively. Find the first eight terms.

18. Find the sum of the first 100 terms of the A.P. of Exercise 9.

19. Find the sum of the first 70 terms of the A.P. of Exercise 10.

20. Find the sum of the first 80 terms of the A.P. of Exercise 11.

21. Find the sum of the first 90 terms of the A.P. of Exercise 12.

22. Find the sum of the first 60 terms of the A.P. of Exercise 13.

23. Find the sum of the first 100 terms of the A.P. of Exercise 14.

24. Find the sum of the first 200 terms of the A.P. of Exercise 15.

25. Find the sum of the first 90 terms of the A.P. of Exercise 16.

26. Find the sum of the first 300 terms of the A.P. of Exercise 17.

27. Office equipment was purchased for $25,000 and is assumed to have a scrap value of $5000 after 8 years. Determine the value of the equipment V_k after k years, $0 < k \le 8$, assuming that the value is depreciated linearly.

28. A truck was purchased for $35,000 and is assumed to have a trade-in value of $8000 after 10 years. Determine the value of the truck V_k after k years, $0 < k \le 10$, assuming that the value is depreciated linearly.

29. The value of a building which cost its owner $300,000 is declining by 5% per year. What will the building be worth at the end of the kth year for $k = 1, 2, 3, 4$, and 5?

30. The value of a machine which cost its owner $60,000 is declining by 8% per year. What will the machine be worth at the end of the kth year for $k = 1, 2, 3, 4$, and 5?

31. An amount of $500 is deposited in a bank that advertises a nominal interest rate of 6% compounded monthly. Find the amount at the end of each of the first six conversion periods.

32. If $300 is deposited in a bank that advertises a nominal interest rate of 8% compounded quarterly, what is the amount at the end of each of the first six conversion periods?

33. An amount of $800 is deposited in a bank that advertises a nominal interest rate of 8% compounded monthly. Find the amount at the end of each of the first six conversion periods.

34. If $1500 is deposited in a bank that advertises a nominal interest rate of 9% compounded quarterly, what is the amount at the end of each of the first six conversion periods?

35. An amount of $600 is deposited in a bank that advertises a nominal interest rate of 7% compounded monthly. Find the amount at the end of each of the first six conversion periods.

36. The first term a_1 of a G.P. is 3 and its common ratio r is 4. Find the second, third, fourth, and fifth terms.

37. The first term b_1 of a G.P. is 2 and its common ratio r is 5. Find the second, third, fourth, and fifth terms.

38. The first term c_1 of a G.P. is -3 and its common ratio r is 2. Find the second, third, fourth, and fifth terms.

39. The first term t_1 of a G.P. is 512 and its common ratio r is $\frac{1}{2}$. Find the second, third, fourth, and fifth terms.

40. The first term a_1 of a G.P. is 243 and its common ratio r is $\frac{-1}{3}$. Find the second, third, fourth, and fifth terms.

41. The fifth and seventh terms of a G.P. are 324 and 2916, respectively. Find its first three terms.

42. The eighth and tenth terms of a G.P. are -384 and -1536, respectively. Find its first three terms.

43. The seventh and tenth terms of a G.P. are -1458 and -39366, respectively. Find its first three terms.

44. The 8th and 11th terms of a G.P. are 8 and 1, respectively. Find its first three terms.

45. The fifth and eighth terms of a G.P. are 9 and $\frac{1}{3}$, respectively. Find its first three terms.

46. Find the sum of the first 20 terms of the G.P. of Exercise 36.

47. Find the sum of the first 20 terms of the G.P. of Exercise 37.

48. Find the sum of the first 50 terms of the G.P. of Exercise 38.

49. Find the sum of the first 30 terms of the G.P. of Exercise 39.

50. Find the sum of the first 15 terms of the G.P. of Exercise 40.

51. Find the sum of the first 25 terms of the G.P. of Exercise 41.

52. Find the sum of the first 42 terms of the G.P. of Exercise 42.

53. Find the sum of the first 27 terms of the G.P. of Exercise 43.

54. Find the sum of the first 17 terms of the G.P. of Exercise 44.

55. Find the sum of the first 28 terms of the G.P. of Exercise 45.

56. A bank advertises a nominal rate of interest of 8% compounded monthly. Kris deposits $300 on the first of each month. How much will be in his account immediately after the tenth deposit?

57. A bank advertises a nominal rate of interest of 6% compounded monthly. Kathy deposits $250 on the first of each month. How much will be in her account immediately after the 15th deposit?

58. A bank advertises a nominal rate of interest of 7% compounded monthly. Steven deposits $150 on the first of each month. How much will be in his account immediately after the fifth deposit?

59. A bank advertises a nominal rate of interest of 6% compounded monthly. Barbara deposits $400 on the first of each

month. How much will be in her account immediately before the tenth deposit?

60. A bank advertises a nominal rate of interest of 7% compounded monthly. Mike deposits $500 on the first of each month. How much will be in his account immediately before the tenth deposit?

61. A bank advertises a nominal rate of interest of 6% compounded monthly. Carol deposits $450 on the first of each month. How much will be in her account immediately before the 23rd deposit?

62. Suppose that each bacterium in a culture divides into three bacteria every hour and that there are five bacteria at the start. How many bacteria will be present after 8 hours?

63. Suppose that a dropped Ping-Pong ball always rebounds half the height it falls. If it is dropped from a height of 64 inches, how far will it have traveled when it hits the floor for the tenth time?

64. One-fourth of a container full of alcohol is replaced by water and the solution is mixed thoroughly. Then one-fourth of that solution is replaced by water and the new solution is mixed thoroughly. If this is done repeatedly, what fraction of the original volume of alcohol is left after the tenth replacement?

65. A man has signed a 12-year contract which gives him a $30,000 yearly salary the first year and a 6% raise each year after that. How much will he have earned when his contract expires?

66. Suppose that a new car is worth $18,000 and that at the end of each year its value depreciates by 15% of the value it had at the beginning of that year. What will the car be worth at the end of 8 years?

67. Suppose that a company made you the following offer. Your monthly salary would either be $15,000 or each month you would be paid 1 cent the first day, 2 cents the second day, 4 cents the third day, 8 cents the fourth day, and so on for each of the month's 21 working days. Which offer would you choose? Justify your answer.

68. Suppose that you are playing blackjack at your favorite casino. You are playing at a table where a minimum bet of $1 is required. You bet $1 on the first hand. If you lose, you bet $2 on the next hand. Each time you lose you double your bet on the next hand. How much will you have won if you lose k straight hands, but win the $(k + 1)$st hand?

69. The casino of Exercise 68 has a table limit of $1 minimum and $500 maximum for bets for each hand. Explain the danger of using the strategy of Exercise 68.

5.2 Annuities

Everyone is familiar with financial transactions involving a sequence of equal payments made periodically. For example, you may have: deposited money in a savings account; or borrowed money and repaid the sum, plus interest, in a certain number of payments; or bought life insurance and made yearly payments. Whenever a transaction involves a sequence of equal payments made periodically, it is called an *annuity*. This section is devoted to defining terms, deriving formulas, and giving examples involving annuities. First, we shall give three examples to illustrate the basic ideas involved. Try to do the problems in this section, applying the techniques used in these examples, until you thoroughly understand the basic ideas. The other examples will use formulas to get the answers more efficiently. In most examples, the computations will be made using a calculator. Review the section "How to Use a Calculator," p. xxvii.

Recall the following two basic formulas.

1. If a sum of P dollars earns compound interest at a rate i (written as a decimal) per compounding period, then the total amount A_n dollars after n compounding periods is given by the formula

$$A_n = P(1 + i)^n$$

2. The sum of the first n terms of a geometric progression is

$$a + ar + ar^2 + ar^3 + \cdots + ar^{n-1} = \frac{a(1 - r^n)}{1 - r}$$

provided that $r \neq 1$. Most of the problems of this section can be solved using these two formulas.

Simple Annuities

EXAMPLE 1   On the first of each month Mary deposits $200 in a bank that pays 6% interest compounded monthly. How much is in Mary's account immediately after the 20th deposit? How much interest is earned?

Solution   The rate of interest per conversion period is .005 since $\frac{.06}{12}$ = .005. Immediately after the 20th deposit, the value of that deposit is still $200. The 19th deposit has been in the bank 1 month, so its value is $200(1 + .005). The 18th deposit has been in the bank 2 months, its value is $200(1 + .005)^2$. In general, the kth deposit has been in the bank $(20 - k)$ months, and its value is $200(1 + .005)^{20-k}$. Thus, the value of the first deposit is $200(1 + .005)^{19}$. (See Figure 5.2.) The total value (deposits plus interest) is given by

$$200 + 200(1.005) + 200(1.005)^2 + \cdots + 200(1.005)^{19}$$

This is the sum of the first 20 terms of a geometric progression with $a = 200$ and $r = 1.005$. It is equal to

$$\frac{200[1 - (1.005)^{20}]}{1 - 1.005} = \frac{200(1 - 1.10489558)}{-.005} = 4195.8232$$

Therefore, Mary has $4195.82 in her savings account. She has made 20 deposits of $200 each for a total of $4000. Since $4195.82 - 4000 = 195.82$, $195.82 is the earned interest.

Do Exercise 17.

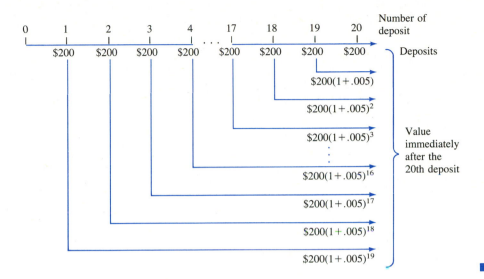

FIGURE 5.2

The foregoing is an example of a *simple annuity* because the interval between payments coincides with the conversion period at which the interest is being paid. The $4195.82 represents the *amount* (or *future value*) of the annuity.

Sinking Fund

EXAMPLE 2   When John entered college, his parents decided to reward him upon graduation with a trip to Europe. They estimate the trip will cost $8000. They wish to make equal deposits on the first of each month, starting October 1, 1992, in a bank that pays a nominal rate of 6% interest compounded monthly. The last deposit will be June 1, 1996, at which time John will graduate. How much should they deposit each month so that the value of the annuity will be $8000?

Solution   John's parents will make 45 deposits of D dollars each. We must find the value of D which will make the sum of all deposits plus interest equal to $8000. As in Example 1, $i = .005$. On June 1, 1991, immediately after the 45th deposit, the value of that deposit will still be D dollars. At that time, the 44th deposit will have earned interest for 1 month, so its value will be $D(1 + .005)$, and the 43rd deposit will have earned interest for 2 months and its value will be $D(1 + .005)^2$. In general, the kth deposit will have earned interest for $(45 - k)$ months and its value will be $D(1 + .005)^{45-k}$. (See Figure 5.3). The total value of all deposits plus interest will be

$$D + D(1.005) + D(1.005)^2 + D(1.005)^3 + \cdots + D(1.005)^{44}$$

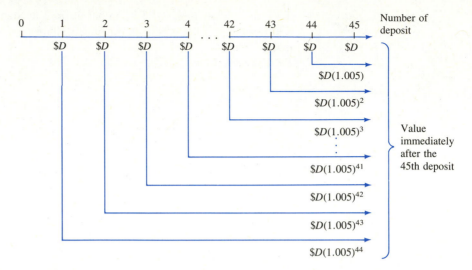

FIGURE 5.3

This is the sum of the first 45 terms of a geometric progression with $a = D$ and $r = 1.005$. Its value is

$$\frac{D[1 - (1.005)^{45}]}{1 - 1.005} \doteq \frac{D(1 - 1.25162082)}{-.005} = (50.324164)D$$

This must equal $8000, so

$$(50.324164)D \doteq 8000$$

and

$$D \doteq 8000/50.324164$$

$$\doteq 158.9693572$$

Thus, John's parents will deposit $158.97 each month.
Do Exercise 19.   ■

Example 2 illustrates what is often called a *sinking fund*. A sinking fund is a fund into which equal payments are made at regular intervals in order to accumulate (deposits plus interest) a definite amount of money by a specified date. Often, sinking funds are established by companies to meet future replacement costs of equipment.

Deferred Annuity Certain

EXAMPLE 3   On the day of their daughter's birth, Mr. and Mrs. Wise decided to set aside a sum of money to provide for the daughter's college education. They wish to make a single deposit in a bank that pays a rate of interest of 12% compounded yearly to provide a payment of $9000 on each of the daughter's 18th, 19th, 20th, and 21st birthdays. How much should they deposit?

Solution It is easier to think of the single deposit as the sum of four deposits A_1, A_2, A_3, and A_4, made at the same time where A_1 plus interest will provide the first $9000, A_2 plus interest will provide the second $9000, and so on. On the daughter's 18th birthday, A_1 will have earned interest for 18 years. Thus, its value will be $A_1(1 + .12)^{18}$. Since A_1 provides the first $9000, we must have $A_1(1 + .12)^{18} = 9000$. Thus,

$$A_1 = 9000(1.12)^{-18}$$

Similarly, we find that

$$A_2 = 9000(1.12)^{-19}$$
$$A_3 = 9000(1.12)^{-20}$$
$$A_4 = 9000(1.12)^{-21} \qquad \text{(See Figure 5.4)}$$

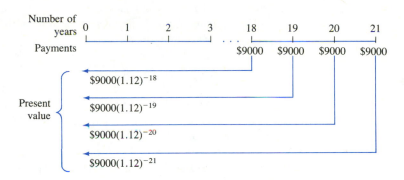

FIGURE 5.4

The total, single deposit is given by

$$9000(1.12)^{-18} + 9000(1.12)^{-19} + 9000(1.12)^{-20} + 9000(1.12)^{-21}$$

This is the sum of the first four terms of a geometric progression with $a = 9000(1.12)^{-18}$ and $r = 1.12^{-1}$. Its value is

$$\frac{9000(1.12)^{-18}[1 - (1.12^{-1})^4]}{1 - 1.12^{-1}} = \frac{9000(1.12)^{-17}(1 - (1.12)^{-4})}{1.12 - 1}$$

$$\doteq \frac{(1310.799068)(0.364481921)}{.12}$$

$$\doteq 3981.35$$

Thus, on the day of their daughter's birth, Mr. and Mrs. Wise should deposit $3981.35 in the bank.

Do Exercise 21. ∎

Example 3 is an illustration of what is called a *deferred annuity certain*. It is "deferred" because the payments start at a future date, and it is "certain" because the payments are not contingent upon some uncertain event, such as the daughter being alive. In case of death, the payments would go to the heirs. The sum $3981.35 is the *present value* of this annuity.

The first three examples illustrate different transactions called annuities. We shall now formally define the basic terms even though some of them already have been introduced. We shall then derive several formulas and give examples to demonstrate how they can be used.

Annuities: Basic Terms Defined

DEFINITION 5.4:   An *ordinary annuity* is a series of equal payments to be paid, or received, at the end of consecutive periods of equal length. If the payments are made at the beginning of each period, the annuity is called an *annuity due*.

An annuity is said to be *simple* if the interval of time between successive payments coincides with the conversion period at which interest is being paid. Otherwise, it is called a *general annuity*.

The *payment period* is the interval of time at the end (or beginning) of which each payment is to be made.

The *term* of an annuity is the time from the beginning of the first payment period to the end of the last payment period.

A *deferred annuity* (or *intercepted annuity*) is an annuity for which the first payment is not made until more than one payment period has elapsed. It is an *immediate annuity* if the term begins immediately.

An annuity is said to be *certain* if its term is fixed.

A *contingent annuity* is an annuity for which the term depends on some uncertain event such as the death of an individual. In particular, a *temporary annuity* is an annuity payable over a given period, provided the recipient continues to live through that period, otherwise terminating at death. A *last survivor annuity* is an annuity to be paid until the last of two or more survivors' lives end. A *perpetuity* is an annuity whose payments continue forever.

A *sinking fund* is an annuity that is established for the purpose of accumulating a sum of money to pay an obligation at a future designated date. Often sinking funds are established by companies to provide money for anticipated replacement costs of equipment.

The *amount* (*future value*) of an annuity is the sum of money which would be accumulated by the end of the term if each of the payments were invested at a given rate of compound interest at the time of payment until the end of the term. The *accumulated value* of an annuity at a certain date is the sum of the compound amounts of all payments to that date.

An amount deposited today that yields a given sum in the future is called the *present value* of the sum.

The *present value of an annuity* is the sum of the present values of all the payments at the beginning of the term of the annuity.

Derivations

We now derive formulas for the amount and present value of an annuity. In these, R, i, and n will denote the periodic payment, the interest rate per period (written as a decimal), and the number of periods, respectively.

Let S dollars be the amount of an ordinary annuity. Since the annuity is ordinary, the payments are made at the end of each period. Therefore the value of the nth payment, at the time of the nth payment, is R dollars; the $(n-1)$st payment has earned interest for one period, so its value is $\$R(1+i)$; the $(n-2)$nd payment has earned interest for two periods, so its value is $\$R(1+i)^2$; and so on. In general, the kth payment has earned interest for $(n-k)$ periods, and its value is $\$R(1+i)^{n-k}$. (See Figure 5.5.) Thus,

$$S = R + R(1+i) + R(1+i)^2 + \cdots + R(1+i)^{n-1}$$

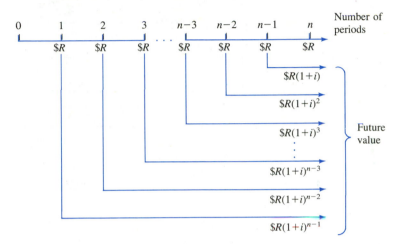

FIGURE 5.5

Since the expression on the right is the sum of the first n terms of a G.P. with $a = R$ and $r = 1 + i$, we write

$$S = \frac{R[1 - (1+i)^n]}{1 - (1+i)} = R \cdot \frac{(1+i)^n - 1}{i} \tag{1}$$

The expression $[(1+i)^n - 1]/i$ is often abbreviated $s_{\overline{n}|i}$ (read: "s angle n at i"). (See Table 5.1, p. 424.) Thus, the amount S of an ordinary annuity is given by

$$S = R \cdot s_{\overline{n}|i} \tag{2}$$

Now let A dollars be the present value of an ordinary annuity. Let A_k dollars be the present value of the kth payment. Since the payments are made at the end of each period, A_1 will have earned interest for one period, hence we have $A_1(1 + i) = R$ and $A_1 = \frac{R}{(1+i)}$. Similarly, A_2 will have earned interest for two periods and we have $A_2(1+i)^2 = R$, so that $A_2 = R/[(1+i)^2]$. In general, it is easy to see that $A_k = R/[(1+i)^k]$. (See Figure 5.6.) Thus,

$$A = \frac{R}{1+i} + \frac{R}{(1+i)^2} + \cdots + \frac{R}{(1+i)^n}$$

This is the sum of the first n terms of a G.P. whose first term and common ratio are $\frac{R}{(1+i)}$ and $\frac{1}{(1+i)}$, respectively. Therefore,

$$A = \frac{\dfrac{R}{1+i}\left[1 - \dfrac{1}{(1+i)^n}\right]}{1 - \dfrac{1}{1+i}}$$

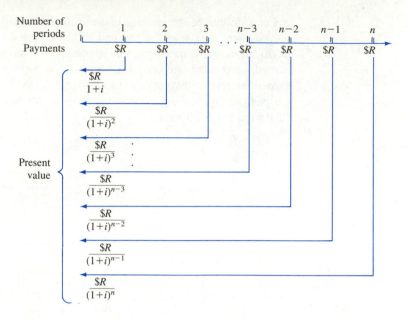

FIGURE 5.6

Multiplying the numerator and denominator of this fraction by $(1 + i)$, we obtain

$$A = R \cdot \frac{1 - (1 + i)^{-n}}{i} \tag{3}$$

The fractional quantity is often abbreviated $a_{\overline{n}|i}$ (read: "a angle n at i"). (See Table 5.2, p. 425.) Thus, we have

$$A = R \cdot a_{\overline{n}|i}{}^{*} \tag{4}$$

---

*We derived the present value of an ordinary annuity by adding $A_1, A_2, \ldots, A_n$ where each A_k was the value of the kth payment at the beginning of the first period. We could have argued instead, that the present value A, deposited as a lump sum together with the compound interest for n periods, should yield the same amount at the end of the nth period. Thus, we should have

$$A(1 + i)^n = S$$

and

$$A = S(1 + i)^{-n}$$

We found earlier that

$$S = R \cdot \frac{(1 + i)^n - 1}{i}$$

Therefore,

$$A = R \cdot \frac{(1 + i)^n - 1}{i}(1 + i)^{-n}$$

$$A = R \cdot \frac{1 - (1 + i)^{-n}}{i}$$

This is the same formula we obtained earlier.

Now we wish to calculate the amount of an annuity due. Rather than derive another formula, we can proceed as follows. Payments are usually made at the beginning of each period, but assume instead that the term of the annuity started one period earlier and make an extra payment at the end of the term. Then, the first payment is made at the end of the first period, the second payment is made at the end of the second period, and so on. The extra payment made at the end will then be the $(n + 1)$st payment made at the end of the $(n + 1)$st period. We, therefore, have changed the annuity to an ordinary annuity with $(n + 1)$ periods. We can calculate the amount of this new annuity using Formula (1) (see the following annuity formulas), then subtract the extra payment which is made at the end of the nth period to obtain the amount of the original annuity.

Similarly, suppose we wish to calculate the present value of an annuity due. Since the payments are made at the beginning of the period, the second payment is made at the beginning of the second period which can also be considered the end of the first period. Thus, if the first payment was not made, we would have an ordinary annuity with $(n - 1)$ periods. We could calculate the present value of that annuity using Formula (3) and add to the result the value of the first payment which we assumed was not made.

Annuity Formulas

Amount of an ordinary annuity:

$$S = R \cdot \frac{(1 + i)^n - 1}{i} \qquad (1)$$

$$S = R \cdot s_{\overline{n}|i} \qquad (2)$$

Present value of an ordinary annuity:

$$A = R \cdot \frac{1 - (1 + i)^{-n}}{i} \qquad (3)$$

$$A = R \cdot a_{\overline{n}|i} \qquad (4)$$

Applications

EXAMPLE 4   John deposits $500 at the end of every 3 months for a period of 4 years in a bank that pays a nominal rate of 12% interest compounded quarterly. What is the amount in his account at the end of the 4 years? How much interest did he earn?

Solution   The rate of interest per period is .03 since $\frac{.12}{4} = .03$. In 4 years, there are 16 periods. The payments are made at the end of each period, so we use Formula (1) for an ordinary annuity. We obtain

$$S = 500 \left[\frac{(1 + .03)^{16} - 1}{.03} \right] \doteq 10{,}078.44$$

Thus, at the end of the fourth year, John will have $10,078.44 in his account. We used a calculator to obtain this result. John made 16 deposits of $500 each for a total of $8000. Therefore, the interest earned was $2,078.44.
Do Exercise 23.

EXAMPLE 5   Betty deposits $300 at the beginning of each month for a period of 4 years in a bank that pays a nominal rate of 12% interest compounded monthly. What is the amount in her account at the end of the 4 years?

Solution   We must find the amount of an annuity due (since the payments are made at the beginning of each month). We have $R = 300$, $n = 48$, and $i = .01$ $\left(\text{since } \frac{.12}{12} = .01\right)$. We cannot use Formula (2) yet because the annuity is not ordinary. We change the problem by assuming that Betty makes an extra payment at the end of the fourth year, thus making 49 payments. We further assume that the term of the annuity started 1 month before she made the first payment, so that all 49 payments are made at the end of the periods. We have changed the problem to that of an ordinary annuity, and now Formula (2) can be used. We get

$$S = 300 \cdot s_{\overline{49}|.01}$$

$$\doteq 300(62.83483)$$

$$\doteq 18,850.449 \quad *$$

Since Betty did not, in fact, make the 49th payment, we must subtract the amount of that payment and conclude that after 4 years the amount in Betty's account is $18,550.45 since $18,850.45 - 300 = 18,550.45$. We have rounded the answer to the nearest cent.

Do Exercise 25. (See Exercise 45.)   ■

EXAMPLE 6   Roy wishes to purchase an annuity that will pay his daughter Lisa $1000 at the end of each month for the next 4 years. If the rate of interest is 12% compounded monthly, how much must he pay for this annuity?

Solution   This annuity is ordinary since the payments are made at the end of each period. Thus, we wish to find the present value of an ordinary annuity and we use Formula (4), where $R = 1000$, $n = 48$, and $i = .01$ (since $\frac{.12}{12} = .01$). We get

$$A = 1000 \cdot a_{\overline{48}|.01}$$

$$\doteq 1000(37.973958)$$

$$\doteq 37,973.96 \quad *$$

Roy will have to pay $37,973.96 for this annuity.

Do Exercise 27.   ■

EXAMPLE 7   Peter won $2,400,000 at a state lottery. The state must pay $120,000 at the beginning of each of the next 20 years and purchases an annuity for that purpose. What is the present value of this annuity if the rate of interest is 8% compounded annually?

---

*The value of $s_{\overline{49}|.01}$ was obtained from Table 5.1 on page 425. We found the row with 49 on the left (since $n = 49$) and the column with .01 at the top (since $i = .01$). The number in that row and that column was 62.83483, the value of $s_{\overline{49}|.01}$.

**The value of $a_{\overline{48}|.01}$ was obtained using Table 5.2 on page 426.

Solution We will use Formula (4). Since the payments are made at the beginning of each period, the annuity is due. If we assume that the first payment is not made, then the second payment becomes the first payment made at the end of the first year, the third payment becomes the second payment made at the end of the second year, and so on. We have changed the annuity to an ordinary annuity where $R = 120{,}000$, $n = 19$, and $i = .08$. The present value of this annuity is

$$S = 120{,}000 \, \frac{1-(1 + .08)^{-19}}{.08}$$

$$\doteq 120{,}000(9.60359920)$$

$$\doteq 1{,}152{,}431.904$$

We must add to this value the first payment which was made at the beginning of the first year. Hence, the present value of the original annuity due is \$1,272,431.90 since \$1,152,431.90 + \$120,000 = \$1,272,431.90.

Do Exercise 29. (See Exercise 46.) ∎

EXAMPLE 8 On her 30th birthday, Norma established a retirement fund by paying \$200 a month to a life insurance company that pays 9% interest compounded monthly. If she retires on her 65th birthday, at which time she makes her last payment, and if she wishes to make equal withdrawals from the retirement fund until she reaches her 80th birthday, the first withdrawal being made 1 month after her 65th birthday, how much will she get each month from the retirement fund?

Solution Norma established an ordinary annuity since the payments are made at the end of each period (the first payment was made 1 month after the beginning of the term of the annuity). The amount of this annuity on her 65th birthday may be calculated using Formula (1) with $R = 200$, $i = .0075$ (since $\frac{.09}{12} = .0075$), and $n = 420$ (since $35 \cdot 12 = 420$). We get

$$S = 200 \cdot \frac{(1 + .0075)^{420} - 1}{.0075}$$

$$\doteq 200(2{,}941.78448)$$

$$\doteq 588{,}356.896$$

When Norma retires, she essentially purchases an ordinary annuity whose present value is \$588,356.90. Since she wishes to make monthly withdrawals until she reaches her 80th birthday, there will be 180 payments of R dollars each. Using Formula (3), we get

$$588{,}356.90 = R \cdot \frac{1 - (1 + .0075)^{-180}}{.0075}$$

$$\doteq R(98.59340886)$$

Therefore,

$$R \doteq \frac{588{,}356.90}{98.59340886}$$

$$\doteq 5967.507431$$

Rounding the answer to the nearest cent, we conclude that Norma will receive $5967.51 a month. While Norma paid only $84,000 into the fund ($200 \cdot 420 = 84,000$) she will withdraw $1,074,151.80 (since $5967.51 \cdot 180 = 1,074,151.80$). The difference represents the interest earned.

Do Exercise 31.   ■

EXAMPLE 9   A roofing company establishes a sinking fund that will provide $200,000 to replace its trucks in 5 years. It purchases an ordinary annuity and agrees to make equal payments every 6 months. How much is each payment if the rate of interest is 8% compounded semiannually?

Solution   Since the company purchased an ordinary annuity, we may use Formula (2) where $S = 200,000$, R is to be found, $n = 10$, and $i = .04$. We get

$$200,000 = R \cdot \frac{(1 + .04)^{10} - 1}{.04}$$

$$\doteq R \cdot (12.006107)$$

Thus,

$$R \doteq \frac{200,000}{12.006107} \doteq 16,658.189$$

Thus, the company should pay $16,658.19 at the end of every 6 months for a period of 5 years.

Do Exercise 33.   ■

EXAMPLE 10   Ralph would like to have $120,000 in his savings account 15 years from now. He is only able to deposit $200 at the end of each month for the next 4 years in a bank that pays a nominal interest rate of 12% compounded monthly. How much must each of his monthly deposits be in the last 11 years if he is to reach his goal?

Solution   Since the payments are made at the end of each month, the amount Ralph will have in his account in 4 years may be obtained using Formula (2). We get

$$S = 200(s\,\overline{_{48|}}._{01})$$

$$\doteq 200(61.222604)$$

$$\doteq 12,244.5208$$

This amount will continue to earn interest in the next 11 years. If Ralph did not make any additional deposits, the amount he would have in 11 more years would be given by

$$A = 12,244.5208(1 + .01)^{132}$$

$$\doteq 12,244.5208(3.71895856)$$

$$\doteq 45,536.86544$$

Rounding to the nearest cent, we see that the additional amount Ralph must accumulate is $74,463.13 since $120,000 - 45,536.87 = 74,463.13$.

Using Formula (1) with $n = 132$, $i = .01$, and $S = 74,463.13$, we get

$$74,463.13 = R \cdot \frac{(1 + .01)^{132} - 1}{.01}$$

$$\doteq R \cdot (271.895856)$$

Therefore,

$$R \doteq \frac{74,463.13}{271.895856}$$

$$\doteq 273.866366$$

Rounding to the nearest cent, Ralph should deposit $273.87 at the end of each month for the next 11 years.
Do Exercise 35.   ■

EXAMPLE 11   Upon retirement, Fran used her life savings of $150,000 to purchase an annuity. The life insurance company from which she purchased the annuity offers 8% interest compounded semiannually and she is to receive one of 40 equal payments every 6 months starting in 6 months. How much will she receive every 6 months?

Solution   The payment being made at the end of each period indicates that the annuity is ordinary. Thus, we may use Formula (3) where $A = 150,000$, R is to be found, $n = 40$, and $i = .04$. We get

$$150,000 = R \cdot \frac{1 - (1 + .04)^{-40}}{.04}$$

$$\doteq R(19.79277389)$$

$$R \doteq \frac{150,000}{19.79277389}$$

$$\doteq 7578.52$$

Fran will receive $7578.52 at the end of each 6-month period for the next 20 years.
Do Exercise 37.   ■

We conclude with an example of a special type of annuity called a *perpetuity*.

Perpetuities

EXAMPLE 12   Suppose someone wishes to establish a fund, earning 8% interest compounded annually, for the purpose of maintaining a family crypt. The maintenance cost is $300 per year with the first payment due immediately. How much should be paid now to purchase an annuity that will take care of the perpetual maintenance of the family crypt?

Solution   The first $300 would have to be paid now. Then, an amount earning $300 in interest each year should be deposited so that each year the interest could be withdrawn to take care of the maintenance of the crypt, keeping the principal amount in the bank.

This principal should be $P where P(.08) = 300. Thus, $P = \frac{300}{.08} = 3750$. Adding the first payment, we obtain 4050 since $3750 + 300 = 4050$. The cost of this perpetuity is $4050.
Do Exercise 41.

∎

Exercise Set 5.2

In Exercises 1–6, find the value of the given symbol using a calculator.

1. $(1 + .02)^{23}$

2. $(1 + .03)^{42}$

3. $(1 + .04)^{26}$

4. $(1 + .05)^{-17}$

5. $(1 + .02)^{-26}$

6. $(1 + .06)^{-9}$

In Exercises 7–15, find the value of the given symbol using the tables on pages 424–426.

7. $s_{\overline{10}|.0025}$

8. $s_{\overline{20}|.005}$

9. $s_{\overline{15}|.0075}$

10. $s_{\overline{32}|.01}$

11. $s_{\overline{20}|.0125}$

12. $a_{\overline{17}|.0025}$

13. $a_{\overline{12}|.005}$

14. $a_{\overline{34}|.0075}$

15. $a_{\overline{17}|.01}$

In Exercises 16–21, use only the formulas that were used in Examples 1, 2, and 3 and do the computations with a calculator.

16. Steven deposits $150 in his savings account on the first of each month. Steven's bank pays 8% interest compounded monthly. How much is in Steven's account immediately after the 25th deposit? How much is earned interest?

17. Julie deposits $300 in her savings account on the first of each month. Julie's bank pays 6% interest compounded monthly. How much is in Julie's account immediately after the 40th deposit? How much is earned interest?

18. When Gwen entered college, her parents decided that they would reward her upon graduation with a trip to Japan. They estimate the trip will cost $6000. They wish to make equal deposits on the first of each month, starting December 1, 1990, in a bank that pays 9% interest compounded monthly. The last deposit will be made on June 1, 1994. How much should they deposit each month?

19. When Mike entered college, his parents decided that they would reward him upon graduation with a trip to India. They estimate the trip will cost $9000. They wish to make equal deposits on the first of each month, starting March 31, 1990, in a bank that pays 12% interest compounded monthly. The last deposit will be made on December 31, 1994. How much should they deposit each month?

20. On the day of their son's birth, Mr. and Mrs. Planer decided to set aside a sum of money to provide for his college education. They wish to make a single deposit in a bank that pays 9% interest compounded yearly in order to provide a payment of $8500 on each of the son's 18th, 19th, 20th, and 21st birthdays. How much should they deposit?

21. On the day of their daughter's birth, Mr. and Mrs. Bernard decided to set aside a sum of money to provide for the daughter's college education. They wish to make a single deposit in a bank that pays 8% interest compounded yearly in order to provide a payment of $9500 on each of the daughter's 18th, 19th, 20th, and 21st birthdays. How much should they deposit?

In Exercises 22–39, you may use Formulas 1–4 and a calculator. If, for a problem, the values of n and i appear in the tables on pages 424–426, you may use the tables for that problem.

22. Lisa deposits $450 at the end of every 6 months for a period of 7 years in a bank that pays 7% interest compounded semiannually. What is the amount in her account at the end of 7 years? How much interest did she earn?

23. Louis deposits $250 at the end of each month for a period of 6 years in a bank that pays 6% interest compounded monthly. What is the amount in his account at the end of 6 years? How much interest did he earn?

24. Shirley deposits $250 at the beginning of each month for a period of 3 years in a bank that pays 9% interest compounded monthly. What is the amount in her account at the end of 3 years? How much interest did she earn?

25. Peter deposits $350 at the beginning of each quarter for a period of 5 years in a bank that pays 8% interest compounded quarterly. What is the amount in his account at the end of 5 years? How much interest did he earn?

26. Ted wishes to purchase an annuity that will pay his son Kris $750 at the end of each month for the next 3 years starting in 1 month. If the rate of interest is 6% compounded monthly, how much must Ted pay for this annuity?

27. Dorothy wishes to purchase an annuity that will pay her son Mark $950 at the end of each quarter for the next 6 years starting in 3 months. If the rate of interest is 8% compounded quarterly, how much must Dorothy pay for this annuity?

28. What is the present value of an annuity due that will pay $15,000 per year for a period of 15 years if the rate of interest is 7% compounded annually?

29. What is the present value of an annuity due that will pay $500 per month for a period of 4 years if the rate of interest is 9% compounded monthly?

30. On his 25th birthday, Charles established a retirement fund by paying $250 a month to a life insurance company that pays 6% interest compounded monthly. If he retires on his 65th birthday, at which time he will make his last payment, and if he wishes to make equal withdrawals from the retirement fund until he reaches his 75th birthday, the first withdrawal being made 1 month after his 65th birthday, how much will he get each month from the fund?

31. On her 35th birthday, Louise established a retirement fund by paying $350 a month to a life insurance company that pays 9% interest compounded monthly. If she retires on her 65th birthday, at which time she will make her last payment, and if she wishes to make equal withdrawals from the retirement fund until she reaches her 75th birthday, the first withdrawal being made 1 month after her 65th birthday, how much will she get each month from the fund?

32. A trucking company wishes to establish a sinking fund that will provide $350,000 to replace its trucks in 4 years. For this purpose, it purchases an ordinary annuity and will make equal payments every quarter. How much is each payment if the rate of interest is 9% compounded quarterly?

33. A manufacturer wishes to establish a sinking fund that will provide $500,000 to replace some of its machinery 5 years from now. For this purpose, it purchases an ordinary annuity and will make equal payments every 6 months. How much is each payment if the rate of interest is 7% compounded semiannually?

34. Steven would like to have $250,000 in his savings account 20 years from now. He will be able to deposit $150 at the end of each month for the next 5 years in a bank that pays 6% interest compounded monthly. How much must each of his monthly deposits be in the last 15 years if he is to reach his goal?

35. Kathy would like to have $150,000 in her savings account 12 years from now. She will be able to deposit $100 at the end of each month for the next 4 years in a bank that pays 12% interest compounded monthly. How much must each of her monthly deposits be in the last 8 years if she is to reach her goal?

36. Upon retirement, George used his life savings of $125,000 to purchase an annuity. The life insurance company from which he purchased the annuity pays 6% interest compounded monthly and he is to receive one of 240 equal payments every month, starting in 1 month. How much will he receive each month?

37. On her 62nd birthday, Gwen used her life savings of $200,000 to purchase an annuity. The life insurance company from which she purchased the annuity pays 9% interest compounded monthly. Gwen is to receive equal monthly payments starting in 1 month, until she receives the last payment on her 75th birthday. How much will she receive each month?

38. John has been depositing $150 on the first of each month over the last 15 years. The interest was 9% compounded monthly for the first 5 years, but it decreased to 6% compounded monthly for the last 10 years. How much was in his account immediately after the 180th deposit?

39. Rose has been depositing $250 on the first of each month over the last 25 years. The interest was 6% compounded monthly for the first 8 years, but it increased to 9% compounded monthly for the last 17 years. How much was in her account immediately after the 300th deposit?

40. Suppose that Mrs. Tradition wishes to establish a fund, earning 7% compounded annually, to maintain a family crypt. The maintenance cost is $250 per year with the first payment due immediately. How much must she pay now to purchase an annuity that will take care of the perpetual maintenance of the crypt?

41. Suppose that Mr. Goodfellow wishes to establish a fund, earning 8% compounded annually, to award a scholarship to the student majoring in business at his alma mater. The maintenance of the scholarship will cost $9500 a year with the first payment due immediately. How much should be paid now to purchase an annuity that will take care of the perpetual maintenance of the scholarship?

42. Prove the formula

$$\frac{1}{a_{\overline{n}|i}} = \frac{1}{s_{\overline{n}|i}} + i$$

43. Prove that the amount of an annuity due is given by

$$S = \frac{R[(1 + i)^{n+1} - 1 - i]}{i} = R(S_{\overline{n+1}|i} - 1)$$

44. Prove that the present value of an annuity due is given by

$$A = \frac{R[1 + i - (1 + i)^{-n+1}]}{i} = R(1 + a_{\overline{n+1}|i})$$

45. Solve Example 5 of this section using the formula of Exercise 43.

46. Solve Example 7 of this section using the formula of Exercise 44.

47. Do Exercise 24 using the formula of Exercise 43.

48. Do Exercise 25 using the formula of Exercise 43.

49. Do Exercise 28 using the formula of Exercise 44.

50. Do Exercise 29 using the formula of Exercise 44.

TABLE 5.1
AMOUNT OF ANNUITY
OF 1 PER PERIOD

$$S_{\overline{n}|i} = \frac{(1 + i)^n - 1}{i}$$

n	$\frac{1}{4}\%$	$\frac{1}{2}\%$	$\frac{3}{4}\%$	1%	$1\frac{1}{4}\%$
1	1.000000	1.000000	1.000000	1.000000	1.000000
2	2.002499	2.005000	2.007500	2.010000	2.012500
3	3.007505	3.015025	3.022556	3.030100	3.037656
4	4.015023	4.030100	4.045225	4.060401	4.075627
5	5.025060	5.050251	5.075564	5.101005	5.126572
6	6.037623	6.075502	6.113631	6.152015	6.190654
7	7.052717	7.105880	7.159483	7.213535	7.268037
8	8.070347	8.141409	8.213179	8.285670	8.358888
9	9.090523	9.182116	9.274778	9.368527	9.463374
10	10.113249	10.228027	10.344339	10.462212	10.581666
11	11.138532	11.279167	11.421921	11.566834	11.713936
12	12.166377	12.335563	12.507586	12.682502	12.860361
13	13.196793	13.397241	13.601393	13.809327	14.021115
14	14.229784	14.464227	14.703403	14.947421	15.196379
15	15.265359	15.536549	15.813679	16.096895	16.386334
16	16.303521	16.614231	16.932281	17.257864	17.591163
17	17.344280	17.697302	18.059273	18.430442	18.811052
18	18.387640	18.785789	19.194717	19.614747	20.046190
19	19.433609	19.879718	20.338678	20.810894	21.296767
20	20.482192	20.979116	21.491218	22.019003	22.562977
21	21.533398	22.084012	22.652402	23.239193	23.845014
22	22.587230	23.194432	23.822295	24.471585	25.143077
23	23.643699	24.310404	25.000962	25.716301	26.457365
24	24.702807	25.431957	26.188469	26.973463	27.788082
25	25.764564	26.559116	27.384883	28.243198	29.135433
26	26.828975	27.691912	28.590269	29.525630	30.499626
27	27.896046	28.830372	29.804696	30.820886	31.880871
28	28.965785	29.974524	31.028231	32.129095	33.279382
29	30.038200	31.124396	32.260943	33.450386	34.695374
30	31.113295	32.280018	33.502900	34.784890	36.129066
31	32.191078	33.441419	34.754172	36.132739	37.580679
32	33.271555	34.608626	36.014828	37.494066	39.050438
33	34.354734	35.781669	37.284939	38.869006	40.538568
34	35.440620	36.960577	38.564576	40.257696	42.045300
35	36.529221	38.145380	39.853810	41.660273	43.570866
36	37.620543	39.336107	41.152714	43.076876	45.115502
37	38.714594	40.532788	42.461359	44.507645	46.679446
38	39.811380	41.735452	43.779819	45.952721	48.262939
39	40.910909	42.944129	45.108168	47.412248	49.866225
40	42.013185	44.158850	46.446479	48.886371	51.489553

TABLE 5.1
AMOUNT OF ANNUITY OF 1 PER PERIOD

$$S_{\overline{n}|i} = \frac{(1+i)^n - 1}{i}$$

n	$\frac{1}{4}\%$	$\frac{1}{2}\%$	$\frac{3}{4}\%$	1%	$1\frac{1}{4}\%$
41	43.118218	45.379644	47.794828	50.375234	53.133172
42	44.226012	46.606543	49.153289	51.878987	54.797337
43	45.336577	47.839575	50.521938	53.397776	56.482304
44	46.449918	49.076773	51.900853	54.931754	58.188332
45	47.566043	50.324167	53.290109	56.481072	59.915686
46	48.684957	51.575788	54.689785	58.045882	61.664632
47	49.806669	52.833667	56.099958	59.626341	63.435440
48	50.931185	54.097835	57.520707	61.222604	65.228383
49	52.058514	55.368324	58.952113	62.834830	67.043738
50	53.188659	56.645166	60.394253	64.463178	68.881784

TABLE 5.2
AMOUNT OF ANNUITY OF 1 PER PERIOD

$$A_{\overline{n}|i} = \frac{1 - (1+i)^{-n}}{i}$$

n	$\frac{1}{4}\%$	$\frac{1}{2}\%$	$\frac{3}{4}\%$	1%	$1\frac{1}{4}\%$
1	0.997506	0.995025	0.992556	0.990099	0.987654
2	1.992524	1.985099	1.977723	1.970395	1.963115
3	2.985061	2.970248	2.955556	2.940985	2.926534
4	3.975123	3.950496	3.926110	3.901965	3.878058
5	4.962716	4.925867	4.889439	4.853431	4.817835
6	5.947846	5.896385	5.845597	5.795476	5.746010
7	6.930519	6.862074	6.794637	6.728194	6.662725
8	7.910741	7.822960	7.736613	7.651677	7.568124
9	8.888520	8.779064	8.671576	8.566017	8.462344
10	9.863860	9.730412	9.599579	9.471304	9.345525
11	10.836767	10.677027	10.520674	10.367628	10.217803
12	11.807249	11.618933	11.434912	11.255077	11.079311
13	12.775310	12.556152	12.342345	12.133740	11.930184
14	13.740957	13.488708	13.243022	13.003702	12.770552
15	14.704197	14.416626	14.136994	13.865052	13.600545
16	15.665033	15.339926	15.024312	14.717873	14.420291
17	16.623475	16.258633	15.905024	15.562251	15.229918
18	17.579525	17.172769	16.779180	16.398268	16.029548
19	18.533192	18.082357	17.646829	17.226008	16.819307
20	19.484480	18.987420	18.508019	18.045552	17.599315
21	20.433396	19.887980	19.362798	18.856982	18.369694
22	21.379946	20.784060	20.211214	19.660379	19.130562
23	22.324136	21.675682	21.053314	20.455820	19.882036
24	23.265970	22.562867	21.889145	21.243386	20.624233
25	24.205456	23.445639	22.718754	22.023155	21.357268
26	25.142599	24.324019	23.542188	22.795203	22.081252
27	26.077405	25.198029	24.359492	23.559607	22.796298
28	27.009879	26.067691	25.170711	24.316442	23.502517
29	27.940030	26.933025	25.975892	25.065784	24.200016
30	28.867859	27.794055	26.775079	25.807707	24.888905

TABLE 5.2
AMOUNT OF ANNUITY OF 1 PER PERIOD

$$A_{\overline{n}|i} = \frac{1 - (1 + i)^{-n}}{i}$$

n	$\frac{1}{4}\%$	$\frac{1}{2}\%$	$\frac{3}{4}\%$	1%	$1\frac{1}{4}\%$
31	29.793376	28.650802	27.568317	26.542284	25.569289
32	30.716584	29.503285	28.355649	27.269588	26.241273
33	31.637490	30.351527	29.137121	27.989691	26.904961
34	32.556099	31.195550	29.912775	28.702665	27.560455
35	33.472417	32.035373	30.682655	29.408579	28.207857
36	34.386451	32.871018	31.446804	30.107504	28.847266
37	35.298205	33.702505	32.205264	30.799509	29.478781
38	36.207685	34.529856	32.958079	31.484662	30.102500
39	37.114898	35.353091	33.705289	32.163032	30.718518
40	38.019848	36.172230	34.446937	32.834685	31.326932
41	38.922541	36.987293	35.183064	33.499688	31.927834
42	39.822983	37.798302	35.913711	34.158107	32.521317
43	40.721180	38.605275	36.638919	34.810007	33.107474
44	41.617136	39.408234	37.358729	35.455452	33.686394
45	42.510859	40.207198	38.073180	36.094507	34.258167
46	43.402353	41.002187	38.782312	36.727235	34.822881
47	44.291624	41.793221	39.486166	37.353698	35.380623
48	45.178676	42.580320	40.184780	37.973958	35.931479
49	46.063518	43.363502	40.878194	38.588077	36.475535
50	46.946152	44.142788	41.566445	39.196116	37.012874

5.3 Mortgages

Installment buying is of great interest since almost everyone has been, or will be, involved in making purchases where it is necessary, because of the size of the original price, to pay only part of the price (the down payment) and pay off the balance by making payments at regular intervals. The purchase of a car or a home are examples of this type of financial transaction. Essentially, when a lending institution lends a sum of money to a person to be repaid with interest in equal payments at regular intervals, the lending institution purchases an annuity from that person. If the payments are made at the end of each period, the annuity is ordinary. Furthermore, if the interval of time between successive payments coincides with the conversion period at which the interest is being paid, the annuity is also simple and we can use Formula (3) of the preceding section.

$$A = R \cdot \frac{1 - (1 + i)^{-n}}{i}$$

where A dollars is the present value of the annuity, R dollars is the amount of each payment, i (written as a decimal) is the rate of interest per conversion period, and n is the number of periods (payments). Thus, A dollars is the amount that the lending institution lends to the purchaser at the time of the purchase. When making such a purchase, the buyer is concerned about the portion of each payment that is interest and the portion that is being repaid on the principal (original loan). The treatment of annuities in the preceding section did not answer the following questions: how much of the loan remains to be paid after k payments? how much of the kth payment is an interest payment? how much is used to repay the loan? We shall derive formulas to answer these and related questions.

Derivation of Formulas

Suppose that a lending institution lends A dollars to an individual who will repay the loan plus interest by making n monthly payments of R dollars each, the first payment being made after 1 month. The rate of interest is i (written as a decimal) compounded monthly on the unpaid balance. Then, the relationships between A, R, n, and i are given by Formula (3) of the preceding section.

Let A_k dollars ($k = 0, 1, 2, \ldots, n$) be the unpaid balance on the loan after the kth payment, where $A_0 = A$ and $A_n = 0$. Let B_k dollars ($k = 1, 2, 3, \ldots, n$) be the amount of the kth payment used to repay the principal. Let I_k dollars ($k = 1, 2, 3, \ldots, n$) be the amount of the kth payment used for interest payment. Then, $B_k + I_k = R$ for each k.

We wish to derive formulas for B_k and I_k. A simple way to derive these is to make use of the formulas we derived in the preceding section. (A more direct approach may be found in the interesting treatment of installment buying in [2], at the end of this section.)

Immediately after the kth payment has been made, $(n - k)$ payments remain to be made. Thus, the principal that remains to be paid is the present value of these $(n - k)$ payments. Using Formula (3) of the previous section, we find the present value to be

$$A_k = R \cdot \frac{1 - (1 + i)^{-(n-k)}}{i} = R \cdot \frac{1 - (1 + i)^{k-n}}{i}$$

It is easy to see that the amount of the kth payment used to repay the principal is the difference between what was left to be paid after the $(k-1)$st payment and what is left to be paid after the kth payment. That is, $B_k = A_{k-1} - A_k$. Therefore,

$$B_k = R \cdot \frac{1 - (1 + i)^{k-1-n}}{i} - R \cdot \frac{1 - (1 + i)^{k-n}}{i}$$

$$= R \cdot \frac{(1 + i)^{k-n} - (1 + i)^{k-1-n}}{i}$$

$$= R \cdot \frac{(1 + i)^{k-1-n}[(1 + i) - 1]}{i} = R \cdot (1 + i)^{k-1-n}$$

and

$$I_k = R - B_k$$

$$= R - R \cdot (1 + i)^{k-1-n} = R[1 - (1 + i)^{k-1-n}]$$

We now derive a formula that will enable us to find the number of payments necessary to pay off a certain loan. Start with the formula

$$A = R \cdot \frac{1 - (1 + i)^{-n}}{i}$$

This yields

$$(1 + i)^n = \frac{R}{R - Ai}$$

Taking the natural logarithm of both sides, we get

$$n\ln(1 + i) = \ln(R) - \ln(R - Ai)$$

and

$$n = \frac{\ln(R) - \ln(R - Ai)}{\ln(1 + i)}$$

Summary of Related Formulas

Total amount of the loan:

$$A = R \cdot \frac{1 - (1 + i)^{-n}}{i} \tag{1}$$

Unpaid balance on the loan after the kth payment:

$$A_k = R \cdot \frac{1 - (1 + i)^{k-n}}{i} \tag{2}$$

Amount of the principal repaid in the kth payment:

$$B_k = R \cdot (1 + i)^{k-1-n} \tag{3}$$

Amount of interest paid in the kth payment:

$$I_k = R[1 - (1 + i)^{k-1-n}] \tag{4}$$

Number of payments necessary to pay off the loan:

$$n = \frac{\ln(R) - \ln(R - Ai)}{\ln(1 + i)} \tag{5}$$

A = Amount of the loan, in dollars.
R = Payment per period, in dollars.
i = Rate of interest per period, written as a decimal.
n = Number of periods (payments).
A_k = Unpaid balance on the loan, after the kth payment, in dollars.
B_k = Amount of the principal paid in the kth payment, in dollars.
I_k = Amount of interest paid in the kth payment, in dollars.

A CLOSER LOOK   When someone borrows money and agrees to repay the loan plus interest by making equal payments at regular intervals, it is understood that the interest is being charged on the unpaid balance of the principal. We have based the derivation of the five formulas on the concept of annuity since the lending institution does, in fact, purchase an annuity from the borrower. We should verify that the results we obtain are consistent with the concept of installment buying.

1. The interest paid in the $(k + 1)$st payment should be the interest earned by the unpaid balance after the kth payment. That is, we should have

$$I_{k+1} = A_k \cdot i$$

Replacing k by $k + 1$ in Formula (4), we get

$$I_{k+1} = R[1 - (1 + i)^{k-n}]$$

Using Formula (2), we obtain

$$A_k \cdot i = R \cdot \frac{1 - (1 + i)^{k-n}}{i} \cdot i$$

$$= R[1 - (1 + i)^{k-n}]$$

Thus, $I_{k+1} = A_k \cdot i$ since both quantities are equal to $R[1 - (1 + i)^{k-n}]$.

2. The unpaid balance on the loan after the kth payment A_k should be equal to the total amount of the loan A, minus the sum of the amounts of the principal paid in each of the first k payments. That is, the following should be true:

$$A_k = A - (B_1 + B_2 + \cdots + B_k) \qquad \text{(See Exercise 19.)}$$

3. B_1, the amount of principal repaid in the first payment, is $R(1 + i)^{-n}$, which is the present value of the nth payment. Similarly, B_2, the amount of principal repaid in the second payment, is $R(1 + i)^{1-n}$, which is the present value of the $(n - 1)$st payment. In general, B_k, the amount of principal repaid in the kth payment, is equal to the present value of the $(n - k + 1)$st payment, since both of these quantities are equal to $R(1 + i)^{k-1-n}$.

4. See Exercise 21.

Applications

EXAMPLE 1   Mr. and Mrs. Beauman bought a house for $150,000. They made a 20% down payment, with the balance amortized by a 25-year mortgage at a rate of 9% compounded monthly.

a. Find the amount of their monthly mortgage payment.
b. Find the amount of interest paid in the 100th payment.
c. Find the amount of principal paid in the 100th payment.
d. Find the amount of the mortgage the Beaumans will have paid in 15 years.
e. Find the total amount of interest the Beaumans will have paid once the mortgage is paid off.

Solution **a.** The down payment was $30,000 since $150,000(0.2) = 30,000$. Thus, the mortgage is $120,000. Use Formula (1) with $A = 120,000$, $i = .0075$ (since $\frac{.09}{12} = .0075$), $n = 300$ (since $25 \cdot 12 = 300$), and R is to be found. We get

$$120,000 = R \cdot \frac{1 - (1 + .0075)^{-300}}{.0075}$$

$$\doteq R(119.1616222)$$

Thus,

$$R \doteq \frac{120,000}{119.1616222}$$

$$\doteq 1007.035636$$

Rounding the answer to the nearest cent, we conclude that the Beaumans will have a $1007.04 monthly mortgage payment.

b. The amount of interest paid in the 100th payment may be found using Formula (4) with $R = 1007.04$, $i = .0075$, $k = 100$, and $n = 300$. We get

$$I_{100} = 1007.04[1 - (1 + .0075)^{100-1-300}]$$

$$\doteq 1007.04(.777287813)$$

$$\doteq 782.7599192$$

Rounding to the nearest cent, the interest paid in the 100th payment is $782.76.

c. Using Formula (3) to find the amount of principal repaid in the 100th payment, with $R = 1007.04$, $i = .0075$, $k = 100$, and $n = 300$. We obtain

$$B_{100} = 1007.04(1 + .0075)^{100-1-300}$$

$$\doteq 1007.04(.222712187)$$

$$\doteq 224.2800809$$

Thus, the amount of interest paid in the 100th payment is $224.28. Note that, as expected, $224.28 + 782.76 = 1007.04$.

d. The amount of principal the Beaumans have paid after 15 years is the total amount of the mortgage minus the unpaid balance after 15 years. To find the unpaid balance after 15 years use Formula (2) with $R = 1007.04$, $i = .0075$, $n = 300$, and $k = 180$ (since $15 \cdot 12 = 180$).

$$A_{180} = 1007.04 \cdot \frac{1 - (1 + .0075)^{180-300}}{.0075}$$

$$\doteq 1007.04(78.94169267)$$

$$\doteq 79,497.44218$$

Rounding to the nearest cent, $A_{180} = 79,497.44$. Thus, after 15 years, the Beaumans will have paid $40,502.56 of the principal since $120,000 - 79,497.44 = 40,502.56$.

e. Since the Beaumans are making monthly payments of $1007.04 for a period of 25 years, when the mortgage is paid off they will have paid a total of $302,112.00

since $12 \cdot 25 \cdot 1007.04 = 302,112$. Therefore, the total amount they paid in interest is \$182,112 since $302,112 - 120,000 = 182,112$.

Do Exercise 3. ◼

EXAMPLE 2 Mr. and Mrs. Dougherty have a combined take-home pay of \$4187 per month. A mortgage company will lend them an amount which will yield a maximum monthly payment of 25% of their take-home pay. If the rate of interest is 12% compounded monthly, the loan is to be amortized over a period of 25 years, and a 20% down payment is required, what is the maximum value of a house the Doughertys can purchase?

Solution The maximum monthly payment is given by

$$\$4187(.25) = \$1046.75$$

Since the loan is to be amortized over a period of 25 years, the number of monthly payments will be 300. The interest rate per month is .01 since $\frac{.12}{12} = .01$. Using Formula (1), we get

$$A = 1046.75 \frac{1 - (1 + .01)^{-300}}{.01}$$

$$\doteq 1046.75(94.94655125)$$

$$\doteq 99,385.30252$$

Rounding to the nearest cent, we conclude that the Doughertys may borrow up to \$99,385.30. But a 20% down payment is required; therefore, if V dollars is the maximum value of a house they can purchase, we must have

$$V - (.2)V = 99,385.30$$

That is,

$$(.8)V = 99,385.30$$

and

$$V = \frac{99,385.3}{.8} \doteq 124,231.625$$

Thus, the maximum value of a house the Doughertys can purchase is approximately \$124,230.

Do Exercise 9. ◼

EXAMPLE 3 Ann bought a new car for \$12,500. She is required to make a 20% down payment. She borrows the balance from her credit union. Her debt is to be amortized over a period of 5 years at 9% interest compounded monthly.

a. How much is each monthly payment?
b. How much of the 30th payment will pay interest?
c. How much of the 30th payment is loan payment?
d. If Ann decides to sell her car after 3 years, how much should she get to be able to pay the balance of her loan?

Solution   **a.** Ann must make a $2500 down payment since $12{,}500(.20) = 2500$. Therefore, she will borrow $10,000 from her credit union. Use Formula (1) with $A = 10{,}000$, $n = 60$ (since $5 \cdot 12 = 60$), and $i = .0075$ (since $\frac{.09}{12} = .0075$).

$$10{,}000 = R \cdot \frac{1 - (1 + .0075)^{-60}}{.0075}$$

$$\doteq R(48.17337347)$$

Therefore,

$$R \doteq \frac{10{,}000}{48.17337347} \doteq 207.5835525$$

Rounding to the nearest cent, Ann's payments are $207.58 per month.

b. To find the amount of interest paid in the 30th payment, we use Formula (4) with $k = 30$, $R = 207.58$, $i = .0075$, and $n = 60$. We get

$$I_{30} = 207.58[1 - (1 + .0075)^{30-1-60}]$$

$$= 207.58(0.206762384)$$

$$= 42.91973567$$

Thus, $42.92 is paid in interest in the 30th payment.

c. Using part **b**, the amount of the loan paid in the 30th payment is $164.66 since $207.58 - 42.92 = 164.66$.

d. In 3 years, Ann will have made 36 payments. The remaining balance is found using Formula (2) with $k = 36$ and the values of R, i, and n as in part **b**.

$$A_{36} = 207.58 \cdot \frac{1 - (1 + .0075)^{36-60}}{.0075}$$

$$= 207.58(21.88914613)$$

$$= 4{,}543.748954$$

Ann should sell her car for at least $4543.75 to be able to pay off the loan.
Do Exercise 13.   ■

EXAMPLE 4   Upon graduation from college, Roy had accumulated a debt of $7600. No interest was charged while he was in college, but the lending institution charges 9% interest compounded monthly starting at graduation. Roy wishes to repay his debt as quickly as possible but can only afford payments of $200 per month. How long will it take him to repay his debt?

Solution   We shall use Formula (5) with $R = 200$, $A = 7600$, and $i = .0075$ since $\frac{.09}{12} = .0075$. We get

$$n = \frac{\ln(200) - \ln[200 - 7600(.0075)]}{\ln(1 + .0075)}$$

$$= 44.89722564$$

Hence, it will take Roy 45 months to repay his student loan. (The last payment will be a bit less than $200.)
Do Exercise 15.   ■

Exercise Set 5.3

1. Mr. and Mrs. Sherman bought a lot for $36,500. They made a 20% down payment, with the balance amortized by a 25-year mortgage at 7.25% interest compounded monthly.

   a. Find the amount of their monthly mortgage payment.
   b. Find the amount of interest paid in the 150th payment.
   c. Find the amount of principal paid in the 150th payment.
   d. Find the amount of the mortgage the Shermans will have paid in 12 years.
   e. Find the total amount of interest the Shermans will have paid once the mortgage is paid off.

2. Mr. and Mrs. Larsen bought a house for $200,000. They made a 40% down payment, with the balance amortized by a 25-year mortgage at 9% interest compounded monthly.

   a. Find the amount of their monthly mortgage payment.
   b. Find the amount of interest paid in the 200th payment.
   c. Find the amount of principal paid in the 200th payment.
   d. Find the amount of the mortgage the Larsens will have paid in 17 years.
   e. Find the total amount of interest the Larsens will have paid once the mortgage is paid off.

3. Mr. and Mrs. Bateau bought a boat for $12,500. They made a 20% down payment, with the balance amortized by a 5-year mortgage at 9% interest compounded monthly.

   a. Find the amount of their monthly mortgage payment.
   b. Find the amount of interest paid in the 25th payment.
   c. Find the amount of principal paid in the 25th payment.
   d. Find the amount of the mortgage the Bateaus will have paid in 3 years.
   e. Find the total amount of interest the Bateaus will have paid once the mortgage is paid off.

4. Charles bought a car for $10,000. He made a 20% down payment, with the balance amortized by a 3-year loan at 12% interest compounded monthly.

   a. Find the amount of his monthly payment.
   b. Find the amount of interest paid in the 22nd payment.
   c. Find the amount of principal paid in the 22nd payment.
   d. Find the amount of the loan Charles will have paid in 2 years.
   e. Find the total amount of interest Charles will have paid once the loan is paid off.

5. Mr. and Mrs. Kelley bought a house for $120,000. They paid 25% down, with the balance amortized by a 20-year mortgage at 12% interest compounded monthly.

   a. Find the amount of their monthly mortgage payment.
   b. Find the amount of interest paid in the 125th payment.
   c. Find the amount of principal paid in the 125th payment.
   d. Find the amount of the mortgage the Kelleys will have paid in 13 years.
   e. Find the total amount of interest the Kelleys will have paid once the mortgage is paid off.

6. Mr. and Mrs. Nelson bought a house for $190,000. They paid 25% down, with the balance amortized by a mortgage at 12% interest compounded monthly. Find the amount of their monthly mortgage payment if the loan is to be paid off in a. 40 years, b. 30 years, c. 25 years, d. 20 years, e. 15 years, and f. 10 years. Compare the monthly payments.

7. Same as Exercise 6 if the house was bought for $210,000 and the rate of interest was 9% compounded monthly.

8. Mr. and Mrs. Leva have a combined take-home pay of $5180 per month. A mortgage company will lend them an amount which will yield a maximum monthly payment of 20% of their take-home pay. If the rate of interest is 11% compounded monthly, the loan is to be amortized over a period of 30 years, and a 20% down payment is required, what is the maximum value of a house the Levas can purchase?

9. Mr. and Mrs. Chang have a combined take-home pay of $5212 per month. A mortgage company will lend them an amount which will yield a maximum monthly payment of 25% of their take-home pay. If the rate of interest is 11.5% compounded monthly, the loan is to be amortized over a period of 20 years, and a 15% down payment is required, what is the maximum value of a house the Changs can purchase?

10. Dave has a take-home pay of $3214 per month. A bank will lend him an amount which will yield a maximum monthly payment of 15% of his take-home pay toward the purchase of a boat. If the rate of interest is 12% compounded monthly, the loan is to be amortized over a period of 10 years, and a 30% down payment is required, what is the maximum value of a boat Dave can purchase?

11. Mr. and Mrs. Udeen have a combined take-home pay of $45,228 per year. A mortgage company will lend them an amount which will yield a maximum monthly payment of 25% of their take-home pay. If the rate of interest is 10% compounded monthly, the loan is to be amortized over a period of 35 years, and a 15% down payment is required, what is the maximum value of a house the Udeens can purchase?

12. Jerry bought a car for $12,500. He is required to make a 20% down payment. He borrows the balance from his credit union. His debt is to be amortized over a period of 5 years at 9% interest compounded monthly.

a. How much is each monthly payment?

b. How much of the 23rd payment will pay interest?

c. How much of the 23rd payment is loan payment?

d. If Jerry decides to trade his car after 2 years, how much must he pay the credit union to pay off his loan, assuming there is no penalty for early payment?

13. Kathy bought a new car for $12,000. She makes a one-third down payment. She borrows the balance from her bank. Her debt is to be amortized over a period of 3 years at 12% interest compounded monthly.

a. How much is each monthly payment?

b. How much of the 19th payment will pay interest?

c. How much of the 19th payment is loan payment?

d. If Kathy decides to sell her car after the 25th payment, how much should she get to be able to pay the balance of her loan from the proceeds?

14. Mr. and Mrs. Galbraith bought a new home for $120,000. They made a 25% down payment and borrowed the balance from a mortgage company. Their debt is to be amortized over a period of 20 years at 12% interest compounded monthly.

a. How much is each monthly payment?

b. How much of the 50th payment will pay interest?

c. How much of the 50th payment is principal payment?

d. If the Galbraiths decide to sell their home after 10 years and succeed in selling it for $130,000, how much cash will be available to them (neglecting the costs of the transaction)?

15. Upon graduation from college, Lisa had accumulated a debt of $9800. No interest was charged while she was in college, but the lending institution charges 9% interest compounded monthly starting at graduation. Lisa wishes to repay her debt as quickly as possible, but can afford payments of only $250 per month. How long will it take to repay her debt?

16. To build a new hot tub in his backyard, Bob borrowed $9850. The lending institution charges 12% interest compounded monthly. Bob wishes to repay his debt as quickly as possible, but can afford payments of only $375 a month. How long will it take to repay his debt?

17. To start a small business, Sharon had accumulated a debt of $57,600. No interest was charged until the business started, but the lending institution charged 12% interest compounded monthly as soon as the business was in operation. Sharon wished to repay her debt as quickly as possible and decided that she could pay $1200 a month. How long will it take to repay her debt?

18. Mr. and Mrs. Esser borrowed $15,000 to remodel their home. The lending institution charges 11% interest compounded monthly. The Essers wish to repay their debt as quickly as possible but can afford payments of only $380 a month. How long will it take to repay their debt?

19. Verify that the unpaid balance on a loan after the kth payment A_k is equal to the total amount of the loan A minus the sum of the amounts of the principal paid in each of the first k payments. That is, verify that

$$A_k = A - (B_1 + B_2 + B_3 + \cdots + B_k)$$

(*Hint:* Use Formula (2) on the left for A_k and Formulas (1) and (3) on the right for A and each of the B_i's; then, simplify using the formula for the sum of the first n terms of a G.P.)

20. Formula (5) gives the number of payments necessary to pay off a loan in terms of $\ln(R)$, $\ln(R - Ai)$, and $\ln(1 + i)$. Clearly R and $1 + i$ are positive, so $\ln(R)$ and $\ln(1 + i)$ are defined. Verify that $R - Ai > 0$. (*Hint:* Use Formula (1) and replace A by $R[1 - (1 + i)^{-n}]/i$.)

21. The total amount of interest paid on a loan is the total amount paid (nR) minus the original amount of the loan (A). It is also the sum of the interest amounts paid with each of the n payments. That is

$$nR - A = I_1 + I_2 + I_3 + \cdots + I_n$$

Verify this fact using Formulas (1) and (4) of this section.

For those interested in pursuing these applications further, the following readings will be helpful.

1. John D. Baildon. Arithmetic Progressions and the Consumer, *The College Mathematics Journal*, Vol. 16, No. 5 (1985), 395–397.

2. Ann D. Holley, A Question of Interest, *The Two-Year College Mathematics Journal*, Vol. 9, No. 2 (1978), 81–83.

3. Morris Morduchow, Discrete and Continuous Compounding, *The American Mathematical Monthly*, Vol. 92, No. 10 (1985), 734–735.

4. Stanley G. Wayment, Another Question of Interest, *The Two-Year College Mathematics Journal*, Vol. 11, No. 4 (1980), 252–254.

5.4 Chapter Review

IMPORTANT SYMBOLS AND TERMS

s_i [5.1]
$[x]$ [5.1]
$s_{\overline{n}|i}$ [5.2]
$a_{\overline{n}|i}$ [5.2]
Accumulated value [5.2]
Amount (future value) [5.2]
Amount of interest paid in the kth payment [5.3]
Amount of the principal repaid in the kth payment [5.3]
Annuity [5.2]
Annuity due [5.2]
A.P. [5.1]
Arithmetic progression [5.1]
Certain annuity [5.2]
Common difference [5.1]
Common ratio [5.1]

Contingent annuity [5.2]
Deferred annuity certain [5.2]
Finite sequence [5.1]
General annuity [5.2]
Geometric progression [5.1]
Immediate annuity [5.2]
Infinite sequence [5.1]
ith term [5.1]
kth term [5.1]
Last survivor annuity [5.2]
Mortgage [5.3]
Nominal rate of interest [5.1]
Number of payments necessary to pay off the loan [5.3]
Ordinary annuity [5.2]

Payment period [5.2]
Perpetuity [5.2]
Present value [5.2]
Simple annuity [5.2]
Sinking fund [5.2]
Sum of the first n terms [5.1]
Temporary annuity [5.2]
Term [5.2]
Total amount of the loan [5.3]
Unpaid balance on the loan after the kth payment [5.3]

SUMMARY

A finite sequence is a function whose domain is the set $\{1, 2, \ldots, n\}$; an infinite sequence is a function whose domain is the set $\{1, 2, 3, \ldots, n, \ldots\}$. The sequence $a, a + d, a + 2d, a + 3d, \ldots$ is called an arithmetic progression. The kth term of an A.P. is

$$s_k = a + (k - 1)d$$

and the sum of the first n terms is

$$S_n = \frac{n[2a + (n - 1)d]}{2} = \frac{n(s_1 + s_n)}{2}$$

The sequence $a, ar, ar^2, ar^3, \ldots$ is called a geometric progression. The kth term of a G.P. is

$$s_k = ar^{k-1}$$

and the sum of the first n terms is

$$S_n = \frac{a(1 - r^n)}{1 - r} \quad \text{whenever } 1 - r \neq 0$$

If interest is compounded at regular time intervals, and if P dollars is the principal (present value), i is the rate of interest per conversion period (written as a decimal), n is the number of conversion periods, and S_n is the compound amount (future value) after n conversion periods, then

$$S_n = P(1 + i)^n \quad \text{and} \quad P = S_n(1 + i)^{-n}$$

Whenever a transaction involves a sequence of equal payments made periodically, this transaction is called an annuity. In an *ordinary annuity*, each payment is made at the end of a payment period and in an *annuity due* each payment is made at the beginning of a payment period.

The basic formulas for ordinary annuities are

$$S = R \cdot \frac{(1 + i)^n - 1}{i} = R \cdot s_{\overline{n}|i}$$

$$A = R \cdot \frac{1 - (1 + i)^{-n}}{i} = R \cdot a_{\overline{n}|i}$$

and for an annuity due:

$$S = R(s_{\overline{n+1}|i} - 1)$$

$$A = R(1 + a_{\overline{n-1}|i})$$

where R = amount of each payment, n = number of payments, i = rate of interest per period (written as a decimal), S = amount (future value) of the annuity, and A = the present value of the annuity.

A sinking fund is an annuity which is established for the express purpose of accumulating a sum of money to pay an obligation at a future designated date.

Suppose that a lending institution lends A dollars, at the periodic rate of interest i (written as a decimal) on the unpaid balance, to be repaid in n periodic payments of R dollars each and each payment is made at the end of each period. Let A_k dollars be the unpaid balance on the loan after the kth payment, B_k dollars be the portion of the kth payment used to repay the principal, and I_k dollars be the portion of the kth payment used to pay interest on the unpaid balance. Then

$$A = R \cdot \frac{1 - (1 + i)^{-n}}{i}$$

$$A_k = R \cdot \frac{1 - (1 + i)^{k-n}}{i}$$

$$B_k = R \cdot (1 + i)^{k-1-n}$$

$$I_k = R[1 - (1 + i)^{k-1-n}]$$

$$n = \frac{\ln(R) - \ln(R - Ai)}{\ln(1 + i)}$$

SAMPLE EXAM QUESTIONS

1. Find the first five terms of each of the sequences defined by

   a. $s_n = \dfrac{4}{3n + 1}$

   b. $s_n = (-1)^n \dfrac{2n}{n + 3}$

   c. $s_n = \dfrac{n^2 + 2n + 3}{2n^2 + n - 2}$

2. If \$700 is deposited in a bank that pays 6% interest compounded monthly, find the amount in that account at the end of each of the first 6 months.

3. Let s be an arithmetic progression whose first term is -3 and whose common difference is 3.

   a. Find the first four terms of this A.P.
   b. Find the sum of the first 200 terms.

4. The 35th and 57th terms of an A.P. are 20 and 31, respectively.

   a. Find the first five terms of this A.P.
   b. Find the sum of the first 1000 terms.

5. Let s be a G.P. whose first term is 4 and whose common ratio is -2.

   a. Find the second, third, and fourth terms.
   b. Find the sum of the first 20 terms.

6. The seventh and tenth terms of a G.P. are 192 and 1536, respectively.

   a. Find the first five terms.
   b. Find the sum of the first 20 terms.

7. A bank pays 6% interest compounded monthly. Wendy deposits $325 the first of each month. How much is in her account immediately after the 54th deposit?

8. Suppose that one-third of a container full of pure alcohol is replaced by water and the solution is mixed thoroughly. Then, one-third of the solution is replaced by water and mixed thoroughly. If this is done repeatedly, what fraction of the original amount of alcohol is left after the 20th replacement?

9. Suppose that a new car is worth $20,000 and that at the end of each year its value depreciates 17% of the value it had at the beginning of that year. What will the car be worth at the end of 10 years?

10. Tim deposits $500 on the first of each month in a bank that pays 7% interest compounded monthly. How much is in his account immediately before the 40th deposit?

11. Jeremy deposits $300 on the first of each month in a bank that pays 6.5% interest compounded monthly. How much is in Jeremy's account immediately after the 30th deposit?

12. When Lisa entered college, her parents decided that they would reward her upon graduation with a trip to France. They estimate the trip will cost $10,000. They wish to make equal deposits on the first of each month, starting October 1, 1990, in a bank that pays 6% interest compounded monthly. The last deposit will be June 1, 1994, at which time Lisa will graduate. How much should they deposit each month?

13. On the day of their son's birth, Mr. and Mrs. Tarte decided to set aside a sum of money to provide for his college education. They wish to make a single deposit in a bank that pays 9% interest compounded yearly in order to provide a payment of $12,000 on each of the son's 18th, 19th, 20th, and 21st birthdays. How much should they deposit?

14. Janet deposits $200 at the beginning of each month in a bank that pays 9% interest compounded monthly. How much will be in her account at the end of 5 years?

15. John deposits $350 at the end of every 3 months for a period of 5 years in a bank that pays 8% interest compounded quarterly. What is the final amount in his account?

16. Mary wishes to purchase an annuity that will pay her son $800 at the end of each month for the next 6 years. If the rate of interest is 8% compounded monthly, how much must she pay for this annuity?

17. What is the present value of an annuity due that will pay $15,000 per year for a period of 20 years if the rate of interest is 9% compounded annually?

18. Nancy won $500,000 at the Washington State Lottery. After taxes, she is to receive $20,000 a year for the next 20 years with the first payment being made immediately. The State of Washington purchases an annuity that will provide these payments. If the rate of interest is 8% compounded annually, what is the cost of the annuity?

19. On his 25th birthday, John established a retirement fund by paying $300 per month to a life insurance company that pays 8% interest compounded monthly, the first payment being made 1 month after his birthday. If he retires on his 65th birthday, at which time he makes the last payment, and if he wishes to make equal withdrawals until he reaches his 75th birthday, the first withdrawal being made 1 month after his 65th birthday, how much will he get each month from the retirement fund?

20. A company establishes a sinking fund that will provide $300,000 to update its machinery in 6 years. For this purpose, it purchases an ordinary annuity and will make equal payments every 3 months. How much is each payment if the rate of interest is 8% compounded quarterly?

21. Judy would like to have $200,000 in her savings account 20 years from now. She will be able to deposit $250 at the end of each month for the next 5 years in a bank that pays 9% interest compounded monthly. How much must she deposit monthly in the last 15 years if she is to reach her goal?

22. Mr. and Mrs. Bernard purchased a house for $240,000. They made a 30% down payment, with the balance amortized by a 20-year mortgage at 11.5% interest compounded monthly.

    a. Find the amount of their monthly mortgage payment.
    b. Find the amount of interest in the 50th payment.
    c. Find the amount of principal in the 50th payment.
    d. Find the amount of the mortgage the Bernards will have paid in 13 years.
    e. Find the total amount of interest the Bernards will have paid once the mortgage is paid off.

23. Mr. and Mrs. Sherman have a combined take-home pay of $5540 per month. A mortgage company will lend them an amount that will yield a maximum monthly payment of 25% of their take-home pay. If the rate of interest is 12% compounded monthly, the loan is to be amortized over a period of 20 years, and a 20% down payment is required, what is the maximum value of a house the Shermans can purchase?

24. Gwen bought a new car for $15,000. She made a 25% down payment and borrowed the balance from her credit union. Her debt is to be amortized over a period of 5 years at 12% interest compounded monthly, the first payment being made one month from the date of purchase.

    a. How much is each monthly payment?
    b. How much of the 20th payment will pay interest?
    c. How much of the 20th payment is loan payment?
    d. If Gwen sells her car immediately after the 24th payment, how much should she get to pay the balance of the loan?

25. Upon graduation from college, Pierre had accumulated a debt of $12,500 in financial aid. No interest was charged while he was in school, but the lending institution charges 12% interest compounded monthly starting at graduation. Pierre wishes to repay his debt as promptly as possible, but can pay only $250 a month. How long will it take to pay his debt?

26. Kathy establishes a fund earning 9% compounded annually to maintain a family crypt. The maintenance cost is $350 per year with the first payment due immediately. How much must be paid to purchase an annuity that will take care of the perpetual maintenance of the family crypt?

Probability

Almost everyone has some idea of the meaning of probability, and the word *probably* is (probably!) one of the most widely used words in English. In fact, because of the various connotations of *probability*, there are differences of opinion about probability theory even among mathematicians and philosophers. Nevertheless, the basic laws and calculations are almost universal. The purpose of this chapter is to familiarize you with the fundamental ideas and to show how the theory may be derived from a few simple axioms. Although our goal is to give you the ability to apply the acquired knowledge in your major field ("probably" business or economics), many of the examples and exercises will involve tossing coins, rolling dice, and drawing cards from a standard deck or marbles from an urn. There are two reasons for this. The first is that such experiments are simple and unambiguous; therefore, we will be able to concentrate on the mathematical ideas we are trying to illustrate. The second is that we are assuming most of you are using this book in a course that is a prerequisite to business and economics courses; as a consequence, you are "probably" not yet ready to handle concepts that come later in these courses.

You are urged to review Sections 1.1 and 1.2 on sets and Venn diagrams before beginning the study of this chapter.

6.1 Introduction

We are often forced to make decisions on probable knowledge and incomplete evidence. Things we often call facts are actually a mixture of physical data and inferences about them. Let us, for example, examine what surgeons mean when they tell a patient who has undergone eye surgery that his chances of recovery are about three out of four. This means that out of a great many patients who have undergone the same type of surgery, approximately 75% have recovered. But can the truth of the statement be tested? Suppose the patient does not recover. Does this mean that the doctor's statement was false? No, because on the average 25% of the patients are not expected to recover. It means only that the present patient was one of the unlucky ones. In spite of the fact that there is no satisfactory way of testing the validity of the statement, we feel it is still significant. In fact, each day most of us are willing to entrust our lives on actions based on probability considerations. The fact that someone can purchase over $100,000 worth of flight insurance for only $5 makes you feel that traveling by air is extremely safe since insurance companies base their rates on probability.

Often, probabilities are assigned to events that cannot be duplicated exactly. These are called *subjective probabilities* and are assigned on the basis of repeated trials and personal interpretation of these trials. In statistical studies, *success* means that the event under consideration actually occurred. The *relative frequency* of success is the ratio of the number of successes to the number of trials. We can now formulate a definition of probability.

DEFINITION 6.1: Suppose that in *n* trials of an experiment there are s_n successes and that the sequence s_n/n of relative frequencies of successes approaches a number as *n* gets larger and larger. Then this number is defined to be the *probability* of success in a single trial.

In practice, the number of trials that can possibly be performed is always finite. Therefore, the probability is approximated by the relative frequency for the largest number of trials that have been performed for the experiment. The following example illustrates this definition of probability.

EXAMPLE 1   A company is making light bulbs. Its quality control team tested light bulbs chosen at random among those produced in the last 6 months. The following table gives the relative frequencies for different numbers of bulbs tested. The results listed in this table are cumulative. For example, 5 of the first 100 bulbs were defective and 6 of the next 100 were defective; thus, the second line shows that 11 of 200 were defective.

Number of bulbs tested n	Number of defective bulbs s_n	Relative frequency $\dfrac{s_n}{n}$
100	5	.05
200	11	.055
300	18	.06
500	29	.058
1000	55	.055
2000	105	.0525
3000	162	.054
4000	214	.0535
5000	266	.0532
6000	318	.053

What is the probability that a light bulb selected at random is defective?

Solution   Since the results are cumulative and the largest number of trials is 6000, we conclude that if a light bulb is chosen randomly, the probability that it is defective is approximately .053.
Do Exercise 1.   ∎

Uniform Sample Space

Often, especially in games of chance, we can define probability without collecting data on relative frequencies. We illustrate with an example.

EXAMPLE 2   Suppose two dice are rolled.* What are the chances that the sum of the faces is five?

Solution   When two dice are rolled, the set of possible outcomes consists of all the different ways that the two dice may come to rest. If we let x and y denote the numbers that come up on the first die and the second die, respectively, then an outcome is represented by the ordered pair (x, y), where x and y are members of the set $\{1, 2, 3, 4, 5, 6\}$.

---

*Unless otherwise stated, we tacitly assume that the dice are "fair." This means only that they are not biased to favor certain outcomes.

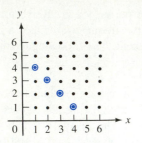

FIGURE 6.1

Since there are 6 choices for x and 6 choices for y, there are 36 ordered pairs. These ordered pairs are represented in Figure 6.1. With "fair" dice, each of the 36 outcomes occurs equally often. Thus, the chance of any one of them occurring is $\frac{1}{36}$. We are interested not in just one ordered pair (x, y), but in those for which $x + y = 5$. There are four such outcomes: $(1, 4)$, $(2, 3)$, $(3, 2)$, and $(4, 1)$. They are the circled points in Figure 6.1. It is reasonable to say that if two fair dice are rolled, the chances that the sum of the faces is 5 is $\frac{1}{9}$ since $\frac{4}{36} = \frac{1}{9}$.

Do Exercise 5.   ■

Since we are often interested in a set of possible outcomes, we must decide what the chances are that one or more of these outcomes actually will occur. We have the following definition.

> **DEFINITION 6.2:** The set of all possible outcomes of an experiment is called the *sample space* for that experiment. Each outcome is a *point* in the sample space and each subset of the sample space is an *event*. If all outcomes are equally likely, we say that they are *equiprobable* and the sample space is *uniform*. In that case, the *probability* of an event is the quotient of the number of elements in the event by the number of elements in the sample space.

In this book, we consider mostly finite sample spaces. Symbolically, if $n(A)$ denotes the number of elements in a set A, we write $p(E) = \frac{n(E)}{n(S)}$ where S is a uniform sample space, E is an event (a subset of S), and $p(E)$ denotes the probability of the event E.*

Since E is a subset of S, we will always have

$$0 \le n(E) \le n(S)$$

from which it follows that

$$0 \le \frac{n(E)}{n(S)} \le \frac{n(S)}{n(S)}$$

and

$$0 \le p(E) \le 1$$

In Example 3, the sample space consisted of 36 points. An event can be a singleton (a subset with just one point) or it can be a subset with several points (for example, the event that the sum of the faces is 5 is the subset $\{(1, 4), (2, 3), (3, 2), (4, 1)\}$). Since the empty set is a subset of any set, it is always an event. It is called the *impossible event*. If S is a sample space and E is a subset of S, then $S \backslash E$ (the complement of E in S, also denoted E') represents the event that E does not occur. If E_1 and E_2

---

*In one sense, the preceding definition of probability is circular because the expression "equally likely" involves the notion of probability. However, this technical difficulty should not cause a problem in understanding the concepts.

are subsets of S, then $E_1 \cup E_2$ is the event that either E_1 or E_2 (or both) occur, and $E_1 \cap E_2$ is the event that both E_1 and E_2 occur. If S is a sample space and E is an event, it is clear that $n(E) + n(E') = n(S)$ since every member of the sample space must be either in E or in E' and no outcome is in both E and E'. Thus,

$$\frac{n(E) + n(E')}{n(S)} = \frac{n(S)}{n(S)}$$

and

$$\frac{n(E)}{n(S)} + \frac{n(E')}{n(S)} = 1$$

Therefore,

$$p(E) + p(E') = 1$$

EXAMPLE 3 Two dice are rolled. Represent the following situations in terms of events and calculate the probability of each event.

a. The sum of the faces is 3.
b. The sum of the faces is 7.
c. The sum of the faces is 7 or 3.
d. The sum of the faces is 7 and 3.
e. The sum of the faces is not 3.

Solution The sample space is the set of all ordered pairs (x, y) where x and y are members of $\{1, 2, 3, 4, 5, 6\}$. Therefore, the sample space has 36 members.

a. The sum of the faces is 3 if, and only if, the outcome is a member of the event $E_3 = \{(1, 2), (2, 1)\}$. Since there are 2 members in E_3 and 36 members in the sample space, the probability that the sum of the numbers is 3 is $\frac{1}{18}$ since $\frac{2}{36} = \frac{1}{18}$. We write

$$p(E_3) = \frac{1}{18}$$

b. The sum of the faces is 7 if, and only if, the outcome is a member of the event $E_7 = \{(1, 6), (2, 5), (3, 4), (4, 3), (5, 2), (6, 1)\}$. Since there are 6 members in E_7, the probability that the sum is 7 is $\frac{1}{6}$ since $\frac{6}{36} = \frac{1}{6}$. We write

$$p(E_7) = \frac{1}{6}$$

c. The sum of the faces is 7 or 3 if, and only if, the outcome is a member of the event $E_7 \cup E_3 = \{(1, 6,), (2, 5), (3, 4), (4, 3), (5, 2), (6, 1), (1, 2), (2, 1)\}$. Since there are 8 members in $E_7 \cup E_3$, the probability that the sum is 7 or 3 is $\frac{2}{9}$ since $\frac{8}{36} = \frac{2}{9}$. We write

$$p(E_7 \cup E_3) = \frac{2}{9}$$

d. The sum of the faces is 7 and 3 if, and only if, the outcome is a member of the event $E_7 \cap E_3$. But $E_7 \cap E_3 = \emptyset$, which means that this event is the impossible event. The number of elements in $\emptyset$ is 0. Hence, the probability that the sum is 7 and 3 is 0 since $\frac{0}{36} = 0$. We write

$$p(E_7 \cap E_3) = 0$$

e. The sum of the faces is not 3 if, and only if, the outcome is a member of the event E_3'. But

$$p(E_3) + p(E_3') = 1$$

Therefore,

$$\frac{1}{18} + p(E_3') = 1$$

and the probability that the sum of the numbers is not 3 is $\frac{17}{18}$. We write

$$p(E_3') = \frac{17}{18}$$

Do Exercise 11.   ∎

Giving Odds

Often, probabilities are given in terms of *odds*, especially in games of chance. The odds in favor of an event E is defined to be the ratio $\frac{p(E)}{p(E')}$. Thus, if the odds in favor of event E are m to n, with $mn \neq 0$, then $\frac{p(E)}{p(E')} = \frac{m}{n}$. Equivalently, $\frac{p(E')}{p(E)} = \frac{n}{m}$. Adding 1 to both sides of this equality we get

$$\frac{p(E')}{p(E)} + 1 = \frac{n}{m} + 1$$

or

$$\frac{p(E) + p(E')}{p(E)} = \frac{n + m}{m}$$

Using the fact that $p(E) + p(E') = 1$, we obtain

$$\frac{1}{p(E)} = \frac{n + m}{m}$$

and so

$$p(E) = \frac{m}{n + m}$$

Also,

$$p(E') = 1 - p(E) = 1 - \frac{m}{n + m} = \frac{n + m - m}{n + m} = \frac{n}{n + m}$$

EXAMPLE 4 If the odds in favor of UCLA going to the next Rose Bowl are 2 to 13, what is the probability that UCLA will go to the next Rose Bowl?

Solution Replacing m by 2 and n by 13 in the formula

$$p(E) = \frac{m}{n + m}$$

we obtain

$$p(E) = \frac{2}{13 + 2} = \frac{2}{15}$$

The probability that UCLA will play in the next Rose Bowl is $\frac{2}{15}$.
Do Exercise 15.   ∎

EXAMPLE 5 Suppose the probability of an event E is 0.21. What are the odds in favor of that event?

Solution When $p(E) = 0.21$, $p(E') = 0.79$ since $1 - 0.21 = 0.79$. Hence, the ratio $\frac{0.21}{0.79} = \frac{21}{79}$ gives the odds in favor of event E. The odds in favor of event E are 21 to 79.
Do Exercise 17.   ∎

Biased Sample Space

In many cases, the outcomes of an experiment are not equally likely; however, we can still assign probabilities to the members of a finite sample space E according to the following axioms:

> **Axiom 1:** If $E = \{O_1, O_2, O_3, \ldots, O_n\}$, then for each i, $0 \leq p(O_i) \leq 1$, where $p(O_i)$ is the probability of outcome O_i.
> **Axiom 2:** $p(O_1) + p(O_2) + p(O_3) + \cdots + p(O_n) = 1$.

EXAMPLE 6 Six teams are playing in a basketball tournament. The probabilities of winning the tournament for each of the first five teams is given in the following table.

Team	Probability of winning the tournament
A	.12
B	.21
C	.08
D	.35
E	.09

What is the probability that team F wins the tournament?

Solution Let $p(F)$ denote the probability that team F wins the tournament. By Axiom 2, we have

$$.12 + .21 + .08 + .35 + .09 + p(F) = 1$$

Thus,

$$.85 + p(F) = 1$$

and

$$p(F) = 1 - .85$$
$$= .15$$

The probability that team F wins the tournament is 0.15.
Do Exercise 21.   ∎

Exercise Set 6.1

*The data given in the tables of Exercises 1–3 are cumulative,
as were the data given in the table of Example 1.*

1. An electronics company is testing new picture tubes that it
manufactured in the last 12 months. Its quality control division randomly chose tubes manufactured during that period
and came up with the following results.

Number of tubes tested n	Number of defective tubes s_n
100	3
200	5
300	8
400	11
500	15
600	17
700	22
800	25
900	29
1000	32
1100	35
1200	38
1300	40
1400	44
1500	47

a. Calculate the relative frequency for each of the values
of n.
b. Estimate the probability that a tube selected at random
will be defective.

2. A manufacturer is testing the reliability of its new electric
razor. Its quality control division randomly chose razors
manufactured during the past year and came up with the
following results.

Number of razors tested n	Number of defective razors s_n
100	4
200	9
300	11
400	17
500	22
600	23
700	29
800	34
900	36
1000	41
1100	47
1200	49
1300	55
1400	57
1500	62

a. Calculate the relative frequency for each of the values of
n.
b. Estimate the probability that a razor selected at random
will be defective.

3. A company is manufacturing and testing a new hair dryer.
Its quality control division randomly tested dryers manufactured during the past 6 months and came up with the
following results.

Number of dryers tested n	Number of defective dryers s_n
200	4
400	9
600	11
800	17
1000	21
1200	26
1400	30
1600	34
1800	35
2000	39

   a. Calculate the relative frequency for each of the values of n.

   b. What is the best estimate of the probability that a dryer selected at random will be defective?

In Exercises 4–7, suppose two fair dice are rolled.

4. What are the chances that the sum of the faces is 4?

5. What are the chances that the sum of the faces is 6?

6. What are the chances that the sum of the faces is 7?

7. What are the chances that the sum of the faces is 8?

8. Three fair dice are rolled. What are the chances that the sum of the faces is 4?

9. Three fair dice are rolled. What are the chances that the sum of the faces is 5?

10. Two fair dice are rolled. Represent the following situations in terms of events and calculate the probability of each event.

   a. The sum of the faces is 4.
   b. The sum of the faces is 6.
   c. The sum of the faces is 4 or 6.
   d. The sum of the faces is 4 and 6.
   e. The sum of the faces is not 4.

11. Two fair dice are rolled. Represent the following situations in terms of events and calculate the probability of each event.

   a. The sum of the faces is 8.
   b. The sum of the faces is 9.
   c. The sum of the faces is 8 or 9.
   d. The sum of the faces is 8 and 9.
   e. The sum of the faces is not 8.

12. A fair coin is tossed twice. Since a coin comes up heads (H) or tails (T), each possible outcome is an ordered pair where each entry is either H or T. Thus the sample space has four members.

   a. Describe the sample space.

   b. Represent the following situations in terms of events and calculate the probability of each event.
     (i) Exactly one toss comes up heads.
     (ii) At least one toss comes up heads.
     (iii) Both are heads.
     (iv) They are different.

13. A fair coin is tossed three times. Since a coin comes up heads (H) or tails (T), each possible outcome is an ordered triple where each entry is either H or T. Thus the sample space has eight members.

   a. Describe the sample space.
   b. Represent each of the following situations in terms of events and calculate the probability of each event.
     (i) Exactly one toss comes up heads.
     (ii) At least one toss comes up heads.
     (iii) All three tosses are heads.
     (iv) At least two of the tosses are heads.

14. If the odds in favor of the Chicago Bears going to the next Super Bowl are 3 to 10, what is the probability that the Bears will go to the next Super Bowl?

15. If the odds in favor of the Boston Celtics winning the next NBA championship are 4 to 20, what is the probability that the Celtics will not win the next NBA championship?

16. If the odds in favor of a stock going up in value tomorrow are 5 to 3, what is the probability that its value will increase tomorrow?

17. If the probability that the Seattle Seahawks will win the next Super Bowl is 0.08, what are the odds in favor of the Seahawks winning the next Super Bowl?

18. If the probability that the Montreal Expos win the next World Series is 0.06, what are the odds in favor of the Expos winning the next World Series?

19. If the probability that John wins the next racquetball championship at his club is 0.35, what are the odds of John winning the championship?

20. If the probability that Betty wins the next golf tournament at her club is 0.24, what are the odds of Betty winning the tournament?

21. Five players are entered in a racquetball tournament. The probabilities that players A, B, C, or D win the tournament are 0.23, 0.13, 0.10, and 0.35, respectively. What is the probability that player E wins the tournament?

22. Eight players are entered in a table tennis tournament. The probabilities that players A, B, C, D, E, F, and G win the tournament are 0.13, 0.03, 0.07, 0.11, 0.06, 0.09, and 0.12, respectively. Show that player H is a heavy favorite to win the tournament.

6.2 Counting Techniques

In the preceding section, we explained how a probability function can be defined when the sample space is finite. That is, in a uniform sample space with n elements, if E is an event with m elements, then the probability of E is $p(E) = \frac{m}{n}$. The examples given were so simple that we could list the elements of the sample space. In many applications, however, the sample spaces and events are large and it is necessary to be able to find the number of elements in these sets without actually listing them. Imagine, for example, that you wanted to know how many people were attending a certain show, given that the theater was full. If the theater had 75 rows with 50 seats each, you would not enumerate all the people but would conclude that 3750 people were attending the show since $75(50) = 3750$. We devote this section to a brief introduction to techniques that allow us to determine the number of elements in sets without actually listing them.

Tree Diagrams

EXAMPLE 1   Suppose we draw one marble from an urn containing a red, a green, and a white marble and then flip a coin. Describe the sample space.

Solution   We will use a scheme called a *tree diagram*. (See Figure 6.2.) Let R, G, and W denote the red, green, and white marbles, respectively; and let H and T denote heads and tails, respectively.

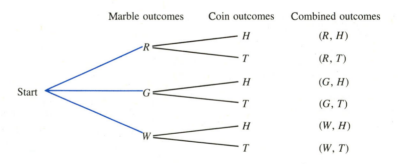

FIGURE 6.2

The set of combined outcomes has six members: (R, H), (R, T), (G, H), (G, T), (W, H), and (W, T). The first element of each pair represents the marble drawn and the second element represents the way the coin landed. There are three possible outcomes when we first draw a marble and two possible outcomes when we flip a coin; and $3 \cdot 2 = 6$.
Do Exercise 17.   ∎

EXAMPLE 2   A coin is tossed three times. Use a tree diagram to find all possible combined outcomes.

Solution   We let H and T represent heads and tails, respectively, and draw the following tree diagram. (See Figure 6.3.) Following the branches of the tree diagram, we get the following eight possible combined outcomes of heads and tails: (H, H, H), $(H, H,$

T), (H, T, H), (H, T, T), (T, H, H), (T, H, T), (T, T, H), (T, T, T). Note that there are two possible outcomes with each of the three tosses and that $2 \cdot 2 \cdot 2 = 8$.
Do Exercise 15.

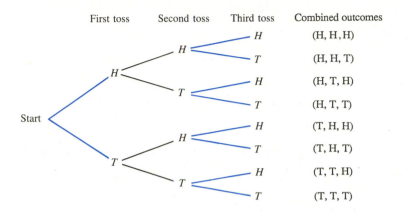

FIGURE 6.3

Although a tree diagram is a convenient way to list combined outcomes when we perform a sequence of experiments in order, it is not practical when the number of outcomes in an experiment is large. In such cases, we use the multiplication principle that we introduced earlier. For convenience, we state this principle again.

Suppose that we perform an experiment in two parts. The first part has m possible outcomes and, regardless of which of these occurs, the second part has n possible outcomes. Then, when the two parts of the experiment are performed, in that order, there are mn possible outcomes. This principle can be generalized as follows.

> **Fundamental Principle of Counting** Suppose that we perform an experiment in k parts. The first part has n_1 possible outcomes; regardless of which of these occurs, the second part has n_2 possible outcomes; regardless of which of these occurs, the third part has n_3 possible outcomes; and so on. Then when the k parts of the experiment are performed, in that order, the number of possible outcomes is $n_1 \cdot n_2 \cdot \cdots \cdot n_k$.

EXAMPLE 3   A license plate is made up of 3 of the 26 letters of the alphabet followed by 3 of the 10 digits. How many different license plates can we have if no letter or digit is repeated?

Solution   There are 26 ways to pick one letter. Then, since we don't allow repetition, there are 25 ways to pick a second letter. The third letter can be picked in 24 ways. Similarly, there are 10 ways to pick the first digit, 9 ways to pick the next digit, and 8 ways to pick the last digit. Therefore we can have 11,232,000 different license plates since

$$26 \cdot 25 \cdot 24 \cdot 10 \cdot 9 \cdot 8 = 11{,}232{,}000$$ ∎

We now derive some formulas that will be useful in calculating the number of elements in a set. For some of these formulas, we will need to recall the meaning of the symbol $n!$.

If n is a positive integer, the product $n(n - 1)(n - 2) \cdots (3)(2)(1)$ is abbreviated $n!$ (read: "n factorial").

For example,

$$3! = 3(3 - 1)(3 - 2) = 3 \cdot 2 \cdot 1 = 6$$

and

$$5! = 5(5 - 1)(5 - 2)(5 - 3)(5 - 4) = 5 \cdot 4 \cdot 3 \cdot 2 \cdot 1 = 120$$

As a consequence of this definition, we see that if k is a positive integer, then

$$(k + 1)! = (k + 1)k!$$

Although we have not defined $0!$, if we replace k by 0 in the foregoing formula, we obtain

$$1! = 1(0!)$$

That is,

$$1 = 0!$$

Therefore, we define $0!$ to be equal to 1. With this value attached to $0!$, the formula

$$(k + 1)! = (k + 1)k!$$

is true for any nonnegative integer k.

Permutations and Combinations

DEFINITION 6.3: Suppose we have a set of n distinct objects from which we select r members. The subset obtained is called a *combination of n things taken r at a time*. If the members of this subset are arranged in a definite order, then the ordered subset obtained is called a *permutation of n things taken r at a time*.

If n and r are integers such that $0 \le r \le n$, the symbol $C(n, r)$ denotes the number of combinations of n things taken r at a time and the symbol $P(n, r)$ denotes the number of permutations of n things taken r at a time. In the next example, we will show how to calculate $P(n, r)$ and $C(n, r)$ for small values of n and r.

EXAMPLE 4   Find the values of $P(7, 3)$ and $C(7, 3)$.

Solution   Given a set with seven members, to form a permutation with three members, we must fill three places without using the same object twice. There are seven ways to fill the first place. Once this has been done, there are six ways to fill the second place. When the first two places have been filled, there are five ways to fill the third place. Using the multiplication principle, the total number of distinct three-member permutations that can be formed is 210 since $7 \cdot 6 \cdot 5 = 210$. Thus,

$$P(7, 3) = 210$$

Note that, multiplying and dividing by 4!, we obtain

$$P(7, 3) = 7 \cdot 6 \cdot 5 = \frac{7 \cdot 6 \cdot 5 \cdot 4 \cdot 3 \cdot 2 \cdot 1}{4 \cdot 3 \cdot 2 \cdot 1} = \frac{7!}{4!} = \frac{7!}{(7 - 3)!}$$

To obtain any one of the 210 permutations of seven things taken three at a time, we do two things in order. We pick a subset with three members. There are $C(7, 3)$ ways to do this. We then must arrange that subset in some order. There are three ways to fill the first place, two ways to fill the second place, and only one way to fill the third place. Thus, for each of the subsets with three members, we can make six ordered subsets since $3 \cdot 2 \cdot 1 = 6$. Therefore, using the multiplication principle, there are $C(7, 3) \cdot 6$ ways to get a permutation of seven things taken three at a time. Since we already know that this number is 210, we can write

$$C(7, 3) \cdot 6 = 210$$

We get

$$C(7, 3) = \frac{210}{6} = 35$$

We could have written

$$C(7, 3) = \frac{P(7, 3)}{3!} = \frac{7!}{(7 - 3)!(3!)} \qquad \blacksquare$$

Using exactly the same reasoning as in the previous example, we get the following formulas.

If n and r are integers with $0 \le r \le n$, then

$$P(n, r) = \frac{n!}{(n - r)!}$$

and

$$C(n, r) = \frac{n!}{(n - r)!r!}$$

If $r = n$, then

$$P(n, n) = \frac{n!}{(n - n)!} = \frac{n!}{0!} = \frac{n!}{1} = n!$$

Thus, there are $n!$ permutations of n things taken n at a time.

EXAMPLE 5   In how many ways can the positions of president, vice president, and secretary be filled, in that order, in a club of 15 members if no member can hold more than one position?

Solution   The number of ways the positions can be filled is the number of permutations of 15 things take 3 at a time, namely

$$P(15, 3) = \frac{15!}{(15 - 3)!} = \frac{15 \cdot 14 \cdot 13 \cdot 12!}{12!} = 15 \cdot 14 \cdot 13 = 2730$$

Do Exercise 27.   ▪

EXAMPLE 6   How many lines are determined by five points, no three of which are collinear?*

Solution   Since two distinct points determine a line, we are interested in finding how many two-point subsets there are in the given set. That number is

$$C(5, 2) = \frac{5!}{(5 - 2)!2!} = \frac{5!}{3!2!} = \frac{5 \cdot 4 \cdot 3!}{3!2!} = \frac{20}{2} = 10$$

Therefore, ten lines are determined by the five points.
Do Exercise 29.   ▪

 It is important to remember that when determining the number of ways a subset of a given set may be selected, we use permutations if the subset is ordered, and if the subset is not ordered, we use combinations.**

EXAMPLE 7   A club membership is composed of 42 women and 37 men. How many different committees of seven members can be formed with the following restrictions:

a. The committee must have four women and three men.
b. Women must have the majority on the committee.

Solution   **a.** In forming a committee of four women and three men, we do two things. We first choose 4 women among the 42 women. Since the members chosen need not be ordered, we use combinations. The number of ways we can choose the four women is $C(42, 4)$. We then choose 3 men among the 37 male members. Again, these need not be ordered. Thus, we use combinations to find that the number of ways we can choose the three men is $C(37, 3)$. By the fundamental principle of counting, the number of ways we can choose the four women and the three men is $C(42, 4) \cdot C(37, 3)$. Using the formula to calculate $C(n, r)$, we get

$$C(42, 4) \cdot C(37, 3) = \frac{42!}{(42 - 4)!4!} \cdot \frac{37!}{(37 - 3)!3!}$$
$$= \frac{42 \cdot 41 \cdot 40 \cdot 39 \cdot 38!}{38!4!} \cdot \frac{37 \cdot 36 \cdot 35 \cdot 34!}{34!3!}$$

---

*Points are collinear if they lie on the same straight line.
**Each time we select a subset A with r elements, given a set S with n elements, we get a corresponding subset, the complement of A in S with $n - r$ elements. Therefore, the number of subsets with r elements is the same as the number of subsets with $n - r$ elements. That is,

$$C(n, r) = C(n, n - r)$$

This equality can be verified using the formula for $C(n, r)$. (See Exercise 19.)

$$= \frac{42 \cdot 41 \cdot 40 \cdot 39}{4!} \cdot \frac{37 \cdot 36 \cdot 35}{3!}$$

$$= 111{,}930 \cdot 7770$$

$$= 869{,}696{,}100$$

Therefore, 869,696,100 different committees of four women and three men can be formed. Whether we choose the women first or the men is irrelevant since

$$C(42, 4) \cdot C(37, 3) = C(37, 3) \cdot C(42, 4)$$

b. In a committee of seven members, the women will have the majority if there are four, five, six, or seven women on that committee. We already know that we can form 869,696,100 committees with four women and three men. In exactly the same manner, we calculate the number of committees with five women and two men. We get

$$C(42, 5) \cdot C(37, 2) = 850{,}668 \cdot 666 = 566{,}544{,}888$$

The number of committees with six women and one man is

$$C(42, 6) \cdot C(37, 1) = 5{,}245{,}786 \cdot 37 = 194{,}094{,}082$$

The number of committees with seven women and no men is, of course, $C(42, 7) = 26{,}978{,}328$. To find the total number of different committees with a majority of women, we simply add the results we just obtained to get

$$869{,}696{,}100 + 566{,}544{,}888 + 194{,}094{,}082 + 26{,}978{,}328 = 1{,}657{,}313{,}398$$

Therefore, 1,657,313,398 committees of seven members with a majority of women can be formed.

Do Exercise 37. ■

In discussing permutations of n things taken r at a time, we have assumed that the n things were distinct. We may encounter situations where we have a set of objects not all of which are distinguishable from one another. Suppose we have a set of n elements consisting of only two types of objects. There are n_1 indistinguishable objects of type I and n_2 indistinguishable objects of type II with $n_1 + n_2 = n$. If all n objects were distinct, there would be $n!$ permutations. However, some of these permutations are indistinguishable. Let $P(n; n_1, n_2)$ denote the number of distinguishable permutations. Assume that we have labeled the objects of type I and type II so that all objects are now distinguishable. Then with each of the original indistinguishable permutations, we can form $n_1!n_2!$ distinguishable permutations because the objects of type I can be arranged in $n_1!$ ways and those of type II can be arranged in $n_2!$ ways. Thus, the total number of distinguishable permutations is $P(n; n_1, n_2) \cdot n_1! \cdot n_2!$. Consequently,

$$P(n; n_1, n_2) \cdot n_1! \cdot n_2! = n!$$

and

$$P(n; n_1, n_2) = \frac{n!}{n_1! \cdot n_2!}$$

This formula may be generalized for cases where we have sets with more than two subsets of like elements.

Suppose we have a set with n elements and k subsets whose union is the given set. The first subset is of type I and has n_1 indistinguishable elements, the second subset is of type II and has n_2 indistinguishable elements, and the kth subset is of type k and has n_k indistinguishable elements. Consequently, $n_1 + n_2 + \cdots + n_k = n$.

> The number $P(n; n_1, n_2, \ldots, n_k)$ of distinct permutations is given by
>
> $$P(n; n_1, n_2, \ldots, n_k) = \frac{n!}{n_1! n_2! \ldots n_k!}$$

EXAMPLE 8   A store has three identical A television sets, four identical B television sets, and two identical C television sets. At the beginning of the Christmas sale, the store manager puts these nine sets on display by arranging them against one of the walls in the store. In how many different ways can he do this?

Solution   Since he has nine sets, of which the three A sets are indistinguishable, the four B sets are indistinguishable, and the two C sets are also indistinguishable, the number of distinct ways that the manager can arrange the nine sets against a wall is 1260 since

$$P(9; 3, 4, 2) = \frac{9!}{3! 4! 2!} = 1260$$

Do Exercise 23.   ■

Set Partitioning

We conclude this section with a brief discussion of set partitioning. For example, when four players are playing bridge and a hand is dealt using all 52 cards, an ordered partition (H_1, H_2, H_3, H_4) of the deck of cards is obtained where each H_i has 13 members. Suppose that you want to know how many different hands can be dealt. To deal with such problems, we need a formula that will allow us to find the number of distinct ordered partitions of a set.

> **DEFINITION 6.4:**   Given a set S with n members, an *ordered partition* of S is an ordered collection of subsets $H_1, H_2, H_3, \ldots, H_k$ of S such that each member of S is in exactly one of the subsets H_i.

Suppose that H_1 has r_1 members, H_2 has r_2 members, H_3 has r_3 members, $\ldots$, and H_k has r_k members, so that $r_1 + r_2 + r_3 + \cdots + r_k = n$. If we let

$C(n; r_1, r_2, \ldots, r_k)$ denote the number of such ordered partitions, then it can be shown that

$$C(n; r_1, r_2, \ldots, r_k = \frac{n!}{r_1! r_2! \cdots r_n!}\ *$$

EXAMPLE 9 In how many different ways can 15 salespersons be assigned to 3 departments if 5 must be assigned to department A, 7 to department B, and 3 to department C?

Solution Any such assignment is an ordered partition (H_A, H_B, H_C) of the set of 15 salespersons where H_A has five elements, H_B has seven elements, and H_C has three elements. The number of ordered partitions is 360,360 since

$$\frac{15!}{5!7!3!} = 360{,}360$$

Do Exercise 41.

Exercise Set 6.2

In Exercises 1–14, evaluate the given expression.

1. 7!

2. $\dfrac{12!}{8!}$

3. $\dfrac{15!}{9!3!3!}$

4. $P(5, 3)$

5. $P(6, 6)$

6. $P(13, 6)$

7. $C(5, 3)$

8. $C(6, 2)$

9. $C(13, 7)$

10. $C(7, 2)$ and $C(7, 5)$

11. $C(10, 3)$ and $C(10, 7)$

12. $C(20, 7)$ and $C(20, 13)$

13. $P(10; 6, 4)$

14. $P(20; 4, 5, 3, 8)$

15. A coin is tossed four times. With the help of a tree diagram, list all possible combined outcomes.

16. A coin is tossed twice, then a die is rolled once. With the help of a tree diagram, list all possible combined outcomes.

17. A coin is tossed three times. Then a marble is drawn from an urn containing a red marble, a green marble, and a blue marble. With the help of a tree diagram, list all possible combined outcomes.

18. An urn contains marbles numbered 1, 2, 3, and 4. A marble is drawn, then a coin is tossed twice. With the help of a tree diagram, list all possible combined outcomes.

19. Using the formula

$$C(n, k) = \frac{n!}{(n - k)!k!}$$

show that if n and r are integers with $0 \le r \le n$, then

$$C(n, r) = C(n, n - r)$$

20. How many permutations of the three letters x, y, and z are there? List all of them.

21. How many permutations of the letters a, b, c, and d are there? List all of them.

22. How many distinct letter arrangements can be made using all the letters in the word *Mississippi*?

---

*Note the similarity of this formula to the formula giving the number of distinguishable permutations of n objects when some of these objects are indistinguishable. Also if $k = 2$, then $r_2 = n - r_1$ and the foregoing formula may be written

$$C(n; r_1, r_2) = \frac{n!}{r_1!(n - r_1)!}$$

This is precisely the formula giving the number of combinations of n things taken r_1 at a time.

23. Repeat Exercise 22 with the word *mathematics*.

24. Repeat Exercise 22 with the word *statistics*.

25. How many different signals, each consisting of seven flags placed in a horizontal line, can be formed using a set of three indistinguishable blue flags and four indistinguishable red flags?

26. Repeat Exercise 25 with a set of four indistinguishable blue flags, five indistinguishable red flags, and six indistinguishable green flags.

27. In how many ways can the positions of president, vice president, and secretary be filled in a club of 13 members if no member can hold more than one position?

28. How many lines are determined by seven points, no three of which are collinear?

29. Repeat Exercise 28 with ten points.

30. How many planes are determined by eight points, no four of which are coplanar* and no three of which are collinear? (*Hint:* Each three points determine a plane.)

31. Repeat Exercise 30 with 12 points.

32. How many permutations of the letters of the word *hypersonic* are there? If you could write one of these every 5 seconds, how long would it take you to list them all?

33. A foreman is in charge of 12 machines and has 12 workers to assign to these machines, one worker per machine. In how many different ways can this be done? Suppose that the foreman wishes to make the assignments in the best possible way and thinks about each of the possible assignments for two seconds. How long would it take him? Does the answer surprise you?

34. Repeat Exercise 33 with 15 machines and 15 workers.

35. Six soldiers are selected from a squad of eight for patrol duty. In how many ways can this be done?

36. How many committees of five can be formed from a group of six Republicans and three Democrats if each committee is to have at least three Republicans?

37. A club membership is composed of 18 women and 20 men. How many committees of five can be formed if **a.** exactly three committee members must be men? **b.** men must have the majority on the committee?

38. A basketball coach has three centers, six forwards, and five guards available. She must field a team made up of one center, two forwards, and two guards. In how many ways can this be done?

39. Repeat Exercise 38 if there are two centers, seven forwards, and six guards available.

40. In how many ways can ten cooks be assigned to three kitchens if five cooks are needed for kitchen I, two cooks are needed for kitchen II, and three cooks are needed for kitchen III?

41. In how many distinct ways can 5-card hands be dealt to 6 poker players from a standard 52-card deck? (*Hint:* Number the players 1, 2, 3, 4, 5, and 6, then each hand corresponds to an ordered partition $(H_1, H_2, H_3, H_4, H_5, H_6, B)$ of the deck into seven subsets where each H_i is the hand player i holds and B is the set of cards left after the hand is dealt.)

42. A publishing company has 25 salespersons. In how many ways can they be assigned to four territories if eight are assigned to the northern territory, seven are assigned to the eastern territory, six are assigned to the western territory, and four are assigned to the southern territory?

6.3 Conditional Probability

When we consider a probability problem for which the sample space is finite and all outcomes are equally likely, we need only to be able to calculate the number of elements in the sample space and the number of elements in an event to find the probability of that event. The preceding section contains formulas that enable us to find these numbers without actually listing all the outcomes. Mathematicians have formulated another method. There are ways to express a complex probability problem in terms of simpler ones. We briefly describe some of these techniques.

Frequently, an event is simply a combination of events. For example, if one die is thrown twice, the experiment is to throw it once, and then, after this is done, to throw

---

*Points that lie on the same plane are said to be coplanar.

it again. Each throw is independent of the other. This is not always the case, however, with sequential events. To best formulate the basic techniques of this section, we must distinguish between events that are dependent, independent, or mutually exclusive. Before giving the formal definitions, we give some examples.

Suppose one card is drawn from a standard deck of cards and then another is drawn without replacing the first. What is the probability that both are spades? Clearly, if the first card is a spade, then the probability of the second card being a spade is $\frac{12}{51}$ (since there are 12 spades left among 51 cards). On the other hand, if the first card is not a spade, then the probability of the second being a spade is $\frac{13}{51}$. We see that the probability of the second card being a spade depends on the outcome of the first draw and therefore the two events are dependent.

However, suppose a card is drawn, replaced in the deck, and then the deck is shuffled before a second card is drawn. Then, the probability that the second card is a spade is $\frac{13}{52}$, regardless of the outcome of the first drawing; the two events are independent.

Suppose a single card is drawn from a deck. The event "it is a heart" and the event "it is a club" are mutually exclusive since one cannot happen if the other does. However, the event "it is a heart" and the event "it is a red card" are alternative outcomes of a single draw that are not mutually exclusive since they can both happen.

Conditional Probability

In the first example, in which two cards were drawn without replacement, we saw that the probability that the second card is a spade given that the first was a spade is $\frac{12}{51}$. For two events A and B, $p(A \mid B)$ denotes the probability that A occurs given that B does occur. Thus, if U is the sample space, $p(A \mid U)$ is synonymous to $p(A)$. Recall that if C is a finite set, $n(C)$ denotes the number of elements in C. For example, if $X = \{2, 4, 5\}$, then $n(X) = 3$. Now suppose that U is a finite sample space in which all outcomes are equally likely and let A and B be events in U. Then, to calculate $p(A \mid B)$, we do not consider points in $A \backslash B$ because, for any of these outcomes, B failed to occur. Thus, we consider only points in $A \cap B$. (See Figure 6.4.) Since B does occur, it plays the role of the sample space and we have

$$p(A \mid B) = \frac{n(A \cap B)}{n(B)}$$

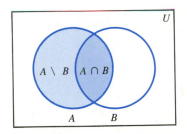

FIGURE 6.4

But,

$$p(B) = \frac{n(B)}{n(U)}$$

Therefore,

$$p(A \mid B) \cdot p(B) = \frac{n(A \cap B)}{n(B)} \cdot \frac{n(B)}{n(U)} = \frac{n(A \cap B)}{n(U)}$$

and

$$p(A \mid B) \cdot p(B) = p(A \cap B)$$

A CLOSER LOOK   This formula was derived when we had a finite uniform sample space. Suppose we have a finite sample space U in which the outcomes are not necessarily equally likely. Let

$$U = \{o_1, o_2, o_3, \ldots, o_n\}$$

If $p(o_i)$ denotes the probability that outcome o_i occurs, we have

$$0 \leq p(o_i) \leq 1$$

for each i, and

$$p(o_1) + p(o_2) + p(o_3) + \cdots + p(o_n) = 1$$

Now suppose that A and B are events in U. Again let $p(A \mid B)$ denote the probability that event A occurs given that event B has occurred. We must assign new probabilities to the original outcomes. We assign the probability 0 to any member of U that is not in B since it is given that B has occurred and, therefore, any outcome that is not in B cannot occur. Consequently, if o_i is not in B, we let $p(o_i \mid B) = 0$. We also treat the event B as our new sample space. It is natural to assign the conditional probabilities to elements of B in the same ratio as their probabilities were assigned in the original sample space. That is, if o_r and o_s are in B, then

$$\frac{p(o_r \mid B)}{p(o_s \mid B)} = \frac{p(o_r)}{p(o_s)}$$

Without loss of generality, we may assume that the elements of U have been labeled so that $B = \{o_1, o_2, o_3, \ldots, o_q\}$ and $U \backslash B = \{o_{q+1}, o_{q+2}, o_{q+3}, \ldots, o_n\}$. Then, for any integer i between 1 and q, inclusive, we have

$$\frac{p(B)}{p(o_i)} = \frac{p(o_1) + p(o_2) + p(o_3) + \cdots + p(o_q)}{p(o_i)}$$

$$= \frac{p(o_1)}{p(o_i)} + \frac{p(o_2)}{p(o_i)} + \frac{p(o_3)}{p(o_i)} + \cdots + \frac{p(o_q)}{p(o_i)}$$

$$= \frac{p(o_1 \mid B)}{p(o_i \mid B)} + \frac{p(o_2 \mid B)}{p(o_i \mid B)} + \frac{p(o_3 \mid B)}{p(o_i \mid B)} + \cdots + \frac{p(o_q \mid B)}{p(o_i \mid B)}$$

$$= \frac{p(o_1 \mid B) + p(o_2 \mid B) + p(o_3 \mid B) + \cdots + p(o_q \mid B)}{p(o_i \mid B)} = \frac{1}{p(o_i \mid B)}$$

We have shown that $p(B)/p(o_i) = 1/p(o_i \mid B)$. Consequently, we have $p(o_i \mid B) = p(o_i)/p(B)$. That is, the conditional probability of any outcome o_i in B is obtained by multiplying the original probability of o_i by $1/p(B)$. Note the following:

1. We have assumed that $p(o_i) \neq 0$. In the case that $p(o_i) = 0$, we let $p(o_i \mid B) = 0$.

2. We have used the fact that since $B = \{o_1, o_2, \ldots, o_q\}$,

$$p(o_1) + p(o_2) + p(o_3) + \cdots + p(o_q) = p(B)$$

3. We have also used the fact that since we treat B as our new sample space, then

$$p(o_1 \mid B) + p(o_2 \mid B) + p(o_3 \mid B) + \cdots + p(o_q \mid B) = 1$$

Again let us assume that the elements of U have been labeled so that $A \cap B = \{o_1, o_2, \ldots, o_m\}$. When we calculate the conditional probability $p(A \mid B)$, we must add the conditional probabilities of those outcomes which are in both A and B (remember B is the new sample space). Thus,

$$p(A \mid B) = p(o_1 \mid B) + p(o_2 \mid B) + p(o_3 \mid B) + \cdots + p(o_m \mid B)$$

$$= \frac{p(o_1)}{p(B)} + \frac{p(o_2)}{p(B)} + \frac{p(o_3)}{p(B)} + \cdots + \frac{p(o_m)}{p(B)}$$

$$= \frac{p(o_1) + p(o_2) + p(o_3) + \cdots + p(o_m)}{p(B)}$$

$$= \frac{p(A \cap B)}{p(B)}$$

and

$$p(A \mid B) \cdot p(B) = p(A \cap B)$$

Note that this is exactly the same relationship as we obtained when all outcomes in U were equally likely. Therefore, we have the following definition.

DEFINITION 6.5: Suppose that A and B are events in a sample space U. Then the *conditional probability* of A given that B has occurred, denoted $p(A \mid B)$, is defined by

$$p(A \mid B) = \frac{p(A \cap B)}{p(B)}, \text{ provided that } p(B) \neq 0$$

Similarly,

$$p(B \mid A) = \frac{p(A \cap B)}{p(A)}, \text{ provided that } p(A) \neq 0$$

EXAMPLE 1   Suppose a bag contains six red marbles and four blue marbles, and a second bag contains four red marbles and five blue marbles. A marble is drawn from the first bag and placed in the second. The second bag is shaken and a marble is drawn from it. What is the probability that a red marble was drawn each time?

Solution   An outcome is drawing one marble and then another. It can be represented by an ordered pair (x, y). Let

$$A = \{(x, y) \mid (x, y) \in U \text{ and } x \text{ is a red marble}\}$$

and

$$B = \{(x, y) \mid (x, y) \in U \text{ and } y \text{ is a red marble}\}$$

Then the probability that a red marble was drawn each time is obtained as follows:

$$p(A \cap B) = p(A) \cdot p(B \mid A)$$
$$= \frac{6}{10} \cdot \frac{5}{10} = \frac{3}{10}$$

We know $p(A) = \frac{6}{10}$ since six of the ten marbles in the first bag were red. Also $p(B \mid A) = \frac{5}{10}$, since if the first marble drawn is red and placed in the second bag, then five of the ten marbles in that bag are red.

Do Exercise 1.   ■

EXAMPLE 2   The probability that a freshman takes an English course is .62. The probability of taking a mathematics course is .35 and the probability of taking both is .23.

a. What is the probability that a freshman will take an English course given that a mathematics course is also taken by that student?
b. What is the probability that a freshman will take a mathematics course given that an English course is also taken by that student?

Solution   Let E denote the event that the student takes an English course and M be the event that the student takes a mathematics course. It is given that $p(E) = .62$, $p(M) = .35$, and $p(E \cap M) = .23$.

a. The probability that the student takes an English course given that a mathematics course is taken is $p(E \mid M)$; the value for which is obtained as follows:

$$p(E \mid M) = \frac{p(E \cap M)}{p(M)} = \frac{.23}{.35} \doteq .6571$$

b. The probability that the student takes a mathematics course given that an English course is taken is $p(M \mid E)$, the value for which is obtained as follows:

$$p(M \mid E) = \frac{p(E \cap M)}{p(E)} = \frac{.23}{.62} \doteq .371.$$

Do Exercise 3a, b.   ■

Independent Events

DEFINITION 6.6:   If A and B are events and $p(A) = p(A \mid B)$, we say that A and B are *independent events*. In that case, we have

$$p(A \cap B) = p(A) \cdot p(B)$$

EXAMPLE 3   An urn contains three white marbles and one black marble. A marble is drawn from the urn and replaced, the urn is shaken, and a marble is drawn again. What is the probability that both marbles drawn are white?

Solution Since before the second drawing the original situation was restored, the two events are independent. Each time, the probability of drawing a white marble is $\frac{3}{4}$. Therefore the probability of drawing a white marble both times is $\frac{3}{4} \cdot \frac{3}{4} = \frac{9}{16}$.

Do Exercise 5. ■

Mutually Exclusive Events

DEFINITION 6.7: If U is a sample space and A and B are disjoint subsets of U, then A and B are said to be *mutually exclusive events*. (See Figure 6.5.)

FIGURE 6.5

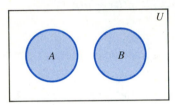

If A and B are mutually exclusive events, then

$$p(A \cup B) = p(A) + p(B)$$

EXAMPLE 4 Suppose a card is drawn at random from a standard deck of cards. What is the probability that it is either a king or a queen?

Solution The probability of drawing a king is $\frac{4}{52} = \frac{1}{13}$. The probability of drawing a queen is also $\frac{1}{13}$. Hence, the probability of drawing a king or a queen is $\frac{1}{13} + \frac{1}{13} = \frac{2}{13}$, since those two events are mutually exclusive.

Do Exercise 9. ■

If two events A and B are not mutually exclusive, and we calculate the probability that one or the other will occur, we use the formula

$$p(A \cup B) = p(A) + p(B) - p(A \cap B)$$

To verify this, suppose the outcomes have been labeled in such a way that

$$A \cap B = \{c_1, c_2, \cdots, c_k\}$$
$$A = \{c_1, c_2, \cdots, c_k, a_1, a_2, \cdots, a_m\}$$
$$B = \{c_1, c_2, \cdots, c_k, b_1, b_2, \cdots, b_n\}$$

Then,

$$A \cup B = \{c_1, c_2, \cdots, c_k, a_1, a_2, \cdots, a_m, b_1, b_2, \cdots, b_n\}$$

Thus,

$$p(A \cup B) = p(c_1) + p(c_2) + \cdots + p(c_k) + p(a_1)$$
$$+ p(a_2) + \cdots + p(a_m) + p(b_1) + p(b_2) + \cdots + p(b_n)$$
$$p(A) = p(c_1) + p(c_2) + \cdots + p(c_k)$$
$$+ p(a_1) + p(a_2) + \cdots + p(a_m)$$
$$p(B) = p(c_1) + p(c_2) + \cdots + p(c_k) + p(b_1) + p(b_2) + \cdots + p(b_n)$$
$$p(A \cap B) = p(c_1) + p(c_2) + \cdots + p(c_k)$$

It is now easy to verify that

$$p(A \cup B) = p(A) + p(B) - p(A \cap B)$$

EXAMPLE 5   A card is drawn from a standard deck of cards. What is the probability that it is either a spade or a king?

Solution   Here the sample space is the set of 52 cards. Let $A = \{x \mid x$ is a spade$\}$ and $B = \{x \mid x$ is a king$\}$. Then clearly $A \cap B \neq \emptyset$. In fact, $A \cap B$ has exactly one member, the king of spades. Thus, $p(A \cap B) = \frac{1}{52}$, $p(A) = \frac{13}{52} = \frac{1}{4}$, and $p(B) = \frac{4}{52} = \frac{1}{13}$. It follows that

$$p(A \cup B) = \frac{1}{4} + \frac{1}{13} - \frac{1}{52} = \frac{13 + 4 - 1}{52} = \frac{16}{52} = \frac{4}{13}$$

Do Exercise 11.   ∎

EXAMPLE 6   What is the probability that the freshman of Example 2 takes both English and mathematics given that at least one of the two is taken?

Solution   We first find the probability that the student takes at least one of the two subjects. We get

$$p(E \cup M) = p(E) + p(M) - p(E \cap M) = .62 + .35 - .23 = .74$$

Thus,

$$p(E \cap M \mid E \cup M) = \frac{p(E \cap M)}{p(E \cup M)} = \frac{.23}{.74} \doteq .3108$$

Do Exercise 3c.   ∎

Sequence of Dependent Events

The following theorem allows us to solve problems in which we have a sequence of k dependent events.

THEOREM 6.1: Suppose we have a sequence $E_1, E_2, \ldots, E_k$ of k dependent events. The probability of the first event E_1 occurring is p_1. After the first event has occurred, the probability of the second event E_2 occurring is p_2; after these two events have occurred, the probability of the third event E_3 occurring is p_3, and so on, then the probability that all k events occur, in that order, is $p_1 p_2 p_3 \cdots p_k$.

When applying this theorem to certain problems, the number of events that are considered failures may be much smaller than the number of events that are considered successes. In such cases, it is simpler to first calculate the probability of failure, then subtract the result from 1 to get the probability of success.

EXAMPLE 7   Suppose 23 people have been randomly selected. What is the probability that at least two of them celebrate their birthdays the same day?

Solution   If "at least two of the people celebrate their birthdays the same day" is success, failure is "no two people celebrate their birthdays the same day." That is, all 23 people have birthdays falling on different days of the year. We assume that each year has 365 days and that it is equally likely for a person to be born on any of these days.

Let us name these people $P_1, P_2, P_3, \ldots, P_{23}$. The 23 people will celebrate their birthdays on different days if the following sequence of events occurs. P_1 celebrates his/her birthday on some day of the year; this has probability 1. Once this happens, P_2 celebrates his/her birthday on any of the other 364 days and this event has probability $\frac{364}{365}$. Now P_3 must celebrate his/her birthday on any day of the year except the days that P_1 and P_2 celebrate their birthdays, and the probability of this event is $\frac{363}{365}$. Continuing in this manner, and using Theorem 6.1, we find that the probability that no 2 of the 23 people celebrate their birthdays the same day is

$$\frac{365}{365} \cdot \frac{364}{365} \cdot \frac{363}{365} \cdot \cdots \cdot \frac{343}{365} \doteq .4927$$

Therefore, the probability that at least 2 of the 23 people celebrate their birthdays the same day is .5073 since $1 - .4927 = .5073$. Does this result surprise you?
Do Exercise 15. ∎

It is often useful in a probability problem to represent the situation pictorially. Venn diagrams and probability trees may be used, as illustrated in the next three examples.

Venn Diagrams

EXAMPLE 8   A sociologist theorized that couples with religious affiliations had a better chance to remain married than those without affiliations. To test his theory, he conducted a study of 1000 randomly selected couples who had married ten years previously. He let U be the universe (the set of married couples he surveyed), R be the set of all married couples who had religious affiliations, and M be the set of all couples who were still married. He then drew the following Venn diagram. (See Figure 6.6.)

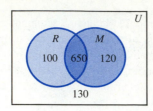

FIGURE 6.6

a. Find the probability that a couple with a religious affiliation is still married.
b. Find the probability that a couple who is still married has a religious affiliation.
c. Find the probability that a couple with a religious affiliation is divorced.

Solution a. We must find the probability that a couple is still married given that it has a religious affiliation. That is, we must find $p(M \mid R)$. From the Venn diagram, we readily see that

$$n(U) = 1000$$
$$n(M \cap R) = 650$$
$$n(M) = 770$$
$$n(R) = 750$$

Thus,

$$p(M \cap R) = \frac{650}{1000} = .65$$

$$p(R) = \frac{75}{1000} = .75$$

It follows that

$$p(M \mid R) = \frac{p(M \cap R)}{p(R)} = \frac{.65}{.75} \doteq .8667$$

Thus, if we know that a couple has a religious affiliation, there is a probability of .8667 that they are still married.

b. We must find $p(R \mid M)$. Since $p(R \mid M) = \frac{p(R \cap M)}{p(M)}$, we first calculate $p(M)$. From the Venn diagram we see that

$$n(M) = 770$$

Hence,

$$p(M) = \frac{770}{1000} = .77$$

Thus,

$$p(R \mid M) = \frac{p(R \cap M)}{p(M)} = \frac{.65}{.77} \doteq .8442$$

Therefore, if we know that a couple is still married, the probability that the couple has a religious affiliation is .8442.

c. Since M represents the set of all couples who are still married, M' represents the set of all couples who are divorced. (Here, we assume that all couples studied are still married or divorced, and that none of the people studied has died.) From the Venn diagram we see that $n(M') = 230$ since $130 + 100 = 230$. Thus,

$$p(M') = \frac{230}{1000} = .23$$

We also see that

$$n(M' \cap R) = 100$$

Thus,

$$p(M' \cap R) = \frac{100}{1000} = .1$$

Hence,

$$p(M' \mid R) = \frac{p(M' \cap R)}{p(R)} = \frac{.1}{.75} \doteq .1333$$

Therefore, if a couple has a religious affiliation, the probability that they are divorced is .1333.

Do Exercise 21. ■

Probability Trees

Another device that is helpful in picturing the possible outcomes in a probability problem is the probability tree. A probability tree is used to represent a multiphase process and consists of chance nodes from which branches are extended. Each chance node represents a phase of the process and each possible outcome of that phase is represented by a branch extending from the node. The probability of each outcome is written by the corresponding branch connecting adjacent nodes. A sequence of branches starting at the first node and ending at a tip is called a *path*. The paths are mutually exclusive and represent all possible outcomes of the process. Thus, the set of all possible paths represents the sample space and the probability of each outcome in the sample space is obtained by multiplying the probabilities written by the connecting branches of the corresponding path. (See Figure 6.7.)

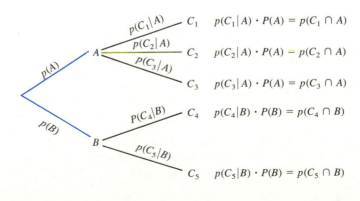

FIGURE 6.7

EXAMPLE 9   An urn contains three red, five white, and two blue marbles. Two marbles are drawn, one after the other, without replacement. What is the probability that one blue and one white marble were drawn?

Solution   A probability tree showing the different possible outcomes is drawn in Figure 6.8. In the tree, R, W, and B represent the events "drawing a red marble," "drawing a white marble," and "drawing a blue marble," respectively. Since there are ten marbles in the urn and all outcomes are equally likely, $p(R) = .3$, $p(W) = .5$, and $p(B) = .2$. Since the first marble is not replaced, there are only nine marbles in the urn when the second marble is drawn. If the first marble drawn is red, only two red marbles remain in the urn, so that $p(R \text{ on second draw} \mid R \text{ on first draw}) = \frac{2}{9}$. The other conditional probabilities are computed similarly and the results are shown on the probability tree.

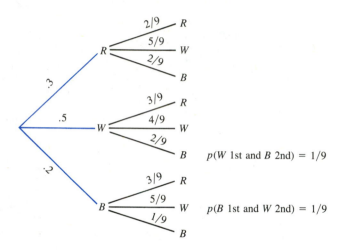

FIGURE 6.8

Since we were asked to find the probability that one blue and one white marble were drawn, we need only show the probability of a white marble being drawn on the first draw and a blue marble being drawn on the second draw, as well as the probability of a blue marble being drawn on the first draw and a white marble being drawn on the second draw. These probabilities are calculated as follows:

$$p(W \text{ 1st and } B \text{ 2nd}) = p(B \text{ 2nd} \mid W \text{ 1st}) \cdot p(W \text{ 1st}) = \frac{2}{9}(.5) = \frac{1}{9}$$

and

$$p(B \text{ 1st and } W \text{ 2nd}) = p(W \text{ 2nd} \mid B \text{ 1st}) \cdot p(B \text{ 1st}) = \frac{5}{9}(.2) = \frac{1}{9}$$

These two events are mutually exclusive, so the final probability is the sum of these two probabilities. That is, $p(\text{one } W \text{ and one } B) = \frac{1}{9} + \frac{1}{9} = \frac{2}{9}$.
Do Exercise 23.   ∎

   Note that in the preceding example, we showed only the necessary probabilities on the probability tree. Often, some of the paths will be shorter than others because we may leave out branches that are not relevant to the particular problem. We illustrate this in the following example.

EXAMPLE 10 Suppose that the Boston Celtics and Los Angeles Lakers are playing in the finals of the NBA championship and that the Lakers win the first two games in Boston. Games 3, 4, and 5 are scheduled in Los Angeles, while games 6 and 7, if necessary, will be played in Boston. The two teams are evenly matched and the only advantage is the home court. For each game, the probability of the home team winning that game is .6. What is the probability that the Celtics will win the championship in spite of losing the first two games at home?

Solution We represent the possible outcomes in a probability tree. An L by a node indicates that the Lakers won that game, while a C indicates that the Celtics were victorious. Also .6 written by a branch indicates the probability of the home team winning, while .4 written by a branch indicates the probability of the visiting team winning. Note that as soon as the Lakers win two more games, the champion is determined and we end the path. Similarly, as soon as the Celtics win four games we end the path. (See Figure 6.9.) For example, the top path has only two branches, indicating that the Lakers won in four straight games. The probability of this is .36 since $(.6)(.6) = 0.36$. (Remember that games 3 and 4 are played in Los Angeles.) The lowest path indicates that the Celtics won games 3, 4, 5, and 6. The occurrence of this outcome has probability 0.0384, since the Celtics are the visiting team for games 3, 4, and 5 and the home team for game 6 and $(.4)(.4)(.4)(.6) = 0.0384$. Since we wish to find the probability of the Celtics winning the championship, we consider only paths that end with a C.

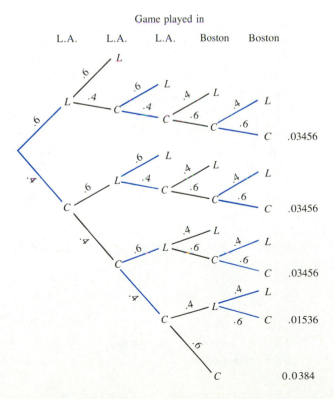

FIGURE 6.9

We calculated the probabilities of all outcomes represented by those paths and showed the results on the probability tree. For example, the first path that ends with a C represents the outcome where the Lakers won game 3 and the Celtics won the last four games. This outcome has probability 0.03456 since $(.6)(.4)(.4)(.6)(.6) = 0.03456$. To find the probability of the Celtics winning the championship, we add the probabilities indicated by the five paths that terminate with a C, since these represent the outcome "the Celtics win" and they are mutually exclusive. We get

$$0.03456 + 0.03456 + 0.03456 + 0.01536 + 0.0384 = 0.15744$$

Therefore, the probability of the Celtics winning the championship after losing the first two games at home is 0.15744.
Do Exercise 29. ∎

Infinite Sample Spaces

Although we have considered so far only sample spaces with a finite number of members, there are instances where we may wish to discuss the possibility of having infinitely many outcomes. We illustrate this situation with the following example.

EXAMPLE 11   Steven, Michael, and Kris take turns tossing a coin, in that order. The first one to throw tails wins. What is the probability that Michael wins?

Solution   The results of the tosses can be represented by ordered n-tuples. A successful outcome ("Michael wins") is represented by an n-tuple where the first $n - 1$ entries are H, and the last entry representing a toss by Michael is a T. For example, (H, T), (H, H, H, H, T), (H, H, H, H, H, H, H, T), represent successful outcomes. The probability of the first of these is $\frac{1}{4}$ since $\frac{1}{2} \cdot \frac{1}{2} = \frac{1}{4}$, the probability of the second is $\frac{1}{32}$ since $\frac{1}{2} \cdot \frac{1}{2} \cdot \frac{1}{2} \cdot \frac{1}{2} \cdot \frac{1}{2} = \frac{1}{32}$, and so on. It is clear that, theoretically, we must consider infinitely many outcomes that are mutually exclusive. Adding the probabilities of all these outcomes we obtain

$$\frac{1}{2^2} + \frac{1}{2^5} + \frac{1}{2^8} + \frac{1}{2^{11}} + \cdots$$

We will see in Chapter 9 that this expression is a geometric series, and that its value can be found by using the formula

$$a + ar + ar^2 + ar^3 + \cdots = \frac{a}{1 - r}$$

whenever $|r| < 1$. In this series, $a = \frac{1}{4}$ and $r = \frac{1}{8}$. Thus,

$$\frac{1}{2^2} + \frac{1}{2^5} + \frac{1}{2^8} + \frac{1}{2^{11}} + \cdots = \frac{\frac{1}{4}}{1 - \frac{1}{8}} = \frac{2}{7}$$

Thus, the probability that Michael wins is $\frac{2}{7}$.
Do Exercise 33. ∎

Exercise Set 6.3

1. A bag contains five red marbles and seven blue marbles. A second bag contains three red marbles and four blue marbles. A marble is drawn from the first bag and placed in the second. The second bag is shaken and a marble is drawn from it. What is the probability that a red marble was drawn each time?

2. An urn contains six white marbles and three green marbles. A second urn contains seven white marbles and six green marbles. A marble is drawn from the first urn and placed in the second. The second urn is shaken and a marble is drawn from it. What is the probability that a red marble was drawn each time?

3. The probability that a freshman takes an English course is .73. The probability of taking a mathematics course is .27 and the probability of taking both is .18. What is the probability that a freshman will take

   a. an English course given that a mathematics course is also taken by that student?
   b. a mathematics course given that an English course is also taken by that student?
   c. both English and mathematics given that at least one of the two is taken by that student?

4. The probability that a sophomore takes a philosophy course is .21. The probability of taking a sociology course is .28 and the probability of taking both is .12. What is the probability that a sophomore will take

   a. a philosophy course given that a sociology course is also taken by that student?
   b. a sociology course given that a philosophy course is also taken by that student?
   c. both philosophy and sociology given that at least one of the two is taken by that student?

5. A bag contains five red marbles and two blue marbles. A marble is drawn from the bag and replaced, the bag is shaken, and a marble is drawn again. What is the probability that both marbles drawn are red?

6. Repeat Exercise 5 without replacement.

7. An urn contains eight white marbles and five black marbles. A marble is drawn from the urn and replaced, the urn is shaken, and a marble is drawn again. What is the probability that both marbles drawn are white?

8. Repeat Exercise 7 without replacement.

9. A card is drawn at random from a standard deck of cards. What is the probability that it is either an ace or a jack?

10. A card is drawn at random from a standard deck of cards. What is the probability that it is either a ten or a face card?

11. A card is drawn from a standard deck of cards. What is the probability that it is either a heart or a queen?

12. A card is drawn from a standard deck of cards. What is the probability that it is either a spade or a face card?

13. Suppose two dice are rolled. What is the probability that the sum of the numbers that come up is either even or less than seven?

14. Suppose two dice are rolled. What is the probability that the sum of the numbers that come up is either odd or less than nine?

15. Suppose that ten people have been randomly selected. What is the probability that at least two of them celebrate their birthdays the same day? (*Hint:* See Example 7 and use a calculator.)

16. Repeat Exercise 15 for 15 people.

17. Repeat Exercise 15 for 17 people.

18. Repeat Exercise 15 for 20 people.

19. Repeat Exercise 15 for 30 people.

20. Repeat Exercise 15 for 40 people.

21. A sociologist theorized that people who do not drink alcoholic beverages have a better chance to complete their college education than those who do. To test her theory, she conducted a study of 2000 randomly selected people in the 23-to-25 age group. She let U be the universe (the set of people she surveyed), A be the set of all people who abstain from drinking alcoholic beverages, and C be the set of all people who completed their college education. She then drew the following Venn diagram.

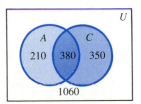

Find the probability that

   a. a person who does not drink alcoholic beverages completes a college education.
   b. a person who completes a college education does drink alcoholic beverages.

c. a person who does not drink alcoholic beverages does not complete a college education.

Did the results of the survey agree with the theory?

22. A sociologist theorized that teenagers who participate in sports have a better chance of not smoking than those who do not participate. To support his theory, he conducted a study of 1500 randomly selected young people who had just turned 20. He let U be the universe (the set of young people he surveyed), A be the set of all people who participate in some sport, and N be the set of all people who do not smoke. He then drew the following Venn diagram.

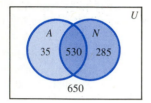

Find the probability that

a. a teenager who participates in sports does not smoke.
b. a teenager who does not smoke participates in sports.
c. a teenager who participates in sports smokes.

Did the results of the survey agree with the theory?

Draw a probability tree to solve Exercises 23–29.

23. An urn contains five red, three white, and seven blue marbles. Two marbles are drawn, one after the other, without replacement. What is the probability that one red and one white marble were drawn?

24. A bag contains four green, seven red, and one blue marble. Two marbles are drawn, one after the other, without replacement. What is the probability that one green and one red marble were drawn?

25. Suppose that the Seattle Supersonics and Milwaukee Bucks are playing in the finals of the NBA championship and that the Bucks win the first two games in Seattle. Games 3, 4, and 5 are scheduled in Milwaukee, and games 6 and 7, if necessary, will be played in Seattle. The two teams are evenly matched and the only advantage is the home court. For each game, the probability of the home team winning that game is .55. What is the probability that the Supersonics will win the championship in spite of losing the first two games at home?

26. Repeat Exercise 25 if the teams split the first two games in Seattle.

27. Suppose that the Montreal Expos and Detroit Tigers are playing in the World Series and that the Expos win the first two games in Detroit. Games 3, 4, and 5 are scheduled in Montreal, and games 6 and 7, if necessary, will be played in Detroit. The probabilities of Montreal winning games 3, 4, 5, 6, and 7 are .65, .55, .6, .35, and .3, respectively. What is the probability that the Tigers will win the World Series in spite of losing the first two games at home?

28. Repeat Exercise 27 if the two teams split the first two games in Detroit.

29. Repeat Exercise 27 if the Tigers win the first two games at home.

30. Suppose that the New York Rangers and Quebec Nordiques are playing for the Stanley Cup and that the Nordiques win the first two games in New York. Games 3, 4, and 5 are scheduled in Quebec, and games 6 and 7, if necessary, will be played in New York. The two teams are evenly matched and the only advantage is the home arena. For each game, the probability of the home team winning that game is .62. What is the probability that the Rangers will win the Stanley Cup in spite of losing the first two games at home?

31. Repeat Exercise 30 if the teams split the first two games in New York.

32. In Example 10 of this section, we found the probability that the Boston Celtics would win the championship after losing the first two games was 0.15744. Using a tree diagram, find the probabilities that

a. the Lakers will win in four games.
b. the Lakers will win in five games.
c. the Lakers will win in six games.
d. the Lakers will win in seven games.
e. the Lakers will win the championship.

Add your answer in part **e** to 0.15744 and verify that this sum is 1. Why?

33. Betty, Shirley, and Fran take turns tossing a coin, in that order. The first one to throw heads wins. What is the probability that Betty wins?

34. In Exercise 33, what is the probability that Fran wins?

35. Jeremy and André take turns rolling a die, in that order. The first to roll a 6 wins. What is the probability that Jeremy wins?

36. In Exercise 35, what is the probability that André wins?

37. Julie, Gwen, and Laura take turns rolling a die. The first one to roll a number less than 3 wins. What is the probability that Julie wins?

38. In Exercise 37, what is the probability that Gwen wins?

39. In Exercise 37, what is the probability that Laura wins?

6.4 Bayes' Formula

We begin by considering a sample space S and n mutually exclusive events E_1, E_2, E_3, . . . , E_n, such that each outcome in the sample space is in one of the events E_i. In other words, $E_i \cap E_j = \emptyset$ if $i \neq j$, and

$$E_1 \cup E_2 \cup E_3 \cup \cdots \cup E_n = U.$$

Let A be an event such that $p(A) \neq 0$. Bayes' formula* provides a simple tool for computing the conditional probability of each event E_i given A, from the conditional probability of A given each event E_i, and the unconditional probability of each E_i.

Derivation of Bayes' Formula

We have

$$P(E_1) + P(E_2) + P(E_3) + \cdots + P(E_n) = 1$$

Let i be any integer between 1 and n, inclusive. Then,

$$p(E_i \mid A) \cdot p(A) = p(E_i \cap A)$$

and

$$p(A \mid E_i) \cdot p(E_i) = p(A \cap E_i)$$

But, $E_i \cap A = A \cap E_i$. Thus, $p(E_i \cap A) = p(A \cap E_i)$, from which it follows that

$$p(E_i \mid A) \cdot p(A) = p(A \mid E_i) \cdot p(E_i)$$

Since $p(A) \neq 0$, we get

$$p(E_i \mid A) = \frac{p(A \mid E_i) \cdot p(E_i)}{p(A)}$$

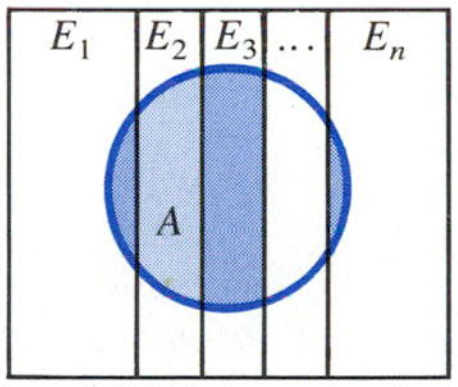

Note that,

$$A = (A \cap E_1) \cup (A \cap E_2) \cup (A \cap E_3) \cup \cdots \cup (A \cap E_n)$$

FIGURE 6.10

(See the Venn diagram in Figure 6.10.)

Since the events $A \cap E_i$, $1 \leq i \leq n$ are mutually exclusive, we have

$$p(A) = p(A \cap E_1) + p(A \cap E_2) + p(A \cap E_3) + \cdots + p(A \cap E_n)$$

It follows that

$$p(E_i \mid A) = \frac{p(A \mid E_i) \cdot p(E_i)}{p(A)}$$

$$= \frac{p(A \mid E_i) \cdot p(E_i)}{p(A \cap E_1) + p(A \cap E_2) + p(A \cap E_3) + \cdots + p(A \cap E_n)}$$

*This formula was first derived by the English Presbyterian minister Thomas Bayes and was first published in 1763 although Bayes died in 1761.

But, for each j we have $p(A \cap E_j) = p(A \mid E_j) \cdot P(E_j)$. Hence,

$$p(E_i \mid A) = \frac{p(A \mid E_i) \cdot p(E_i)}{p(A \mid E_1) \cdot p(E_1) + p(A \mid E_2) \cdot p(E_2) + \cdots + p(A \mid E_n) \cdot p(E_n)}$$

Applications

Bayes' formula may appear difficult to memorize. In actuality, problems involving this formula can be done using a probability tree as we shall illustrate in the following example.

EXAMPLE 1   A novice golfer must hit a certain shot. He has eight different clubs in his bag, only one of which is the right club for this particular shot. The probability that he hits a good shot is .25 if he uses the right club, but only .12 if he uses a wrong club. Suppose he picks a club randomly from his bag and hits a good shot. What is the probability that he selected the right club?

Solution   Let G and B represent the events "hit a good shot" and "hit a bad shot," respectively. Let R and W represent the events "selecting the right club" and "selecting the wrong club," respectively. Since he is just as likely to select one club as any other (remember he is a novice), we have $p(R) = \frac{1}{8} = .125$ and $p(W) = \frac{7}{8} = .875$. It is given that $p(G \mid R) = .25$ and that $p(G \mid W) = .12$. Hence, $p(B \mid R) = .75$ since $1 - .25 = .75$, and $p(B \mid W) = .88$ since $1 - .12 = .88$. We are asked to find $p(R \mid G)$.

Using Bayes' formula, we know that

$$p(R \mid G) = \frac{p(G \mid R) \cdot p(R)}{p(G \mid R) \cdot p(R) + p(G \mid W) \cdot p(W)}$$

Thus,

$$p(R \mid G) = \frac{.25(.125)}{.25(.125) + .12(.875)} \doteq .2294$$

Thus, if it is known that the novice hit a good shot, the probability that he had selected the right club for that shot is approximately .2294. We draw a probability tree for this problem. (See Figure 6.11.)

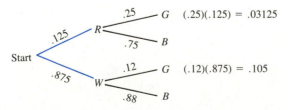

FIGURE 6.11

We see from the diagram that the numerator of the expression giving $p(R \mid G)$ is the product $p(G \mid R)p(R) = .25(.125) = .03125$. This is the product of the probabilities on the branches in the path to G through R. On the other hand, the denominator of that expression is $p(G \mid R)p(R) + p(G \mid W)p(W) = .25(.125) + .12(.875) = .13625$. This is the sum of all products of the probabilities on the branches of the paths leading to G.

Do Exercise 1. ∎

Therefore, instead of using the formula

$$p(E_i \mid A) = \frac{p(A \mid E_i) \cdot p(E_i)}{p(A \mid E_1) \cdot p(E_1) + p(A \mid E_2) \cdot p(E_2) + \cdots + p(A \mid E_n) \cdot p(E_n)}$$

to solve this type of problem, we can use the following procedure.

STEP 1. Draw a probability tree for the problem, indicating by each branch of the tree $p(E_i)$ and $p(A \mid E_i)$ for $i = 1, 2, \ldots, n$.

STEP 2. Calculate the product of the probabilities on the branches of each path leading to A.

STEP 3. Add all products obtained in Step 2.

STEP 4. To calculate $p(E_i \mid A)$, divide the product of the probabilities on the branches of the path leading to A through E_i (this is one of the products calculated in Step 2) by the sum obtained in Step 3.

EXAMPLE 2 Three machines are used in manufacturing a certain gadget. Machines A, B, and C are used to produce 30%, 45%, and 25% of the total output, respectively. It is known that 2% of the gadgets made using machine A are defective, 4% of the gadgets made using machine B are defective, and 3% of the gadgets made using machine C are defective. A gadget is selected at random from the total output. If it is defective, what is the probability that it was made using machine B?

Solution **Step 1.** We begin by drawing a probability tree (see Figure 6.12), where A, B, and C represent the events "produced using machine A," "produced using machine B," and "produced using machine C," respectively; D represents the event "the gadget is defective" and D' represents the event "the gadget is not defective."

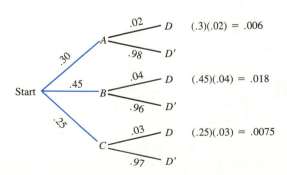

FIGURE 6.12

Step 2. Since it is given that the gadget chosen at random is defective, we calculate each of the products of the probabilities on the branches of each path leading to D.

For the first path we get .3(.02) = .006.
For the second path we get .45(.04) = .018
For the third path we get .25(.03) = .0075.

Step 3. Now we add all products obtained in Step 2.

$$.006 + .018 + .0075 = .0315$$

Step 4. Since we wish to find the probability that the defective gadget was produced using machine B, we consider the second path since it leads to D through B. The product of the probabilities of the branches in that path was obtained in Step 2. It is .018. To find the desired probability, we divide .018 by the sum obtained in Step 3. We get

$$\frac{.018}{.0315} \doteq .57143$$

where the result has been rounded off to five decimal places. Thus, the probability that the gadget was made using machine B, given that it is defective, is approximately .57143.

Do Exercise 9.   ■

Sometimes we are faced with the problem of finding probabilities of experiments that have several stages. We illustrate this with an example.

EXAMPLE 3   Suppose that an urn contains three fair coins and a coin with tails on both sides. A coin is selected randomly from the urn and it is tossed.

a. Tails comes up. What is the probability that the coin selected was fair?
b. The coin is tossed again, and again tails comes up. Calculate the probability that the coin selected was fair.
c. The coin is tossed a third time and this time heads comes up. What is the probability that the coin selected was a fair coin?

Solution   **a. Step 1.** We first draw a probability tree where F represents the event "the coin selected was fair," F' represents the event "the coin selected has two tails," H represents the event "heads came up" and T represents the event "tails came up." We indicate the probability of each of these by the corresponding branch on the tree. (See Figure 6.13.)

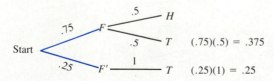

FIGURE 6.13

Step 2. We calculate the product of the probabilities on the branches of each path leading to T. We get, for the first path leading to T, .75(.5) = .375, and for the second path leading to T, .25(1) = .25.

Step 3. We now add the products obtained in Step 2. We get .375 + .25 = .625.

Step 4. We divide .375, which is the product of the probabilities on the branches of the path leading to T through F, by the sum obtained in Step 3. We get $\frac{.375}{.625} = .6$.

Therefore, the probability that the coin selected was fair, given that the toss came up tails, is .6.

b. We use the result of part **a**. The probability $p(F \mid T)$ was found to be .6. Consequently, $p(F' \mid T) = .4$. These probabilities can now be used for the second stage. We go through the same steps as in part **a**.

Step 1. We first draw a probability tree and indicate the new probabilities on the branches of the tree. (See Figure 6.14.)

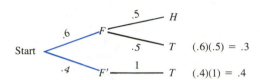

FIGURE 6.14

Step 2. We calculate the product of the probabilities on the branches of each path leading to T. We get, for the first path leading to T, .6(.5) = .3, and for the second path leading to T, .4(1) = .4.

Step 3. We now add the products obtained in Step 2. We get .3 + .4 = .7.

Step 4. We divide .3, which is the product of the probabilities on the branches of the path leading to T through F, by the sum obtained in Step 3. We get $\frac{.3}{.7} = \frac{3}{7}$. The probability that the coin selected was fair, given that the tosses came up tails both times, is $\frac{3}{7}$.

c. If at any stage the coin is tossed and heads comes up, then we know that a fair coin was selected since it could not be the coin with tails on both sides. In that case, the probability that a fair coin was selected is 1.

Do Exercise 19.

Exercise Set 6.4

1. A novice golfer must hit a certain shot. He has six different clubs in his bag, only one of which is the right club for this particular shot. The probability that he hits a good shot is .23 if he uses the right club, but only .1 if he uses a wrong club. Suppose that he picks a club randomly from his bag and hits a good shot. What is the probability that he selected the right club?

2. Repeat Exercise 1 if the golfer has five clubs in his bag and the probability that he hits a good shot is .26 if he uses the right club and .11 if he uses a wrong club.

3. In Exercise 1, what is the probability that the golfer selected the right club given that the shot he hit was a bad shot?

4. Urn 1 contains four green marbles and three red marbles, and urn 2 contains five red marbles and seven red marbles.

A coin is tossed. If heads comes up, urn 1 is selected and a marble is drawn from that urn. If tails comes up, urn 2 is selected and a marble is drawn. Suppose the marble drawn is red.

a. What is the probability that the first urn was selected?
b. What is the probability that the second urn was selected?

5. Repeat Exercise 4 if urn 1 has 3 green marbles and 10 red marbles, and urn 2 has 13 green marbles and 9 red marbles.

6. Urn 1 contains five black marbles and seven white marbles, and urn 2 contains six black marbles and ten white marbles. A die is rolled. If a 2 comes up, urn 1 is selected and a marble is drawn from that urn. If any other number comes up, urn 2 is selected and a marble is drawn. Suppose the marble drawn is white.

a. What is the probability that the first urn was selected?

b. What is the probability that the second urn was selected?

7. Urn 1 contains four yellow marbles and six blue marbles, and urn 2 contains seven yellow marbles and nine blue marbles. A die is rolled. If a number greater than 4 comes up, urn 1 is selected and a marble is drawn from that urn. If any other number comes up, urn 2 is selected and a marble is drawn. Suppose the marble drawn is blue.

a. What is the probability that the first urn was selected?

b. What is the probability that the second urn was selected?

8. Urn 1 contains five red marbles and six white marbles, urn 2 contains seven red marbles and nine blue marbles, and urn 3 contains ten red marbles and three blue marbles. A die is rolled. If an even number comes up, urn 1 is selected and a marble is drawn from that urn. If a 1 or a 3 comes up, urn 2 is selected and a marble is drawn. If a 5 comes up, urn 3 is selected and a marble is drawn. Suppose the marble drawn is red.

a. What is the probability that the first urn was selected?

b. What is the probability that the second urn was selected?

c. What is the probability that the third urn was selected?

d. If a white marble is drawn, what is the probability that urn 1 was selected?

9. Three machines are used in manufacturing a certain gadget. Machines A, B, and C are used to produce 20%, 65%, and 15% of the total output, respectively. It is known that 3% of the gadgets made using machine A are defective, 5% of the gadgets made using machine B are defective, and 4% of the gadgets made using machine C are defective. A gadget is selected at random from the total output. If it is defective, what is the probability that it was made using machine B?

10. A company manufactures car batteries at three plants. Plant A produces 850 batteries a day, plant B produces 630 batteries a day, and plant C produces 520 batteries a day. Experience indicates that 4% of the batteries produced at plant A are defective, 5% of the batteries produced at plant B are defective, and 2% of those produced at plant C are defective. Each day, all batteries produced are shipped to the company's warehouse. Suppose that a battery randomly selected in the warehouse is found to be defective. What is the probability that it was produced at **a.** plant A? **b.** plant B? **c.** plant C?

11. A company manufactures VCRs at four plants. Plant A produces 350 VCRs a week, plant B produces 420 VCRs a week, plant C produces 330 VCRs a week, and plant D produces 400 VCRs a week. Experience indicates that 2% of the VCRs produced at plant A are defective, 4% of the VCRs produced at plant B are defective, 3% of those pro-

duced at plant C are defective, and 1.5% of those produced at plant D are defective. Each week, all VCRs produced are shipped to the company's warehouse. Suppose that a VCR randomly selected in the warehouse is found to be defective. What is the probability that it was produced at **a.** plant A? **b.** plant B? **c.** plant C? **d.** plant D?

12. In Exercise 10, suppose that a battery randomly selected in the warehouse is not defective. What is the probability that it was produced at **a.** plant A? **b.** plant B? **c.** plant C?

13. In Exercise 11, suppose that a VCR randomly selected in the warehouse is not defective. What is the probability that it was produced at **a.** plant A? **b.** plant B? **c.** plant C? **d.** plant D?

14. The director of an MBA program decided that applicants to the program will be required to take a qualifying exam in mathematics. After several years of administering the qualifying test, it was found that 70% of all applicants completed the program and that 85% of those who completed the program had passed the test, while only 12% of those who did not complete the program had passed the test. Suppose an applicant has passed the test. What is the probability that this applicant will complete the program?

15. Repeat Exercise 14 with 60% of all applicants completing the program, 82% of those completing the program passed the test, and only 14% of those not completing the program passed the test.

16. It is known that .02% of the people in a city have AIDS. A new test is tried and it is found to give a positive reaction to 45% of the people who have the disease and to 15% of the people who do not have the disease. A certain person's test is positive. What is the probability that this person has the disease?

17. It is known that 2.6% of the people in the United States are diabetic. A new glucose test is tried and it is found to give a positive reaction to 90% of the people who are diabetic and to 5% of the people who are not diabetic. A certain person's test is positive. What is the probability that this person is diabetic?

18. Suppose that an urn contains four fair coins and a coin with tails on both sides. A coin is selected randomly from the urn and it is tossed.

a. Tails comes up. What is the probability that the coin selected was fair?

b. The coin is tossed again, and again tails comes up. Calculate the probability that the coin selected was fair.

c. The coin is tossed a third time and this time heads comes up. What is the probability that the coin selected was a fair coin?

19. Suppose that an urn contains six fair coins and two coins with heads on both sides. A coin is selected randomly from the urn and it is tossed.

a. Heads comes up. What is the probability that the coin selected was fair?

b. The coin is tossed again, and again heads comes up. Calculate the probability that the coin selected was fair.

c. The coin is tossed a third time and this time tails comes up. What is the probability that the coin selected was a fair coin?

6.5 Expected Value

In the 17th century, the Chevalier de Méré, a gambling enthusiast, asked a great French mathematician, Blaise Pascal (1623–1662), how the stakes should be divided in an unfinished game of chance. Subsequently, Pascal corresponded with another great French mathematician, Pierre de Fermat (1601–1665), about the problem raised by the Chevalier's request. The resulting series of letters laid the foundation for probability theory. It is therefore natural that we use simple games of chance to illustrate some of the ideas involved. However, you should keep in mind that probability theory is an important subject with wide applications both in applied and pure mathematics.

Expectation

We begin with the following example.

EXAMPLE 1 In drawing a single card from a standard deck, you win $5 if you draw an ace, $2 if you draw a face card, and 25 cents if any other card is drawn. It costs $1 to be allowed to draw a card. Would you expect to win money in the long run by playing this game?

Solution Suppose you play the game 520 times. Since the probability of drawing an ace is $\frac{4}{52}$, you would expect to draw an ace approximately 40 times, and win approximately $200 doing so. Similarly, there are 12 face cards in the deck; thus, the probability of drawing a face card is $\frac{12}{52}$ and you would expect to draw a face card approximately 120 times, winning approximately $240 doing so. Finally, you would expect to draw some other card approximately 360 times (since $520 - (40 + 120) = 360$), and win approximately $90 dollars doing so. Thus, if you played 520 times, you would win approximately $530. However, it would cost you $520 dollars to play 520 games. Thus, you would expect to win about $10 dollars if you played 520 games or an average of $\frac{1}{52}$ dollar per game (slightly less than 2 cents per game).

Do Exercise 1. ∎

Of course, we know that things will not always happen exactly as expected. In the preceding example, when the game is played 520 times, we may not have exactly 40 aces drawn or 120 face cards drawn. However, if the game is played a great number of times, it is likely that the average gain per game will be slightly less than 2 cents. We used the 520 games to illustrate the basic idea.

Suppose we present the example in a different way. Each time we draw an ace we win $5, but we pay $1 to play, so the net payoff is only $4. Each time we draw a face card we win $2, but it cost $1; thus, the net payoff is only $1. Each time we draw any other card we win 25 cents, but the cost is still $1, so we lose 75 cents and the net payoff is $-\$.75$. If we play 520 games, we can calculate the average payoff as follows:

$$\frac{40(4) + 120(1) + 360(-.75)}{520} = 4 \cdot \frac{40}{520} + 1 \cdot \frac{120}{520} + (-.75) \cdot \frac{360}{520}$$

$$= 4 \cdot \frac{1}{13} + 1 \cdot \frac{3}{13} + (-.75) \cdot \frac{9}{13}$$

$$= 4p(A) + 1p(F) + (-.75)p(T)$$

where A is the event "drawing an ace," F is the event "drawing a face card," and T is the event "drawing any other card."

> **DEFINITION 6.8:**   If a game has n outcomes, the probabilities of which are $p_1, p_2, \ldots, p_n$, respectively, and if the net payoff is the amount A_i if the ith outcome occurs, where $(1 \le i \le n)$, then the *expectation* for the game is the sum
>
> $$p_1 A_1 + p_2 A_2 + \cdots + p_n A_n.$$

This concept is useful when deciding whether or not to enter a gambling game. Since the expectation represents approximately the average winnings per trial, provided the game is played many times, then the decision to enter the game should be made only if the expectation is positive. This, of course, does not apply to the purchase of insurance, where it is customary for the insured to have a negative expectation. The purpose of insurance is not to win money, but to spread the cost of accidents among the insured.

Probability Distribution

To expand on this, we will introduce the concept of random variables. Basically, a random variable is a variable whose values are real numbers that depend on a random process.

> **DEFINITION 6.9:**   A *random variable* is a function X whose domain is a sample space and whose range is a set of real numbers. If the range is finite, or if it can be put in a one-to-one correspondence with the set of positive integers, then X is said to be a *discrete random variable*. If the range is an interval of real numbers, the random variable is said to be *continuous*.

Suppose, for example, a die is rolled twice and we represent an outcome by an ordered pair (x, y), where x is the number that comes up on the first roll and y is the number that comes up on the second roll. Then, the sample space S has 36 members.

We may define a random variable X by $X(x, y) = x + y$. For example, $X(1, 3) = 4$, $X(2, 5) = 7$, and so on. Then, the range of X is the set $\{2, 3, 4, 5, 6, 7, 8, 9, 10, 11, 12\}$ and X is a discrete random variable. Now suppose we keep tossing a fair coin until tails comes up and X is the number of tosses it takes to get tails. In this case, the sample space is $\{T, HT, HHT, HHHT, HHHHT, HHHHHT, \ldots\}$. Then, $X(T) = 1, X(HT) = 2, X(HHT) = 3$, and so on. Again, X is a discrete random variable. On the other hand, if a traffic light changes every 90 seconds and X seconds is the time a randomly selected motorist must wait at a red light, then the value of X is in the interval $[0, 90]$ and therefore X is a continuous random variable.

Note that for each value of a random variable we have a corresponding event in the sample space. For example, if $X(x, y) = x + y$, where x and y represent the numbers that came up on two rolls of a die, then

$$\{(x, y) \mid X(x, y) = 5\} = \{(1, 4), (2, 3), (3, 2), (4, 1)\}$$

so that the event $E = \{(1, 4), (2, 3), (3, 2), (4, 1)\}$ corresponds to the value 5 of the random variable. Consequently, we can assign to each value of a random variable a probability that the random variable will take that value. In this example, since the outcomes in a roll of a fair die are equally likely, $p(E) = \frac{4}{36} = \frac{1}{9}$. So, $p(X = 5) = p(E) = \frac{1}{9}$, where $p(X = 5)$ denotes the probability that the random variable assumes the value 5. In general, the following definition applies.

> **DEFINITION 6.10:** Suppose S is a sample space with a probability function p and X is a discrete random variable with domain S. For each value of x in the range of X, let
>
> $$p(X = x) = p(E), \text{ where } E = \{s \mid s \in S \text{ and } X(s) = x\}$$
>
> The *probability function* (or *probability distribution*) of X is the function f, defined by $f(x) = p(X = x)$ for each x in the range of X.

EXAMPLE 2 A fair coin is tossed three times and X is the number of tails that come up. Determine the probability distribution of X.

Solution In this experiment, the sample space has eight elements. Each of these is an ordered triple. For example, (T, T, H) represents the outcome "tails came up the first two tosses, then heads came up." The number of tails that can come up is 0, 1, 2, or 3. Thus, the range of the random variable X is the set $\{0, 1, 2, 3\}$. This set is the domain of the probability distribution.

The event corresponding to $X = 0$ is $\{(H, H, H)\}$. Since all outcomes are equally likely, the probability of that event is $\frac{1}{8}$, and $f(0) = P(X = 0) = \frac{1}{8}$. The event corresponding to $X = 1$ is $\{(T, H, H), (H, T, H), (H, H, T)\}$. The probability of that event is $\frac{3}{8}$, and $f(1) = P(X = 1) = \frac{3}{8}$. The event corresponding to $X = 2$ is $\{(T, T, H), (T, H, T), (H, T, T)\}$. The probability of that event is $\frac{3}{8}$, and $f(2) = P(X = 2) = \frac{3}{8}$. The event corresponding to $X = 3$ is $\{(T, T, T)\}$. The probability of that event is $\frac{1}{8}$, and $f(3) = P(X = 3) = \frac{1}{8}$.

The results are summarized in the following *probability table*.

x	$f(x) = P(X = x)$
0	$\frac{1}{8}$
1	$\frac{3}{8}$
2	$\frac{3}{8}$
3	$\frac{1}{8}$

It should be noted that

$$f(0) + f(1) + f(2) + f(3) = \frac{1}{8} + \frac{3}{8} + \frac{3}{8} + \frac{1}{8} = 1$$

and that $0 \le f(x) \le 1$ for each $x \in \{0, 1, 2, 3\}$. This follows from the fact that the events corresponding to $X = 0$, $X = 1$, $X = 2$, and $X = 3$ are mutually exclusive and their union is the sample space.
Do Exercise 5.　　　　　　　　　　　　　　　　　　　　　　　　　　■

In general, if X is a discrete random variable with domain S and range D, and f is the probability distribution of X, then **a.** The domain of f is D, **b.** $0 \le f(x) \le 1$, for all x in D, and **c.** $\displaystyle\sum_{x \in D} f(x) = 1$.

EXAMPLE 3　A fair coin is tossed until heads comes up. Let X be the number of tosses it takes to get heads. Determine the probability distribution of X.

Solution　The sample space is {H, TH, TTH, TTTH, TTTTH, . . .}. Since the coin is fair, we have $p(\text{H}) = \frac{1}{2}$, $p(\text{TH}) = \left(\frac{1}{2}\right)\left(\frac{1}{2}\right) = \frac{1}{4}$, $p(\text{TTH}) = \left(\frac{1}{2}\right)\left(\frac{1}{2}\right)\left(\frac{1}{2}\right) = \frac{1}{8}$, $p(\text{TTTH}) = \left(\frac{1}{2}\right)\left(\frac{1}{2}\right)\left(\frac{1}{2}\right)\left(\frac{1}{2}\right) = \frac{1}{16}$; and so on. If X is the number of tosses required to get heads, the range of X is $\{1, 2, 3, 4, \ldots\}$. Thus, the domain of the probability distribution f is also $\{1, 2, 3, 4, \ldots\}$. We have

$$f(1) = P(X = 1) = p(\text{H}) = \frac{1}{2}$$

$$f(2) = P(X = 2) = p(\text{TH}) = \frac{1}{4}$$

$$f(3) = P(X = 3) = p(\text{TTH}) = \frac{1}{8}$$

$$f(4) = P(X = 4) = p(\text{TTTH}) = \frac{1}{16}$$

$$\vdots$$

In general, $f(n) = 1/2^n$.

We see that

$$f(1) + f(2) + f(3) + f(4) + \cdots = \frac{1}{2} + \frac{1}{4} + \frac{1}{8} + \frac{1}{16} + \cdots$$

$$= \frac{\dfrac{1}{2}}{1 - \dfrac{1}{2}} = 1$$

which agrees with our earlier statement about probability distributions. ■

Histogram

The probability distribution of a discrete random variable is often represented pictorially by a *probability histogram*. The probability histogram for the random variable of Example 2 is shown in Figure 6.15.

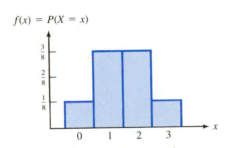

FIGURE 6.15

The rectangle over each value of x has width 1 and height $f(x) = P(X = x)$. The sum of the areas of the rectangles is 1 square unit.

Mathematical Expectation

Returning to Example 1, we can define the value of a discrete random variable to be the net payoff (winnings − cost). The range of X is $\{-.75, 1, 4\}$. The probability distribution function is defined as follows:

$$f(-.75) = P(X = -.75) = p(\text{drawing a card other than an ace or a face card})$$

$$= \frac{9}{13}$$

$$f(1) = P(X = 1) = p(\text{drawing a face card}) = \frac{3}{13}$$

$$f(4) = P(X = 4) = p(\text{drawing an ace}) = \frac{1}{13}$$

The expectation for that game was calculated as follows:

$$(-.75)\left(\frac{9}{13}\right) + (1)\left(\frac{3}{13}\right) + (4)\left(\frac{1}{13}\right) = \frac{.25}{13} = \frac{1}{52} \tag{*}$$

But, $\frac{9}{13} = f(-.75)$, $\frac{3}{13} = f(1)$, and $\frac{1}{13} = f(4)$. So, equality (*) may be written

$$(-.75)f(-.75) + (1)f(1) + (4)f(4) = \frac{1}{52}$$

This equality suggests the following definition.

> **DEFINITION 6.11:** Suppose S is a finite sample space and X is a discrete random variable with probability distribution f defined on the range $\{x_1, x_2, x_3, \ldots, x_n\}$ of X. The *mathematical expectation* of X is given by
>
> $$E(X) = x_1 f(x_1) + x_2 f(x_2) + x_3 f(x_3) + \cdots + x_n f(x_n)$$

EXAMPLE 4   Calculate the mathematical expectation of the random variable of Example 2.

Solution   The values of X were the numbers of tails that can come up when a fair coin is tossed three times. Therefore, the range of X is $\{0, 1, 2, 3\}$. We found in Example 2 that $f(0) = \frac{1}{8}$, $f(1) = \frac{3}{8}$, $f(2) = \frac{3}{8}$, and $f(3) = \frac{1}{8}$. Thus, the mathematical expectation of the random variable is

$$E(X) = 0\left(\frac{1}{8}\right) + 1\left(\frac{3}{8}\right) + 2\left(\frac{3}{8}\right) + 3\left(\frac{1}{8}\right) = \frac{12}{8} = 1.5$$

Do Exercise 11.   ■

This result should not surprise you since we would expect tails to come up about half of the time; thus for three tosses, we expect the average number of tails to be 1.5. Note that for no outcome of the experiment do we have 1.5 tails coming up. The number 1.5 is simply an average, just as when an instructor announces that the class average on a certain test is 73.52, probably no student actually scored 73.52 on that exam.

EXAMPLE 5   The annual premium that a Seattle homeowner must pay an insurance company to protect her house, worth $150,000, against total destruction by fire is $125. From experience, the company has determined that the probability of a house being totally destroyed by fire in that neighborhood is 0.00025. Calculate the expected gain per policy to the company.

Solution   Let X be the gain to the company per policy. If the house is not destroyed by fire, the company gains $125. On the other hand, if the house is destroyed by fire, the company loses $149,875 since $150,000 - 125 = 149,875$. Therefore, the value of X is 125 in the first case and $-149,875$ in the second case. If f is the probability distribution of X, we have

$$f(-149,875) = P(X = -149,875) = 0.00025$$

and

$$f(125) = P(X = 125) = 0.99975$$

The expected value of X is found as follows:

$$E(X) = -149,875f(-149,875) + 125f(125)$$
$$= -149,875(0.00025) + 125(0.99975) = 87.5$$

On the average, the insurance company will gain $87.50 per policy.
Do Exercise 13. ■

Variance and Standard Deviation

Since the mathematical expectation $E(X)$ of a random variable X is the average value of X in the long run, it is a measure of the central tendency of X. You have undoubtedly seen cases of two exams where the averages were identical, but in one exam all scores were close to the average and in the other exam there were a few high grades and a few low grades, but not many near the average. Thus, the average alone does not tell us anything about the dispersion of the grades. We need to have some way to measure the dispersion of the values of a random variable. The variance, often denoted $VAR(X)$ or σ^2 (read: "sigma squared"), and the standard deviation are two such measures. Basically, the variance of a discrete random variable X is the mathematical expectation of the squares of the deviations of the values of the random variable X from its expected value $E(X)$.

DEFINITION 6.12: If X is a random variable whose range is $\{x_1, x_2, x_3, \ldots, x_n\}$, f is the probability distribution of X, and $E(X)$ is the mathematical expecation of X, then the *variance* of X is

$$\sigma^2 = [x_1 - E(X)]^2 f(x_1) + [x_2 - E(X)]^2 f(x_2) + \cdots + [x_n - E(X)]^2 f(x_n)$$

In applications, it is sometimes difficult to use variance as a measure of dispersion of a random variable because the values of the variance and the variable are expressed in different units. For example, if X represents a number of feet, the variance would represent a number of square feet. To alleviate this difficulty, we use the standard deviation.

DEFINITION 6.13: The *standard deviation* of a random variable is the non-negative square root of its variance.

EXAMPLE 6   Suppose both random variables X and Y have range $\{1, 2, 3, 4\}$. Suppose f and g are the probability distributions of X and Y, respectively, defined by the following probability tables.

x	$f(x) = P_1(X = x)$
1	.3
2	.4
3	.2
4	.1

y	$g(y) = P_2(Y = y)$
1	.55
2	.1
3	.05
4	.3

a. Show that both X and Y have the same mathematical expectation.
b. Find the variance and standard deviation of X.
c. Find the variance and standard deviation of Y.
d. Draw a probability histogram for X and for Y.

Solution   **a.** We first find $E(X)$.

$$E(X) = 1 \cdot f(1) + 2 \cdot f(2) + 3 \cdot f(3) + 4 \cdot f(4)$$
$$= 1(.3) + 2(.4) + 3(.2) + 4(.1) = 2.1$$

We now find $E(Y)$.

$$E(Y) = 1 \cdot g(1) + 2 \cdot g(2) + 3 \cdot g(3) + 4 \cdot g(4)$$
$$= 1(.55) + 2(.1) + 3(.05) + 4(.3) = 2.1$$

Thus, $E(X) = E(Y)$.

b. The variance of X is found as follows:

$$VAR(X) = (1 - 2.1)^2(.3) + (2 - 2.1)^2(.4) + (3 - 2.1)^2(.2)$$
$$+ (4 - 2.1)^2(.1) = .89$$

Therefore, the standard deviation of X is

$$\sigma = \sqrt{VAR(X)} = \sqrt{.89} \doteq .943398$$

c. The variance of Y is found as follows:

$$VAR(Y) = (1 - 2.1)^2(.55) + (2 - 2.1)^2(.1) + (3 - 2.1)^2(.05)$$
$$+ (4 - 2.1)^2(.3) = 1.79$$

The standard deviation of Y is

$$\sigma = \sqrt{VAR(Y)} = \sqrt{1.79} \doteq 1.337909$$

d. The histograms for X and Y are shown in Figure 6.16.
Do Exercise 21.

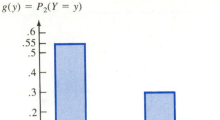

$f(x) = P_1(X = x)$

$g(y) = P_2(Y = y)$

FIGURE 6.16

As we pointed out earlier, the variance is a rough measure of the spread of values of the random variable. In fact, if VAR(X) is small and each of the terms used to calculate its value is nonnegative, then each of these terms must be small. Therefore if for any i, $|x_i - E(X)|$ is large, then the corresponding $f(x_i)$ must be small. In other words, when the variance is small, large deviations of X from the mean $E(X)$ are not likely. On the other hand, when the variance is large, it is likely that not all values of X lie near the mean.

A CLOSER LOOK   The formula

$$\sigma^2 = [x_1 - E(X)]^2 f(x_1) + [x_2 - E(X)]^2 f(x_2) + \cdots + [x_n - E(X)]^2 f(x_n)$$

which defines the variance, can be simplified to a form that is often easier to use. This simplified formula is given in the following theorem.

> **THEOREM 6.2:**   Suppose X is a random variable whose range is $\{x_1, x_2, x_3, \ldots, x_n\}$, f is the probability distribution of X, and $E(X)$ is the mathematical expected value of X (the mean). Then the variance of X is given by
>
> $$\text{VAR}(X) = \sigma^2 = \left[\sum_{i=1}^{n} x_i^2 f(x_i) \right] - [E(X)]^2 = E(X^2) - [E(X)]^2$$

Proof: (Optional) If you are reading this proof, you should be familiar with the sigma notation introduced in Section 1.7. If you are not familiar with this notation but you want to know how the formula is obtained, you should work Exercises *24, *25, and *26 at the end of this section. In the proof, we shall make use of the following facts.

$$E(X) = \sum_{i=1}^{n} x_i f(x_i) \qquad \text{(by definition)}$$

and

$$\sum_{i=1}^{n} f(x_i) = 1 \qquad \text{(since } f \text{ is a probability density function)}$$

We proceed as follows.

$$\text{VAR}(X) = \sum_{i=1}^{n} [x_i - E(X)]^2 f(x_i)$$

$$= \sum_{i=1}^{n} \{x_i^2 - 2x_i E(X) + [E(X)]^2\} f(x_i)$$

$$= \sum_{i=1}^{n} \{x_i^2 f(x_i) - 2x_i E(X) f(x_i) + [E(X)]^2 f(x_i)\}$$

$$= \sum_{i=1}^{n} x_i^2 f(x_i) - \sum_{i=1}^{n} 2x_i E(X) f(x_i) + \sum_{i=1}^{n} [E(X)]^2 f(x_i)$$

$$= \sum_{i=1}^{n} x_i^2 f(x_i) - 2E(X) \sum_{i=1}^{n} x_i f(x_i) + [E(X)]^2 \sum_{i=1}^{n} f(x_i)$$

$$= \left[\sum_{i=1}^{n} x_i^2 f(x_i)\right] - 2[E(X)]^2 + [E(X)]^2$$

$$= \left[\sum_{i=1}^{n} x_i^2 f(x_i)\right] - [E(X)]^2 = E(X^2) - [E(X)]^2$$

EXAMPLE 7   Find the variance of the random variable X of Example 6 using the preceding formula.

Solution   It was found in Example 6 that $E(X) = 2.1$. Using this formula and the values of x_i and $f(x_i)$ given in the probability table of Example 6, we obtain

$$\text{VAR}(X) = [1^2(.3) + 2^2(.4) + 3^2(.2) + 4^2(.1)] - (2.1)^2$$
$$= 5.3 - 4.41 = 0.89$$

As expected, this is the result obtained in Example 6.   ■

Exercise Set 6.5

1. In drawing a single card from a standard deck, you win $10 if you draw an ace, $5 if you draw a face card, and 50 cents if any other card is drawn. It costs $2 to be allowed to draw a card. Would you expect to win money in the long run by playing this game? Approximately how much would you expect to win or lose if you played 200 times?

2. In a throw of two dice, you win $3 if the sum of the numbers that come up is 7 or 11, and lose $1 if the sum is anything else. Would you expect to win money in the long run by playing this game? Approximately how much would you expect to win or lose if you played 100 times?

3. You are given a bag containing six black coins and three white coins and you are offered the following choice.

   a. Draw one coin. If it is white, you win $5; if it is black, you lose $1.
   b. Draw two coins. If both are white, you win $10; if they are of different color, you lose $2.

   Assuming that you want to play one of the two games, which game should you play?

4. A fair coin was tossed four times and X is the number of heads that came up. Determine the probability distribution of X.

5. Repeat Exercise 4 if the coin is tossed five times.

6. Two fair dice are rolled. Let X be the sum of the two numbers that come up. Determine the probability distribution of X.

7. A fair die is rolled until a six comes up. Let X represent the number of rolls it takes for a six to come up. Determine the probability distribution of X.

8. A bag contains five black marbles and eight white marbles. You draw a marble randomly. If it is black, you stop; otherwise, you replace the marble in the bag, shake the bag, and draw a marble again. If it is black, you stop; otherwise, you repeat the process until you draw a black marble. Let X be the number of times you must draw a marble to get a black one. Determine the probability distribution of X.

9. Calculate the mathematical expectation of the random variable of Exercise 4.

10. Calculate the mathematical expectation of the random variable of Exercise 5.

11. Calculate the mathematical expectation of the random variable of Exercise 6.

12. The annual premium that a San Francisco homeowner must pay an insurance company to protect his house, worth $275,000, against total destruction by fire is $325. From experience, the company has determined that the probability of a house being totally destroyed by fire in that neighborhood is 0.00018. Calculate the expected gain per policy to the company.

13. Repeat Exercise 12 if the annual premium is $360, the house is worth $325,000, and the probability of the house being totally destroyed by fire is 0.00025.

14. Repeat Exercise 12 if the annual premium is $280, the house is worth $275,000, and the probability of the house being totally destroyed by fire is 0.0006.

15. An insurance company offers the following medical plan to university employees. It will pay $150 a day for hospitalization up to a maximum of 7 days per year. The annual premium for this policy is $50. From experience with similar groups, the company has estimated that the probabilities that an employee is hospitalized for exactly 1 day, exactly 2 days, exactly 3 days, exactly 4 days, exactly 5 days, exactly 6 days, and more than 6 days are 0.0015, 0.0025, 0.0035, 0.005, 0.0065, 0.008, and 0.02, respectively. Calculate the expected gain per policy to the company.

16. The insurance company of Exercise 15 wishes to have an expected gain of $25 per policy and has hired an actuary to calculate the necessary annual premium to reach that goal. What should the annual premium be?

17. The Accounting Club at a small university is holding a raffle to raise money for its activities. Five hundred tickets at $1.50 each have been sold. The first prize is $200, the second prize is $100, and the third prize is $50. Let X be the random variable that represents the net gain on one ticket. Determine the expected value of X.

18. Suppose a game consists of tossing a fair coin three times. The number of dollars you receive is equal to the square of the number of tails that came up. If you must pay $3 to play the game, is the game a fair game? (A fair game is defined to be a game for which the expected value per play is $0.)

19. Calculate the variance of the random variable Y of Example 6 of this section using the formula

$$\text{VAR}(X) = \sigma^2 = \left[\sum_{i=1}^{n} x_i^2 f(x_i)\right] - [E(X)]^2$$

20. Suppose both random variables X and Y have range $\{2, 4, 6, 8\}$. Suppose f and g are probability distributions of X and Y, respectively, defined by the following probability tables.

x	$f(x) = P_1(X = x)$	y	$g(y) = P_2(Y = y)$
2	.4	2	.5
4	.3	4	.2
6	.2	6	.1
8	.1	8	.2

a. Show that both X and Y have the same mathematical expectation.
b. Find the variance and standard deviation of X.
c. Find the variance and standard deviation of Y.
d. Draw a probability histogram for X and Y.

21. Suppose both random variables X and Y have range $\{1, 3, 5, 7, 9\}$. Suppose f and g are probability distributions of X and Y, respectively, defined by the following probability tables.

x	$f(x) = P_1(X = x)$	y	$g(y) = P_2(Y = y)$
1	.1	1	.15
3	.1	3	.25
5	.3	5	.05
7	.2	7	.05
9	.3	9	.5

a. Show that both X and Y have the same mathematical expectation.
b. Find the variance and standard deviation of X.
c. Find the variance and standard deviation of Y.
d. Draw a probability histogram for X and for Y.

22. Calculate the variance and standard deviation of the random variable X of Exercise 4.

23. Repeat Exercise 22 for the random variable of Exercise 6.

*24. The range of a random variable X is $\{x_1, x_2\}$ and f is the probability distribution of X. Show that $\text{VAR}(X) = x_1^2 f(x_1) + x_2^2 f(x_2) - [E(X)]^2$. (Hint: By definition $\text{VAR}(X) = [x_1 - E(X)]^2 f(x_1) + [x_2 - E(X)]^2 f(x_2)$. Expand $[x_1 - E(X)]^2$ and $[x_2 - E(X)]^2$, rearrange the terms and use the fact that $E(X) = x_1 f(x_1) + x_2 f(x_2)$ and that $f(x_1) + f(x_2) = 1$.)

*25. The range of a random variable X is $\{x_1, x_2, x_3\}$ and f is the probability distribution of X. Show that VAR$(X) = x_1^2f(x_1) + x_2^2f(x_2) + x_3^2f(x_3) - [E(X)]^2$. (*Hint:* See Exercise *24.)

*26. The range of a random variable X is $\{x_1, x_2, x_3, x_4\}$ and f is the probability distribution of X. Show that VAR$(X) = x_1^2f(x_1) + x_2^2f(x_2) + x_3^2f(x_3) + x_4^2f(x_4) - [E(X)]^2$. (*Hint:* See Exercise *24.)

6.6 The Binomial Distribution

In this section, we shall discuss a discrete probability distribution that is connected with repeated, but independent, trials of an experiment where only two outcomes are possible. There are many applications in which such experiments take place. When a television network conducts a survey, either a person is or is not watching a certain program; when the quality control of a manufacturing company tests an item randomly selected, it either is or is not defective, and so on.

The Binomial Theorem

The terms in the expansion of a power of a binomial are useful in describing the probability distributions of random variables. We begin with a brief discussion of the binomial theorem. You undoubtedly recall, from elementary algebra, at least the first few of the following formulas. (See also the "Getting Started" section.)

$$(a + b)^1 = a + b$$
$$(a + b)^2 = a^2 + 2ab + b^2$$
$$(a + b)^3 = a^3 + 3a^2b + 3ab^2 + b^3$$
$$(a + b)^4 = a^4 + 4a^3b + 6a^2b^2 + 4ab^3 + b^4$$
$$(a + b)^5 = a^5 + 5a^4b + 10a^3b^2 + 10a^2b^3 + 5ab^4 + b^5$$

and so on.

We observe the following for $n = 1, 2, 3, 4,$ and 5:

1. The expansion of $(a + b)^n$ has $n + 1$ terms.
2. The first term is a^n and the last is b^n.
3. After the first term of the expansion, in each successive term, the exponent assigned to b increases by one from 0 to n, while that assigned to a decreases by one, from n to 0. Thus the sum of the exponents for each term is n.
4. If we multiply the coefficient of any term, except the last, by the exponent assigned to a in that term and then divide by one more than the exponent assigned to b, we get the coefficient of the next term.
5. The coefficients of terms equidistant from the ends of the expansion are equal.
6. Recalling that $C(n, r) = \frac{n!}{(n - r)!r!}$, we can verify that the coefficient of each of the terms involving $a^{n-r}b^r$ is $C(n, r)$.

The observations we just made for $n = 1, 2, 3, 4,$ and 5 are special cases of the following general theorem. (See Exercise *41.)

> **THEOREM 6.3 (Binomial Theorem):** If n is a positive integer and a, b are real (or complex) numbers, then $(a + b)^n = C(n, 0)a^n + C(n, 1)a^{n-1}b^1 + C(n, 2)a^{n-2}b^2 + \cdots + C(n, n-1)a^1b^{n-1} + C(n, n)b^n$. Using the sigma notation,
>
> $$(a + b)^n = \sum_{r=0}^{n} C(n, r)a^{n-r}b^r$$

EXAMPLE 1   Write the expansion of $(a + b)^7$.

Solution   The expansion will have eight terms. The first and last terms are a^7 and b^7, respectively. Without coefficients, the terms are a^7, a^6b^1, a^5b^2, a^4b^3, a^3b^4, a^2b^5, a^1b^6, and b^7. The coefficients are $C(7, 0)$, $C(7, 1)$, $C(7, 2)$, $C(7, 3)$, $C(7, 4)$, $C(7, 5)$, $C(7, 6)$, and $C(7, 7)$. These coefficients are easily calculated using the formula

$$C(n, r) = \frac{n!}{(n - r)!r!}$$

They are 1, 7, 21, 35, 35, 21, 7, 1. As an example, we show one of the calculations.

$$C(7, 3) = \frac{7!}{(7 - 3)!3!} = \frac{7(6)(5)(4!)}{4!3!} = \frac{7(6)(5)}{3(2)(1)} = 35$$

Thus,

$$(a + b)^7 = a^7 + 7a^6b + 21a^5b^2 + 35a^4b^3 + 35a^3b^4 + 21a^2b^5 + 7ab^6 + b^7$$

Do Exercise 1.

Bernoulli Trials

We are now ready to begin the discussion of the main topic of this section. Repeated independent trials of an experiment are said to be *Bernoulli trials* if there are only two possible outcomes for each trial and if their probabilities remain the same throughout the n trials of the experiment. Jakob (Jacques) Bernoulli (1654–1705) was a member of a Swiss family of mathematicians and scientists. He was one of the first people who undertook the systematic study of probability problems related to experiments with only two possible outcomes.

In working with Bernoulli experiments, it is customary to refer to one of the two outcomes as a *success*, denoted S, and to the other as a *failure*, denoted F. If we let the probability of success be p and the probability of failure be q, we have

$$P(S) + P(F) = p + q = 1$$

For example, if we roll a fair die and ask for the probability of a 5 showing, we may consider the outcome "a 5 showing" a success and the outcome "any of the numbers 1, 2, 3, 4, or 6 showing" a failure, then $P(S) = \frac{1}{6}$ and $P(F) = \frac{5}{6}$.

Suppose a Bernoulli trial is repeated n times. We may want to calculate the probability that exactly r of these trials were successes, where $0 \leq r \leq n$.

In applying our results to real-life situations, there will be very few cases where the conditions described above are perfectly satisfied. However, in most cases, the assumption that the conditions are satisfied will not affect the results. For example, suppose 30% of a population of 15,000 is watching a football game on a Sunday afternoon. If the television station calls one of these people, the probability that this person is watching the game will be $\frac{3}{10}$, but the probability that the second person called is watching will be $\frac{4500}{14,999}$, or $\frac{4499}{14,999}$, depending on whether the first person called was, or was not, watching the game. Clearly, these numbers are near, but not equal, to .3. For all practical purposes, we may assume that the probability remains .3 as long as the number of people called remains small. This again illustrates that when mathematics is used in real-life applications, the mathematical model only approximates the situation. This creates no difficulty as long as we are careful in choosing a model that best fits the problem at hand.

Binomial Distribution

In working with Bernoulli trials, we shall use the following rule.

Suppose that there are n independent trials of an experiment and that the probability of outcome o_i in the ith trial is $p(o_i)$. Then, the probability of the event $(o_1, o_2, \ldots, o_n)$ occurring is given by

$$p(o_1, o_2, \ldots, o_n) = p(o_1)p(o_2) \cdots p(o_n)$$

The next two examples illustrate the basic idea behind a binomial distribution.

EXAMPLE 2 Suppose a coin is weighted so that the probability that tails comes up on any toss is $\frac{3}{4}$. The coin is tossed five times. What is the probability that heads will come up on the second and fourth tosses and tails will come up on the first, third, and fifth tosses?

Solution For convenience, success means the result of the toss is tails and failure means the coin came up heads. Then $p(S) = \frac{3}{4}$ and $p(F) = \frac{1}{4}$ since $1 - \frac{3}{4} = \frac{1}{4}$. Furthermore, let S_i and F_j denote success and failure in the ith and jth toss, respectively. We want to find the probability of $(S_1, F_2, S_3, F_4, S_5)$. This probability is found as follows:

$$p(S_1, F_2, S_3, F_4, S_5) = p(S_1)p(F_2)p(S_3)p(F_4)p(S_5)$$
$$= \left(\frac{3}{4}\right)\left(\frac{1}{4}\right)\left(\frac{3}{4}\right)\left(\frac{1}{4}\right)\left(\frac{3}{4}\right) = \left(\frac{3}{4}\right)^3\left(\frac{1}{4}\right)^2 = \frac{27}{1024}$$

Do Exercise 3. ∎

Before considering the next example, we should note that since multiplication is commutative, we would have obtained exactly the same result had we been asked to calculate the probability that tails will come up on the second, fourth, and fifth tosses and heads will come up on the first and third tosses.

EXAMPLE 3   If we use the coin of Example 2 and toss it five times, what is the probability that tails will come up exactly three times?

Solution   For convenience, success means that tails came up exactly three times. Then, the outcome $(S_1, F_2, S_3, F_4, S_5)$ is a success, but so are the outcomes $(F_1, S_2, S_3, F_4, S_5)$ and $(S_1, F_2, F_3, S_4, S_5)$. Each of these outcomes has probability $\left(\frac{3}{4}\right)^3\left(\frac{1}{4}\right)^2$, as was found in Example 2. We must consider all possible outcomes for which we have exactly three S's and two F's. Note that in $(S_1, F_2, S_3, F_4, S_5)$ the S's have subscripts 1, 3, and 5, while in $(F_1, S_2, S_3, F_4, S_5)$ they have subscripts 2, 3, and 5, and in $(S_1, F_2, F_3, S_4, S_5)$, the subscripts for S are 1, 4, and 5. Therefore, with each successful outcome, we can associate a subset of three elements chosen from the set $\{1, 2, 3, 4, 5\}$. We know, however, that the number of such subsets is $C(5, 3)$, which is easily found to be 10. Thus, we have exactly ten outcomes in a successful event and each has probability $\left(\frac{3}{4}\right)^3\left(\frac{1}{4}\right)^2$. We could add the probabilities, but the ten terms of that sum are equal, so we obtain the same result by multiplying the common term by 10. Therefore, the probability that exactly three tails come up when the weighted coin is tossed five times is $\frac{135}{512}$, since $10\left(\frac{3}{4}\right)^3\left(\frac{1}{4}\right)^2 = C(5, 3)\left(\frac{3}{4}\right)^3\left(\frac{1}{4}\right)^2 = \frac{135}{512}$.
Do Exercise 9.   ∎

Because the number of outcomes in a successful event corresponds to a coefficient of the expansion of a power of a binomial, an experiment such as the one in the preceding example is called a *binomial experiment* and the probability distribution of the number of successes is called a *binomial distribution*.

The result of the preceding example can be generalized to any number n of Bernoulli trials. This generalization is stated in the following theorem.

> **THEOREM 6.4:**   Suppose that in a binomial experiment the probabilities of success and failure in any trial are p and q, respectively. Let X be the number of successes in n independent trials. Then, the probability distribution f of X is given by
>
> $$f(x) = P(X = x) = C(n, x)p^x q^{n-x}$$
>
> where $p + q = 1$ and x is an integer such that $0 \le x \le n$.

Any random variable with this probability distribution is called a *binomial random variable* and is said to have a *binomial distribution*.*

EXAMPLE 4   Evaluate $b\left(3; 7, \frac{1}{3}\right)$.

---

*In many textbooks on probability, the symbol $b(x; n, p)$ is used to denote $C(n, x)p^x q^{n-x}$.

Solution   We have

$$b\left(3; 7, \frac{1}{3}\right) = C(7, 3)\left(\frac{1}{3}\right)^3\left(\frac{2}{3}\right)^{7-3}$$

$$= \frac{7!}{4!3!}\left(\frac{1}{3}\right)^3\left(\frac{2}{3}\right)^4 = 35 \cdot \frac{1}{27} \cdot \frac{16}{81} = \frac{560}{2187}$$

Do Exercise 15.

EXAMPLE 5   Suppose X is a binomial random variable with $p = \frac{1}{4}$ and $n = 4$. Find the probability distribution of X.

Solution   We have $q = 1 - \frac{1}{4} = \frac{3}{4}$. We shall use the formula

$$f(x) = P(X = x) = C(n, x)p^x q^{n-x}$$

with $n = 4$ and $x = 0, 1, 2, 3, 4$.

We get

$$f(0) = P(X = 0) = C(4, 0)\left(\frac{1}{4}\right)^0\left(\frac{3}{4}\right)^4 = \frac{4!}{4!0!} \cdot 1 \cdot \frac{81}{256} = \frac{81}{256}$$

$$f(1) = P(X = 1) = C(4, 1)\left(\frac{1}{4}\right)^1\left(\frac{3}{4}\right)^3 = \frac{4!}{3!1!} \cdot \frac{1}{4} \cdot \frac{27}{64} = \frac{27}{64}$$

$$f(2) = P(X = 2) = C(4, 2)\left(\frac{1}{4}\right)^2\left(\frac{3}{4}\right)^2 = \frac{4!}{2!2!} \cdot \frac{1}{16} \cdot \frac{9}{16} = \frac{27}{128}$$

$$f(3) = P(X = 3) = C(4, 3)\left(\frac{1}{4}\right)^3\left(\frac{3}{4}\right)^1 = \frac{4!}{1!3!} \cdot \frac{1}{64} \cdot \frac{3}{4} = \frac{3}{64}$$

$$f(4) = P(X = 4) = C(4, 4)\left(\frac{1}{4}\right)^4\left(\frac{3}{4}\right)^0 = \frac{4!}{0!4!} \cdot \frac{1}{256} \cdot 1 = \frac{1}{256}$$

Note that

$$f(0) + f(1) + f(2) + f(3) + f(4) = \frac{81}{256} + \frac{27}{64} + \frac{27}{128} + \frac{3}{64} + \frac{1}{256}$$

$$= \frac{81 + 108 + 54 + 12 + 1}{256} = \frac{256}{256} = 1$$

This result is expected since f is a probability distribution.
Do Exercise 17.

As we have seen earlier, it is sometimes advantageous when we wish to find the probability of an event E, to compute the probability of the complement E' and then use the fact that $P(E) + P(E') = 1$, as in the following example.

EXAMPLE 6   A fair die is rolled ten times. Find the probability of a five coming up at least twice.

Solution   Let X be the number of times that a five comes up. Then X is a discrete random variable with a binomial distribution. Here $n = 10$, $p = \frac{1}{6}$, and $q = \frac{5}{6}$. The computations will be simplified if we first find the probability that the number of times a five came up is less than two.

$$P(X < 2) = P(X = 0) + P(X = 1)$$

$$= C(10, 0)\left(\frac{1}{6}\right)^0\left(\frac{5}{6}\right)^{10} + C(10, 1)\left(\frac{1}{6}\right)^1\left(\frac{5}{6}\right)^9$$

$$= \frac{10!}{10!0!}\left(\frac{1}{6}\right)^0\left(\frac{5}{6}\right)^{10} + \frac{10!}{9!1!}\left(\frac{1}{6}\right)^1\left(\frac{5}{6}\right)^9$$

$$= 1 \cdot 1 \cdot \frac{5^{10}}{6^{10}} + 10 \cdot \frac{1}{6} \cdot \frac{5^9}{6^9} = \left(\frac{5}{6}\right)^{10}(1 + 2) \doteq .48452$$

Therefore, the probability that a five comes up at least twice in ten rolls of a fair die is .51548 since $1 - .48452 = .51548$. The answer has been rounded off to five decimal places.

Do Exercise 21. ■

The formulas used to calculate the mathematical expectation and variance of a binomial random variable are simple.

> **THEOREM 6.5:** Suppose that in a binomial experiment the probabilities of success and failure in any trial are p and q, respectively. Let X be the number of successes in n independent trials. Then, the mathematical expectation (mean), variance, and standard deviation of X are given by $E(X) = np$, $\text{VAR}(X) = npq$, $\sigma = \sqrt{npq}$, respectively.

A CLOSER LOOK The proof of the first of these formulas is given for those of you who feel at ease with manipulations involving the sigma notation and factorials. It may be omitted without any loss of understanding of the concepts. The proof of the formula for the variance is more involved and is left as an exercise. (See Exercise **40.)

Proof: (Optional) The possible values of X are $0, 1, 2, \cdots, n$. Thus, if f is the probability distribution of X, we have

$$E(X) = 0 \cdot f(0) + 1 \cdot f(1) + 2 \cdot f(2) + \cdots + n \cdot f(n)$$

But in this case, $f(k) = b(k; n, p) = C(n, k)p^kq^{n-k}$. Thus,

$$E(X) = 0 + \sum_{k=1}^{n} kb(k; n, p) = \sum_{k=1}^{n} kC(n, k)p^kq^{n-k}$$

$$= \sum_{k=1}^{n} k\frac{n!}{(n-k)!k!}p^kq^{n-k}$$

$$= np \sum_{k=1}^{n} \frac{(n-1)!}{(n-k)!(k-1)!}p^{k-1}q^{n-k}$$

$$= np(C(n-1, 0)q^{n-1} + C(n-1, 1)q^{n-2}p^1 + C(n-1, 2)q^{n-3}p^2$$
$$+ \cdots + C(n-1, n-2)q^1p^{n-2} + C(n-1, n-1)p^{n-1})$$
$$= np(q + p)^{n-1} = np(1)^{n-1} = np$$

EXAMPLE 7   Find the mean, variance, and standard deviation for the binomial random variable of Example 5.

Solution   Using the formula given in Theorem 6.5 with $n = 4$, $p = \frac{1}{4}$, and $q = \frac{3}{4}$, we get

$$E(X) = 4\left(\frac{1}{4}\right) = 1$$

$$\text{VAR}(X) = 4\left(\frac{1}{4}\right)\left(\frac{3}{4}\right) = \frac{3}{4}$$

and

$$\sigma = \sqrt{\text{VAR}(X)} = \sqrt{\frac{3}{4}} = \frac{\sqrt{3}}{2}$$

Do Exercise 33.   ■

Although it is not necessary to do so, since we already have obtained the desired results, we shall calculate these quantities using the results of Example 5 to compare the efficiency of the two methods.

In Example 5, the probability distribution was $f(0) = \frac{81}{256}$, $f(1) = \frac{27}{64}$, $f(2) = \frac{27}{128}$, $f(3) = \frac{3}{64}$, and $f(4) = \frac{1}{256}$. Thus,

$$E(X) = 0 \cdot \frac{81}{256} + 1 \cdot \frac{27}{64} + 2 \cdot \frac{27}{128} + 3 \cdot \frac{3}{64} + 4 \cdot \frac{1}{256}$$

$$= \frac{108 + 108 + 36 + 4}{256} = 1$$

$$\text{VAR}(X) = 0^2 f(0) + 1^2 f(1) + 2^2 f(2) + 3^2 f(3) + 4^2 f(4) - E(X)^2$$

$$= 0^2 \cdot \frac{81}{256} + 1^2 \cdot \frac{27}{64} + 2^2 \cdot \frac{27}{128} + 3^2 \cdot \frac{3}{64} + 4^2 \cdot \frac{1}{256} - 1^2$$

$$= \frac{108 + 216 + 108 + 16 - 256}{256} = \frac{3}{4}$$

As we expect, the results are the same.

Testing Small Samples

One of the important applications of binomial experiments is in making decisions about whether to accept or reject a large number of items on the basis of evidence on a small sample. We conclude this section with an example to illustrate this type of problem.

EXAMPLE 8   An auto supply company has a sampling scheme to inspect large lots of incoming car batteries. Twenty batteries are to be tested and the lot rejected if three or more are found to be defective. Suppose that a lot contains exactly 4% defective batteries. What is the probability that the lot will be accepted?

Solution   Let X be the number of defective batteries found. Then the lot will be accepted if $X < 3$. Thus, we must find $P(X < 3)$. But, $P(X < 3) = P(X = 0) + P(X = 1) + P(X = 2)$. Note that the probability of the first battery tested being defective is .04. Strictly speaking, the probability that the second battery tested is defective depends on whether or not the first battery tested was defective. However, since we were told that the lot of batteries was large, we may assume that for each of the 20 batteries tested, the probability that it is defective remains .04. Hence, $p = .04$, $q = .96$ (since $1 - .04 = .96$), and $n = 20$. It follows that

$$P(X < 3) = P(X = 0) + P(X = 1) + P(X = 2)$$
$$= C(20, 0)(.04)^0(.96)^{20} + C(20, 1)(.04)^1(.96)^{19}$$
$$+ C(20, 2)(.04)^2(.96)^{18}$$
$$= 1 \cdot 1 \cdot (.96)^{20} + 20(.04)(.96)^{19} + 190(.04)^2(.96)^{18}$$
$$= (.96)^{18}[(.96)^2 + 20(.04)(.96) + 190(.04)^2]$$
$$\doteq .4796(1.9936) \doteq .95614$$

Therefore, the probability that a lot is accepted is .95614. Consequently, the probability that a lot is rejected is .04386 since $1 - .95614 = .04386$.
Do Exercise 37.   ■

Exercise Set 6.6

1. Write the expansion of $(a + b)^6$.

2. Write the expansion of $(2x + 3y)^8$.

In Exercises 3–14, assume that the trials are independent, although in some cases this is not actually true. However, in each case, the values of p and q remain nearly constant so that this assumption is justified.

3. Suppose a coin is weighted so that the probability that heads comes up on any toss is $\frac{5}{7}$. The coin is tossed five times. What is the probability that heads will come up on the second and fifth tosses, and tails will come up on the first, third, and fourth tosses?

4. Suppose a coin is weighted so that the probability that tails comes up on any toss is $\frac{8}{13}$. The coin is tossed six times. What is the probability that heads will come up on the second and fourth tosses, and tails will come up on the first, third, fifth, and sixth tosses?

5. Suppose a die is weighted so that the probability that a six comes up on any roll is $\frac{1}{4}$. The die is rolled five times. What is the probability that a six comes up only on the second and fourth rolls?

6. Suppose a die is weighted so that the probability that a two comes up on any roll is $\frac{4}{13}$. The die is rolled seven times. What is the probability that a two comes up only on the first, second, and fourth rolls?

7. Suppose the coin of Exercise 3 is tossed five times. What is the probability that exactly three tails will come up?

8. Repeat Exercise 7 if the coin is tossed eight times and we want exactly five tails to come up.

9. Suppose the coin of Exercise 4 is tossed eight times. What is the probability that exactly three heads will come up?

10. Repeat Exercise 9 if the coin is tossed ten times and we want exactly six tails to come up.

11. The die of Exercise 5 is rolled seven times. What is the probability that a six comes up exactly four times?

12. Repeat Exercise 11 if the die is rolled nine times and we want the probability of a six coming up exactly five times.

13. The die of Exercise 6 is rolled eight times. What is the probability that a two comes up exactly four times?

14. Repeat Exercise 13 if the die is rolled 11 times and we want the probability of a two coming up exactly four times.

15. Evaluate the following:

    a. $b(2; 8, .2)$ **b.** $b(3; 10, .3)$ **c.** $b(9; 13, .4)$

16. Evaluate the following:

    a. $b(3; 9, .6)$ **b.** $b(6; 11, .2)$ **c.** $b(7; 14, .1)$

17. Suppose X is a random variable with $p = \frac{1}{5}$ and $n = 4$.

    a. Find the probability distribution of X.
    b. Draw a probability histogram for X.

18. Repeat Exercise 17 with $p = \frac{2}{3}$ and $n = 5$.

19. Repeat Exercise 17 with $p = \frac{3}{5}$ and $n = 6$.

20. A fair die is rolled 12 times. Find the probability that a six comes up at least twice.

21. A fair die is rolled 20 times. Find the probability that a one comes up at least three times.

22. Suppose a basketball player hits 76% of the free throws he attempts. If he attempts eight free throws in a game, what is the probability that **a.** he will hit exactly six free throws? **b.** he will hit at least six free throws?

23. Suppose a basketball player hits 80% of the free throws she attempts. If she attempts eight free throws in a game, what is the probability that **a.** she will hit exactly seven free throws? **b.** she will hit at least seven free throws?

24. In a field goal kicking contest, John is successful on 55% of his attempts and he needs at least six successful kicks in order to win. He has ten attempts left in the contest. What is the probability that he will win?

25. If there are six children in a family, what is the probability that at least two are boys? (Assume that the probability that a child is a boy is $\frac{1}{2}$.)

26. If there are seven children in a family, what is the probability that at least four are girls? (Assume that the probability that a child is a girl is $\frac{1}{2}$.)

27. A company wins a contract on 65% of the bids it submits. If the company plans to bid on ten projects next year and if it must win at least four of the contracts to have a successful year, what is the probability that next year will be successful?

28. Mr. Nonbright, who is a student in a mathematics course, is taking a true-false examination which has 20 questions.

If he guesses the answer for each question, what is the probability that he gets **a.** all twenty correct? **b.** at least eighteen correct? **c.** at least fifteen correct?

29. Repeat Exercise 28 if the test is multiple choice and five answers are given for each question, only one of which is correct.

30. What is the expected score of the student in Exercise 28?

31. What is the expected score of the student in Exercise 29?

32. Compute the expected value (mean), variance, and standard deviation of the binomial random variable with $n = 8$ and $p = .3$.

33. Repeat Exercise 32 with $n = 12$ and $p = .4$.

34. Repeat Exercise 32 with $n = 15$ and $p = .2$.

35. Repeat Exercise 32 with $n = 18$ and $p = .3$.

36. Repeat Exercise 32 with $n = 20$ and $p = .35$.

37. A hardware store has a sampling scheme to inspect large lots of incoming light bulbs. Fifteen light bulbs are to be tested and the lot rejected if two or more are found to be defective. Suppose that a lot contains exactly 3% defective light bulbs. What is the probability that the lot will be accepted?

38. The quality control department of a large chain of computer stores established the following sampling scheme to inspect large lots of incoming computers. Fifteen computers would be tested and the lot rejected if two or more were found to be defective. Suppose that a lot contains exactly 3% defective computers. What is the probability that the lot will be accepted?

*39. The lines in the proof of Theorem 6.4 have been numbered. Give a justification for claiming equality between

    a. lines (1) and (2)   **b.** lines (2) and (3)
    c. lines (3) and (4)   **d.** lines (4) and (5)

$$E(X) = 0 + \sum_{k=1}^{n} kb(k; n, p) = \sum_{k=1}^{n} kC(n, k)p^k q^{n-k} \quad (1)$$

$$= \sum_{k=1}^{n} k\frac{n!}{(n - k)!k!}p^k q^{n-k} \quad (2)$$

$$= np \sum_{k=1}^{n} \frac{(n - 1)!}{(n - k)!(k - 1)!}p^{k-1}q^{n-k} \quad (3)$$

$$= np(C(n - 1, 0)q^{n-1} +$$
$$C(n - 1, 1)q^{n-2}p^1$$
$$+ C(n - 1, 2)q^{n-3}p^2 +$$
$$\cdots + C(n - 1, n - 2)q^1 p^{n-2}$$
$$+ C(n - 1, n - 1)p^{n-1}) \quad (4)$$

$$= np(q + p)^{n-1} = np(1)^{n-1} = np \quad (5)$$

**40. Prove the formula $\text{VAR}(X) = npq$ which appeared in Theorem 6.5. (*Hint:* Start with $\text{VAR}(X) = [0^2 f(0) + 1^2 f(1) + 2^2 f(2) + \cdots + n^2 f(n)] - E(X)^2 = [0^2 f(0) + 1^2 f(1) + 2^2 f(2) + \cdots + n^2 f(n)] - (np)^2$. Then replace each $f(k)$ by $C(n, k)p^k q^{n-k}$. Factor np out of the sum in brackets and simplify each term. The sum in brackets should reduce to $[1 \cdot b(0; n - 1, p) + 2 \cdot b(1; n - 1, p) + 3 \cdot b(2; n - 1, p) + \cdots + n \cdot b(n - 1; n - 1, p)]$. Each term can be written as the sum of two terms. For example, $3 \cdot b(2; n - 1, p)$ can be written $2 \cdot b(2; n - 1, p) +$ $b(2; n - 1, p)$. After doing this, rearrange the terms and group them into two sums so that one sum is the mean of a binomial random variable for an experiment with $n - 1$ trials each with probability of success p (thus equal to $(n - 1)p$), and the other sum is the sum of all values of a probability distribution (thus equal to 1). Then simplify and use the fact that $1 - p = q$.)

*41. Using mathematical induction, prove the binomial theorem.

6.7 More on Markov Chains

Markov chains were introduced in Section 3.8 because they provided an interesting application of matrix multiplication. However, this introduction was limited by the fact that we had not discussed probability prior to that introduction. In this section, we give a few more examples involving Markov chains and we shall use some of the results we obtained elsewhere in this chapter. You are urged to read Section 3.8 before proceeding. However, we shall state again the main concepts.

Review of Definitions

We have seen that if we have a finite sequence of independent trials with possible outcomes $O_1, O_2, \ldots, O_n$, where for each i the probability of O_i is p_i, then the probability of $(O_1, O_2, \ldots, O_n)$ is the product $p_1 p_2 \cdots p_n$. In the theory of Markov chains, we generalize this result slightly by making the outcome of any trial, except the first, dependent on the outcome of the immediately preceding trial, and only on that outcome. Thus, to every ordered pair of outcomes (O_i, O_j), there corresponds a conditional probability p_{ij}. That is, given that O_i has occurred at some trial, the probability that O_j will occur at the next trial is p_{ij}. In addition, we must be given the probability p_k of the outcome O_k occurring at the initial trial. The vector $[\, p_1 \quad p_2 \quad \cdots \quad p_n \,]$ is called the *initial-state vector*. We must have $p_k \geq 0$ for each k, and $p_1 + p_2 + \cdots + p_n = 1$, since the p_k's are probabilities and one of the outcomes $O_1, O_2, \ldots, O_n$ must occur. The matrix

$$\begin{bmatrix} p_{11} & p_{12} & p_{13} & \cdots & p_{1n} \\ p_{21} & p_{22} & p_{23} & \cdots & p_{2n} \\ p_{31} & p_{32} & p_{33} & \cdots & p_{3n} \\ \vdots & \vdots & \vdots & & \vdots \\ p_{n1} & p_{n2} & p_{n3} & \cdots & p_{nn} \end{bmatrix}$$

is called the *transition matrix*. If the possible outcomes of any trial are $O_1, O_2, \ldots, O_n$, and if during the experiment O_i has just occurred in the latest trial, we say that the experiment is in the state O_i. If outcome O_j occurs in the next trial, we say that the experiment has gone through the *transition* from state O_i to state O_j. In the transition matrix, all p_{ij}'s must be nonnegative since they are probabilities, and for any i, the sum $p_{i1} + p_{i2} + \cdots + p_{in} = 1$, since when the experiment is in state O_i, it must be in one of the states $O_1, O_2, \ldots, O_n$ in the next trial. A *state vector* is a vector $[\begin{array}{cccc} p_1 & p_2 & p_3 & \cdots & p_n \end{array}]$ where each entry p_i gives the probability of the system being in state O_i. Again, we must have $p_k \geq 0$ for each k, and $p_1 + p_2 + \cdots + p_n = 1$.

The examples that follow are given to clarify the preceding definitions.

EXAMPLE 1   Professor Meany, who teaches a mathematics course for business students, gives frequent pop quizzes. Members of a fraternity have studied the pattern he follows. They have established that if he gives a quiz on a certain day, the probability that he will give a quiz the next class day is 0.1. On the other hand, if he does not give a quiz, then the probability that he will give a quiz the next class day is 0.3. He never gives a quiz the first week of the semester, but the probability that he gives a quiz on the first day of the second week is 0.4.

a. Find the probability that he will give a quiz on the second day of the second week.
b. Find the probability that he will not give a quiz the second day of the second week.

Solution   **a.** We first solve the problem using a probability tree. We let Q_i represent the outcome "he gives a quiz on the ith day of the second week" and Q_i' the outcome "he does not give a quiz on the ith day of the second week." We were given $p(Q_1) = 0.4$, hence $p(Q_1') = .6$ since $1 - .4 = .6$. Similarly, it was given that $p(Q_2 \mid Q_1) = 0.1$ and $p(Q_2 \mid Q_1') = 0.3$. Consequently, $p(Q_2' \mid Q_1) = 0.9$ and $p(Q_2' \mid Q_1') = 0.7$. We have drawn a probability tree showing this information in Figure 6.17.

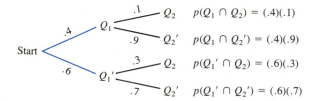

FIGURE 6.17

Therefore,

$$p(Q_2) = p(Q_1 \cap Q_2) + p(Q_1' \cap Q_2)$$
$$= (0.4)(0.1) + (0.6)(0.3) = 0.22$$

The probability that he will give a quiz the second day of the second week is 0.22.

b. We can obtain the probability that he will not give a quiz the second day of the second week immediately since $1 - .22 = .78$. However, using the probability tree, we get

$$p(Q_2') = p(Q_1 \cap Q_2') + p(Q_1' \cap Q_2') = (0.4)(0.9) + (0.6)(0.7) = 0.78$$

as expected.   ■

If we were asked to find the probability that Professor Meany gives a quiz on the fifth day of the second week, using a probability tree would be inefficient because the number of branches that would have to be drawn would be too large. In this case, it is best to use matrix multiplication as we now illustrate.

EXAMPLE 2   Repeat Example 1 by first writing the appropriate transition matrix and initial-state vector and then showing that the first-state vector is the product of the initial-state vector by the transition matrix.

Solution   If we think of "he gives a quiz" as outcome O_1 and "he does not give a quiz" as outcome O_2, then the probability p_{11} is 0.1. We know this because the probability that he gives a quiz one day if he gave a quiz the previous day is 0.1. Also, $p_{12} = 0.9$, since we must have $p_{11} + p_{12} = 1$. Similarly, $p_{21} = 0.3$ since we know that if he does not give a quiz one day, the probability that he gives a quiz the next day is 0.3. Thus, p_{22} is 0.7 because $p_{21} + p_{22} = 1$. So, the transition matrix is

$$\begin{bmatrix} .1 & .9 \\ .3 & .7 \end{bmatrix}$$

For the initial-state vector, it was given that on the first day of the second week, the probability that "he gives a quiz" is 0.4. Thus, $p_1 = 0.4$. Consequently, $p_2 = 0.6$ since we must have $p_1 + p_2 = 1$. Therefore, the initial-state vector is [.4   .6]. If we multiply the initial-state vector (which is a 1×2 matrix) by the 2×2 transition matrix, we will get a 1×2 matrix (a vector). The first entry of that vector is the dot product of [.4   .6] by the first column of the transition matrix. It is calculated as follows:

$$(.4)(.1) + (.6)(.3) = .22$$

Compare this calculation to that of $p(Q_2)$ in Example 1. Similarly, the second entry of the product is the dot product of [.4   .6] by the second column of the transition matrix. It is calculated as follows:

$$(.4)(.9) + (.6)(.7) = .78$$

Compare this calculation to that of $p(Q_2')$ in Example 1. We can summarize the calculations as follows:

$$[.4 \quad .6] \begin{bmatrix} .1 & .9 \\ .3 & .7 \end{bmatrix} = [.22 \quad .78]$$

The second-state vector is [.22   .78]. ∎

EXAMPLE 3   What is the probability that Professor Meany of Example 1 will give a pop quiz on the following days.

a. On the third day of the second week?
b. On the fourth day of the second week?
c. On the fifth day of the second week?

Solution   **a.** We wish to get a third-state vector. We simply multiply the second-state vector by the transition matrix.

$$[.22 \quad .78] \begin{bmatrix} .1 & .9 \\ .3 & .7 \end{bmatrix} = [.256 \quad .744]$$

Thus, the probability that Professor Meany will give a pop quiz on the third day of the second week is 0.256.

b. We now multiply the third-state vector by the transition matrix to get the fourth-state vector. We get

$$[\ .256 \quad .744 \ ] \begin{bmatrix} .1 & .9 \\ .3 & .7 \end{bmatrix} = [\ .2488 \quad .7512 \ ]$$

We see that the probability that the professor will give a pop quiz on the fourth day of the second week is 0.2488.

c. We get the fifth-state vector by multiplying the fourth-state vector by the transition matrix.

$$[\ .2488 \quad .7512 \ ] \begin{bmatrix} .1 & .9 \\ .3 & .7 \end{bmatrix} = [\ .25024 \quad .74976 \ ]$$

The probability that he gives a quiz on the fifth day of the second week is 0.25024. ∎

Steady-State Vector

In the preceding example, the fourth- and fifth-state vectors seem to be approaching the vector [.25   .75]. In fact, calculating the sixth- and seventh-state vectors, we get [.249952   .750048] and [.2500096   .7499904], respectively. Multiplying the vector [.25   .75] by the transition matrix, we get

$$[\ .25 \quad .75 \ ] \begin{bmatrix} .1 & .9 \\ .3 & .7 \end{bmatrix} = [\ .25 \quad .75 \ ]$$

We see that the transition of the state vector [.25   .75] leads to the same state vector. This vector is called the *steady-state vector*. In the preceding example, the fact that the steady-state vector is [.25   .75] means that after a few days, the probability that Professor Meany gives a pop quiz on an arbitrary day will be 0.25.
Do Exercise 15.

Recall that a transition matrix **T** is said to be regular if for some positive integer n, all entries of the nth power of **T** are positive. As we observed in Section 3.8, if **T** is a regular transition matrix, then the successive state vectors approach some fixed state vector **v** regardless of the initial-state vector. (See also Theorem 6.6.)

EXAMPLE 4   Let $\mathbf{T} = \begin{bmatrix} .1 & 0 & .9 \\ .2 & .3 & .5 \\ .6 & .2 & .2 \end{bmatrix}$ be a transition matrix.

a. Show that **T** is regular.
b. Find the steady-state vector.

Solution **a.**

$$\mathbf{T}^2 = \begin{bmatrix} .1 & 0 & .9 \\ .2 & .3 & .5 \\ .6 & .2 & .2 \end{bmatrix} \cdot \begin{bmatrix} .1 & 0 & .9 \\ .2 & .3 & .5 \\ .6 & .2 & .2 \end{bmatrix}$$

$$= \begin{bmatrix} .55 & .18 & .27 \\ .38 & .19 & .43 \\ .22 & .10 & .68 \end{bmatrix}$$

All entries of $\mathbf{T}^2$ are positive. Hence $\mathbf{T}$ is regular.

b. Since the transition matrix is regular, we know that a steady-state vector exists. Let $[\, x \quad y \quad z \,]$ be this vector. Then, we must have

$$[\, x \quad y \quad z \,] \cdot \begin{bmatrix} .1 & 0 & .9 \\ .2 & .3 & .5 \\ .6 & .2 & .2 \end{bmatrix} = [\, x \quad y \quad z \,]$$

Performing the multiplication on the left, we get a three-dimensional row vector that is equal to the vector on the right. Because corresponding entries of these vectors are equal, we obtain the system

$$\begin{cases} .1x + .2y + .6z = x \\ \quad\quad .3y + .2z = y \\ .9x + .5y + .2z = z \end{cases}$$

which may be written

$$\begin{cases} -.9x + .2y + .6z = 0 & \quad (1) \\ \quad\quad -.7y + .2z = 0 & \quad (2) \\ .9x + .5y - .8z = 0 & \quad (3) \end{cases}$$

Adding equations (1) and (2) and multiplying the resulting equation by -1 gives equation (3). We may omit that equation. Since the steady-state vector is a probability vector, we have $x + y + z = 1$. Therefore, we must solve the system

$$\begin{cases} -.9x + .2y + .6z = 0 \\ \quad\quad -.7y + .2z = 0 \\ x + y + z = 1 \end{cases}$$

The solution is $x = \frac{230}{635}$, $y = \frac{90}{635}$, and $z = \frac{315}{635}$. The steady-state vector is

$$\mathbf{v} = \begin{bmatrix} \dfrac{230}{635} & \dfrac{90}{635} & \dfrac{315}{635} \end{bmatrix} \doteq [\, .3622 \quad .1417 \quad .4961 \,]$$

Note that the three entries are between 0 and 1, as they should be since they are probabilities. You are encouraged to verify that the sum of the three entries in the vector $\mathbf{v}$ is 1, and that the product $\mathbf{vT}$ is $\mathbf{v}$.

Do Exercise 5. ■

In many cases, we have a fixed transition matrix $\mathbf{T}$ for a Markov chain and an initial-state vector $\mathbf{v}_0$. We are then asked to find a state vector $\mathbf{v}_k$ for the kth trial of the experiment. In such cases, we proceed to find $\mathbf{v}_1$ by calculating the product $\mathbf{v}_0\mathbf{T}$, then we find v_2 by calculating the product v_1T, and so on, until we find v_k by calculating the product $v_{k-1}T$. In other cases, we have a fixed transition matrix but we wish to find state vectors v_k using different initial-state vectors. For this, it is best to first find the kth power T^k of T and use the fact that $v_k = v_0T^k$. Once T^k has been found, we need only multiply any initial-state vector v_0 by T^k to find the corresponding state vector v_k. The matrix T^k is often called the *k-step transition matrix* for the Markov chain.

EXAMPLE 5   The transition matrix for a Markov process is

$$T = \begin{bmatrix} .2 & .8 \\ .6 & .4 \end{bmatrix}$$

Find the state vector for the fifth trial if the initial state vector is

a. [1   0]   **b.** [0   1]   **c.** [.4   .6]

d. [.2   .8]   **e.** [.3   .7]   **f.** [.9   .1]

Solution   **a.** If we let v_0 be the initial-state vector for the first trial, then $v_1 = v_0T$ is the state vector for the second trial, $v_2 = v_0T^2$ is the state vector for the third trial, $v_3 = v_0T^3$ is the state vector for the fourth trial, and $v_4 = v_0T^4$ is the state vector for the fifth trial. We first find T^4 as follows:

$$T^2 = \begin{bmatrix} .2 & .8 \\ .6 & .4 \end{bmatrix} \cdot \begin{bmatrix} .2 & .8 \\ .6 & .4 \end{bmatrix} = \begin{bmatrix} .52 & .48 \\ .36 & .64 \end{bmatrix}$$

$$T^4 = \begin{bmatrix} .52 & .48 \\ .36 & .64 \end{bmatrix} \cdot \begin{bmatrix} .52 & .48 \\ .36 & .64 \end{bmatrix} = \begin{bmatrix} .4432 & .5568 \\ .4176 & .5824 \end{bmatrix}$$

(We used the fact that $T^4 = T^2 \cdot T^2$.) Now, when $v_0 = [1 \quad 0]$,

$$v_4 = [1 \quad 0] \cdot \begin{bmatrix} .4432 & .5568 \\ .4176 & .5824 \end{bmatrix} = [.4432 \quad .5568]$$

b. When $v_0 = [0 \quad 1]$,

$$v_4 = [0 \quad 1] \cdot \begin{bmatrix} .4432 & .5568 \\ .4176 & .5824 \end{bmatrix} = [.4176 \quad .5824]$$

c. When $v_0 = [.4 \quad .6]$,

$$v_4 = [.4 \quad .6] \cdot \begin{bmatrix} .4432 & .5568 \\ .4176 & .5824 \end{bmatrix} = [.42784 \quad .57216]$$

d. When $v_0 = [.2 \quad .8]$,

$$v_4 = [.2 \quad .8] \cdot \begin{bmatrix} .4432 & .5568 \\ .4176 & .5824 \end{bmatrix} = [.42272 \quad .57728]$$

e. When $v_0 = [.3 \quad .7]$,

$$v_4 = [.3 \quad .7] \cdot \begin{bmatrix} .4432 & .5568 \\ .4176 & .5824 \end{bmatrix} = [.42528 \quad .57472]$$

f. When $v_0 = [\ .9 \quad .1\ ]$,

$$v_4 = [\ .9 \quad .1\ ] \cdot \begin{bmatrix} .4432 & .5568 \\ .4176 & .5824 \end{bmatrix} = [\ .44064 \quad .55936\ ]$$ ■

EXAMPLE 6   Let **T** be the transition matrix of Example 5.

a. Find T^7.
b. What do you notice?
c. Find the steady-state vector for the matrix **T**.
d. Comment.

Solution   **a.** We have already found T^2 and T^4 in the previous example. Thus,

$$T^7 = T^4 \cdot T^2 \cdot T = \begin{bmatrix} .4276352 & .5723648 \\ .4292736 & .5707264 \end{bmatrix}$$

b. Note that if we round off the entries to two decimal places, the two rows of the matrix T^7 would be identical.

c. Let the steady-state vector be $[\ x \quad y\ ]$. Then,

$$[\ x \quad y\ ] \cdot \begin{bmatrix} .2 & .8 \\ .6 & .4 \end{bmatrix} = [\ x \quad y\ ]$$

and

$$[\ .2x + .6y \quad .8x + .4y\ ] = [\ x \quad y\ ]$$

This matrix equation is equivalent to the system

$$\begin{cases} .2x + .6y = x \\ .8x + .4y = y \end{cases}$$

which may be written

$$\begin{cases} -.8x + .6y = 0 \\ .8x - .6y = 0 \end{cases}$$

Clearly, these two equations are equivalent, we need to keep only one of them. We must also have $x + y = 1$ since $[\ x \quad y\ ]$ is a probability vector. We now solve the system

$$\begin{cases} .8x - .6y = 0 \\ x + y = 1 \end{cases}$$

The solution is easily found to be $x = \frac{3}{7}$ and $y = \frac{4}{7}$. Thus, the steady-state vector is $\left(\frac{3}{7} \quad \frac{4}{7} \right) \doteq [\ .42857 \quad .57143\ ]$.

d. We see that the steady-state vector is nearly equal to each of the two rows of the matrix T^7.

Do Exercise 27. ■

The result of the preceding example is a special case of what always happens with regular transition matrices. We state the following theorem without proof.

> **THEOREM 6.6:** Suppose **T** is a regular transition matrix. Then,
>
> 1. **T** has a unique steady-state vector **v** all of whose entries are positive.
> 2. If **W** is a square matrix, all of whose rows are equal to **v**, the powers **T**, $\mathbf{T}^2$, $\mathbf{T}^3$, $\mathbf{T}^4$, . . . approach **W**.
> 3. If **u** is an arbitrary state vector, the sequence of vectors $\mathbf{uT}$, $\mathbf{uT}^2$, $\mathbf{uT}^3$, $\mathbf{uT}^4$, . . . approaches the steady-state vector **v**.

Exercise Set 6.7

In Exercises 1–6, a transition matrix for a Markov chain is given. In each exercise, show that the given matrix is regular and find the steady-state vector.

1. $\begin{bmatrix} .3 & .7 \\ 1 & 0 \end{bmatrix}$ **2.** $\begin{bmatrix} .2 & .8 \\ 1 & 0 \end{bmatrix}$

3. $\begin{bmatrix} .75 & .25 \\ 1 & 0 \end{bmatrix}$ **4.** $\begin{bmatrix} .3 & .2 & .5 \\ .1 & .9 & 0 \\ .4 & .5 & .1 \end{bmatrix}$

5. $\begin{bmatrix} .3 & 0 & .7 \\ 0 & .2 & .8 \\ .4 & .1 & .5 \end{bmatrix}$ **6.** $\begin{bmatrix} .2 & 0 & .8 \\ 0 & .3 & .7 \\ .4 & .6 & 0 \end{bmatrix}$

In Exercises 7 and 8, show that the given transition matrix is not regular.

7. $\begin{bmatrix} .3 & .7 \\ 0 & 1 \end{bmatrix}$ **8.** $\begin{bmatrix} 1 & 0 & 0 \\ .2 & .7 & .1 \\ .3 & .6 & .1 \end{bmatrix}$

In Exercises 9, 10, and 11, determine the values of the variables that will transform the given matrix into a transition matrix for a Markov chain.

9. $\begin{bmatrix} .2 & .7 & x \\ .3 & y & .5 \\ z & .2 & .4 \end{bmatrix}$ **10.** $\begin{bmatrix} .6 & x & 2y \\ y & 3x & .3 \\ x & 2y & z \end{bmatrix}$

11. $\begin{bmatrix} 2x & 3y & .2 \\ z & .4 & 3x \\ .4 & 4x & y \end{bmatrix}$

12. The transition matrix for a Markov process is

$$\mathbf{T} = \begin{bmatrix} .7 & .3 \\ .6 & .4 \end{bmatrix}$$

and $\mathbf{v}_0 = [\ .4 \quad .6\ ]$, $\mathbf{w}_0 = [\ .2 \quad .8\ ]$ are initial-state vectors. Do the following:

a. Perform the multiplication $\mathbf{v}_0\mathbf{T}$ and interpret the result with the help of a probability tree.
b. Perform the multiplication $\mathbf{w}_0\mathbf{T}$ and interpret the result with the help of a probability tree.
c. Find $\mathbf{w}_0\mathbf{T}^2$ and interpret the result.
d. Find $\mathbf{w}_0\mathbf{T}^3$ and interpret the result.

13. Repeat Exercise 12 with

$$\mathbf{T} = \begin{bmatrix} .3 & .2 & .5 \\ .1 & .7 & .2 \\ .4 & .5 & .1 \end{bmatrix}$$

$\mathbf{v}_0 = [\ .2 \quad .4 \quad .4\ ]$ and $\mathbf{w}_0 = [\ .5 \quad .3 \quad .2\ ]$.

14. Professor Tuffy, who teaches a mathematics course for business students, gives frequent pop quizzes. Members of a sorority have studied the pattern she follows. They have established that if she gives a quiz on a certain day, the probability that she will give a quiz the next day is 0.08. On the other hand, if she does not give a quiz, then the probability that she will give a quiz the next day is 0.27. She never gives a quiz the first three days of the quarter, but the probability that she gives a quiz on the fourth day of the quarter is 0.35. Find the following probabilities.

a. She will give a quiz on the fifth day of the quarter.
b. She will not give a quiz on the fifth day of the quarter.

c. She will give a quiz on the sixth day of the quarter.

d. She will not give a quiz on the sixth day of the quarter.

e. She will give a quiz on the tenth day of the quarter.

f. She will not give a quiz on the tenth day of the quarter.

15. A group attempting to fight cancer in a large city began an active campaign against smoking. For each of several months a survey was conducted. On the basis of their study, they found they could predict that 85% of adults who smoke in any given month will continue to do so in the next month, and only 2.5% of adults who do not smoke in any given month will do so the next month.

 a. If initially 25% of the adult population smoked, what percentage will still be smoking after 3 months?

 b. If the campaign is continued over a long period of time, what percentage of the adult population will continue smoking?

16. Assume that a nation is controlled by three political parties: X, Y, and Z. The results at each election depend somewhat on which party is in control at the time of the election. If party X is in control now, the probabilities that parties X, Y, Z will be in control after the next elections are 0.4, 0.35, and 0.25, respectively. If party Y is in control now, the probabilities that parties X, Y, Z will be in control after the next elections are 0.2, 0.45, and 0.35, respectively. If party Z is in control now, the probabilities that parties X, Y, Z will be in control after the next elections are 0.3, 0.35, and 0.35, respectively. Given that the elections take place every four years and that party Y won in 1988, what is the probability that **a.** party X will take control in 1996? **b.** party Y will take control in 2000? **c.** party Z will take control in 2004?

17. A new transportation system has been established in Seattle and the city is conducting a survey over several months to establish the worth of that system. The survey found that the people traveling to work either use the new system, drive their own car, or car pool. If a person uses the new system now, the probabilities that he/she continues to use the new system, uses his/her own car, or car pools the next month are 0.8, 0.15, and 0.05, respectively. If a person uses his/her own car now, the probabilities that he/she switches to the new system, continues to use his/her own car, or car pools the next month are 0.3, 0.6, and 0.1, respectively. Finally, if a person car pools now, the probabilities that he/she uses the new system, uses his/her own car, or continues to car pool the next month are 0.3, 0.05, and 0.65, respectively. The first month the new system went into operation, 20% of the working people used the new system, 75% used their own cars, and 5% car pooled.

 a. What percentages of the work force will be using each type of transportation after 3 months?

 b. After the system has been in operation for a long time, what percentages of the work force will be using each type of transportation?

18. Sally is taking a mathematics class at 8:00 A.M. and has difficulty arriving on time. Suppose the events that she is on time or late on consecutive days form a Markov chain. If she is on time on a given day, then the probability that she is on time the next day is 0.4. On the other hand, if she is late on a given day, then the probability that she will be late again on the next day is 0.1.

 a. Suppose Sally was on time the first day of class, what is the probability that she will be on time on the fifth day of class?

 b. Suppose Sally was late the first day of class, what is the probability that she will be on time on the fifth day of class?

19. Repeat Exercise 18**a** if we know only that the probability she is late on the first day of class is 0.2.

20. Repeat Exercise 18**a** if we know only that the probability she is on time the first day of class is 0.65.

21. A car insurance company classifies its insured motorists into three categories. Those who have not had an accident during the previous year are in the A_0 category, those who had exactly one accident during the previous year are in the A_1 category, while those who had more than one accident during the previous year are in the A_+ category. At the end of each year, all insured motorists are reclassified based only on their record during the preceding year. Long studies by the company have shown that if an insured motorist is in A_0 in any given year, then the probabilities that he/she will be in A_0, A_1, or A_+ the following year are 0.85, 0.1, and 0.05, respectively. If an insured motorist is in A_1 in any given year, then the probabilities that he/she will be in A_0, A_1, or A_+ the following year are 0.9, 0.08, and 0.02, respectively. Finally, if an insured motorist is in A_+ in any given year, then the probabilities that he/she will be in A_0, A_1, or A_+ the following year are 0.5, 0.3, and 0.2, respectively. Suppose that initially the company insured only motorists who belonged in A_0.

 a. Find the probability that after 3 years, an insured motorist randomly selected belongs to A_0.

 b. Find the probability that after 4 years, an insured motorist randomly selected belongs to A_1.

 c. Find the probability that after 5 years, an insured motorist randomly selected belongs to A_+.

22. Suppose we play the following game. We have three dice which are weighted. When the blue die is rolled, the probability that a 1 or 2 comes up is 0.2, the probability that a 3 or 4 comes up is 0.3, and the probability that a 5 or 6

comes up is 0.5. For the white die, the probabilities are 0.3, 0.4, and 0.3, respectively; for the red die, they are 0.1, 0.5, and 0.4, respectively. The color die that is rolled is determined by the number that came up on the previous roll. That is, if a 1 or 2 comes up, use the blue die on the next roll; if a 3 or 4 comes up, use the white die on the next roll; and if a 5 or 6 comes up, use the red die on the next roll.

a. What is the probability that we roll the blue die on the fifth roll if the blue die was used on the first roll?

b. What is the probability that we roll the red die on the fifth roll if the white die was used on the first roll?

c. What is the probability that we roll the white die on the fifth roll if the first roll was determined as follows: Draw a card from a standard deck. If it is an ace, roll the blue die; if it is a face card, roll the white die; if it is any other card, roll the red die.

23. Let **T** be the regular matrix of Exercise 1. Find the matrix **W** that the sequence **T**, **T**^{2}, **T**^{3}, **T**^{4}, . . . approaches.

24. Repeat Exercise 23 for the matrix of Exercise 2.

25. Repeat Exercise 23 for the matrix of Exercise 3.

26. Repeat Exercise 23 for the matrix of Exercise 4.

27. Repeat Exercise 23 for the matrix of Exercise 5.

28. Repeat Exercise 23 for the matrix of Exercise 6.

29. The matrix of Exercise 7 is not regular. Does it have a steady-state vector?

30. The matrix of Exercise 8 is not regular. Does it have a steady-state vector?

*31. Let $\mathbf{v} = [\ a_1 \quad a_2 \quad a_3 \quad \ldots \quad a_n\ ]$ be an n-dimensional row vector such that $a_1 + a_2 + a_3 + \cdots + a_n = 1$. (The a_i's need not be nonnegative since they are not necessarily probabilities.) Let **W** be the $n \times n$ matrix with all rows equal to **v**. Show that the product **vW** is equal to **v**.

6.8 Chapter Review

IMPORTANT SYMBOLS AND TERMS

$b(x; n, p)$ [6.6]
$C(n, r)$ [6.2]
$C(n; r_1, r_2, \ldots, r_k)$ [6.2]
$n!$ [6.2]
$p(A \mid B)$ [6.3]
$P(n, r)$ [6.2]
$P(n; n_1, n_2, \ldots, n_k)$ [6.2]
$p(E)$ [6.1]
$P(X = x)$ [6.5]
σ [6.5]
$\sigma = \sqrt{\text{VAR}(X)}$ [6.5]
Bayes' formula [6.4]
Bernoulli trials [6.6]
Binomial distribution [6.6]
Binomial experiment [6.6]
Binomial random variable [6.6]
Binomial theorem [6.6]
Combination of n things taken r at a time [6.2]
Conditional probability [6.3]
Continuous random variable [6.5]
Discrete random variable [6.5]

Equiprobable [6.1]
Event [6.1]
Expectation [6.5]
Fundamental principle of counting [6.2]
Impossible event [6.1]
Independent events [6.3]
Infinite sample space [6.3]
Initial-state vector [6.7]
k-step transition matrix [6.7]
Markov chain [6.7]
Mathematical expectation (mean) [6.5], [6.6]
Multiplication principle [6.2]
Mutually exclusive events [6.3]
n factorial [6.2]
Odds [6.1]
Ordered partition [6.2]
Outcome [6.1]
Permutation of n things taken r at a time [6.2]
Probability function (probability distribution) [6.5]

Probability histogram [6.5]
Probability of success [6.1]
Probability table [6.5]
Probability tree [6.3]
Random variable [6.5]
Regular transition matrix [6.7]
Relative frequency [6.1]
Sample space [6.1]
Standard deviation [6.5], [6.6]
State vector [6.7]
Steady-state vector [6.7]
Subjective probabilities [6.1]
Success [6.1]
Transition matrix [6.7]
Tree diagram [6.2]
Uniform sample space [6.1]
Variance [6.6]
Variance of a discrete random variable [6.5]

SUMMARY   An event is a subset of a sample space (the set of all possible outcomes of an experiment). If S, E, $n(S)$, and $n(E)$ denote a sample space, an event, the number of elements in the sample space, and the number of elements in the event, respectively, and if the outcomes are equally likely, then the probability of the event E is

$$p(E) = \frac{n(E)}{n(S)}$$

The odds in favor of an event E is the ratio $\frac{p(E)}{p(E')}$.

Suppose we perform k experiments, in order, and there are n_1, n_2, . . . , n_k possible outcomes, respectively, for these k experiments. Then, the number of combined outcomes performed in the given order is $n_1 \cdot n_2 \cdot \cdots \cdot n_k$. A subset with r elements chosen from a set of n distinct objects is called a combination of n things taken r at a time. If the members of this subset are arranged in a definite order, then the ordered subset is called a permutation of n things taken r at a time. If n and r are integers such that $0 \le r \le n$, the symbols $C(n, r)$ and $P(n, r)$ denote the numbers of combinations and permutations, respectively, of n things taken r at a time. We can calculate these numbers using the formulas

$$C(n, r) = \frac{n!}{r!(n - r)!} \text{ and } P(n, r) = \frac{n!}{(n - r)!}$$

If we have a set with n elements composed of k subsets, the first subset being type 1 and having n_1 indistinguishable elements, the second subset being type 2 and having n_2 indistinguishable elements, and the kth subset being type k and having n_k indistinguishable elements, then the number of distinct permutations of these n elements (taken n at a time) is denoted $P(n; n_1, n_2, \ldots, n_k)$ and is given by $P(n; n_1, n_2, \ldots, n_k) = n!/[n_1!n_2! \ldots n_k!]$.

Given a set S with n members, an ordered partition of S is an ordered collection H_1, H_2, . . . , H_n of subsets of S such that each member of S is in exactly one of the subsets H_i. Suppose H_1 has r_1 members, H_2 has r_2 members, . . . , H_k has r_k members, and let $C(n; r_1, r_2, \ldots, r_k)$ denote the number of ordered partitions, then $C(n; r_1, r_2, \ldots, r_k) = n!/[r_1!r_2! \ldots r_k!]$.

If A and B are events, the conditional probability of A, given that B has occurred, is denoted $p(A \mid B)$ and is

$$p(A \mid B) = \frac{p(A \cap B)}{p(B)}, \text{ provided that } p(B) \ne 0.$$

If A and B are events and $p(A) = p(A \mid B)$, we say that A and B are independent. If A and B are disjoint, we say that they are mutually exclusive. In this case, $p(A \cup B) = p(A) + p(B)$. If two events are not mutually exclusive, we may use the formula $p(A \cup B) = p(A) + p(B) - p(A \cap B)$.

If we have a sequence E_1, E_2, . . . , E_k of k dependent events, where the probability of the first event E_1 occurring is p_1 and, after the first event has occurred, the probability of the second event occurring is p_2, and so forth, then the probability that all k events occur, in that order, is $p_1 \cdot p_2 \cdot p_3 \cdot \cdots \cdot p_k$.

If E_1, E_2, . . . , E_n are mutually exclusive events whose union is U and A is an event such that $p(A) \ne 0$, then Bayes' formula states that

$$p(E_i \mid A) = \frac{p(A \mid E_i) \cdot p(E_i)}{p(A \mid E_1) \cdot p(E_1) + p(A \mid E_2) \cdot p(E_2) + \cdots + p(A \mid E_n) \cdot p(E_n)}$$

If a game has n outcomes, the probabilities of which are p_1, p_2, . . . , and p_n, respectively, and if the net payoff to a person is the amount A_i if the ith outcome occurs, where $1 \le i \le n$, then the person's expectation for that game is the sum $p_1 \cdot A_1 + p_2 \cdot A_2 + \cdots + p_n \cdot A_n$.

If S is a sample space with a probability function P and X is a discrete random variable with domain S, the probability distribution of X is the function f, whose domain is the range of X, defined by

$$f(x) = P(X = x) = P(E) \text{ where } E = \{s \mid s \in S \text{ and } X(s) = x\}$$

A probability distribution is often represented by a probability histogram. If S is a finite sample space and X is a discrete random variable with probability distribution f defined on the range $\{x_1, x_2, \ldots, x_n\}$ of X, then the mathematical expectation of X is

$$E(x) = x_1 \cdot f(x_1) + x_2 \cdot f(x_2) + x_3 \cdot f(x_3) + \cdots + x_n \cdot f(x_n)$$

Its variance (denoted $\text{VAR}(X)$ or σ^2) is

$$\sigma^2 = [x_1 - E(X)]^2 f(x_1) + [x_2 - E(X)]^2 f(x_2) + \cdots + [x_n - E(X)]^2 f(x_n)$$

$$= \left[\sum_{i=1}^{n} x_i^2 f(x_i) \right] - [E(X)]^2 = E(X^2) - [E(X)]^2$$

and its standard deviation is $\sigma = \sqrt{\text{VAR}(X)}$.

The Binomial Theorem states that if n is a positive integer and a and b are real (or complex) numbers, then

$$(a + b)^n = \sum_{i=0}^{n} C(n, r) a^{n-r} b^r$$

If X is the number of successes in n independent trials of a binomial experiment, then the probability distribution of X is given by

$$f(x) = P(X = x) = C(n, x) p^x q^{n-x}$$

where $p + q = 1$ and x is an integer such that $0 \leq x \leq n$. The mathematical expectation (mean), variance, and standard deviation of X are given by

$$E(X) = np, \quad \text{VAR}(X) = npq, \quad \text{and} \quad \sigma = \sqrt{npq}$$

respectively.

If in a Markov chain it is given that outcome O_i has occurred at some trial and that the probability that outcome O_j will occur at the next trial is p_{ij}, then the $n \times n$ marix $[p_{ij}]$ is called the transition matrix. If it is given that at the initial trial the probability of outcome O_i occurring is p_i, then the vector $[\, p_1 \quad p_2 \quad \cdots \quad p_n \,]$ is called the initial-state vector.

If we let $\mathbf{v}_0$ and $\mathbf{T}$ denote the initial-state vector and transition matrix, respectively, then $\mathbf{v}_1 = \mathbf{v}_0 \mathbf{T}$ is the state vector for the first trial, $\mathbf{v}_2 = \mathbf{v}_1 \mathbf{T}$ is the state vector for the second trial, and so on. For the kth trial, the jth entry of the vector $\mathbf{v}_k$ gives the probability that the jth outcome will occur. A transition matrix $\mathbf{T}$ is said to be regular if for some positive integer n all entries of the nth power of $\mathbf{T}$ are positive. If $\mathbf{T}$ is a regular transition matrix, then the successive state vectors approach some fixed vector $\mathbf{v}$ regardless of the initial vector. This vector has the property that $\mathbf{v}\mathbf{T} = \mathbf{v}$.

SAMPLE EXAM QUESTIONS

1. Evaluate the expressions

   a. $5!$ b. $\dfrac{12!}{10!}$ c. $\dfrac{17!}{14!5!3!}$ d. $P(7, 3)$

   e. $P(8, 8)$ f. $C(12, 3)$ g. $P(15; 3, 5, 7)$ h. $C(18; 4, 6, 8)$

2. A die is rolled and then a coin is tossed twice. With the help of a tree diagram, list all the possible outcomes.

3. How many distinct letter arrangements can be made using all the letters of the word *Massachusetts*?

4. A basketball team has three centers, three power forwards, three small forwards, and four guards. The coach must choose one center, one power forward, one small forward, and two guards to form his starting team. In how many ways can he do it?

5. How many committees of six can be formed from a group of seven Republicans and six Democrats if each committee is to have at least three Democrats?

6. In how many ways can 5-card hands be dealt to five poker players from a standard 52-card deck?

7. A store has two identical A television sets, three identical B television sets, and five identical C television sets. At the end of a sale, the store manager displays these sets by arranging them against one of the walls in her store. In how many distinguishable ways can she do this?

8. A foreman is in charge of eight machines and has eight workers, one worker per machine. In how many ways can this be done? Suppose the foreman thinks about each of the possible ways for one second. How long will it take him to reach a decision?

9. A fair coin is tossed four times.

   a. Describe the sample space. (It is a set of ordered quadruples.)
   b. Represent the following situations in terms of events and calculate the probability of each event: (i) exactly one toss results in tails; (ii) at least one toss results in tails; (iii) at least two of the tosses result in tails; (iv) all four tosses result in tails.

10. The odds in favor of the Seattle Supersonics winning the next NBA championship are 2 to 488. What is the probability that Seattle will win the next championship?

11. The probability that Lendl wins the next U.S. Open Tennis Championship is .32. What are the odds in favor of Lendl winning?

12. Shirley, Betty, Norma, Fran, Barbara, Carol, and Kathy are the only entries in a golf tournament. The probabilities that Shirley, Betty, Norma, Fran, and Kathy win are .11, .05, .09, .23, and .32, respectively. What is the probability that Barbara or Carol wins the tournament?

13. Suppose an urn contains five black and four white marbles, and a second urn contains three black and six white marbles. A marble is drawn from the first urn and placed in the second urn. The second urn is shaken and a marble is drawn from it. What is the probability that a black marble was drawn each time?

14. The probability that a freshman takes an English course is .7, the probability that she takes a mathematics course is .25, and the probability that she takes both is .18. What is the probability that a freshman takes **a.** an English course given that she takes a mathematics course? **b.** a mathematics course given that she takes an English course? **c.** both English and mathematics given that she takes at least one of the two subjects?

15. Suppose that a bag contains five red chips and three blue chips. A chip is drawn from the bag and replaced, the bag is shaken, and a chip is drawn again. What is the probability that both chips drawn were blue?

16. A card is drawn from a standard deck of cards.
   a. What is the probability that it is either a king or an ace?
   b. What is the probability that it is either a diamond or a queen?

17. Suppose that five people have been asked to randomly choose a number between 1 and 50, inclusive. What is the probability that no two of them chose the same number?

18. An urn contains five black, three white, and four yellow marbles. Two marbles are drawn, one after the other, without replacing the first before the second is drawn. With the help of a probability tree, find the probability that one black and one yellow marble were drawn.

19. Gwen, Julie, and Marian take turns rolling a fair die, in that order. The first one to roll a 6 wins. What is the probability that Marian wins?

20. A sociologist theorized that people who do not drink alcohol have a better chance to be nonsmokers than those who do. To support his theory, he conducted a study of 1000 randomly selected people in the 18-to-23 age group. He let U be the set of people he studied (the universe), A be the set of people who abstain from drinking alcohol, and N be the set of nonsmokers. He then drew the following Venn diagram.

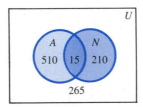

a. Find the probability that a person who does not drink alcohol is a nonsmoker.
b. Find the probability that a person who is a nonsmoker does drink alcohol.
c. Find the probability that a person who does not drink alcohol is a smoker.

21. A novice enters a marksmanship competition and has the choice of six different rifles, only one of which is the right one for a shot from a certain distance. The probability that he hits a bull's-eye is .12 if he uses the right rifle and only .03 if he uses a wrong rifle. Suppose he picks a rifle randomly and hits a bull's-eye. What is the probability that he selected the right rifle?

22. Three machines are used in manufacturing a certain golf ball. Machines I, II, and III are used to manufacture 32%, 43%, and 25% of the total output, respectively. It is known that 1.5% of the balls manufactured using machine I are defective, 2.1% of the balls manufactured using machine II are defective, and 3.7% of the balls manufactured using machine III are defective. A ball is selected at random from the total output and is found to be defective. What is the probability that it was made using machine III?

23. It costs $2 to roll a pair of fair dice. If a sum of 2 or 12 comes up, you are paid $10; if a sum of 3 or 11 comes up, you are paid $5; and if a sum of 4 or 10 comes up, you are paid $3. You lose if any other sum comes up. If you played this game many times, how much would you expect to win or lose per game, on the average?

24. Suppose that a fair coin is tossed four times and that X is the number of heads that came up.

a. Determine the probability distribution of X.
b. Draw a probability histogram for X.
c. Calculate the mathematical expection of X.

25. The annual premium that a Boston homeowner must pay to an insurance company to protect her house, worth $280,000, against total destruction by fire is $300. From experience, the company has determined that the probability of a house being totally destroyed by fire in any one year is .00021. Calculate the company's expected gain.

26. Suppose that a random variable X has range $\{1, 2, 3, 4, 5\}$ and that its probability distribution is defined by the following probability table.

x	$f(x) = P(X = x)$
1	.2
2	.25
3	.1
4	.3
5	.15

a. Find the variance and standard deviation of X.

b. Draw a probability histogram for X.

27. Suppose a coin is weighted so that the probability that heads comes up on any toss is .6. The coin is tossed four times. What is the probability that heads comes up exactly two times?

28. The coin of Exercise 27 is tossed five times. What is the probability that heads comes up
a. exactly three times? b. exactly four times? c. at least three times?

29. Suppose that X is a binomial random variable with $p = .3$ and $n = 3$.

a. Find the probability distribution of X.

b. Find the mean, variance, and standard deviation of X.

30. A fair die is rolled 12 times. Find the probability that a three came up at least twice.

31. A hardware store has a sampling scheme to inspect large lots of incoming light bulbs. Twenty-five light bulbs are to be tested and the lot is rejected if four or more are found to be defective. Suppose a lot contains exactly 5% defective light bulbs. What is the probability that the lot will be rejected?

32. Show that the given transition matrix is not regular.

$$\begin{bmatrix} .2 & .8 \\ 0 & 1 \end{bmatrix}$$

33. Determine the values of the variables that will transform the given matrix into a transition matrix for a Markov chain.

$$\begin{bmatrix} .1 & .3 & x \\ .4 & y & .2 \\ z & .5 & .3 \end{bmatrix}$$

34. Let $A = \begin{bmatrix} .6 & .4 \\ 1 & 0 \end{bmatrix}$.

a. Show that A is a regular transition matrix for a Markov chain.

b. Find the steady-state vector.

35. An organization attempting to fight heart disease in a large city began an active campaign against smoking. For each of several months a survey is conducted. The survey found that 78% of those who smoke in any given month will continue to do so the next month, and only 2% of those who do not smoke in any given month will smoke the next month.

a. If initially 20% of the adult population of that city smoked, what percentage will still be smoking after 4 months?

b. If the campaign is continued over a long period, what percentage of the adult population will be smoking after many months?

Elementary Theory of Games and Graphs

7

Society is constantly confronted with the task of making decisions. For example, following the 1989 San Francisco earthquake, homeowners who were not carrying earthquake insurance had to decide whether to buy the expensive insurance based on the probability of an earthquake reoccuring. The mathematical models used in solving this type of problem are introduced in a branch of mathematics called game theory.* In this chapter, we give a brief introduction to this topic along with a brief discussion of graphs and digraphs.

7.1   Matrix Games

A book entitled "Game Theory" may be mistakenly bought by someone aspiring to be a football coach, or by someone hoping to make a fortune in Las Vegas. Although game theory does have applications in football and in games of chance, it was developed to analyze decision making in competitive situations in economics, business, and warfare.

The games treated in this chapter are rather elementary, because of the nature of this course. You should realize, however, that the objective of game theory is to arrive at an accurate analysis of more important "games." For example, the competition for the computer market may be regarded as a game in which there are many players (the companies manufacturing computers) and where the stakes are high.

Matrix Games

Many games can be put into matrix form. An $m \times n$ matrix **A** represents a game if the game has the following features.

1. There are only two players. However, each player may represent a group of people such as a corporation or a nation.
2. On each play of the game, one player (usually called R) makes any one of m choices, while the other player (usually named C) makes any one of n choices.
3. If R makes choice R_i and C makes choice C_j, then C pays R the amount a_{ij}, where a_{ij} is the i-j entry of matrix **A**, and the payment is made in the appropriate units (not necessarily money). If a_{ij} is positive, then R receives a_{ij} from C. On the other hand, if a_{ij} is negative, then C receives $|a_{ij}|$ units from R.

EXAMPLE 1   The matrix $\begin{bmatrix} 2 & 4 & -3 & 6 \\ -1 & 3 & 7 & 0 \\ 3 & -6 & 5 & 1 \end{bmatrix}$ represents a game where the payoffs are in dollars.

---

*The French mathematician Emile Borel gave a first mathematical analysis of games in 1921. However, the foundation of game theory first appeared in 1944 in the book, *Games and Economic Behavior*, published by mathematician John von Neumann and economist Oskar Morgenstern who immigrated to this country from Hungary and Germany, respectively.

    a. How many choices does each player have?

    b. If player R chooses R_2 and player C chooses C_3, what is the payment?

    c. Same as **b** if player R chooses R_3 and player C chooses C_2.

Solution   **a.** Since the matrix has 3 rows and 4 columns, player R has three choices and player C has four choices.

    b. If player R chooses R_2 and player C chooses C_3, the payment will be the 2-3 entry of the matrix. But $a_{23} = 7$. Thus player C will pay $7 to player R since 7 is positive.

    c. Since player R chooses R_3 and player C chooses C_2, the payment will be the 3-2 entry of the matrix. But $a_{32} = -6$. Thus, the payment will be $6 from player R to player C. The payment is from R to C because -6 is negative.

Do Exercise 1. ■

EXAMPLE 2   Players R and C may each select one of three numbers: 1, 2, or 3. They simultaneously select a number. If they selected the same number, there is no payment. If they selected different numbers and the sum of these numbers is even, player R wins; otherwise, player C wins. The winner receives $p from the other player where p is the sum of the two numbers selected. Give the matrix representation of this game.

Solution   We let $R_i = i$ and $C_j = j$. Then $a_{ii} = 0$ for $i = 1, 2,$ and 3 since in this case they selected the same number.

$$R_1 + C_2 = 1 + 2 = 3$$

Since 3 is odd, player C wins and receives $3 from R. So $a_{12} = -3$.

$$R_1 + C_3 = 1 + 3 = 4$$

Since 4 is even, player R wins and receives $4 from C. So $a_{13} = 4$.

    Similarly, we find that $a_{21} = -3$, $a_{23} = -5$, $a_{31} = 4$, and $a_{32} = -5$. Therefore, the matrix representation of the game is

$$\mathbf{A} = \begin{bmatrix} 0 & -3 & 4 \\ -3 & 0 & -5 \\ 4 & -5 & 0 \end{bmatrix}$$

Do Exercise 3. ■

DEFINITION 7.1: A *two-person zero-sum game* is a game involving only two players. After each play a payoff of some amount is made so that one player's win is the other player's loss.

    If a two-person zero-sum game is represented by an $m \times n$ matrix, the convention is to call R the player who chooses one of the m rows and C the player who chooses one of the n columns. The matrix is called the *game matrix* or *payoff matrix*. In general, a positive entry in the payoff matrix indicates a win for R while a negative entry indicates a win for C. We assume that the game may be played repeatedly and that the main motivation for R is to maximize his winnings, while that for C is to minimize his losses.

Strictly Determined Games

In this and the following three sections we consider the following questions. How should R play to maximize his winnings and how should C play to minimize his losses? We begin with an example.

EXAMPLE 3   Consider the following game matrix

$$A = \begin{bmatrix} 4 & 1 & 0 & 2 \\ 3 & 2 & 5 & 7 \\ 1 & -3 & 6 & -4 \end{bmatrix}$$

The payoffs are in dollars. How should R and C play to achieve their goals?

Solution   R may be tempted to play the second row to try to win \$7. However, R knows that C is intelligent and will certainly avoid choosing the fourth column. So R should instead analyze each move to see how much can be won even if C makes the best possible countermove. Therefore, R should consider the worst payoff for each of the three choices. We circle the smallest entry in each of the rows to obtain

$$A = \begin{bmatrix} 6 & 1 & ⓪ & 2 \\ 5 & ③ & 5 & 7 \\ 1 & -3 & 6 & ⓸ \end{bmatrix}$$

Obviously, R will choose the largest of these encircled numbers, 3 in this case, and will always play the second row.

On the other hand, C tries to minimize losses. So C considers the worst payoff for each move and finds the largest number for each of the four choices. Thus we box the largest element in each column to obtain

$$A = \begin{bmatrix} \boxed{6} & 1 & 0 & 2 \\ 5 & \boxed{3} & 5 & \boxed{7} \\ 1 & -3 & \boxed{6} & -4 \end{bmatrix}$$

Since C is trying to minimize losses, C will choose the smallest of these boxed numbers, which is 3, and will always play column two.
Do Exercise 7.   ■

Observe that in the preceding example the conclusions are the best strategies for both players. That is, if R continues to play the second row and C deviates from the second column, then C's losses will not decrease. Also if C continues to play the second column and R deviates from the second row, then R's winnings will not increase. The game is said to be strictly determined.

DEFINITION 7.2:   If in a game matrix there is an entry which is both the smallest entry in its row and the largest entry in its column, then this entry is called a *saddle point*. A game matrix is said to be *strictly determined* if it has a saddle point.

The *optimum strategies* for the players is for R to always choose a row containing a saddle point and for C to always choose a column containing a saddle point.

THEOREM 7.1:   If a game matrix has two or more saddle points, then they are equal.

Proof: Suppose that a_{ij} and a_{rs} are saddle points. Then we must have

$$a_{ij} \leq a_{is}, \qquad \text{since the saddle point is row minimum}$$

and

$$a_{ij} \geq a_{rj}, \qquad \text{since the saddle point is column maximum}$$

Similarly,

$$a_{rs} \leq a_{rj} \text{ and } a_{rs} \geq a_{is}$$

Combining these inequalities, we obtain

$$a_{is} \leq a_{rs} \leq a_{rj} \leq a_{ij} \leq a_{is}$$

Therefore,

$$a_{is} = a_{rs} = a_{rj} = a_{ij} = a_{is}$$ ■

DEFINITION 7.3:   A saddle point is called the *value* of the strictly determined game. A game is *fair* if its value is zero. In a game matrix the process of finding optimum strategies for R and C and the value of the game is called *solving* the game.

In general, optimum strategies need not be unique as is shown in the following example.

EXAMPLE 4   Consider the following game matrix

$$\mathbf{A} = \begin{bmatrix} 6 & 5 & 8 & 5 \\ 1 & 2 & 4 & 3 \\ 7 & 5 & 9 & 5 \end{bmatrix}$$

a. Show that the game is strictly determined and find its value.
b. Find optimum strategies for each player.

Solution   **a.** We first circle the minimum entry in each row to obtain

$$A = \begin{bmatrix} 6 & ⑤ & 8 & ⑤ \\ ① & 2 & 4 & 3 \\ 7 & ⑤ & 9 & ⑤ \end{bmatrix}$$

Next we box the maximum entry in each column and get

$$A = \begin{bmatrix} 6 & ⑤ & 8 & ⑤ \\ ① & 2 & ⑨ & 3 \\ ⑦ & ⑤ & 6 & ⑤ \end{bmatrix}$$

The 5 appears both in boxes and circles. So each of these four 5s is both a row minimum and column maximum. Thus each is a saddle point. Therefore the game is strictly determined.

b. The optimal strategies for R are to choose either the first or third row. Those for C are to select either the second or fourth column.

Do Exercise 9.   ■

> **DEFINITION 7.4:**   If a strategy for a player is to consistently choose a certain row or column, then such a strategy is called a *pure strategy*.

Nonstrictly Determined Games

Not all game matrices are strictly determined as is shown in the next example.

EXAMPLE 5   Consider the following game matrix

$$A = \begin{bmatrix} -1 & 2 & 3 \\ 3 & 4 & 2 \\ 5 & -2 & 6 \end{bmatrix}$$

Show that this game is not strictly determined.

Solution   Circle the minimum value in each row and box the maximum value in each column to obtain

$$A = \begin{bmatrix} ⊖1 & 2 & 3 \\ 3 & ④ & ② \\ ⑤ & ⊖2 & ⑥ \end{bmatrix}$$

No entry is both circled and boxed. So there is no saddle point. The game is not strictly determined.

Do Exercise 11.   ■

In the next three sections we will learn how to solve nonstrictly determined games.

Recessive Rows and Columns

The dimensions of some game matrices may be lowered because certain rows or columns would never be chosen by intelligent players.

EXAMPLE 6   Consider the following game matrix

$$A = \begin{bmatrix} 0 & -1 & -2 & 4 \\ 1 & 2 & 4 & 3 \\ 6 & 5 & 9 & 3 \end{bmatrix}$$

Delete any row or column that would never be chosen by an intelligent player.

Solution   Recalling that R's goal is to maximize winnings, we see that R would never choose the second row over the third row because every entry of the third row is greater than or equal to the corresponding entry of the second row. So regardless of C's move, R has more to gain by choosing the third row over the second row. We eliminate the second row to obtain

$$B = \begin{bmatrix} 0 & -1 & -2 & 4 \\ 6 & 5 & 9 & 3 \end{bmatrix}$$

Since C is trying to minimize losses, C would never choose the first column over the second column. To see this, observe that each entry of the first column is larger than the corresponding entry in the second column; hence, regardless of R's choice, C would lose more by choosing the first over the second column. So we delete the first column to obtain

$$C = \begin{bmatrix} -1 & -2 & 4 \\ 5 & 9 & 3 \end{bmatrix}$$

Do Exercise 13.   ∎

DEFINITION 7.5:   Let **A** be an $m \times n$ game matrix. If each entry in the ith row of **A** is less than or equal to the corresponding entry in the kth row; that is

$$a_{ij} \leq a_{kj}, \qquad j = 1, 2, \ldots, n$$

then the ith row is said to be *recessive* and the kth row is said to be *dominant*.

If each entry in the jth column of **A** is less than or equal to the corresponding entry in the kth column; that is

$$a_{ij} \leq a_{ik}, \qquad i = 1, 2, \ldots, m$$

then the kth column is said to be *recessive* and the jth column is said to be *dominant*.

Optimal strategies will never require the selection of rows or columns that are recessive. Thus, such rows and columns may be deleted.

EXAMPLE 7 Consider the following game matrix

$$A = \begin{bmatrix} 1 & 2 & 7 & -1 & -3 \\ 6 & 5 & 3 & 2 & 4 \\ 4 & 2 & 1 & 0 & 3 \\ -7 & 3 & 0 & 1 & 15 \\ 0 & 7 & 5 & -3 & -4 \end{bmatrix}$$

Reduce its dimensions if appropriate. If it is strictly determined (if it is not, say why), find the following.

a. All saddle points.
b. Optimal strategies for R and C.
c. The value of the game.

Solution Observe that every entry in the third row is less than the corresponding entry of the second row. So the third row is recessive and may be omitted. We obtain

$$B = \begin{bmatrix} 1 & 2 & 7 & -1 & -3 \\ 6 & 5 & 3 & 2 & 4 \\ -7 & 3 & 0 & 1 & 15 \\ 0 & 7 & 5 & -3 & -4 \end{bmatrix}$$

Every entry of the second column is greater than the corresponding entry of the fourth column. Thus the second column is recessive and may be omitted. We get

$$C = \begin{bmatrix} 1 & 7 & -1 & -3 \\ 6 & 3 & 2 & 4 \\ -7 & 0 & 1 & 15 \\ 0 & 5 & -3 & -4 \end{bmatrix}$$

Observe now that the fourth row of matrix **C** is recessive and may be omitted. We obtain

$$D = \begin{bmatrix} 1 & 7 & -1 & -3 \\ 6 & 3 & 2 & 4 \\ -7 & 0 & 1 & 15 \end{bmatrix}$$

a. To find the saddle points, we circle the minimum values in each row and box the maximum values in each column. We obtain

$$D = \begin{bmatrix} 1 & \boxed{7} & -1 & \boxed{-3} \\ \boxed{6} & 3 & \boxed{2} & 4 \\ \boxed{-7} & 0 & 1 & \boxed{15} \end{bmatrix}$$

The 2 has been both circled and boxed. So it is the saddle point.

b. The optimal strategy for R is to choose row 2 of matrix **D**, and the optimal strategy for C is to choose column 3 of matrix **D**. It should be obvious to you that to obtain optimal strategies for the original game matrix **A**, R needs only to choose row 2 and C should choose column 4 of matrix **A**.

c. The value of the game is 2.

Do Exercise 15.

Applications

As we stated earlier, game theory was developed in an attempt to systematically analyze problems involving decision making in economics, business, the social sciences, and warfare. We now give two examples of such applications. You will find several more in the exercises and in the following sections.

EXAMPLE 8   Each of two athletic clubs, say S and T, is planning to build a new facility. Each club has two locations to choose from, labeled A and B for S, and C and D for T. A consulting firm analyzed the situation as follows. If S builds in A and T in C, then S will get 62% of the total membership. If S builds in A and T in D, then T will get 60% of the total membership. If S builds in B and T in C, then S will get 65% of the total membership. Finally, if S builds in B and T in D, then S will get 75% of the total membership. Write this consulting firm's report as a game matrix where payoffs are measured by the number of percentage points above 50%. Then solve the game.

Solution   Let club S be the row player and club T be the column player. The first and second rows represent locations A and B for S, respectively. Also, the first and second columns represent locations C and D for T. If S builds in A and T in C, then S will get 62% of the total membership, which is 12 percentage points above 50%. So $a_{11} = 12$. If S builds in A and T in D, then T will get 60% of the total membership, which is 10 percentage points above 50%. However, a gain by T is a loss for S, so $a_{12} = -10$. Similarly, we find $a_{12} = 15$ and $a_{22} = 25$. Thus

$$A = \begin{bmatrix} 12 & -10 \\ 15 & 25 \end{bmatrix}$$

Now circle the smallest entry in each row and box the largest entry in each column to obtain

$$A = \begin{bmatrix} 12 & \boxed{-10} \\ \boxed{\textcircled{15}} & \boxed{25} \end{bmatrix}$$

The saddle point is 15 since it is both circled and boxed. So the value of the game is 15. The optimum strategies are for club S to build in location B and for club T to build in location C. Club S will get 65% of the available membership.
Do Exercise 23.   ■

EXAMPLE 9   A county has three small cities A, B, and C. It is known that 35% of the population shops in A, 40% shops in B, and 25% shops in C. Each of the two competing grocery chains X and Y is planning to build a store in one of the three cities. A consulting firm provided the following information. If both stores are built in the same city, X will get 58% of the total business. If the stores are built in different cities, each will get 70% of the business of the city it is in, and X will get 40% of the business from the third city. Translate this information into a game matrix where payoffs are measured by the number of percentage points above 50%. Is this game strictly determined?

Solution   Let us agree that chain X will be player R and chain Y will be player C. Also, a store being built in city A, B, or C by X will be represented by rows 1, 2, and 3, respectively. Similarly, a store being build by Y in city A, B, or C will be represented by columns 1, 2, and 3, respectively.

Observe first that if both chains build in the same city, X gets 58% of the total business. This is 8 percentage points above 50%. So $a_{11} = a_{22} = a_{33} = 8$. Now, if X builds in A and Y builds in B, then X gets 70% of the business in A, 30% of the business in B (since Y gets 70% of the business in B) and 40% of the business in C. Recalling that 35% of the people shop in A, 40% shop in B, and 25% shop in C, we calculate

$$(.7)(.35) + (.3)(.4) + (.4)(.25) = .465$$

So X gets 46.5% of the total business. But this is 3.5 percentage points below 50%. So $a_{12} = -3.5$.

If X builds in A and Y builds in C, then X gets 70% of the business in A, 30% of the business in C (since Y gets 70% of the business in C) and 40% of the business in B. We calculate

$$(.7)(.35) + (.3)(.25) + (.4)(.4) = .48$$

So X gets 48% of the total business. But this is 2 percentage points below 50%. So $a_{13} = -2$. Similarly, we obtain $a_{21} = -1.5$, $a_{23} = -.5$, $a_{31} = -6$, and $a_{32} = -6.5$. The game matrix is

$$A = \begin{bmatrix} 8 & -3.5 & -2 \\ -1.5 & 8 & -.5 \\ -6 & -6.5 & 8 \end{bmatrix}$$

Clearly, no entry of this matrix is both a row minimum and column maximum. Thus the game is not strictly determined.

Do Exercise 25. ∎

Exercise Set 7.1

1. The matrix

$$\begin{bmatrix} 3 & 6 & -2 & 5 \\ -3 & 6 & 4 & 1 \\ 5 & -4 & 9 & 2 \end{bmatrix}$$

represents a game where the payoffs are in dollars.

   a. How many choices does each player have?
   b. If player R chooses R_2 and player C chooses C_1, what is the payment?
   c. Same as **b** if player R chooses R_3 and player C chooses C_3.

2. The matrix

$$\begin{bmatrix} 3 & 9 & -2 & 4 \\ 7 & -5 & 8 & 2 \\ 5 & -4 & 6 & 3 \end{bmatrix}$$

represents a game where the payoffs are in dollars.

   a. How many choices does each player have?
   b. If player R chooses R_2 and player C chooses C_3, what is the payment?
   c. Same as **b** if player R chooses R_3 and player C chooses C_2.

3. Players R and C may each select one of three numbers: 3, 4, or 5. They simultaneously select a number. If they selected the same number, there is no payment. If they selected different numbers and the sum of these numbers is odd, player R wins; otherwise, player C wins. The winner receives $\$p$ from the other player where p is the sum of the two numbers selected. Give the matrix representation of this game.

4. Ron and Charles each have a quarter. They simultaneously flip the coins. If the two coins match (both heads or both tails), Ron wins; otherwise, Charles wins. For each play, the loser pays the other player $2. Give the matrix representation of this game.

5. Rosemary and Cindy choose, independently, one of three numbers: 1, 3, or 5. If they choose identical numbers, then Rosemary pays Cindy $x where x is the chosen number; otherwise, Cindy pays Rosemary $y where y is the number chosen by Rosemary. Give the matrix representation of this game.

6. Consider the following game matrix

$$A = \begin{bmatrix} 2 & 3 & 4 & 1 \\ 4 & -2 & 3 & -1 \\ 2 & 3 & -1 & 0 \end{bmatrix}$$

The payoffs are in dollars. How should R and C play to achieve their goals?

7. Consider the following game matrix

$$A = \begin{bmatrix} 3 & 2 & 1 & 4 & 5 \\ -2 & -3 & 0 & 3 & -2 \\ 4 & 5 & -5 & 1 & 2 \end{bmatrix}$$

The payoffs are in dollars. How should R and C play to achieve their goals?

8. Consider the following game matrix

$$A = \begin{bmatrix} 3 & 5 & 3 & 7 \\ 2 & 4 & 1 & -1 \\ 3 & 6 & 3 & 8 \end{bmatrix}$$

a. Show that the game is strictly determined and find its value.
b. Find optimum strategies for each player.

9. Consider the following game matrix

$$A = \begin{bmatrix} 5 & 2 & 6 & 2 \\ 4 & 2 & 8 & 2 \\ 0 & -1 & -4 & 1 \end{bmatrix}$$

a. Show that the game is strictly determined and find its value.
b. Find optimum strategies for each player.

10. Consider the following game matrix

$$A = \begin{bmatrix} -2 & 3 & 5 \\ 3 & 1 & 4 \\ 6 & 5 & -2 \end{bmatrix}$$

Show that this game is not strictly determined.

11. Consider the following game matrix

$$A = \begin{bmatrix} -1 & 3 & 6 & 0 \\ 2 & 5 & 3 & -1 \\ -2 & 0 & 3 & -3 \end{bmatrix}$$

Show that this game is not strictly determined.

12. Consider the following game matrix

$$A = \begin{bmatrix} 1 & -3 & -4 & 2 \\ 2 & 1 & -2 & 3 \\ 6 & 0 & 3 & 4 \end{bmatrix}$$

Delete any row or column that would never be chosen by an intelligent player.

13. Same as Exercise 12 for the following game matrix

$$A = \begin{bmatrix} 3 & -2 & 4 & 1 & 6 \\ 1 & -4 & 3 & 0 & 2 \\ 4 & -3 & 5 & -4 & -5 \end{bmatrix}$$

In Exercises 14–22, reduce the dimensions of the given game matrices if appropriate. For each matrix that is strictly determined (if it is not, say why), find the following.

a. All saddle points.
b. Optimal strategies for R and C.
c. The value of the game.

14. $\begin{bmatrix} 2 & 6 \\ 5 & 4 \end{bmatrix}$

15. $\begin{bmatrix} -2 & 1 \\ -4 & -4 \end{bmatrix}$ 16. $\begin{bmatrix} 2 & 3 & 0 \\ 4 & 1 & 5 \end{bmatrix}$

17. $\begin{bmatrix} 1 & 0 & -3 & -1 \\ 3 & 1 & -2 & 0 \\ -1 & -4 & 0 & -1 \\ 0 & -3 & 4 & 1 \end{bmatrix}$

18. $\begin{bmatrix} 1 & 3 & -1 \\ 2 & 3 & 2 \\ 1 & 2 & 3 \\ -1 & 0 & 2 \end{bmatrix}$

19. $\begin{bmatrix} 1 & 0 & -1 & 2 \\ 0 & 5 & -2 & -4 \\ 2 & 2 & 1 & 3 \\ 3 & -1 & 0 & 18 \end{bmatrix}$

20. $\begin{bmatrix} -1 & 2 & -5 & 2 \\ 0 & 5 & -3 & 6 \\ 6 & 3 & 2 & 4 \\ -8 & -1 & 10 & 0 \end{bmatrix}$

21. $\begin{bmatrix} 1 & 4 & 0 & -1 & 3 \\ 4 & 5 & 2 & 1 & 4 \\ 0 & 0 & 6 & -2 & -4 \\ -9 & 17 & -2 & 0 & 15 \\ -10 & 2 & -3 & -1 & 2 \end{bmatrix}$

22. $\begin{bmatrix} -3 & 5 & 4 & -3 & -7 \\ -1 & 6 & 5 & -3 & -5 \\ 4 & 2 & 1 & 0 & 2 \\ 3 & 1 & 0 & -1 & 1 \\ -9 & 0 & -2 & -1 & 19 \end{bmatrix}$

23. Each of two athletic clubs, say W and T, is planning to build a new facility. Each club has two locations to choose from, labeled *I* and *II* for W, and *III* and *IV* for T. A consulting firm analyzed the situation as follows. If W builds in *I* and T in *III*, then W will get 60% of the total membership. If W builds in *I* and T in *IV*, then T will get 55% of the total membership. If W builds in *II* and T in *III*, then W will get 62% of the total membership. Finally, if W builds in *II* and T in *IV*, then W will get 70% of the total membership. Write this consulting firm's report as a game matrix where payoffs are measured by the number of percentage points above 50%. Then solve the game.

24. Same as Exercise 23 but with the consulting firm's report as follows. If W builds in *I* and T in *III*, then W will get 55% of the total membership. If W builds in *I* and T in *IV*, then T will get 65% of the total membership. If W builds in *II* and T in *III*, then W will get 58% of the total membership. Finally, if W builds in *II* and T in *IV*, then W will get 60% of the total membership.

25. A county has three small cities 1, 2, and 3. It is known that 25% of the population shops in 1, 35% shops in 2, and 40% shops in 3. Each of the two competing grocery chains A and B is planning to build a store in one of the three cities. A consulting firm provided the following information. If both stores are built in the same city, A will get 60% of the total business. If the stores are built in different cities, each will get 75% of the business of the city it is in, and A will get 45% of the business from the other city. Translate this information into a game matrix where payoffs are measured by the number of percentage points above 50%. Is this game strictly determined?

26. Same as Exercise 25 but with the report from the consulting firm as follows. If both stores are built in the same city, A will get 70% of the total business. If the stores are built in different cities, each will get 90% of the business of the city it is in, and A will get 55% of the business from the other city.

27. A county has three small cities X, Y, and Z. The distance between X and Y is 24 miles, while Z is 17 miles from X and 15 miles from Y. Each of two department store chains R and C plans to build a store in one of the three cities. Each of the three cities provides the same amount of business. If the stores are built in the same cities, R and C will split the business evenly; however, if they are built in different cities, then the store closer to the third town will get all of that town's business. Write the game matrix and solve the game. (*Hint:* Measure the payoffs in the amount of business R has over C. For example, if they build in the same city, the payoff is 0 to both. If they build in different cities and R is closer to the third city than C, the payoff is 1, because R has one city's business more than C. If C is closer to the third city than R, the payoff is -1.)

28. Nations R and C are at war. The commanders of their respective navies are faced with the following problem. Nation C must have a convoy go from port X to port Y. In doing so, its navy may sail east of a certain island or west of that island. Either route would take 5 days. The weather forecast is fog and poor visibility for the eastern route and clear skies and good visibility for the western route. The commander of R's navy must decide where to concentrate most of the air reconnaissance. If R concentrates the reconnaissance on the eastern route, then if C takes the eastern route, the convoy will not be sighted the first two days, allowing three days of bombing. However, if C takes the western route, the convoy will be sighted the second day, allowing four days of bombing. If R concentrates the reconnaissance on the western route, then if C takes the eastern route, the convoy will not be sighted until the third day, allowing only two days of bombing. But if C takes the western route, the convoy will be sighted immediately, allowing five days of bombing. What are the optimum strategies for both commanders?

29. Two football coaches from universities R and C know that a certain number of blue-chip players from regions A and B will enroll at one of the two universities. If R concentrates his recruiting efforts in region A, then R will get 65% of the players if C concentrates in region A also, and 70% of the players if C concentrates in region B. On the other hand, if R concentrates in region B, then he will get 60% of the players if C concentrates in region A, but only 40% of the players if C concentrates in region B. What are the optimum strategies for both coaches?

30. During a football game between the Rats and the Crocodiles, the Rats have the ball at midfield and it is first and ten. The Rats quarterback has four plays he can call, the Crocodiles defense has five possible defensive alignments it can use. The number of yards gained by the Rats for each possible play called and defensive alignment is given in the following game matrix.

$$\mathbf{A} = \begin{bmatrix} -3 & 6 & 12 & -3 & -7 \\ 9 & 7 & 6 & 4 & 5 \\ 1 & -2 & 1 & 2 & 6 \\ 10 & -7 & 2 & 0 & 1 \end{bmatrix}$$

What are the optimum strategies for both teams?

7.2 Mixed Strategy Games

We now consider how to find optimum strategies for nonstrictly determined games. We must first define carefully the meaning of the word *strategy*. In the preceding section, we considered strictly determined games and saw that if a game matrix had a saddle point, the optimum strategies for R and C were to consistently play a row and column containing a saddle point. These strategies were called *pure strategies*. John von Neumann established that if a game is not strictly determined then it is best for each player to mix his selection of choices using a probability distribution and a chance device that would generate this distribution. In this section we begin to illustrate how this can be done.

Pure and Mixed Strategies

> **DEFINITION 7.6:** Let $\mathbf{A}$ be an $m \times n$ game matrix. A *strategy for R* is an m-dimensional probability row vector
>
> $$\mathbf{P} = [\, p_1 \quad p_2 \quad \cdots \quad p_m \,]$$
>
> where $p_i \geq 0$ for $i = 1, 2, \ldots, m$ and $p_1 + p_2 + \ldots + p_m = 1$.
> A *strategy for C* is an n-dimensional probability column vector
>
> $$\mathbf{Q} = \begin{bmatrix} q_1 \\ q_2 \\ \vdots \\ q_n \end{bmatrix}$$
>
> where $q_i \geq 0$ for $i = 1, 2, \ldots, n$ and $q_1 + q_2 + \cdots + q_n = 1$.
> If one of the entries in $\mathbf{P}$ (or $\mathbf{Q}$) is 1 and the others are 0, the strategy is called a *pure strategy*. Otherwise, it is called a *mixed strategy*.

In general p_i is the probability that R will choose the ith row and q_j is the probability that C will choose the jth column. The reason for choosing row and column vectors for strategies will soon be apparent.

Observe that although we are using the term strategy, we are not implying that $\mathbf{P}$ and $\mathbf{Q}$ are the best strategies for players R and C, respectively.

Expected Value of a Game Matrix

We begin with the following example.

EXAMPLE 1   Consider the following game matrix

$$\mathbf{A} = \begin{bmatrix} 30 & -45 \\ -60 & 90 \end{bmatrix}$$

where the payoffs are in dollars. Let

$$\mathbf{P} = \begin{bmatrix} \frac{1}{3} & \frac{2}{3} \end{bmatrix} \text{ and } \mathbf{Q} = \begin{bmatrix} \frac{1}{5} \\ \frac{4}{5} \end{bmatrix}.$$

be strategies for R and C, respectively.

a. What is the probability that the payoff is \$30 to R?
b. What is the probability that the payoff is \$45 to C?
c. What is the probability that the payoff is \$60 to C?
d. What is the probability that the payoff is \$90 to R?
e. What is the expected value of the game for R?
f. Find the product **PAQ** and compare your answer to that of part **e.**

Solution   **a.** The payoff will be \$30 to R if R chooses row 1 and C chooses column 1. But the probabilities of these choices are $\frac{1}{3}$ and $\frac{1}{5}$, respectively. So the probability of both choices being made is $\left(\frac{1}{3}\right)\left(\frac{1}{5}\right) = \frac{1}{15}$.

b. The payoff will be \$45 to C if R chooses row 1 and C chooses column 2. The probabilities of these choices are $\frac{1}{3}$ and $\frac{4}{5}$, respectively. So the probability of both choices being made is $\left(\frac{1}{3}\right)\left(\frac{4}{5}\right) = \frac{4}{15}$.

c. Similarly, we find the probability that the payoff will be \$60 to C is $\left(\frac{2}{3}\right)\left(\frac{1}{5}\right) = \frac{2}{15}$.

d. Furthermore, the probability that the payoff will be \$90 to R is $\left(\frac{2}{3}\right)\left(\frac{4}{5}\right) = \frac{8}{15}$.

e. Consequently, the expected value of the game is

$$E = 30\left(\frac{1}{15}\right) + (-45)\left(\frac{4}{15}\right) + (-60)\left(\frac{2}{15}\right) + 90\left(\frac{8}{15}\right) = 30$$

The expected value of the game (for player R) is \$30.

f.
$$\mathbf{PAQ} = \begin{bmatrix} \frac{1}{3} & \frac{2}{3} \end{bmatrix} \begin{bmatrix} 30 & -45 \\ -60 & 90 \end{bmatrix} \begin{bmatrix} \frac{1}{5} \\ \frac{4}{5} \end{bmatrix}$$

$$= \begin{bmatrix} -30 & 45 \end{bmatrix} \begin{bmatrix} \frac{1}{5} \\ \frac{4}{5} \end{bmatrix} = \begin{bmatrix} (-30)\frac{1}{5} + (45)\frac{4}{5} \end{bmatrix} = \begin{bmatrix} 30 \end{bmatrix}$$

In general, a 1×1 matrix is written without brackets. Therefore, we write **PAQ** $= 30$. Note that this is the expected value of the game.
Do Exercise 1.

In general, if $\mathbf{A} = [\, a_{ij}\,]$ is an $m \times n$ game matrix,

$$\mathbf{P} = [\, p_1 \quad p_2 \quad \cdots \quad p_m \,] \text{ and } \mathbf{Q} = \begin{bmatrix} q_1 \\ q_2 \\ \vdots \\ q_n \end{bmatrix}$$

are strategies for R and C, respectively, then the payoff a_{ij} will occur if R chooses row i and C chooses column j. The probabilities of these choices are p_i and q_j, respectively. Therefore, the probability of the payoff a_{ij} is $p_i q_j$. To find the expectation of the game, we add all products $p_i q_j a_{ij}$ for $i = 1, 2, \ldots, m$ and $j = 1, 2, \ldots, n$. It is easy to verify that this sum is the product $\mathbf{PAQ}$.

DEFINITION 7.7:   If $\mathbf{A}$ is an $m \times n$ game matrix,

$$\mathbf{P} = [\, p_1 \quad p_2 \quad \cdots \quad p_m \,], \text{ and } \mathbf{Q} = \begin{bmatrix} q_1 \\ q_2 \\ \vdots \\ q_n \end{bmatrix}$$

are strategies for R and C, respectively, then the *expectation* of R when R uses strategy $\mathbf{P}$ and C uses strategy $\mathbf{Q}$ is $E(\mathbf{P}, \mathbf{Q})$ where

$$E(\mathbf{P}, \mathbf{Q}) = \mathbf{PAQ}$$

EXAMPLE 2   Consider the following game matrix

$$\mathbf{A} = \begin{bmatrix} 21 & -42 & 105 \\ 63 & -84 & 168 \end{bmatrix}$$

where the payoffs are in dollars. Let

$$\mathbf{P} = \begin{bmatrix} \frac{2}{3} & \frac{1}{3} \end{bmatrix} \text{ and } \mathbf{Q} = \begin{bmatrix} \frac{1}{7} \\ \frac{2}{7} \\ \frac{4}{7} \end{bmatrix}$$

be strategies for R and C, respectively. Find the expectation of R.

Solution   We first calculate the product $\mathbf{PA}$. We get

$$\mathbf{PA} = \begin{bmatrix} \frac{2}{3} & \frac{1}{3} \end{bmatrix} \begin{bmatrix} 21 & -42 & 105 \\ 63 & -84 & 168 \end{bmatrix} = [\, 35 \quad -56 \quad 126 \,]$$

Therefore

$$\mathbf{PAQ} = [\, 35 \quad -56 \quad 126 \,] \begin{bmatrix} \frac{1}{7} \\ \frac{2}{7} \\ \frac{4}{7} \end{bmatrix} = 61$$

Note that we wrote the product without brackets. If R and C use strategies **P** and **Q**, respectively, the expectation of R is $61.
Do Exercise 3. ■

The Value of a Game Matrix

The fundamental theorem of game theory is stated in Theorem 7.2.

> **THEOREM 7.2:** Let **A** be an $m \times n$ game matrix. There exist strategies **P*** and **Q*** for R and C, respectively, and a number v such that
>
> **P*****AQ** $\geq v$ for every strategy **Q** of C
>
> and
>
> **PAQ*** $\leq v$ for every strategy **P** of R

The proof of this theorem is beyond the scope of this book and is omitted.

> **DEFINITION 7.8:** The strategies **P*** and **Q*** described in the statement of the fundamental theorem are called *optimum (optimal) strategies* for R and C, respectively, and v is the value of the game. The game is said to be *fair* in the case $v = 0$.

We have seen in the preceding section that each player may have more than one optimum strategy. This was illustrated in the case where the game matrix has more than one saddle point. However, the value of a game matrix is unique. We shall prove this at the end of this section.

In this book, unless otherwise specified,

$$\mathbf{P}^* = [\, p_1^* \quad p_2^* \quad \cdots \quad p_m^* \,] \text{ and } \mathbf{Q}^* = \begin{bmatrix} q_1^* \\ q_2^* \\ \vdots \\ q_n^* \end{bmatrix}$$

denote optimum strategies for R and C, respectively.

> **THEOREM 7.3:** If **A** is a game matrix, then
>
> $E(\mathbf{P}^*, \mathbf{Q}^*) = \mathbf{P}^*\mathbf{AQ}^* = v$

Proof: By the fundamental theorem, **P*****AQ** $\geq v$ for every strategy **Q** of C. In particular, it must be true for the strategy **Q***. Thus

$$\mathbf{P}^*\mathbf{AQ}^* \geq v \tag{1}$$

Also $\mathbf{PAQ^*} \leq v$ for every strategy $\mathbf{P}$ of R and therefore it must be true for the strategy $\mathbf{P^*}$. Thus

$$\mathbf{P^*AQ^*} \leq v \tag{2}$$

Inequalities (1) and (2) yield

$$\mathbf{P^*AQ^*} = v \qquad \blacksquare$$

A Solution to a 2 × 2 Game Matrix

If a nonstrictly determined game matrix is 2×2, there exists a set of formulas for the solution of the game in terms of the entries of the game matrix. We do not have formulas for the solution of the more general case. However, in Sections 7.3 and 7.4 we will show how to convert the problem of solving a game matrix to an equivalent problem in linear programming that can be solved using the methods of Chapter 4.

First we must establish conditions under which a 2×2 game matrix is nonstrictly determined. We begin with the following example.

EXAMPLE 3   Complete the following matrix to obtain an example of a game matrix that is nonstrictly determined.

$$\begin{bmatrix} 2 & \\ & 7 \end{bmatrix}$$

Solution   First observe that we cannot have

$$2 \leq a_{12} \leq 7$$

To see this, assume that the above double inequality holds and write

$$\mathbf{A} = \begin{bmatrix} 2 & a_{12} \\ a_{21} & 7 \end{bmatrix}$$

Then, regardless of the value of a_{21}, $\mathbf{A}$ is strictly determined because

1. $a_{21} \leq 2$ implies 2 is a saddle point,
2. $2 < a_{21} < 7$ implies a_{21} is a saddle point, and
3. $7 \leq a_{21}$ implies 7 is a saddle point.

Therefore a_{12} must be less than 2 or greater than 7. Let $a_{12} = 1$ to obtain

$$\mathbf{A} = \begin{bmatrix} 2 & 1 \\ a_{21} & 7 \end{bmatrix}$$

It is easy to see that we cannot have $7 \leq a_{21}$ (otherwise 7 would be a saddle point) and by the same argument, we cannot have $2 \leq a_{21} \leq 7$. So we must have $a_{21} < 2$. Let $a_{21} = 0$ to obtain

$$\mathbf{A} = \begin{bmatrix} 2 & 1 \\ 0 & 7 \end{bmatrix}$$

Observe that if we had chosen a number greater than 7 for a_{12} we would have had to choose a number greater than 7 for a_{21} also. For example, we could have obtained

$$\mathbf{A} = \begin{bmatrix} 2 & 8 \\ 9 & 7 \end{bmatrix}$$

Do Exercise 5.　　　　　　　　　　　　　　　　　　　　　　　　　　　　■

The ideas used in the preceding example may be used to prove the following theorem.

> **THEOREM 7.4:**　The game matrix
>
> $$\mathbf{A} = \begin{bmatrix} a & b \\ c & d \end{bmatrix}$$
>
> is nonstrictly determined if, and only if,
>
> $$\max\{b, c\} < \min\{a, d\} \tag{1}$$
>
> or
>
> $$\max\{a, d\} < \min\{b, c\} \tag{2}$$

Proof:　Clearly if either (1) or (2) holds, then none of the entries a, b, c, or d is simultaneously a row minimum and a column maximum. So there is no saddle point and the game is nonstrictly determined.

Conversely, suppose that **A** is nonstrictly determined. Then using the same argument as in Example 3, we must have either

$$b < \min\{a, d\} \text{ and } c < \min\{a, d\}$$

which implies

$$\max\{b, c\} < \min\{a, d\}$$

or

$$b > \max\{a, d\} \text{ and } c > \max\{a, d\}$$

which implies

$$\max\{a, d\} < \min\{b, c\} \qquad\qquad ■$$

(See also Exercise *34.)

We are now ready to discuss a procedure to find optimal strategies for R and C in a 2×2 nonstrictly determined game matrix. The underlying principle is that R wants to maximize expected earnings by maximizing the minimum expected gain. Similarly, C wants to minimize expected losses by minimizing the maximum expected loss.

EXAMPLE 4   Consider the nonstrictly determined game matrix

$$\mathbf{A} = \begin{bmatrix} 3 & 1 \\ -2 & 5 \end{bmatrix}$$

Determine the optimal strategy for each player.

Solution   Suppose the probability that R chooses row 1 is p. Then R will choose row 2 with probability $1 - p$. If C chooses column 1, R's expected earning will be

$$E_R = 3p + (-2)(1 - p) = 5p - 2 \tag{1}$$

Similarly, if C chooses column 2, R's expected earning will be

$$E_R = p + 5(1 - p) = -4p + 5 \tag{2}$$

We have graphed both equations in Figure 7.1.

Let (p_0, E_0) be the point of intersection. Then if $0 < p < p_0$, the graph of equation 1 is below that of equation 2. So R's expected gain will be less if C chooses column 1. On the other hand, if $p_0 < p$, the graph of equation 2 is below that of equation 1. So R's expected gain will be less if C chooses column 2. It is now clear that the point (p_0, E_0) displays the maximum of the minimum expected earnings. We solve the system

$$\begin{cases} E_R = 5p - 2 \\ E_R = -4p + 5 \end{cases}$$

We get

$$p = \tfrac{7}{9} \quad \text{and} \quad E_R = \tfrac{17}{9}$$

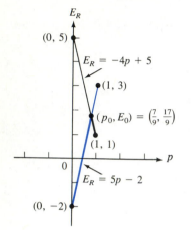

FIGURE 7.1

Therefore an optimal strategy for player R is $\mathbf{P^*} = \begin{bmatrix} \tfrac{7}{9} & \tfrac{2}{9} \end{bmatrix}$. Similarly, suppose q is the probability that player C chooses column 1. Then, C will choose column 2 with probability $1 - q$.

If R chooses row 1, R's expected earning will be

$$E_R = 3q + (1 - q) = 2q + 1 \tag{1}$$

Similarly, if R chooses row 2, R's expected earning will be

$$E_R = (-2)q + 5(1 - q) = -7q + 5 \tag{2}$$

We have graphed both equations in Figure 7.2.

Let (q_0, E_0) be the point of intersection. If $0 < q < q_0$, the graph of equation (2) is above that of equation (1). In this case R wins more (C loses more) if R chooses row 2. If $q > q_0$, the graph of equation (1) is above that of equation (2) and therefore R wins more (C loses more) if R chooses row 1. Since C is trying to minimize losses, C's purpose will be accomplished at the point of intersection.

We solve the system

$$\begin{cases} E_R = 2q + 1 \\ E_R = -7q + 5 \end{cases}$$

We get

$$q = \tfrac{4}{9} \quad \text{and} \quad E_R = \tfrac{17}{9}$$

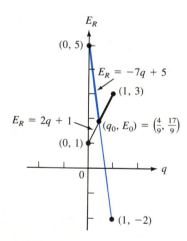

FIGURE 7.2

Hence, an optimum strategy for C is

$$\mathbf{Q^*} = \begin{bmatrix} \frac{4}{9} \\ \frac{5}{9} \end{bmatrix}$$

The value of the game is

$$E(\mathbf{P^*}, \mathbf{Q^*}) = \mathbf{P^*AQ^*} = \begin{bmatrix} \frac{7}{9} & \frac{2}{9} \end{bmatrix} \begin{bmatrix} 3 & 1 \\ -2 & 5 \end{bmatrix} \begin{bmatrix} \frac{4}{9} \\ \frac{5}{9} \end{bmatrix}$$

$$= \begin{bmatrix} \frac{17}{9} & \frac{17}{9} \end{bmatrix} \begin{bmatrix} \frac{4}{9} \\ \frac{5}{9} \end{bmatrix} = \frac{17}{9}$$

Observe that this is the value of E_R found in solving the preceding systems of equations. **Do Exercise 9.** ■

Using the same argument as in Example 4, the following theorem may be proved.

THEOREM 7.5: Suppose the game matrix

$$\mathbf{A} = \begin{bmatrix} a & b \\ c & d \end{bmatrix}$$

is nonstrictly determined and let

$$D = (a + d) - (b + c)$$

$$p_1^* = \frac{d - c}{D}, \quad p_2^* = 1 - p_1^*$$

$$q_1^* = \frac{d - b}{D}, \quad q_2^* = 1 - q_1^*$$

and

$$v = \frac{ad - bc}{D}$$

Then

$$\mathbf{P^*} = \begin{bmatrix} p_1^* & p_2^* \end{bmatrix} \quad \text{and} \quad \mathbf{Q^*} = \begin{bmatrix} q_1^* \\ q_2^* \end{bmatrix}$$

are optimum strategies for R and C, respectively, and v is the value of the game.

Proof: The details are analogous to the reasoning in Example 4 and are omitted here. However, you should verify the following:

1. $D \neq 0$ (see Exercise 28).

2. $0 < \dfrac{d - c}{D} < 1$ (see Exercise 29).

3. $0 < \dfrac{d - b}{D} < 1$ (see Exercise 30). ■

EXAMPLE 5   Solve the following game matrix

$$\mathbf{A} = \begin{bmatrix} 4 & 3 & 7 & 5 \\ 0 & -2 & 4 & -1 \\ 8 & 6 & 2 & -1 \end{bmatrix}$$

Solution   Row 1 dominates row 2, so we delete row 2 to obtain

$$\mathbf{B} = \begin{bmatrix} 4 & 3 & 7 & 5 \\ 8 & 6 & 2 & -1 \end{bmatrix}$$

Column 1 and column 3 are recessive. We delete them to obtain

$$\mathbf{C} = \begin{bmatrix} 3 & 5 \\ 6 & -1 \end{bmatrix}$$

The 2×2 matrix $\mathbf{C}$ is nonstrictly determined. So we can use the formulas from Theorem 7.5. Here $a = 3$, $b = 5$, $c = 6$, and $d = -1$. So $D = [3 + (-1)] - (5 + 6)] = -9$. Now

$$p_1{}^* = \frac{d - c}{D} = \frac{-1 - 6}{-9} = \frac{7}{9}, \quad p_2{}^* = 1 - \frac{7}{9} = \frac{2}{9}$$

$$q_1{}^* = \frac{d - b}{D} = \frac{-1 - 5}{-9} = \frac{2}{3}, \quad q_2{}^* = 1 - \frac{2}{3} = \frac{1}{3}$$

and

$$v = \frac{ad - bc}{D} = \frac{3(-1) - 5(6)}{-9} = \frac{11}{3}$$

Then

$$\mathbf{P}^* = \begin{bmatrix} \frac{7}{9} & \frac{2}{9} \end{bmatrix} \quad \text{and} \quad \mathbf{Q}^* = \begin{bmatrix} \frac{2}{3} \\ \frac{1}{3} \end{bmatrix}$$

are optimum strategies for R and C, respectively, and $\frac{11}{3}$ is the value of the game. In terms of the original game matrix, we would have

$$\mathbf{P}^* = \begin{bmatrix} \frac{7}{9} & 0 & \frac{2}{9} \end{bmatrix} \quad \text{and} \quad \mathbf{Q}^* = \begin{bmatrix} 0 \\ \frac{2}{3} \\ 0 \\ \frac{1}{3} \end{bmatrix}$$

Observe that

$$\begin{bmatrix} \frac{7}{9} & 0 & \frac{2}{9} \end{bmatrix} \begin{bmatrix} 4 & 3 & 7 & 5 \\ 0 & -2 & 4 & -1 \\ 8 & 6 & 2 & -1 \end{bmatrix} \begin{bmatrix} 0 \\ \frac{2}{3} \\ 0 \\ \frac{1}{3} \end{bmatrix} = \frac{11}{3}$$

which is the value of the game.
Do Exercise 11.

In the preceding example player R would never choose row 2 of the original matrix. In this game, player R could place seven red marbles and two green marbles in an urn and draw one at random. If a red marble is drawn (with probability $\frac{7}{9}$), R would choose row 1. If a green marble is drawn (with probability $\frac{2}{9}$), R would choose row 3. Player C would proceed similarly.

Having found the optimum strategies **P*** and **Q*** and the value of the game v for a 2×2 game matrix using the formulas of Theorem 7.5, the solution can be checked by showing that **P*AQ*** $= v$. If this does not check and you wish to locate your error, a further check can be done using the fact that **P*A** $= [v \quad v]$ and **AQ*** $= \begin{bmatrix} v \\ v \end{bmatrix}$. (See Exercises 27 and *33.)

Proof of the Uniqueness of the Value of a Game Matrix (Optional)

A CLOSER LOOK

> **THEOREM 7.6:** Let **A** be an $m \times n$ game matrix, **P*** and **Q*** be optimal strategies for R and C, respectively, and v be the corresponding value of the game. Then each entry of **P*A** is greater than or equal to v, and each entry of **AQ*** is less than or equal to v.

Proof: Write **P*A** $= [x_1 \quad x_2 \quad \ldots \quad x_n]$. Since v is the value of the game corresponding to **P*** and **Q***,

$$\mathbf{P^*AQ} \geq v$$

for every strategy **Q** of C. In particular, for the pure strategy $\mathbf{Q}_j$ which has a 1 in its jth row and 0's elsewhere, we have

$$v \leq \mathbf{P^*AQ}_i = [x_1 \quad x_2 \quad \ldots \quad x_n] \begin{bmatrix} 0 \\ \vdots \\ 1 \\ \vdots \\ 0 \end{bmatrix} = x_j$$

So $x_j \geq v$ for every $j = 1, 2, \ldots, n$.

The proof that each entry of **AQ*** is less than or equal to v is analogous and is omitted. ■

THEOREM 7.7:   Let **A** be an $m \times n$ game matrix, and **P*** and **Q*** be optimal strategies for R and C, respectively. If there is an number v such that each entry of **P*****A** is greater than or equal to v and each entry of **A****Q*** is less than or equal to v, then v is the value of the game corresponding to **P*** and **Q***.

Proof: Let $\mathbf{P}^*\mathbf{A} = [\, x_1 \quad x_2 \quad \cdots \quad x_n \,]$. Then

$$\mathbf{P}^*\mathbf{A}\mathbf{Q}^* = [\, x_1 \quad x_2 \quad \cdots \quad x_n \,]\begin{bmatrix} q_1^* \\ q_2^* \\ \cdot \\ \cdot \\ \cdot \\ q_n^* \end{bmatrix}$$

$$= x_1q_1^* + x_2q_2^* + \cdots + x_nq_n^* \geq vq_1^* + vq_2^* + \cdots + vq_n^*$$
$$= v(q_1^* + q_2^* + \cdots + q_n^*) = v$$

To justify the steps above, observe that $x_i \geq v$ for each i (given), $q_i^* \geq 0$ for each i and $q_1^* = q_2^* + \cdots + q_n^* = 1$, since **Q*** is a probability vector. So

$$\mathbf{P}^*\mathbf{A}\mathbf{Q}^* \geq v$$

Similarly, let

$$\mathbf{A}\mathbf{Q}^* = \begin{bmatrix} y_1 \\ y_2 \\ \cdot \\ \cdot \\ y_m \end{bmatrix}$$

Then

$$\mathbf{P}^*\mathbf{A}\mathbf{Q}^* = [\, p_1^* \quad p_2^* \quad \cdots \quad p_m^* \,]\begin{bmatrix} y_1 \\ y_2 \\ \cdot \\ \cdot \\ \cdot \\ y_m \end{bmatrix}$$

$$= p_1^*y_1 + p_2^*y_2 + \cdots + p_m^*y_m \leq p_1^*v + p_2^*v + \cdots + p_m^*v$$
$$= (p_1^* + p_2^* + \cdots + p_m^*)v = v$$

We have shown that $\mathbf{P}^*\mathbf{A}\mathbf{Q}^* \geq v$ and $\mathbf{P}^*\mathbf{A}\mathbf{Q}^* \leq v$. Therefore $\mathbf{P}^*\mathbf{A}\mathbf{Q}^* = v$. ∎

THEOREM 7.8:   Every matrix game has a unique value.

Proof: Suppose that **A** is an $m \times n$ game matrix, **P*** and **Q*** are optimal strategies for R and C, respectively, with u the corresponding value of the game, and **P****

and $\mathbf{Q}^{**}$ are optimal strategies for R and C, respectively, with v the corresponding value of the game. We wish to prove that $u = v$.

Let

$$\mathbf{P}^*\mathbf{A} = [\ x_1 \quad x_2 \quad \cdots \quad x_n\ ], \ \mathbf{AQ}^* = \begin{bmatrix} y_1 \\ y_2 \\ \cdot \\ \cdot \\ \cdot \\ y_m \end{bmatrix}$$

$$\mathbf{P}^{**}\mathbf{A} = [\ \overline{x_1} \quad \overline{x_2} \quad \cdots \quad \overline{x_n}\ ]$$

and

$$\mathbf{AQ}^{**} = \begin{bmatrix} \overline{y_1} \\ \overline{y_2} \\ \cdot \\ \cdot \\ \cdot \\ \overline{y_m} \end{bmatrix}$$

Then, by Theorem 7.6, we have

$$x_i \geq u \text{ and } \overline{x_i} \geq v \text{ for } i = 1, 2, \ldots, n$$

and

$$y_j \leq u \text{ and } \overline{y_j} \leq v \text{ for } j = 1, 2, \ldots, m$$

It follows that

$$\begin{aligned}
\mathbf{P}^*\mathbf{AQ}^{**} &= [\ x_1 \quad x_2 \quad \cdots \quad x_n\ ]\mathbf{Q}^{**} \\
&= x_1 q_1^{**} + x_2 q_2^{**} + \cdots + x_n q_n^{**} \\
&\geq u q_1^{**} + u q_2^{**} + \cdots + u q_n^{**} \\
&= u(q_1^{**} + q_2^{**} + \cdots + q_n^{**}) = u
\end{aligned}$$

In exactly the same way we can show that

$$\mathbf{P}^*\mathbf{AQ}^{**} \leq v$$

Thus we have shown

$$u \leq \mathbf{P}^*\mathbf{AQ}^{**} \leq v$$

It follows that

$$u \leq v \tag{1}$$

In the same manner it can be shown that

$$v \leq \mathbf{P}^{**}\mathbf{AQ}^* \leq u, \text{ so that}$$

$$v \leq u \tag{2}$$

Combining inequalities (1) and (2) we get $u = v$. ■

Exercise Set 7.2

1. Consider the following game matrix

$$A = \begin{bmatrix} 35 & -70 \\ -105 & 140 \end{bmatrix}$$

where the payoffs are in dollars.
Let

$$P = \begin{bmatrix} \frac{2}{5} & \frac{3}{5} \end{bmatrix} \text{ and } Q = \begin{bmatrix} \frac{3}{7} \\ \frac{4}{7} \end{bmatrix}$$

be strategies for R and C, respectively.

a. What is the probability that the payoff is $35 to R?
b. What is the probability that the payoff is $70 to C?
c. What is the probability that the payoff is $105 to C?
d. What is the probability that the payoff is $140 to R?
e. What is the expected value of the game for R?
f. Perform the matrix multiplication **PAQ** and compare your answer to that of part **e**.

2. Same as Exercise 1 with

$$A = \begin{bmatrix} -21 & 42 & -63 \\ 105 & -63 & 21 \end{bmatrix}, P = \begin{bmatrix} \frac{2}{3} & \frac{1}{3} \end{bmatrix},$$

$$\text{and } Q = \begin{bmatrix} \frac{2}{7} \\ \frac{1}{7} \\ \frac{4}{7} \end{bmatrix}$$

and appropriate payoffs to players R and C.

3. Consider the following game matrix

$$A = \begin{bmatrix} 40 & -60 & 120 \\ 80 & -40 & 160 \end{bmatrix}$$

where the payoffs are in dollars. Let

$$P = \begin{bmatrix} \frac{1}{4} & \frac{3}{4} \end{bmatrix} \text{ and } Q = \begin{bmatrix} \frac{1}{5} \\ \frac{3}{5} \\ \frac{1}{5} \end{bmatrix}$$

be strategies for R and C, respectively. Find the expectation of R.

4. Same as Exercise 3 with

$$A = \begin{bmatrix} 60 & -60 & 120 \\ 90 & -90 & 120 \\ 30 & 150 & 180 \end{bmatrix}, P = \begin{bmatrix} \frac{1}{5} & \frac{2}{5} & \frac{2}{5} \end{bmatrix},$$

$$\text{and } Q = \begin{bmatrix} \frac{1}{2} \\ \frac{1}{3} \\ \frac{1}{6} \end{bmatrix}$$

5. Complete the following to obtain an example of a game matrix that is nonstrictly determined.

$$\begin{bmatrix} 3 & \\ & 5 \end{bmatrix}$$

6. Same as Exercise 5 with

$$\begin{bmatrix} -2 & \\ & 8 \end{bmatrix}$$

7. Same as Exercise 5 with

$$\begin{bmatrix} & -3 \\ 5 & \end{bmatrix}$$

8. Same as Exercise 5 with

$$\begin{bmatrix} & 9 \\ -4 & \end{bmatrix}$$

9. Consider the nonstrictly determined game matrix

$$A = \begin{bmatrix} 5 & 2 \\ -3 & 7 \end{bmatrix}$$

Determine the optimal strategy for each player without using the formulas of Theorem 7.5 (see Example 4).

10. Same as Exercise 9 with

$$A = \begin{bmatrix} 4 & 3 \\ -5 & 6 \end{bmatrix}$$

In Exercises 11–17 do the following:

a. *Check if the game is strictly determined. If it is, solve it.*
b. *If the game is not strictly determined, reduce its dimensions, if possible.*
c. *Find the optimal strategies* **P*** *and* **Q*** *for R and C, respectively (in terms of the original game matrix), and the value of the game.*

11. A $= \begin{bmatrix} 2 & 3 & 5 & 7 \\ 3 & 6 & -1 & 0 \\ 1 & 5 & -2 & -3 \end{bmatrix}$

12. $A = \begin{bmatrix} -2 & 0 & 1 & 5 \\ 8 & 3 & 5 & 7 \\ 0 & 2 & -4 & 3 \end{bmatrix}$

13. $A = \begin{bmatrix} 3 & -2 & 1 & 2 \\ 2 & -4 & 3 & 2 \\ 4 & -6 & 1 & -8 \end{bmatrix}$

14. $A = \begin{bmatrix} 3 & 1 & 0 & -2 \\ 5 & 3 & 2 & 1 \\ 2 & 2 & 0 & -1 \\ 0 & -2 & 6 & 5 \end{bmatrix}$

15. $A = \begin{bmatrix} 0 & 1 & 2 & -1 \\ 1 & 2 & 7 & 3 \\ 5 & 3 & 2 & -2 \\ 9 & 11 & 5 & 10 \end{bmatrix}$

16. $A = \begin{bmatrix} 4 & -4 & -5 \\ 2 & -3 & 0 \\ 5 & -6 & -8 \\ 0 & -9 & 4 \end{bmatrix}$

17. $A = \begin{bmatrix} 2 & -1 & 0 & 5 & 7 \\ 1 & 2 & 1 & 0 & 3 \\ 4 & 3 & 4 & 2 & 4 \\ 2 & 0 & 1 & 1 & 3 \end{bmatrix}$

18. Roy and Cherie each flip a coin. If both coins match, Cherie pays Roy $1; otherwise, Roy pays Cherie $1. Find the optimum strategy for each player and the value of the game.

19. Suppose Ron and Carol play the following game. They simultaneously show one or two fingers. If the total number of fingers shown is T, then Carol pays Ron $$T$ if T is even; otherwise, Ron pays Carol $$T$. Determine the optimal strategy for each player. What is the value of the game?

20. Same as Exercise 19, but with each player showing two or three fingers.

21. Rosemary and Charlie play the following game. For each-play each draws a marble from an urn (the marble is replaced after each play). Rosemary's urn contains 3 blue, 4 red, and 2 white marbles; Charlie's urn contains 1 blue, 3 red, 4 white, and 3 green marbles. The payoff matrix is

	Blue	Red	White	Green
Blue	2	1	-3	0
Red	4	2	5	1
White	2	-6	1	4

For example if Rosemary draws a red marble and Charlie draws a blue marble, Charlie pays Rosemary $4. However, if Rosemary draws a white marble and Charlie draws a red marble, then Rosemary pays Charlie $6.

a. What is Rosemary's expected value of the game?

b. Suppose Charlie changes to his optimal strategy and Rosemary continues to play by drawing marbles; what is Rosemary's new expected value of the game?

c. Suppose Rosemary changes to her optimal strategy and Charlie continues to play by drawing marbles; what is Rosemary's new expected value of the game?

d. What is the value of the game if they both switch to their optimal strategies?

22. Same as Exercise 21 with this payoff matrix

	Blue	Red	White	Green
Blue	1	2	-4	1
Red	6	5	2	1
White	1	-9	3	2

23. There are only two sporting goods stores in town X, say store R and store C. They will between them get all of the town's business. Each month each must decide to use one, and only one, of the following forms of advertising: radio, television, mail, or newspaper. A consulting firm provided the following information, where each entry in the matrix indicates the number of percentage points above (below) 50% of the total business gained by store R.

	Radio	TV	Mail	Paper
Radio	3	6	-10	0
TV	4	8	-6	10
Mail	-8	2	1	-4
Paper	3	5	7	9

a. Find the optimum strategy for each store and the value of the game.

b. What is the expected value of the game for store R if R always chooses the newspaper and C uses its optimum strategy?

c. What is the expected value of the game for store R if R uses its optimum strategy and C always chooses television?

24. Same as Exercise 23 with the following matrix.

	Radio	TV	Mail	Paper
Radio	1	3	-12	2
TV	6	5	-3	15
Mail	-5	1	4	-7
Paper	3	2	6	8

25. Politicians R and C are campaigning for the governship of state M. They are giving speeches each day in different cities of the state. In each speech, each will make one, and only one, of two promises. The League of Women Voters conducted a survey and provided the followng information.

If R promises X and C promises Z, R will get 15 points.
If R promises X and C promises W, C will get 18 points.
If R promises Y and C promises Z, R will get 10 points.
If R promises Y and C promises W, R will get 13 points.

Each point represents a percentage point above 50% of the voting population. Write this information as a game matrix, determine the optimum strategies and the value of the game.

26. Repeat Exercise 25 with the following information.

If R promises X and C promises Z, C will get 10 points.
If R promises X and C promises W, R will get 15 points.
If R promises Y and C promises Z, R will get 12 points.
If R promises Y and C promises W, C will get 5 points.

In Exercises 27–32,

$$A = \begin{bmatrix} a & b \\ c & d \end{bmatrix}$$

is a nonstrictly determined game matrix. Use Theorem 7.4 and the formulas of Theorem 7.5 to do each exercise.

27. Let P^* and Q^* be the optimal strategies for R and C, respectively, and let v be the value of the game. Show that

$$P^*A = [\, v \quad v \,] \text{ and } AQ^* = \begin{bmatrix} v \\ v \end{bmatrix}$$

28. Verify that $D \neq 0$.

29. Verify that $0 < \dfrac{d - c}{D} < 1$.

30. Verify that $0 < \dfrac{d - b}{D} < 1$.

31. Show that if each entry of matrix A is increased by k to obtain matrix B, then the optimal strategies for R and C of game B are the same as those of game A but the value of the game has been increased by k.

32. Show that if each entry of matrix A is multiplied by the positive number k to obtain matrix C, then the optimal strategies for R and C of game C are the same as those of game A but the value of the game has been multiplied by k.

***33.** If A is a 2×2 game matrix, P^* and Q^* are optimal strategies for R and C, respectively, and v is the value of the game, then $E(P^*, Q^*) = P^*AQ^* = v$. Suppose that P and Q are strategies for R and C, respectively, and that

$$E(P, Q) = PAQ = v.$$

Must P and Q be optimum strategies? (*Hint:* To show that the answer is no, consider a nonstrictly determined game matrix A, find the optimal strategy P^* for R and the value v of the game. Let $Q = \begin{bmatrix} 1 \\ 0 \end{bmatrix}$ and show that $P^*AQ = v$ and that Q is not an optimum strategy for C. (See also Exercise 27.))

***34.** Prove that the game matrix $A = \begin{bmatrix} a & b \\ c & d \end{bmatrix}$ is strictly determined if, and only if, $\max\{b, c\} \geq \min\{a, d\}$ and $\max\{a, d\} \geq \min\{b, c\}$. (See Theorem 7.4.)

7.3 Solution of a 2 × 2 Game Matrix Using Linear Programming

We do not have formulas to solve a nonstrictly determined $m \times n$ game matrix when either m, n, or both are greater than 2. However, each such game has a corresponding linear programming problem that can be solved using the methods described in Chapter 4. Although a nonstrictly determined 2×2 game matrix may be solved using the formulas of Theorem 7.5 from the preceding section, we will use it to begin our presentation of linear programming. In this manner, the details will not get too involved and you will be able to concentrate on the ideas behind the method.

Preliminary Results

First, we must establish some preliminary facts needed to justify the steps that will be taken.

> **THEOREM 7.9:** Let **A** be an $m \times n$ game matrix, **P*** and **Q*** be optimal strategies, and v be the value of the game. Let **B** be the $m \times n$ matrix obtained by adding the constant k to each entry of **A**. Then **P*** and **Q*** are optimal strategies of the game matrix **B** and $v + k$ is the value of the new game.

Proof: For the special 2×2 case, see Exercise 31 of the preceding section. The (optional) proof of the general case is given at the end of this section. ∎

Now let **A** be a 2×2 nonstrictly determined game matrix. If necessary, add a positive constant k to each entry of **A** to obtain a matrix **B** all of whose entries are positive. We shall describe how to obtain optimal strategies **P*** and **Q***, and the value v of the game matrix **B**. Then **P*** and **Q*** will also be optimal strategies of the game matrix **A**, and $v - k$ will be its value.

Recall that R's goal is to find a strategy **P*** and the *largest* possible number v such that

$$\mathbf{P^*BQ} \geq v$$

for every strategy **Q** of C.

The following theorem is useful in translating this problem to a linear programming problem.

> **THEOREM 7.10:** Let **B** be a 2×2 game matrix; $\mathbf{P} = [\, p_1 \quad p_2 \,]$, $\mathbf{Q}_1 = \begin{bmatrix} 1 \\ 0 \end{bmatrix}$, and $\mathbf{Q}_2 = \begin{bmatrix} 0 \\ 1 \end{bmatrix}$ be strategies (probability vectors); and v be a number. Then $\mathbf{PBQ} \geq v$ for any choice of strategy **Q** by C if, and only if, $\mathbf{PBQ}_1 \geq v$ and $\mathbf{PBQ}_2 \geq v$.

Proof: Suppose first that $\mathbf{PBQ} \geq v$ for any choice of strategy **Q** by C. Then, since C may choose either $\mathbf{Q}_1$ or $\mathbf{Q}_2$, we have $\mathbf{PBQ}_1 \geq v$ and $\mathbf{PBQ}_2 \geq v$. Conversely, suppose that $\mathbf{PBQ}_1 \geq v$ and $\mathbf{PBQ}_2 \geq v$ and let

$$\mathbf{Q} = \begin{bmatrix} q_1 \\ q_2 \end{bmatrix}$$

Then

$$\mathbf{Q} = \begin{bmatrix} q_1 \\ q_2 \end{bmatrix} = q_1 \begin{bmatrix} 1 \\ 0 \end{bmatrix} + q_2 \begin{bmatrix} 0 \\ 1 \end{bmatrix} = q_1\mathbf{Q}_1 + q_2\mathbf{Q}_2$$

So

$$\mathbf{PBQ} = \mathbf{PB}(q_1\mathbf{Q}_1 + q_2\mathbf{Q}_2) = \mathbf{PB}q_1\mathbf{Q}_1 + \mathbf{PB}q_2\mathbf{Q}_2$$
$$= q_1\mathbf{PBQ}_1 + q_2\mathbf{PBQ}_2 \geq q_1 v + q_2 v = (q_1 + q_2)v = v$$

Therefore

$$\mathbf{PBQ} \geq v$$

You should be able to justify these steps using the properties of matrix multiplication, inequalities, and the definition of strategy (probability vector). ∎

We will also need the following theorem.

> **THEOREM 7.11:** If each entry of a game matrix is positive, then the value of the game is positive.

Proof: See Exercise 16. ∎

Translation of the Problem

We are now ready to translate the game matrix problem to a linear programming problem. Let

$$\mathbf{B} = \begin{bmatrix} a & b \\ c & d \end{bmatrix}$$

where a, b, c, and d are positive. This requirement is necessary, as we shall soon see, but it is not too restrictive because of Theorem 7.9. (See also Exercise 15.) We wish to find a strategy $\mathbf{P}^* = [\, p_1^* \quad p_2^* \,]$ and the *largest* number v such that

$$\mathbf{P}^*\mathbf{BQ}_1 \geq v \quad (1) \quad \text{and} \quad \mathbf{P}^*\mathbf{BQ}_2 \geq v \quad (2)$$

Note that

$$\mathbf{P}^*\mathbf{BQ}_1 = [\, p_1^* \quad p_2^* \,] \begin{bmatrix} a & b \\ c & d \end{bmatrix} \begin{bmatrix} 1 \\ 0 \end{bmatrix} = p_1^* a + p_2^* c$$

and

$$\mathbf{P}^*\mathbf{BQ}_2 = [\, p_1^* \quad p_2^* \,] \begin{bmatrix} a & b \\ c & d \end{bmatrix} \begin{bmatrix} 0 \\ 1 \end{bmatrix} = p_1^* b + p_2^* d$$

Inequalities (1) and (2) are equivalent to the following system.

$$\begin{cases} p_1^* a + p_2^* c \geq v & (3) \\ p_1^* b + p_2^* d \geq v & (4) \end{cases}$$

By Theorem 7.11, we know that the value of v that we seek is positive. Therefore we may divide both sides of inequalities (3) and (4) by v and let $\dfrac{p_1^*}{v} = x_1$ and $\dfrac{p_2^*}{v} = x_2$ to obtain

$$\begin{cases} ax_1 + cx_2 \geq 1 & (5) \\ bx_1 + dx_2 \geq 1 & (6) \end{cases}$$

where $x_1 \geq 0$ and $x_2 \geq 0$.

The problem is to maximize v subject to the preceding constraints.

Observe now that

$$x_1 + x_2 = \frac{p_1{}^*}{v} + \frac{p_2{}^*}{v} = \frac{p_1{}^* + p_2{}^*}{v} = \frac{1}{v} \qquad \text{(since } p_1{}^* + p_2{}^* = 1\text{)}$$

Thus to maximize v, it is sufficient to minimize $z = \frac{1}{v}$.

We have shown that the game matrix problem is equivalent to the following linear programming problem:

Minimize $z = \frac{1}{v} = x_1 + x_2$ subject to

$$\begin{cases} ax_1 + cx_2 \geq 1 \\ bx_1 + dx_2 \geq 1 \\ x_1 \geq 0,\ x_2 \geq 0 \end{cases}$$

Using the methods presented in Chapter 4, we solve this linear programming problem for x_1, x_2, and the minimum value of z. Then using

$$z = \frac{1}{v},\ \frac{p_1{}^*}{v} = x_1 \text{ and } \frac{p_2{}^*}{v} = x_2$$

we can find v, $p_1{}^*$, and $p_2{}^*$.

Now C's goal is to find a strategy $\mathbf{Q}^*$ and the *smallest* possible number v' such that $\mathbf{PBQ}^* \leq v'$ for any choice of strategy $\mathbf{P}$ by R. Therefore we want to find a strategy

$$\mathbf{Q}^* = \begin{bmatrix} q_1{}^* \\ q_2{}^* \end{bmatrix}$$

and the *smallest* number v' such that

$$\mathbf{P_1BQ}^* \leq v' \quad (7) \quad \text{and} \quad \mathbf{P_2BQ}^* \leq v' \quad (8)$$

(See Exercise 17.)

But

$$\mathbf{P_1BQ}^* = [\ 1 \quad 0\ ] \begin{bmatrix} a & b \\ c & d \end{bmatrix} \begin{bmatrix} q_1{}^* \\ q_2{}^* \end{bmatrix} = q_1{}^*a + q_2{}^*b$$

Similarly,

$$\mathbf{P_2BQ}^* = [\ 0 \quad 1\ ] \begin{bmatrix} a & b \\ c & d \end{bmatrix} \begin{bmatrix} q_1{}^* \\ q_2{}^* \end{bmatrix} = q_1{}^*c + q_2{}^*d$$

So inequalitites (7) and (8) are equivalent to the following system.

$$\begin{cases} q_1{}^*a + q_2{}^*b \leq v' & \qquad (9) \\ q_1{}^*c + q_2{}^*d \leq v' & \qquad (10) \end{cases}$$

By Theorem 7.11, we know that the value of v' that we seek is positive. So, we may divide both sides of inequalities (9) and (10) by v', then let $\frac{q_1{}^*}{v'} = y_1$ and $\frac{q_2{}^*}{v'} = y_2$ to obtain

$$\begin{cases} ay_1 + by_2 \leq 1 & \qquad (11) \\ cy_1 + dy_2 \leq 1 & \qquad (12) \end{cases}$$

where $y_1 \geq 0$, $y_2 \geq 0$.

The problem is to minimize v' subject to the preceding constraints.
Observe now that

$$y_1 + y_2 = \frac{q_1{}^*}{v'} + \frac{q_2{}^*}{v'} = \frac{q_1{}^* + q_2{}^*}{v'} = \frac{1}{v'} \qquad \text{(since } q_1{}^* + q_2{}^* = 1\text{)}$$

Thus to minimize v', it is sufficient to maximize $Z = \dfrac{1}{v'}$.

We have shown that the game matrix problem is equivalent to the following linear programming problem.

Maximize $Z = \dfrac{1}{v'} = y_1 + y_2$ subject to

$$\begin{cases} ay_1 + by_2 \le 1 \\ cy_1 + dy_2 \le 1 \\ y_1 \ge 0, \, y_2 \ge 0 \end{cases}$$

Using the methods presented in Chapter 4, we solve this linear programming problem for y_1, y_2, and the maximum value of Z. Then using

$$Z = \frac{1}{v'}, \, \frac{q_1{}^*}{v'} = y_1 \text{ and } \frac{q_2{}^*}{v'} = y_2$$

we find v', $q_1{}^*$, and $q_2{}^*$.

Note that the second linear programming problem we obtained is the dual of the first. Thus we should obtain the same optimal values. That is, min z = max Z. So $v = v'$. For convenience, we summarize the method just described.

Procedure to Solve a 2 × 2 Matrix Game Using Linear Programming

Let **A** be a 2 × 2 nonstrictly determined game. To find optimum strategies **P***
and **Q*** for R and C, respectively, and the value of the game v proceed as follows:

Step 1. If any entry in **A** is not positive, add to each entry of matrix **A** any constant k large enough to obtain

$$\mathbf{B} = \begin{bmatrix} a & b \\ c & d \end{bmatrix}$$

where a, b, c, and d are positive. Otherwise, proceed directly to Step 2. In that case, $\mathbf{A} = \mathbf{B}$ and $k = 0$.

Step 2. Set up the following two linear programming problems.
I. Minimize $z = x_1 + x_2$ subject to

$$\begin{cases} ax_1 + cx_2 \ge 1 \\ bx_1 + dx_2 \ge 1 \\ x_1 \ge 0, \, x_2 \ge 0 \end{cases}$$

II. Maximize $Z = y_1 + y_2$ subject to

$$\begin{cases} ay_1 + by_2 \leq 1 \\ cy_1 + dy_2 \leq 1 \\ y_1 \geq 0, \ y_2 \geq 0 \end{cases}$$

Step 3. Solve each linear programming problem. Since there are only two decision variables in each problem, geometric solutions are simplest. Since the second problem is the dual of the first, the minimum value of z must equal the maximum value of Z.

Step 4. Substitute the values of x_1, x_2, z, y_1, y_2, and Z found in Step 3 in

$$v = \frac{1}{z} \quad \left(\text{or in } v = \frac{1}{z}\right)$$

$$\mathbf{P^*} = [\ vx_1 \quad vx_2\ ] \text{ and } \mathbf{Q^*} = \begin{bmatrix} vy_1 \\ vy_2 \end{bmatrix}$$

to find the optimal strategies $\mathbf{P^*}$ and $\mathbf{Q^*}$ and the value v of the game matrix $\mathbf{B}$.

Step 5. The optimal strategies of the original game matrix $\mathbf{A}$ are also $\mathbf{P^*}$ and $\mathbf{Q^*}$. Its value is $V = v - k$.

Step 6. Check your answer verifying that $\mathbf{P^*AQ^*} = v$. You can also check your answers by solving the original problem using the formulas of Theorem 7.5 of the preceding section.

EXAMPLE 1 Consider the following game matrix

$$\mathbf{A} = \begin{bmatrix} 2 & -2 \\ -1 & 3 \end{bmatrix}$$

a. Show that it is nonstrictly determined.
b. Write two corresponding linear programming problems.
c. Solve these problems geometrically.
d. What are the optimal strategies and the value of the game matrix $\mathbf{A}$?

Solution **a.** Observe that

$$\max\{-1, -2\} < \min\{2, 3\}$$

So by Theorem 7.4, Section 7.2, the game is nonstrictly determined.
b. We obtain the corresponding linear progrmming problems as follows:
Add 3 to every entry of matrix $\mathbf{A}$ to obtain

$$\mathbf{B} = \begin{bmatrix} 5 & 1 \\ 2 & 6 \end{bmatrix}$$

Note that $k = 3$.

Here $a = 5$, $b = 1$, $c = 2$, and $d = 6$. Therefore the linear programming problems are

I. Minimize $z = x_1 + x_2$ subject to

$$\begin{cases} 5x_1 + 2x_2 \geq 1 \\ x_1 + 6x_2 \geq 1 \\ x_1 \geq 0,\ x_2 \geq 0 \end{cases}$$

II. Maximize $Z = y_1 + y_2$ subject to

$$\begin{cases} 5y_1 + y_2 \leq 1 \\ 2y_1 + 6y_2 \leq 1 \\ y_1 \geq 0,\ y_2 \geq 0 \end{cases}$$

We solve each linear programming problem geometrically. The regions of feasible solutions are shown in Figures 7.3 and 7.4.

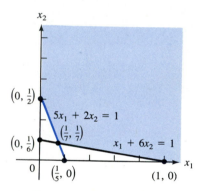

FIGURE 7.3   **FIGURE 7.4**

Since in each case the extremum must occur at a vertex, we evaluate $z = x_1 + x_2$ at the three vertices of the region of Figure 7.3. The vertices were easily obtained solving systems of linear equations.

We have

At $\left(0, \frac{1}{2}\right)$, $z = 0 + \frac{1}{2} = \frac{1}{2}$

At $\left(\frac{1}{7}, \frac{1}{7}\right)$, $z = \frac{1}{7} + \frac{1}{7} = \frac{2}{7}$

At $(1, 0)$, $z = 1 + 0 = 1$

The region of Figure 7.3 is unbounded. The graph of any line $x_1 + x_2 = c$, where $c < \frac{2}{7}$, does not intersect the region. Therefore the minimum value of z is $\frac{2}{7}$. It occurs at $x_1 = \frac{1}{7}$, $x_2 = \frac{1}{7}$.

We also evaluate Z at each vertex of the region of Figure 7.4.

At $(0, 0)$, $Z = 0 + 0 = 0$

At $\left(0, \frac{1}{6}\right)$, $Z = 0 + \frac{1}{6} = \frac{1}{6}$

At $\left(\frac{5}{28}, \frac{3}{28}\right)$, $Z = \frac{5}{28} + \frac{3}{28} = \frac{2}{7}$

At $\left(\frac{1}{5}, 0\right)$, $Z = \frac{1}{5} + 0 = \frac{1}{5}$

The maximum value of Z is $\frac{2}{7}$. It occurs at $y_1 = \frac{5}{28}$, $y_2 = \frac{3}{28}$. As expected, min z = max Z. Since $\frac{1}{v} = z$ and $z = \frac{2}{7}$, we have $v = \frac{7}{2}$. Furthermore

$$p_1{}^* = vx_1 = \frac{7}{2} \cdot \frac{1}{7} = \frac{1}{2} \quad \text{and} \quad p_2{}^* = vx_2 = \frac{7}{2} \cdot \frac{1}{7} = \frac{1}{2}$$

So $\mathbf{P}^* = \begin{bmatrix} \frac{1}{2} & \frac{1}{2} \end{bmatrix}$.

Also

$$q_1{}^* = vy_1 = \frac{7}{2} \cdot \frac{5}{28} = \frac{5}{8} \quad \text{and} \quad q_2{}^* = vy_2 = \frac{7}{2} \cdot \frac{3}{28} = \frac{3}{8}$$

Therefore

$$\mathbf{Q}^* = \begin{bmatrix} \frac{5}{8} \\ \frac{3}{8} \end{bmatrix}$$

The value of the original game is $V = v - k = \frac{7}{2} - 3 = \frac{1}{2}$. Performing the multiplication $\mathbf{P}^*\mathbf{A}\mathbf{Q}^*$, we obtain

$$\begin{bmatrix} \frac{1}{2} & \frac{1}{2} \end{bmatrix} \begin{bmatrix} 2 & -2 \\ -1 & 3 \end{bmatrix} \begin{bmatrix} \frac{5}{8} \\ \frac{3}{8} \end{bmatrix} = \begin{bmatrix} \frac{1}{2} & \frac{1}{2} \end{bmatrix} \begin{bmatrix} \frac{5}{8} \\ \frac{3}{8} \end{bmatrix} = \frac{1}{2}$$

This is, as expected, the value V of the game found earlier. You should also check that the same optimum strategies and the same value of the game would be obtained using the formulas of Theorem 7.5 of the preceding section.
Do Exercise 1.   ■

Proof of Theorem 7.9 (Optional)

The statement of the theorem is given again for convenience.

> Let $\mathbf{A}$ be an $m \times n$ game matrix, $\mathbf{P}^*$ and $\mathbf{Q}^*$ be optimal strategies, and v be the value of the game. Let $\mathbf{B}$ be the $m \times n$ matrix obtained by adding the constant k to each entry of $\mathbf{A}$. Then $\mathbf{P}^*$ and $\mathbf{Q}^*$ are optimal strategies of the game matrix $\mathbf{B}$, and $v + k$ is the value of the new game.

Proof: Let $\mathbf{K}$ be the $m \times n$ matrix all of whose entries are k. Then,

$$\mathbf{B} = \mathbf{A} + \mathbf{K}$$

Since $\mathbf{P}^*$ and $\mathbf{Q}^*$ are optimal strategies of the game matrix $\mathbf{A}$ and v is its value by Definition 7.8, we know that

$$\mathbf{P}^*\mathbf{A}\mathbf{Q} \geq v \text{ for any choice of } \mathbf{Q} \text{ by C} \tag{13}$$

and

$$\mathbf{P}\mathbf{A}\mathbf{Q}^* \leq v \text{ for any choice of } \mathbf{P} \text{ by C} \tag{14}$$

We want to prove

$$\mathbf{P}^*\mathbf{BQ} \geq v + k \text{ for any choice of } \mathbf{Q} \text{ by C}$$

and

$$\mathbf{PBQ}^* \leq v + k \text{ for any choice of } \mathbf{P} \text{ by C.}$$

By Exercise 19 at the end of this section, we have

$$\mathbf{P}^*\mathbf{KQ} = k \tag{15}$$

Adding equation (15) to inequality (13), we obtain

$$\mathbf{P}^*\mathbf{AQ} + \mathbf{P}^*\mathbf{KQ} \geq v + k \text{ for any choice of } \mathbf{Q} \text{ by C} \tag{16}$$

Using the distributive law of multiplication of matrices over addition (twice), we get

$$\mathbf{P}^*(\mathbf{A} + \mathbf{K})\mathbf{Q} \geq v + k \text{ for any choice of } \mathbf{Q} \text{ by C} \tag{17}$$

That is

$$\mathbf{P}^*\mathbf{BQ} \geq v + k \text{ for any choice of } \mathbf{Q} \text{ by C} \tag{18}$$

In exactly the same manner, but using inequality (14), we can show

$$\mathbf{PBQ}^* \leq v + k \text{ for any choice of } \mathbf{P} \text{ by C} \tag{19}$$

Thus, $\mathbf{P}^*$ and $\mathbf{Q}^*$ are optimum strategies for the game matrix $\mathbf{B}$ and $v + k$ is its value. ∎

Exercise Set 7.3

In Exercises 1–6, do the following:

a. Show that the game matrix is nonstrictly determined.
b. If appropriate, delete recessive rows and columns to obtain a 2 × 2 game matrix.
c. Write two corresponding linear programming problems.
d. Solve these problems geometrically.
e. What are the optimal strategies and the value of the original game matrix?

1. Use matrix $\mathbf{A}$ of Exercise 9, Exercise Set 7.2.

2. Use matrix $\mathbf{A}$ of Exercise 10, Exercise Set 7.2.

3. Use matrix $\mathbf{A}$ of Exercise 11, Exercise Set 7.2.

4. Use matrix $\mathbf{A}$ of Exercise 14, Exercise Set 7.2.

5. Use matrix $\mathbf{A}$ of Exercise 15, Exercise Set 7.2.

6. Use matrix $\mathbf{A}$ of Exercise 17, Exercise Set 7.2.

Solve Exercises 7–14 using the linear programming method.

7. Do Exercise 19, Exercise Set 7.2.

8. Do Exercise 20, Exercise Set 7.2

9. Do Exercise 21, Exercise Set 7.2

10. Do Exercise 22, Exercise Set 7.2

11. Do Exercise 23, Exercise Set 7.2

12. Do Exercise 24, Exercise Set 7.2

13. Do Exercise 25, Exercise Set 7.2

14. Do Exercise 26, Exercise Set 7.2

15. Let $\mathbf{A} = \begin{bmatrix} a & b \\ c & d \end{bmatrix}$ be a game matrix and suppose that at least one of its entries is not positive. Let $e = \min\{a, b, c, d\}$ and $k = |e| + 1$. Show that each entry of $\mathbf{B} = \begin{bmatrix} a + k & b + k \\ c + k & d + k \end{bmatrix}$ is positive.

16. Let $\mathbf{A}$ be an $m \times n$ game matrix all of whose entries are positive. Show that the value of the game is positive. (*Hint:* Use the fact that in a strategy vector no entry is negative and at least one entry is positive.)

*17. Let $\mathbf{B}$ be a 2×2 game matrix; $\mathbf{P}_1 = [\,1 \quad 0\,]$, $\mathbf{P}_2 = [\,0 \quad 1\,]$, and $\mathbf{Q} = \begin{bmatrix} q_1 \\ q_2 \end{bmatrix}$ be strategies (probability vectors); and v be a number. Prove that

$PBQ \leq v$ for any choice of strategy P by R

if, and only if,

$P_1BQ \leq v$ and $P_2BQ \leq v$

(*Hint:* See the proof of Theorem 7.10, this section.)

18. Let $K = \begin{bmatrix} k & k & k \\ k & k & k \end{bmatrix}$, $P = [\, p_1 \quad p_2 \,]$, and

$Q = \begin{bmatrix} q_1 \\ q_2 \\ q_3 \end{bmatrix}$ be strategies. Show that **a. PK =**

$[\, k \quad k \quad k \,]$ and **b.** $PKQ = k$. (*Hint:* Recall that $p_1 + p_2 = q_1 + q_2 + q_3 = 1$.)

19. Let K be an $m \times n$ game matrix all entries of which are k.

Let $P = [\, p_1 \quad p_2 \quad \cdots \quad p_m \,]$ and $Q = \begin{bmatrix} q_1 \\ q_2 \\ \vdots \\ q_n \end{bmatrix}$ be

strategies. Show that $PKQ = k$.

7.4 Solution of an *m* × *n* Game Matrix Using Linear Programming

In this section, we generalize the methods of Section 7.3 to apply to $m \times n$ game matrices. The simplex method will be used to solve the corresponding linear programming problems. A generalization of the procedure to convert a 2×2 game matrix into two linear programming problems, each the dual of the other, may be used for $m \times n$ game matrices. We will omit the details and instead give a summary of the procedure for the general case.

Procedure

Procedure to Solve an *m* × *n* Matrix Game Using Linear Programming

Step 1. If there are recessive rows or columns, delete them to obtain an $m \times n$ matrix **A** free of recessive rows and columns.

Step 2. If any entry in **A** is not positive, add to each entry of matrix **A**, a constant k large enough to obtain

$$B = \begin{bmatrix} a_{11} & a_{12} & \cdots & a_{1n} \\ a_{21} & a_{22} & \cdots & a_{2n} \\ \vdots & \vdots & \vdots & \vdots \\ a_{m1} & a_{m2} & \cdots & a_{mn} \end{bmatrix}$$

where $a_{ij} > 0$, for $i = 1, 2, \ldots, m$ and $j = 1, 2, \ldots, n$. Otherwise, proceed directly to Step 3. In that case, $A = B$ and $k = 0$.

Step 3. Set up the following two linear programming problems.

1. Minimize $z = x_1 + x_2 + \cdots + x_m$
   subject to

$$\begin{cases} a_{11}x_1 + a_{21}x_2 + \cdots + a_{m1}x_m \geq 1 \\ a_{12}x_1 + a_{22}x_2 + \cdots + a_{m2}x_m \geq 1 \\ \qquad \vdots \qquad\quad \vdots \qquad\quad \vdots \\ a_{1n}x_1 + a_{2n}x_2 + \cdots + a_{mn}x_m \geq 1 \\ \quad x_1 \geq 0, x_2 \geq 0, \ldots, x_m \geq 0 \end{cases}$$

2. Maximize $Z = y_1 + y_2 + \cdots + y_n$
   subject to

$$\begin{cases} a_{11}y_1 + a_{12}y_2 + \cdots + a_{1n}y_n \leq 1 \\ a_{21}y_1 + a_{22}y_2 + \cdots + a_{2n}y_n \leq 1 \\ \qquad \vdots \qquad\quad \vdots \qquad\quad \vdots \\ a_{m1}y_1 + a_{m2}y_2 + \cdots + a_{mn}y_n \leq 1 \\ \quad y_1 \geq 0, y_2 \geq 0, \ldots, y_n \geq 0 \end{cases}$$

Step 4. Observe that 2 is a linear programming problem in standard form and is the dual of 1. So we solve 2 by the simplex method. In this manner, we get the solution to problem 2, but we automatically get the solution of the minimization problem (part 1) since part 2 is the dual of part 1. In the final tableau, we get the values of the variables that will yield the optimum strategies for R in the objective row and in the slack variables columns. That is, x_1 is found in the objective row in the first slack variable column, x_2 is found in the objective row in the second slack variable column, and so on. So in solving the second linear programming problem, it is convenient to name the first slack variable x_1, the second slack variable x_2, and so on.

Note that we are using the same names for the slack variables of the second problem and the decision variable of the first. This should not create confusion since we get the solution of the first problem from the final tableau of the second problem.

Step 5. Substitute the values of $x_1, x_2, \ldots, x_m, y_1, y_2, \ldots, y_n$, and Z found in Step 4 in

$$v = \frac{1}{Z}, \mathbf{P}^* = [\, vx_1 \quad vx_2 \quad \cdots \quad vx_m \,], \mathbf{Q}^* = \begin{bmatrix} vy_1 \\ vy_2 \\ \vdots \\ vy_n \end{bmatrix}$$

Step 6. $\mathbf{P}^*$ and $\mathbf{Q}^*$ are optimal strategies for the original game and $V = v - k$ is the value of the original game.

Step 7. Check your solution by verifying that $\mathbf{P}^*\mathbf{A}\mathbf{Q}^* = V$.

Before reading the examples, you are urged to review the summary of Chapter 4 in Section 4.8. In particular, you should be familiar with the procedures of selecting the pivot element in a tableau and of pivoting.

EXAMPLE 1 Solve the game matrix

$$\begin{bmatrix} 2 & -3 & 5 \\ 0 & -5 & 2 \\ -1 & 3 & -2 \end{bmatrix}$$

Solution Step 1. Observe that the second row is recessive. We delete it to get

$$\mathbf{A} = \begin{bmatrix} 2 & -3 & 5 \\ -1 & 3 & -2 \end{bmatrix}$$

Step 2. Add 4 to each entry of matrix $\mathbf{A}$. Note that $k = 4$.

$$\mathbf{B} = \begin{bmatrix} 6 & 1 & 9 \\ 3 & 7 & 2 \end{bmatrix}$$

Step 3. Set up the following two linear programming problems.
1. Minimize $z = x_1 + x_2$
   subject to

$$\begin{cases} 6x_1 + 3x_2 \geq 1 \\ x_1 + 7x_2 \geq 1 \\ 9x_1 + 2x_2 \geq 1 \\ x_1 \geq 0, x_2 \geq 0 \end{cases}$$

2. Maximize $Z = y_1 + y_2 + y_3$
   subject to

$$\begin{cases} 6y_1 + y_2 + 9y_3 \leq 1 \\ 3y_1 + 7y_2 + 2y_3 \leq 1 \\ y_1 \geq 0, y_2 \geq 0, y_3 \geq 0 \end{cases}$$

Step 4. To solve the second linear programming problem, we introduce the slack variables x_1 and x_2 and write the equation of the objective function as follows.

$$\begin{cases} 6y_1 + y_2 + 9y_3 + x_1 \quad\quad = 1 \\ 3y_1 + 7y_2 + 2y_3 \quad\quad + x_2 = 1 \\ Z - y_1 - y_2 - y_3 \quad\quad\quad = 0 \end{cases}$$

The initial tableau is

Tableau I

Basis	y_1	y_2	y_3	x_1	x_2	Values of basic variables
x_1 ←	⑥	1	9	1	0	1
x_2	3	7	2	0	1	1
Z	−1	−1	−1	0	0	0

The most negative indicators in the third row are in the first three columns. So we may choose any one of these as the pivot column. We choose the first. Since $\frac{1}{6} < \frac{1}{3}$, 6 is the pivot. Pivoting, we get the next tableau.

Tableau II

Basis	y_1	y_2	y_3	x_1	x_2	Values of basic variables
y_1	1	$\frac{1}{6}$	$\frac{3}{2}$	$\frac{1}{6}$	0	$\frac{1}{6}$
x_2 ←	0	$\frac{13}{2}$	$\frac{-5}{2}$	$\frac{-1}{2}$	1	$\frac{1}{2}$
Z	0	$\frac{-5}{6}$	$\frac{1}{2}$	$\frac{1}{6}$	0	$\frac{1}{6}$

The only negative indicator in the third row is $\frac{-5}{6}$. Thus the second column is the pivot column. Since $\frac{1}{2}/\frac{13}{2} < \frac{1}{6}/\frac{1}{6}$, the pivot is $\frac{13}{2}$. Pivoting, we get

Tableau III

Basis	y_1	y_2	y_3	x_1	x_2	Values of basic variables
y_1	1	0	$\frac{61}{39}$	$\frac{7}{39}$	$\frac{-1}{39}$	$\frac{2}{13}$
y_2	0	1	$\frac{-5}{13}$	$\frac{-1}{13}$	$\frac{2}{13}$	$\frac{1}{13}$
Z	0	0	$\frac{7}{39}$	$\frac{4}{39}$	$\frac{5}{39}$	$\frac{3}{13}$

There are no negative indicators in the third row. So the maximum value of Z is $\frac{3}{13}$. It occurs at $y_1 = \frac{2}{13}$, $y_2 = \frac{1}{13}$, $y_3 = 0$. The solution to problem I may be read in the third row of the final tableau for the dual problem above. The minimum value of z is $\frac{3}{13}$ and it occurs at $x_1 = \frac{4}{39}$, $x_2 = \frac{5}{39}$.

Step 5. Using the value of x_1, x_2, y_1, y_2, y_3, and Z found in Step 4, we calculate

$$v = \frac{1}{Z} = \frac{1}{\frac{3}{13}} = \frac{13}{3}$$

$$\mathbf{P}^* = [\, p_1{}^* \quad p_2{}^* \,] = [\, vx_1 \quad vx_2 \,] = \left[\, \frac{13}{3}\left(\frac{4}{39}\right) \quad \frac{13}{3}\left(\frac{5}{39}\right) \,\right] = \left[\, \frac{4}{9} \quad \frac{5}{9} \,\right]$$

$$\mathbf{Q}^* = \begin{bmatrix} q_1{}^* \\ q_2{}^* \\ q_3{}^* \end{bmatrix} = \begin{bmatrix} vy_1 \\ vy_2 \\ vy_3 \end{bmatrix} = \begin{bmatrix} \frac{13}{3}\left(\frac{2}{13}\right) \\ \frac{13}{3}\left(\frac{1}{13}\right) \\ \frac{13}{3}(0) \end{bmatrix} = \begin{bmatrix} \frac{2}{3} \\ \frac{1}{3} \\ 0 \end{bmatrix}$$

Step 6. The optimum strategies for the original game matrix are also

$$\mathbf{P^*} = \begin{bmatrix} \frac{4}{9} & \frac{5}{9} \end{bmatrix} \text{ and } \mathbf{Q^*} = \begin{bmatrix} \frac{2}{3} \\ \frac{1}{3} \\ 0 \end{bmatrix}$$

The value of the original game is $V = v - k = \frac{13}{3} - 4 = \frac{1}{3}$.

Step 7. We check our solution by verifying that

$$\mathbf{P^*AQ^*} = v$$

We find

$$\begin{bmatrix} \frac{4}{9} & \frac{5}{9} \end{bmatrix} \begin{bmatrix} 2 & -3 & 5 \\ -1 & 3 & -2 \end{bmatrix} \begin{bmatrix} \frac{2}{3} \\ \frac{1}{3} \\ 0 \end{bmatrix} = \begin{bmatrix} \frac{1}{3} & \frac{1}{3} & \frac{10}{9} \end{bmatrix} \begin{bmatrix} \frac{2}{3} \\ \frac{1}{3} \\ 0 \end{bmatrix} = \frac{1}{3}$$

as expected. (See Exercise 16.)
Do Exercise 7. ■

Applications

EXAMPLE 2   Candidates R and C are campaigning for the governorship of state W. They are giving speeches each day in different cities of the state. In each speech, each will make one, and only one, of three promises. A reliable publication surveyed the voting population and provided the following information, where each point represents a percentage point over 50% of the voting population.
  If R promises I, then

  R will get 10 points if C promises X,
  C will get 3 points if she promises Y, and
  R will get 4 points if C promises Z.

If R promises II, then

  C will get 5 points if she promises X,
  R will get 7 points if C promises Y, and
  C will get 6 points if she promises Z.

If R promises III, then

  R will get 2 points if C promises X,
  C will get 4 points if she promises Y, and
  R will get 5 points if C promises Z.

a. Write this information as a game matrix.
b. Find the optimum strategy for each candidate and the value of the game.
c. If R faithfully uses his optimum strategy but C changes to pure strategies, how will the expected values of the game be affected?

Solution   **a.** Let rows 1, 2, and 3 represent promises I, II, and III by candidate R and columns 1, 2, and 3 represent promises X, Y, and Z by candidate C. We obtain the following game matrix

$$\mathbf{A} = \begin{bmatrix} 10 & -3 & 4 \\ -5 & 7 & -6 \\ 2 & -4 & 5 \end{bmatrix}$$

b. To solve the game matrix, we proceed as follows.

Step 1. There are no recessive rows or columns. So we go to Step 2.

Step 2. The smallest negative entry is -6, so we add 7 to each entry of matrix $\mathbf{A}$ to obtain

$$\mathbf{B} = \begin{bmatrix} 17 & 4 & 11 \\ 2 & 14 & 1 \\ 9 & 3 & 12 \end{bmatrix}$$

Step 3. We set up the following two linear programming problems.

1. Minimize $z = x_1 + x_2 + x_3$
   subject to

$$\begin{cases} 17x_1 + 2x_2 + 9x_3 \geq 1 \\ 4x_1 + 14x_2 + 3x_3 \geq 1 \\ 11x_1 + x_2 + 12x_3 \geq 1 \\ x_1 \geq 0, x_2 \geq 0, x_3 \geq 0 \end{cases}$$

2. Maximize $Z = y_1 + y_2 + y_3$
   subject to

$$\begin{cases} 17y_1 + 4y_2 + 11y_3 \leq 1 \\ 2y_1 + 14y_2 + y_3 \leq 1 \\ 9y_1 + 3y_2 + 12y_3 \leq 1 \\ y_1 \geq 0, y_2 \geq 0, y_3 \geq 0 \end{cases}$$

Step 4. Since we will obtain the solutions of the first problem from the final tableau of the second problem, we let the slack variables for the second problem be x_1, x_2, and x_3. We obtain the initial tableau for the second problem.

Tableau I

Basis	y_1	y_2	y_3	x_1	x_2	x_3	Values of basic variables
x_1	17	4	11	1	0	0	1
x_2	2	14	1	0	1	0	1
x_3	9	3	⑫	0	0	1	1
Z	-1	-1	-1	0	0	0	0

The most negative indicators in the bottom row are in the first three columns. So we may choose any of these as the pivot column. We choose the third. Since $\frac{1}{12}$ is less than both 1 and $\frac{1}{11}$, 12 is the pivot. Pivoting, we obtain

Tableau II

Basis	y_1	y_2	y_3	x_1	x_2	x_3	Values of basic variables
← x_1	$\frac{35}{4}$	$\left(\frac{5}{4}\right)$	0	1	0	$\frac{-11}{12}$	$\frac{1}{12}$
x_2	$\frac{5}{4}$	$\frac{55}{4}$	0	0	1	$\frac{-1}{12}$	$\frac{11}{12}$
y_3	$\frac{3}{4}$	$\frac{1}{4}$	1	0	0	$\frac{1}{12}$	$\frac{1}{12}$
Z	$\frac{-1}{4}$	$\frac{-3}{4}$	0	0	0	$\frac{1}{12}$	$\frac{1}{12}$

Since $\frac{-3}{4}$ is the most negative indicator in the bottom row, the second column is the pivot column. Observe that $\frac{1}{12}/\frac{5}{4} = \frac{1}{15}$ is less than or equal to both $\frac{11}{12}/\frac{55}{4} = \frac{1}{15}$ and $\frac{1}{12}/\frac{1}{4} = \frac{1}{3}$. Thus we choose the first row as the pivot row. Pivoting, we get

Tableau III

Basis	y_1	y_2	y_3	x_1	x_2	x_3	Values of basic variables
y_2	7	1	0	$\frac{4}{5}$	0	$\frac{-11}{15}$	$\frac{1}{15}$
← x_2	-95	0	0	-11	1	$\left(10\right)$	0
y_3	-1	0	1	$\frac{-1}{5}$	0	$\frac{4}{15}$	$\frac{1}{15}$
Z	5	0	0	$\frac{3}{5}$	0	$\frac{-7}{15}$	$\frac{2}{15}$

Clearly the sixth column must be the pivot column and the second row must be the pivot row. Pivoting, we obtain

Tableau IV

Basis	y_1	y_2	y_3	x_1	x_2	x_3	Values of basic variables
y_2	$\frac{1}{30}$	1	0	$\frac{-1}{150}$	$\frac{11}{150}$	0	$\frac{1}{15}$
x_3	$\frac{-19}{2}$	0	0	$\frac{-11}{10}$	$\frac{1}{10}$	1	0
y_3	$\frac{23}{15}$	0	1	$\frac{7}{75}$	$\frac{-2}{75}$	0	$\frac{1}{15}$
Z	$\frac{17}{30}$	0	0	$\frac{13}{150}$	$\frac{7}{150}$	0	$\frac{2}{15}$

There are no negative indicators in the bottom row. So the maximum value of Z has been attained. We have max $Z = \frac{2}{15}$. It occurs at $y_1 = 0$, $y_2 = \frac{1}{15}$, $y_3 = \frac{1}{15}$.

The solution to problem I may be read in the fourth row of the final tableau for the dual problem. The minimum value of z is $\frac{2}{15}$. It occurs at $x_1 = \frac{13}{150}$, $x_2 = \frac{7}{150}$, $x_3 = 0$.

Step 5. Using the values of x_1, x_2, x_3, y_1, y_2, y_3, and Z found in Tableau IV, we calculate

$$v = \frac{1}{Z} = \frac{1}{\frac{2}{15}} = \frac{15}{2}$$

$$\mathbf{P}^* = [\, p_1^* \quad p_2^* \quad p_3^* \,] = [\, vx_1 \quad vx_2 \quad vx_3 \,]$$

$$= \left[\, \tfrac{15}{2}\!\left(\tfrac{13}{150}\right) \quad \tfrac{15}{2}\!\left(\tfrac{7}{150}\right) \quad 0 \,\right] = \left[\, \tfrac{13}{20} \quad \tfrac{7}{20} \quad 0 \,\right]$$

$$\mathbf{Q}^* = \begin{bmatrix} q_1^* \\ q_2^* \\ q_3^* \end{bmatrix} = \begin{bmatrix} vy_1 \\ vy_2 \\ vy_3 \end{bmatrix} = \begin{bmatrix} 0 \\ \tfrac{15}{2}\!\left(\tfrac{1}{15}\right) \\ \tfrac{15}{2}\!\left(\tfrac{1}{15}\right) \end{bmatrix} = \begin{bmatrix} 0 \\ \tfrac{1}{2} \\ \tfrac{1}{2} \end{bmatrix}$$

Step 6. The optimum strategies for the original game matrix are also

$$\mathbf{P}^* = \left[\, \tfrac{13}{20} \quad \tfrac{7}{20} \quad 0 \,\right] \text{ and } \mathbf{Q}^* = \begin{bmatrix} 0 \\ \tfrac{1}{2} \\ \tfrac{1}{2} \end{bmatrix}$$

Step 7. We check our solution by verifying that

$$\mathbf{P}^*\mathbf{A}\mathbf{Q}^* = v$$

We find

$$\mathbf{P}^*\mathbf{A}\mathbf{Q}^* = \left[\, \tfrac{13}{20} \quad \tfrac{7}{20} \quad 0 \,\right] \begin{bmatrix} 10 & -3 & 4 \\ -5 & 7 & -6 \\ 2 & -4 & 5 \end{bmatrix} \begin{bmatrix} 0 \\ \tfrac{1}{2} \\ \tfrac{1}{2} \end{bmatrix}$$

$$= \left[\, \tfrac{19}{4} \quad \tfrac{1}{2} \quad \tfrac{1}{2} \,\right] \begin{bmatrix} 0 \\ \tfrac{1}{2} \\ \tfrac{1}{2} \end{bmatrix} = \tfrac{1}{2}$$

as expected.

Therefore R should promise I 65% of the time (since $\tfrac{13}{20} = .65$), II 35% of the time (since $\tfrac{7}{20} = .35$), and never promise III. Similarly, C should never promise X, promise Y 50% of the time, and promise Z 50% of the time.

c. If R faithfully uses his optimum strategy but C alters her strategy, the expected value of the game will depend on which pure strategy C adopts. If she consistently promises X, then her strategy will be $\mathbf{Q}_1$, and the value of the game will be

$$E(\mathbf{P}^*, \mathbf{Q}_1) = \left[\, \tfrac{13}{20} \quad \tfrac{7}{20} \quad 0 \,\right] \begin{bmatrix} 10 & -3 & 4 \\ -5 & 7 & -6 \\ 2 & -4 & 5 \end{bmatrix} \begin{bmatrix} 1 \\ 0 \\ 0 \end{bmatrix}$$

$$= \left[\, \tfrac{19}{4} \quad \tfrac{1}{2} \quad \tfrac{1}{2} \,\right] \begin{bmatrix} 1 \\ 0 \\ 0 \end{bmatrix} = \tfrac{19}{4}$$

If C consistently promises Y, then her strategy will be $\mathbf{Q}_2$. In that case, the value of the game will be

$$E(\mathbf{P}^*, \mathbf{Q}_2) = \begin{bmatrix} \frac{13}{20} & \frac{7}{20} & 0 \end{bmatrix} \begin{bmatrix} 10 & -3 & 4 \\ -5 & 7 & -6 \\ 2 & -4 & 5 \end{bmatrix} \begin{bmatrix} 0 \\ 1 \\ 0 \end{bmatrix}$$

$$= \begin{bmatrix} \frac{19}{4} & \frac{1}{2} & \frac{1}{2} \end{bmatrix} \begin{bmatrix} 0 \\ 1 \\ 0 \end{bmatrix} = \frac{1}{2}$$

If C consistently promises Z, then her strategy will be $\mathbf{Q}_3$ and the value of the game will be

$$E(\mathbf{P}^*, \mathbf{Q}_3) = \begin{bmatrix} \frac{13}{20} & \frac{7}{20} & 0 \end{bmatrix} \begin{bmatrix} 10 & -3 & 4 \\ -5 & 7 & -6 \\ 2 & -4 & 5 \end{bmatrix} \begin{bmatrix} 0 \\ 0 \\ 1 \end{bmatrix}$$

$$= \begin{bmatrix} \frac{19}{4} & \frac{1}{2} & \frac{1}{2} \end{bmatrix} \begin{bmatrix} 0 \\ 0 \\ 1 \end{bmatrix} = \frac{1}{2}$$

Do Exercise 9. ■

EXAMPLE 3   A real estate development company purchased a large tract of land near Kent. The company plans to use the land for a shopping center, condominiums, and a sports arena. The state is considering building a highway nearby, but no decision has been made yet. A consulting firm provided estimates of the profit percentages to be made in the different cases. If the state builds the highway, the percentages will be 25%, 10%, and 18% on the shopping center, the condominiums, and the sports arena, respectively. If the state does not build the highway, the percentages will be 6%, 20%, and 12% on the shopping center, the condominiums, and the sports arena, respectively. If it is assumed that the state plays the role of an opponent to the company, what portions of the land should be used for the shopping center, the condominiums, and the sports arena?

Solution   Let the company be player R and the state be player C. Also the first, second, and third rows represent the shopping center, the condominiums, and the sports arena, respectively. The first and second columns represent the highway being built and the highway not being built, respectively. Then we obtain the following game matrix.

$$\mathbf{B} = \begin{bmatrix} .25 & .06 \\ .10 & .20 \\ .18 & .12 \end{bmatrix}$$

There are no recessive rows or columns and all entries are positive. So we go directly to Step 3.

Step 3. We set up the following two linear programming problems.

1. Minimize $z = x_1 + x_2 + x_3$
   subject to

$$\begin{cases} .25x_1 + .1x_2 + .18x_3 \geq 1 \\ .06x_1 + .2x_2 + .12x_3 \geq 1 \\ x_1 \geq 0, \, x_2 \geq 0, \, x_3 \geq 0 \end{cases}$$

2. Maximize $Z = y_1 + y_2$
   subject to

$$\begin{cases} .25y_1 + .06y_2 \leq 1 \\ .1y_1 + .2y_2 \leq 1 \\ .18y_1 + .12y_2 \leq 1 \\ y_1 \geq 0, \, y_2 \geq 0 \end{cases}$$

As usual, we solve the second problem which is the dual of the first. We will read the solution of the first problem from the bottom row of the final tableau in the solution of the second problem.

Step 4. To solve the second linear programming problem, we introduce the slack variables x, x_2, and x_3 and write the equation of the objective function as follows:

$$\begin{cases} .25y_1 + .06y_2 + x_1 \qquad\qquad = 1 \\ .1y_1 + .2y_2 \qquad + x_2 \qquad = 1 \\ .18y_1 + .12y_2 \qquad\qquad + x_3 = 1 \\ Z - \quad y_1 - \quad y_2 \qquad\qquad\quad = 0 \end{cases}$$

The initial tableau is

Tableau I

Basis	y_1	y_2	x_1	x_2	x_3	Values of basic variables
x_1	.25	.06	1	0	0	1
x_2	.1	(.2)	0	1	0	1
x_3	.18	.12	0	0	1	1
Z	-1	-1	0	0	0	0

The most negative indicators in the bottom row are in the first two columns. So we may choose either of these as the pivot column. We choose the second. Since $\frac{1}{.2} = 5$ is less than both $\frac{1}{.06} = \frac{50}{3}$ and $\frac{1}{.12} = \frac{25}{3}$, the second row is the pivot row. Therefore .2 is the pivot. Pivoting, we get the next tableau.

Tableau II

Basis	y_1	y_2	x_1	x_2	x_3	Values of basic variables
← x_1	(22)	0	1	$-.3$	0	.7
y_2	.5	1	0	5	0	5
x_3	.12	0	0	$-.6$	1	.4
Z	$-.5$	0	0	5	0	5

↑

The first column is the pivot column since it contains the only negative indicator in the bottom row. It is easy to verify that $\frac{.7}{.22}$ is less than both $\frac{5}{.5}$ and $\frac{.4}{.12}$. So the first row is the pivot row. Pivoting, we get

Tableau III

Basis	y_1	y_2	x_1	x_2	x_3	Values of basic variables
y_1	1	0	$\frac{50}{11}$	$\frac{-15}{11}$	0	$\frac{35}{11}$
y_2	0	1	$\frac{-25}{11}$	$\frac{125}{22}$	0	$\frac{75}{22}$
x_3	0	0	$\frac{-6}{11}$	$\frac{-24}{55}$	1	$\frac{1}{55}$
Z	0	0	$\frac{25}{11}$	$\frac{95}{22}$	0	$\frac{145}{22}$

There are no negative indicators in the bottom row. So the maximum value of Z is $\frac{145}{22}$. It occurs at $y_1 = \frac{35}{11}$, $y_2 = \frac{75}{22}$. The solution to problem I may be read in the bottom row of the final tableau for the dual problem. The minimum value of z is $\frac{145}{22}$, and it occurs at $x_1 = \frac{25}{11}$, $x_2 = \frac{95}{22}$, $x_3 = 0$.

Step 5. Using the values of x_1, x_2, x_3, y_1, y_2, and max Z found in Tableau III, we calculate

$$v = \frac{1}{Z} = \frac{1}{\frac{145}{22}} = \frac{22}{145}$$

$$\mathbf{P^*} = [\; p_1^* \quad p_2^* \quad p_3^* \;] = [\; vx_1 \quad vx_2 \quad vx_3 \;]$$
$$= \left[\; \frac{22}{145}\left(\frac{25}{11}\right) \quad \frac{22}{145}\left(\frac{95}{22}\right) \quad 0 \;\right] = \left[\; \frac{10}{29} \quad \frac{19}{29} \quad 0 \;\right]$$

$$\mathbf{Q^*} = \begin{bmatrix} q_1^* \\ q_2^* \end{bmatrix} = \begin{bmatrix} vy_1 \\ vy_2 \end{bmatrix} = \begin{bmatrix} \frac{22}{145}\left(\frac{35}{11}\right) \\ \frac{22}{145}\left(\frac{75}{22}\right) \end{bmatrix} = \begin{bmatrix} \frac{14}{29} \\ \frac{15}{29} \end{bmatrix}$$

We check our solution by verifying that

$$\mathbf{P^*BQ^*} = v$$

We find

$$\begin{bmatrix} \frac{10}{29} & \frac{19}{29} & 0 \end{bmatrix} \begin{bmatrix} .25 & .06 \\ .10 & .20 \\ .18 & .12 \end{bmatrix} \begin{bmatrix} \frac{14}{29} \\ \frac{15}{29} \end{bmatrix} = \begin{bmatrix} \frac{22}{145} & \frac{22}{145} \end{bmatrix} \begin{bmatrix} \frac{14}{29} \\ \frac{15}{29} \end{bmatrix} = \frac{22}{145}$$

as expected.

The company should use $\frac{10}{29}$ (approximately 34.5%) of the land for a shopping center, $\frac{19}{29}$ (approximately 65.5%) of the land for condominiums, and no land for a sports arena.

Do Exercise 11. ∎

Exercise Set 7.4

In Exercises 1–8, solve the given game matrix.

1. $\begin{bmatrix} 0 & -1 & 3 \\ -1 & 1 & -2 \end{bmatrix}$

2. $\begin{bmatrix} -3 & 4 \\ 1 & 0 \\ 2 & -1 \end{bmatrix}$

3. $\begin{bmatrix} -1 & 2 & -1 \\ -1 & -2 & 0 \\ 0 & -1 & -1 \end{bmatrix}$

4. $\begin{bmatrix} -3 & 4 & -2 \\ 1 & 0 & 3 \\ 2 & -1 & 3 \end{bmatrix}$

5. $\begin{bmatrix} -1 & 3 & 0 \\ 0 & 1 & -2 \\ 1 & 4 & -3 \end{bmatrix}$

6. $\begin{bmatrix} 0 & 3 & 2 \\ 4 & 5 & -2 \\ -2 & 1 & 3 \end{bmatrix}$

7. $\begin{bmatrix} 2 & 3 & 0 & 7 \\ -1 & 5 & -2 & 5 \\ 0 & 2 & 3 & -2 \end{bmatrix}$

8. $\begin{bmatrix} -1 & 1 & 2 & -3 \\ 0 & 3 & -1 & 4 \\ 2 & 4 & 0 & -7 \\ 3 & 6 & 5 & -5 \end{bmatrix}$

9. Candidates Roberta and Charlie are campaigning for the governorship of state N. They are giving speeches each day in different cities of the state. In each speech, each will make one, and only one, of three promises. The League of Women Voters surveyed the voting population and provided the following information, where each point represents a percentage point over 50% of the voting population. If Roberta promises I, then

   Roberta will get 3 points if Charlie promises A,
   Roberta will get 8 points if Charlie promises B, and
   Charlie will get 7 points if he promises C.

   If Roberta promises II, then

   Charlie will get 2 points if he promises A,
   Roberta will get 3 points if Charlie promises B, and
   Roberta will get 18 points if Charlie promises C.

   If Roberta promises III, then

   Roberta will get 13 points if Charlie promises A,
   Charlie will get 12 points if he promises B, and
   Roberta will get 3 points if Charlie promises C.

   a. Write this information as a game matrix.
   b. Find the optimum strategy for each candidate and the value of the game.
   c. If Roberta faithfully uses her optimum strategy but Charlie changes to pure strategies, how will the expected values of the game be affected?

10. Repeat Exercise 9 if the information provided by the League of Women Voters is as follows:

    If Roberta promises I, then

    Roberta will get 19 points if Charlie promises A,
    Roberta will get 4 points if Charlie promises B, and
    Charlie will get 1 point if he promises C.

    If Roberta promises II, then

    Charlie will get 6 points if he promises A,
    Roberta will get 9 points if Charlie promises B, and
    Roberta will get 4 points if Charlie promises C.

    If Roberta promises III, then

    Roberta will get 4 points if Charlie promises A,
    Charlie will get 11 points if he promises B, and
    Roberta will get 14 points if Charlie promises C.

11. A real estate development company purchased a large tract of land near Monterey. The company plans to use the land for a shopping center, restaurants, and a sports arena. The state is considering building a highway nearby but no decision has been made yet. A consulting firm provided estimates of the profit percentages to be made in the different cases. If the state builds the highway, the percentages will be 63%, 48%, and 18% on the shopping center, the restaurants, and the sports arena, respectively. If the state does not build the highway, the percentages will be 3%, 33%, and 48% on the shopping center, the restaurants, and the sports arena,

respectively. If it is assumed that the state plays the role of an opponent to the company, what portions of the land should be used for the shopping center, restaurants, and the sports arena?

12. Repeat Exercise 11 if the information provided by the consulting firm is as follows: If the state builds the highway, the percentages will be 17%, 23%, and 21% on the shopping center, the restaurants, and the sports area, respectively. If the state does not build the highway, the percentages will be 25%, 19%, and 23% on the shopping center, the restaurants, and the sports arena, respectively.

13. Two large competing grocery stores R and C are located in the same city. Each store chooses to use one and only one form of advertising at any one time: radio, television, or the newspapers; however, they may change from week to week. A consulting firm provided the following information to the stores. If R uses the radio, then R will get 2 points if C uses the radio and 14 points if C uses the newspapers; however, C will get 10 points if it uses television. If R uses television, then R will get 26 points if C uses television and 2 points if C uses the newspapers; but C will get 10 points if it uses the radio. If R uses the newspapers, then R will get 14 points if C uses the radio; and C will get 22 points if it uses television and 10 points if it uses the newspaper. Each point represents a percentage point over 50% of the total population that shops regularly at one of the two stores. Write this information as a game matrix and solve the game.

14. Repeat Exercise 13 if the information provided by the consulting firm is as follows: If R uses the radio, then R will

get 4 points if C uses the radio and 14 points if C uses television; however, C will get 6 points if it uses the newspapers. If R uses television, then R will get 4 points if C uses television and 14 points if C uses the newspapers; but C will get 6 points if it uses the radio. If R uses the newspapers, then R will get 14 points if C uses the radio and 4 points if C uses the newspaper; however, C will get 6 points if it uses television.

15. An investor wishes to invest $12,000 during a period when the state legislature is in session. The legislature will pass one, and only one, of three propositions which it is considering. This will affect the returns from the three stocks the investor is considering. An investment firm provided the following information. If proposition I passes, the returns to the investor from stocks A, B, and C will be 7%, 1%, and 13%, respectively. If proposition II passes, the returns to the investor from stocks A, B, and C will be 1%, 19%, and −5%, respectively. Finally, if proposition III passes, the returns to the investor from stocks A, B, and C will be 13%, 7%, and 1%, respectively. What is the investor's optimum strategy?

16. Suppose that in Example 1 of this section R continues to use his optimum strategy but C switches to a pure strategy

$$\mathbf{Q}_1 = \begin{bmatrix} 1 \\ 0 \\ 0 \end{bmatrix}, \mathbf{Q}_2 = \begin{bmatrix} 0 \\ 1 \\ 0 \end{bmatrix}, \text{ or } \mathbf{Q}_3 = \begin{bmatrix} 0 \\ 0 \\ 1 \end{bmatrix}.$$

How would the expected value of the game change?

7.5 Graphs

Venn diagrams were used to solve certain problems pictorially. In particular, they were useful in illustrating relationships between sets and in solving certain survey problems. There are other types of problems that can be better understood using geometric models. A branch of mathematics that is particularly well suited for providing geometric models to a great variety of problems is *graph theory*. As you will soon see, the meaning of the word *graph* here is quite different from its meaning when we worked with graphs of functions.

Introduction of Problems

Here are a few examples of problems that can be solved using graph theory.

1. An ice cream vendor wishes to cover every street in a certain neighborhood, but does not want to travel any street more than once. Is this possible? If so, how should he do it?

2. Suppose a complex task must be completed. It may be useful to divide the task into a number of smaller projects and to make up a schedule for the completion of each of the small projects. However, some of these may require the completion of others before they can be initiated. The problem is to make up a schedule that will enable the completion of the whole task in the shortest possible time.

3. Suppose an infectious disease is discovered on a certain campus and the health department must stop it from spreading. Through interviews with students who have contracted the disease, it has isolated a certain number of students. Can the health department check whether some student who could spread the disease has not been isolated?

4. A large company wishes to support a candidate during a political campaign. It would like to provide this support to the candidate who will be the most influential if elected. How should the company choose whom to support?

We will solve these problems and more in this and the next two sections.

A Graph as a Geometric Model

We begin with one of the oldest problems involving graphs. The town of Königsberg (now Kaliningrad) is crossed by the Pregel River. There are two islands in the river. The islands are connected to each other and to the opposite banks by seven bridges. (See Figure 7.5.)

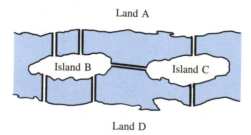

Land A

Island B   Island C

Land D

FIGURE 7.5

In the 18th century the townsfolk amused themselves with the following problem: Is it possible to cross the seven bridges in a continuous walk without going over any of them more than once?

We make up a geometric model for this problem as follows. Let the four masses of land (the two banks and the two islands) be represented by the points A, B, C, and D. (See Figures 7.5 and 7.6.) Now connect two of these points if, and only if, the corresponding masses of land are connected by a bridge. (See Figure 7.6.) A curve connecting two points is called an *edge*. The Königsberg bridge problem can now be phrased as follows: Can someone start at any of the points A, B, C, or D, traverse each edge exactly once, and return to the starting point? The Swiss mathematician Leonhard Euler (1707–1783) was able to solve this problem in 1736. We will provide the solution at the end of this section.

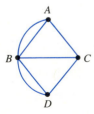

FIGURE 7.6

Terminology

The geometric representation of a graph in Figure 7.6 helps us intuitively, but it is not a definition. We must introduce the definition of a graph as well as the definitions of other terms which we will frequently use in the remainder of this chapter.

DEFINITION 7.9: Let V be a finite set of points called *vertices* (plural of *vertex*). An *edge* is a curve joining two distinct vertices. Two vertices joined by an edge are called the *endpoints* of that edge. A *graph* consists of the set V of vertices together with a finite set E of edges. The endpoints of each edge in E must be in V. If two vertices u and v are endpoints of the same edge e, then we say that u and v are *adjacent vertices*. Further, we say that e is *incident with u and v*, and that u and v are *incident with e*. Two edges that have the same endpoints are said to be *parallel*. The *order* of the graph is the number of vertices in V. A vertex that is not the endpoint of any edge in E is said to be *isolated*. A graph with no parallel edges is called a *simple graph*.

Observe that a loop (a curve that joins a point to itself) is not an edge. In Figure 7.7a, A is an isolated vertex, e_4 and e_5 are parallel edges; Figure 7.7b does not illustrate a graph because there is a loop; and Figure 7.7c represents a simple graph. In Figure 7.7c, observe that AD and BC intersect. However, the point of intersection is not a vertex because it is not labeled.

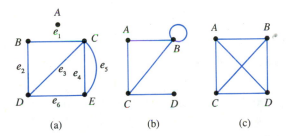

FIGURE 7.7

To give a satisfactory solution to the Königsberg bridge problem we need several more definitions. If you were trying to solve the problem by trial and error, you might consider the graph in Figure 7.6—put your pencil on one of the vertices and trace the graph, trying to pass through every vertex and going over each edge once, and only once, until you returned to the vertex where you started.

DEFINITION 7.10: A *path* between the vertices v_1 and v_k is a sequence v_1, e_1, v_2, e_2, . . . , v_{k-1}, e_{k-1}, v_k where each v_i is a vertex, each e_i is an edge, and for each $i = 1, 2, 3, \ldots, k - 1$, e_i has endpoints v_i and v_{i+1}. The *length* of the path is $k - 1$. Note that the length of the path is the number of edges in the path.

If no confusion arises, we may identify a path by simply naming the vertices. However, care must be used if there are parallel edges in the graph. For example, in Figure 7.8, A, e_1, B, e_2, C and A, e_1, B, e_3, C are different paths, although the corresponding sequences of vertices are the same ($A, B,$ and C).

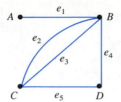

FIGURE 7.8

DEFINITION 7.11: A graph G is *connected* if for every pair of distinct vertices v_i and v_j in G, there is a path between v_i and v_j.

The graph in Figure 7.9**a** is connected, while the graph in Figure 7.9**b** is not connected.

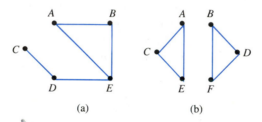

FIGURE 7.9

           (a)   (b)

Traceable Graphs

DEFINITION 7.12: A graph G is *traceable* if there is a path with no repeated edges that contains every vertex and every edge of G.

EXAMPLE 1   Show that the graph pictured in Figure 7.10 is traceable.

Solution   The path $A, e_1, B, e_2, C, e_3, D, e_4, B, e_5, E, e_6, D, e_7, F$ contains every vertex and every edge of the path; furthermore, no edge is repeated. Therefore the graph is traceable.

Do Exercise 1. ■

It is easy to see that the answer to the Königsberg bridge problem will be "yes" if the corresponding graph is traceable. Therefore, we must establish some conditions to tell us whether a graph is traceable.

We first observe that a graph G that is not connected is not traceable because there are at least two distinct vertices u and v in G with no path between u and v.

Next we observe that if a graph G is traceable, then there are two vertices u and v, not necessarily distinct, and a path between u and v containing all vertices and edges

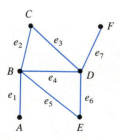

FIGURE 7.10

of G and having no repeated edges. So if w is any other vertex, each time w appears in the path it must be incident with two edges—one to enter the vertex and one to leave it. Consequently, if w appears k times in the path, it must be incident with $2k$ vertices. That is, except possibly for the vertices u and v where the path started and ended, each other vertex in the path is incident with an even number of vertices.

DEFINITION 7.13: Let G be a graph and v be a vertex of G. The *degree of v in G* is the number of edges of G incident with v. It is denoted $\deg_G v$, or simply $\deg v$ if no confusion arises. A vertex is said to be *even* if its degree is even, otherwise it is said to be *odd*.

We now state the following theorem.

THEOREM 7.12: A graph G can be traced if, and only if, the following conditions are satisfied:

1. The graph is connected.
2. It has exactly two odd vertices or no odd vertices.
3. If all vertices are even, then the tracing may begin at any vertex, but will end at the same vertex.
4. If it has exactly two odd vertices, then the tracing must begin at one of them and end at the other.

In Example 1, we showed that the given graph was traceable. Observe that B, C, D, and E were even vertices, while A and F were odd.

EXAMPLE 2 Solve the Königsberg bridge problem.

Solution The graph in Figure 7.6 is a model of the problem. We observed earlier that it would be possible to start on a piece of land, walk and visit each piece of land, cross each bridge once—and only once—and return to the starting point if, and only if, the graph of Figure 7.6 was traceable. We see that $\deg B = 5$ and $\deg A = \deg C = \deg D = 3$. Since the graph has more than two vertices of odd degrees, the graph is not traceable.
Do Exercise 9. ■

EXAMPLE 3 An ice cream vendor wishes to drive down each street of the neighborhood that has been mapped below, but does not want to travel over any street more than once. Can he do it? If so, describe the path he could use.

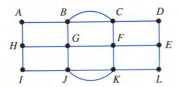

Solution Observe that the figure represents a connected graph where deg B = deg C = deg G = deg F = deg J = deg K = 4, deg A = deg D = deg I = deg L = 2, and deg H = deg E = 3. Since the graph is connected and has exactly two odd vertices, it is traceable. The required path must start at either H or E. Label the edges as done in Figure 7.11. Then the path H, e_1, G, e_2, F, e_3, E, e_4, D, e_5, C, e_6, B, e_8, C, e_{14}, F, e_{16}, K, e_{12}, J, e_{11}, I, e_{10}, H, e_9, A, e_7, B, e_{19}, G, e_{15}, J, e_{13}, K, e_{18}, L, e_{17}, E is a trace of the graph.

The ice cream vendor should follow the route described by that path.
Do Exercise 11. ■

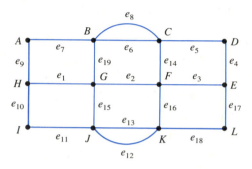

FIGURE 7.11

As an exercise, you should find two other tracings of the graph of Example 3.

Networks

Perhaps one of the most useful applications of graph theory stems from the study of networks. A familiar example of a network is the interstate highway system. A model of this network may be found in a road atlas of the United States. Observe that the distance between any two adjacent cities is often shown near the highway connecting these cities.

> **DEFINITION 7.14:** A *network* is a connected graph each of whose edges has a real number associated with it. The *value* of a path in a network is the sum of the numbers associated with the edges that appear in the path.

EXAMPLE 4 Figure 7.12 represents part of a road atlas of the United States. What is the value of the path Seattle-Ellengsburg-Pendleton-Missoula?

Solution The distances (in miles) indicated on the map by the highways connecting Seattle-Ellengsburg, Ellengsburg-Pendleton, and Pendleton-Missoula are 107, 168, and 399, respectively. So the value of the path is 674 since 107 + 168 + 399 = 674.
Do Exercise 13.

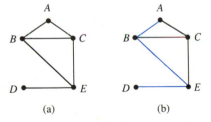

FIGURE 7.12

Often, as we shall see in the next example, the numbers associated with the edges represent costs. A common problem is that of linking certain units together so that they are able to communicate with each other. It is usually too expensive, or not practical, to link each unit directly to every other unit. (See Exercise 28.) So it is necessary to determine paths from one unit to another, which may include intermediate units, to efficiently (economically) link all units together.

We need a few more definitions.

DEFINITION 7.15: If G and H are graphs, every vertex of H is a vertex of G, and every edge of H is an edge of G, then H is said to be a *subgraph* of G.

In Figure 7.13**b** the solid colored edges and vertices illustrate a subgraph of the graph of Figure 7.13**a**.

FIGURE 7.13

(a)   (b)

DEFINITION 7.16: A path in a graph that starts and ends at the same vertex is called a *cycle*. A *tree* is a connected graph that does not contain any cycles. A *spanning tree* for a connected graph G is a subgraph of G, say T, such that T is a tree and the set of vertices of G and of T are equal.

In Figure 7.14**a** the path A, e_1, B, e_2, C, e_3, A is a cycle. So the graph is not a tree. The graph of Figure 7.14**b** is a tree. In Figure 7.14**c**, the solid blue curves illustrate a spanning tree of the graph of Figure 7.14**a**.

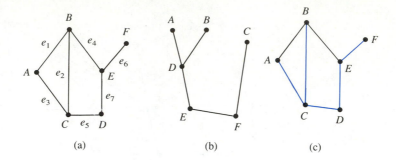

FIGURE 7.14

(a)   (b)   (c)

DEFINITION 7.17: *A minimum spanning tree* in a network is a spanning tree for which the sum of the numbers assigned to the edges is less than or equal to the sum of the numbers assigned to the edges of any other spanning tree in the graph.

We must have an algorithm to select a minimum spanning tree in a network. The idea is quite simple and is described as follows.

Procedure to Select a Minimum Spanning Tree

Let *G* be a network. For convenience, call the number associated with an edge the cost of that edge.

Step 1. Start with a least-cost edge. (If there is more than one, select any of them.)

Step 2. Choose a least-cost edge that has not yet been selected, that shares a vertex with an edge already selected, and whose other vertex is not on any edge already selected.

Repeat Step 2 until all vertices are on edges that have been selected. Observe that this will happen when the number of selected edges is one less than the number of vertices in the graph.

EXAMPLE 5 In Figure 7.15, the vertices represent cities and the edges represent pipelines that currently link some of these cities together. The government decided that some of the older lines must be replaced by new ones. The cost to replace an old link by a new one is shown in Figure 7.15 as an integer by the edge representing that link, where each integer represents a number of $100,000 units. Find a set of lines which are to be replaced by new lines such that the following two conditions are satisfied:

1. Any two cities are connected by a set of new pipelines.
2. The total cost of replacement is not greater than the total cost of any other replacement that would yield a system satisfying condition 1.

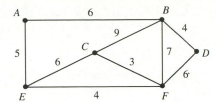

FIGURE 7.15

Solution   The graph of Figure 7.15 represents a network. The least-cost edge is *CF*. So we select it. (See Figure 7.16a.) The least-cost edge that shares a vertex with *CF* is *EF*. We select it. (See Figure 7.16b. The edges selected in the next three steps will be shown in Figures 7.16c, 7.16d, and 7.16e.) The least-cost edge that has not yet been selected, which shares a vertex with *CF* and *EF* and whose other vertex is not on *CF* or *EF*, is *AE*. We select it. A least-cost edge distinct from *CF*, *EF*, and *AE*, which shares a vertex with one of these vertices and whose other vertex is not on *CF*, *EF*, or *AE*, is *FD*. We select it. (Note that we could have chosen *AB* as well.) The least-cost edge that has not yet been selected, which shares a vertex with an edge already selected and whose other vertex is not on an edge already selected is *BD*. We select it. All the vertices have now been used.

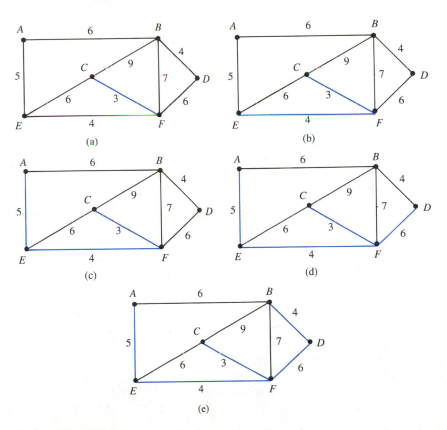

FIGURE 7.16

Observe that the original graph had six vertices and we have selected five edges. We have obtained a minimum spanning tree. The government should replace the

pipeline between C and F, F and E, E and A, F and D, and D and B. The total cost will be \$2,200,000 since $3 + 4 + 5 + 6 + 4 = 22$.
Do Exercise 27.

In general, there will be more than one minimum spanning tree. (See Exercise 29.)

Exercise Set 7.5

In Exercises 1–8, determine whether the given graph is traceable. If it is not traceable, state why. If it is, trace it.

1.
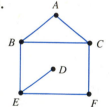

2.
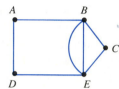

3.
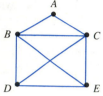

4.
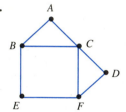

5.
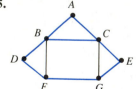

6.
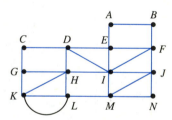

7.

8.
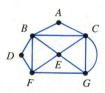

9. A river divides a small town into five distinct sections connected by bridges as shown in the following figure.

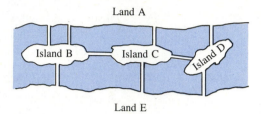

Is it possible for a jogger to start at one section of town, traverse each bridge once, and only once, and return to her starting point? If it is possible, describe her path.

10. An ice cream vendor wishes to drive down each street of the mapped neighborhood, but does not want to travel over any street more than once. Can he do it? If so, describe the path he could use.

11. A law firm has offices on the first floor of a building. The floor plan is shown below. Assuming that each door is left open during the day, is it possible for a security guard to enter one of the suites from the hallway, lock each door after passing through it, and exit without having to open a door that he just locked? If it is possible, describe his path.

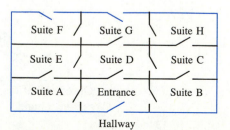

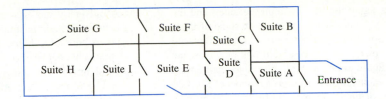

FIGURE FOR EXERCISE 12

12. Repeat Exercise 11 with the floor plan above.

13. The figure at the bottom of the page represents part of a road atlas of the United States. What is the value of the path Chicago-Toledo-Cincinnati-Indianapolis-Springfield?

In Exercises 14–19, determine whether the given graph is a tree. Justify your answers. If a graph is not a tree, find a spanning tree.

14.

15.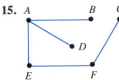

16.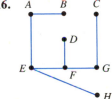

17.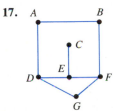

18.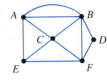

19.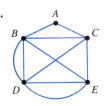

In Exercises 20–23, find a minimum spanning tree.

20.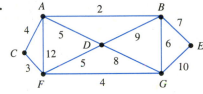

21.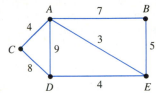

22.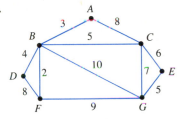

23.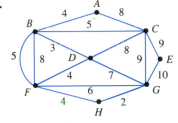

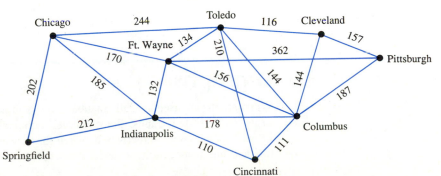

FIGURE FOR EXERCISE 13

24. In the following figure the vertices represent cities and the edges represent pipelines that currently link some of these cities together. Some of these older lines must be replaced by new ones. The cost to replace an old link by a new one is shown in the figure as an integer by the edge representing that link, where each integer represents a number of $10,000 units. Find a set of lines which are to be replaced by new lines such that the following two conditions are satisfied:

1. Any two cities are connected by a set of new pipelines.
2. The total cost of replacement is not greater than the total cost of any other replacement that would yield a system satisfying condition 1.

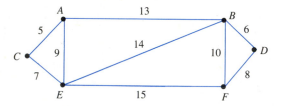

25. A company wishes to install private communication lines between its branch offices located in San Antonio, Victoria, Austin, Houston, Dallas, Ft. Worth, Cisco, and Brownwood. The distances along the possible routes where these lines could be built are shown on the following diagram. Find a line of minimum total length which connects any two of the branch offices.

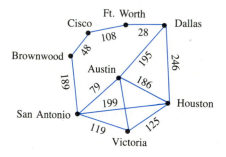

26. A university has provided a number of computers for its faculty. The administration has decided to connect all computers with a local area network. The cost of installing fiber optic cable between certain pairs of computers is shown on the diagram. The integer by each edge represents the cost in $1000 units. Determine how to install the network at a minimum cost but still have any two computers connected by a fiber optic path.

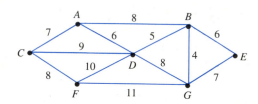

27. Repeat Exercise 26 with the following diagram.

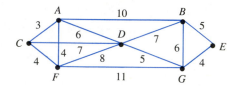

28. If in a network connecting 100 computers each computer is directly linked to every other computer in the network, how many connections are there? (*Hint:* See Section 6.2.)

29. Find a minimum spanning tree different from that obtained in the solution of Example 5.

7.6 Digraphs

When graphs are used as mathematical models, it is often necessary to assign direction to the edges of the graphs. For example, a company may have an organization chart in which the flow of directives between the officers of the company is displayed. In this and the next section, we briefly discuss directed graphs.

Digraphs

> **DEFINITION 7.18:**   A *digraph* (abbreviation of *directed graph*) G is a finite
> set V of *vertices* and a set E of ordered pairs (u, v) where $u \in V$, $v \in V$, and
> $u \neq v$. Each element of E is called a *directed edge* of G. If (u, v) is a member
> of E, then u and v are called the *initial point* and the *terminal point* of the
> directed edge, respectively.

In the geometric representation of a digraph, a directed edge (A, B) is represented
by an arrow pointing from A to B.

EXAMPLE 1   Let G be a digraph where $V = \{A, B, C, D, F\}$ and $E = \{(A, C), (A, F), (C, B), (F, D), (F, C), (D, A)\}$ are the sets of vertices and directed
edges, respectively. Represent this digraph geometrically.

Solution   The digraph is shown in Figure 7.17. We have drawn arrows from A to C and from
A to F because the directed edges (A, C) and (A, F) are in E. The other arrows were
drawn for similar reason.
Do Exercise 3.

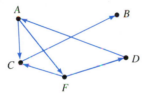

FIGURE 7.17

EXAMPLE 2   The following is the organization chart of a small university where P, V_1, V_2, and V_3
represent the president, vice president for finance, academic vice president, and vice
president for student life, respectively (see Figure 7.18). Also, X and Y represent the
director of personnel and the director of admissions, respectively; D_1, D_2, D_3, D_4, and
D_5 represent deans of the different schools; and S and I represent the athletic director
and the director of intramural sports. Represent this chart as a digraph.

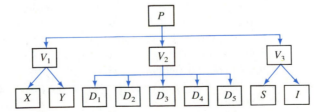

FIGURE 7.18

Solution   The set of vertices is

$$V = \{P, V_1, V_2, V_3, X, Y, D_1, D_2, D_3, D_4, D_5, S, I\}$$

The set of directed edges is

$$E = \{(P, V_1), (P, V_2), (P, V_3), (V_1, X), (V_1, Y), (V_2, D_1),$$
$$(V_2, D_2), (V_2, D_3), (V_2, D_4), (V_2, D_5), (V_3, S), (V_3, I)\}$$

Do Exercise 5.   ■

> **DEFINITION 7.19:** A *directed path* in a digraph G is a sequence of directed edges $(v_1, v_2), (v_2, v_3), (v_3, v_4), \ldots, (v_{n-2}, v_{n-1}), (v_{n-1}, v_n)$. The number of directed edges in a directed path is called the *length* of the path.

Observe that except for the last edge, the terminal point of each edge is the initial point of the next.

> **DEFINITION 7.20:** If in a digraph a directed path exists from u to v, then we say that v is *reachable* from u. If v is reachable from u, then the minimum of the lengths of all paths from u to v is called the *distance between u and v* and is denoted $d(u, v)$. If v is not reachable from u, we let $d(u, v) = \infty$ (read "infinity").

EXAMPLE 3   Consider the digraph in Figure 7.19.

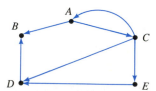

FIGURE 7.19

a. Find two directed paths of length 2 and a directed path of length 3 from C to B.
b. Is E reachable from A?
c. Is E reachable from B?
d. Find $d(A, B)$ and $d(B, E)$.

Solution
a. $(C, A), (A, B)$ is a directed path of length 2 from C to B. Another is $(C, D),$ (D, B). Also $(C, E), (E, D), (D, B)$ is a directed path of length 3 from C to B.
b. E is reachable from A because we have the directed path $(A, C), (C, E)$.
c. Since no directed edge starts at B, E is not reachable from B.
d. The only directed paths from A to D are $(A, C), (C, D)$ (length 2) and $(A, C),$ $(C, E), (E, D)$ (length 3). Thus $d(A, D) = 2$. Since E is not reachable from B, we write $d(B, E) = \infty$. Observe that $d(E, B) = 2$.

Do Exercise 7.   ■

Matrix Representation of Digraphs

Although the digraphs we use as examples in this book are simple, in the applications it is often difficult to obtain meaningful information simply by examining a digraph.

For example, it may be necessary to determine the number of directed paths of a certain length and this can be difficult in cases where there are a large number of vertices and edges. There are several matrices that are associated with a digraph. These matrices can be effectively used to analyze the situation represented by a digraph. To describe the first of these matrices, we label the vertices $v_1, v_2, \ldots, v_n$, where n is the number of vertices.

> **DEFINITION 7.21:** Let G be a digraph where E is the set of directed edges and $V = \{v_1, v_2, \ldots, v_n\}$ is the set of vertices. The *adjacency matrix* of G is the $n \times n$ matrix $\mathbf{A} = [a_{ij}]$ where
>
> $$a_{ij} = \begin{cases} 1 & \text{if } (v_i, v_j) \in E \\ 0 & \text{if } (v_i, v_j) \notin E \end{cases}$$
>
> for $i, j = 1, 2, \ldots, n$.

EXAMPLE 4   Obtain the adjacency matrix of the digraph of Example 3.

Solution   The vertices are $A, B, C, D,$ and E. We label them $v_1, v_2, v_3, v_4,$ and v_5. Thus A will be written by the first row and above the first column of the adjacency matrix, B will be written by the second row and above the second column, and so on. The directed edges are $(A, C), (A, B), (C, D), (C, E), (C, A), (E, D),$ and (D, B). Since (A, B) and (A, C) are directed edges of the digraph, we write 1 in the 1-2 and 1-3 positions of the matrix and 0s elsewhere in the first row. The reason for the other entries should be clear.

$$\begin{array}{c} \\ A \\ B \\ C \\ D \\ E \end{array} \begin{array}{cc} \begin{array}{ccccc} A & B & C & D & E \end{array} \\ \left[\begin{array}{ccccc} 0 & 1 & 1 & 0 & 0 \\ 0 & 0 & 0 & 0 & 0 \\ 1 & 0 & 0 & 1 & 1 \\ 0 & 1 & 0 & 0 & 0 \\ 0 & 0 & 0 & 1 & 0 \end{array} \right] \end{array}$$

Observe that each entry in the second row is 0 since no directed edge in the given digraph starts at B.
Do Exercise 11. ∎

EXAMPLE 5   Draw a digraph whose associated adjacency matrix is

$$\begin{array}{c} \\ A \\ B \\ C \\ D \\ E \end{array} \begin{array}{cc} \begin{array}{ccccc} A & B & C & D & E \end{array} \\ \left[\begin{array}{ccccc} 0 & 1 & 0 & 1 & 0 \\ 0 & 0 & 0 & 1 & 0 \\ 0 & 0 & 0 & 1 & 0 \\ 0 & 0 & 1 & 0 & 0 \\ 0 & 1 & 0 & 1 & 0 \end{array} \right] \end{array}$$

Solution   Since the 1-2 entry is 1, we draw an arrow from A to B. The 1-4 entry is also 1, so we draw an arrow from A to D, and so on. The digraph is drawn in Figure 7.20. **Do Exercise 15.**   ■

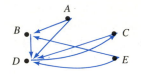

FIGURE 7.20

As we shall soon see, the product of an adjacency matrix with itself provides information about the digraph. Let $v_1, v_2, \ldots, v_n$ be the vertices of a digraph G and $\mathbf{A} = [a_{ij}]$ be the associated adjacency matrix. Let $\mathbf{A}^2 = [b_{ij}]$. Then b_{ij} is the dot product of the ith row of $\mathbf{A}$ by its jth column. That is

$$b_{ij} = a_{i1}a_{1j} + a_{i2}a_{2j} + \cdots + a_{in}a_{nj}$$

Observe that each term of this sum is either 0 or 1. Furthermore, $a_{ik}a_{kj} = 1$ if, and only if, $a_{ik} = 1$ and $a_{kj} = 1$. But this is so if, and only if, (v_i, v_k) and (v_k, v_j) are directed edges in the digraph. That is, there is a directed path of length 2 with initial point v_i, terminal point v_j, and passing through v_k. Therefore

$$b_{ij} = \text{number of directed paths of length 2 from } v_i \text{ to } v_j$$

In exactly the same manner, the following theorem can be proved.

THEOREM 7.13: Suppose the vertices of a digraph have been labeled $v_1, v_2, \ldots, v_n$ and let $\mathbf{A}$ be the corresponding adjacency matrix. Then the i-j entry of the matrix $\mathbf{A}^k$ is the number of directed paths of length k from v_i to v_j.

EXAMPLE 6   Let $\mathbf{M}$ be the adjacency matrix obtained in Example 4. Calculate $\mathbf{M}^2$ and $\mathbf{M}^3$ and interpret the results.

Solution

$$\mathbf{M}^2 = \begin{bmatrix} 0 & 1 & 1 & 0 & 0 \\ 0 & 0 & 0 & 0 & 0 \\ 1 & 0 & 0 & 1 & 1 \\ 0 & 1 & 0 & 0 & 0 \\ 0 & 0 & 0 & 1 & 0 \end{bmatrix} \cdot \begin{bmatrix} 0 & 1 & 1 & 0 & 0 \\ 0 & 0 & 0 & 0 & 0 \\ 1 & 0 & 0 & 1 & 1 \\ 0 & 1 & 0 & 0 & 0 \\ 0 & 0 & 0 & 1 & 0 \end{bmatrix}$$

$$= \begin{bmatrix} 1 & 0 & 0 & 1 & 1 \\ 0 & 0 & 0 & 0 & 0 \\ 0 & 2 & 1 & 1 & 0 \\ 0 & 0 & 0 & 0 & 0 \\ 0 & 1 & 0 & 0 & 0 \end{bmatrix}$$

Since row 1 corresponds to vertex A and columns 1, 4, and 5 correspond to vertices A, D, and E, respectively, the entries in the first row of $\mathbf{M}^2$ indicate that there is a path of length 2 from A to A, a path of length 2 from A to D, and a path of length 2 from A to E. The entries in the second row (all 0s) show that there is no path of length

2 starting at B. The entries in row 3 indicate that there are two paths of length 2 from C to B, and so on.

$$\mathbf{M}^3 = \mathbf{M}^2\mathbf{M} = \begin{bmatrix} 1 & 0 & 0 & 1 & 1 \\ 0 & 0 & 0 & 0 & 0 \\ 0 & 2 & 1 & 1 & 0 \\ 0 & 0 & 0 & 0 & 0 \\ 0 & 1 & 0 & 0 & 0 \end{bmatrix} \cdot \begin{bmatrix} 0 & 1 & 1 & 0 & 0 \\ 0 & 0 & 0 & 0 & 0 \\ 1 & 0 & 0 & 1 & 1 \\ 0 & 1 & 0 & 0 & 0 \\ 0 & 0 & 0 & 1 & 0 \end{bmatrix}$$

$$= \begin{bmatrix} 0 & 2 & 1 & 1 & 0 \\ 0 & 0 & 0 & 0 & 0 \\ 1 & 1 & 0 & 1 & 1 \\ 0 & 0 & 0 & 0 & 0 \\ 0 & 0 & 0 & 0 & 0 \end{bmatrix}$$

The entries in the first row of $\mathbf{M}^3$ indicate that there are two paths of length 3 from A to B, a path of length 3 from A to C, and a path of length 3 from A to D. The entries in the second row (all 0s) show that there is no path of length 3 starting at B. The entries in row 3 indicate that there is a path of length 3 from C to A, and so on.
Do Exercise 19. ∎

Dominance

Digraphs may be used as models in situations where certain people, or teams, have influence or dominance over others.

EXAMPLE 7   A lobbyist wants a committee to introduce a bill that would favor his company. He wishes to obtain the support of the most influential member of the committee. From experience, he knows that Albert has influence over Betty, Charles, and Darlene; Betty has influence over Darlene and Frank; Charles has influence over Betty, Elaine, and Frank; Darlene has influence over Charles and Frank; Elaine has influence over Albert and Darlene; and Frank has influence over Albert. Determine the most influential member of the committee.

Solution   Since influence is exerted in only one direction, we may use a directed graph as a model of the relationships between committee members. The directed graph is shown in Figure 7.21.

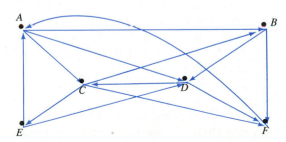

FIGURE 7.21

Observe that Albert and Charles each exert influence over three committee members. So we must break the tie. We look for directed paths of length 2 because each of these indicates that a committee member exerts influence over another through a third member. For example, Albert exerts influence over Betty, who in turn influences Frank. Thus Albert exerts two-stage influence over Frank. This is indicated by the directed path of length 2 $(A, B), (B, F)$. For convenience, we have recorded the number of directed paths of length 2 between each pair of vertices.

$$
\begin{array}{c c c c c c c}
 & A & B & C & D & E & F \\
A & 0 & 1 & 1 & 1 & 1 & 3 \\
B & 1 & 0 & 1 & 0 & 0 & 1 \\
C & 2 & 0 & 0 & 2 & 0 & 1 \\
D & 1 & 1 & 0 & 0 & 1 & 1 \\
E & 0 & 1 & 2 & 1 & 0 & 1 \\
F & 0 & 1 & 1 & 1 & 0 & 0
\end{array}
$$

The sum of the entries in each row of this matrix gives the total number of two-stage influence that each committee member exerts over other committee members. Adding each of these sums to the total number of one-stage influence for each of the committee members, we obtain

Committee member	Number of one- or two-stage influence
Albert	10
Betty	5
Charles	8
Darlene	6
Elaine	7
Frank	4

Albert is the most influential committee member and the lobbyist should try to gain his support.
Do Exercise 23. ■

Digraphs can also be used in the study of *dominance*. Given two distinct elements A and B of a set (for example A and B could represent sports teams or individuals) either A dominates B, or B dominates A, but not both. If X dominates Y, we say that X has a *one-stage dominance* over Y. If in turn Y has a one-stage dominance over Z, then we say that X has a *two-stage dominance* over Z. If we wish to rank the members of a set according to the total number of one-stage dominances that each member has, it may be that the result is a tie. In that case, we can use the total number of two-stage dominances to break the tie.

EXAMPLE 8   Bob, Jeremy, Lou, Tim, Andre, and Steve were assigned the numbers 1, 2, 3, 4, 5, and 6, respectively, for the purpose of recording the results of a round-robin racquetball tournament. The results are tabulated in the following matrix, where a 1 in the *i-j*

position signifies that player i defeated player j, while a 0 indicates that player i lost to player j, or $i = j$.

$$\begin{bmatrix} 0 & 1 & 0 & 1 & 1 & 1 \\ 0 & 0 & 1 & 0 & 1 & 0 \\ 1 & 0 & 0 & 1 & 1 & 1 \\ 0 & 1 & 0 & 0 & 1 & 0 \\ 0 & 0 & 0 & 0 & 0 & 1 \\ 0 & 1 & 0 & 1 & 0 & 0 \end{bmatrix}$$

a. Draw a digraph that illustrates the one-stage dominances of each player over the others.

b. Show that the tournament ended in a tie.

c. Use two-stage dominances to break the tie.

Solution **a.** {1, 2, 3, 4, 5, 6} is the set of vertices of the digraph. We draw an arrow from i to j if, and only if, player i beat player j. The digraph is sketched in Figure 7.22.

b. Adding the entries in each row, we see that Bob has four wins (4 one-stage dominances over the other players), Jeremy has two wins, Lou has four wins, Tim has two wins, Andre has 1 win, and Steve has two wins. So there is a tie between Bob and Lou.

c. Observe that the given matrix is the adjacency matrix of the digraph. To find the number of two-stage dominances that each player has over the others, we let $\mathbf{T}$ be the matrix that gives the one-stage dominances and calculate $\mathbf{T}^2$. We find

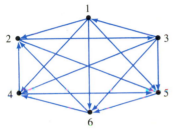

FIGURE 7.22

$$\mathbf{T}^2 = \begin{bmatrix} 0 & 1 & 0 & 1 & 1 & 1 \\ 0 & 0 & 1 & 0 & 1 & 0 \\ 1 & 0 & 0 & 1 & 1 & 1 \\ 0 & 1 & 0 & 0 & 1 & 0 \\ 0 & 0 & 0 & 0 & 0 & 1 \\ 0 & 1 & 0 & 1 & 0 & 0 \end{bmatrix} \cdot \begin{bmatrix} 0 & 1 & 0 & 1 & 1 & 1 \\ 0 & 0 & 1 & 0 & 1 & 0 \\ 1 & 0 & 0 & 1 & 1 & 1 \\ 0 & 1 & 0 & 0 & 1 & 0 \\ 0 & 0 & 0 & 0 & 0 & 1 \\ 0 & 1 & 0 & 1 & 0 & 0 \end{bmatrix}$$

$$= \begin{bmatrix} 0 & 2 & 1 & 1 & 2 & 1 \\ 1 & 0 & 0 & 1 & 1 & 2 \\ 0 & 3 & 0 & 2 & 2 & 2 \\ 0 & 0 & 1 & 0 & 1 & 1 \\ 0 & 1 & 0 & 1 & 0 & 0 \\ 0 & 1 & 1 & 0 & 2 & 0 \end{bmatrix}$$

To obtain the total numbers of one-stage and two-stage dominances each player has over the others, we calculate

$$\mathbf{T} + \mathbf{T}^2 = \begin{bmatrix} 0 & 3 & 1 & 2 & 3 & 2 \\ 1 & 0 & 1 & 1 & 2 & 2 \\ 1 & 3 & 0 & 3 & 3 & 3 \\ 0 & 1 & 1 & 0 & 2 & 1 \\ 0 & 1 & 0 & 1 & 0 & 1 \\ 0 & 2 & 1 & 1 & 2 & 0 \end{bmatrix}$$

The 3 in the 1-2 position indicates that Bob (player 1) has 3 one-stage or two-stage dominances over Jeremy (player 2). The meaning of the other entries should be clear. The sums of the entries in the 1st, 2nd, 3rd, 4th, 5th, and 6th rows are 11, 7, 13, 5, 3, and 6, respectively. Thus Lou has a total of 13 dominances, Bob has 11, Jeremy 7, Steve 6, Tim 5, and Andre 2. Since Lou has the greatest total number of one-stage and two-stage dominances over the others, he is the winner of the tournament.

Do Exercise 25.   ∎

Exercise Set 7.6

In Exercises 1–4, draw the given digraphs.

1. G is the digraph where $V = \{A, B, C, D, F\}$ and $E = \{(A, D), (B, F), (C, D), (C, A), (F, D), (D, B)\}$ are the sets of vertices and directed edges, respectively.

2. G is the digraph where $V = \{R, S, T, U, W\}$ and $E = \{(R, T), (R, W), (S, R), (S, W), (T, R), (W, U)\}$ are the sets of vertices and directed edges, respectively.

3. G is the digraph where $V = \{A, B, C, D, F, H\}$ and $E = \{(A, F), (B, A), (B, H), (C, F), (D, H), (F, B), (H, C)\}$ are the sets of vertices and directed edges, respectively.

4. G is the digraph where $V = \{R, S, T, U, W, X\}$ and $E = \{(R, X), (S, R), (S, U), (T, W), (U, R), (W, X), (X, T), (X, W)\}$ are the sets of vertices and directed edges, respectively.

5. The following is the organization chart of an insurance company where P, V_1, V_2, V_3 represent the president, vice president for finance, vice president of personnel, and actuarial vice president, respectively. A, B, and C represent three chief accountants, respectively; D_1, D_2, and D_3 represent three department heads; and X, Y, and Z represent three actuaries. Represent this chart as a digraph.

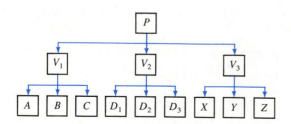

6. Consider the following digraph.

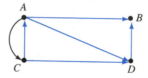

a. Find two directed paths of length 2 from C to B.
b. Is D reachable from C?
c. Is A reachable from B?
d. Find $d(A, D)$ and $d(B, C)$.

7. Consider the following digraph.

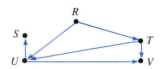

a. Find two directed paths of length 2 and a directed path of length 3 from R to V.
b. Is U reachable from R?
c. Is R reachable from T?
d. Find $d(R, V)$ and $d(T, U)$.

8. Consider the following digraph.

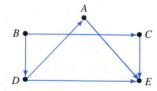

a. Find two directed paths of length 2 and a directed path of length 3 from B to E.
b. Is E reachable from C?
c. Is C reachable from E?
d. Find $d(B, E)$ and $d(C, E)$.

9. Consider the following digraph.

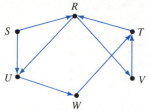

a. Find two directed paths of length 3 and a directed path of length 4 from S to T.

b. Is R reachable from V?

c. Is V reachable from R?

d. Find $d(S, T)$ and $d(T, S)$.

10. Consider the following digraph.

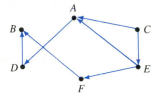

a. Find two directed paths of length 3 and a directed path of length 4 from C to B.

b. Is E reachable from F?

c. Is F reachable from E?

d. Find $d(C, B)$ and $d(B, C)$.

11. Obtain the adjacency matrix of the graph of Exercise 7.

12. Obtain the adjacency matrix of the graph of Exercise 8.

13. Obtain the adjacency matrix of the graph of Exercise 9.

14. Obtain the adjacency matrix of the graph of Exercise 10.

15. Draw a digraph whose associated adjacency matrix is

	A	B	C	D
A	0	0	1	0
B	1	0	1	0
C	0	1	0	0
D	1	1	0	0

16. Draw a digraph whose associated adjacency matrix is

	A	B	C	D	E
A	0	0	1	0	1
B	1	0	0	0	0
C	0	1	0	0	1
D	0	1	0	0	1
E	0	0	1	0	0

17. Draw a digraph whose associated adjacency matrix is

	R	S	T	U	V
R	0	0	1	0	1
S	1	0	1	0	0
T	0	1	0	0	1
U	1	1	0	0	0
V	0	0	1	0	0

18. Draw a digraph whose associated adjacency matrix is

	A	B	C	D	E	F
A	0	0	1	0	1	0
B	1	0	0	0	1	0
C	0	0	0	1	0	0
D	1	0	1	0	1	0
E	0	1	0	1	0	0
F	1	0	1	0	0	0

19. Let $\mathbf{M}$ be the adjacency matrix of Exercise 15. Calculate $\mathbf{M}^2$ and $\mathbf{M}^3$ and interpret the results.

20. Let $\mathbf{M}$ be the adjacency matrix of Exercise 16. Calculate $\mathbf{M}^2$ and $\mathbf{M}^3$ and interpret the results.

21. Let $\mathbf{M}$ be the adjacency matrix of Exercise 17. Calculate $\mathbf{M}^2$ and $\mathbf{M}^3$ and interpret the results.

22. Let $\mathbf{M}$ be the adjacency matrix of Exercise 18. Calculate $\mathbf{M}^2$ and $\mathbf{M}^3$ and interpret the results.

23. A lobbyist wants a committee to introduce a bill that would favor her company. She wishes to obtain the support of the most influential member of the committee. From experience, she knows that Allen has influence over Carl and Doug; Bob has influence over Allen and Fran; Carl has influence over Bob, Earl, and Fran; Doug has influence over Carl, Earl, and Fran; Earl has influence over Allen and Fran; and Fran has influence over Allen. Determine the most influential member of the committee.

24. Gwen, Anay, Laura, Julie, Judy, and Marian were assigned the numbers 1, 2, 3, 4, 5 and 6, respectively, for the purpose of recording the results of a round-robin tennis tournament. The results are tabulated in matrix form, where a 1 in the i-j position signifies that player i defeated player j, while a 0 indicates that player i lost to player j, or $i = j$.

$$
\begin{bmatrix}
0 & 1 & 1 & 1 & 0 & 1 \\
0 & 0 & 1 & 1 & 0 & 0 \\
0 & 0 & 0 & 1 & 0 & 1 \\
0 & 0 & 0 & 0 & 1 & 0 \\
1 & 1 & 1 & 0 & 0 & 1 \\
0 & 1 & 0 & 1 & 0 & 0
\end{bmatrix}
$$

a. Draw a digraph that illustrates the one-stage dominances of each player over the others.

b. Show that the tournament ended in a tie.

c. Use two-stage dominances to break the tie.

25. Phillip, Gregory, Lee Michael, Timothy, Katie, and Beth were assigned the numbers 1, 2, 3, 4, 5, and 6, respectively, for the purpose of recording the results of a round-robin golf tournament. The results are tabulated in matrix form, where a 1 in the i-j position signifies that player i defeated player j, while a 0 indicates that player i lost to player j, or $i = j$.

$$\begin{bmatrix} 0 & 0 & 0 & 0 & 1 & 1 \\ 1 & 0 & 0 & 1 & 0 & 1 \\ 1 & 1 & 0 & 0 & 1 & 1 \\ 1 & 0 & 1 & 0 & 1 & 1 \\ 0 & 1 & 0 & 0 & 0 & 0 \\ 0 & 0 & 0 & 0 & 1 & 0 \end{bmatrix}$$

a. Draw a digraph that illustrates the one-stage dominances of each player over the others.

b. Show that the tournament ended in a tie.

c. Use two-stage dominances to break the tie.

7.7 More Applications

In this section, we use a digraph to measure the status of an individual within an organization. We also discuss a scheduling problem, and give an example for controlling disease propagation.

Status

In a digraph where the vertices represent individuals within an organization, a directed edge from u to v may mean that u directly supervises v or that u gives orders to v. It is clear that if w is reachable from u, then u has some status over w, and the greater the distance $d(u, w)$, the greater the status. To see this, assume that Janet is a dean at some college. She directly supervises Ron who is a chairperson, and Ron supervises Margaret who is a professor. Then Janet has more status over Margaret than Ron. In the applications, the following definition has been given.

> **DEFINITION 7.22:**   Suppose that a digraph G represents an organization where (u, v) is a directed edge of G if, and only if, u supervises v directly. *The status of v_i in G is denoted $S(v_i)$ and is given by*
>
> $$S(v_i) = \sum d(v_i, v_j)$$
>
> where the sum is over all v_j's that are reachable from v_i.

EXAMPLE 1   The structures of two organizations are described in Figure 7.23. Find the status of v_1, v_2, v_3, v_4, v_5, v_6, and w_1, w_2, w_3, w_4.

Solution   In organization A, the vertices v_2, v_3, v_4, v_5, v_7, v_8, and v_9 can be reached from v_1. We get

$$S(v_1) = d(v_1, v_2) + d(v_1, v_3) + d(v_1, v_4) + d(v_1, v_5) + d(v_1, v_7) + d(v_1, v_8) + d(v_1, v_9)$$
$$= 1 + 1 + 2 + 2 + 3 + 3 + 3 = 15$$

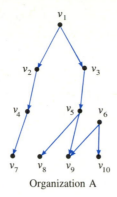
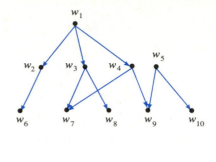

Organization A   Organization B

FIGURE 7.23

Similar computations give the following:

$$S(v_2) = 1 + 2 = 3, \ S(v_3) = 1 + 2 + 2 = 5$$
$$S(v_4) = 1, \ S(v_5) = 1 + 1 = 2, \ S(v_6) = 1 + 1 = 2$$
$$S(w_1) = 1 + 1 + 1 + 2 + 2 + 2 + 2 = 11$$
$$S(w_2) = 1, \ S(w_3) = 1 + 1 = 2$$
$$S(w_4) = 1 + 1 = 2, \text{ and } S(w_5) = 1 + 1 = 2$$

Observe that v_1 in organiztaion A has higher status than w_1 in organization B although v_1 and w_1 are heads of organizations with the same number of employees.
Do Exercise 1. ■

Distance Matrices

If the organizational structure of a company is complex, it is difficult to compute the status of its members from the corresponding digraph. It is useful, as we shall soon see, to make use of another matrix associated with a digraph.

DEFINITION 7.23: If $V = \{v_1, v_2, \ldots, v_n\}$ is the set of vertices of a digraph G, the *distance matrix* **D** of G is the $n \times n$ matrix $[d_{ij}]$ where

$$d_{ij} = \begin{cases} 0 & \text{if } i = j \\ d(v_i, v_j) & \text{if } i \neq j \end{cases}$$

EXAMPLE 2 Find the distance matrix for the following digraph.

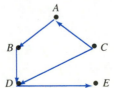

FIGURE 7.24

Solution There are five vertices in the digraph. So the distance matrix is 5×5. The vertices A, B, C, D, and E are labeled v_1, v_2, v_3, v_4, and v_5, respectively. We enter 0 in the

i-i position for $i = 1, 2, 3, 4$, and 5. Recall that $d(u, w)$ is the length of the shortest path from u to w if w is reachable from u, and $d(u, w) = \infty$ if w is not reachable from u. We have

$$d(A, B) = 1 \text{ so } d_{12} = 1, \quad d(A, D) = 2 \text{ so } d_{14} = 2$$

$$d(A, E) = 3 \text{ so } d_{15} = 3, \quad d(B, D) = 1 \text{ so } d_{24} = 1$$

$$d(B, E) = 2 \text{ so } d_{25} = 2, \quad d(C, A) = 1 \text{ so } d_{31} = 1$$

$$d(C, B) = 2 \text{ so } d_{32} = 2, \quad d(C, D) = 1 \text{ so } d_{34} = 1$$

$$d(C, E) = 2 \text{ so } d_{35} = 2, \quad d(D, E) = 1 \text{ so } d_{45} = 1$$

We enter these results to form the distance matrix **D**. We then enter ∞ in the remaining positions.

$$
\begin{array}{c}
\begin{array}{ccccc} A & B & C & D & E \end{array} \\
\begin{array}{c} A \\ B \\ C \\ D \\ E \end{array}
\left[
\begin{array}{ccccc}
0 & 1 & \infty & 2 & 3 \\
\infty & 0 & \infty & 1 & 2 \\
1 & 2 & 0 & 1 & 2 \\
\infty & \infty & \infty & 0 & 1 \\
\infty & \infty & \infty & \infty & 0
\end{array}
\right]
\end{array}
$$

Note that the ∞ entries in the bottom row indicate that no vertex other than E can be reached from E.
Do Exercise 7. ■

We were able to find the distance matrix for the digraph of Example 2 because this digraph was simple. In digraphs with many vertices and edges it is difficult to find the lengths of the shortest paths between some of the vertices. It is useful to have a procedure for computing the entries of the distance matrix **D** of a digraph from its adjacency matrix **A**.

Procedure to Find the Distance Matrix D of a Digraph

Step 1. Label the vertices $v_1, v_2, \ldots, v_n$.

Step 2. Find the adjacency matrix $\mathbf{A} = [a_{ij}]$.

Step 3. Enter 0 in the i-i position of the distance matrix **D** for $i = 1, 2, 3, \ldots, n$.

Step 4. For each 1 in the adjacency matrix, enter a 1 in the corresponding position of the distance matrix **D**.

Step 5. Compute $\mathbf{A}^2$. For each positive i-j entry of $\mathbf{A}^2$ for which there is no entry in the corresponding i-j position of matrix **D**, enter a 2 in that position for **D**.

Step 6. Compute $\mathbf{A}^3$. For each positive i-j entry of $\mathbf{A}^3$ for which there is no entry in the corresponding i-j position of matrix **D**, enter a 3 in that position for **D**.

Continue this process. In general, enter k in the i-j position of matrix **D** if k is the smallest power to which **A** must be raised so that the i-j entry of $\mathbf{A}^k$ is positive. As soon as a computation of $\mathbf{A}^m$ does not contribute a new entry in matrix **D**, stop the process.*

Step 7. Complete matrix **D** by entering ∞ in every position that has not been filled.

EXAMPLE 3 Use the procedure described above to find the distance matrix for the digraph of Example 2.

Solution Labeling the vertices A, B, C, D, and E as v_1, v_2, v_3, v_4, and v_5, respectively, the adjacency matrix is

$$\mathbf{A} = \begin{bmatrix} 0 & 1 & 0 & 0 & 0 \\ 0 & 0 & 0 & 1 & 0 \\ 1 & 0 & 0 & 1 & 0 \\ 0 & 0 & 0 & 0 & 1 \\ 0 & 0 & 0 & 0 & 0 \end{bmatrix}$$

We start a 5×5 matrix with a 0 in the i-i position for $i = 1, 2, 3, 4,$ and 5. We also enter 1s in the 1-2, 2-4, 3-1, 3-4, and 4-5 positions since these entries are 1 in the adjacency matrix. We get

$$\mathbf{D} = \begin{bmatrix} 0 & 1 & & & \\ & 0 & & 1 & \\ 1 & & 0 & 1 & \\ & & & 0 & 1 \\ & & & & 0 \end{bmatrix}$$

We compute $\mathbf{A}^2$ and get

$$\mathbf{A}^2 = \begin{bmatrix} 0 & 0 & 0 & 1 & 0 \\ 0 & 0 & 0 & 0 & 1 \\ 0 & 1 & 0 & 0 & 1 \\ 0 & 0 & 0 & 0 & 0 \\ 0 & 0 & 0 & 0 & 0 \end{bmatrix}$$

---

*Observe that if $d(u, v) = k$ where $k > 1$, then there must be a vertex w such that $d(u, w) = k - 1$. Therefore if a power $\mathbf{A}^k$ of the adjacency matrix does not contribute a new entry in **D**, we have no two vertices u and w such that $d(u, w) = k$. Consequently, there are no two vertices u and v such that $d(u, v) > k$. Thus, we need not continue to calculate powers of **A**.

Observe that the 1-4, 2-5, 3-2, and 3-5 entries are positive and that the corresponding positions in matrix **D** are still blank. So we enter 2s in these positions in matrix **D**, and get

$$\mathbf{D} = \begin{bmatrix} 0 & 1 & & 2 & \\ & 0 & & 1 & 2 \\ 1 & 2 & 0 & 1 & 2 \\ & & & 0 & 1 \\ & & & & 0 \end{bmatrix}$$

We compute $\mathbf{A}^3$, and get

$$\mathbf{A}^3 = \begin{bmatrix} 0 & 0 & 0 & 0 & 1 \\ 0 & 0 & 0 & 0 & 0 \\ 0 & 0 & 0 & 1 & 0 \\ 0 & 0 & 0 & 0 & 0 \\ 0 & 0 & 0 & 0 & 0 \end{bmatrix}$$

The 1-5 and 3-4 entries are the only positive entries. The 1-5 position in matrix **D** is still blank. So we enter a 3 there. However, the 3-4 position of matrix **D** has already been filled. So we proceed.

$$\mathbf{A}^4 = \begin{bmatrix} 0 & 0 & 0 & 0 & 0 \\ 0 & 0 & 0 & 0 & 0 \\ 0 & 0 & 0 & 0 & 1 \\ 0 & 0 & 0 & 0 & 0 \\ 0 & 0 & 0 & 0 & 0 \end{bmatrix}$$

The 3-5 entry is the only positive entry. But the 3-5 entry of matrix **D** has already been filled. So the computation of $\mathbf{A}^4$ did not yield any new entry for the distance matrix. We need not continue. Since no vertex in the digraph is at a distance 4 from another vertex, no vertex will be at a distance 5 (or larger) from another vertex. We now enter ∞ in each position of matrix **D** that is still blank. We get

$$\mathbf{D} = \begin{bmatrix} 0 & 1 & \infty & 2 & 3 \\ \infty & 0 & \infty & 1 & 2 \\ 1 & 2 & 0 & 1 & 2 \\ \infty & \infty & \infty & 0 & 1 \\ \infty & \infty & \infty & \infty & 0 \end{bmatrix}$$

As expected this is the same result obtained in Example 2.
Do Exercise 9.   ■

You may think that it was simpler to obtain the distance matrix directly from the digraph (Example 2), than it was using matrix multiplication (Example 3). However, in a case where a digraph is complex and a computer is used to perform matrix multiplication, the second method is more efficient. The following definition is useful in using matrices to determine the status of individuals in an organization.

> **DEFINITION 7.24:** Let $\{v_1, v_2, \ldots, v_n\}$ be the vertices of a digraph G and let $\mathbf{D}$ be the distance matrix of G. The *distance sum of vertex v_i* is the sum of the finite entries in the ith row of matrix $\mathbf{D}$.

EXAMPLE 4   Find the distance sum of each of the vertices of the digraph of Example 2.

Solution   The distance matrix for this digraph was found in both Example 2 and in Example 3. It is

$$\mathbf{D} = \begin{bmatrix} 0 & 1 & \infty & 2 & 3 \\ \infty & 0 & \infty & 1 & 2 \\ 1 & 2 & 0 & 1 & 2 \\ \infty & \infty & \infty & 0 & 1 \\ \infty & \infty & \infty & \infty & 0 \end{bmatrix}$$

So the distance sum of v_1 is 6 since $1 + 2 + 3 = 6$. Similarly, we find that the distance sums of v_2, v_3, v_4, and v_5 are 3, 6, 1, and 0, respectively. **Do Exercise 13.**   ∎

Comparing the definitions of status and of distance sum, it is clear that if the vertices of a digraph represent individuals in an organization and the directed edge (u, v) is in the digraph if, and only if, u directly supervises v, then the status of any v_i is equal to the distance sum of v_i.

EXAMPLE 5   Determine the status of each individual in an organization where the supervising responsibilities are given by finding the distance sum of each vertex of the corresponding digraph.

Individual	Directly supervises
Ann	Bob and Charles
Bob	Dennis
Charles	Dennis and Elaine
Dennis	Fran and Gwen
Elaine	Gwen and Henri

Solution   We first draw the digraph where we use the first letter of an individual's name to label the corresponding vertex. (See Figure 7.25.)

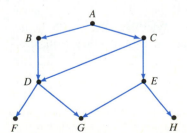

FIGURE 7.25

If we label the vertices A, B, C, D, E, F, G, and H as v_1, v_2, v_3, v_4, v_5, v_6, v_7, and v_8, respectively, we can write the distance matrix directly from the digraph. We obtain

$$
\mathbf{D} = \begin{bmatrix}
0 & 1 & 1 & 2 & 2 & 3 & 3 & 3 \\
\infty & 0 & \infty & 1 & \infty & 2 & 2 & \infty \\
\infty & \infty & 0 & 1 & 1 & 2 & 2 & 2 \\
\infty & \infty & \infty & 0 & \infty & 1 & 1 & \infty \\
\infty & \infty & \infty & \infty & 0 & \infty & 1 & 1 \\
\infty & \infty & \infty & \infty & \infty & 0 & \infty & \infty \\
\infty & \infty & \infty & \infty & \infty & \infty & 0 & \infty \\
\infty & \infty & \infty & \infty & \infty & \infty & \infty & 0
\end{bmatrix}
$$

Since $0 + 1 + 1 + 2 + 2 + 3 + 3 + 3 = 15$, the distance sum of v_1 is 15. Similarly, the distance sums of v_2, v_3, v_4, v_5, v_6, v_7, and v_8 are 5, 8, 2, 2, 0, 0, and 0, respectively. Thus the status of the individuals are $S(\text{Ann}) = 15$, $S(\text{Charles}) = 8$, $S(\text{Bob}) = 5$, $S(\text{Dennis}) = S(\text{Elaine}) = 2$, and $S(\text{Fran}) = S(\text{Gwen}) = S(\text{Henri}) = 0$. If the organization pays its employees according to their status, Fran, Gwen, and Henri should have equal and lowest salaries in the organization. Next highest should be Dennis and Elaine with equal salaries, next highest should be Bob, then Charles, then Ann with the highest salary in the organization.*
Do Exercise 17.   ■

Activity Digraph

Suppose that you are planning to have your home built. You wish to do the contracting yourself, but you would like to have the house completed as soon as possible. How should you schedule the different tasks to accomplish your goal?

Given a project composed of tasks, we construct an *activity digraph* as follows. Each task is assigned a vertex and with each vertex we associate the number t if the task corresponding to that vertex requires t units of time to complete. We introduce a vertex S (start) and a vertex E (end) and assign the number 0 to each of these vertices. The directed edge (S, W) is in the digraph if the task W may start before the completion of any other task. Also the directed edge (U, E) is in the digraph if no other task requires U to be completed before that task begins. Further, (U, V) will be a directed edge of the digraph if, and only if, task V may begin immediately after completion of task U. We now give the following definition.

DEFINITION 7.25: Suppose G is an activity digraph and u is a vertex in G. For each directed path from S to u, add the numbers assigned to the vertices along that path. Any of these paths with the largest sum is called a *critical path* for u. (There may be more than one critical path.)

---

*If we used matrix multiplication to find the distance matrix, we would have to compute $\mathbf{A}^2$, $\mathbf{A}^3$, and $\mathbf{A}^4$ where $\mathbf{A}$ is the adjacency matrix of the digraph. Since $\mathbf{A}$ is an 8×8 matrix, the computations—without the use of a computer—would be tedious.

EXAMPLE 6   A project consists of tasks A, B, C, D, F, G, H, and I. Its activity digraph is shown in Figure 7.26.

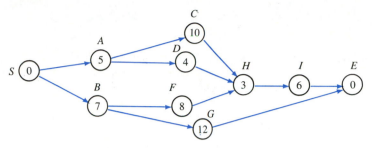

FIGURE 7.26

a. Find a critical path for H.
b. Find a critical path for the project.

Solution   a. The directed paths from S to H are

$$S, A, C, H; \quad S, A, D, H; \quad \text{and} \quad S, B, F, H$$

Adding the numbers assigned to the vertices we get

$$0 + 5 + 10 + 3 = 18 \quad \text{for the first path}$$
$$0 + 5 + 4 + 3 = 12 \quad \text{for the second path}$$
$$0 + 7 + 8 + 3 = 18 \quad \text{for the third path}$$

There are two critical paths for H. They are S, A, C, H and S, B, F, H.

b. The directed paths from S to E are

$$S, A, C, H, I, E; \quad S, A, D, H, I, E; \quad S, B, F, H, I, E; \quad \text{and} \quad S, B, G, E.$$

Adding the numbers assigned to the vertices we get

$$0 + 5 + 10 + 3 + 6 + 0 = 24 \quad \text{for the first path}$$
$$0 + 5 + 4 + 3 + 6 + 0 = 18 \quad \text{for the second path}$$
$$0 + 7 + 8 + 3 + 6 + 0 = 24 \quad \text{for the third path}$$
$$0 + 7 + 12 + 0 = 19 \quad \text{for the fourth path}$$

So there are two critical paths for the project. They are S, A, C, H, I, E and S, B, F, H, I, E.
It is now easy to schedule each task.
Do Exercise 21. ∎

Procedure to Schedule Tasks for a Project

Step 1.   Draw the activity digraph for the project.

Step 2.   For each task u, find a critical path for u.

Step 3.   Add all the numbers assigned to the vertices on that critical path except that assigned to u. This sum gives us the number of units of time to allow before scheduling u.*

EXAMPLE 7   Mr. and Mrs. Entrepreneur decided to have their house built. They consulted sub-contractors and found that each task will require the amount of time shown in the following table.

Task	Time (in weeks)
Clear land	1
Build foundation	2
Build structure	5
Plumbing	1
Electrical work	2
Finish interior	3
Finish exterior	2
Landscape	1

a. When should they schedule each task?
b. How long will it take to complete the house?

Solution   Let us label each task as follows:

A:	Clear land
B:	Build foundation
C:	Build structure
D:	Plumbing
F:	Electrical work
G:	Finish interior
H:	Finish exterior
I:	Landscape

A possible relationship between the tasks is shown in the next table.

Task	Preceding tasks
A	None
B	A
C	A, B
D	A, B, C
F	A, B, C
G	A, B, C, D, F
H	A, B, C
I	A, B, C, H

---

*In the applications, techniques such as PERT (Program Evaluation and Review Technique) make use of activity digraphs.

Using this information, we draw the activity digraph.

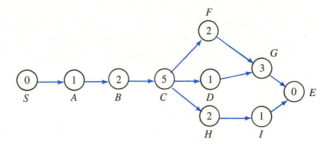

FIGURE 7.27

It is clear now that the following schedule should be used.

Begin A (clear land) immediately. Start B (build foundation) in 1 week, since the only path leading to B is S, A, B and $0 + 1 = 1$. Start C (build structure) in 3 weeks, since the only path leading to C is S, A, B, C and $0 + 1 + 2 = 3$. For similar reasons, we should start D (plumbing) in 8 weeks, F (electrical work) in 8 weeks, H (exterior work) in 8 weeks, I (landscaping) in 10 weeks, and G (interior work) in 10 weeks.

There are two paths leading to G. However, S, A, B, C, F, G is a critical path and $0 + 1 + 2 + 5 + 2 = 10$. There are three paths leading to E (end). However, S, A, B, C, F, G, E is a critical path. Since $0 + 1 + 2 + 5 + 2 + 3 = 13$, the house should be completed in 13 weeks.
Do Exercise 23.   ■

Control of Disease Propagation

When an infectious disease begins to spread, it is useful to isolate those who are already infected. However, is there a way to check whether someone has been infected and has not yet been identified? We conclude this chapter with an example of this situation.

EXAMPLE 8   Seven students suffering from an infectious disease have reported to the clinic of a small university. Health officials believe that the disease was introduced by one person and spread to others. They immediately isolate the seven students and interview them trying to determine how the disease could have spread within the group. They determine the following:

Student	Could have given the disease to
Albert	Frank
Betty	Albert, Frank
Colleen	Genevieve
Doug	Colleen
Earl	Betty, Albert
Frank	Earl
Genevieve	Doug

Use a digraph to determine whether some other student on campus has the disease but has not yet been identified.

Solution The digraph will have seven vertices. Use the first letter of each student's name to identify a vertex with a student. We draw an edge from S to T if S could have given the disease to T. The digraph is drawn in Figure 7.28.

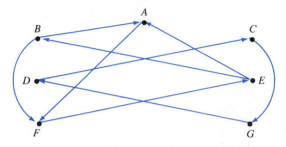

FIGURE 7.28

It is clear that if the disease spread within the group from S to T, then T must be reachable from S. That is, there must be a directed path from S to T. There is no directed path from A to C. Thus the disease could not have started with Albert and spread within the group to Colleen.

Similarly, the disease could not have started with Betty since there is no directed path from B to C. It could not have started with Colleen since there is no directed path from C to A. Checking the other four vertices, we find that the disease could not have started with one of the isolated students and spread within the group. Therefore there must be at least one other individual on campus who has the disease but who has not been isolated yet.

Do Exercise 27. ■

Exercise Set 7.7

In Exercises 1–4, the structures of organizations are described by digraphs. Find the status of each member.

1.

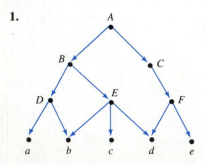

2.

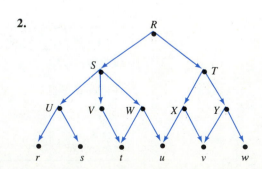

3.

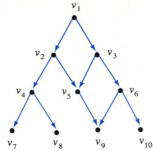

4.

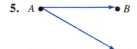

In Exercises 5–8, find the distance matrix for the given digraph.

5. 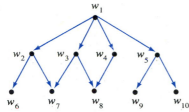   **6.**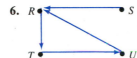

7. 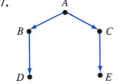   **8.**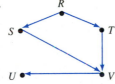

In Exercises 9–12, use the procedure described on page 522.

9. Find the distance matrix of the digraph of Exercise 5.

10. Find the distance matrix of the digraph of Exercise 6.

11. Find the distance matrix of the digraph of Exercise 7.

12. Find the distance matrix of the digraph of Exercise 8.

13. Find the distance sum of each of the vertices of the digraph of Exercise 5.

14. Find the distance sum of each of the vertices of the digraph of Exercise 6.

15. Find the distance sum of each of the vertices of the digraph of Exercise 7.

16. Find the distance sum of each of the vertices of the digraph of Exercise 8.

17. Determine the status of each individual in an organization, where the supervising responsibilities are given, by finding the distance sum of each vertex of the corresponding digraph.

Individual	Directly supervises
Bob	Gwen and Julie
Gwen	Peter and Marc
Julie	Beth and Paul
Peter	Katie and Marie
Marc	Wendy and Bertha
Beth	Shirley and Steve
Paul	Lisa

18. Repeat Exercise 17 with

Individual	Directly supervises
Bill	Carol and Mike
Carol	Ted
Mike	Ted and Judy
Ted	Nancy and Terry
Judy	Terry, Louis, and John

19. Repeat Exercise 17 with

Individual	Directly supervises
Janet	Carl and Kathie
Carl	Ahmad and Wynne
Kathie	Burnett and Andre
Ahmad	Allen and Marie
Wynne	Ralph and Frank
Burnett	Betty and Norma
Andre	Janine, Tim, and Gregory

20. Repeat Exercise 17 with

Individual	Directly supervises
Jean	Pierre, Yvette, and Arlette
Pierre	Paul and Claude
Arlette	Louise and Henri
Yvette	Solange and Ursule
Paul	Diane and Marie
Claude	Rose and Desiree
Louise	Juliette
Henri	Yolande, Marcel, Deedee, and Annie

FIGURE FOR EXERCISE 21

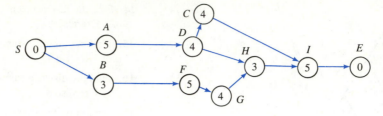

FIGURE FOR EXERCISE 22

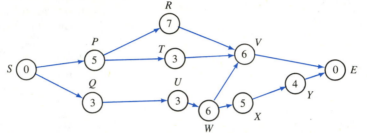

21. A project consists of tasks A, B, C, D, F, G, H, and I. Its activity digraph is shown above.

   a. Find a critical path for H.
   b. Find a critical path for the project.

22. A project consists of tasks P, Q, R, T, U, V, W, X, and Y. Its activity digraph is shown above.

   a. Find a critical path for V.
   b. Find a critical path for the project.

-23. A couple wanted to have their house built. They consulted subcontractors and found that each task will require the amount of time shown in the following table.

   a. When should they schedule each task?
   b. How long will it take to complete the house?

Task	Time (in weeks)
Purchase land	3
Clear land	2
Build foundation	3
Build structure	7
Plumbing	2
Electrical work	2
Finish interior	4
Finish exterior	3
Landscape	2

24. A book has been accepted for publication. The different tasks for the project are listed in the following table together with the time required for each and the interdependency of the tasks.

   a. When should each task be scheduled?
   b. How long will it take to complete the book?

Task label	Task	Time required (in weeks)	Tasks that must preceed
A	Manuscript review	4	None
B	Author revises manuscript	5	A
C	Editor checks manuscript	3	A, B
D	Editor marks manuscript	6	A, B, C
F	Art work	5	A, B, C, D
G	Manuscript to compositor	11	A, B, C, D
H	Author checks galley proofs	5	A, B, C, D, G
I	Author checks art work	3	A, B, C, D, F
J	Setting of page proofs	4	A, B, C, D, F, G, H, I
K	Author checks page proofs	5	A, B, C, D, F, G, H, I, J
L	Printing and binding	8	A, B, C, D, F, G, H, I, J, K
M	Marketing	12	A, B, C, D

25. A project consists of eight tasks A, B, C, D, F, G, H, and I. The activity digraph for this project is given where the units of time are months. (See page 533)

   a. When should each task be scheduled?
   b. When will the project be completed?

26. A project consists of nine tasks A, B, C, D, F, G, H, I and J. The activity digraph for this project is given where the units of time are weeks. (See page 533)

   a. When should each task be scheduled?
   b. When will the project be completed?

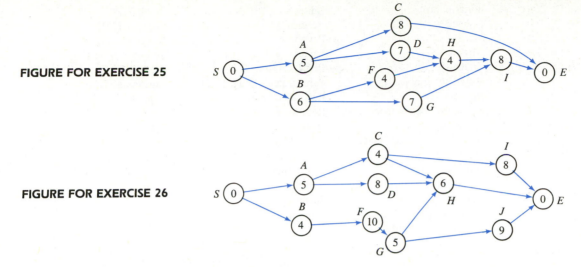

FIGURE FOR EXERCISE 25

FIGURE FOR EXERCISE 26

27. Seven students of a small university have reported to the clinic suffering from an infectious disease. Health officials believe that the disease was introduced by one person and spread to others. They immediately isolate the seven students and interview them trying to determine how the disease could have spread within the group. They determine the following:

Student	Could have given the disease to
Alvin	Ed
Bob	Calvin
Calvin	Fern
Darlene	Alvin
Ed	Darlene
Fern	Bob
Gary	Bob and Calvin

Use a digraph to determine whether some other student on campus has the disease but has not yet been identified.

28. Repeat Exercise 27 with eight students and the following information.

Student	Could have given the disease to
Antoine	Beatrice and Charlotte
Beatrice	Charlotte
Denise	Irene, Françoise, and George
Françoise	George
George	Irene
Hubert	Françoise
Irene	Hubert

7.8 Chapter Review

IMPORTANT SYMBOLS AND TERMS

$E(\mathbf{P}, \mathbf{Q})$ [7.2]
$d(u, v)$ [7.6]
$d(u, v) = \infty$ [7.6]
$S(v_i)$ [7.6]
Activity digraph [7.7]
Adjacency matrix [7.6]
Adjacent vertices [7.5]
Connected graph [7.5]
Critical path [7.7]
Cycle [7.5]
Digraph [7.6]
  activity [7.7]
Directed path [7.6]

Distance
  between vertices [7.6]
  sum of a vertex [7.7]
Dominance [7.6]
  one-stage [7.6]
  two-stage [7.6]
Dominant
  row [7.1]
  column [7.1]
Edge [7.5]
  directed [7.6]
Endpoint [7.5]
Expectation [7.2]

Expected value of a game [7.2]
Game
  fair [7.1], [7.2]
  matrix [7.1]
  strictly determined [7.1]
  two-person [7.1]
  value of [7.1], [7.2]
  zero-sum, [7.1]
Graph [7.5]
  simple [7.5]
  traceable [7.5]
Incident [7.5]

SUMMARY   An $m \times n$ matrix **A** represents a game if the game has the following features:

1. There are only two players.
2. On each play of the game, one player (usually called R) makes any one of m choices, while the other player (usually named C) makes any one of n choices.
3. If R makes choice R_i and C makes choice C_j, then C pays R the amount a_{ij}, where a_{ij} is the i-j entry of matrix **A**.

A *two-person zero-sum game* is a game involving only two players. After each play a payoff of some amount is made so that one player's win is the other player's loss. The matrix representing a game is called a *game matrix* or *payoff matrix*. If in a game matrix there is an entry which is both the smallest entry in its row and the largest entry in its column, then this entry is called a *saddle point*. A game matrix is said to be *strictly determined* if it has a saddle point. The *optimum strategies* for the players is for R to always choose a row containing a saddle point and for C to always choose a column containing a saddle point. A saddle point is called the *value* of the strictly determined game. A game is *fair* if its value is zero. If a strategy for a player is to consistently choose a certain row, or column, then such a strategy is called a *pure strategy*.

Let **A** be an $m \times n$ matrix. If each entry in the ith row of **A** is less than or equal to the corresponding entry in the kth row, then the ith row is said to be *recessive* and the kth row is said to be *dominant*. If each entry in the jth column of **A** is less than or equal to the corresponding entry in the kth column, then the kth column is said to be *recessive* and the jth column is said to be *dominant*. Optimal strategies will never require the selection of rows or columns that are recessive. Therefore such rows and columns may be deleted.

If a game is not strictly determined, then it is best for each player to mix his or her selection of choices using a probability distribution and a chance device to generate the distribution.

Let **A** be an $m \times n$ game matrix. A *strategy for R* is an m-dimensional probability row vector $\mathbf{P} = [\ p_1 \quad p_2 \quad \cdots \quad p_m\ ]$ and a *strategy for C* is an n-dimensional probability column vector $\mathbf{Q} = \begin{bmatrix} q_1 \\ q_2 \\ \vdots \\ q_n \end{bmatrix}$. If one of the entries in **P** (or **Q**) is 1 and the others are 0, the strategy is called a *pure strategy*. Otherwise, it is called a *mixed strategy*.

In general p_i is the probability that R will choose the ith row and q_j is the probability that C will choose the jth column.

If $\mathbf{A} = [a_{ij}]$ is an $m \times n$ game matrix, $\mathbf{P} = [\, p_1 \quad p_2 \quad \cdots \quad p_m \,]$ and

$$\mathbf{Q} = \begin{bmatrix} q_1 \\ q_2 \\ \vdots \\ q_n \end{bmatrix}$$ are strategies for R and C, respectively, then the *expectation* of R when R uses

strategy $\mathbf{P}$ and C uses strategy $\mathbf{Q}$ is $\mathbf{PAQ}$.

The fundamental theorem of game theory states that if $\mathbf{A}$ is an $m \times n$ game matrix, then there exist strategies $\mathbf{P}^*$ and $\mathbf{Q}^*$ for R and C, respectively, and a number v such that

$$\mathbf{P}^*\mathbf{AQ} \geq v \text{ for every strategy } \mathbf{Q} \text{ of C}$$

and

$$\mathbf{PAQ}^* \leq v \text{ for every strategy } \mathbf{P} \text{ of R}$$

The strategies $\mathbf{P}^*$ and $\mathbf{Q}^*$ described in the statement of the fundamental theorem are called *optimum (optimal) strategies* for R and C, respectively, and v is the value of the game. The game is said to be *fair* in the case $v = 0$. If $\mathbf{A}$ is a game matrix, then $\mathbf{P}^*\mathbf{AQ}^* = v$.

The game matrix $\mathbf{A} = \begin{bmatrix} a & b \\ c & d \end{bmatrix}$ is nonstrictly determined if, and only if,

(1) $\max\{b, c\} < \min\{a, d\}$

or

(2) $\max\{a, d\} < \min\{b, c\}$

Suppose now that the game matrix

$$\mathbf{A} = \begin{bmatrix} a & b \\ c & d \end{bmatrix}$$

is nonstrictly determined and let

$$D = (a + d) - (b + c)$$
$$p_1^* = \frac{d - c}{D}, \quad p_2^* = 1 - p_1^*$$
$$q_1^* = \frac{d - b}{D}, \quad q_2^* = 1 - q_1^*$$
$$v = \frac{ad - bc}{D}$$

Then $\mathbf{P}^* = [\, p_1^* \quad p_2^* \,]$ and $\mathbf{Q}^* = \begin{bmatrix} q_1^* \\ q_2^* \end{bmatrix}$ are optimum strategies for R and C, respectively, and v is the value of the game. For each $m \times n$ game matrix where either m, n, or both are greater than 2 there is a corresponding linear programming problem that may be solved using the methods described in Chapter 4. See the seven-step procedure in Section 7.4.

Let V be a finite set of points called *vertices* (plural of *vertex*). An *edge* is a curve joining two distinct vertices. Two vertices joined by an edge are called the *endpoints* of that edge. A *graph* consists of the set V of vertices together with a finite set E of edges. The end points of each edge in E must be in V. If two vertices u and v are endpoints of the same edge e, then we say that u and v are *adjacent vertices*. Further, we say that e is *incident with u and v*, and that u and v are *incident with e*. Two edges that have the same endpoints are said to be *parallel*. The *order* of the graph is the number of vertices in V. A vertex that is not the endpoint of any edge in E is said to be *isolated*. A graph with no parallel edges is called a *simple graph*. A loop (a curve that joins a point to itself) is not an edge. A *path* between the vertices v_1 and v_k is a sequence $v_1, e_1, v_2, e_2, \ldots, v_{k-1}, e_{k-1}, v_k$, where each v_i is a vertex, each e_i is an edge, and for each $i = 1, 2, 3, \ldots, k - 1$, e_i has endpoints v_i and v_{i+1}.

The *length* of the path is the number of edges in the path. A graph G is *connected* if for every pair of distinct vertices v_i and v_j in G, there is a path between v_i and v_j. A graph G is *traceable* if there is a path with no repeated edges that contains every vertex and every edge of G. Let G be a graph and V be a vertex of G. The *degree of v in G* is the number of edges of G incident with v. It is denoted $\deg_G v$, or simply $\deg v$. A vertex is said to be *even* if its degree is even, otherwise it is said to be *odd*. A graph G can be traced if, and only if, the following conditions are satisfied:

1. The graph is connected.
2. It has exactly two odd vertices, or no odd vertices.
3. If all vertices are even, then the tracing may begin at any vertex, but will end at the same vertex.
4. If it has exactly two odd vertices, then the tracing must begin at one of them and end at the other.

One of the most useful applications of graph theory stems from the study of networks. A familiar example of a network is the interstate highway system. *A network* is a connected graph each of whose edges has a real number associated with it. The *value* of a path in a network is the sum of the numbers associated with the edges that appear in the path. A path in a graph which starts and ends at the same vertex is called a *cycle*. A *tree* is a connected graph that does not contain any cycles. A *spanning tree* for a connected graph G is a subgraph of G, say T, such that T is a tree and the set of vertices of G and of T are equal. *A minimum spanning tree* in a network is a spanning tree for which the sum of the numbers assigned to the edges is less than or equal to the sum of the numbers assigned to the edges of any other spanning tree in the graph.

We select a minimum spanning tree as follows. Let G be a network. For convenience, call the number associated with an edge the cost of that edge.

Step 1. Start with a least-cost edge. (If there is more than one, select any of them.)
Step 2. Choose a least-cost edge which has not yet been selected, which shares a vertex with an edge already selected, and whose other vertex is not on any edge already selected.

Repeat Step 2 until all vertices are on edges that have been selected.

A *digraph* (abbreviation of *directed graph*) G is a finite set V of *vertices* and a set E of ordered pairs (u, v) where $u \in V$, $v \in V$, and $u \neq v$. Each element of E is called a *directed edge* of G. If (u, v) is a member of E, then u and v are called the *initial point* and the *terminal point* of the directed edge, respectively.

In the geometric representation of a digraph, a directed edge (A, B) is represented by an arrow pointing from A to B. A *directed path* in a digraph G is a sequence of directed edges $(v_1, v_2), (v_2, v_3), (v_3, v_4), \ldots, (v_{n-2}, v_{n-1}), (v_{n-1}, v_n)$. The number of directed edges in a directed path is called the *length* of the path. If in a digraph a directed path exists from u to v, then we say that v is *reachable* from u. If v is reachable from u, then the minimum of the lengths of all paths from u to v is called the *distance between u and v* and is denoted $d(u, v)$. If v is not reachable from u, we let $d(u, v) = \infty$ (read "infinity").

Let G be a digraph where E is the set of directed edges and $V = \{v_1, v_2, \ldots, v_n\}$ is the set of vertices. The *adjacency matrix* of G is the $n \times n$ matrix $\mathbf{A} = [a_{ij}]$ where

$$a_{ij} = \begin{cases} 1 & \text{if } (v_i, v_j) \in E \\ 0 & \text{if } (v_i, v_j) \notin E \end{cases}$$

for $i, j = 1, 2, \ldots, n$.

If the vertices of a digraph are labeled $v_1, v_2, \ldots, v_n$ and $\mathbf{A}$ is the corresponding adjacency matrix, then the i-j entry of the matrix $\mathbf{A}^k$ is the number of directed paths of length k from v_i to v_j.

Digraphs can also be used in the study of *dominance*. Given two distinct elements A and B of a set (for example A and B could represent sports teams or people) either A dominates B, or B dominates A, but not both. If X dominates Y, we say that X has a *one-stage dominance* over Y. If in turn Y has a one-stage dominance over Z, then we say that X has a *two-stage dominance* over Z. If we wish to rank the members of a set according to the total amount of one-stage dominance that each member has, a tie may result. In that case, we use the total amount of two-stage dominance to break the tie.

In a digraph where the vertices represent individuals within an organization, a directed edge from u to v may mean that u directly supervises v, or that u gives orders to v. If w is reachable from u, then u has some status over w, and the greater the distance $d(u, w)$, the greater the status. If a digraph G represents an organization where (u, v) is a directed edge of G if, and only if, u supervises v directly, then the *status of v_i* in G is $S(v_i) = \Sigma\, d(v_i, v_j)$ where the sum is over all v_j's that are reachable from v_i. If the organizational structure of a company is complex, use another matrix associated with a digraph.

If $V = \{v_1, v_2, \ldots, v_n\}$ is the set of vertices of a digraph G, the *distance matrix* $\mathbf{D}$ of G is the $n \times n$ matrix $[d_{ij}]$ where

$$d_{ij} = \begin{cases} 0 & \text{if } i = j \\ d(v_i, v_j) & \text{if } i \neq j \end{cases}$$

In digraphs with many vertices and edges it is difficult to find the lengths of the shortest paths between some of the vertices. In such cases use the seven-step procedure described in Section 7.7.

Let $\{v_1, v_2, \ldots, v_n\}$ be the vertices of a digraph G and let $\mathbf{D}$ be the distance matrix of G. The *distance sum of a vertex v_i* is the sum of the finite entries in the ith row of matrix $\mathbf{D}$. If the vertices of a digraph represent individuals in an organization and the directed edge (u, v) is in the digraph if, and only if, u directly supervises v, then the status of any v_i is equal to the distance sum of v_i.

Given a project composed of tasks, we can construct an *activity digraph* as follows. Each task is assigned a vertex and with each vertex we associate the number t if the task corresponding to that vertex requires t units of time to complete. We introduce a vertex S (start) and a vertex E (end) and assign the number 0 to each of these vertices. The directed edge (S, w) is in the digraph if the task w starts before the completion of any other task. Also, the directed edge (u, E) is in the digraph if no other task requires u to be completed before that task begins. Further, (u, v) is a directed edge of the digraph if, and only if, task v begins immediately after completion of task u.

Suppose that G is an activity digraph and u is a vertex in G. For each directed path from S to u, add the numbers assigned to the vertices along that path. Any of these paths with the largest sum is called a *critical path* for u. (There may be more than one critical path.) In scheduling tasks for a project use the following procedure.

Step 1. Draw the activity digraph for the project.
Step 2. For each task u find a critical path for u.

Step 3. Add all the numbers assigned to the vertices on that critical path except that assigned to u. This sum gives the number of units of time to allow before scheduling u.

In application techniques such as PERT (Program Evaluation and Review Technique), use activity digraphs.

When an infectious disease begins to spread in a group it is useful to isolate those who are already infected. A digraph may be used to determine whether someone in the group has the disease but has not yet been identified.

SAMPLE EXAM QUESTIONS

1. The matrix

$$\begin{bmatrix} 5 & 7 & -6 & 3 \\ -4 & 8 & 2 & 5 \\ 9 & -7 & 6 & 3 \end{bmatrix}$$

   represents a game where the payoffs are in dollars.

   a. How many choices does each player have?
   b. If player R chooses R_2 and player C chooses C_1, what is the payment?
   c. Same as **b** if player R chooses R_3 and player C chooses C_3.

2. Players R and C simultaneously select one of the numbers 4, 5, or 6. If they selected the same number, there is no payment. If they selected different numbers and the sum of these numbers is odd, player R wins; otherwise, player C wins. The winner receives p from the other player, where p is the sum of the two numbers selected. Give the matrix representation of this game.

3. Consider the following game matrix

$$A = \begin{bmatrix} 1 & 2 & 3 & 0 \\ 3 & -3 & 2 & -2 \\ 1 & 2 & -2 & -1 \end{bmatrix}$$

   The payoffs are in dollars. How should R and C play to achieve their goals?

4. Consider the following game matrix

$$A = \begin{bmatrix} 2 & 4 & 2 & 6 \\ 1 & 3 & 0 & -2 \\ 2 & 5 & 2 & 7 \end{bmatrix}$$

   a. Show that the game is strictly determined and find its value.
   b. Find optimum strategies for each player.

5. Consider the following game matrix

$$A = \begin{bmatrix} -3 & 2 & 4 \\ 2 & 0 & 3 \\ 5 & 4 & -3 \end{bmatrix}$$

   Show that this game is not strictly determined.

6. Consider the following game matrix

$$\mathbf{A} = \begin{bmatrix} 3 & -1 & -2 & 4 \\ 4 & 3 & 0 & 5 \\ 8 & 2 & 5 & 6 \end{bmatrix}$$

Delete any row or column which would never be chosen by an intelligent player.

7. Reduce the dimensions of the given game matrices if appropriate. For each matrix that is strictly determined (if it is not, say why), find: (a) all saddle points; (b) optimal strategies for R and C; and (c) the value of the game.

(a) $\begin{bmatrix} 2 & 3 & 0 \\ 4 & 1 & 5 \end{bmatrix}$

(b) $\begin{bmatrix} 1 & 3 & -1 \\ 2 & 3 & 2 \\ 1 & 2 & 3 \\ -1 & 0 & 2 \end{bmatrix}$

8. Each of two athletic clubs, R and C, is planning to build a new facility. Each club has two locations to choose from labeled I and II for R and III and IV for C. A consulting firm analyzed the situation as follows. If R builds in I and C in III, then R will get 62% of the total membership. If R builds in I and C in IV, then C will get 57% of the total membership. If R builds in II and C in III, then R will get 64% of the total membership. Finally, if R builds in II and C in IV, then R will get 72% of the total membership. Write this as a game matrix where payoffs are measured by the number of percentage points above 50%. Then solve the game.

9. A county has three small cities A, B, and C. It is known that 23% of the population shops in A, 33% shops in B, and 44% shops in C. Each of the two competing grocery chains X and Y is planning to build a store in one of the three cities. A consulting firm provided the following information. If both stores are built in the same city, X will get 63% of the total business. If the stores are built in different cities, each will get 80% of the business of the city it is in, and X will get 35% of the business from the other city. Translate this information into a game matrix where payoffs are measured by the number of percentage points above 50%. Is this game strictly determined?

10. Two football coaches from universities R and C know that a certain number of blue-chip players from regions I and II will enroll at one of the two universities. If R concentrates recruiting efforts in region I, then R will get 67% of the players if C concentrates in region I also, and 65% of the players if C concentrates in region II. On the other hand, if R concentrates in region II, then he will get 55% of the players if C concentrates in region I, but only 35% of the players if C concentrates in region II. What are the optimum strategies for both coaches?

11. During a football game between the Raccoons and the Cats, the Raccoons have the ball on their 40-yard line and it is first and ten. The Raccoon quarterback has four plays he can call, the Cats' defense has five possible defensive alignments it can use. The number of yards gained by the Raccoons for each possible play and defensive alignment is given in the following game matrix.

$$A = \begin{bmatrix} -4 & 5 & 11 & -4 & -8 \\ 8 & 6 & 5 & 3 & 4 \\ 0 & -3 & 0 & 1 & 5 \\ 9 & -8 & 1 & -1 & 0 \end{bmatrix}$$

What are the optimum strategies for both teams?

12. Consider the following game matrix

$$A = \begin{bmatrix} 20 & -80 \\ -100 & 160 \end{bmatrix}$$

where the payoffs are in dollars. Let

$$P = \begin{bmatrix} \frac{1}{4} & \frac{3}{4} \end{bmatrix} \text{ and } Q = \begin{bmatrix} \frac{2}{5} \\ \frac{3}{5} \end{bmatrix}$$

be strategies for R and C, respectively.

a. What is the probability that the payoff to R is $20?
b. What is the probability that the payoff to C is $80?
c. What is the probability that the payoff to C is $100?
d. What is the probability that the payoff to R is $160?
e. What is the expected value of the game for R?
f. Perform the matrix multiplication **PAQ** and compare your answer to that of part **e**.

13. Consider the following game matrix

$$A = \begin{bmatrix} 45 & -90 & 150 \\ 60 & -75 & 180 \end{bmatrix}$$

where the payoffs are in dollars. Let

$$P = \begin{bmatrix} \frac{1}{3} & \frac{2}{3} \end{bmatrix} \text{ and } Q = \begin{bmatrix} \frac{2}{5} \\ \frac{1}{5} \\ \frac{2}{5} \end{bmatrix}$$

be strategies for R and C, respectively. Find the expectation of R.

14. Complete the following matrix to create an example of a game matrix that is nonstrictly determined.

$$\begin{bmatrix} 5 & \\ & 7 \end{bmatrix}$$

15. Consider the nonstrictly determined game matrix

$$A = \begin{bmatrix} 6 & 3 \\ -2 & 8 \end{bmatrix}$$

Determine the optimal strategy for each player.

In Exercises 16–18, do the following:

a. Check if the game is strictly determined. If it is, solve it.
b. If the game is not strictly determined, reduce its dimensions, if possible.

c. Find optimal strategies P* and Q* for R and C, respectively (in terms of the original game matrix), and the value of the game.

16. $A = \begin{bmatrix} 1 & 2 & 4 & 6 \\ 2 & 5 & -2 & -1 \\ 0 & 4 & -3 & -4 \end{bmatrix}$

17. $A = \begin{bmatrix} -4 & -2 & -1 & 3 \\ 6 & 1 & 3 & 5 \\ -2 & 0 & -6 & 1 \end{bmatrix}$

18. $A = \begin{bmatrix} 4 & -1 & 2 & 3 \\ 3 & -3 & 4 & 3 \\ 5 & -5 & 2 & -7 \end{bmatrix}$

19. Ron and Carl each flip a coin. If both coins match, Carl pays Ron $2; otherwise, Ron pays Carl $2. Find the optimum strategy for each player and the value of the game.

20. Rose and Chuck play the following game. For each play, each draws a marble from an urn (the marble is replaced after each play). Rose's urn contains 2 blue, 5 red, and 3 white marbles; Chuck's urn contains 3 blue, 2 red, 5 white, and 4 green marbles. The payoff matrix is

	Blue	Red	White	Green
Blue	3	2	-2	1
Red	5	3	6	2
White	3	-5	2	5

For example if Rose draws a red marble and Chuck draws a blue marble, Chuck pays Rose $5. However, if Rose draws a white marble and Chuck draws a red marble, then Rose pays Chuck $5.

a. What is Rose's expected value of the game?

b. Suppose that Chuck changes to his optimal strategy and Rose continues to play by drawing marbles, what is Rose's new expected value of the game?

c. Suppose that Rose changes to her optimal strategy and Chuck continues to play by drawing marbles, what is Rose's new expected value of the game?

d. What is the value of the game if they both switch to their optimal strategies?

21. There are only two sporting goods stores in a town, R and C. They will between them get all of the town's business. Each month each must decide to use one, and only one, of the following forms of advertising: radio, television, mail, or newspaper. A consulting firm provided the following information, where each entry in the matrix indicates the number of percentage points above (below) 50% of the total business gained by store R.

	Radio	TV	Mail	Paper
Radio	5	8	-8	2
TV	6	6	-4	12
Mail	-6	4	2	-2
Paper	5	7	9	11

a. Find the optimum strategy for each store and the value of the game.

b. What is the expected value of the game for store R if R always chooses the newspaper and C uses its optimum strategy?

c. What is the expected value of the game for store R if R uses its optimum strategy and C always chooses television?

22. For the given game matrix do the following:

   a. Show that it is nonstrictly determined.
   b. If appropriate, delete recessive rows and columns to obtain a 2×2 game matrix.
   c. Write two corresponding linear programming problems.
   d. Solve these problems geometrically.
   e. What are the optimal strategies and the value of the original game matrix?

$$\mathbf{A} = \begin{bmatrix} 2 & 4 & 5 & 8 \\ 3 & 7 & -1 & 2 \\ 1 & 6 & -3 & -1 \end{bmatrix}$$

23. Solve the following using the linear programming method. Ron and Carol simultaneously show two or three fingers. If the total number of fingers shown is T, then Carol pays Ron $\$T$ if T is even; otherwise, Ron pays Carol $\$T$. Determine the optimal strategies for each player. What is the value of the game?

24. Solve the following problem using the linear programming method.

Rosemary chooses one of the colors blue, red, or white. Simultaneously, Charlies chooses one of the colors blue, red, white, or green. The payoff matrix is

	Blue	Red	White	Green
Blue	2	1	−3	0
Red	4	2	5	1
White	2	−6	1	4

For example, if Rosemary chooses red and Charlie chooses blue, he pays her $\$4$. However, if Rosemary chooses white and Charlie chooses red, then she pays him $\$6$.

   a. What are their optimal strategies?
   b. What is the value of the game?

25. There are only two sporting goods stores in town X, store R and store C. They will between them get all of the town's business. Each month each must decide to use one, and only one, of the following forms of advertising: radio, television, mail, or newspaper. A consulting firm provided the following information, where each entry in the matrix indicates the number of percentage points above (below) 50% of the total business gained by store R.

	Radio	TV	Mail	Paper
Radio	6	9	−7	3
TV	7	11	−3	13
Mail	−5	5	4	−1
Paper	6	8	10	12

Using the linear programming method find the optimum strategy for each store and the value of the game.

26. Solve the given game matrix using the simplex method.

$$\mathbf{A} = \begin{bmatrix} 0 & -1 & 3 \\ -1 & 1 & -2 \end{bmatrix}$$

27. Same as Exercise 26 with

$$A = \begin{bmatrix} 1 & 2 & -1 & 6 \\ -2 & 4 & -3 & 4 \\ -1 & 1 & 2 & -3 \end{bmatrix}$$

28. Two large competing grocery stores R and C are located in the same city. Each store chooses to use one and only one form of advertising at any one time—radio, television, or newspaper—however, they may change from week to week. A consulting firm provided the following information to the stores. If R uses the radio, then R will get 4 points if C uses the radio and 16 points if C uses the newspapers; however, C will get 8 points if it uses television. If R uses television, then R will get 28 points if C uses television and 4 points if C uses the newspapers; but C will get 8 points if it uses the radio. If R uses the newspapers, then R will get 16 points if C uses the radio and C will get 20 points if it uses television and 8 points if it uses the newspapers. Each point represents a percentage point over 50% of the total population that shops regularly at one of the two stores. Write this information as a game matrix and solve the game using linear programming.

29. Determine whether each graph is traceable. If it is not traceable, state why. If it is, trace it.

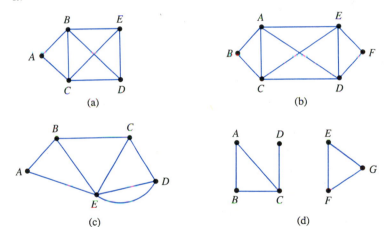

30. A river divides a small town into five distinct sections connected by bridges as shown in the following figure.

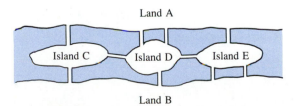

Is it possible for a jogger to start at one section of town, cross each bridge once, and only once, and return to her starting point? If it is possible, describe her path.

31. The following figure represents part of a road atlas of the United States. What is the value of the path Seattle-Ellengsburg-Spokane-Pendleton-Missoula?

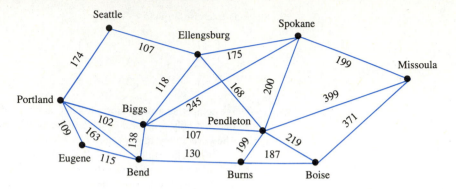

32. Determine whether the given graphs are trees. Justify your answers. If a graph is not a tree, find a spanning tree.

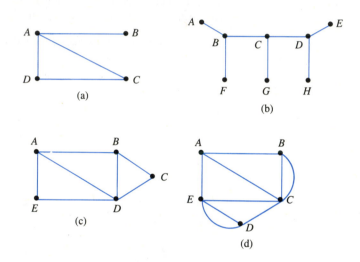

(a)

(b)

(c)

(d)

33. Find a minimum spanning tree for the following graphs.

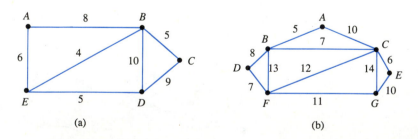

(a)

(b)

34. A university has provided a number of computers for its faculty. The administration has decided to connect all computers with a local area network. The cost of installing fiber optic cable between certain pairs of computers is shown on the diagram. The integer by each edge represents the cost in $1000 units. Determine how to install the network at a minimum cost and still have any two computers connected by a fiber optic path.

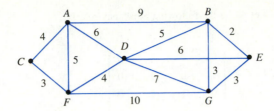

35. Draw the digraph G where $V = \{A, B, C, D, F\}$ and $E = \{(A, B)$ (B, D), (C, F), (C, D), (F, D), $(D, B)\}$ are the sets of vertices and directed edges, respectively.

36. The following is the organization chart of a publishing company where P, V_1, V_2, and V_3 represent the president, editorial vice president, production vice president, and sales vice president, respectively. A, B, and C represent editors, D_1, D_2, and D_3 represent production department heads; and X, Y, and Z represent sales managers. Represent this chart as a digraph.

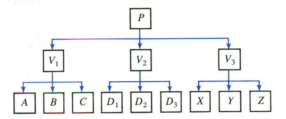

37. Consider the following digraph.

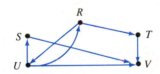

a. Find two directed paths of length 2 and a directed path of length 3 from R to V.
b. Is S reachable from R?
c. Is T reachable from S?
d. Find $d(R, V)$ and $d(T, U)$.

38. Obtain the adjacency matrix of the graph of Exercise 37.

39. Draw a digraph whose associated adjacency matrix is

$$
\begin{array}{c c c c c c}
 & A & B & C & D & E \\
A & 0 & 1 & 0 & 0 & 1 \\
B & 1 & 0 & 0 & 1 & 0 \\
C & 0 & 1 & 0 & 0 & 1 \\
D & 1 & 0 & 1 & 0 & 1 \\
E & 1 & 0 & 1 & 0 & 0
\end{array}
$$

40. Let **M** be the following adjacency matrix.

$$
\begin{array}{c c}
 & \begin{array}{cccc} A & B & C & D \end{array} \\
\begin{array}{c} A \\ B \\ C \\ D \end{array} &
\left[\begin{array}{cccc}
0 & 0 & 1 & 0 \\
1 & 0 & 1 & 0 \\
0 & 1 & 0 & 0 \\
1 & 1 & 0 & 0
\end{array}\right]
\end{array}
$$

Calculate $\mathbf{M}^2$ and $\mathbf{M}^3$ and interpret the results.

41. A lobbyist wants a committee to introduce a bill that favors his company. He wishes to obtain the support of the most influential member of the committee. From experience, he knows that André has influence over Charles and Debbie; Betty has influence over André and Frank; Charles has influence over Betty, Ed, and Frank; Debbie has influence over Charles, Ed, and Frank; Ed has influence over André and Frank; and Frank has influence over André. Determine the most influential member of the committee.

42. The structure of an organization is described by the following digraph. Find the status of each member.

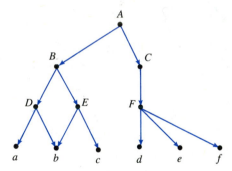

43. Find the distance matrix for the given digraph.

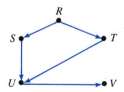

44. Find the distance matrix of the following digraph after finding the adjacency matrix and using matrix multiplication.

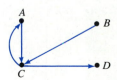

45. Find the distance sum of each of the vertices of the digraph of Exercise 44.

46. Determine the status of each individual in an organization, where the supervising responsibilities are tabulated, by finding the distance sum of each vertex of the corresponding digraph.

Individual	Directly supervises
Bob	Carl and Michelle
Carl	Tanya
Michelle	Tanya and John
Tanya	Nicholle and Teresa
John	Teresa, Laura, and Julie

47. A project consists of tasks P, Q, R, T, U, V, W, X, and Y. Its activity digraph is given.

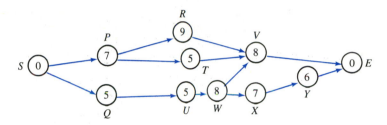

a. Find a critical path for V.
b. Find a critical path for the project.

48. A couple wants to have a house built. They consult subcontractors and find that each task will require the amount of time shown in the following table.

Task	Time (in weeks)
Purchase land	4
Clear land	3
Build foundation	4
Build structure	8
Plumbing	3
Electrical work	3
Finish interior	5
Finish exterior	4
Landscape	3

a. When should they schedule each task?
b. How long will it take to complete the house?

49. A project consists of nine tasks A, B, C, D, F, G, H, I, J. The activity digraph for this project is given on page 548, where the units of time are weeks.

a. When should each task be scheduled?
b. When will the project be completed?

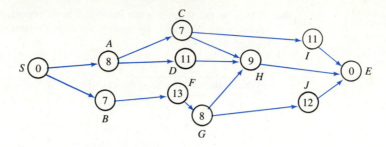

50. Seven students suffering from an infectious disease have reported to the clinic of a small university. Health officials believe that the disease was introduced by one person and spread to others. They immediately isolated the seven students and interviewed them trying to determine how the disease could have spread within the group. They determined the following.

Student	Could have given the disease to
Allen	Elaine
Betty	Colette
Colette	Frank
Doug	Allen
Elaine	Doug
Frank	Bob
Gill	Betty and Colette

Use a digraph to determine whether some other student on campus has the disease but has not yet been identified.

Mathematical Induction

The Principle of Mathematical Induction

In Chapter 5, Section 5.1, you learned that if $a_1, a_2, a_3, \ldots$ is an arithmetic progression, then the sum of the first n terms is given by

$$a_1 + a_2 + \cdots + a_n = \frac{(a_1 + a_n)n}{2}$$

and, if the first term and common ratio of a geometric progression are a and r, respectively, then

$$a + ar + ar^2 + \cdots + ar^{n-1} = \frac{a(1 - r^n)}{1 - r}$$

provided that $r \neq 1$. These two formulas are statements involving an integer n. It was proved that both are true for all positive integers. Often, however, there is a case where we wish to show that a statement $S(n)$ involving an integer is true for all positive integers. The principle of mathematical induction is the tool that we can use in such a case.

In the experimental sciences we might test a formula with a variable n for $n = 1$, $2, 3, \ldots, k$ and, if it is true for these tests, we would conclude that the formula is probably true for all positive integers n, especially if the value of k is large. This procedure is *not* valid in mathematics. Consider the statement "$n^2 - n + 41$ is a prime number." If we test this statement for $n = 1, 2, 3, \ldots, 40$, it is true in all 40 cases. We might be tempted to conclude it is true for all positive integers! However, let us try one more value and replace n by 41. We get "$41^2 - 41 + 41$ is a prime number" and that statement is false since $41^2 - 41 + 41 = 41^2$ which is not prime because it is divisible by 41.* An example that might be even more convincing is the equality

$$1 + 2 + \cdots + n = \frac{n(n + 1)}{2} + (n - 1)(n - 2)(n - 3) \ldots (n - 10{,}000)$$

---

*Numbers in the form $2^n - 1$, where n is a positive integer, are called *Mersenne numbers* after the French monk, Marin Mersenne (1588–1648). Mersenne primes are Mersenne numbers that are prime. It has been verified that $2^p - 1$ is prime for $p = 2, 3, 5, 7, 13, 17, 19, 31, 61, 89, 107,$ and 127 and is composite for all other $p < 257$. (See Exercises 28 and 29.)

which becomes a true statement if we replace n successively by the first 10,000 positive integers, but it is false thereafter.

The following example might help make the principle of mathematical induction intuitively plausible. Suppose a person wishes to climb a ladder with infinitely many rungs. If that person can step on the first rung of the ladder and after having stepped on any rung he or she can step on the next rung, then it will be possible for that person to climb the ladder. This idea is similar to the idea now formally stated as an axiom.

PRINCIPLE OF MATHEMATICAL INDUCTION: Let $S(n)$ be a proposition involving an integer. If

1. *The proposition is true for $n = 1$ and*
2. *The truth of the proposition for $n = k$ implies the truth of the proposition for the next value $n = k + 1$,*

then the proposition is true for all positive integers.*

Proofs by Induction

EXAMPLE 1   In Chapter 5, Section 5.1, we derived a formula for the sum of the first n terms of a geometric progression. Use the principle of mathematical induction to prove that formula.

Solution   The first term of the G.P. is a and the common ratio is r, where $r \neq 1$. We must prove that

$$a + ar + ar^2 + \cdots + ar^{n-1} = \frac{a(1 - r^n)}{1 - r}$$

is true for all positive integral values of n.

1. First, check the formula for $n = 1$. When we have only one term on the left side the result is

$$a = \frac{a(1 - r^1)}{1 - r}$$

which is true. The first step of the induction proof is complete.
2. Suppose now that the formula is true for $n = k$. That is, suppose that

$$a + ar + ar^2 + \cdots + ar^{k-1} = \frac{a(1 - r^k)}{1 - r} \tag{1}$$

is true.

---

*In some instances, a statement $S(n)$ may be true for all integers $n \geq n_1$ where n_1 is some integer. In such cases, induction starts at n_1. In other words, the first step is to verify that the statement is true for $n = n_1$, the smallest integral value of n for which the statement is to be proven true. The second step consists of verifying that if we assume that the statement is true for $n = k$ where $k \geq n_1$, then the statement is true for $n = k + 1$. (See, for example, Exercises 21 and 22.)

We must therefore show that the formula is true for $n = k + 1$. That is, we must prove that

$$a + ar + ar^2 + \cdots + ar^{(k+1)-1} = \frac{a(1 - r^{k+1})}{1 - r} \tag{2}$$

is also true.

Adding ar^k to both sides of the true equality (1), we obtain

$$a + ar + ar^2 + \cdots + ar^{k-1} + ar^k = \frac{a(1 - r^k)}{1 - r} + ar^k$$

$$= \frac{a(1 - r^k) + ar^k(1 - r)}{1 - r}$$

$$= \frac{a - ar^k + ar^k - ar^{k+1}}{1 - r}$$

$$= \frac{a(1 - r^{k+1})}{1 - r}$$

Therefore,

$$a + ar + ar^2 + \cdots + ar^{(k+1)-1} = \frac{a(1 - r^{k+1})}{1 - r}$$

is true. This completes the second step of the induction proof and we conclude that if $r \neq 1$, the formula

$$a + ar + ar^2 + \cdots + ar^{n-1} = \frac{a(1 - r^n)}{1 - r}$$

is true for all positive integral values of n.
Do Exercise 1.   ■

EXAMPLE 2   Prove that the sum of the squares of the first n positive integers is equal to $\frac{n(n + 1)(2n + 1)}{6}$.

Solution   1. First, verify that the statement is true for $n = 1$. We have

$$1^2 = \frac{1(1 + 1)(2 + 1)}{6}$$

which yields the true statement $1 = 1$. The first step of the induction proof is complete.

*2. Next, suppose that the statement is true for $n = k$. That is, suppose that

$$\sum_{i=1}^{k} i^2 = \frac{k(k + 1)(2k + 1)}{6} \tag{1}$$

is true.

---

*See Section 1.8 for the meaning and use of the symbol Σ.

We must prove that the statement is also true for $n = k + 1$. That is, we must show that

$$\sum_{i=1}^{k+1} i^2 = \frac{(k+1)[(k+1)+1][2(k+1)+1]}{6} \qquad (2)$$

is also true.

Add $(k+1)^2$ to both sides of the true equality (1) to get

$$\left(\sum_{i=1}^{k} i^2\right) + (k+1)^2 = \frac{k(k+1)(2k+1)}{6} + (k+1)^2$$

This can be written

$$\sum_{i=1}^{k+1} i^2 = \frac{k(k+1)(2k+1) + 6(k+1)^2}{6}$$

$$= \frac{(k+1)[k(2k+1) + 6(k+1)]}{6}$$

$$= \frac{(k+1)(2k^2 + 7k + 6)}{6}$$

$$= \frac{(k+1)(k+2)(2k+3)}{6}$$

$$= \frac{(k+1)[(k+1)+1][2(k+1)+1]}{6}$$

This completes the second step of the induction proof. We conclude that the formula

$$\sum_{i=1}^{n} i^2 = \frac{n(n+1)(2n+1)}{6}$$

is true for all positive integral values of n.

Do Exercise 3.   ■

EXAMPLE 3   Prove that for each positive integer n, the following is true:

$$1 + \frac{1}{2} + \frac{1}{4} + \frac{1}{8} + \cdots + \frac{1}{2^n} < 2$$

Solution   1. Let $n = 1$ to obtain the inequality

$$1 + \frac{1}{2} < 2$$

which is true since $1.5 < 2$. The first step of the induction proof is complete.

2. Assume that the statement is true for $n = k$. That is, assume that

$$1 + \frac{1}{2} + \frac{1}{4} + \frac{1}{8} + \cdots + \frac{1}{2^k} < 2 \qquad (1)$$

is true.

We must therefore show that the statement is true for $n = k + 1$ also. That is, that

$$1 + \frac{1}{2} + \frac{1}{4} + \frac{1}{8} + \cdots + \frac{1}{2^{k+1}} < 2 \tag{2}$$

is also true.

Multiplying both sides of the true inequality (1) by $\frac{1}{2}$, we get

$$\frac{1}{2} + \frac{1}{4} + \frac{1}{8} + \frac{1}{16} + \cdots + \frac{1}{2^{k+1}} < 1$$

Adding 1 to both sides of the inequality we obtain

$$1 + \frac{1}{2} + \frac{1}{4} + \frac{1}{8} + \frac{1}{16} + \cdots + \frac{1}{2^{k+1}} < 2$$

This completes the second step of the induction proof. Therefore the inequality

$$1 + \frac{1}{2} + \frac{1}{4} + \frac{1}{8} + \cdots + \frac{1}{2^n} < 2$$

is true for all positive integral values of n.
Do Exercise 11.   ■

The following two examples illustrate that both parts of the principle of mathematical induction must be verified to confirm that both are true.*

EXAMPLE 4   Consider the formula

$$5 + 10 + 15 + \cdots + 5n = \frac{5n(n + 1)}{2} + (n - 1)$$

Show that the formula is true for $n = 1$, but not true for another positive integer.

Solution   Replace n by 1 to have only one term on the left side of the formula. The equality

$$5 = \frac{5 \cdot 1 \cdot (1 + 1)}{2} + (1 - 1)$$

is true. However, if we replace n by 2, we have two terms on the left side.

$$5 + 10 = \frac{5 \cdot 2 \cdot (2 + 1)}{2} + (2 - 1)$$

This can be simplified to the **false** statement $15 = 16$.   ■

---

*The analogy of the infinite ladder should make it clear that the truth of *both* parts of the principle of mathematical induction must be verified if this principle is to be used. Knowing that a person can step on the first rung of a ladder does not guarantee that the whole ladder can be climbed unless we are sure that after a certain rung is stepped on, the next one can also be stepped on. However, knowing that a person can indeed step on any rung of a ladder, whenever he or she has stepped on the previous rung, also requires that the first rung be stepped on before we can actually be sure that the person can climb the whole ladder.

EXAMPLE 5   Show that the statement "$1 + 2 + 3 + \cdots + n = \frac{n(n + 1)}{2} + 1$ holds for all positive integers" is false although it satisfies the second condition of the principle of mathematical induction.

Solution   If we replace n by some positive integer, say 2, we get

$$1 + 2 = \frac{2(2 + 1)}{2} + 1$$

which yields the false statement $3 = 4$.

We now show that the second condition of the principle of mathematical induction is satisfied. Assume that the formula is true for $n = k$. Then the equality

$$1 + 2 + 3 + \cdots + k = \frac{k(k + 1)}{2} + 1$$

is true. Adding $(k + 1)$ to both sides of this equality, we obtain

$$1 + 2 + 3 + \cdots + k + (k + 1) = \frac{k(k + 1)}{2} + 1 + (k + 1)$$

$$= \frac{k(k + 1) + 2(k + 1)}{2} + 1$$

$$= \frac{(k + 1)(k + 2)}{2} + 1$$

$$= \frac{(k + 1)[(k + 1) + 1]}{2} + 1$$

Thus, we have shown that if we assume the truth of the formula for $n = k$, the truth of the formula for $n = k + 1$ will follow. Thus, the second condition of the principle of mathematical induction is satisfied, but the statement "$1 + 2 + 3 + \cdots + n = \frac{n(n + 1)}{2} + 1$ holds for all positive integers" is false.

Do Exercise 27.   ■

Exercise Set A

In Exercises 1–22, use mathematical induction to prove that the given equation, or statement, is true for all positive integral values of n.

1. $1 + 2 + 3 + \cdots + n = \frac{n(n + 1)}{2}$

2. $1^3 + 2^3 + 3^3 + \cdots + n^3 = \frac{n^2(n + 1)^2}{4}$

3. $1^4 + 2^4 + 3^4 + \cdots + n^4 = \frac{n(n + 1)(2n + 1)(3n^2 + 3n - 1)}{30}$

4. $1 \cdot 2 + 2 \cdot 3 + 3 \cdot 4 + \cdots + n(n + 1) = \frac{n(n + 1)(n + 2)}{3}$

5. $2 + 2^2 + 2^3 + \cdots + 2^n = 2^{n+1} - 2$

6. $1 + 3 + 5 + \cdots + (2n - 1) = n^2$

7. $\frac{1}{1 \cdot 2} + \frac{1}{2 \cdot 3} + \frac{1}{3 \cdot 4} + \cdots + \frac{1}{n(n + 1)} = \frac{n}{n + 1}$

8. $\frac{1}{1 \cdot 3} + \frac{1}{2 \cdot 4} + \frac{1}{3 \cdot 5} + \cdots + \frac{1}{n(n + 2)} = \frac{n(3n + 5)}{4(n + 1)(n + 2)}$

9. $\frac{5}{1 \cdot 2} \cdot \frac{1}{3} + \frac{7}{2 \cdot 3} \cdot \frac{1}{3^2} + \cdots + \frac{2n + 3}{n(n + 1)} \cdot \frac{1}{3^n} = 1 - \frac{1}{n + 1} \cdot \frac{1}{3^n}$

10. $1 + \dfrac{1}{3} + \dfrac{1}{9} + \dfrac{1}{27} + \cdots + \dfrac{1}{3^n} < \dfrac{3}{2}$

11. $1 + \dfrac{1}{4} + \dfrac{1}{16} + \dfrac{1}{64} + \cdots + \dfrac{1}{4^n} < \dfrac{4}{3}$

12. $2^{2n} - 1$ is a multiple of 3.

13. $2n^3 + 3n^2 + n$ is a multiple of 6.

14. $\displaystyle\sum_{i=1}^{n} (4i - 3) = n(2n - 1)$

15. $\displaystyle\sum_{i=1}^{n} (3i + 2) = \dfrac{n(3n + 7)}{2}$

16. $\displaystyle\sum_{i=1}^{n} i(i + 2) = \dfrac{n(n + 1)(2n + 7)}{6}$

17. $\displaystyle\sum_{i=1}^{n} (7i - 5) = \dfrac{n(7n - 3)}{2}$

18. $\displaystyle\sum_{i=1}^{n} (2i - 3)^2 = \dfrac{n(4n^2 - 12n + 11)}{3}$

19. $(a - b)$ is a factor of $a^n - b^n$. (*Hint:* $a^{k+1} - b^{k+1} = (a^{k+1} - ab^k) + (ab^k - b^{k+1})$.)

20. $(a + b)$ is a factor of $a^{2n-1} + b^{2n-1}$. (*Hint:* See Exercise 19.)

21. $2^n < n!$ for all integers $n > 3$. (*Hint:* Recall that $n! = n(n - 1)(n - 2) \cdot \ldots \cdot 2 \cdot 1$. Show that the statement is true for $n = 4$ and that the truth of the statement for $n = k$, where $k > 3$, implies the truth of the statement for $n = k + 1$.)

22. Prove that if $a \neq 0$ and $a > -1$, then

$$(1 + a)^n > 1 + na$$

is true for each integer $n > 1$.

23. Note that
$$1 = 1$$
$$1 - 4 = -(1 + 2)$$
$$1 - 4 + 9 = 1 + 2 + 3$$
$$1 - 4 + 9 - 16 = -(1 + 2 + 3 + 4)$$
$$1 - 4 + 9 - 16 + 25 = 1 + 2 + 3 + 4 + 5$$

Guess a general statement involving the variable n that the foregoing five equalities suggest. Then prove that your statement is true for all positive integral values of n.

24. Repeat Exercise 23 using the following six equalities:

$$1 = 1$$
$$1 - 3 = -2$$
$$1 - 3 + 5 = 3$$
$$1 - 3 + 5 - 7 = -4$$
$$1 - 3 + 5 - 7 + 9 = 5$$
$$1 - 3 + 5 - 7 + 9 - 11 = -6$$

25. Repeat Exercise 23 using the following five equalities:

$$1 - 1/2 = 1/2$$
$$(1 - 1/2)(1 - 1/3) = 1/3$$
$$(1 - 1/2)(1 - 1/3)(1 - 1/4) = 1/4$$
$$(1 - 1/2)(1 - 1/3)(1 - 1/4)(1 - 1/5) = 1/5$$
$$(1 - 1/2)(1 - 1/3)(1 - 1/4)(1 - 1/5)(1 - 1/6) = 1/6$$

26. Consider the statement

$$2 + 4 + 6 + \cdots + 2n = n^2 + n + 1$$

Prove that the truth of this equality for $n = k$ implies its truth for $n = k + 1$. There are, however, positive integers that are not solutions of the sentence. Does this contradict the induction principle? Are there any positive integers that are solutions of the given equality?

27. Repeat Exercise 26 with the statement

$$1 + 2 + 3 + \cdots + n = \dfrac{(2n + 1)^2}{8}$$

28. Show that $2^{11} - 1$ is not prime.

29. Show that $2^n - 1$ is not prime whenever $n = n_1 \cdot n_2$, where n_1 and n_2 are integers greater than 1. (*Hint:* Use the fact that $a - b$ is a factor of $a^k - b^k$, for each positive integer k, and $c^{pq} = (c^p)^q$.)

Answers to Odd-Numbered Exercises

If you wish to see detailed solutions of the odd-numbered exercises, you may want to obtain a copy of the Student Solutions Manual. For further help, a Study Guide is available where you will find suggested strategies to study the chapters, additional solved examples, detailed solutions to the even-numbered sample exam questions from the chapter summaries, and a complete set of 3×5 study cards. Your instructor may be able to order these supplements from your college bookstore.

Getting Started (Pages xvii–xxx)

1. $(x - y)(x^6 + x^5y + x^4y^2 + x^3y^3 + x^2y^4 + xy^5 + y^6)$ **3.** $(2x - 5y)(4x^2 + 10xy + 25y^2)$

5. $(2a - 3c)(16a^4 + 24a^3c + 36a^2c^2 + 54ac^3 + 81c^4)$ **7.** $(x + y)(x^2 - xy + y^2)$ **9.** $\left(\dfrac{x}{2} + 3y\right)\left(\dfrac{x^2}{4} - \dfrac{3}{2}xy + 9y^2\right)$

11. 1 6 15 20 15 6 1 **13.** $x^6 + 6x^5y + 15x^4y^2 + 20x^3y^3 + 15x^2y^4 + 6xy^5 + y^6$
15. $x^7 + 7x^6y + 21x^5y^2 + 35x^4y^3 + 35x^3y^4 + 21x^2y^5 + 7xy^6 + y^7$ **17.** $8x^3 + 36x^2y + 54xy^2 + 27y^3$
19. $81a^4 - 216a^3b + 216a^2b^2 - 96ab^3 + 16b^4$

21. $256x^8 + 512x^7y + 448x^6y^2 + 224x^5y^3 + 70x^4y^4 + 14x^3y^5 + \dfrac{7}{4}x^2y^6 + \dfrac{xy^7}{8} + \dfrac{y^8}{256}$

23. a. 12 **b.** $792a^7b^5$ **c.** $220a^9b^3$ **25. a.** 14 **b.** $364a^3b^{11}$ **c.** $2002a^5b^9$
27. a. 17 **b.** $2380a^4b^{13}$ **c.** $12{,}376a^6b^{11}$ **29. a.** 20 **b.** $15{,}504a^{15}b^5$ **c.** $1140a^{17}b^3$
31. a. 15 **b.** $5005a^6b^9$ **c.** $6435a^8b^7$ **33.** 2600 **35.** 1275

37. a. 9, 3 **b.** 36, 6 **c.** 100, 10 **d.** 225, 15 **e.** 441, 21 **f.** 784, 28, $\dfrac{n^2(n + 1)^2}{4}$ **39.** 5083

41. 7269.29 **43.** 25.832571 **45.** 30.36845238 **47.** 947 **49.** 3833 **51.** -4059 **53.** 803.3827
55. 989,052.3819 **57.** 41,260,314,020 **59.** $-21{,}000.93812$ **61.** 62.25143833 **63.** 74.96647626
65. $-1{,}592{,}648.342$ **67.** -5.23401763 **69. a.** \$11,102.59 **b.** \$11,351.89 **c.** \$11,483.61 **d.** \$11,574.23
71. \$871.11 **73.** \$37,460.61 **75.** \$300.00

Exercise Set 1.1 (Pages 7–8) Elementary Introduction to Sets

1. {1, 2, 3, 4, 5, 6} **3.** {1} **5.** {Wisconsin, Washington, West Virginia, Wyoming} **7.** {c, d, e}
9. {Johnson, Nixon, Ford, Carter, Reagan, Bush} **11.** {7} **13.** {6} **15.** {2} **17.** ∅ **19.** {1}
21. Lincoln, Jefferson, Truman **23.** 3 and 4 are the only members of the set. **25.** 5, 9, 233 **27.** 1, 2, 4
29. Harry Reasoner, Barbara Walters, Phil Donahue **31.** false **33.** true **35.** false
37. 4, ∅, {1, 2}, {1}, {2}
39. 32, ∅, {1}, {2}, {3}, {4}, {5}, {1, 2}, {1, 3}, {1, 4}, {1, 5}, {2, 3}, {2, 4}, {2, 5}, {3, 4}, {3, 5}, {4, 5}, {3, 4, 5}, {2, 4, 5}, {2, 3, 5}, {2, 3, 4}, {1, 4, 5}, {1, 3, 5}, {1, 3, 4}, {1, 2, 5}, {1, 2, 4}, {1, 2, 3}, {1, 2, 3, 4}, {1, 2, 3, 5}, {1, 2, 4, 5}, {1, 3, 4, 5}, {2, 3, 4, 5}, {1, 2, 3, 4, 5} **41.** 3 **43.** 15 **45.** $2^n - 1$

47. a. 1   **b.** 6   **c.** 15   **d.** 20   **e.** 15   **f.** 6   **g.** 1   **49.** {0, 7, 8, 9}   **51.** {0, 1, 2, 5, 6, 7, 9}
53. {1, 2, 3, 4, 5, 6}, {2, 5}   **55.** {2, 3, 4, 5, 8}, ∅   **57.** {0, 1, 3, 4, 6, 7, 8, 9}, {0, 1, 3, 4, 6, 7, 8, 9}
59. {0, 1, 2, 5, 6, 7, 8, 9}, {0, 1, 2, 5, 6, 7, 8, 9}

Exercise Set 1.2 (Pages 12–13) Venn Diagrams

1.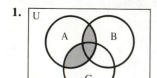

$A \cap (B \cup C)$

3.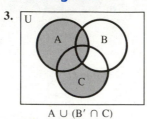

$A \cup (B' \cap C)$

5.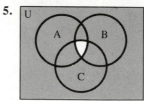

$(A \cap B)' \cup C'$

7.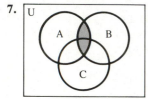

$A \cap B$

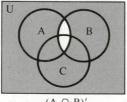

$(A \cap B)'$

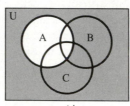

A'

9.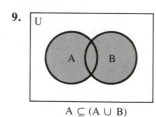

$A \subseteq (A \cup B)$

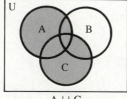

B'

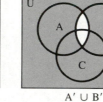

$A' \cup B'$

11.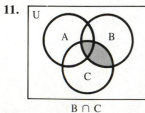

$B \cap C$

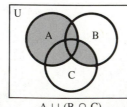

$A \cup (B \cap C)$

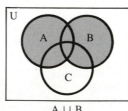

$A \cup B$

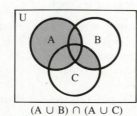

$A \cup C$ $(A \cup B) \cap (A \cup C)$

13. a. 6   **b.** 8   **c.** 281   **15. a.** 15   **b.** 55   **c.** 115   **d.** 100   **e.** 60   **17. a.** 13   **b.** 20   **c.** 134

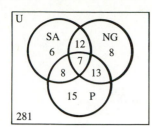

19. a. 50   **b.** 60   **c.** 50   **d.** 20

Exercise Set 1.3 (Pages 23–25) Equations in One Variable

1. all real numbers except -2, 3, and -5 **3.** all real numbers except -1, 1, and -6
5. all real numbers greater than or equal to 4   **7.** $\{-3\}$ **9.** $\{2\}$ **11.** $\{-2\}$ **13.** $\{2\}$ **15.** $\{1, 5\}$ **17.** $\{2, -9\}$
19. $\{5, -15\}$ **21.** $\{6\}$ **23.** $\{-5, 2\}$ **25.** $\left\{\frac{-28}{5}, 7\right\}$ **27.** $\left\{\frac{-20}{3}, 3\right\}$ **29.** $\left\{2, \frac{-7}{3}\right\}$ **31.** $\{8\}$ **33.** $\{-2\}$
35. $\{2\}$ **37.** \$10,000 at 12% and \$30,000 at 8%   **39.** \$44   **41.** 41   **43.** 62 hours, \$27
45. 80 pounds of the first kind and 40 pounds of the second   **47.** $3\frac{1}{3}$ quarts   **49.** 300   **51.** 12 gallons
53. 70 cm by 70 cm   **55.** 20 ft (along existing wall) by 25 ft

Exercise Set 1.4 (Pages 35–36) Inequalities

1. $(-5, +\infty)$ **3.** $(-\infty, 7]$ **5.** $\left(\frac{130}{21}, +\infty\right)$ **7.** $(-1, 3)$ **9.** no solutions

11. $[-10, 2]$ **13.** $[-2, 5]$ **15.** $(-\infty, 1.5)$ **17.** $[-1.5, 1.5]$

19. $(-\infty, -\sqrt{19} - 4) \cup (\sqrt{19} - 4, +\infty)$ **21.** $(-5, -4) \cup (1, +\infty)$ **23.** $(-\infty, 2) \cup (3, 5)$

25. $(-4, 2) \cup (4, +\infty)$ **27.** $(-\infty, -5)$ **29.** $(-\infty, -\sqrt{3} - 1) \cup (-1, \sqrt{3} - 1)$

31. $\left(\frac{-3 - \sqrt{33}}{2}, -3\right) \cup \left(\frac{-3 + \sqrt{33}}{2}, +\infty\right)$ **33.** $\left(-\sqrt{5}, \frac{3 - \sqrt{3}}{3}\right] \cup \left[\frac{3 + \sqrt{3}}{3}, \sqrt{5}\right)$

35. $(-\infty, -1 - 3\sqrt{2}) \cup (-5, -1 + 3\sqrt{2})$ **37.** $(-2/3, 2) \cup (3, 4)$ **39.** $(-\infty, -2) \cup (-1, +\infty)$

45. 25   **47.** $10 < p < 30$ **49.** $25 < p < 45$ **51.** at least 40,000 units but no more than 60,000 units
53. between 2 and 4 quarts   **55.** $(-\infty, -5) \cup (-3, 5)$

Exercise Set 1.5 (Pages 45–46) Function

1. a. $\{0, 1, 2, 3, 4\}$ **b.** 2, 6, 8   **3. a.** $\{0, 1, 2, 3, 4\}$ **b.** $f(0) = 0, f(2) = 8, f(4) = 64$
5. a. S **b.** 5, 5, 37   **7.** $\{x \mid x \neq 4\}$ **9.** $\{x \mid x \neq 4 \text{ or } -4\}$ **11.** The domain is $(4, \infty)$ **13.** $2, -.5, -\frac{2}{7}$
15. $\frac{1}{3}, \frac{1}{15}, \frac{1}{3}$ **17.** 1, 2, undefined
19.

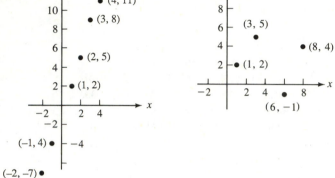

21.

23. $y = 4x + 18$ **25.** $y = 2$ **27.** $y = -x + 11$ **29.** $y = 9$ **31.** $x + 2y + 7 = 0$
33.

35.

37.

39. b.

41.

43. a. $p = -\frac{1}{3}x + 75$ **b.** $r = -\frac{1}{3}x^2 + 75x$
c. 81, $1,987

45.

26.25 feet

20 feet   20 feet

26.25 feet

47. $10 per credit hour   **49. a.** $p = -\frac{1}{2}x + 20$ **b.** $r = -\frac{1}{2}x^2 + 20x$ **c.** 16

Exercise Set 1.6 (Pages 57–58) The Exponential Function

1. 540   **3.** $\frac{81}{49}$ **5.** 972   **7.** 2304   **9.** $3^{-88/15}$ **11.**

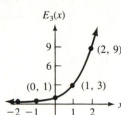

13.

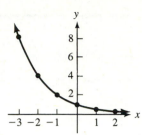

15.

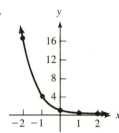

17. a.

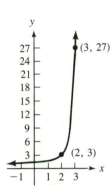

b.

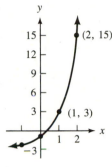

19. $10,199.44, $5,199.44   **21.** $9,861.69, $1,861.69   **23.** Bank C   **25.** 8.24%   **27.** 7.25%   **29.** $15,429.70
31. $96,006   **33.** $464,372   **35.** 4,661,528   **37.** .625 gm   **39.** .6931 gm
41. a. 49.63   **b.** 76.87   **c.** 100.02   **43. a.** $21,675   **b.** $15,660   **c.** $8,175
45. a. $6,782   **b.** $4,560   **c.** $3,066

Exercise Set 1.7 (Pages 64–65) The Logarithmic Function

1. a. 1.8060   **b.** 1.5562   **c.** 1.7323   **3. a.** .1204   **b.** .357825   **c.** −.0178428
5. a. .6309   **b.** 2.3776   **c.** 1.8927   **7. a.** .4055   **b.** −.2876   **c.** .5234
9. a. $3 + \ln 5$ **b.** $\frac{2}{3}$ **c.** 2.2958   **11.** {39}   **13.** $\left\{-\frac{20}{3}\right\}$ **15.** ∅   **17.** {6}   **19.** {−3}   **21.** {−6, 1}
23. {4}   **25.** ∅   **27.** −8.4603   **29.** 56 quarters   **31.** 221 months   **33.** 13.73 years   **35.** March of 1991
37. $6\frac{1}{2}$ weeks   **39.** In the year 1993   **41.** 701.4 years   **47.** 13.1 years

Exercise Set 1.8 (Pages 68–69) The Sigma Notation

1. 67   **3.** 55   **5.** 862   **7.** 0   **9.** 85   **11.** 244   **13.** 15   **15.** 1,136,275   **17.** 12,060
19. 207,917,325   **21.** 102,541,820   **23.** −1674   **25.** $\sum_{i=1}^{9} (2i - 1)$ **27.** $\sum_{i=1}^{7} (2 + 3i)$ **29.** $\sum_{i=1}^{8} i^2$ **31.** $\sum_{i=2}^{6} i^3$
33. $\sum_{i=2}^{11} (-1)^i$ **35.** $\sum_{i=1}^{9} i^2(-1)^{i+1}$

Chapter Review 1.9 (Pages 72–75) Sample Exam Questions

1. a. false   **b.** true   **c.** false   **d.** false   **e.** true   **f.** true   **g.** true   **h.** false   **i.** false   **3.** 31
5. not be paid   **7. a.** {5}   **b.** {2, −3}
9. a. $(4, \infty)$ **b.** $[5, \infty)$ **c.** $(-3, 7)$ **d.** $[-5, -1]$

e. $(-\infty, -3] \cup [7, \infty)$ **f.** $\varnothing$ **g.** $\left(\frac{7}{8}, \frac{17}{8}\right)$ **h.** $(-14, 1)$

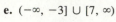

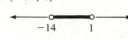

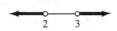

i. $(-3, -2] \cup [2, +\infty)$ **j.** $(-\infty, -3) \cup (-2, 1) \cup (5, +\infty)$ **k.** $(-\infty, 2) \cup (3, +\infty)$ **l.** $(2, +\infty)$

11. $[2, +\infty)$ **13.**

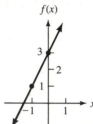

15. $h(x)$

17. a. $y = \frac{1}{2}x + \frac{5}{2}$
   b. $y - 5 = -\frac{2}{3}(x - 2)$
   c. $y = 2x + 5$
   d. $x = 1$ **e.** $y = 4$

19.

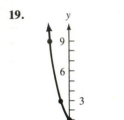

21. \$837.68, \$137.68
23. a. .7781   **b.** .9030   **c.** .2385   **d.** $-.8293$ **e.** 2.398   **f.** 3.3222
25. 23.2 years   **27. a.** 2870   **b.** 150   **c.** $\frac{100}{101}$ **29.** 20 feet
31. a. $p = -\frac{1}{2}q + 110$ **b.** $R(q) = -\frac{1}{2}q^2 + 110q$ **c.** 50 racquets, \$85, \$1,000
33. \$9,000 at 6% and \$3,000 at 8%
35. at least 4.48 but not more than 9.184

Exercise Set 2.1 (Pages 87–88) Systems of Two Linear Equations in Two Variables

1. $\{(1, -2)\}$ **3.** $\{(3, -1)\}$ **5.** $\varnothing$ **7.** $\{(4, -3)\}$ **9.** $\{(-2, -3)\}$ **11.** $\left\{\left(c, \frac{5-c}{3}\right) \mid c \text{ is a real number}\right\}$
13. $\{(7, -1)\}$ **15.** $\left\{\left(\frac{-66}{41}, \frac{277}{41}\right)\right\}$ **17.** $\left\{\left(\frac{1}{3}, \frac{1}{2}\right)\right\}$ **19.** $\{(-1, 2)\}$ **21.** $\{(-3, 5)\}$ **23.** $\{(5, -2)\}$
25. $\left\{\left(\frac{3c-2}{4}, c\right) \mid c \text{ is an arbitrary real number}\right\}$ **27.** $\left\{\left(\frac{1}{3}, \frac{1}{2}\right)\right\}$ **29.** $\{(2, 5)\}$ **31.** $\varnothing$ **33.** $\{(-1, 5)\}$ **35.** $\{(1, -2)\}$
37. $\{(3, -1)\}$ **39.** $\varnothing$ **41.** $\{(4, -3)\}$ **43.** $\{(-2, -3)\}$ **45.** $\left\{\left(c, \frac{5-c}{3}\right) \mid c \text{ is a real number}\right\}$ **47.** $\{(7, -1)\}$
49. $\left\{\left(\frac{-66}{41}, \frac{277}{41}\right)\right\}$ **51.** $\left\{\left(\frac{1}{3}, \frac{1}{2}\right)\right\}$ **53.** $\{(2, 5), (2, -5), (-2, 5), (-2, -5)\}$ **55.** $\left\{\left(-\frac{1}{2}, -\frac{1}{3}\right)\right\}$ **57.** $\left\{\left(\frac{1}{3}, -1\right)\right\}$
59. \$110 with the Dallas fan and \$130 with the L.A. fan   **61.** Mix 30 pounds of Brand A with 70 pounds of Brand B.
63. \$32,000 must be invested at 6.5% and \$18,000 at 9.5%.

Exercise Set 2.2 (Pages 97–99) Systems of Three Linear Equations in Three Variables

1. $\{(-1, 2, 1)\}$ **3.** $\{(-2, 2, 3)\}$ **5.** $\varnothing$ **7.** $\{(2, 1, 5)\}$ **9.** $\{(-3, 1, -1)\}$ **11.** $\{(2, 2, 2)\}$
13. $\{(-1, -3, 2)\}$ **15.** $\{(-1, 2, 1)\}$ **17.** $\{(-2, 2, 3)\}$ **19.** $\{(-3, 2, 0)\}$
21. $\left\{\left(\frac{5}{7} + \frac{9}{7}c, \frac{13}{7}c + \frac{8}{7}, c\right) \mid c \text{ is a real number}\right\}$ **23.** $\{(2, 2, 2)\}$ **25.** $\{(-1, -3, 2)\}$ **27.** $\{(-1, 2, 1)\}$
29. 300 pounds of kind A, 80 pounds of B, and 20 pounds of C
31. \$25,000 in bond A, \$15,000 in bond B and \$10,000 in bond C   **33.** X costs \$9.25 per unit, Y \$8.50 and Z \$13.75.
35. 5 houses of type A, 3 of type B, and 6 of type C   **37.** The wife's age is 40, the professor's 44, and the son's 16.
39. 0 lbs of A, $\frac{2600}{17}$ lbs of B, and $\frac{800}{17}$ lbs of C

Exercise Set 2.3 (Page 104) Graphical Solutions of Systems of Equations

1. $\{(1, 4), (-5, -8)\}$ **3.** $\{(2, 5), (-2, -3)\}$ **5.** $\{(-2, -1), (1, 5)\}$ **7.** $\{(1, 9), (-2, -6)\}$
9. $\{(-1, 5), (2, 2.11)\}$ **11.** $\{(1, 2), (3, 8)\}$ **13.** $\{(.5, .5)\}$ **15.** $\{(-.98, 1.40)\}$

Exercise Set 2.4 (Pages 107–109) Systems of Linear Inequalities in Two Variables

1.

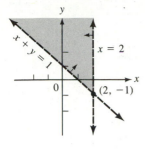

3.

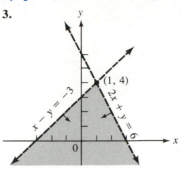

5.

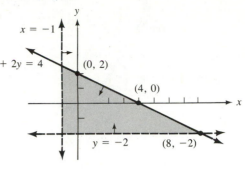

7.

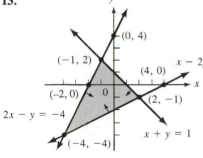

9.

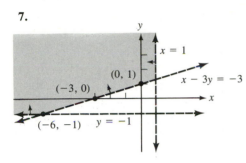

11.

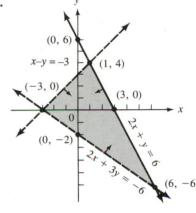

13.

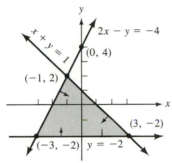

15.

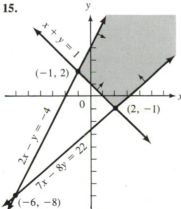

17.

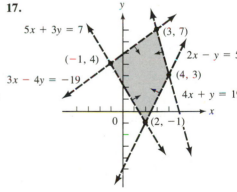

19.

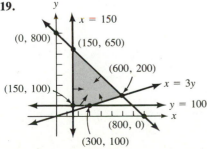

21.

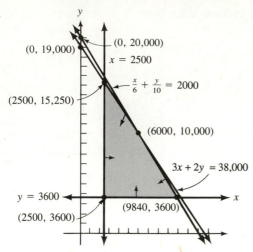

23.

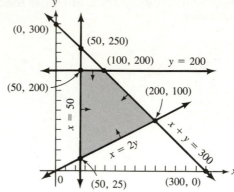

25.

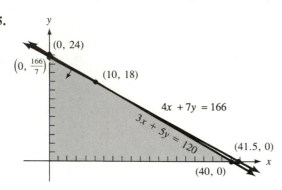

27.

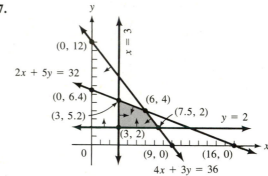

29.

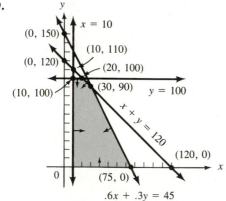

Exercise Set 2.5 (Pages 120–124) Applications to Business and Economics

1. a. \$25,000   **b.** 12,500   **3. a.** 300   **b.** $300 < q < 700$ **5. a.** 50   **b.** $50 < q < 80$
7. \$350, 33,000 units

9. a. $30, 1200 thousand units
   b. $36, 1452 thousand units

c.

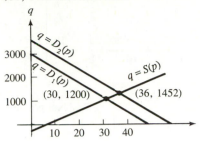

11. a. $45, 2775 thousand units
   b. $48, 3138 thousand units

c.

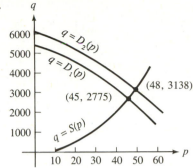

13. a. $21, 54 thousand units   **b.** $24, 39 thousand units

15. a. $D(p) = -32p + 1862$ **b.** $S(p) = 24p - 98$ **c.** $35, 742 units   **d.** $38, 646 units

17. a. $D(p) = -10p + 560$ **b.** $S(p) = 15p - 40$ **c.** $24, 320 units   **d.** $21, 350 units

19. a. $S(p) = 12.5p - 16$ **b.** 8 francs/liter, 84 million liters   **c.** 4.4 francs/liter

21. a. $S(p) = 30p - 69$ **b.** $14/unit, 351 thousand units   **c.** $5.70/unit

23. a. $3.20, $2.50 are the equilibrium prices   **b.** The equilibrium quantities are 7.1 thousand units and 16.1 thousand units.

25. The equilibrium prices are $4.50, $3.20, and $5.10 per unit. The equilibrium quantities are 551.9, 209.4, and 452.1 thousand units.

27. The equilibrium prices are $9.80, $10.20, and $8.75 per unit. The equilibrium quantities are 1386.1, 993.85, and 1465.4 thousand units.

Chapter Review 2.6 (Pages 127–129) Sample Exam Questions

1. $\{(2, -3)\}$ **3.** $\{(2, -5)\}$ **5.** $\{(2, -1)\}$ **7.** $\{(-3, 2)\}$ **9.** $\{(2, 3), (2, -3), (-2, 3), (-2, -3)\}$

11. $\left\{\left(-\frac{1}{3}, \frac{1}{2}\right)\right\}$ **13.** $\{(2, 21), (-1, 3)\}$ **15.** $\{(1, 12), (-2, 0)\}$ **17.** $\{(9, 10, -12)\}$ **19.** $\{(14, -15, -13)\}$

21. $\varnothing$ **23.** $\{(2, 3, 1)\}$ **25.** $\left\{\left(\frac{1}{2}, \frac{-3}{2}, 0\right)\right\}$ **27.** $\left\{\left(c, \frac{-17}{3} + \frac{7}{3}c, \frac{-5c}{3} + \frac{16}{3}\right) \mid c \text{ is a real number}\right\}$ **29.** $\{(3, 2, -1)\}$

31.

33.

35. The value of the first store is $750,000 and that of the second store is $550,000.

37. The adults' wage is $72 and the teens' is $32.

39. Kris can do the job in 16 days and Pete can do it in 20 days.

41.

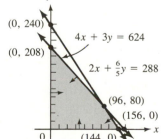

43. a. $4.95 per pound and $2.50 per pound   **b.** 202,500 pounds and 499,000 pounds

Exercise Set 3.1 (Pages 136–139) Matrix Notation and Matrix Arithmetic

1. $2 \times 3, 3 \times 2, 3 \times 4$ **3.** $\begin{bmatrix} 5 & 7 & 9 \\ 8 & 10 & 12 \end{bmatrix}$ **5.** $\begin{bmatrix} 3 & 5 & 7 \\ 6 & 8 & 10 \\ 11 & 13 & 15 \end{bmatrix}$ **7.** $\begin{bmatrix} 0 & 4 & 5 & 6 \\ 5 & 0 & 7 & 8 \\ 7 & 8 & 0 & 10 \\ 9 & 10 & 11 & 0 \end{bmatrix}$

9. $\begin{bmatrix} 1 & 5 & 6 & 7 \\ 7 & 4 & 9 & 10 \\ 10 & 11 & 9 & 13 \end{bmatrix}$ **11.** $\begin{bmatrix} -1 & 2 & 2 & 2 \\ 4 & 1 & 4 & 4 \\ 6 & 6 & -1 & 6 \\ 8 & 8 & 8 & 1 \end{bmatrix}$ **13.** $\begin{bmatrix} -9 & -12 & 6 \\ -6 & 12 & 0 \\ -21 & -18 & -36 \end{bmatrix}$ **15.** Not possible

17. $\begin{bmatrix} 30 & 19 \\ 15 & 9 \\ 17 & -2 \\ 3 & 10 \end{bmatrix}$ **19.** $\begin{bmatrix} 11 & 58 & -3 \\ 25 & -1 & -24 \\ 35 & 10 & 11 \end{bmatrix}$ **21.** $y = 3, x = 5, z = 7$ **23.** $x = 1, y = 2, z = 4, w = 3$

25. $x = -2, y = 3$ **27.** $y = 2, x = 1, z = 1$ **29.** $\begin{bmatrix} 1 & 2 & 9 & 9 \\ 8 & 2 & 5 & 6 \\ 6 & 5 & 2 & 11 \\ 0 & 6 & 6 & 7 \\ 7 & 7 & 2 & 0 \end{bmatrix}$

31. c. $\begin{bmatrix} 0 & 5 & 3 & 3 & 4 \\ 1 & 0 & 5 & 2 & 5 \\ 3 & 1 & 2 & 5 & 0 \\ 4 & 4 & 1 & 0 & 3 \\ 2 & 1 & 6 & 3 & 0 \end{bmatrix}$ **d.** Player 1   **33. a.** $\begin{bmatrix} 60 & 39 & 57 \\ 71 & 46 & 40 \end{bmatrix}$ **b.** $\begin{bmatrix} 74 & 48 & 70 \\ 87 & 56 & 49 \end{bmatrix}$

Exercise Set 3.2 (Pages 151–154) Matrix Multiplication

1. $\begin{bmatrix} -3 & 18 \\ 41 & 20 \\ 16 & 34 \end{bmatrix}$, undefined, undefined. $\begin{bmatrix} 1 & 5 & 12 \\ 15 & 39 & -18 \end{bmatrix}$, $\begin{bmatrix} 61 & 4 & 23 \\ 42 & 6 & 18 \\ 11 & 14 & 13 \end{bmatrix}$, $\begin{bmatrix} 11 & 5 \\ 39 & 69 \end{bmatrix}$

3. Undefined, undefined, undefined, undefined, undefined, $\begin{bmatrix} 51 & 24 & -1 \end{bmatrix}$

5. Undefined, $\begin{bmatrix} 70 & 1 & 62 & 5 \\ 15 & 7 & 10 & 20 \end{bmatrix}$, undefined, undefined, $\begin{bmatrix} 48 \\ 21 \end{bmatrix}$, undefined   **7.** $\begin{bmatrix} 1356 & 960 & 1360 \\ -72 & -192 & -152 \end{bmatrix}$

9. $\begin{bmatrix} 46 & 43 & 62 \\ 72 & 76 & 109 \\ 6 & 23 & 32 \end{bmatrix}$ **11.** $\begin{bmatrix} 1 & 15 \\ 47 & -1 \end{bmatrix}$, $\begin{bmatrix} -2 & 15 & -17 \\ 16 & 0 & 28 \\ 14 & 25 & 2 \end{bmatrix}$

13. $\begin{bmatrix} -24 & 52 \\ 24 & 5 \end{bmatrix}$, $\begin{bmatrix} 9 & 23 & 31 & -12 \\ 0 & -12 & -6 & -9 \\ 9 & 31 & 35 & -6 \\ 18 & 10 & 44 & -51 \end{bmatrix}$ **15.** $\begin{bmatrix} 1 & 0 \\ 0 & 1 \end{bmatrix}$, $\begin{bmatrix} 1 & 0 \\ 0 & 1 \end{bmatrix}$

17.
$$\begin{bmatrix} 30 & 24 & 18 \\ 84 & 69 & 54 \\ 138 & 114 & 90 \end{bmatrix}, \begin{bmatrix} 90 & 114 & 138 \\ 54 & 69 & 84 \\ 18 & 24 & 30 \end{bmatrix}$$
19.
$$\begin{bmatrix} 7 & 14 & -11 \\ 17 & 40 & 19 \\ 9 & 15 & -17 \end{bmatrix}, \begin{bmatrix} 0 & 14 & -4 \\ 12 & 46 & -3 \\ -12 & 3 & -16 \end{bmatrix}$$

21. $A = \begin{bmatrix} 1 & 2 \\ -1 & 3 \end{bmatrix}, B = \begin{bmatrix} -1 & 8 \\ -4 & 7 \end{bmatrix}$ **23.** $\begin{bmatrix} 1 & -2 \\ 3 & -6 \end{bmatrix}$ **27.** $\begin{bmatrix} 1 & 0 \\ 0 & 1 \end{bmatrix}, \begin{bmatrix} 1 & 0 \\ 0 & 1 \end{bmatrix}$

29. $\begin{bmatrix} 1 & 0 \\ 0 & 1 \end{bmatrix}$ and $\begin{bmatrix} 1 & 0 & 121 \\ 0 & 1 & 71 \\ 0 & 0 & 0 \end{bmatrix}$ **31.** $\begin{bmatrix} 5 & 2 \\ 7 & 3 \end{bmatrix}\begin{bmatrix} x \\ y \end{bmatrix} = \begin{bmatrix} -3 \\ 5 \end{bmatrix}$ **33.** $\begin{bmatrix} 11 & 6 \\ 9 & 5 \end{bmatrix}\begin{bmatrix} x \\ y \end{bmatrix} = \begin{bmatrix} 7 \\ -5 \end{bmatrix}$

35.
$$\begin{bmatrix} 1 & -4 & 0 & 1 \\ 5 & 0 & -5 & 4 \\ 0 & 1 & 3 & 1 \\ 2 & 2 & -1 & 2 \end{bmatrix}\begin{bmatrix} x \\ y \\ z \\ w \end{bmatrix} = \begin{bmatrix} -2 \\ 3 \\ 5 \\ -1 \end{bmatrix}$$

37. a. $\begin{bmatrix} 33,700 & 36,700 & 36,300 & 34,575 \\ 30,400 & 33,300 & 32,900 & 31,560 \\ 29,500 & 33,450 & 32,525 & 31,575 \end{bmatrix}$ The number in the ith row and jth column is the cost of project j using supplier i.

b. Use supplier C for project I, supplier B for project II, supplier C for project III, and supplier B for project IV.

39.

From To → ↓	Vancouver	Victoria
Quebec	37	19
Montreal	37	16
Toronto	34	19

41. $\begin{bmatrix} 83.65 & 96.25 & 91.25 & 90.35 & 95.95 & 42.4 \\ 87.4 & 94.70 & 90.2 & 87.6 & 92.2 & 35.7 \\ 85.45 & 95.40 & 90.85 & 88.7 & 94.55 & 38.55 \end{bmatrix}$ The number in the ith row and jth column is the final grade given to student j by professor i.

43. a. $AB = BA = \begin{bmatrix} 12 & 6 \\ 36 & 18 \end{bmatrix}$

b. det $A = 0$ and det $B = -24$. So, there does not exist a 2×2 matrix C such that A and B are powers of C, otherwise 0 and -24 would be powers of det$C$, and this is impossible.

Exercise Set 3.3 (Pages 171–175) Solution of Linear Systems by Row Reduction

1. $\begin{bmatrix} 2 & 3 & | & -5 \\ 3 & -5 & | & 17 \end{bmatrix}$ **3.** $\begin{bmatrix} -7 & 2 & 5 & | & 4 \\ 3 & 0 & 7 & | & 0 \\ 0 & -5 & 2 & | & 7 \end{bmatrix}$ **5.** $\begin{cases} 2x & = 3 \\ 5x + 3y & = -2 \end{cases}$ **7.** $\begin{cases} 2x & + 3z + 4w = 7 \\ -3x + 2y + 5z & = 2 \\ 2x - y & + 4w = -3 \end{cases}$

9. $\varnothing$ **11.** $\{(\frac{1}{2}(4 - 3c), c) \mid c \text{ is a real number}\}$ **13.** $\varnothing$ **15.** $\{(1, 1, 2)\}$ **17.** $\varnothing$ **19.** $\{(2, 2, 1)\}$
21. $\{(3 + 10a, -4 - 17a, a) \mid a \text{ is a real number}\}$ **23.** $\{(3, 4, -2)\}$ **25.** $\{(5, 2, 3)\}$
27. $w = 1, x = -1, y = 0, \text{ and } z = 2$ **29.** $\{(3, 0, 1, -2)\}$ **31.** $\varnothing$ **33.** $\varnothing$
35. $\{(2 - \frac{a}{2}, 1, a, 3) \mid a \text{ is a real number}\}$ **37.** $\{(2 - a, -2 + 4a, a, 3) \mid a \text{ is a real number}\}$
39. $\{(7 - 3a, a - 3, a, -3) \mid a \text{ is a real number}\}$ **41.** $\{(-\frac{a}{2}, 0, a, 0) \mid a \text{ is a real number}\}$
43. $\{(-a, 4a, a, 0) \mid a \text{ is a real number}\}$ **45.** $\{(-3a, a, a, 0) \mid a \text{ is a real number}\}$
47. a. $\{(45, 19)\}$ **b.** $\{(29, 13)\}$ **c.** $\{(57, 24)\}$ **49. a.** $\{(38, 7, 18)\}$ **b.** $\{(-10, -1, -5)\}$ **c.** $\{(71, 8, 32)\}$
51. a. $\{(76, 23, -64)\}$ **b.** $\{(-71, -21, 64)\}$ **c.** $\{(177, 55, -152)\}$
53. b. $\{(5, 3)\}, \{(6, 2)\}, \{(4, 5)\}, \{(6, 6)\}, \{(7, 4)\}$ **55. b.** $\{(60, 70, 90)\}, \{(50, 75, 100)\}, \text{ no feasible solution}$

57. 0, 27, 37.5, \$3,120   **59.** 56, 0, 28, \$4,200   **61.** 50, 60, 40, 30, \$10,700
63. The equilibrium price for commodity X is \$4.50, for Y is \$3.20, and for Z is \$5.10. The equilibrium quantity for X is 551,900, for Y is 209,400, and for Z is 452,100.
65. The equilibrium prices for commodities X, Y, and Z are \$9.80, \$10.20, and \$8.75, respectively. The corresponding equilibrium quantities are 1,386,100,   993,850, and 1,465,400.

Exercise Set 3.4 (Pages 186–187) Multiplicative Inverse

5. $\begin{bmatrix} 3 & -7 \\ -2 & 5 \end{bmatrix}$ **7.** $\begin{bmatrix} 3 & -13 \\ -2 & 9 \end{bmatrix}$ **9.** $\begin{bmatrix} -3 & 2 \\ \frac{5}{4} & -\frac{3}{4} \end{bmatrix}$ **11.** $\begin{bmatrix} 7 & 6 & -17 \\ 2 & 2 & -5 \\ -6 & -5 & 15 \end{bmatrix}$

13. $\begin{bmatrix} 5 & -5 & 4 \\ 0 & 3 & 1 \\ 2 & -1 & 2 \end{bmatrix}$ **15.** No inverse **17.** $\frac{1}{5}\begin{bmatrix} -1 & 1 & -1 \\ -9 & 19 & -24 \\ 7 & -12 & 17 \end{bmatrix}$ **19.** $\frac{1}{273}\begin{bmatrix} -18 & -2 & 63 \\ 6 & 31 & -21 \\ 45 & 5 & -21 \end{bmatrix}$

21. $\begin{bmatrix} 35 & -54 & 34 & -73 \\ 1 & -2 & 1 & -2 \\ 10 & -16 & 10 & -21 \\ -31 & 47 & -30 & 65 \end{bmatrix}$ **23.** $\begin{bmatrix} 128 & 35 & -28 & -73 \\ 4 & 1 & -1 & -2 \\ 37 & 10 & -8 & -21 \\ -114 & -31 & 25 & 65 \end{bmatrix}$ **25.** $\{(-29, 21)\}$

27. $\{(-74, 51)\}$ **29.** $\{(37, -14)\}$ **31.** $\{(-58, -17, 52)\}$ **33.** $\{(13, 16, 11)\}$ **35.** No solutions
37. $\left\{\left(\frac{-7}{5}, \frac{-128}{5}, \frac{94}{5}\right)\right\}$ **39.** $\left\{\left(\frac{-95}{273}, \frac{244}{273}, \frac{374}{273}\right)\right\}$ **41.** $\{(-604, -18, -175, 535)\}$ **43.** $\{(478, 16, 139, -425)\}$
45. At the central store 160 pounds of kind A are mixed with 40 pounds of kind B. At the northern store 180 pounds of kind A are mixed with 120 pounds of kind B. At the southern store 100 pounds of kind A are mixed with 150 pounds of kind B.
47. On weekdays sell 4000 seats in section A and 2000 in B. On Saturday sell 1000 seats in section A and 5000 in B. On Sunday sell 3000 seats in section A and 3000 in B.   **49.** $\{(10, 6)\}, \{(9, 7)\}, \{(11, 5)\}, \{(10, 7)\}, \{(9, 8)\}$
51. $\{(20, 15)\}, \{(22, 16)\}, \{(21, 20)\}, \{(25, 18)\}$ **53.** $\{(3, 5, 4)\}, \{(4, 6, 3)\}, \{(6, 3, 5)\}$

Exercise Set 3.5 (Pages 201–202) Determinants

1. -7 **3.** 2   **5.** -21 **7.** -35 **9.** 18   **11.** 10   **13.** -1 **15.** -3 **17.** $x = 3$
19. $x = 6$ or $x = -1$ **21.** $\{(-16, 23)\}$ **23.** $\{(2, 1)\}$ **25.** $\{(-5, 3)\}$ **27.** $\{(-3, 1)\}$ **29.** $\{(-3, 7)\}$
31. $\{(-1, 2, 1)\}$ **33.** $\{(-2, 2, 3)\}$ **35.** Cramer's rule cannot be used to solve the system.   **37.** $\{(2, 1, 5)\}$
39. $\{(-3, 1, -1)\}$ **41.** $\{(2, 2, 2)\}$ **43.** $\{(-1, -3, 2)\}$ **45.** $\{(-1, 2, 1)\}$
47. 300 pounds of A, 80 pounds of B, and 20 pounds of C
49. \$25,000 in Bond A, \$15,000 in Bond B, and \$10,000 in Bond C
51. X costs \$9.25 per unit, Y costs \$8.50 per unit, and Z costs \$13.75 per unit.
53. 5 type A houses, 3 type B houses, and 6 type C houses
55. The wife's age is 40, the husband's is 44, and the son's is 16.   **57.** $-x + y = 1$ **59.** $13x + 2y = 27$
61. $x - 9y = 22$ **63.** 75   **65.** -162 **67.** -6 **69.** 12 sq. units   **71.** 28 sq. units

Exercise Set 3.6 (Page 208) Inverse of a Square Matrix Using Adjoints

5. $\begin{bmatrix} 3 & -7 \\ -2 & 5 \end{bmatrix}$ **7.** $\begin{bmatrix} 3 & -13 \\ -2 & 9 \end{bmatrix}$ **9.** $\begin{bmatrix} -3 & 2 \\ \frac{5}{4} & \frac{-3}{4} \end{bmatrix}$ **11.** $\begin{bmatrix} 7 & 6 & -17 \\ 2 & 2 & -5 \\ -6 & -5 & 15 \end{bmatrix}$ **13.** $\begin{bmatrix} 5 & -5 & 4 \\ 0 & 3 & 1 \\ 2 & -1 & 2 \end{bmatrix}$

15. A^{-1} does not exist.   **17.** $.2\begin{bmatrix} -1 & 1 & -1 \\ -9 & 19 & -24 \\ 7 & -12 & 17 \end{bmatrix}$ **19.** $\frac{1}{273}\begin{bmatrix} -18 & -2 & 63 \\ 6 & 31 & -21 \\ 45 & 5 & -21 \end{bmatrix}$

21. $\begin{bmatrix} 35 & -54 & 34 & -73 \\ 1 & -2 & 1 & -2 \\ 10 & -16 & 10 & -21 \\ -31 & 47 & -30 & 65 \end{bmatrix}$ **23.** $\begin{bmatrix} 128 & 35 & -28 & -73 \\ 4 & 1 & -1 & -2 \\ 37 & 10 & -8 & -21 \\ -114 & -31 & 25 & 65 \end{bmatrix}$

Exercise Set 3.7 (Pages 220–222) Applications in Economics and Cryptography: Input-Output Analysis and Coded Messages

1. a. Industry I should produce 1296 units and Industry II should produce 1464 units.
b. Industry I should produce 1668 units and Industry II should produce 1862 units.
c. Industry I should produce 1492 units and Industry II should produce 1478 units.
3. a. Industry I should produce 5440 units, Industry II should produce 3990 units.
b. Industry I should produce 6120 units, Industry II should produce 4570 units.
c. Industry I should produce 7660 units, Industry II should produce 5885 units.
5. a. Industry I should produce 3380.8 units, Industry II should produce 4184 units.
b. Industry I should produce 3220 units, Industry II should produce 4100 units.
c. Industry I should produce 3512.8 units, Industry II should produce 4044 units.
7. a. Industries I, II, and III must produce approximately 3966, 4743, and 4076 units, respectively.
b. Industries I, II, and III must produce approximately 4155, 4019, and 4178 units, respectively.
c. Industries I, II, and III must produce approximately 4223, 4242, and 4752 units, respectively.
9. a. The units required of Industries I, II, and III are 4392, 5034, and 5198, respectively.
b. 5660, 6560, 6000 units are required of the respective industries.
c. 5006, 5795, and 5291 units are required of the respective industries.
11. a. Industries I and II must produce 2360 and 2640 millions of dollars per year, respectively.
b. The industries must produce 3120 and 3380 millions of dollars per year, respectively.
c. The industries must produce 3544 and 3796 millions of dollars per year, respectively.
13. a. Industries I and II must produce 2860 and 2235 millions of dollars per year, respectively.
b. The industries must produce 4600 and 3650 millions of dollars per year, respectively.
c. The industries must produce 4440 and 3290 millions of dollars per year, respectively.
15. a. Industries I and II must produce 3401.6 and 4068 millions of dollars per year, respectively.
b. The industries must produce 3514.4 and 4112 millions of dollars per year, respectively.
c. The industries must produce 3153.6 and 4028 millions of dollars per year, respectively.
17. a. Industries I, II, and III must produce approximately 5286, 4231, and 4981 million dollars per year, respectively.
b. The industries must produce approximately 5633, 5181, and 5659 million dollars per year, respectively.
19. a. In millions of dollars per year, Industries I, II, and III must produce approximately 4614, 4675, and 4235, respectively.
b. The industries must produce approximately 6811, 8107, and 7124 millions of dollars per year, respectively.
21. DRUGS KILL   **23.** STUDY EVERY DAY
25. 5, 61, 31, 3, 50, 26, 9, 70, 32, 27, 128, 73, 27, 136, 81, 16, 84, 41, 22, 99, 56, 5, 32, 15

Exercise Set 3.8 (Pages 229–231) Markov Chains

1. Not a state vector   **3.** Not a state vector   **5.** A state vector   **7.** Not a transition matrix
9. Not a transition matrix   **11.** A transition matrix   **13.** T is regular.   **15.** T is not regular.   **17.** T is regular.
19. [.3   .7]   **21.** [.45   .55]   **23.** [.25   .35   .4]   **25.** [.25   .75]   **27.** [.17   .83]
29. [.2   .6   .2]
31. a. The stock has approximately a 36.8% chance of increasing, a 29.6% chance of staying the same, and a 33.7% chance of decreasing on January 9.
b. The stock has approximately a 36.6% chance of increasing, 33.8% chance of decreasing, and 29.6% chance of staying the same on March 28.
c. The stock has approximately a 36.6% chance of increasing, a 29.6% chance of staying the same, and a 33.8% chance of decreasing on September 29.
33. a. 28.75%   **b.** 47.875%   **c.** 19.9%   **35. a.** 31.95%   **b.** 35.0925%   **c.** 38.41925%

Chapter Review 3.9 (Pages 234–239) Sample Exam Questions

1. $\begin{bmatrix} -1 & -4 \\ 1 & -2 \end{bmatrix}$ **3.** $\begin{bmatrix} 0 & 5 & 8 & 11 \\ -1 & 0 & 5 & 8 \\ -6 & -3 & 0 & 3 \\ -13 & -10 & -7 & 0 \end{bmatrix}$

5. a. $x = 3, y = 4, z = 7$ **b.** $x = 6, y = 2, z = 3$ **c.** $x = 3, y = -4, z = 1$

7. a. $\mathbf{A} = \begin{bmatrix} 1 & 2 \\ 3 & 4 \end{bmatrix}$ and $\mathbf{B} = \begin{bmatrix} 5 & 6 \\ 7 & 8 \end{bmatrix}$ **b.** $\mathbf{A} = \begin{bmatrix} 2 & 0 \\ 0 & 2 \end{bmatrix}$ and $\mathbf{B} = \begin{bmatrix} \frac{1}{2} & 0 \\ 0 & \frac{1}{2} \end{bmatrix}$

9. a. $\{(a - 3.2, 3.4 + .5a, a) \mid a \text{ is a real number}\}$ **b.** $\{(2, -1, 3)\}$
11. a. $\{(72, -29, 11)\}$ **b.** $\{(54, -23, 10)\}$ **c.** $\{(-25, 8, -1)\}$
13. a. $\{(37, -30)\}$ **b.** $\{(-24, -18, 39)\}$ **c.** $\{(43, -28, 23)\}$ **15. a.** $\{(2, -3)\}$ **b.** $\{(-1, 2, 4)\}$
17. a. Industries I, II, and III must produce approximately 4759, 4817, and 4891 million dollars per year, respectively.
 b. The industries must produce approximately 5817, 5627, and 5850 million dollars per year, respectively.
 c. The industries must produce approximately 8446, 8483, and 9504 million dollars per year, respectively.
19. GOOD LUCK
21. a. Not a transition matrix **b.** A transition matrix **c.** Not a transition matrix **d.** A transition matrix
23. [.48 .52] **25. a.** 10.25% **b.** 37.875% **c.** 47.83625%

27. a. $\begin{cases} 8x + 5y = 70 \\ 3x + 2y = 20 \end{cases} \begin{cases} 8x + 5y = 75 \\ 3x + 2y = 30 \end{cases} \begin{cases} 8x + 5y = 64 \\ 3x + 2y = 28 \end{cases} \begin{cases} 8x + 5y = 75 \\ 3x + 2y = 33 \end{cases} \begin{cases} 8x + 5y = 62 \\ 3x + 2y = 25 \end{cases}$

b. $\{(40, -50)\}, \{(0, 15)\}, \{(-12, 32)\}, \{(-15, 39)\},$ and $\{(-1, 14\}.$ However, only on Tuesday is the solution feasible.

Exercise Set 4.1 (Pages 248–250) The Geometric Approach

1. 59, 3 **3.** 68, −7 **5.** 22.65, 7.5 **7.** 250, 0 **9.** 47, −239 **11.** 427, 187
13. No maximum value, minimum value of 38 **15.** No maximum or minimum **17.** No maximum, no minimum
19. $z = 370x + 290y$, 600 acres of wheat and 200 acres of barley
21. $z = 350x + 275y$, 2500 units of the deluxe model and 15,250 units of the standard model
23. $z = 80x + 70y$, 50 shares of stock A and 25 shares of stock B **25.** 23 **27.** 6 baskets and 4 vases
29. 30 first-class rooms and 90 regular rooms

Exercise Set 4.2 (Page 255) The Three-Dimensional Coordinate System

1. a.

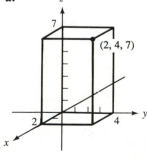

b.

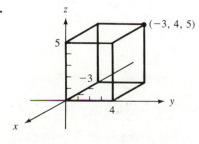

c.

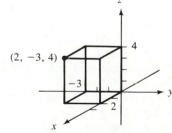

3. a. 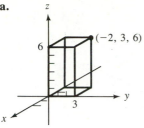 (−2, 3, 6)

b. 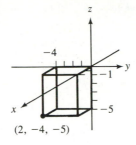 (2, −4, −5)

c. 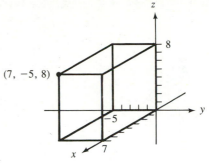 (7, −5, 8)

5.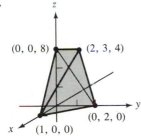

(0, 0, 20) (12, 0, 0) (0, 30, 0)

7.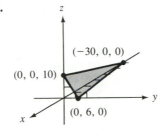

(−30, 0, 0) (0, 0, 10) (0, 6, 0)

9. a. (2, 3, 4)
b.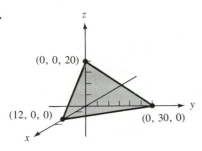

(0, 0, 8) (2, 3, 4) (1, 0, 0) (0, 2, 0)

11. a. (7, 5, 6)
b.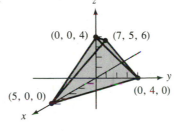

(0, 0, 4) (7, 5, 6) (5, 0, 0) (0, 4, 0)

13. a. (8, 7, 4)
b.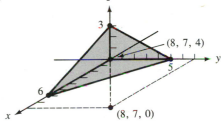

3 (8, 7, 4) 6 (8, 7, 0)

15. 15   **17.** 28   **19.** 10   **21.** 84

Exercise Set 4.3 (Pages 264–265) Introduction to the Simplex Method

1. $3x + y + s = 15$ **3.** $2x + 3y + s = 12$ **5.** $2x + y + s = 14$

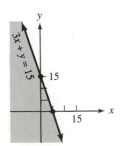

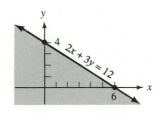

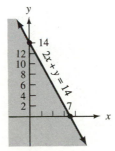

7. $4x + 5y + 4z + s = 40$

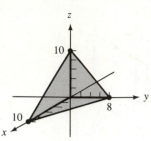

9. $4x + y + 4z + s = 16$

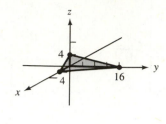

11. $2x + 2y + 3z + s = 12$

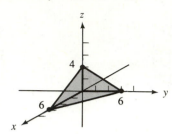

13. a.

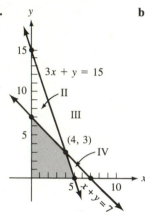

b. $x + y + r = 7$
$3x + y + s = 15$

15. a.

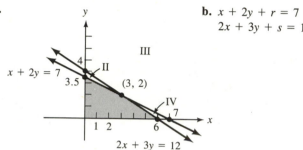

b. $x + 2y + r = 7$
$2x + 3y + s = 12$

17. a.

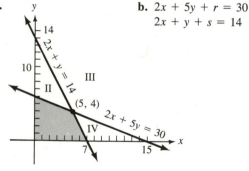

b. $2x + 5y + r = 30$
$2x + y + s = 14$

19. a.

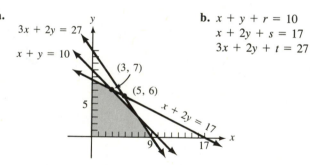

b. $x + y + r = 10$
$x + 2y + s = 17$
$3x + 2y + t = 27$

21. a.

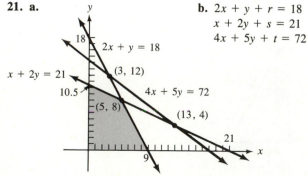

b. $2x + y + r = 18$
$x + 2y + s = 21$
$4x + 5y + t = 72$

23. a.

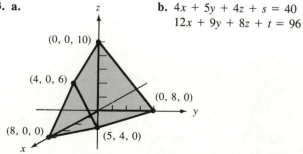

b. $4x + 5y + 4z + s = 40$
$12x + 9y + 8z + t = 96$

25. a.

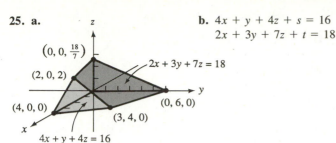

b. $4x + y + 4z + s = 16$
$2x + 3y + 7z + t = 18$

27. a.

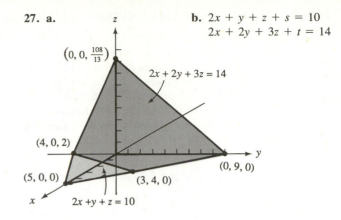

b. $2x + y + z + s = 10$
$2x + 2y + 3z + t = 14$

Exercise Set 4.4 (Page 278) Moving from Corner to Corner

1. $(0, -3, 0, 5, 0)$ **3.** $(13, 0, -2, 0, 17, 0)$ **5.** $(-2, 3, 0, 5, 0, 13)$ **7.** $(0, 0), (0, 7), (4, 3)$
9. $(0, 0), \left(0, \frac{7}{2}\right), (3, 2)$ **11.** $(0, 0), (7, 0), (5, 4)$ **13.** $(0, 0), (9, 0), (7, 3), (3, 7)$
15. $(0, 0), (9, 0), (5, 8), (0, 10.5)$ **17.** $(0, 0, 0), (8, 0, 0), (5, 4, 0), (4, 0, 6)$
19. $(0, 0, 0), (4, 0, 0), (2, 0, 2), (3, 4, 0)$ **21.** $(0, 0, 0), (5, 0, 0), (3, 4, 0), (4, 0, 2)$
23. $(0, 0, 0), (0, 0, 15), (3, 0, 12), (3, 0, 0), (6, 4, 5), (6, 4, 0)$

Exercise Set 4.5 (Pages 294–297) The Simplex Method

1. 82   **3.** No maximum   **5.** 400   **7.** 672   **9.** 684   **11.** 52   **13.** 158   **15.** 1925   **17.** 80   **19.** 195
21. 116   **23.** 125
25. The maximum profit is $15,000, which occurs when the complete 100 acres is planted in tomatoes.
27. The maximum profit is $50,000, which occurs when all 100 acres are planted in oranges and none in lemons.
29. The maximum profit is $7,000, which occurs when 100 units of product C is produced, 30 units of product B, and none of product A.
31. The maximum profit is $14,000, which occurs when 20 acres are planted in wheat, 20 in barley, and none in peanuts.
33. The maximum revenue is $11,600 and occurs for 800 pounds of variety I, 1800 pounds of variety II, and 400 pounds of variety III.
35. The maximum profit is $550, which occurs when 35 units of A, 0 units of B, and 5 units of C are produced.
37. The maximum profit is $9,000,000, which occurs when $(1 - t)40$ type A houses, $(1 - t)172 + 180t$ type B houses, and $(1 - t)22 + 30t$ type C houses are constructed for $0 \le t \le 1$.

Exercise Set 4.6 (Pages 313–315) Minimization and Duality

1. 1539   **3.** 5700   **5.** 1260   **7.** 98   **9.** 684   **11.** 270   **13.** 3850   **15.** 600   **17.** 1960   **19.** 80
21. 58
23. The minimum cost is $47,500, which occurs when $25,000 is spent on radio advertising and $22,500 is spent on TV advertising.
25. The minimum cost is $1,818,000. This occurs when plants A, B, and C are operated for 30, 12, and 18 days, respectively.
27. The minimum cholesterol is 88 units, which occurs for 0, 6, and 10 ounces of foods A, B, and C respectively.
29. The minimum cost is $774,000, which occurs when the plants at Arlington, Bellingham, and Mount Vernon operate for 0, 9, and 27 days respectively.
31. The minimum cost is $9,900, and this occurs when 450 units are sent from Site A to Distributor I, 150 units from Site A to Distributor II, no units from Site B to Distributor I, and 600 units from Site B to Distributor II.
33. The minimum cost is $9,625, which occurs when 20, 5, 0, and 25 units are shipped from Factory A to Distributors I and II and from Factory B to Distributors I and II respectively.

Exercise Set 4.7 (Pages 333–335) The Big *M* Method

1. 34 **3.** No maximum **5.** 90 **7.** 52 **9.** 39 **11.** 54 **13.** 45 **15.** 42

17. $P = 400x + 300y$, 5 acres of peanuts and 13 acres of corn

19. $P = 120x + 95y$, 27,000 units of model A and 2000 units of model B

21. $R = 1.05x + 1.15y$, 500 gallons of regular and 1100 gallons of premium

23. $C = 20x + 35y + 25z$, 175 shares of stock A, 75 shares of stock B, and 250 shares of stock C

25. $C = 3x + 3.75y$, 10 pounds of Type I and 10 pounds of Type II

27. $C = 8x_1 + 6x_2 + 7x_3 + 5x_4$, 100 motors are shipped from site A to plant I, 700 motors are shipped from site A to plant II, 500 motors are shipped from site B to plant I, and 0 motors are shipped from site B to plant II.

Chapter Review 4.8 (Pages 336–339) Sample Exam Questions

1. The maximum value of z is 25. The minimum value is 8. **3.** The maximum value of w is 52. The minimum value is 10.

5. The maximum value of z is 50.62$\overline{6}$. The minimum value is 17.

7. a.

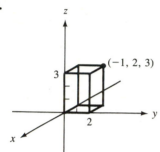

b.

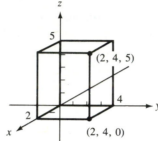

c.

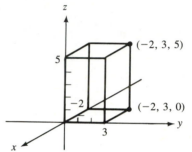

9. 10

11. a. (3, 4, 6)

b.

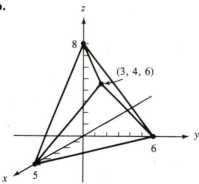

c. 44

13. a. (0, 0), (12, 0), (10, 4) **b.** (0, 0), (0, 15), (5, 10), (8, 5), (10, 0)

c. (0, 0, 0), (5, 0, 0), (0, 6, 0), (0, 10, 15)

15. No maximum

17. 44

19. 2430

21. 90

23. The maximum profit is $15,600, which occurs for 40 gadgets A and 100 gadgets B.

25. The Arlington plant operates for 18 hours and the Marysville plant operates for 3 hours.

27. 39

29. 51

Exercise Set 5.1 (Pages 348–349) Sequences, Arithmetic and Geometric Progressions

1. $2, \frac{2}{3}, \frac{2}{5}, \frac{2}{7}, \frac{2}{9}$ **3.** $-1, -\frac{2}{5}, -\frac{2}{10}, -\frac{2}{17}, -\frac{2}{26}$ **5.** $\frac{1}{2}, -\frac{2}{3}, \frac{3}{4}, -\frac{4}{5}, \frac{5}{6}$ **7.** $2, 1, -\frac{3}{4}, -\frac{7}{11}, -\frac{8}{14}$

9. $-4, -1, 2, 5, 8, 11, 14, 17, 20, 23$ **11.** $-5, -7, -9, -11, -13, -15, -17, -19$ **13.** $-3, 1, 5, 9, 13$

15. 10, 8, 6, 4, 2 **17.** $-8, -5, -2, 1, 4, 7, 10, 13$ **19.** 12,915 **21.** 16,020 **23.** 15,050 **25.** 19,665

27. $22,500, $20,000, $17,500, $15,000, $12,500, $10,000, $7500, $5000

29. $285,000, $270,750, $257,212.50, $244,351.875, $232,134.2813

31. $502.50, $505.01, $507.54, $510.08, $512.62, $515.19 **33.** 805.33, 810.70, 816.11, 821.55, 827.02, 832.54

35. $603.50, $607.02, $610.56, $614.12, $617.71, $621.31   **37.** 10, 50, 250, 1250   **39.** 256, 128, 64, 32
41. If $r = 3$, $a = 4$, $a_2 = 12$, $a_3 = 36$. If $r = -3$, $a = 4$, $a_2 = -12$, $a_3 = 36$ **43.** $-2, -6, -18$ **45.** 729, 243, 81
47. $\dfrac{5^{20} - 1}{2}$ **49.** 1023.9999   **51.** $2(3^{25} - 1)$ or $1 + 3^{27}$ **53.** $1 - 3^{27}$ **55.** $\dfrac{3^7 - 3^{-21}}{2}$ **57.** $3884.14
59. $3691.21   **61.** $10,489.68   **63.** 191.75 inches   **65.** $506,098.24   **67.** Second option
69. If the player loses 9 straight hands, he could not double his last bet of $256 because $572 is over the maximum limit.

Exercise Set 5.2 (Pages 362–364) Annuities

1. 1.576899264   **3.** 2.772469785   **5.** .591457366   **7.** 10.113249   **9.** 15.813679   **11.** 22.562977
13. 11.618933   **15.** 15.562251   **17.** $13,247.65, $1247.65   **19.** $115.25   **21.** $8504.07
23. $21,602.21, $3602.21   **25.** $8674.16, $1674.16   **27.** $17,968.23   **29.** $20,243.08   **31.** $8116.88
33. $42,620.68   **35.** $838.42   **37.** $2179.36   **39.** $260,383.36   **41.** $128,250   **45.** $18,550.45
47. $10,365.34   **49.** $146,182.02

Exercise Set 5.3 (Pages 373–374) Mortgages

1. a. $211.06   **b.** $126.06   **c.** $85.00   **d.** $7,916.73   **e.** $34,118
3. a. $207.58   **b.** $48.96   **c.** $158.62   **d.** $5456.25   **e.** $2,454.80
5. a. $990.98   **b.** $678.52   **c.** $312.45   **d.** $33,862.52   **e.** $147,835.20
7. a. $1214.89   **b.** $1267.28   **c.** $1321.73   **d.** $1417.07   **e.** $1597.47   **f.** $1995.14   **9.** $143,745.18
11. $128,947.89   **13. a.** $265.71   **b.** $43.57   **c.** $222.14   **d.** $2,754.78   **15.** 47 months   **17.** 66 months

Chapter Review 5.4 (Pages 376–378) Sample Exam Questions

1. a. $1, \frac{4}{7}, \frac{4}{10}, \frac{4}{13}, \frac{4}{16}$ **b.** $-\frac{2}{4}, \frac{4}{5}, -\frac{6}{6}, \frac{8}{7}, -\frac{10}{8}$ **c.** $6, \frac{11}{8}, \frac{18}{19}, \frac{27}{34}, \frac{38}{53}$ **3. a.** $-3, 0, 3, 6$ **b.** 59,100
5. a. $-8, 16, -32$ **b.** $-1,398,100$ **7.** $20,090.42   **9.** $3103.21   **11.** $9743.96   **13.** $8983.35
15. $8,504.08   **17.** $149,251.72   **19.** $12,706.66   **21.** $337.28   **23.** $157,231.11   **25.** 70 months

Exercise Set 6.1 (Pages 386–387) Introduction

1. a.

Relative frequency $\dfrac{s}{n}$
.03
.025
.02$\overline{6}$
.0275
.03
.028$\overline{3}$
.031428571
.03125
.03$\overline{2}$
.032
.031$\overline{8}$
.031$\overline{6}$
.03076923
.03142857
.031$\overline{3}$

b. .031$\overline{3}$ **3. a.**

Relative frequency $\dfrac{s}{n}$
.02
.0225
.018$\overline{3}$
.02125
.021
.021$\overline{6}$
.021428571
.02125
.0194
.0195

b. .0195

5. $\frac{5}{36}$ **7.** $\frac{5}{36}$ **9.** $\frac{1}{36}$
11. a. $\{(2, 6), (3, 5), (4, 4), (5, 3), 6, 2)\}$, $\frac{5}{36}$ **b.** $\{(3, 6), (4, 5), (5, 4), (6, 3)\}$, $\frac{1}{9}$
   c. $\{(2, 6), (3, 5), (4, 4), (5, 3), (6, 2), (3, 6), (4, 5), (5, 4), (6, 3)\}$, $\frac{1}{4}$ **d.** $\varnothing$, 0   **e.** $\frac{31}{36}$

13. a. {(T, T, T), (T, T, H), (T, H, T), (H, T, T), (H, H, T), (H, T, H), (T, H, H), (H, H, H)}
  b. i. {(T, T, H), (T, H, T), (H, T, T)}, $\frac{3}{8}$
    ii. {(T, T, H), (T, H, T), (H, T, T), (T, H, H), (H, T, H), (H, H, T), (H, H, H)}, $\frac{7}{8}$ **iii.** {(H, H, H)}, $\frac{1}{8}$
    iv. {(H, H, T), (H, T, H), (T, H, H), (H, H, H)}, $\frac{1}{2}$ **15.** $\frac{5}{6}$ **17.** 2 to 23   **19.** 7 to 13   **21.** .19

Exercise Set 6.2 (Pages 395–396) Counting Techniques

1. 5,040   **3.** 100,100   **5.** 720   **7.** 10   **9.** 1,716   **11.** 120   **13.** 210
15. (T, T, T, T), (T, T, T, H), (T, T, H, T), (T, T, H, H), (T, H, T, T),
(T, H, T, H), (T, H, H, T), (T, H, H, H), (H, T, T, T), (H, T, T, H),
(H, T, H, T), (H, T, H, H), (H, H, T, T), (H, H, T, H), (H, H, H, T),
(H, H, H, H)
17. (T, T, T, R), (T, T, T, G), (T, T, T, B), (T, T, H, R), (T, T, H, G),
(T, T, H, B), (T, H, T, R), (T, H, T, G), (T, H, T, B), (T, H, H, R),
(T, H, H, G), (T, H, H, B), (H, H, H, R), (H, H, H, G), (H, H, H, B),
(H, H, T, R), (H, H, T, G), (H, H, T, B), (H, T, H, R), (H, T, H, G),
(H, T, H, B), (H, T, T, R), (H, T, T, G), (H, T, T, B)

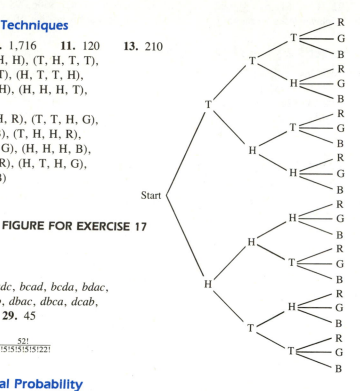

FIGURE FOR EXERCISE 17

21. 24, *abcd, abdc, acbd, acdb, adcb, adbc, bacd, badc, bcad, bcda, bdac,
bdca, cabd, cadb, cbad, cbda, cdab, cdba, dabc, dacb, dbac, dbca, dcab,
dcba*   **23.** 4,989,600   **25.** 35   **27.** 1,716   **29.** 45
31. 220   **33.** 479,001,600, 30.4 years   **35.** 28
37. a. 174,420   **b.** 277,134   **39.** 630   **41.** $\frac{52!}{5!5!5!5!5!5!22!}$

Exercise Set 6.3 (Pages 409–410) Conditional Probability

1. $\frac{5}{24}$ **3. a.** $\frac{2}{3}$ **b.** $\frac{18}{73}$ **c.** $\frac{9}{41}$ **5.** $\frac{25}{49}$ **7.** $\frac{64}{169}$ **9.** $\frac{2}{13}$ **11.** $\frac{4}{13}$ **13.** $\frac{2}{3}$ **15.** .116948178
17. .31500766   **19.** .706317891   **21. a.** $\frac{38}{59}$ **b.** $\frac{35}{73}$ **c.** $\frac{21}{59}$ **23.** $\frac{1}{7}$ **25.** .1738   **27.** .1876525
29. .8375275   **31.** .453122696   **33.** $\frac{4}{7}$ **35.** $\frac{6}{11}$ **37.** $\frac{9}{19}$ **39.** $\frac{4}{19}$

Exercise Set 6.4 (Pages 415–417) Bayes' Formula

1. .3151   **3.** .4611   **5. a.** .6528   **b.** .3472   **7. a.** .3478   **b.** .6522   **9.** .7303
11. a. .1763   **b.** .4232   **c.** .2494   **d.** .1511   **13. a.** .2349   **b.** .2761   **c.** .2698   **15.** .8978
17. .3245   **19. a.** .6   **b.** .4286   **c.** one

Exercise Set 6.5 (Pages 426–428) Expected Value

1. $53.80   **3.** Play game 1.   **5.**

x	$f(x)$
0	$\frac{1}{32}$
1	$\frac{5}{32}$
2	$\frac{10}{32}$
3	$\frac{10}{32}$
4	$\frac{5}{32}$
5	$\frac{1}{32}$

7. $f(k) = \left(\frac{5}{6}\right)^{k-1}\left(\frac{1}{6}\right)$, $k = 1, 2, 3, \ldots$
9. 2
11. 7
13. $278.75
15. $11.375
17. 80¢
19. 1.79

21. a. $E(X) = E(Y) = 6$ **b.** 6.6, 2.569   **c.** 10.6, 3.256

23. $5.8\overline{3}$, 2.415

d.

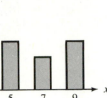

$f(x) = P_1(X = x)$

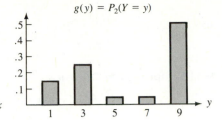

$g(y) = P_2(Y = y)$

Exercise Set 6.6 (Pages 435–437) The Binomial Distribution

1. $a^6 + 6a^5b + 15a^4b^2 + 20a^3b^3 + 15a^2b^4 + 6ab^5 + b^6$ **3.** $\frac{200}{16,807}$ **5.** $\frac{27}{1024}$ **7.** .119   **9.** .281   **11.** .0577

13. .144   **15. a.** .2936   **b.** .2668   **c.** .0243

17.

x	$f(x)$
0	.4096
1	.4096
2	.1536
3	.0256
4	.0016

$f(x) = P(X = x)$

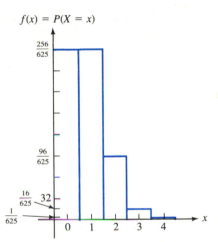

19.

x	$f(x)$
0	.004096
1	.036864
2	.13824
3	.27648
4	.31104
5	.186624
6	.046656

21. .671   **23. a.** .2013   **b.** .8791   **25.** .890625   **27.** .97398

29. a. $(.2)^{20}$ **b.** $(.2)^{18}(124.84)$ **c.** .00000018   **31.** 4   **33.** 4.8, 2.88, 1.697   **35.** 5.4, 3.78, 1.944   **37.** .9270

Exercise Set 6.7 (Pages 444–446) More on Markov Chains

1. The steady-state vector is $\begin{bmatrix} \frac{10}{17} & \frac{7}{17} \end{bmatrix}$.   **3.** The steady-state vector is [.8   .2].

5. The steady-state vector is $\begin{bmatrix} \frac{32}{95} & \frac{7}{95} & \frac{56}{95} \end{bmatrix}$.   **9.** $x = .1$, $y = .2$, and $z = .4$ **11.** $x = .1$, $y = .2$, and $z = .3$

13. a. [.26   .52   .22]   **b.** [.26   .41   .33]   **c.** [.251   .504   .245]

d. [.2237   .5255   .2508]

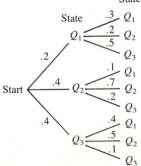

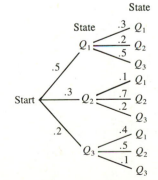

15. a. 20%   **b.** 14.3%   **17. a.** 55%, 29.5%, and 15.5%   **b.** 60%, 24.4%, and 15.6%   **19.** .6125

21. a. .836325   **b.** .1088122   **c.** .055   **23.** $\begin{bmatrix} \frac{10}{17} & \frac{7}{17} \\ \frac{10}{17} & \frac{7}{17} \end{bmatrix}$ **25.** $\begin{bmatrix} .8 & .2 \\ .8 & .2 \end{bmatrix}$ **27.** $\begin{bmatrix} \frac{32}{95} & \frac{7}{95} & \frac{56}{95} \\ \frac{32}{95} & \frac{7}{95} & \frac{56}{95} \\ \frac{32}{95} & \frac{7}{95} & \frac{56}{95} \end{bmatrix}$ **29.** Yes

Chapter Review 6.8 (Pages 448–451) Sample Exam Questions

1. a. 120   **b.** 132   **c.** $5.\overline{6}$ **d.** 210   **e.** 40,320   **f.** 220   **g.** 360,360   **h.** 9,189,180   **3.** 64,864,800
5. 1,058   **7.** 2,520
9. a. The sample space consists of all ordered 4-tuples (x, y, z, w) where x, y, z, and w are members of the set {H, T}.
  b. The probabilities are $\frac{1}{4}$, $\frac{15}{16}$, $\frac{11}{16}$, and $\frac{1}{16}$ **11.** 8 to 17   **13.** $\frac{2}{9}$ **15.** $\frac{9}{64}$ **17.** .8136   **19.** $\frac{25}{91}$ **21.** $\frac{4}{9}$
23. lose approximately 39¢   **25.** $241.20   **27.** .3456

29. a.

x	$f(x)$
0	.343
1	.441
2	.189
3	.027

  b. .9, .63, .794   **31.** .0341   **33.** $x = .6$, $y = .4$, and $z = .2$ **35. a.** 12.2%   **b.** $8\frac{1}{3}$%

Exercise Set 7.1 (Pages 461–464) Matrix Games

1. a. R has 3 choices and C has 4 choices.   **b.** R pays $3 to C   **c.** C pays $9 to R   **3.** $\begin{bmatrix} 0 & 7 & -8 \\ 7 & 0 & 9 \\ -8 & 9 & 0 \end{bmatrix}$

5. $\begin{bmatrix} -1 & 1 & 1 \\ 3 & -3 & 3 \\ 5 & 5 & -5 \end{bmatrix}$ **7.** R plays row 1 and C plays column 3

9. a. The 2s in row 1, columns 2 and 4, and row 2, columns 2 and 4 are saddle points. The value of the game is 2
  b. R should play either row 1 or row 2, C should play either column 2 or column 4
11. No entry is smallest in its row and largest in its column.
13. Delete first row 2, then columns 1 and 3, then row 2, to obtain $[\ -2\quad 1\quad 6\ ]$ **15. a.** -2 is the saddle point
  b. R should always play row 1 and C should always play column 1   **c.** The value of the game is -2

17. First delete rows 1 and 3, then column 1 to obtain $\begin{bmatrix} 1 & -2 & 0 \\ -3 & 4 & 1 \end{bmatrix}$. No entry is both row minimum and column maximum. The game is not strictly determined.

19. Delete row 1 and column 1 to obtain $\begin{bmatrix} 5 & -2 & -4 \\ 2 & 1 & 3 \\ -1 & 0 & 18 \end{bmatrix}$

a. 1 is the saddle point   **b.** R should play row 2 and C should play column 2   **c.** The value of the game is 1

21. First delete rows 1 and 5, then column 2 to obtain $\begin{bmatrix} 4 & 2 & 1 & 4 \\ 0 & 6 & -2 & -4 \\ -9 & -2 & 0 & 15 \end{bmatrix}$ **a.** 1 is the saddle point

  b. R should play row 1 and C should play column 3 (or row 2, column 4, of the original matrix)
  c. the value of the game is 1.

23. The game matrix is $\begin{bmatrix} 10 & -5 \\ 12 & 20 \end{bmatrix}$ 12 is the saddle point. W should build in location 2 and T in location 3. W will get 62% of the membership.

25. The game matrix is $\begin{bmatrix} 10 & -4.5 & -5.5 \\ .5 & 10 & -2.5 \\ 2 & 0 & 10 \end{bmatrix}$. No entry is both a row minimum and column maximum. The game is not

strictly determined.

27. The game matrix is $\begin{bmatrix} 0 & -1 & -1 \\ 1 & 0 & -1 \\ 1 & 1 & 0 \end{bmatrix}$. The saddle point is 0 in row 3 column 3. Both stores should build in Z and they will

split the business evenly.   **29.** Both coaches should concentrate in A and R will get 65% of the players.

Exercise Set 7.2 (Pages 476–478) Mixed Strategy Games

1. a. $\frac{6}{35}$ **b.** $\frac{8}{35}$ **c.** $\frac{9}{35}$ **d.** $\frac{12}{35}$ **e.** $11   **f.** 11   **3.** 7   **5.** $\begin{bmatrix} 3 & 8 \\ 10 & 5 \end{bmatrix}$ **7.** $\begin{bmatrix} 6 & -3 \\ 5 & 7 \end{bmatrix}$

9. The optimal strategy for **R** is $\begin{bmatrix} \frac{10}{13} & \frac{3}{13} \end{bmatrix}$ and that for **Q** is $\begin{bmatrix} \frac{5}{13} \\ \frac{8}{13} \end{bmatrix}$.

11. a. The game is nonstrictly determined.   **b.** Deleting recessive rows and columns we obtain $\begin{bmatrix} 2 & 5 \\ 3 & -1 \end{bmatrix}$.

c. In terms of the original matrix the optimal strategy for **R** is $\begin{bmatrix} \frac{4}{7} & \frac{3}{7} & 0 \end{bmatrix}$ and for **C** it is $\begin{bmatrix} \frac{6}{7} \\ 0 \\ \frac{1}{7} \\ 0 \end{bmatrix}$. The value of the game is $\frac{17}{7}$.

13. The game is strictly determined. Its value is -2. The optimal strategies are for **R** to play row 1 and for C to play column 2.

15. b. delete recessive rows and columns to obtain $\begin{bmatrix} 1 & 7 \\ 9 & 5 \end{bmatrix}$. The optimal strategies (in terms of the original matrix) are

$\begin{bmatrix} 0 & \frac{2}{5} & 0 & \frac{3}{5} \end{bmatrix}$ for **R** and $\begin{bmatrix} \frac{1}{5} \\ 0 \\ \frac{4}{5} \\ 0 \end{bmatrix}$ for **C**.

17. b. delete recessive rows and columns to obtain $\begin{bmatrix} -1 & 5 \\ 3 & 2 \end{bmatrix}$

c. $p^* = \begin{bmatrix} \frac{1}{7} & 0 & \frac{6}{7} & 0 \end{bmatrix}$ and $Q^* = \begin{bmatrix} 0 \\ \frac{3}{7} \\ 0 \\ \frac{4}{7} \\ 0 \end{bmatrix}$. The value of the game is $\frac{17}{7}$.

19. The value of the game is $\frac{-1}{12}$. The optimal strategies are $\begin{bmatrix} \frac{7}{12} & \frac{5}{12} \end{bmatrix}$ for Ron and $\begin{bmatrix} \frac{7}{12} \\ \frac{5}{12} \end{bmatrix}$ for Carol.

21. a. $\$\frac{16}{9}$ **b.** $\$\frac{93}{99}$ **c.** $\$\frac{30}{11}$ **d.** $\$\frac{14}{11}$

23. a. $P^* = \begin{bmatrix} 0 & \frac{2}{7} & 0 & \frac{5}{7} \end{bmatrix}$ and $Q^* = \begin{bmatrix} \frac{13}{14} \\ 0 \\ \frac{1}{14} \\ 0 \end{bmatrix}$. The value of the game is $\frac{23}{7}$.   **b.** $\frac{23}{7}$ **c.** $\frac{41}{7}$

25. The game matrix is $\begin{bmatrix} 15 & -18 \\ 10 & 13 \end{bmatrix}$. The optimum strategies are $P^* = \begin{bmatrix} \frac{1}{12} & \frac{11}{12} \end{bmatrix}$ and $Q^* = \begin{bmatrix} \frac{31}{36} \\ \frac{5}{36} \end{bmatrix}$. The value of the game is $\frac{125}{12}$.

Exercise Set 7.3 (Pages 486–487) Solution of a 2 × 2 Game Matrix Using Linear Programming

1. a. min $\{5, 7\}$ > max $\{-3, 2\}$. So the game is nonstrictly determined.   **c.** Minimize $z = x_1 + x_2$ subject to $9x_1 + x_2 \geq 1$; $6x_1 + 11x_2 \geq 1$; $x_1 \geq 0$; $x_2 \geq 0$ and maximize $Z = y_1 + y_2$ subject to $9y_1 + 6y_2 \leq 1$; $y_1 + 11y_2 \leq 1$; $y_1 \geq 0$; $y_2 \geq 0$.

d.

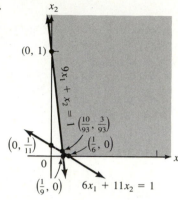

Z attains a minimum $\frac{13}{19}$
where $x_1 = \frac{10}{93}$, $x_2 = \frac{3}{93}$

Z attains a maximum $\frac{13}{19}$
where $y_1 = \frac{5}{93}$, $y_2 = \frac{8}{93}$

e. $\mathbf{P}^* = \begin{bmatrix} \frac{10}{13} & \frac{3}{13} \end{bmatrix}$ $\mathbf{Q}^* = \begin{bmatrix} \frac{5}{13} \\ \frac{8}{13} \end{bmatrix}$. The value of the original game is $\frac{41}{13}$.

3. b. Delete row 3, then columns 2 and 4 to obtain $\begin{bmatrix} 2 & 5 \\ 3 & -1 \end{bmatrix}$. Let $k = 2$, to obtain $B = \begin{bmatrix} 4 & 7 \\ 5 & 1 \end{bmatrix}$ **c.** Minimize $z = x_1 + x_2$ subject to $4x_1 + 5x_2 \geq 1$; $7x_1 + x_2 \geq 1$; $x_1 \geq 0$; $x_2 \geq 0$ and maximize $Z = y_1 + y_2$ subject to $4y_1 + 7y_2 \leq 1$; $5y_1 + y_2 \leq 1$; $y_1 \geq 0$; $y_2 \geq 0$.

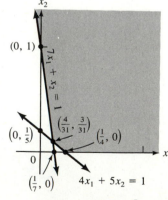

Z attains a minimum $\frac{7}{31}$
where $x_1 = \frac{4}{31}$, $x_2 = \frac{3}{31}$

Z attains a maximum $\frac{7}{31}$
where $y_1 = \frac{6}{31}$, $y_2 = \frac{1}{31}$

e. $\mathbf{P}^* = \begin{bmatrix} \frac{4}{7} & \frac{3}{7} \end{bmatrix}$, $\mathbf{Q}^* = \begin{bmatrix} \frac{6}{7} \\ \frac{1}{7} \end{bmatrix}$, the value of the original game is $\frac{17}{7}$.

5. b. Delete rows 1 and 3, columns 1 and 4, and obtain $\begin{bmatrix} 1 & 7 \\ 9 & 5 \end{bmatrix}$.   **c.** Minimize $z = x_1 + x_2$ subject to $x_1 + 9x_2 \geq 1$;

$7x_1 + 5x_2 \geq 1$; $x_1 \geq 0$; $x_2 \geq 0$ and maximize $Z = y_1 + y_2$ subject to $y_1 + 7y_2 \leq 1$; $9y_1 + 5y_2 \leq 1$; $y_1 \geq 0$; $y_2 \geq 0$.

e. $\mathbf{P^*} = \begin{bmatrix} \frac{2}{5} & \frac{3}{5} \end{bmatrix}$ $\mathbf{Q^*} = \begin{bmatrix} \frac{1}{5} \\ \frac{4}{5} \end{bmatrix}$. The value of the game is $\frac{29}{5}$.

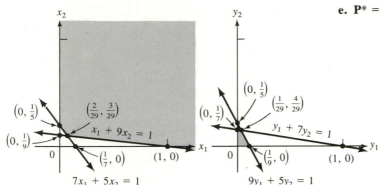

$7x_1 + 5x_2 = 1$

Z attains a minimum $\frac{5}{29}$
where $x_1 = \frac{2}{29}$, $x_2 = \frac{3}{29}$

$9y_1 + 5y_2 = 1$

Z attains a maximum $\frac{5}{29}$
where $y_1 = \frac{1}{29}$, $y_2 = \frac{4}{29}$

7. $\mathbf{A} = \begin{bmatrix} 2 & -3 \\ -3 & 4 \end{bmatrix}$ $\mathbf{P^*} = \begin{bmatrix} \frac{7}{12} & \frac{5}{12} \end{bmatrix}$ $\mathbf{Q^*} = \begin{bmatrix} \frac{7}{12} \\ \frac{5}{12} \end{bmatrix}$. The value of the game is $\frac{-1}{12}$.

9. $\mathbf{A} = \begin{bmatrix} 2 & 1 \\ -6 & 4 \end{bmatrix}$ $\mathbf{P^*} = \begin{bmatrix} \frac{10}{11} & \frac{1}{11} \end{bmatrix}$ $\mathbf{Q^*} = \begin{bmatrix} \frac{3}{11} \\ \frac{8}{11} \end{bmatrix}$. The value of the game is $\frac{14}{11}$.

11. $\mathbf{P^*} = \begin{bmatrix} 0 & \frac{2}{7} & 0 & \frac{5}{7} \end{bmatrix}$, $\mathbf{Q^*} = \begin{bmatrix} \frac{13}{14} \\ 0 \\ \frac{1}{14} \\ 0 \end{bmatrix}$. The value of the game is $\frac{23}{7}$.

13. $\mathbf{P^*} = \begin{bmatrix} \frac{1}{12} & \frac{11}{12} \end{bmatrix}$, $\mathbf{Q^*} = \begin{bmatrix} \frac{31}{36} \\ \frac{5}{36} \end{bmatrix}$, the value of the game is $\frac{125}{12}$.

Exercise Set 7.4 (Pages 498–499) Solution of an $m \times n$ Game Matrix Using Linear Programming

1. $\mathbf{P^*} = \begin{bmatrix} \frac{2}{3} & \frac{1}{3} \end{bmatrix}$, $\mathbf{Q^*} = \begin{bmatrix} \frac{2}{3} \\ \frac{1}{3} \\ 0 \end{bmatrix}$. The value of the game is $\frac{-1}{3}$.

3. $\mathbf{P^*} = \begin{bmatrix} \frac{1}{4} & \frac{3}{8} & \frac{3}{8} \end{bmatrix}$, $\mathbf{Q^*} = \begin{bmatrix} \frac{3}{8} \\ \frac{1}{8} \\ \frac{1}{2} \end{bmatrix}$. The value of the game is $\frac{-5}{8}$.

5. $\mathbf{P^*} = \begin{bmatrix} \frac{4}{5} & 0 & \frac{1}{5} \end{bmatrix}$, $\mathbf{Q^*} = \begin{bmatrix} \frac{3}{5} \\ 0 \\ \frac{2}{5} \end{bmatrix}$. The value of the game is $\frac{-3}{5}$.

7. $P^* = \begin{bmatrix} \frac{3}{5} & 0 & \frac{2}{5} \end{bmatrix}$, $Q^* = \begin{bmatrix} \frac{3}{5} \\ 0 \\ \frac{2}{5} \\ 0 \end{bmatrix}$. The value of the game is $\frac{6}{5}$.

9. a. $A = \begin{bmatrix} 3 & 8 & -7 \\ -2 & 3 & 18 \\ 13 & -12 & 3 \end{bmatrix}$ **b.** $P^* = \begin{bmatrix} \frac{1}{2} & \frac{1}{3} & \frac{1}{6} \end{bmatrix}$, $Q^* = \begin{bmatrix} \frac{1}{2} \\ \frac{1}{3} \\ \frac{1}{6} \end{bmatrix}$, the value of the game is 3.   **c.** 3

11. $A = \begin{bmatrix} .63 & .03 \\ .48 & .33 \\ .18 & .48 \end{bmatrix}$, $P^* = \begin{bmatrix} 0 & \frac{2}{3} & \frac{1}{3} \end{bmatrix}$ $Q^* = \begin{bmatrix} \frac{1}{3} \\ \frac{2}{3} \end{bmatrix}$. Thus, no land should be used for the shopping center, $\frac{2}{3}$ of the land
should be used for restaurants, and $\frac{1}{3}$ for the sports arena.

13.

	Radio	Television	Newspaper
Radio	2	-10	14
Television	-10	26	2
Newspaper	14	-22	-10

$P^* = \begin{bmatrix} \frac{1}{4} & \frac{5}{12} & \frac{1}{3} \end{bmatrix}$, $Q^* = \begin{bmatrix} \frac{7}{12} \\ \frac{1}{4} \\ \frac{1}{6} \end{bmatrix}$, the value of the game is 1. So **R**
should use the radio $\frac{1}{4}$ of the time, television $\frac{5}{12}$ of the time and the newspaper $\frac{1}{3}$ of the time. Also **C** should use the radio $\frac{7}{12}$ of the
time, television $\frac{1}{4}$ of the time and the newspaper $\frac{1}{6}$ of the time.

15.

	Proposition		
	1	2	3
Stock A	7	1	13
Stock B	1	19	7
Stock C	13	-5	1

$P^* = \begin{bmatrix} \frac{1}{4} & \frac{5}{12} & \frac{1}{3} \end{bmatrix}$, $Q^* = \begin{bmatrix} \frac{7}{12} \\ \frac{1}{4} \\ \frac{1}{6} \end{bmatrix}$. The value of the game is 6.5. The investor should invest
$3,000 in stock A, $5,000 in stock B, and $4,000 in stock C.

Exercise Set 7.5 (Pages 508–510) Graphs

1. Traceable; D, C, B, E, D, B, A, C, E   **3.** Not traceable   **5.** Traceable; F, D, B, A, C, B, F, G, C, E, G

7. Traceable; A, ε_1, C, ε_2, A, ε_3, D, ε_4, C, ε_5, D, ε_6, B, ε_7, A, ε_8, B   **9.** Not possible

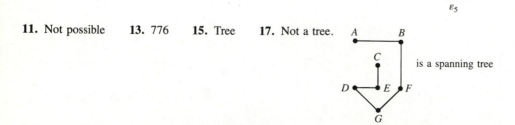

11. Not possible   **13.** 776   **15.** Tree   **17.** Not a tree.   is a spanning tree

19. Not a tree.

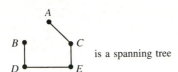

is a spanning tree

21.

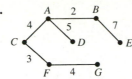

23.

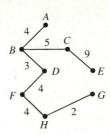

25. A line of minimum length connects Dallas and Fort Worth, Fort Worth and Cisco, Cisco and Brownwood, Brownwood and San Antonio, San Antonio and Austin, San Antonio and Victoria, and Victoria and Houston.

27. Install the network from F to C, from C to A, from A to D, from D to B, from B to E and from B to G.

29.

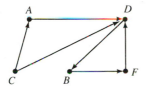

Exercise Set 7.6 (Pages 518–520) Digraphs

1.

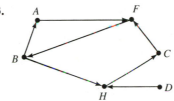

3.

5. $V = \{P, V_1, V_2, V_3, A, B, C, D_1, D_2, D_3, X, Y, Z\}$
$E = \{(P, V_1), (P, V_2), (P, V_3), (V_1, A), (V_1, B), (V_1, C), (V_2, D_1), (V_2, D_2), (V_2, D_3), (V_3, X), (V_3, Y), (V_3, Z)\}$

7. a. $(R, T), (T, V); (R, U), (U, V); (R, T), (T, U), (U, V)$ **b.** yes **c.** yes **d.** 2, 1

9. a. $(S, U), (U, W), (W, T); (S, R), (R, V), (V, T); (S, R), (R, U), (U, W), (W, T)$ **b.** yes **c.** yes **d.** 3, ∞

11.

	R	S	T	U	V
R	0	0	1	1	0
S	0	0	0	0	0
T	0	0	0	1	1
U	0	1	0	0	1
V	0	0	0	0	0

13.

	R	S	T	U	V	W
R	0	0	0	1	1	0
S	1	0	0	1	0	0
T	1	0	0	0	0	0
U	0	0	0	0	0	1
V	0	0	1	0	0	0
W	0	0	1	0	0	0

15.

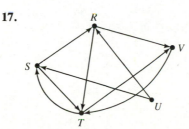

F12

17.

19. $M^2 = \begin{bmatrix} 0 & 1 & 0 & 0 \\ 0 & 1 & 1 & 0 \\ 1 & 0 & 1 & 0 \\ 1 & 0 & 2 & 0 \end{bmatrix}$. There is one path of length 2 from A to B, B to B, B to C, C to A, C to C, and D to A. There

are two paths of length 2 from D to C. There are no other paths of length 2.

$M^3 = \begin{bmatrix} 1 & 0 & 1 & 0 \\ 1 & 1 & 1 & 0 \\ 0 & 1 & 1 & 0 \\ 0 & 2 & 1 & 0 \end{bmatrix}$. There is one path of length 3 from A to A, A to C, B to A, B to B, B to C, and C to B, C to C, and

D to C. There are two paths of length 3 from D to B. There are no other paths of length 3.

21. $M^2 = \begin{bmatrix} 0 & 1 & 1 & 0 & 1 \\ 0 & 1 & 1 & 0 & 2 \\ 1 & 0 & 2 & 0 & 0 \\ 1 & 0 & 2 & 0 & 1 \\ 0 & 1 & 0 & 0 & 1 \end{bmatrix}$. There are exactly two directed paths of length 2 from S to V, T to T, and U to T. There

is exactly one directed path of length 2 from R to S, R to T, R to V, S to S, S to T, T to R, U to R, U to V, V to S, and V to V. There are no other directed paths of length two.

$M^3 = \begin{bmatrix} 1 & 1 & 2 & 0 & 1 \\ 1 & 1 & 3 & 0 & 1 \\ 0 & 2 & 1 & 0 & 3 \\ 0 & 2 & 2 & 0 & 3 \\ 1 & 0 & 2 & 0 & 0 \end{bmatrix}$. There is exactly one directed path of length three from R to R, R to S, R to V, S to R, S

to S, S to V, T to T, and V to R. There are exactly two directed paths of length three from R to T, T to S, U to S, V to T, and U to T. There are exactly three directed paths of length three from S to T, T to V, and U to V. There are no other directed paths of length three.

23. Doug

25. a.

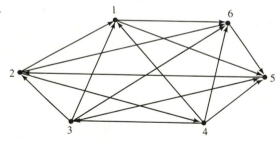

c. Timothy is the winner

Exercise Set 7.7 (Pages 530–533) More Applications

1. $S(A) = 23$, $S(B) = 10$, $S(C) = 5$, $S(D) = 2$, $S(E) = 3$, $S(F) = 2$, $S(a) = S(b) = S(c) = S(d) = S(e) = 0$

3. $S(v_1) = 20$, $S(v_2) = 8$, $S(v_3) = 6$, $S(v_4) = 2$, $S(v_5) = 1$, $S(v_6) = 2$, $S(v_7) = S(v_8) = S(v_9) = S(v_{10}) = 0$

5.

	A	B	C	D
A	0	1	2	1
B	∞	0	∞	∞
C	∞	∞	0	∞
D	∞	∞	1	0

7.

	A	B	C	D	E
A	0	1	1	2	2
B	∞	0	∞	1	∞
C	∞	∞	0	∞	1
D	∞	∞	∞	0	∞
E	∞	∞	∞	∞	0

9.

	A	B	C	D
A	0	1	2	1
B	∞	0	∞	∞
C	∞	∞	0	∞
D	∞	∞	1	0

11.

	A	B	C	D	E
A	0	1	1	2	2
B	∞	0	∞	1	∞
C	∞	∞	0	∞	1
D	∞	∞	∞	0	∞
E	∞	∞	∞	∞	0

13. The distance sum of A is 4, of B is 0, of C is 0, and of D is 1.

15. The distance sum of A is 6, of B is 1, of C is 1, of D is 0, and of E is 0.

17. The status of Bob is 31, of Gwen is 10, of Julie is 8, of Peter, Beth, and Marc is 2, of Paul is 1, and of the other seven individuals is 0.

19. The status of Janet is 37, of Carl is 10, of Kathie is 12, of Ahmad, Wynne and Burnett is 2, of Andre is 3, and of the other nine individuals is 0.

21. a. S, B, F, G, H **b.** S, B, F, G, H, I, E

23. Purchase land immediately, clear land in 3 weeks, build foundation in 5 weeks, build structure in 8 weeks, plumbing done in 15 weeks, electrical work in 15 weeks, finish interior in 17 weeks, finish exterior in 15 weeks, landscape in 18 weeks.
b. 21 weeks

25. a.

Task	Schedule
A	Immediately
B	Immediately
C	5 months hence
D	5 months hence
F	6 months hence
G	6 months hence
H	12 months hence
I	16 months hence

b. 24 months

27. Another person with the disease has not been identified.

Chapter Review 7.8 (Pages 538–548) Sample Exam Questions

1. a. R has 3 choices and C has 4 choices   **b.** R pays \$4 to C **c.** C pays \$6 to R

3. R should play row 1 and C should play column 4.

5. No entry is the smallest in its row and largest in its column.

7. a. No entry is the smallest in its row and largest in its column. The game is not strictly determined.

b. First remove rows 1 and 4, then remove columns 3 and 4. We get $\begin{bmatrix} 2 \\ 1 \end{bmatrix}$. 2 is the saddle point. With respect to the original matrix, R should play row 2 and C should play column 1. The value of the game is 2.

9. X is player R and Y is player C.

$$
\begin{array}{c}
 \\ A \\ B \\ C
\end{array}
\begin{array}{ccc}
A & B & C \\
\begin{bmatrix} 13 & -16.134 & -11.25 \\ -3.6 & 13 & -6.75 \\ 1.35 & -.15 & 13 \end{bmatrix}
\end{array}
$$

This game is not strictly determined.

11. The Racoons should use play number 2 and the Cats should use defensive alignment number 4.

13. \$74   **15.** $\mathbf{P}^* = \begin{bmatrix} \frac{10}{13} & \frac{3}{13} \end{bmatrix}$ and $\mathbf{Q}^* = \begin{bmatrix} \frac{5}{13} \\ \frac{8}{13} \end{bmatrix}$.

17. a. It is strictly determined. Its value is 1. R should always choose row 2 and C should always choose column 2.

19. $\mathbf{A} = \begin{bmatrix} 2 & -2 \\ -2 & 2 \end{bmatrix}$, $\mathbf{P}^* = \begin{bmatrix} \frac{1}{2} & \frac{1}{2} \end{bmatrix}$, $\mathbf{Q}^* = \begin{bmatrix} \frac{1}{2} \\ \frac{1}{2} \end{bmatrix}$, the value of the game is 0.

21. a. Delete recessive rows and columns to get $\begin{bmatrix} 6 & -4 \\ 5 & 9 \end{bmatrix}$, $\mathbf{P}^* = \begin{bmatrix} \frac{2}{7} & \frac{5}{7} \end{bmatrix}$, $\mathbf{Q}^* = \begin{bmatrix} \frac{13}{14} \\ \frac{1}{14} \end{bmatrix}$. In terms of the original matrix, the

optimal strategies are $\begin{bmatrix} 0 & \frac{2}{7} & 0 & \frac{5}{7} \end{bmatrix}$ for R and $\begin{bmatrix} \frac{13}{14} \\ 0 \\ \frac{1}{14} \\ 0 \end{bmatrix}$ for C. The value of the game is $\frac{37}{7}$.   **b.** $\frac{37}{7}$ **c.** $\frac{32}{7}$

23. Ron should show 2 fingers 55% of the time and 3 fingers 45% of the time. Carol should use the same strategy. The expected value of the game for Ron is $\frac{-1}{20}$ dollars.

25. Store R should not use the radio or mail, use TV $\frac{2}{7}$ of the time and the paper $\frac{5}{7}$ of the time. Store C should use the radio $\frac{13}{14}$ of the time, mail $\frac{1}{14}$ of the time, and never use TV or paper. The value of the game is $\frac{44}{7}$.

27. $\mathbf{P}^* = \begin{bmatrix} \frac{3}{5} & 0 & \frac{2}{5} \end{bmatrix}$, $\mathbf{Q}^* = \begin{bmatrix} \frac{3}{5} \\ 0 \\ \frac{2}{5} \\ 0 \end{bmatrix}$. The value of the game is $\frac{1}{5}$.

29. a. Traceable; $E, B, A, C, B, D, C, E, D$ **b.** Traceable; $A, B, C, E, A, D, F, E, D, C, A$
   c. Not traceable; more than two vertices are odd.   **d.** Not traceable; not connected.   **31.** 881

33. a.   **b.**   **35.**

37. a. $(R, T), (T, V); (R, U), (U, V); (R, U), (U, S), (S, V)$.   **b.** yes   **c.** no   **d.** 2, ∞

39.

41. Debbie   **43.**

	R	S	T	U	V
R	0	1	1	2	2
S	∞	0	∞	1	2
T	∞	∞	0	1	2
U	∞	∞	∞	0	1
V	∞	∞	∞	∞	0

49. a.

Task	Schedule
A	Immediately
B	Immediately
C	8 weeks hence
D	8 weeks hence
F	7 weeks hence
G	20 weeks hence
H	28 weeks hence
I	15 weeks hence
J	28 weeks hence

b. 40 weeks

45. The distance sum of A is 3, of B is 5, of C is 2, and of D is 0.
47. a. S, Q, U, W, V **b.** S, Q, U, W, X, Y, E

Index